Y0-AAH-145

PROCEEDINGS OF THE SUMMER INSTITUTE ON BIOLOGICAL CONTROL OF PLANT INSECTS AND DISEASES

EDITED BY
FOWDEN G. MAXWELL
and
F. A. HARRIS

UNIVERSITY PRESS OF MISSISSIPPI
JACKSON

Library of Congress Catalog Card Number 74-77456
ISBN 0-87805-060-4
Manufactured in the United States of America

This volume is authorized
and sponsored by

Southern Regional Education Board
U.S. Department of Health, Education and Welfare,
Office of Environmental Education
Mississippi Agricultural and Forestry
Experiment Station
Mississippi State University
Mississippi State, Mississippi 39762

Table of Contents

Page

Page

ACKNOWLEDGMENTS

The editors wish to thank the Southern Regional Education Board and Dr. T. J. Horne for sponsoring and supporting in part this important Institute on the Biological Control of Plant Insects and Diseases. We wish also to acknowledge the support of the U.S. Department of Health, Education and Welfare, Office of Environmental Education, for their support in helping to publish these proceedings.

Deep appreciation is extended to each of the participants and to each of the faculty of the Department of Entomology, Mississippi State University, for their hard work which contributed to the success of the Institute.

Special thanks are given to Mrs. Henry Green for typing of this manuscript and for other invaluable services in preparation and editing. Appreciation is expressed to Professor C. A. Wilson and Dr. L. W. Hepner for their assistance in proofreading.

Fowden G. Maxwell
F. A. Harris
Editors

FOREWORD

Since the formation of the Council of Higher Education in the Agricultural Sciences in 1956, the Southern Regional Education Board has been concerned with planning and implementing programs designed to strengthen and expand the opportunities in high quality education in agriculture and its related sciences at the college level. The Council has served effectively in formulating policy and providing general guidance for further program development.

Currently, the Council is guiding a five-year Southern Regional Education Board project, supported by the W. K. Kellogg Foundation, designed to advance land grant institutions, agriculture and agricultural sciences in the region. To more effectively guide the planning activities involved in this regional effort the Council membership was organized into four subcommittees, with each assigned a major objective of the project as a particular area of responsibility. The Council subcommittee No. 3, headed by Dr. John A. Ewing, Dean of the Agricultural Experiment Station, University of Tennessee, studied the needs and opportunities for advancing scientific knowledge in the land grant institutions in the region and recommended programs for implementation.

Conducting the Institute on Biological Control of Plant Insects and Diseases was a recommendation of this subcommittee, approved by the Council for implementation. The concept of the institute was supported by the deans of agriculture and directors of the experiment stations of the region as an effective means of further developing research and teaching competence of faculty involved in this relevant area of agriculture. From nominations by the deans a regional planning committee of renowned scientists was formed to develop the institute program and nominate the institute faculty. Following the work of the regional committee the proposed program with recommended institute faculty was given to the host committee at Mississippi State University under the leadership of Dr. F. G. Maxwell, Head, Department of Entomology to revise as necessary in conducting the institute.

This publication of the proceedings of the institute entitled Biological Control of Plant Insects and Diseases is being made available to participants and other scientists and teachers to use in further developing their programs concerning "Biological Control of Plant Insects and Diseases," in the southern region.

T. J. Horne, Project Director
Agricultural Sciences

PEST MANAGEMENT: HISTORY, CURRENT STATUS AND FUTURE PROGRESS

L. D. Newsom
Professor and Head
Department of Entomology
Louisiana State University

Pest management is a term that has come into considerable prominence during the last few years. The general public and even a great many entomologists use the term as though it were descriptive of a new concept, something almost magical, a panacea that will solve all of the problems posed by insects and related pests. Like "the new morality" is a recently accepted term for the old immorality, pest management is a new term for methods of regulating pest populations based on the principles of applied ecology that have for long been practiced by applied entomologists in this country and others. New features involved in the concept are some tools and techniques that have only recently become available.

In treating the history, current status, and future progress of pest management, I have chosen to concentrate my discussion on problems restricted for the most part to the southern United States, principally control of cotton insects and related pests. Firstly, I know of no other area in the country where the problems associated with control of insects and related pests are so severe, complex, and difficult; secondly, cotton insect control especially furnishes an appropriate model to illustrate the mistakes, successes, failures, challenges, and opportunities involved in pest management; and thirdly, I am more familiar with it than any other.

<u>History of Pest Management in the Cotton Growing South</u>: A relatively mild climate, long growing season, intensive cultivation of cotton in a virtual monoculture involving extensive areas, and a large complex of pest species have all contributed to the complexity and severity of the problems of pest control in cotton production in the southern United States. Logically, a discussion of the history of pest management in the area may be divided into four periods: 1) pre-boll weevil invasion; 2) pre-calcium arsenate period; 3) calcium arsenate period; 4) synthetic organic insecticide period.

<u>Pre-boll weevil period</u>: Prior to 1892 when the boll weevil, <u>Anthonomus grandis</u> Boheman crossed the Rio Grande River at Brownsville, Texas (Townsend 1895), insect damage to cotton posed relatively minor problems. There were several species that attacked cotton occasionally but none was a key pest. The bollworm, <u>Heliothis zea</u> (Boddie) had been recognized as a pest of cotton as early as 1820 (Quaintance and Brues, 1905) and the cotton leafworm, <u>Alabama argillacea</u> Hubner, had been an annual immigrant that sporadically caused severe injury in some areas. Of the papers published in the Journal of Economic Entomology during the period 1908-23, only 12 dealt with cotton insect pests. Eight of these were concerned with the boll weevil and the remaining four with the cotton leafworm, <u>Alabama</u> argillacea Hubner.

Pre-calcium arsenate period: Pierce (1922) described the insects that affected cotton at that time and gave recommendations for their control. Species considered to be pests are listed in Table 1.

Table 1.--Insect pests of cotton according to Pierce (1922).

Pest status	
Important	Occasional, of minor importance
Agrotis ipsilon	*Melanoplus differentialis*
Peridroma margaritosa	*Laphygma frugiperda*
Alabama argillacea	*Phyllophaga* spp.
Tetranychus telarius	*Aphis gossypii*
Anthonomus grandis	*Chalcodermus aneus*
Heliothis zea	*Platynota* spp.
	Loxostege sp.
	Estigmene acrea
	Strymon melinus
	Prodenia ornithogalli
	Adelphocoris rapidus
	Leptoglossus spp.
	Aphis maidiradicis
	Monocrepidius vespertinus

Only three of these species listed by Pierce (1922) are considered to be of major importance in cotton producing areas of the southern United States today, viz., the boll weevil, bollworm and twospotted spider mite.

Species that have become major pests of cotton in the area since that time that were not mentioned by Pierce are given in Table 2.

Table 2.--Species considered to be major pests of cotton that were not listed by Pierce (1922).

Heliothis virescens	*Trialeurodes abutilonea*
Pectinophora gossypiella	**Frankliniella* spp.
Lygus lineolaris	*Trichoplusia ni*
Pseudatomoscelis seriatus	

*Thrips are not important pests of cotton but they are so considered by many entomologists.

Pierce's (1922) summary of control measures provides an excellent insight into a philosophy of pest management that can furnish contemporary entomologists with useful suggestions. It is worth quoting in its entirety.

> In summary of the preceding paragraphs a *single system* [Italics mine] may be devised for cotton-insect control.
>
> Best measures for the Early Spring
> 1. Keep down weed growth around the farm.
> 2. Plow in the winter to break up the winter cells in the ground.
> 3. Where necessary set out poison baits to trap cutworms, grasshoppers, May beetles, etc.

4. Plant as early as it can be done safely and yet avoid killing frosts. Plant the variety which is found to be the best producer in your own locality, and which has the qualities of rapid and prolific fruiting.
5. Space the rows in accordance with local experience.
6. Cultivate frequently, but not too deeply.

Best measures to follow during the summer
1. Continue cultivation until the crop is made, or as long as possible.
2. Watch for the first appearance of worms.
3. Dust the cotton with powdered arsenate of lead as soon as grasshoppers or "worms" begin to attack, unless "worm" attack starts late and would hasten ripening.
4. Keep down the weeds.

What to do in the fall
1. Pick the cotton out as soon as possible.
2. Destroy the plants by plowing under or grazing as long before frost as possible.
3. Where practicable plow the fields and plant a cover crop.
4. Where feasible follow a three-year rotation with cotton following some crop other than corn.

Practical measures for the winter
1. Clear up all turn rows and fence rows.
2. Cut and burn all weeds.
3. Plow under all stubble fields that are not to be used otherwise.
4. Grub up old stumps.

Two points about Pierce's system deserve emphasis. Firstly, an insecticide, arsenate of lead, was recommended for the control of "worms" and grasshoppers only. Presumably, "worms" includes the leafworm and bollworm. Secondly, no insecticide was recommended for control of the boll weevil, although it was pointed out in the text that effective control of the pest had been obtained in the Louisiana Delta by dusting with calcium arsenate. It was stated that this new method had not been tested carefully over the entire cotton belt but the thought was expressed that with perhaps "slight modifications" the method might be equally effective over the entire belt.

Thirty years were required for the boll weevil to make its way from its crossing the Rio Grande at Brownsville, Texas to the Atlantic Ocean in North Carolina. During this period losses to the pest had often been catastrophic. However, by follwoing the system outlined by Pierce (1922) growers in most areas were able to continue the profitable production of cotton. Production practices involving early planting and proper fertilization and cultivation of early fruiting, prolific, rapidly maturing varieties allowed the major portion of the crop to be produced before boll weevil populations could reach economic injury levels during most years. Crop residue destruction after harvest but before killing frost substantially reduced the population of overwintering boll weevils. Results obtained by adoption of these methods were surprisingly good considering the destructive potential of the boll weevil. They provided the foundation of a pest management system that enabled the industry to survive for more than three decades. Indeed, production practices involving early planting and proper fertilization of early fruiting, rapidly maturing varieties plus stalk destruction continued to be the most widely practiced and effective method for control of the boll weevil until introduction of the synthetic organic insecticides after World War II.

Although it has been generally overlooked, use of early fruiting, rapidly maturing varieties plus cultural practices that contributed to earliness for control of the boll weevil provides one of the most outstanding successes of varietal resistance for control of a major pest. Early production of the crop to escape periods of heaviest boll weevil infestation had the added advantage of helping to avoid the hazard posed by the cotton leafworm.

The calcium-arsenate period: The effectiveness of powdered lead arsenate for control of the boll weevil was demonstrated by Newell (1908). However, his findings attracted little attention and it was not until after World War I that Coad (1918) reported success in the use of calcium arsenate dusts for boll weevil control. His experiments were done in the vicinity of Tallulah, Louisiana and the results were not demonstrated to be generally applicable to all areas of the South for a period of several years. The demonstration that proper use of calcium arsenate dusts gave excellent control of the boll weevil with highly profitable increases in yields set the stage for a trend in applied entomology, best characterized as excessive reliance upon use of insecticides, that has continued until the present. Research during the pre-calcium arsenate period was concentrated on the biology and ecology of various pest species. These studies set standards of excellence that have not been exceeded since (Hunter and Hinds, 1904; Hunter and Pierce, 1912; Quaintance and Brues, 1905; and Garman and Jewett, 1914).

Grower experience with calcium arsenate was often disappointing as has generally proved to be the case with excessive reliance upon any single component system of pest management. Having an effective new tool, the principles of control summarized by Pierce (1922) in his single system of pest control were neglected and often completely ignored. When calcium arsenate was used on large acreages, populations of predators such as ladybeetles, ground beetles, and lacewings were severely affected. Decimation of predator populations were usually followed by serious outbreaks of the cotton aphid, *Aphis gossypii* Glover, and often by increased damage by *Heliothis* spp. In fact, losses from these induced pests often exceeded the increases resulting from control of the boll weevil.

Most growers failed to acquire the necessary information required to time applications properly. Treatments were frequently made when infestations were below economic injury levels or delayed until the crop had suffered serious damage. Thus, control of the boll weevil by use of calcium arsenate dusts was erratic. Most growers failed to develop confidence in their ability to control the boll weevil effectively and economically with it.

During the period 1920-45 a high percentage of research was devoted to the evaluation of calcium arsenate for boll weevil control and various additives as a means of controlling the infestations of cotton aphid and *Heliothis* spp. that occurred with increasing frequency.

Opportunities for developing an effective program of pest management for cotton insects during the period 1925-1945: For more than four decades after demonstration that calcium arsenate provided an effective and economical means of controlling the boll weevil, entomologists failed to take advantage of their opportunities to develop an effective system of pest management for cotton insects in the southern United States. They ignored the large body of information made available by early studies of the biology and ecology of the boll weevil and bollworm. Furthermore, additional information developed by Isely (1924, 1934) could have been utilized very effectively in a pest management program.

Isely (1924) pointed out a number of features about the biology and ecology of the boll weevil that demonstrated that he was "ahead of the times" in his concept of pest management. He recognized that hibernation was the most critical period in the seasonal history of the pest. He also pointed out that the later in the season the boll weevil matures, to a point, the greater its chances of surviving hibernation. This feature of boll weevil biology is illustrated well in Gaines' (1959) summary of ecological investigations of the boll weevil at Tallulah, Louisiana for the period 1915-1958.

Isely demonstrated that controlled burning and clearing of favorable hibernation quarters were effective measures for reducing the populations of overwintered weevils. He recognized the extremely local nature of the boll weevil problem during most years, especially during early season, and stressed the great differences in abundance, destructiveness, and in control measures required on the same plantation and often in different parts of the same field. It was in his work that the suppression pressures of severe winters and hot, dry summers on boll weevil populations were clearly recognized.

Isely (1934) demonstrated the value of using a trap crop, comprised of an extra early variety planted on a very small percentage of the total acreage, as a means of concentrating populations of overwintered boll weevils. His observations showed that weevils emerging from winter quarters usually concentrated in the earliest fruiting cotton on a farm. By planting an early fruiting variety early on less than 5% of the total acreage he was able to attract most of the overwintering population that found its way into a field. By treating such restricted areas with calcium arsenate dust 3 to 4 times at 5-day intervals damaging populations could often be prevented from developing anywhere on the farm for an entire season. Restricting treatment to such limited areas helped to conserve populations of predators and parasites and hold environmental pollution to minimal levels.

Thus, there was available as early as the mid-1920's sufficient information to form the core of a sound, multifaceted pest management program for cotton insects based on the principles of applied ecology. Such a program would have included the following elements most of which are recognizable as important components of the pest management program for cotton insects that is presently attracting such widespread attention and substantial support.

1. Using varietal resistance in the form of early fruiting, prolific, rapidly maturing varieties with appropriate cultural practices such as early planting, proper fertilization, and appropriate cultural practices required to encourage early production of the crop.

2. Using an early planted trap crop of the earliest maturing varieties available to concentrate the overwintered boll weevil populations on a very small percentage of the total acreage where they could be destroyed by applications of calcium arsenate before having an opportunity to reproduce.

3. Conserving populations of predators and parasites by delaying as long as possible applications of insecticides to any acreage except that planted to trap crops.

4. Inspecting all acreage at weekly intervals to determine when and where insecticide applications were required to regulate pest populations below economic injury thresholds. (A system of using commercial cotton insect scouts for this purpose was initiated in Arkansas by Isely in the late 1920's).

5. Harvesting the crop and destroying crop residues as soon after harvest as possible to eliminate food for the boll weevil and thus decrease its chances for overwintering successfully.

6. Reducing the extent of favorable hibernation quarters by controlled burning and land clearing.

A pest management program having these components would have been effective, economical, ecologically sound, and environmentally safe. Its adoption and use would have undoubtedly reduced greatly, if not have avoided completely most of the problems that have arisen during the last two decades to bring us to the point of crisis in our management of cotton insect pests. It seems quite probable that the current pest management program for cotton insect pests may eventually evolve to the extent that the methods available 45 years ago will, with minor modifications, form the core of the pest management system.

The synthetic organic insecticide period: The synthetic organic insecticides that came into general use after World War II quickly revolutionized prevailing attitudes and practices of growers and entomologists toward cotton insect control. Discovery of the effectiveness of calcium arsenate dust for boll weevil control had started most entomologists along the trail of "treat and count" insecticide evaluation. Growers for the most part, however, had not been able to obtain consistently good results from the use of calcium arsenate. Many, perhaps a substantial majority, were not convinced of its usefulness for cotton insect control. With the introduction of DDT followed by benzene hexachloride, toxaphene, chlordane, aldrin, heptachlor, dieldrin and endrin, growers, for the first time in history, had available cheap, effective insecticides whose use produced results that were immediately apparent. Growers who had been unconvinced that it was possible to control with calcium arsenate and nicotine sulfate or rotenone any cotton insects except the leafworm were quickly convinced that the new insecticides provided the answer to their cotton insect control problems. This set the stage for the growers to join the entomologists in excessive reliance upon use of insecticides for pest control. Ecological principles of animal population regulation were generally forgotten or completely ignored for almost two decades.

Initial successes with the new insecticides were responsible for changes in other areas of cotton production. Cultural practices were changed drastically. Boll weevil control was so effective that early maturing varieties were no longer thought to be necessary. Varieties were developed for extended fruiting periods and later maturity. Application of heavier rates of fertilizers, especially nitrogen, became a standard practice. Irrigation was more frequently and widely used. The result of these changes was to extend the fruiting period of cotton long enough to provide food for the development of one additional generation of the boll weevil in many areas.

Ewing and Parencia (1948) revived the concept of controlling overwintered boll weevils before they could reproduce that had been tested by the early workers by square removal and weevil picking (Coad 1916) and "mopping" with calcium-arsenate molasses mixtures. Unlike Isely (1926) who had proved by his "spot dusting" method that the concept of controlling the overwintered boll weevil population was ecologically sound and effective, Ewing and his co-workers undertook control of thrips, plant bugs, and other so-called "early season" insects as well. Their method consisted of treating all of the cotton acreage in a community, regardless of insect population, with the first application being made at the "pinhead" squaring stage followed by a second 8 to 10 days later. Their initial experiment in Wharton County, Texas in 1948 involved treating all of the cotton acreage on 69 farms in 4 communities, a total of 6,214 acres. Nearby farms were untreated and served as controls. Cotton in the treated area outyielded that in the control area by 44%.

This experiment probably attracted more attention than any other single experiment ever conducted on cotton insects. It was immediately repeated by entomologists throughout the area with different results. Two applications were found to be inadequate to provide the season-long control obtained in the first experiment. Unfortunately, growers had been so excited by publicity given to the initial experiment that many put the method into practice. Even more unfortunately, most of the entomologists in the area had also accepted and strongly recommended the "early season control" method.

The program quickly evolved to one of applying insecticides to cotton on a regularly scheduled basis without regard to pest populations or economic injury thresholds. The idea was so attractive to growers because it required little attention to determining economic injury thresholds and so attractive to the insecticide industry because it provided a large and dependable market for their products that it swept across the Cotton Belt. By the early 1950's most of the growers throughout the South had adopted a "womb to tomb" program of insecticide applications to cotton beginning with emergence of the seedling plants and ending when the crop had matured beyond the point of susceptibility to insect injury.

Thus, the stage was set by an experiment conducted in Wharton County, Texas during 1948 that is responsible for our being here today. Excessive reliance upon use of insecticides for cotton insect control, begun 50 years ago with the finding that calcium arsenate was effective for boll weevil control increased immeasurably with use of the much greater insecticidal and broad spectrum activity of the synthetic organic insecticides. Now this philosophy of pest control had added to it the ultimate in entomological irresponsibility--complete disregard for economic injury thresholds. The expression, "The only good bug is a dead bug," became almost a slogan as all of the principles of applied ecology and common sense were violated by this concept of pest control. It was inevitable that such a program would lead to the crisis in entomology in which we find ourselves today. The consequences may be summarized as follows:

1. The most effective suppression of pest populations that had ever been experienced with substantial increases in yields.

2. Decimation of predator and parasite populations leading to:

   a. Resurgence of treated pests.
   b. Elevation of minor pests, or species of previously unrecognized pest potential, to the status of major pests.

3. Decline of the pest status of some species, for example the cotton aphid and cotton leafworm, to positions of insignificance.

4. Heavy mortality of nontarget species especially pollinating insects and fish.

5. Development of resistance to one or more insecticides by virtually all of the cotton insect pests.

6. Massive environmental pollution with residues of persistent insecticides.

7. Excessive costs of insect control.

8. Decline of confidence by the general public in the entomological profession.

9. Stimulation of the most imaginative research that has ever been done on cotton insects.

10. Convincing applied entomologists of the necessity for developing pest management systems based on principles of applied ecology.

These are too well-known to require further discussion. Suffice it to say, not all of the aspects of the current crisis in entomology are negative. As Shakespeare expressed it, "Sweet are the uses of adversity; which like the toad, ugly and venomous, wears yet a precious jewel in his head." The "jewel" in the head of the applied entomologist is the opportunity now available as never before to develop pest management systems for the control of cotton insect pests that will be effective, economical, environmentally acceptable, and possess a degree of stability and permanence never before achieved.

This opportunity has come about because of the disastrous situation that now exists with regard to the cotton insect problem in the southern United States. The situation may be described as follows:

1. Resistance to all classes of insecticides registered for use on cotton occurs in populations of the tobacco budworm and banded-wing whitefly. The former has been responsible for destruction of a more than 500,000-acre cotton industry in northeastern Mexico and threatens profitable production of cotton in the Rio Grande Valley of Texas with similar ruin. Problems posed by the banded-wing whitefly are rapidly increasing in size and intensity in mid-South states.

2. Effectiveness of natural control agents in regulating populations of _Heliothis_ spp., _Tetranychus_ spp. and _Trialeurodes abutilonea_ is nullified by present insecticide use patterns.

3. Environmental pollution with DDT residues to such an extent that it has been banned for use on cotton after 1972 by order of the Food and Drug Administration. This will necessitate the substitution of heavy dosages of methyl parathion at shorter intervals between applications in areas where the boll weevil is a key pest with a consequent increase in hazard to humans and other animals from acute intoxication.

The gravity of these problems forces entomologists to turn to a system of pest management that could have been developed 45 years ago. Such a system must deal for the present with far more serious problems than existed in the mid-1920's, with one less component to use, viz., rapidly fruiting, early maturing varieties, and with very little additional information for immediate integration into the system. The challenge posed is to initiate a system that will not only halt but reverse the rapidly accelerating deterioration of cotton insect control toward the disaster that has occurred in northeastern Mexico. Such a system must accomplish the following:

1. Maintain satisfactory control of the boll weevil.
2. Prevent higher levels of resistance than now occur from developing in populations of pest species by relaxing the amount of selective pressure being applied. This is critically important in the case of the tobacco budworm and banded-wing whitefly.
3. Make the most effective use of indigenous predators and parasites of _Heliothis_ spp., _Tetranychus_ spp. and _Trialeurodes abutilonea_.

The "reproduction-diapause" system of boll weevil control more nearly meets these requirements than any other available at this time (Brazzel, 1961; Lloyd et al., 1966; Bottrell and Almand, 1968). This system is based on denying the portion of the boll weevil population in which diapause has been induced access to the amount of food required to accumulate enough fat to overwinter successfully. Such a system requires destruction of a high percentage of that portion of the population destined to become adult during the period September 15-October 31 for most of the area. A combination of insecticide application, defoliation and rapid harvest and stalk destruction is employed to achieve the objective of starving or killing outright the weevils that otherwise would accumulate enough fat to overwinter.

Employing this system usually results in the destruction of 80 to 95% of the potential overwintering population. Such levels of suppression often delay the development of economically damaging populations during the following year until the third generation or later. It allows maximum utilization of the indigenous predator-parasite population to be exerted on _Heliothis_ spp., _Tetranychus_ spp. and _Trialeurodes abutilonea_. This usually delays substantially, or prevents, need for insecticide applications to control these pests.

This system will result in a substantial decrease, one-third to one-half, in the number of insecticide applications now being made routinely by a majority of cotton growers in most areas. It will reduce from three to one the number of generations of _Heliothis_ spp. being exposed to dosages high enough to select for increased levels of resistance. It will also result in a substantial decrease in numbers of generations of banded-wing whitefly exposed to the selective pressures of heavy applications of insecticides. By making possible a delay in beginning insecticide applications until mid-season, or later, prey populations of aphids, thrips, spider mites, whiteflies, and lepidopterous eggs and larvae can be maintained in sufficient numbers to hold and increase populations of the parasite-predator complex in cotton fields. In situations where control of early-season pests such as the plant bugs, Lygus spp. or Pseudatomoscelis seriatus becomes mandatory only the most highly selective insecticides, dimethoate or trichlorfon, for example, will be used and then only at the minimum effective dosages.

This relatively simple pest management system is being adopted rapidly and practiced widely in many areas where the boll weevil is a key pest. It should prevent further deterioration of the cotton insect pest problem and serve as a foundation upon which a more sophisticated pest management system quickly can be constructed. Viewing the system pessimistically, it is a vast improvement over what has been practiced throughout the southern United States for about two decades. More optimistically viewed, it will reverse the trend toward higher levels of resistance in populations of several pest species and "buy the time" required for developing more effective and permanent systems of managing cotton insect pests.

The "reproductive-diapause" control system has two very serious flaws. 1) It is based on "automatic" applications of insecticide to all of the acreage in an area without regard to pest population density and economic injury thresholds. 2) It exerts the heaviest pressure on the boll weevil of any system yet devised and thus could be responsible for the development of populations resistant to the O-P insecticides. This has not yet occurred and the fact that it has not is somewhat surprising. Resistance to the point of practical immunity to the O Cl insecticides developed in boll weevil populations after exposure of no more than 15 generations. About 45 generations of boll weevils have been exposed to the O-P insecticides with no measurable level of resistance to these insecticides having been detected. It is possible, though unlikely, that the boll weevil does not possess the necessary mechanism(s) for resisting the O-P compounds. Adopting a system of pest management that increases the probability that a resistant line may be selected is a calculated risk, justified only in a crisis situation. Such a situation now exists.

There is another, quite different but very serious weakness of this system of pest management. It is a system that the insecticide industry and commercial applicators, for the most part, have been unwilling to accept and support. The reasons are simple. Its acceptance by growers will result in substantial reductions in the amount of insecticides used for cotton insect control. Entomologists should immediately initiate a vigorous educational program aimed at overcoming such a narrow, short-sighted attitude. The program cannot be completely successful without cooperation of all segments of the cotton industry. The profligate use of insecticides for control of cotton insect pests has been largely responsible for the decision by EPA officials to ban further use of DDT. Similar action on other chemicals is certain to occur unless the amount of insecticide applied to cotton is reduced substantially.

Another extremely serious threat to the success of the "reproduction-diapause" system of boll weevil control is currently posed by those who advocate use of aldicarb in pre-plant or sidedress applications, or both, on a field scale or to larger areas. Practice of such a technique on a large scale could have disastrous consequences of a magnitude exceeded only by the early-season program of Ewing and Parencia (1948). It has been amply demonstrated that the adverse effects on predators of aldicarb used in this manner may induce outbreaks of Heliothis spp. (Ridgway and Lingren 1967, Ridgway 1969). Destruction of the predators of Heliothis spp., Tetranychus spp. and Trialeurodes abutilonea is casued directly by the toxic effects of aldicarb and indirectly by destroying prey populations required to support predator populations and hold them in the cotton fields. Regardless of how the effects are exerted, use of aldicarb on a large scale would negate the effectiveness of the indigenous parasite-predator complex that is an essential component of a successful pest management system for cotton insect pests.

Future progress: Fortunately, it should not be necessary to rely for long on a pest management system for cotton insects based on the "reproduction-diapause" method of control of the key pest, the boll weevil. Excellent progress has been made during the last decade in the development of some exciting new techniques for control of insects. Some of these are in the advanced stages of field testing and should be ready for inclusion as new components of pest management systems during the next few years. The manipulative methods that appear most promising for early availability as a result of their performance in large-scale field tests include: 1) varietal resistance; 2) biological control; 3) selective insecticides and discriminative use of non-selective insecticides; 4) cultural control; 5) pheromones; and 6) microbial insecticides. Possibilities also exist among other methods of manipulating the boll weevil and other major pests of cotton. Among these are: 1) genetic methods; 2) juvenile hormones; 3) autocidal methods; and 4) electromagnetic energy.

Varietal resistance: The possibility of having useful levels of varietal resistance to the boll weevil available in agronomically acceptable varieties within two or three years appears to be excellent. Frego bract, a mutant form with rolled, strap-like bracts in contrast to the large flat, leaf-like bracts in the normal condition confers useful levels of resistance to the boll weevil (Lincoln, et al., 1971). Since the Frego bract character apparently came from DPL Smoothleaf, a good commercial variety, it has not been difficult to develop agronomically acceptable lines possessing the character. Unfortunately, lines to which the Frego bract character have been transferred are more heavily damaged by *Lygus* spp., *Pseudatomoscelis seriatus*, and *Neurocolpus nubilis* than those that do not possess the character. This poses a serious problem in areas where these plant bugs are important pests and would preclude the effective use of varieties possessing the Frego bract character in such areas unless much more highly selective insecticides for control of the plant bugs than now exist could be discovered. Application of insecticides currently available for control of plant bugs is far too disruptive of populations of the predator-parasite complex that growers must conserve.

Frego bract appears to confer measurable levels of both nonpreference and antibiosis to the boll weevil. Clower et al. (1970) have had promising results by making use of the nonpreference effect and Jenkins and his associates (personal communication) have demonstrated convincing levels of antibiosis as well.

Effective use of varietal resistance currently available that could be included in a pest management system involves nonpreference or antibiosis, or both. The boll weevil is strongly attracted to fruiting cotton. Clower and Jones and their associates have showed that Super okraleaf, a prolific, rapidly fruiting, early variety when planted on a relatively small part of the total acreage planted to cotton attracts and holds a very high percentage of the overwintered and first generation boll weevils in the area. This response makes available a technique that can be used to control a very high percentage of the overwintering and first generation boll weevils with minimum use of insecticides. Combining the nonpreference in Frego bract with red leaf from AK Djura Red or North Carolina Margin sources made it possible to delay the beginning of insecticide applications for more than one month during experiments for two successive years under conditions of extreme boll weevil pressure.

Plant breeders once more have undertaken the development of early maturing, rapidly fruiting varieties capable of producing and maturing maximum yields before boll weevil populations reach economically damaging levels.

Substantial progress has been made. The most exciting results in this area of activity have come from the research of plant breeders in Texas who have selected genetic stocks of cotton that mature very early. They have found that reduction in yields often associated with extreme earliness may be offset by growing such varieties in high density culture (Walker and Niles, 1971).

Biological control: One of the important results of using the synthetic organic insecticides in such a senseless manner during the last 25 years has been the demonstration of the importance of indigenous predators and parasites. One of the greatest advantages of the "diapause" control program is that it conserves these useful insects. Their conservation can be further effected by making use of the varietal resistance discussed above to further reduce insecticide applications and to use them in a more discriminating manner.

DeBach (1971) has stated that importation of new natural enemies from abroad should receive first priority in development of pest management programs. Perhaps many will disagree with this being the top priority item in pest management programs for cotton insects. None, I believe, will deny that it is a seriously neglected method that has great promise.

Recent successes with mass-rearing and properly timed releases of parasites and predators offer considerable promise for exploitation in control of *Heliothis* spp. (Ridgway and Lingren, 1972). (Now that Mr. Carpenter of Manned Space Flight fame has undertaken production of *Trichogramma* as a business venture, perhaps this technique will receive the attention it has long deserved). Use of mass-reared predators and parasites in inundative releases, essentially as living insecticides, has not been properly evaluated as a technique of pest management. Theoretically, it should be very useful.

Selective insecticides and discriminative use of non-selective insecticides: The ultimate weapon for pest management is an insecticide that will affect nothing but the target species, one that the target species cannot overcome by developing resistance to it, and that is economical to use. A number of insecticides meet the latter requirement. None possesses the first two. Numerous strides have been made to evaluate relative toxicity of various insecticides to important insect predators. Examples of substantial differences in toxicity to the various species have been found as well as specificity of response. Unfortunately, with the exception of Ripper et al. (1948; 1949; 1951), results of most of these studies have been confounded by the possibility that populations of some species may have developed substantial levels of resistance to one or more classes of insecticides before the studies were done. Other than the work of Ripper and his associates, the studies reported on reaction of predators to the synthetic organic insecticides appear to have been done after 1954.

With such a large complex of major pest species and predators as is the case with cotton, it is not unexpected that no compound has been found that possesses a useful degree of selectivity. Trichlorfon probably has come closest to being useful in this regard. Even it has not lived up to early claims. Lingren and Ridgway (1967) showed that it was substantially less toxic to *Collops balteata*, *Chrysopa carnea*, and *Hippodamia convergens* than four other insecticides tested but that it was highly toxic to the important hemipterous predators *Geocoris punctipes*, *Nabis americoferus* and *Orius insidiosus*.

Thus, the degree of selectivity available at present does not appear to be of much usefulness in pest management systems for cotton insects. Good results can be obtained, however, by the applications of non-selective pesticides in such ways that they act in a discriminating manner. Seed treatment with low rates of application of disyston or thimet for thrips control illustrates a technique of using pesticides in a discriminating way. Only insects that attack the seedling plants directly are affected. Indirectly, however, seed treatments of this sort can have severe effects on populations of predators that depend on thrips, aphids, spider mites, and whiteflies as a source of food. Destruction of prey will result in starvation or force the predators to leave cotton fields in search of food.

Probably the technique that offers the greatest immediate promise of effective use of insecticides in a highly discriminatory manner is the use of varieties that possess enough earliness to serve as a trap plot for boll weevils when planted on a very small percentage of the total acreage; or, planting a small percentage of the total acreage to a preferred variety and the remainder to a non-preferred variety. In either case boll weevil control could be obtained by treating only a small percentage of the total acreage, leaving the remainder to serve as a refuge for the beneficial species.

Cultural Control: Early harvest and destruction of stalks as advocated by Pierce (1922) have once again assumed major importance as techniques for controlling cotton insect pests. Growers have available some additional tools that the early advocates of early harvest and stalk destruction did not have. Growers now have available harvest aid chemicals such as defoliants and desiccants, mechanical cotton pickers, and more effective mechanical equipment for stalk destruction than was available 50 years ago. Thus, they are now able to accomplish stalk destruction far more rapidly and efficiently than formerly was the case. Consequently, this technique may be expected to play an increasingly important role for control of cotton insect pests.

Pheromones: Isolation, identification and synthesis of the boll weevil sex pheromone, "Grandlure" (Tumlinson, et al. 1969) has added a powerful new component to pest management systems for cotton insect pests. It has been amply demonstrated that traps baited with male boll weevils attract large enough numbers of overwintered boll weevils (Hardee, et al., 1970; Lloyd, et al., 1972) to suppress population buildup. Use of this potent pheromone fits very well into the "diapause" control for boll weevil. Lloyd, et al. (1972) reported that the efficiency of the method was more than 90% if populations of the overwintered weevils were 5/A or fewer. However, when more than 300 overwintered weevils per acre were present efficiency was reduced to about 20%. Traps located around field margins at a density of one per acre appeared to be as effective as a greater density of traps.

A formulation of the pheromone has recently been developed that has about twice the residual life of previous formulations with no loss of effectiveness.

Microbial Pathogens: Insect pathogens appear to promise more for cotton insect control and to give less than any other of the relatively new techniques being investigated for their potential role in pest management. It is highly desirable that some means be found for using them effectively, especially for control of larvae of *Heliothis* spp. Thus far, results have been disappointing. *Bacillus thuringiensis* and a nuclear polyhedrosis virus of Heliothis have been tested extensively. Neither has been able to provide reliable, economical control of *Heliothis* spp. on cotton. However, the prospects of improving the effectiveness of both pathogens appear to be good (Dulmage, 1972).

The microbial pathogens, like predators and parasites, furnish another example where foreign exploration and introduction might be a very profitable research venture.

Autocidal Methods: Treatment of laboratory-reared boll weevils with busulfan (1, 4-butanediol dimethanesulfonate) has been demonstrated to induce permanent sterility in the insects without reducing production of the sex pheromone. It has been suggested (Klassen and Earle, 1970) that this chemosterilant may be useful in producing sterile boll weevils for release in an eradication program.

Recently, all of the proven techniques and promising new methods of boll weevil control have been integrated into a pest management system designed to eradicate the boll weevil from a large area of southwestern Alabama, southeastern Louisiana and southern Mississippi. This ambitious program is supported by several agencies of State and Federal Governments and Cotton Incorporated. The system includes the following components:

1. Reproductive-diapause treatments with O-P insecticides in late summer and early fall.
2. Defoliation, harvesting as early as possible, followed by stalk destruction.
3. Trap plots of an early planted, extra early, cold resistant variety planted in 4 to 6 rows around the margins of all fields in the "eradication zone" and treated with a heavy rate of the systemic insecticide aldicarb, in-furrow followed by sidedress, to kill overwintered weevils attracted to the trap plots.
4. Planting a part of the acreage to the boll weevil resistant variety Frego bract.
5. Trapping overwintered weevils by use of the synthetic pheromone grandlure.
6. Treating with O-P insecticides when squaring begins those areas where overwintered boll weevil numbers are excessively high in order to suppress the population to levels at which pheromone traps perform most effectively.
7. Overflooding the "eradication zone" with laboratory-reared weevils sterilized by treatment with busulfan in programmed releases of numbers large enough to prevent survivors of previous treatments from reproducing.

Although it is highly probable that the experiment will fail, it should provide a rigid test of some of the potential components of a pest management system for cotton insects in the southern United States. With the ban of DDT for use on cotton this becomes more imperative than ever.

There appears to be much reason for optimism that an effective, economical, multifaceted pest management system can be developed that will reverse the trend toward disaster and put cotton insect control on the soundest basis of its entire history.

One of the great Welsh hymns has a line that goes "Watchman tell me of the night, what its signs of promise are." I cannot conclude the discussion of a subject that includes an assessment of the future progress in pest management without "telling you of the night and what its signs of promise are" as I see them.

For about 15 years entomologists have been stressing the necessity for developing pest management programs that do not rely so heavily on use of conventional pesticides. Even more to the point, many have been contending that all we need to develop effective programs of pest management is financial support for work of this sort. Regrettably, a great many of us have oversold the proven and potential effectiveness of the alternatives to use of conventional pesticides. Substantial numbers of entomologists have jumped on the environmental bandwagon in vicious attacks on the use of DDT and related insecticides, seemingly convinced that once DDT is banned all the environmental ills related to pest control will be corrected. Others have stood supinely by without making any effort to help defend the sensible use of all pesticides, DDT included, apparently unaware that effective pest management systems for the vast majority of our major pests require the intelligent use of these chemicals.

Now we find that we have "painted ourselves into a corner" so to speak. Suddenly, we find that we have been handed very substantial sums for the immediate initiation of action and research programs in pest management. In effect, we have been told to "put up or shut up." Gentlemen, these programs must not fail. Should they do so our profession will suffer enormous loss of prestige, but even more unfortunately, the concept of pest management which has been so difficult to sell will be set back for decades.

Signs of potential roadblocks to the successful development of the pest management concept are apparent. Those that appear most significant to me are the following:

1. Lack of sufficient numbers of trained personnel: Personnel properly trained in pest management are few when measured against the magnitude of need. Obviously, this fact is being recognized or we would not be here today. I think it significant that Mr. Ruckelshaus stated in his announcement banning use of DDT on cotton that methyl parathion would be the substitute. But, he also stated that people would have to be trained to use this acutely toxic insecticide. How is this to be done between now and the beginning of the next crop season? Who is to be trained and upon whom do we depend to do the training? What levels of training will be required?

2. Attitudes of the insecticide industry, commercial applicators and growers: Many members of industry and commercial applicators view pest management as inevitably contributing to a substantial decrease in the amount of insecticides used. They are correct. If adoption of pest management systems do not result in such a decrease, they are failures. How do we convince these people that pest management is the route that must be followed when their livelihood is being affected? Their cooperation is essential to the success of this approach to pest control. How do we convince Union Carbide, for example, to abandon their attempts to persuade cotton growers to treat every acre of cotton with rates of aldicarb that are unacceptably hazardous to predator complexes? This is a compound that can be quite useful for application to trap plots of early planted cotton to control overwintered boll weevils. However, costs of developing aldicarb can never be recovered if its use is limited to treatment of trap plots.

There are still substantial numbers of growers, county agents and unfortunately entomologists, who remain unconvinced that they should allow cotton to go untreated when thrips, fleahoppers, lygus bugs, bollworms, tobacco budworms, and miscellaneous species are attacking their crops and causing readily observable injury. This sort of injury is occurring throughout the area now. How many of us here today have yielded to the temptation to recommend treatment for situations of this sort? How many are convinced that having low populations of these pest species present in cotton fields at this time is necessary for the maintenance of effective predator-parasite populations?

4. Cooperation between members of all disciplines involved in crop production: Frego bract cotton has been demonstrated to have great potential for boll weevil control. It has also been demonstrated to be considerably more susceptible to injury by plant bugs. In areas where plant bugs are not a serious problem entomologists and plant breeders have no problem in cooperating effectively. Where plant bugs are a serious problem the question is asked, "Do we proceed with Frego bract because of its importance in boll weevil control and use insecticides to control plant bugs, or do we delay taking advantage of Frego bract while we attempt to get plant bug resistance into the variety?" This is where the cooperation between plant breeder and entomologist may become a bit tenuous.

Clearly, entomolgists and plant breeders must work in the closest sort of harmonious relationship. In the past, plant breeders have had a tendency to concentrate on desirable agronomic qualities, edaphic and climatic adaptability and have tended to pay too little attention to varietal adaptability to pests.

Entomologists can no longer tolerate development and release of varieties possessing levels of susceptibility to pests that will disrupt the pest management system.

The overhead application of arsenical herbicides, DSMA and MSMA, has become an increasingly common practice for weed control in some areas. These herbicides also exert substantial insecticidal action when applied in this manner, so much so that populations of some predator species, especially the coccinellids, are seriously reduced. This is another example where the cooperation between disciplines is likely to grow a bit tenuous.

Obviously, the answer to these problems is the team approach to pest management. It cannot operate successfully without such an approach. The day of the entomologist, agronomist, plant breeder, pathologist, nematologist, plant physiologist, agricultural engineer, and economist each working alone in his narrow little world has gone the way of DDT. Major administrative officers have the obligation and responsibility to organize the various disciplines into effectively cooperative working groups at all levels of research and extension.

Difficulties in developing and registering for use insecticides that are especially needed in pest management: Until very recently, and probably it has not yet changed appreciably, the insecticide industry has had the philosophy of developing compounds effective for control of everything from the aardvark to the zoril. Compounds with such wide spectra of activity are not compatible with pest management programs. On the contrary, compounds that affect nothing but the target species are desirable. Entomologists cannot tolerate the former and insecticide chemists cannot produce the latter. So, here again is a situation where cooperation and accommodation must be achieved.

Unterstenhofer (1970) has discussed, in a very illuminating way, the question of insecticide selectivity from the aspect of industrial research. He called attention to a statement by Mathys and Baggiolini (1965) that "the prospects of integrated pest control in Switzerland are very favorable, but there are still many problems to be solved before the method really is completely ready for practical use." He suggested that this viewpoint is still that of the chemical industry.

He summarized his views as follows:

1. Strictly selective monotoxic compounds cannot yet be developed systematically. The probability of discovering such compounds is extremely slight.
2. Provided expenditure for research development and production is kept at a reasonable level, active ingredients of the oligotoxic and polytoxic categories can be regarded as realistic targets. It must be taken into consideration that the breadth and nature of the spectrum of activity cannot be influenced in the sense of selectivity as required by integrated pest control. Selectivity in this sense is an anthropocentric and not a physiological toxicological criterion. Therefore, selectivity will always be a compromise.

Instead of the sort of selectivity that would be provided by "monotoxic" compounds, he suggests that special attention be devoted to "ecological" selectivity. He suggested the following as possible ways and means of achieving selective action.

a) Choice of the lowest dose rates for the suppression of a pest population.
b) Application of the product at the most suitable time.
c) Use of attractants for the purpose of increasing the probability

of contact between active ingredient and pest with reduced probability of contact between active ingredient and beneficial insect.

d) Production of suitable formulations.

Although it may be argued that production of the highly selective ("monotoxic"), compounds is a difficult task that industry is not willing to tackle at current levels of research funding, Unterstenhofer (1970) may be overly pessimistic in his conviction that such compounds cannot be developed. Nevertheless, we must accept the fact that they are not now available and that it will be some time in the future, if indeed ever, that they may become available. There are tremendous possibilities of achieving the degree of selectivity in effect of insecticides required in pest management systems by the discriminating use of these chemicals. It is work toward achieving ecological selectivity that is likely to be most productive for the immediate future.

Obtaining clearance for use of new chemicals, or new uses of old chemicals, has become a frustratingly difficult problem that must be alleviated. As we move toward increasing usage of microbial pathogens, pheromones, chemosterilants, attractants, repellents, and hormones, it will be necessary that mechanisms for moving these materials through the necessary registration procedures be simplified and accelerated.

Incompatibility of the pest management and eradication philosophies: Development of new techniques of insect control during the last two decades has given new life to the philosophy of eradicating pests rather than attempting to regulate their populations at sub-economic levels. Obviously, eradicating a pest species is the ultimate in pest management. The screwworm eradication program has undoubtedly done more to advance the eradication philosophy than all other eradication efforts combined. It has been a remarkably successful program when measured in terms of degree of control of a species over a large area of its range. However, it has failed to achieve the objective announced more than 10 years ago of eradicating the species from the southwestern U.S. Unbelievably, eradication of the screwworm has been claimed in spite of the fact that 4,236 cases of infestation were reported from the area during May 1972 compared with 6,372 during the same month in 1962.

We are now involved in a very large and expensive effort to eradicate the boll weevil from large areas of three states. I supported this undertaking with the understanding that it would be conducted for a two-year period after which it would be discontinued if eradication from the area was not achieved. I do not believe that it will be achieved. Based on past performance, I do not believe it will be discontinued if its objective is not achieved.

If never before, the time has come for a serious evaluation to be made of the eradication concept. Programs like barberry eradication that was initiated in 1918 and continues to operate with the destruction of barberry bushes during 1964 and again during 1969 exceeding the annual average for the period 1918-1945, the imported fire ant program started during 1957 and continued since with an expenditure of more than $150,000,000, and others of the sort obviously should be discontinued. Entirely too much of the funds available for plant protection research and action programs is being expended on eradication programs. Available information suggests that these funds could more appropriately be expended on pest management.

References

Bottrell, D. G. and L. K. Almand. 1968. Evaluation of the 1967 reproductive-diapause boll weevil control program of the Texas High Plains. Texas Agr. Exp. Sta. M.P. 904.

Brazzel, J. R. 1961. Destruction of diapause boll weevils as a means of boll weevil control. Texas Agr. Exp. Sta. Mim. Publ. No. 511.

Clower, D. F., J. E. Jones, K. B. Benkwith, Jr. and L. W. Sloane. 1970. "Non-preference"--a new approach to boll weevil control. Louisiana Agric. 13, No. 4, 10-11.

Coad, B. R. 1916. Cotton boll weevil control in the Mississippi Delta, with special reference to square picking and weevil picking. USDA Bul. 382.

Coad, B. R. 1918. Recent experimental work on poisoning cotton boll weevils. U.S. Dept. Agr. Bul. 731.

DeBach, P. 1971. The use of imported natural enemies in insect pest management ecology. Proc. Tall Timbers Conf. on Ecological Animal Control by Habitat Mgmt. No. 3.

Dulmage, H. T. 1972. Pathogens, pp. 57-64. In: Southern Coop. Series Bul. 169.

Ewing, K. P. and C. R. Parencia, Jr. 1949. Experiments in early season applications of insecticides for cotton insect control in Wharton County, Texas during 1948. U.S. Bur. Entomol. and Plant Quar. E-772.

Gaines, R. C. 1959. Ecological investigations of the boll weevil, Tallulah, Louisiana, 1915-1958. USDA Tech. Bul. 1208.

Garman, H. and H. H. Jewett. 1914. The life history and habits of the corn earworm. Kentucky Agr. Exp. Sta. Bul. 187.

Hardee, D. D., W. H. Cross, P. M. Huddleston and T. B. Davich. 1970. Survey and control of the boll weevil in West Texas with traps baited with males. J. Econ. Entomol. 63: 1041-1048.

Hunter, W. D. and W. E. Hinds. 1905. The Mexican cotton boll weevil. USDA Bur. Entomol. Bul. 45.

Hunter, W. D. and W. D. Pierce. 1912. The Mexican cotton boll weevil, a summary of the investigations of this insect up to December 31, 1911. U.S. Sen. Doc. 306.

Isely, D. 1924. The boll weevil problem in Arkansas. Arkansas Agr. Exp. Sta. Bul. 190.

Isely, D. 1926. Early summer dispersion of the boll weevil. J. Econ. Entomol. 19: 109-110.

Isely, D. 1943. Relationship between early varieties of cotton and boll weevil injury. J. Econ. Entomol. 27: 762-766.

Klaasen, W. and N. W. Earle. 1970. Permanent sterility induced in boll weevils with busulfan without reducing pheromone production. J. Econ. Entomol. 63: 1195-1198.

Lincoln, C. G., G. Dean, B. A. Waddle, W. C. Yearian, J. R. Phillips and L. Roberts. 1971. Resistance of Frego-type cotton to boll weevil and bollworm. J. Econ. Entomol. 64: 1326-1327.

Lingren, P. D. and R. L. Ridgway. 1967. Toxicity of five insecticides to several insect predators. J. Econ. Entomol. 60: 1639-1641.

Lloyd, E. P., F. C. Tingle, J. R. McCoy and T. B. Davich. 1966. The reproductive-diapause approach to population control of the boll weevil. J. Econ. Entomol. 59: 13-16.

Lloyd, E. P., M. E. Merkl, F. C. Tingle, W. P. Scott, D. D. Hardee and T. B. Davich. 1972. Evaluation of male-baited traps for control of boll weevils following a reproduction-diapause program in Monroe County, Mississippi. J. Econ. Entomol. 65: 522-525.

Mathys, G. and M. Baggiolini. 1965. Praktische Anwendung der integrierten Schadlingsbekampfung in Obstanlagen der Westschweiz. Mitt. Biol. Bundesanstalt 115: 21-30.

Newell, W. and T. C. Barber. 1908. Preliminary report upon experiments with powdered arsenate of lead as a boll weevil poison. Louisiana State Crop Pest Comm. Circ. 23.

Pierce, W. D. 1922. How insects affect the cotton plant and means of combating them. USDA Farm. Bul. 890.

Quaintance, A. L. and C. T. Brues. 1905. The cotton bollworm. USDA Bur. Entomol. Bul. 50.

Ridgway, R. L. 1969. Control of the bollworm and tobacco budworm through conservation and augmentation of predaceous insects. Proc. Tall Timbers Conf. on Ecological Animal Control by Habitat Mgmt. No. 1, pp. 127-144.

Ridgway, R. L., P. D. Lingren, C. B. Cowan, Jr. and J. W. Davis. 1967. Populations of arthropod predators and Heliothis spp. after application of systemic insecticides. J. Econ. Entomol. 60: 1012-1016.

Ridgway, R. L. and P. D. Lingren. 1972. Predaceous and parasitic arthropods as regulators of Heliothis populations, pp. 48-56. In: Southern Coop. Series Bul. 169.

Ripper, W. E., R. M. Greenslade, J. Heath and C. H. Barker. 1948. New formulations of DDT with selective properties. Nature 161: 484-485.

Ripper, W. E., R. M. Greenslade and L. A. Lickerish. 1949. Combined chemical and biological control of insects by means of a systemic insecticide. Nature 163: 787-789.

Ripper, W. E., R. M. Greenslade and G. S. Hartley. 1951. Selective insecticides and biological control. J. Econ. Entomol. 44: 448-459.

Townsend, C. H. T. 1895. Insect life. 7, No. 4.

Tumlinson, J. H., D. D. Hardee, R. C. Gueldner, A. C. Thompson, P. A. Hedin and J. P. Minyard. 1969. Sex pheromones produced by boll weevil: isolation, identification and synthesis. Science 166: 1010-1012.

Unterstenhofer, G. 1970. Integrated pest control from the aspect of industrial research on crop protection chemicals. Pflanzenschutz-Nachrichten 23: 264-272.

Walker, J. K., Jr. and G. A. Niles. 1971. Population dynamics of the boll weevil and modified cotton types. Implications for pest management. B-1109. Texas A&M Univ.

ECOLOGICAL PRINCIPLES AS A BASIS FOR PEST MANAGEMENT IN THE AGROECOSYSTEM

R. L. Rabb[1/], R. E. Stinner[1/], and G. A. Carlson[2/]
North Carolina State University
Raleigh, North Carolina

"Pest management is the selection, integration, and implementation of pest control actions on the basis of predicted economic, ecological, and sociological consequences" (Rabb, 1972).

The concepts and principles of pest management have evolved over many years and may be studied in publications by Beirne (1967); Chant (1964, 1966); Clark et al. (1967); F. A. O. (1966); Geier (1966); Harcourt (1970); Knipling (1966, 1970); Lawson (1969); N. A. S. (1969); Newsom (1970); Pimentel, et al. (1965); Rabb and Guthrie (1970); Smith (1971); Smith, et al. (1964); Stern (1966); Stern, et al. (1959); and many others. In preparing this lecture, we have borrowed heavily from these papers and a recent paper by the senior author (Rabb, 1972). In spite of recognizing the critical need for a comprehensive approach to crop production and pest problems, our presentation will be more restricted than desirable. For pest management to evolve most usefully as a facet of resource management, it is essential for the plant pathologists, weed control specialists, vertebrate pest experts, and entomologists to integrate their activities more effectively. At this time, however, we shall speak primarily in relation to agricultural insect pests, although the same basic ecological principles apply in varying degrees to other types of pest problems.

It is important to clearly visualize pest management in relation to the broad spectrum of pest control strategies and tactics, and perhaps we can accomplish this best by avoiding such terms as pest management, integrated control, chemical and biological control for a few minutes and consider the evolution of pest control programs.

Four principle types of control efforts may develop after the initial recognition of a pest problem (Figure 1). If the preliminary assessment indicates a real or potential problem, emergency use of a control method usually follows. Such methods are designed for the use of individual growers and for temporary alleviation of pest population outbreaks. (For clarity, I shall use "outbreak" to mean any pest population fluctuation above the economic threshold.)

If the pest problem persists, emergency methods may be refined as the pest's interactions with the crop and associated organisms become better understood. Where a temporary suppression method, such as the application of a

1/Professor and Visiting Assistant Professor, Entomology Department
2/Assistant Professor, Economics Department

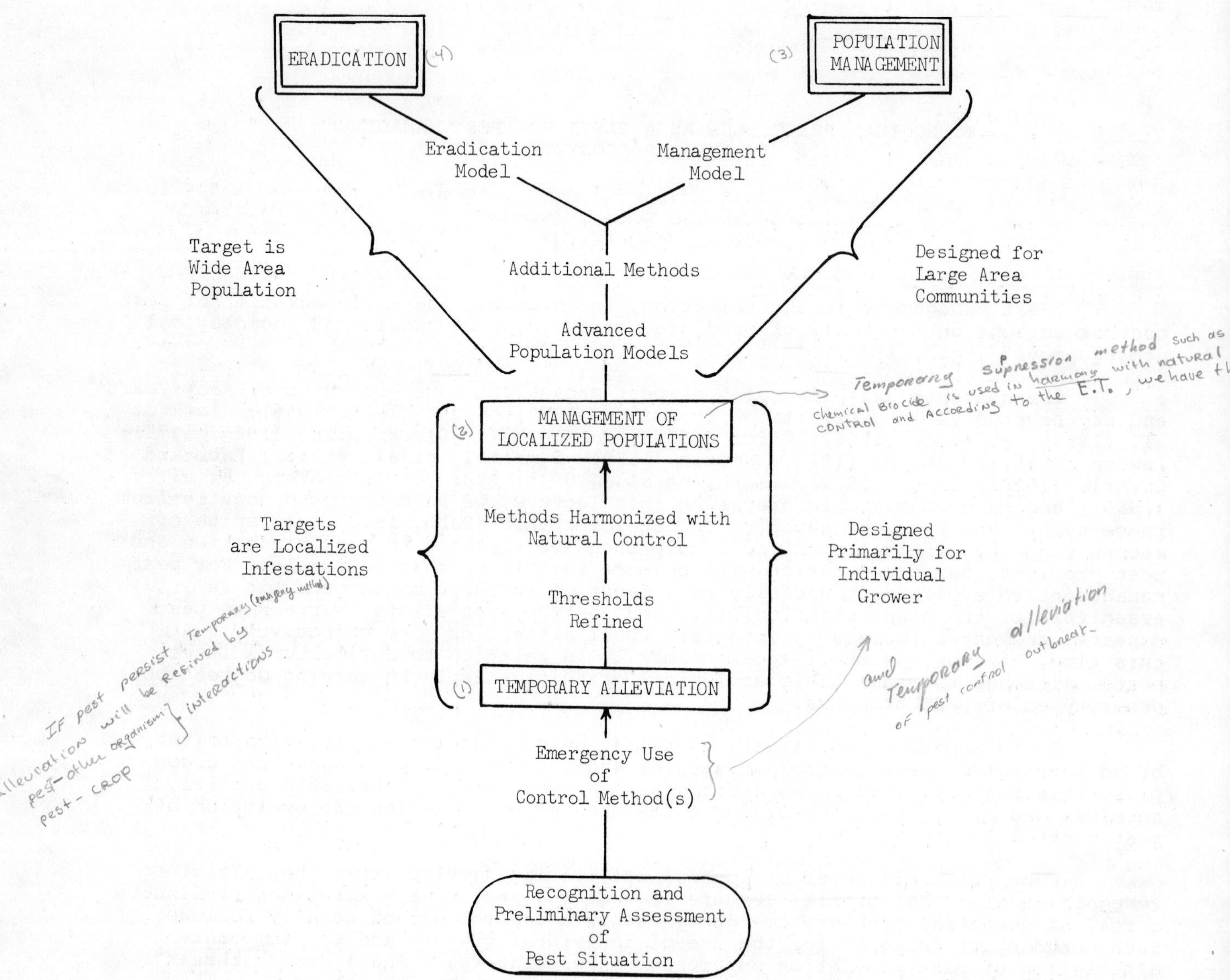

Figure 1.--Evolution of insect pest control actions (from Rabb, 1972).

chemical biocide, is used in harmony with natural control and according to the economic threshold principle, we have the second major category of control activities--i.e., the management of localized populations. In some situations, however, a single method has not been sufficient to achieve control in harmony with environmental quality. In such cases, we attempt to achieve desired results by combining two or more methods of control. Since each pest situation is unique, there is a great variation in the methods which can be combined to best advantage; however, deliberate manipulations of cultural practices and breeding crop varieties for resistance to pest attack are among the most commonly encountered.

Considerable progress has been made in developing means for managing localized pest populations with techniques designed for the independent use of individual growers. However, the more complex and persistent pest problems have not been effectively solved by limited population management. The solution of such problems requires the uniform application of control procedures over wide areas. This shifting of targets from localized infestations to wide area or species populations theoretically provides options: (1) eradication, or (2) more satisfactory population management than possible through a mosaic of independent and unpredictable grower actions.

Eradication and population management not only share the same orientation to the entire pest population, but they also share the same proximate objective--i.e., to lower the mean level of pest abundance so that the frequency of infestaitons above the economic threshold is reduced or eliminated. Eradication and management programs also may require some of the same basic ecological information and may utilize some of the same techniques; however, when applied to a specific pest species, these two goals are obviously incompatible. And, by the same token, population management is incompatible with the poorly conceived idea of a pest-free environment. The public finds it hard to accept the fact that insects as a group, including actual and potential competitors, are absolutely essential to a viable and productive environment.

Eradication is a grave responsibility and should be attempted only after careful study involving diverse perspectives has produced convincing evidence that the benefits to be accrued more than balance the resulting ecological impoverishment caused by removal of the pest species. Eradication of a key pest may make management of a pest complex more practical; however, the step by step eradication of our competitors cannot be accomplished without serious impairment to ecosystem stability and productivity.

Since the population to be eradicated or managed is not contiguous to an individual's property or state or national boundary, a high degree of cooperation among people and governmental units is necessary for success. Certain of the methods required are not practical for the independent use of individual growers and must be the responsibility of specialized personnel. For example, the application of the sterile male technique could not be achieved by the individual citizen.

As in an earlier paper (Rabb, 1972), I deliberately avoided direct reference to chemical, biological and cultural control as well as to integrated control and pest management. I have found that the use of these terms at times tends to segregate us into camps. Although I have been emphasizing the need for multifactorial approaches to our serious pest problems, an objective consideration of the past history and present status of pest control will not fail to reveal invaluable successes of single factor approaches, such as chemical, biological, and cultural methods. Where a unilateral approach has resulted in population management, however, it has been in harmony with natural control processes, regardless of whether the harmony was fortuitous or deliberately achieved.

The terms biological control and natural control are sources of confusion, so I should like to repeat definitions given in a previous paper (Rabb, 1972).

"Biological control is the deliberate use by man of a biotic agent to suppress and/or regulate a pest population. It may be achieved through the deliberate manipulation of environmental factors to enhance the activities of indigenous natural enemies, by the establishment of imported natural enemies, or by inundative releases of biotic agents for temporary alleviation. In contrast, natural control is not a technique, but is the total effect on a given population of all naturally occurring factors, including indigenous natural enemies, which limit and regulate it. These naturally occurring factors seem best considered as agents of natural control unless deliberately used in methods of conservation, importation and establishment, or 'inundative release'."

Biological control specialists can point to many impressive successes (DeBach, 1964 and C.I.B.C., 1971), some of which will surely be discussed in subsequent papers in this Institute. Plant breeders, in cooperation with plant pathologists and entomologists, also have found satisfactory solutions to certain pest problems. For example: the use of varieties of wheat resistant to the Hessian fly not only protected the commodity but also drastically lowered the Hessian fly populations (Painter, 1968 and Pathak, 1970).

These successes attest to the value of traditional single factor methods of pest control; however, all unilateral attempts to solve certain of our pest problems have failed and more comprehensive approaches are indicated. These have commonly been referred to as integrated control or pest management, terms which have been contrasted in "Insect-Pest Management and Control" (N.A.S., 1969) and in a previous paper (Rabb, 1972) as follows:

> There are various views as to how the terms "integrated control" and "pest management" should be used and there is considerable support for synonymizing them. However, "integrated control" at times has been restricted to the modification of insecticidal control in order to protect beneficial insects, and also has been used for the fortuitous combination of practices which, after the fact, were found to harmonize for limited population management. Pest management connotes the deliberate acquisition of ecological knowledge and the application of ecological principles in devising and implementing a program of insect population management. It is based on the recognition of insects as an essential resource; and, thus, is an integral part of resource management. Since management of a pest population sometimes may be accomplished by single factor manipulations, such as the release of an enemy or the use of a resistant variety, and since even eradication may be an acceptable technique in managing a pest complex, pest management seems more appropriate than integrated control to connote the essence of applied entomology (N.A.S., 1969).

Since pest management is an applied science it has no unique basic principles comparable to those characteristic of basic disciplines. However, pest management is characterized by the particular combination of principles it applies. The more important of these may be juxtaposed with three major steps in developing pest management programs: (1) developing an ecological framework, (2) manipulating the ecosystem, and (3) pest management decision making.

Developing an Ecological Framework

The ecosystem:--To fully appreciate the problems associated with the various concepts of pests requires a much broader perspective than we usually have in our day to day specialized activities. These concepts have deep and complex ecological roots, and, of all natural sciences, none requires a more comprehensive approach than ecology.

As we consider an "unwanted" organism or complex of organisms and lay plans for suppression, eradication, or management, we should have the mental agility to easily shift our view in space and time--from the microscopic, we should be able to zoom to the macroscopic, and from the past through the present to the future. It must become second nature to adjust our mental focus from a localized pest outbreak to an entire wide-area pest population, from immediate, ontarget effects of control actions to time-lag effects on nontarget organisms which reverberate throughout the ecosystem.

As a brief preconditioning exercise before discussing specific ecological principles of pest management, let us divest ourselves of our subjective considerations, and view our earth from an astronaut's vantage point. From this detached perspective, as so realistically represented by the color photographs of our Apollo missions, we see the atmosphere, land masses, oceans, and seas of our planet; and, in finer focus, recognize the mountains, plains, forests, grasslands, deserts, lakes, rivers, and other topographic and edaphic characteristics of our earth. Higher magnification exposes the tremendous variety and abundance of plant and animal life involved in the myriad of dynamic ecological and evolutionary processes.

From this hypothetical satellite view, we also can appreciate the energetics of this intricate and complex world ecosystem which involves a continuing series of transformations of the incoming solar radiation-energy transformations which drive our weather systems, biogeochemical cycles, and the various specific ecological processes of central concern to us today.

By extending our hypothetical and detached perspective through time we can see many temporal patterns of change. Geochemical and climatic forces modify the topographical and physiochemical aspects of the earth's crust. Genetic and selective factors modify species composition, with some species becoming extinct and some new ones evolving. Thus, from pole to pole, from desert to tropical rainforest, from the cold mountain spring to the warm, salty depths of tropical seas, characteristic communities of plants and animals evolve and continue to adapt to the everchanging environment.

One of the most characteristic features of this continuing community change is that it is not uniform in rate. Change proceeds much more rapidly at certain places and times than others. Catastrophic events such as earthquakes, severe floods, fire and outbreaks of pests dramatically disrupt segments of our world ecosystem, in some cases to the extent of removing the living components and greatly altering the physiochemical substrate. In such cases, reinvasion of affected areas occurs and a succession of species complexes is set in motion. This succession proceeds rapidly initially but decreases in rate as community homeostasis is approached.

Where do our concepts of pests fit into this picture of our world ecosystem? What criteria should we use for their definition? If, from our satellite perspective we were to designate as pests those species which impair ecosystem productivity, diversity, and stability, we might agree with Corbet (1970) that "the world's major pest is man--the earthpest--whose numbers and activities are threatening the stability of the biosphere."

As Corbet (1970) stated, those engaged in pest management must recognize that the key pest involved is man himself because all other pest situations are direct or indirect consequences of "the need to adjust man's numbers (primarily) and activity (secondarily)--in relation to food, space, the capacity of the environment to recover from exploitation, and time." Time to make these critical adjustments in population and technology is running out, but can be increased to some extent through more efficient crop production. One route to increased production is through more satisfactory control of our competitors--certain species of insects, disease organisms and plants. Pest management is suggested as a valid and useful approach to this complex problem. According to Ray Smith of Berkeley, "consider the ecosystem" is one of the basic principles of pest management. In applying it, we recognize that populations or organisms designated as pests, as well as man, are dynamic subsystems of the world ecosystem which must not be impaired by management practices. As pointed out by Newsom (1967) the effects of practices may reverberate from the site of their application; therefore, both local and general effects must be considered.

The agroecosystem:--Throughout the world, man has greatly altered the ecosystem of which he is a part, and those areas devoted to agriculture have been labeled "agroecosystems" and may be classified according to location and cropping systems.

In each agroecosystem, man uses a variety of energy inputs (Figure 2) to channel as much of the available energy and nutrients as possible into population explosions of prized plants and animals for harvest. Although highly modified, the agroecosystem still maintains its basic trophic structure of producers, herbivores, carnivores, and decomposers, and the agricultural specialists attempt to manage ecological processes somewhat differently at each level. The weed control specialist attempts to prevent invasion and interspecific competition at the producer level, whereas the entomologist seeks to encourage invasion and predation at the carnivore level in an effort to reduce damage from herbivorous insects.

Thus, agriculture is an extremely complex example of applied ecology which cannot be appreciated unless one considers the actions and interactions of the various inputs in terms of the ecological processes affected. Some of these interactions are positive and others are negative in terms of maximum production and environmental quality. In pest management, we seek to influence the choice of production practices so that ecological processes such as invasion, succession, competition, symbiosis, parasitism, and predation are managed optimally at each trophic level.

The life system:--The life system concept is to be discussed in another chapter, so we shall have little to say about it. However, it cannot be ignored in any general discussion of ecological principles. Although the life system concept is considered by some to be a truism (Varley, 1968), it is still a focal point in the study of population dynamics. Clark, et al. (1967) consider a life system as "composed of a subject population and its effective environment which includes the totality of external agencies influencing the population, including man. . . ." Thus, for those of us involved in pest management, where our emphasis is usually directed to the behavior of a single population, the life system approach is invaluable in the development of a conceptual framework.

In complex systems, such as those often observed in biology, it is impossible to grasp the pathways and relative importance of even the major factors without such a conceptual framework. In fact, the results of such relationships are often counterintuitive (Forrester, 1969). Even though the development of high speed digital computers and the adoption of systems analysis techniques as tools for analyzing biological systems have greatly enhanced our knowledge, such techniques are still dependent on the life system approach.

The unfortunate reluctance of many entomologists in accepting the usefulness of systems analysis is, to a great extent, we feel, due to misconceptions as to what systems analysis entails. One might call this the "black box syndrome." We feel this is unfortunate because, although techniques utilized may be highly complex, the concept itself is not. Simply, systems analysis has two basic components: conceptualization and quantification.

Conceptualization of the system involves the development of a pictorial representation of the variables and interactions believed to be involved in the

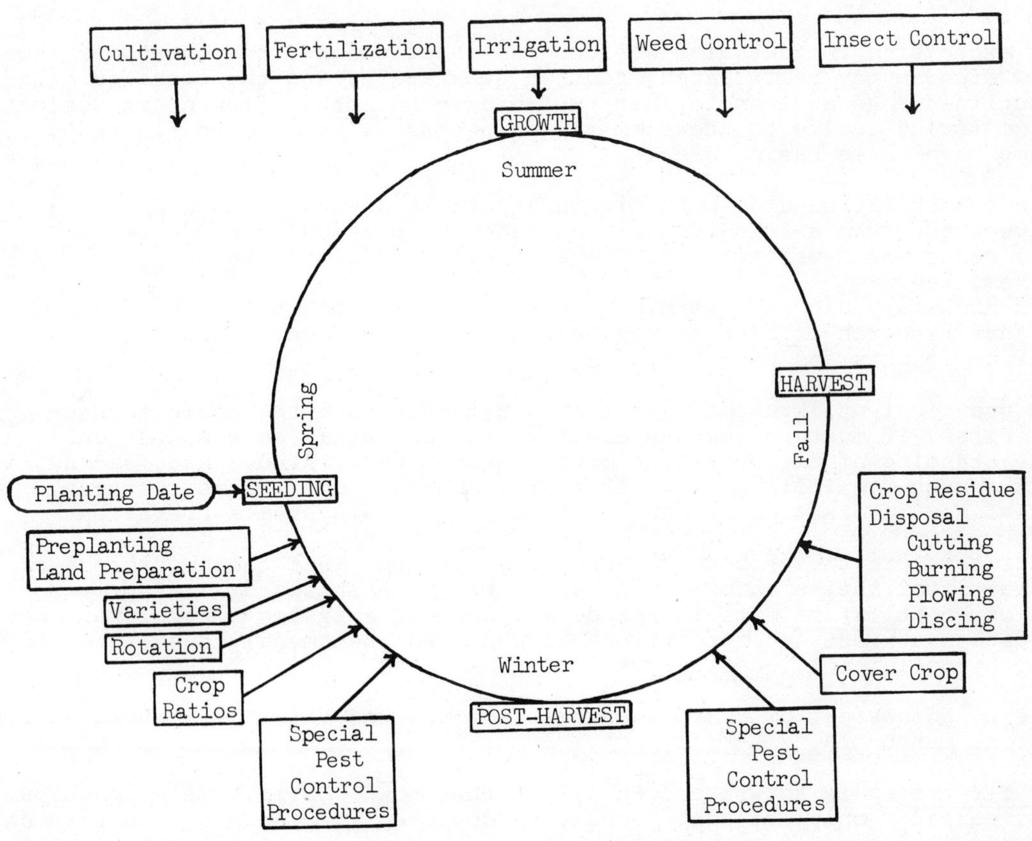

Figure 2.--Inputs of crop production.

system. Initially, this often takes the form of pictures of the variables and arrows between them to designate interactions (Figure 3). In pest management, we must start with the biology of the insect. The conceptualization process then continues to a more formalized flow chart or diagram (Figure 4) in order to more clearly depict the interactions.

There are several ways of approaching the analysis of a system. Ideally, flow diagrams should be set up with the different stages of the population as the major stage variables and then all other factors as modifying parameters. However, certain stages may be difficult to monitor or the modifying parameters may vary so greatly seasonally, that it can often be of more practical use to modify such flow diagrams to seasonal populations flows. For example, the initial flow chart utilized in our studies of Heliothis zea is shown in Figure 5. Of course this represents only the state variable, population size, at different times when it can be accurately estimated. For simplicity, the various modifying factors have not been included.

For expressing the detailed actions and interactions, components of the overall flow diagram may be isolated, redrawn separately, and the detailed relationships outlined. As an example, our preliminary Heliothis flow chart depicting the flow from spring adults to the summer populations is modeled in Figure 6. Such flow diagrams have three basic functions:

(1) as a communication aid, both within a pest management team and between the team and individuals outside the particular project;
(2) as a basis for development of a quantitative model for the system; and
(3) as a necessary aid in determining where information is lacking, so that research priorities may be continually reordered for optimum results and progress.

For practical utilization, any flow chart must have two basic characteristics. First, it must reflect as clearly and accurately as possible our present understanding of the operation of the system under study; and, second, it must contain enough flexibility, such that it is easily modified as research progresses and new information alters our concept of the system's operation.

It is here, in the process of conceptualization, that the importance of highly accurate qualitative information cannot be overemphasized. The proper utilization of quantitative data in the development of a system's model must have as its foundation correct interpretations of observed phenomena; these interpretations are, by definition, qualitative.

Once an adequate flow chart is developed, the 2nd phase of systems analysis quantification, can begin.

The first step is to write down all of the known possible relationships. For example, consider only the adult longevity depicted in Figure 6. In this case, adult longevity is dependent on weather, natural enemies, genetic factors, and nutrition.

The next step is to examine these relationships and separate out those for which biological mechanisms are known. Constants in these relationships can then be determined from data by iterative or least-squares techniques. Next are separated out those for which information is available from regressions or correlations. What remains are the relationships about which little or nothing is known. It is these relationships which must receive the greatest research attention if we are to optimize the progress in our understanding of a particular life system. It is not until these unknown relationships are somewhat defined that we can understand the life system well enough to obtain at least crude predictions of population trends.

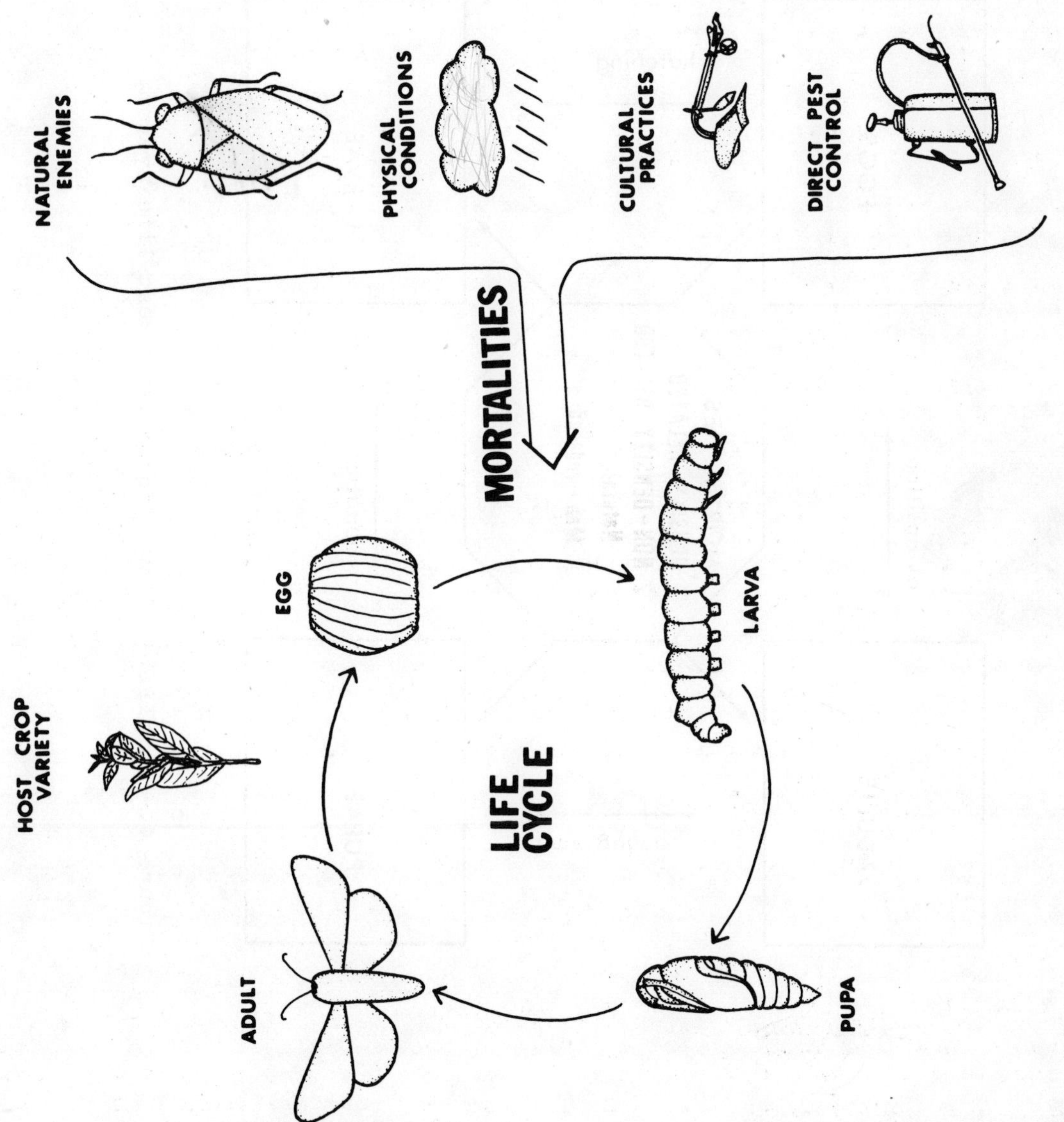

Figure 3.--Initial conceptualization of a pest life system.

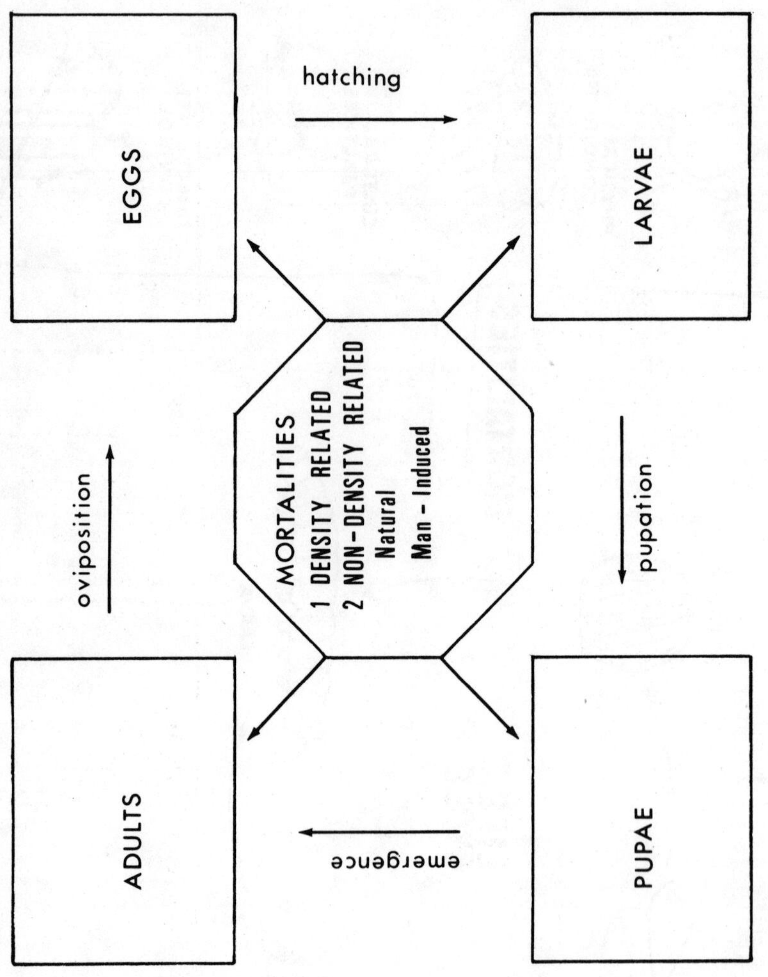

Figure 4.--Simplified flow diagram of a pest life system.

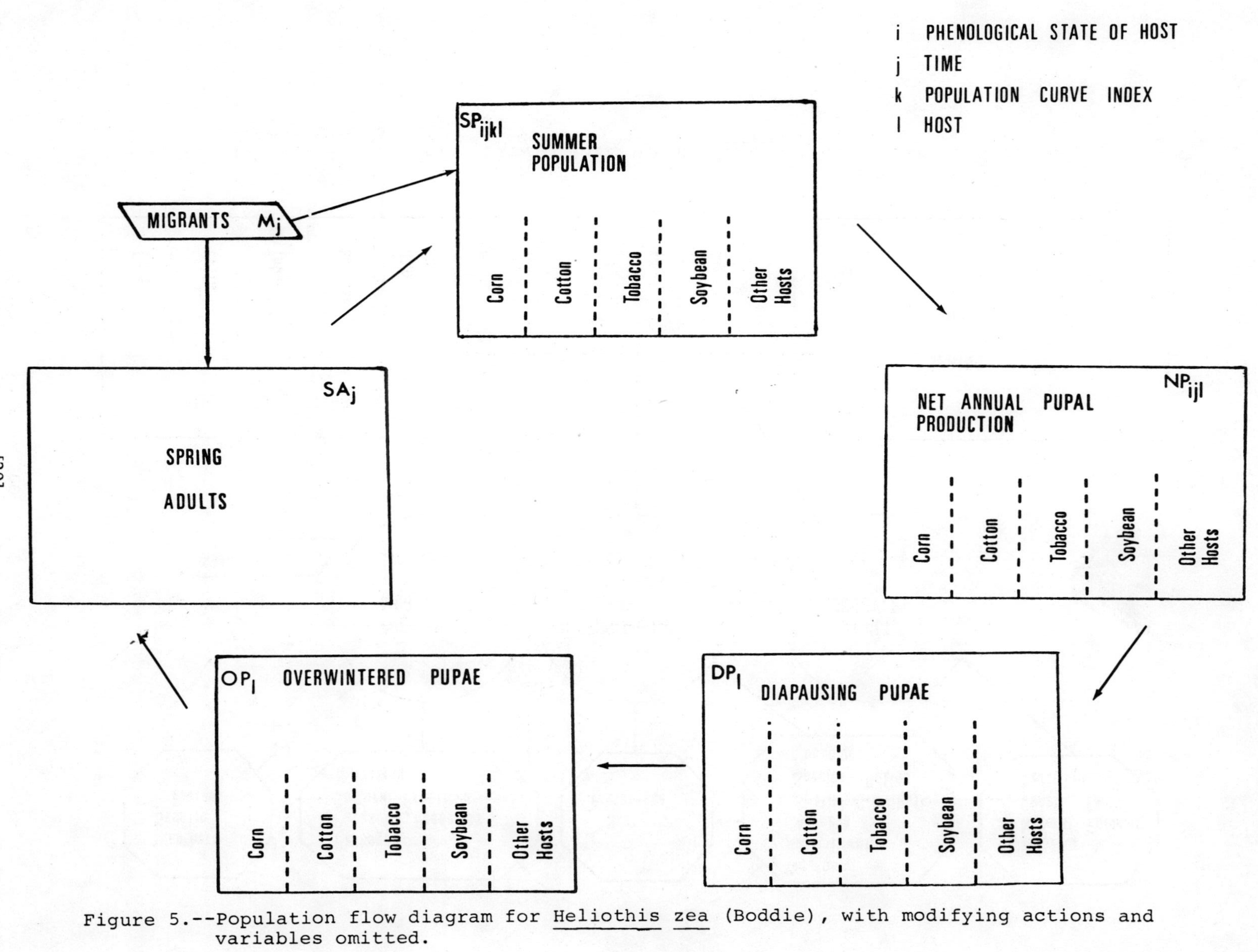

Figure 5.--Population flow diagram for Heliothis zea (Boddie), with modifying actions and variables omitted.

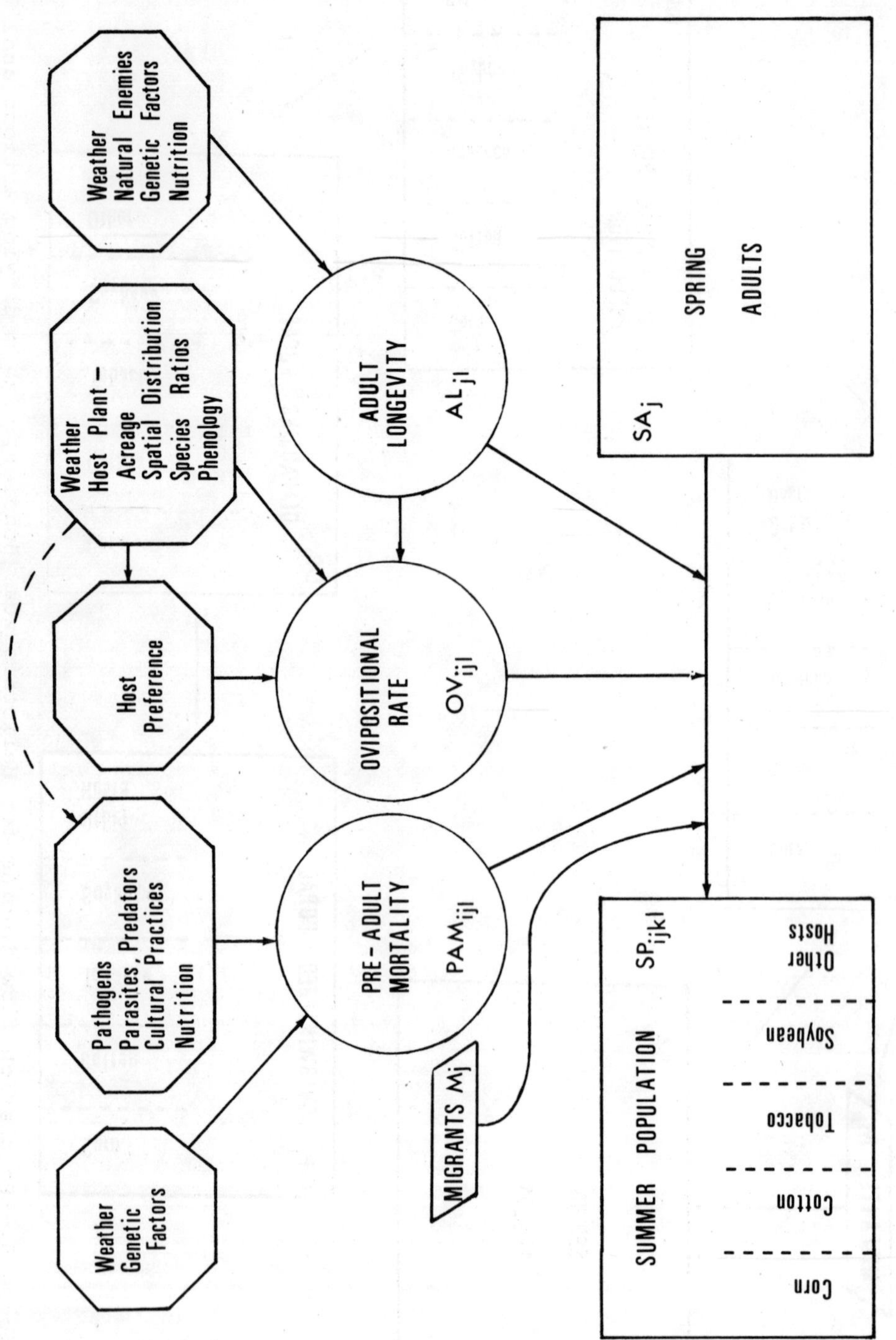

Figure 6.--A partial population flow diagram of _Heliothis zea_. spring adults to summer population.

In the opinion of some scientists, no model is completed until it is totally mechanistic. Although the level of detail at which one works depends initially on the present state of knowledge, from a practical viewpoint, the level of detail of a completed model must be determined by the relative importance of the various relationships to the overall population's system, the degree of accuracy necessary for simulation and prediction, and the funding available. From an economic viewpoint, there is little justification for refinement beyond that required for the accuracy of population prediction necessary in the economic decision-making (Scientifically, of course, there is much justification for as great a degree of understanding of the system as possible).

Although obvious, it is worthwhile to point out that no mathematical or statistical manipulations can increase the accuracy of data. Since any model or prediction will be at most only as accurate as the data upon which it is based, careful consideration should be given to planning research and sampling such that information be as unbiased as possible.

In short, the concepts of the life system and the use of systems analysis are merely extensions of the research process which allow us to more formally define the particular research problem, to determine the objectives of optimum benefit, and to obtain out objectives with a minimum of funding and effort. One could view it as procedures for optimizing the cost/benefit ratio of research.

Population dispersal:--Feasibility of a management system is strongly influenced by the dispersal characteristics of pest organisms and their natural enemies. Population dispersal is so important that it will receive special treatment in a subsequent chapter. I only wish to call attention to its significance in establishing boundaries for a management area.

If management is to lower the mean level of pest abundance, the management practices must be applied uniformly throughout a designated area. Defining the size and shape of this area is one of the most difficult problems confronting the pest management specialist and is directly related to dispersal.

Pests differ markedly in dispersal characteristics as well as in their ecological and geographic distributions. Ideally, the management area might coincide with the pest's distribution; however, this will rarely be practiced because of political and sociological constraints.

When the total land area occupied by a pest species is viewed in broad perspective, its distribution may be divided roughly into population "pockets." Movement between pockets is restricted to various degrees by ecological factors, and, in extreme cases, ecological barriers may provide a high degree of isolation between these pockets or demes of a species population. In such cases, management areas may be defined by ecological barriers to dispersal. The coastal valleys of Peru comprise outstanding examples of ecologically fragmented distributions of many organisms including some well-known pest species. Each valley is essentially isolated by surrounding desert and can be managed with a minimum of attention to invaders dispersing from adjacent valleys. The ecological characteristics of these isolated valleys were ideal for some of the first practical demonstrations of pest population management by Wille (1951) and associates.

Dispersal is of particular significance in biological control where the lack of vagility might be a factor limiting the practical effectiveness of a particular natural enemy. In such situations, a natural enemy with a low rate of dispersal might be artificially inoculated into a field infested with its prey to enhance biological control. Huffaker and Kennett (1956) demonstrated this technique by inoculating strawberry plantings with predatory mites and thus obtaining control of certain phytophagous mite species.

In some parts of the world, i.e., in the eastern United States ecological barriers do not fragment the species distributions of many of our pests into areas of convenient size and shape for management programs. As a result, we are now grappling with the problems of achieving the wide-scale cooperation among communities and states necessary for success. Some pilot pest management programs have been tried or are underway where the management area has been arbitrarily set without regard to ecological barriers. In such cases, it should be clearly understood that the full effect of management practices will be felt only toward the center of the area and the degree of effect will be in proportion to the rate of pest dispersal.

Population levels and damage:--In simplest terms, the objective of pest management is the maintenance of pest populations below levels causing economic damage, discomfort or inconvenience, through the manipulation of various factors. The relationship between physical damage or discomfort and various density levels of pests must be determined in evaluating a pest problem and as a part of the basic ecological information required for devising a management program. Such information is essential for the economic considerations to be discussed presently under "Pest Management Decision Making."

Where a pest is associated with a crop throughout the growing season, the significance of a given pest density varies as the plant grows, flowers, and matures; therefore, a given pest density may be innocuous at one time during the season but damaging at some different stage of the plant phenology. The density-damage relationship of a pest and its host crop also varies geographically and is affected by variations in weather, host plant factors, and natural enemy activity. Additionally, the problem of evaluating density-damage relationships becomes even more complex when several species of pests are involved contemporaneously or sequentially.

Thus, studies of population levels and damage must comprise a continuing part of a pest management program. Initial rough approximations must be refined and modified as both the ecological and economic dimensions of the problem become more fully understood and change with time.

Manipulating the Agroecosystem

Many individuals have contributed to our knowledge of the principles involved in manipulating various factors in the agroecosystem for pest suppression and regulation, and many texts are available which discuss these principles and give examples of their application. Dwight Iseley of the University of Arkansas, however, deserves special recognition, and his text (Iseley, 1948) is still an excellent guide to ecologically sound management practices. The National Academy of Science has provided a more recent text (N.A.S. 1969) in which references to specific examples of the application of ecological principles in insect-pest management can be found.

Within a properly defined and understood ecological framework, the pest management specialist can devise various scientifically feasible management programs. For convenience, the chief principles of value in developing such programs may be discussed under the following eight headings: (1) identification of real and induced pests, (2) use of indigenous natural enemies and maintenance of subeconomic pest populations, (3) controlling diversity, (4) manipulating food, (5) manipulating soil and water, (6) introduction of specific control agents, (7) prevention of new pest invasions, and (8) integration of management inputs.

Identification of real and induced pests:--In many serious pest situations, the pest complex has been highly modified by ongoing control actions which disrupt natural control. The well-known phenomena of resurgence and outbreaks of secondary pests have been observed many times following applications of certain wide-spectrum toxicants. In such situations, induced pests may appear more important than real pests, and an early step in developing pest management is to identify accurately the real and induced pest species in the pest complex (DeBach and Huffaker, 1971).

The real pests are those against which suppression methods were directed initially and which cause unacceptable loss or discomfort on a continuing basis. In contrast, the induced pests are those whose populations rise to unacceptable levels as a result of their natural enemies being reduced by control actions directed toward the real pests. In many cotton growing areas, for example, the cotton boll weevil is a real pest and the cotton bollworm an indirect pest whose populations become damaging as a result of early-season applications of insecticides directed toward the weevil.

Use of indigenous natural enemies and maintenance of subeconomic pest populations:--Most species of plant feeding insects are held at relatively low levels, particularly in undisturbed ecosystems, by a complex of natural control factors. In the disturbed agroecosystem, the pest management specialist seeks to manipulate these factors to lower and stabilize pest populations. Certain of the natural factors, particularly physical ones, can be manipulated to lower pest densities but do not function well for stabilizing the populations at low levels. On the other hand, natural enemies do provide a degree of stability to pest populations because they are geared to their hosts' densities through feed-back mechanisms (Huffaker, 1971). The protection and enhancement of natural enemies is therefore an important management principle.

One of the most obvious means of protecting natural enemies is by the use of selective agents and techniques for suppressing pest outbreaks. Another method is the "dirty-field technique," which is difficult for many farmers and agricultural specialists to accept, and which can, in some situations, cause other problems. This technique, however, is based on the ecologically sound concept that we must manage our insect competitors on a continuing basis rather than attempt to eliminate them entirely. Since natural enemies are among the factors making it possible for us to manage our pests and are themselves dependent upon the pests for their existence, we must maintain subeconomic host populations as food for parasites and predators.

Controlling diversity:--Perhaps the greatest contrast between agroecosystems and undisturbed ecosystems is in diversity, for much of our farmlands is routinely stripped of all vegetation and planted to a single variety of a crop species. The pros and cons of monoculture and polyculture and the relationship between diversity and ecosystem stability continue to be of interest to agriculturists and theoretical ecologists (B.N.L. 1969). There is considerable support for the concept that a shift toward more complex polycultures would be advantageous in developing more satisfactory pest management programs. However, Southwood and Way (1970) remind us that the deliberate addition of diversity may or may not increase pest population stability, depending on the type of diversity added. They make the important distinction between diversity among species at different trophic levels. For example, adding an alternate host for a pest might be advantageous to the pest and thus a counter-productive management practice. On the other hand, adding an alternate host for a natural enemy might be an excellent management practice. Thus, any manipulation of diversity should be assessed separately for the pest and its natural enemies. This is true both when manipulating diversity spatially by the deliberate choice of crop mixtures and by controlling diversity temporally through rotations.

Manipulating food:--Some of our most effective management tools are techniques for altering the quality and availability of food. Two full days of this

Institute will be devoted largely to genetic methods of altering food quality, and speakers will discuss specific accomplishments in breeding varieties of various crops for insect resistance. Food quality, however, also can be manipulated through fertilization, growth regulators, and irrigation.

The availability of food for pests can be regulated to some extent by varying planting dates, as illustrated by the classic example of using fly-free planting dates for winter wheat in areas infested by Hessian flies. Additionally, the timely removal or alteration of crop residues may comprise an effective management practice. For example, the overwintering population of the tobacco hornworm can be substantially reduced by the immediate and uniform destruction of tobacco stalks after harvest.

Pruning and defoliation, the destruction of alternate hosts of pests or in some cases the use of alternate hosts as trap crops, the provision of food plants in fence rows for alternate hosts of important natural enemies, and strip-cropping are other techniques for manipulating food quality and availability.

Manipulating soil and water:--In certain cases, direct mortality can be inflicted on pest populations through the manipulation of soil and water, provided such actions are implemented on the basis of intimate knowledge of the pest's ecology. For example, fall plowing causes heavy mortality of the potentially overwintering populations of corn earworms, and irrigating or withholding irrigation at specific times can be effective in controlling certain species of wireworms in the western United States. Additionally, soil and water manipulations can be used in certain cases to enhance natural enemy activity.

Introduction of chemicals, natural enemies, and other specialized agents and techniques:--At least the major cause and effect pathways among the chief ecosystem elements must be understood before using a chemical, introducing a natural enemy, or utilizing some specialized technique such as light traps or sterile male releases. Since pest management deals chiefly with the integration of control agents and techniques, a discussion of principles involved in developing the specialized single factor control methods is beyond the constraints of our topic. It should be noted, however, that the application of basic information from chemistry, physics, toxicology, physiology, behavior, genetics, microbiology, ecology and other disciplines in developing specific control agents is essential to effective pest management. Those principles of particular relevance in developing and utilizing parasites, predators, and pathogens will be discussed at length by other contributors to this Institute and have been reviewed recently by Huffaker (1971), DeBach (1971) and Cameron (1971).

Prevention of new pest invasions:--A particular pest situation may be accentuated and complicated by a new pest invasion. In addition to direct damage problems, in certain cases, a new pest might negate largely a previously satisfactory management program if its arrival triggers actions which interact negatively with existing practices. Therefore, the exclusion of pests from new areas is both practically and theoretically a sound management principle.

Many of our worst pests originated in other lands, and our quarantine specialists face the difficult task of preventing the introduction and establishment of additional, unwanted species. Their problems have intensified with the exponential expansion of international transportation, and the effectiveness of their efforts is difficult to assess. There is some rationale to suggest that our pest problems might be even more severe than they are if quarantine measures had not been initiated and were not continuing. Nevertheless, perhaps more of the quarantine resources might be used to advantage in learning the ecological relationships of potential invaders so that more satisfactory control and containment practices might be initiated when invasion occurs (N.A.S., 1969).

Integration of management inputs:--For certain pest problems one or more control actions may be integrated into the total crop production system with ease and with little evidence of negative interactions. Fortunately, there are examples of control where this integration has been accomplished either by trial and error, fortuitously or deliberately. Our more difficult and complex problems, however, particularly those involving several different crops and large acreages stretching throughout widely differing geographic areas may require more sophisticated analytical techniques to optimize the various inputs into the management system. Here, again, the techniques of systems analysis may help (Adkisson, 1970; Campbell, 1971; and Watt, 1963, 1968, 1970), both in developing scientifically feasible management systems and in making decisions as to their implementation.

Pest Management Decision Making

For this discussion, let's focus on the economic factors which are coupled with ecological considerations in making two major types of pest management decisions: (1) decisions made by individual producers, and (2) group decisions.

First, consider individual producer pest management decisions: In the simplest case, a producer can be expected to expend resources for pest control until a dollar expended will return a dollar of additional crop revenue.

Figure 7 illustrates one way of viewing how the economic and ecological factors are combined to determine economic injury thresholds, which are population levels calling for appropriate pest control measures to be initiated. Obviously pest density information (the left triangle) is essential in calculating both crop revenue and control costs. As shown in the upper left of the figure, the relationship between pest density and plant damage (quality and quantity losses) is strongly influenced by climate and biological factors such as host susceptibility, natural enemies, etc. This relationship also was mentioned earlier.

Now let's look more closely at the three major factors influencing the action threshold depicted by the triangle to the right.

First, the producer's attitude toward pest damage risks--shown in the flow as Producer Risk Considerations--varies among producers and will be most evident for producers who have limited off-farm employment opportunities, little crop diversity, and limited credit possibilities. The producer's aversion to risk is an important factor involved in decisions and may account for much of the observed difference between advised treatment thresholds and actual pest control practices (Carlson, 1970).

Secondly, we see that the crop income--pest density relationship (center of figure)--is the route through which quality and quantity losses become inputs to the action threshold.

Finally, at the bottom of the figure, control cost becomes another input to the action threshold through the control cost--pest density relationship.

These last two relationships between pest density and crop income and pest density and control costs are depicted in Figure 8. Although this is a hypothetical representation, the shapes of the curves follow from observation and logic.

Net crop income, as depicted by the sold line in the graph, usually will fall faster than proportionally with pest density above some crop tolerance level represented by N_1. This is due to increased commodity quality and synergistic

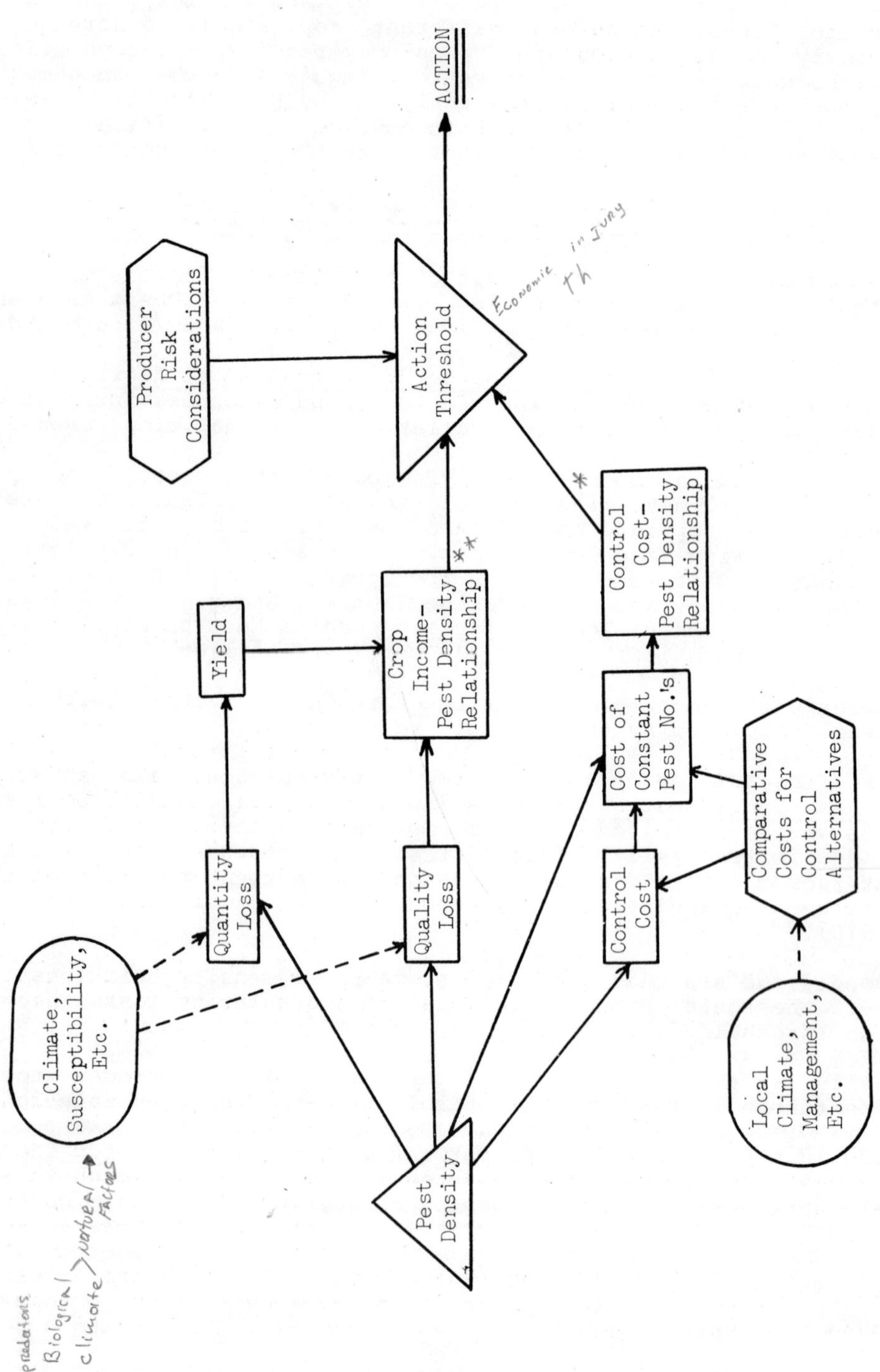

Figure 7.--Economic and biological components of individual producer pest management decisions.

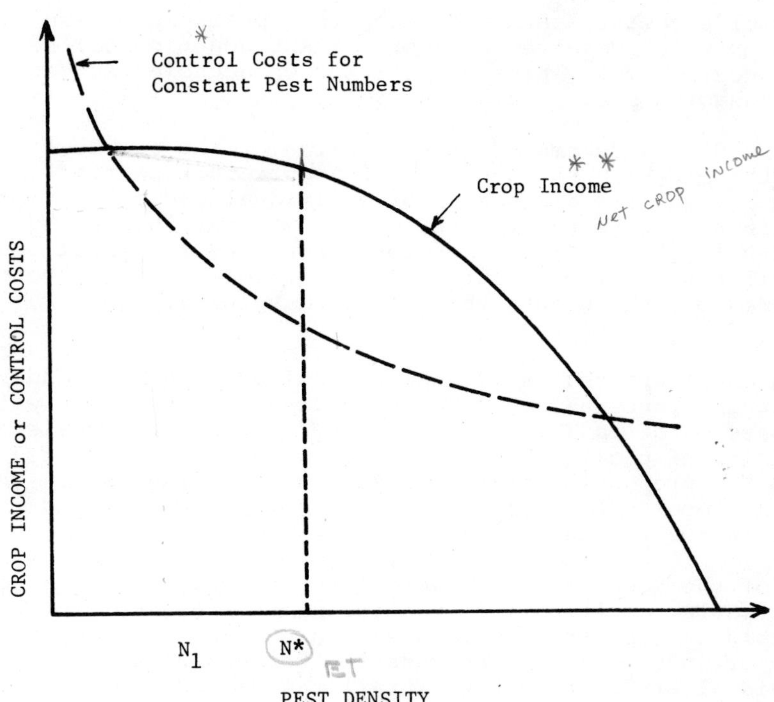

Figure 8.--Crop income and control cost as related to pest density.

damages as pest density increases. On the other hand, control costs, represented by the dashed line, will fall as fixed pest management resources are spread over more pests in a given area. However, control costs will usually decrease less rapidly at higher pest densities as it becomes more and more difficult to suppress higher and higher proportions of a pest population, as shown by the leveling out of the dashed line. Another way to view the same thing is to say that pest management resources yield diminishing returns when more and more resources are used while other productive inputs are held constant.

These cost and benefit relationships when tempered by the first factor mentioned, that is the producer's particular attitude toward risk, will determine the economic threshold, N*. This is the pest density at which the slopes of these two lines are equal (Headley, 1972a).

Earlier, the importance of continuous study of pest density--damage relationships was stressed. Pest densities in later periods depend upon the amount of suppression in earlier periods and side effects of suppression methods. Over time, control costs to reach a given pest density may rise as secondary pests emerge and pesticide resistance develops. For a producer to maximize the returns from pest management assets, he must consider their future as well as their current productivity. For example, it may be more profitable to sustain slight crop losses in current periods in order to slow the rate of resistance development. Thus, the long run action threshold is influenced by current control costs, time costs as reflected by interest rates and the costs of substitute pest management resources (Carlson and Castle, 1972, Headley, 1972b).

Pesticide resistance may develop in a pest population even if a given producer attempts to preserve the nonresistant genetic pool because of the actions of other producers. This, plus other interdependencies suggests a regional approach to managing pest populations.

Figure 9 is a schematic that illustrates the complexities of group pest management--the block in the center. The upper left part of the figure is brought forward from our earlier discussion of individual decisions. But individual producer's willingness to pay for pest control may vary widely due to differences in local pest density, control costs, crops grown and financial postures. In general, the more similar the cropping pattern and the more uniform the threat of pest damage in a region, the higher the chance of voluntary cooperation and group action.

With mobile pests, individual control efforts are rendered less effective. Climate, cropping patterns and natural geographical barriers will influence the rate of dispersion of an infectious disease or pesticide resistant pest variety. The productivity of the total pest control resources of an area may also be quite sensitive to the proportion of the producers participating in the regional program. This may be an important feedback in the decision process and is indicated by the dashed arrow in the figure.

One of the most common arguments for regional pest management is the economies of scale resulting from monitoring or treating many producers' crops jointly. (This will be the major way in which regional control costs will differ from the sum of individual expenditures). For example, it would be very expensive for an individual producer to have a tailored weather forecast to help predict pest population levels, but the same service to many producers may be quite profitable (Carlson, 1970). When pest management services are difficult to exclude from producers, these are known as public goods and there is an economic argument for their collective provision. (Examples are weather information, quarantine protection and controlling mobile pests). Many of these economies of scale and exclusion characteristics depend upon geographical contiguity--hence the emphasis on regions for control (Tullock, 1969).

The Regional Control Program Group (center of Figure 9) will have to deal with other groups of people interested in environmental quality, agribusiness sales, tax, and regional income and employment, because when a group of producers act collectively, they can influence product prices and many other things of interest to the general public. Consequently, bargaining among and within these groups may take place on the various administrative features of the program such as:

(1) revenue generation procedures (taxes, user fees, revenue sharing),
(2) restrictive production practices (seeding and harvest dates, production moratoria, crop residue treatment, pesticide use permits),
(3) voluntary or compulsory participation (which commodities or pests are excluded),
(4) enforcement procedures (penalties, liability bonds, bounty system), and
(5) size of control region.

When pests are very mobile and there are significant economies of scale, larger regions will be favored. However, the larger the region, the more likely that individual producer demands will differ and the more likely mandatory compliance will be necessary. Compulsory participation is also more important when control effectiveness increases at an increasing rate with included acreage.

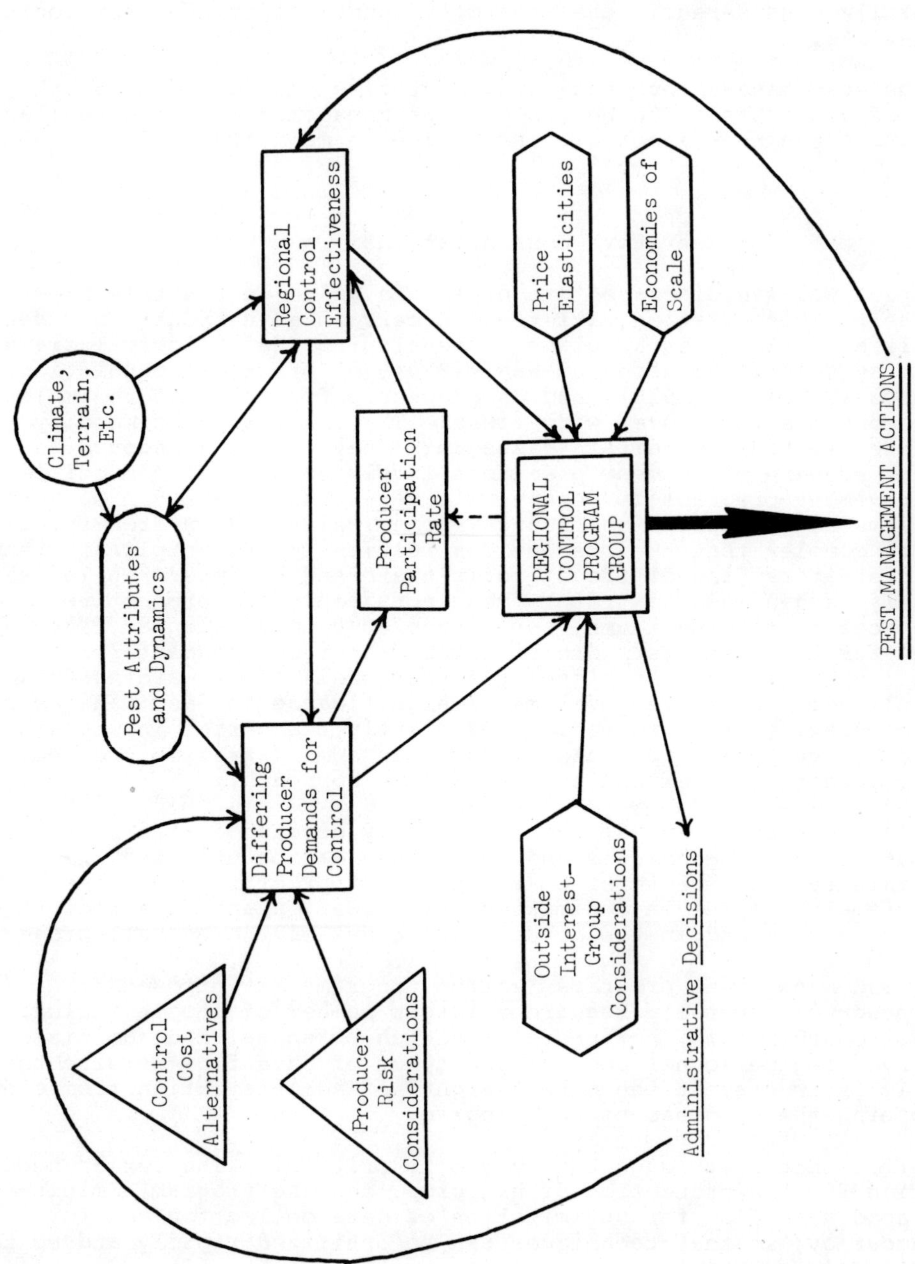

Figure 9.--Regional pest management decisions.

As Figure 9 indicates, the administrative considerations will influence individual producer demands. The control actions taken will affect control effectiveness. Thus, it is evident that regional pest management raises many questions that were absent in working with individual producers. There must be closer and more effective collaboration among entomologists, economists, producers and agribusiness personnel in arriving at socially desirable decisions.

The problems of group decisions, people management and value judgments have not been generally considered in the area of responsibility of entomologists as biological scientists.

However, the pest management specialist must give them increasing attention, because many of the costs and benefits of pest management are indirect and shared, and those which are non-market are increasing in magnitude and significance.

Administrative Considerations

The principles we have discussed are of little value unless they are implemented, and their implementation will require certain adjustments in research and extension. More attention must be given to developing the ecological framework for management systems (Southwood and Way, 1970), on practical population monitoring techniques (Gonzales, 1970), and on producers for the uniform implementation of management practices over wide areas. The traditional commodity approach, while quite fruitful in certain cases, will have to be abandoned for pest complexes which require management practices involving several commodities. Full-scale pest management programs obviously will be expensive and can be justified only for problems of greatest significance economically and environmentally. Such problems are so complex that their solutions require interdisciplinary team efforts which administrators find difficult to initiate and execute. While team projects are urgently needed, administrators must not jeopardize productive individualized projects in forming teams, for many of our most talented researchers are not well-suited for team efforts. Creating the proper environment for the formation of productive teams and establishing a fair basis for rewarding team members are difficult administrative problems of significance to pest management. Additionally, the complexity of coordinating efficiently the varied inputs and outputs of a comprehensive team effort also present an administrative problem, but it can be alleviated to some extent by the systems approach.

The use of systems analysis has been previously discussed with regard to actual research, but the same techniques and approaches can be utilized in developing the administrative aspects of pest management programs. Systems analysis can be extremely helpful in optimizing communication and data flow through the administrative hierarchy and in the actual management of the overall project.

Area-wide pest management programs (action programs, as opposed to a strictly research program) generally require a larger number of people including scouts, supervisors, coordinators, peripheral research personnel and administrators. Complex problems involving personnel and various types of data flow require the use of systems analysis to prevent the mere "weight" of administration from slowing or totally stopping the progress of the program.

Rapid information flow through the system is critical. The farmer needs immediate information for the protection of his crop, and the program administration requires rapid data flow for optimization of data collection and for determining data accuracy, so that techniques can be optimized rapidly and so that personnel may be given instructions or corrections with as little delay as possible. Furthermore, formal lines of communication must be provided between the extension and research personnel involved. These exchanges are of utmost importance if such programs are to evolve in a meaningful and economically practical manner.

The first stages of administrative systems analysis have been undertaken in the Tobacco Pest Management Program presently in operation in North Carolina. As an example, the data flow diagram from their annual report (Ganyard, et al., 1972) is given in Figure 10. It has been slightly modified by the addition of the flow from state extension specialists to research personnel, since the original diagram was concerned only with information compilation. In any viable pest management program, a reciprocal flow of information between research and extension must be maintained with a minimum time lag. The total time for data flow process throughout the system is only one week; and, as can be seen, the grower receives a scout report of his fields on the day of the survey. The very important area of data accuracy is included in quality control listings and error listings, which are returned to the scout supervisor since he has direct responsibility for correcting personnel and data collection problems at the local level (thus avoiding delay in any feedback through the administrative hierarchy).

This particular program has not as yet gone deeply into an analysis of the administrative system, but even the use of flow diagrams has greatly aided the program personnel in determining and correcting for many of the weaknesses and bottlenecks in their data flow system.

Each individual within the program receives the information which he requires in its most useful form and in a minimum of time; the flow of administrative information is optimized such that the funding and effort can be spent towards realizing the goal of the program rather than in administrative aspects.

Tax supported research, extension programs and private enterprise exist in a dynamic relationship which is particularly unsettled now with respect to pest control because of changing economic and environmental priorities. The coordination of the various governmental agencies, business organizations, and private groups with vested interests in pest control is a formidable task, but must be achieved.

Man faces the apparent paradox that his survival depends both on his ability to compete successfully with other species and his ability to do so without eliminating them. We cannot eliminate pest problems in the context of a viable environment, but we can learn to handle them satisfactorily as an integral part of effective resource management.

Acknowledgments

Discussions with various colleagues in the Department of Entomology, North Carolina State University, comprised an influential background for the preparation of this paper. Conversations with Drs. J. R. Bradley, W. M. Brooks, W. V. Campbell, F. E. Guthrie, H. H. Neunzig, and W. J. Mistric were of special value although this acknowledgment does not imply their complete endorsement of the paper's contents. In addition, the authors wish to recognize the support of the North Carolina Agricultural Experiment Station and the National Science Foundation (Grants GB-28855 and GZ-1899).

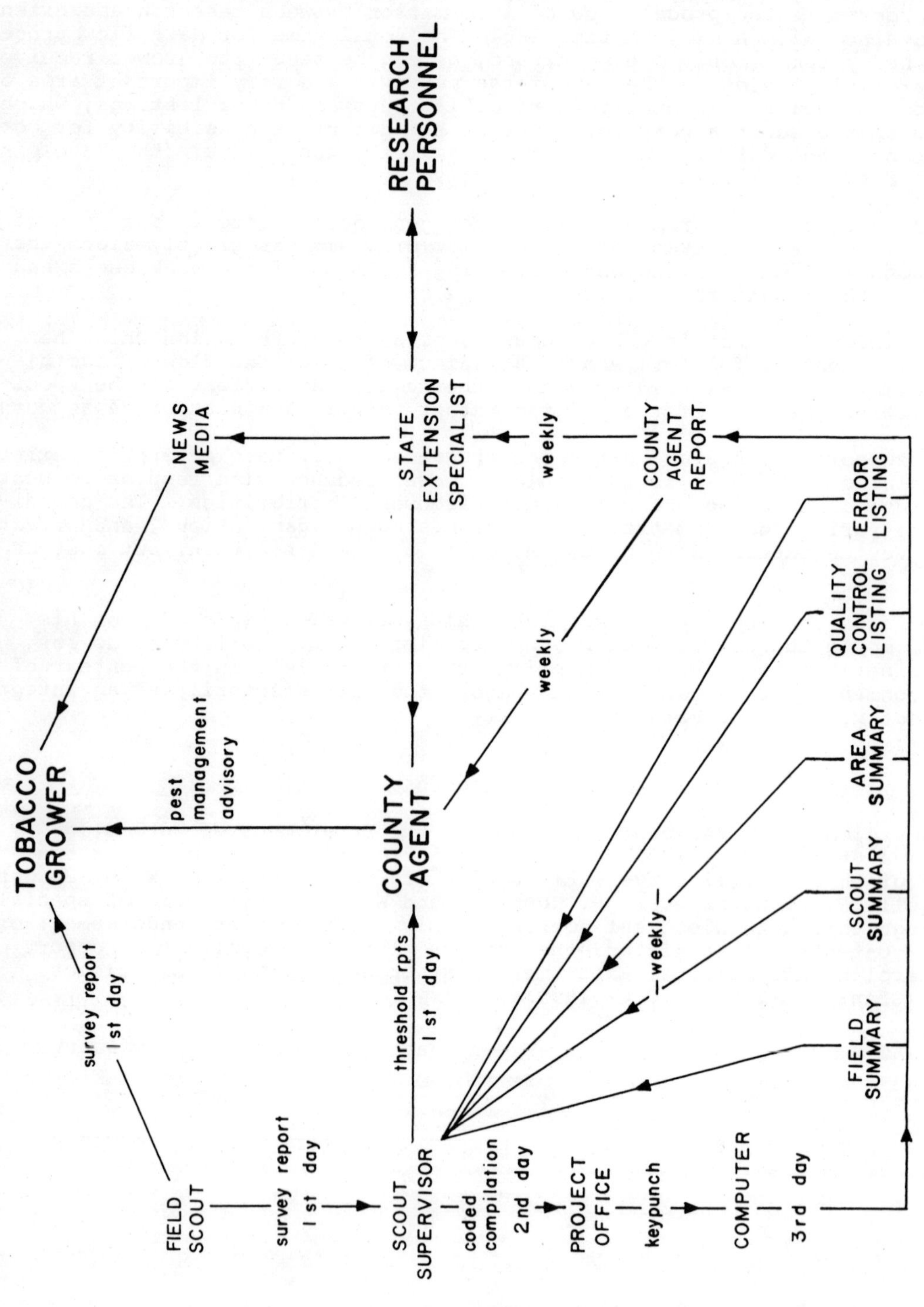

Figure 10.--Tobacco Pest Management data flow system. (Modified from Ganyard, et al. (1972).

References

Adkisson, P. L. 1970. A systems approach to cotton insect control. Proc. Beltwide Cotton Production and Mechanization Conf., National Cotton Council. Houston, Texas. Jan. 8-9, 1970.

Beirne, B. P. 1967. Pest Management. Leonard Hill Books, London.

Brookhaven National Laboratory. 1969. Diversity and Stability in Ecological Systems. Brookhaven Symposia in Biology No. 22: 1-264.

Cameron, J. W. M. 1971. Insect pathogens and their potential for regulation of insect populations. Proc. Tall Timbers Conf. on Ecological Animal Control by Habitat Management, No. 3. Tallahassee, Florida. Feb. 25-27, 1971. pp. 267-277.

Campbell, R. W. 1971. Developing a pest population management system. Proc. Tall Timbers Conf. on Ecological Animal Control by Habitat Management, No. 3. Tallahassee, Florida. Feb. 25-27, 1971.

Carlson, G. A. 1970. A decision theoretic approach to crop disease prediction and control. Amer. J. Agr. Economics 52: 216. May.

Carlson, G. A. and E. N. Castle. 1972. Economics of pest control. In Pest Control Strategies for the Future. Natl. Acad. Sci. Washington, D.C.

Chant, D. A. 1964. Strategy and tactics of insect control. Canadian Entomol. 96: 182-201.

Chant, D. A. 1966. Integrated control systems. Scientific aspects of pest control. pp. 193-218. A symposium of the Agricultural Research Institute of the National Research Council.

Clark, L. R., P. W. Geier, R. D. Hughes and R. F. Morris. 1967. The Ecology of Insect Populations in Theory and Practice. Methuen & Co., Ltd., 232 pp.

Commonwealth Institute of Biological Control, Trinidad. 1971. Biological Control Programmes Against Insects and Weeds in Canada, 1959-1968. Technical Communication No. 4: 1-266.

Corbet, P. S. 1970. Pest management: objectives and prospects on a global scale. In R. L. Rabb and F. E. Guthrie (eds.) Concepts of Pest Management. North Carolina State University, Raleigh, pp. 191-208.

DeBach, P. (ed.) 1964. Biological Control of Insect Pests and Weeds. Reinhold Publ. Co., New York. 844 pp.

DeBach, P. 1971. The use of imported natural enemies in insect pest management ecology. Proc. Tall Timbers Conf. on Ecological Animal Control by Habitat Management, No. 3. Tallahassee, Florida, Feb. 25-27, 1971. pp. 211-233.

DeBach, P. and C. B. Huffaker. 1971. Experimental techniques for evaluation of the effectiveness of natural enemies. In C. B. Huffaker (ed.) Biological Control. Plenum Press, New York. pp. 113-139.

FAO of United Nations. 1966. Proc. FAO Symposium on Integrated Pest Control. October 11-15, 1965, Rome.

Forrester, J. W. 1969. Urban Dynamics. The M.I.T. Press, Cambridge, Mass. 285 pp.

Ganyard, M. C., H. C. Ellis and H. M. Singletary. 1972. North Carolina Tobacco Pest-Management First Annual Report: 1971. 36 pp.

Geier, P. W. 1966. Management of insect pests. Ann. Rev. Entomol. 11: 471-490.

Gonzalez, D. 1970. Sampling as a basis for pest management strategies. Proc. Tall Timbers Conf. on Ecological Animal Control by Habitat Management, Number 2. Tallahassee, Florida, Feb. 26-28, 1970.

Harcourt, D. G. 1970. Crop life tables as a pest management tool. Canadian Entomol. 102(8): 950-955.

Headley, J. C. 1968. Estimating the productivity of agricultural pesticides. Amer. J. Agr. Economics 50: 13 (February).

Headley, J. C. 1972a. Defining the economic threshold. In Pest Control Strategies for the Future. Natl. Acad. Sci., Washington, D.C.

Headley, J. C. 1972b. Economics of agricultural pest control. Ann. Rev. Entomol. 17: 273.

Huffaker, C. B. (ed.). 1971. Biological Control. Plenum Press. New York - London. 1971. 511 pp.

Huffaker, C. B. and C. E. Kennett. 1956. Experimental studies on predation: predation and cyclamen-mite populations on strawberries in California. Hilgardia 26: 191-222.

Iseley, D. 1948. Methods of Insect Control. Parts I and II. 3rd. ed. Burgess Publ. Co., Minneapolis.

Knipling, E. F. 1966. Some basic principles in insect population suppression. Bull. Entomol. Soc. Amer. 12: 7-15.

Knipling, h. F. 1970. Use of organisms to control insect pests. American Assoc. Adv. Sci. Meeting, Chicago, Dec., 1970.

Lawson, F. R. 1969. The relation of insect control to increased food production. Proc. Tall Timbers Habitat Management Conf. 1: 145-173.

National Academy of Sciences. 1969. Insect-pest management and control. Vol. 3 of Principles of Plant and Animal Pest Control. U.S. Natl. Acad. Sci. Publ. 1695, 508 pp.

Newsom, L. D. 1967. Consequences of insecticide use on nontarget organisms. Ann. Rev. Entomol. 12: 257-286.

Newsom, L. D. 1970. The end of an era and future prospects for insect control. Proc. Tall Timbers Conf. on Ecological Animal Control by Habitat Management, Number 2. Tallahassee, Florida, Feb. 26-28, 1970.

Ordish, G. and D. Dufour. 1969. Economic bases for protection against plant diseases. Ann. Rev. Phytopathology 7: 31.

Painter, R. H. 1968. Crops that resist insects provide a way to increase world food supply. Kansas State Agr. Exp. Sta. Bull. 520.

Pathak, M. D. 1970. Genetics of plants in pest management. In R. L. Rabb and F. E. Guthrie (eds.) Concepts of Pest Management. North Carolina State Univ., Raleigh, pp. 138-157.

Pimentel, D., D. Chant, A. Kelman, R. L. Metcalf, L. D. Newsom and C. Smith. 1965. Appendix Yll. Improved pest control practices in restoring the quality of our environment. Rpt. of Environ. Pollution Panel, pp. 227-291. Presidents Science Advisory Committee.

Rabb, R. L. 1972. Principles and concepts of pest management. Proc. Natl. Extension Insect-Pest Management Workshop. (In press).

Rabb, R. L. and F. E. Guthrie. 1970. Concepts of pest management. Proc. Conf. held at North Carolina State University, Raleigh, Mar. 25-27, 1970.

Smith, R. F. 1971. Economic aspects of pest control. Proc. Tall Timbers Conf. on Ecological Animal Control by Habitat Management, No. 3. Tallahassee, Florida.

Smith, R. F., M. Bates, E. J. LeRoux, R. J. Anderson, J. L. George and R. L. Dout. 1964. Bull. Entomol. Soc. Amer. 10: 67-88.

Southwood, T. R. E. and M. J. Way. 1970. Ecological background to pest management. In R. L. Rabb and F. E. Guthrie (eds.). Concepts of Pest Management. North Carolina State University, Raleigh, pp. 6-29.

Stern, V. M. 1966. Significance of the economic threshold in integrated pest control. Proc. FAO Symposium on Integrated Pest Control 2: 41-56. Rome.

Stern, V. M., R. F. Smith, R. van den Boschand K. S. Hagen. 1959. The integrated control concept. Hilgardia 29: 81-101.

Tullock, G. 1969. Problems in the theory of public choice, social cost and government action. American Economic Review LIX: 189 (May).

Varley, G. C. 1968. Review of "The ecology of insect populations in theory and practice" by L. R. Clark, P. W. Geier, R. D. Hughes, and R. F. Morris. J. Ani. Ecol. 37: 275-276.

Watt, K. E. F. 1963. Mathematical populations models for five agricultural crop pests. Mem. Entomol. Soc. Canada 32: 83-91.

Watt, K. E. F. 1968. Ecology and resource management, a quantitative approach. McGraw-Hill Book Co., New York, 450 pp.

Watt, K. E. F. 1970. The systems point of view in pest management. In R. L. Rabb and F. E. Guthrie (eds.) Concepts of Pest Management. North Carolina State University, Raleigh, pp. 71-83.

Wille, J. E. 1951. Biological control of certain cotton insects and the application of new organic insecticides in Peru. J. Econ. Entomol. 44: 13-18.

POPULATIONS DEFINED AND APPROACHES TO MEASURING POPULATION DENSITY, DISPERSAL, AND DISPERSION

R. E. Fye
Cotton Insects Biological Control Investigations
Agricultural Research Service, USDA
Tucson, Arizona

As entomologists seek alternatives to the application of insecticides and apply control measures with more subtle effects, methods for more accurate assessments of the specific insect populations become necessary. The sledge-hammer effect of pesticides is relatively simple to assess; but the effects of biological and cultural controls, which may be postponed appreciably, are more difficult to evaluate. As biological control and more elaborate pest management systems are developed, adequate assessments of population will become increasingly essential.

To develop assessment methods, we must first define the population with which we are going to work. In the broad definition, a population of a given species is comprised of all living individuals of that species at a certain time and in a particular space. This population will have several characteristics of concern to the research worker in his assessment, three important ones being: (1) the age stratification; (2) the tendency for dispersal; and (3) the potential for limiting the population in time and space.

Figure 1 is an exaggerated figurative concept of an insect population. The main segment occupies the central zone; smaller segments penetrate outward as arms and subarms where favorable interactions of the limiting factors of the population permit. "A" represents a segment that was severed from the main population when the limiting interactions placed ecological limits between the main population and the segment. Students of evolution consider these segments as potential new species. The oversimplification in Figure 1 allows a mental superimposition of similar figures representing several species occupying the same overall climatic area. The mental glimpse of the overlapping populations in a single geographic or climatic area reveals the complexity of the ecosystem and demonstrates why the research worker or the manipulator of the ecosystem, must strike some happy medium in his considerations.

Too often in the past the entomologist has placed tight limits on his considerations and has become overfascinated with a single facet, i.e., the density in a delineated area. However, density may vary with time within the ecological limits shown in Figure 1. Population density is a dynamic multivariate phenomenon and the acceptance of an oversimplified numerical expression that ignores the dynamics of the underlying factors which result in a density has led the entomologist to erroneous theoretical and practical approaches to insect population dynamics and control.

The univariate application of population density must also be attributed, at least in part, to the commodity approach to research. The research worker is frequently restricted in his considerations to a given insect on a given crop. At

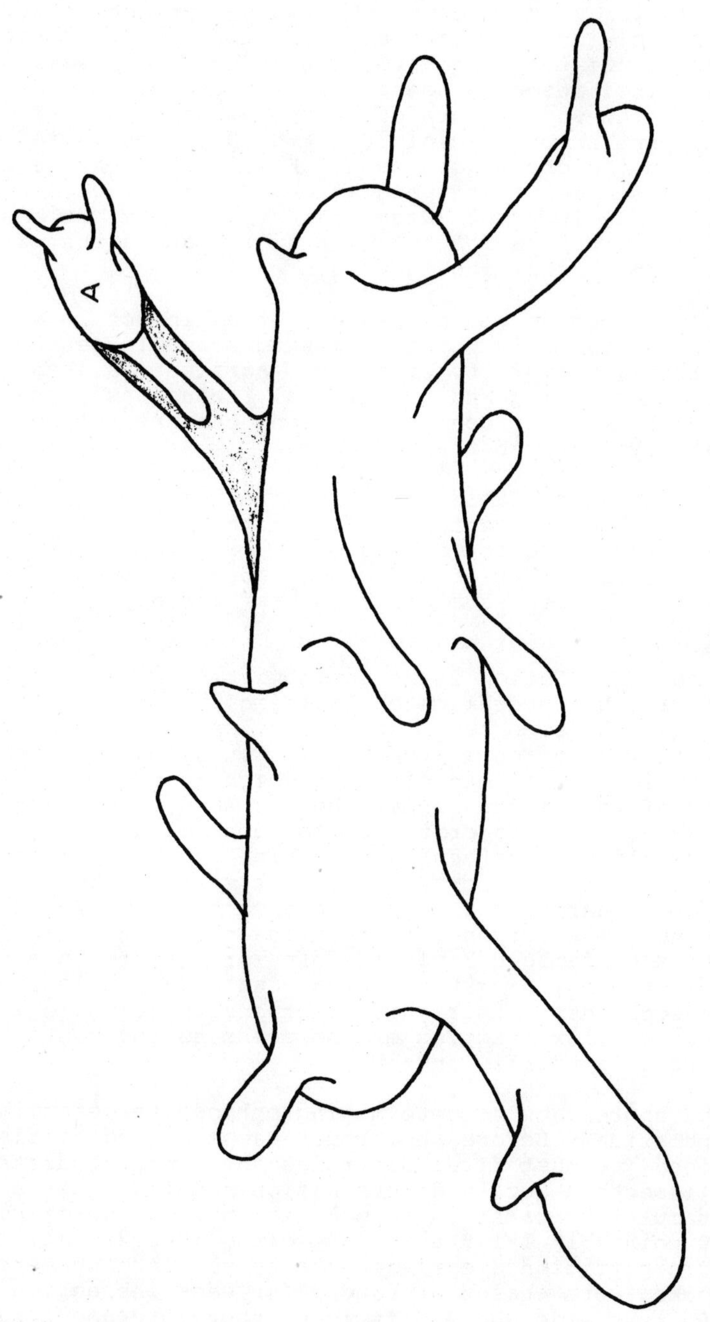

Figure 1.--Schematic drawing of an animal population showing the main central population with arms extending into areas where ecological factors interact favorably to allow penetration.

worst, he may be further restricted to a specific stage of the insect on the specific crop. Thus, the first change necessary to a consideration of population dynamics must be the broadening of the crop base. For example, many of our insect pests, such as the bollworm, *Heliothis zea* (Boddie), have cosmopolitan tastes which result in the presence of larvae on a large variety of crops, weeds, and ornamentals in the same area. Therefore, in the development of biological control and pest management systems for such insects, the total approach is absolutely necessary if effective control introductions or manipulations are to be made. In some cases considerations may be restricted to only one or two stages of an insect though this is a difficult limitation to accept because the numbers of a given stage are directly dependent on the numbers of the previous stage. Time may be another limitation if the pest or beneficial insect need be considered only during a limited period, e.g., a cropping season. Feasibility may be the final criterion for limiting an assessment of a population: (1) in space, (2) to one age segment of a population, or (3) in time. Regardless of the limitations, density must necessarily be a first consideration but with an awareness that density is dynamic and the result of many underlying factors.

The entomologist approaching the assessment of an insect population should first consult a statistician. At this point I wish to acknowledge the long-time assistance of Dr. R. O. Kuehl, University of Arizona Agricultural Experiment Station Statistician, in the sampling and statistical techniques presented. Only with the assistance of statisticians such as Dr. Kuehl can the entomologist acquire the insight necessary for proper assessment.

Perhaps one of the greatest fallacies in controlling insects with insecticides has been the utilization of percentage infestation as a criterion of determining when a crop should be treated for a pest. The practice is particularly abused when the damage level or number of insects has been allowed to remain constant throughout a growing season as a criterion for the application of insecticides.

Let us consider cotton. Cotton is an agronomically and physiologically dynamic crop that is attacked by several major biologically and physiologically dynamic insects. Certainly, the same recommended level of percentage fruit damage is not applicable to the cotton at the initiation of squaring and also later when the cotton has set a full load of fruit which is the maximum that the photosynthetic capability of the plants will support. When a constant percentage is applied to both situations a static criterion for control has been applied to a dynamic situation. In reality, X percent infestation at the commencement of squaring is far more pertinent than X percent at the time when the plant will be setting a limited number of squares as a consequence of having reached its physiological potential under the agronomic situation in a particular field. In addition the use of numbers of damaged fruiting forms or insects on a plant are "after the fact" criteria that may or may not be adequate indications of the need for an application of insecticides. Certainly, in the case of biological control which involves the relatively slow reaction of the pests to the control, the percentage infestation of fruit is totally unusable.

What then are the approaches to determining methods to determine the density of an insect population? Before the proper sampling and statistical techniques can be developed, we must first determine the sampling distributions of the pest and beneficial insects. Such a determination requires that a relatively large number of random samples be taken on some basic unit of the crop. In cotton data were taken from 100 points in 1-1.5 acres of cotton (Fye, et al., 1969a). Initially, the fields were partitioned early in the season into numbered blocks, and the blocks to be checked were chosen at random for each inspection date. The blocks were usually 64-65 rows wide and 200 ft long, therefore about square, but the dimensions were changed slightly if the field configuration did not divide into the standard block. Sample points within the blocks were obtained by numbering the rows and ascertaining the required number of samples per row from random numbers tables. The distance into the row of each sample point was determined by arbitrarily assigning one end as the starting point and then determining the distance into the row from the random numbers tables. In practice, a nylon line with 5-ft lengths delineated by color-coded ribbons was stretched between 2 stakes in the center of the 1-acre block so the checkers could quickly establish their positions; and large placards were placed at 5-row intervals to facilitate the positioning of the checkers at the sample points. At the sample points, the first plant was inspected, and the number of insects and damaged fruit was recorded. Similar data were recorded after inspecting 5, 10, and 20 plants.

Generally, the variance associated with the mean populations determined by the samples has been very large and at least equal to the mean, but in some cases, particularly those in which some type of clustered distribution is involved, the variance was extremely large. Many of the samples of insects that affect cotton have been found to agree with a negative binomial sampling distribution, numerous others have been found in a Poisson distribution, and some, probably those involving a clustering of some nature, do not fit either of these basic sampling distributions. Table 1(Kuehl and Fye, 1972) presents data from Arizona cotton. Note the numbers of distributions which agree with either negative binomial or Poisson configurations. Certain of the distributions fall into both categories since, from a statistical standpoint, the Poisson sampling distribution cannot be distinguished from the negative binomial distribution when insect infestations are low and sample sizes relatively small.

Table 1.--Summary of sample distributions for cotton insect pests in 1967, 1968 and 1969 (Kuehl and Fye, 1972).

Species	Plants/ unit	Number of samples			
		Total	Species present	Poisson[1/]	Negative binomial[1/]
Bollworm larvae	1	152	86	77	17
	5	112	77	67	26
Cabbage looper larvae	1	169	81	66	29
	5	125	82	52	33
Saltmarsh caterpillar larvae	1	91	24	21	1
	5	76	29	16	11
Beet armyworm larvae	1	123	19	19	0
	5	91	33	29	3
Pink bollworm larvae	1	96	12	10	3
	5	83	16	12	2
Boll weevil adult	1	37	8	7	2
	5	38	16	13	3
Lygus adult	1	165	79	71	5
	5	114	82	60	20
Lygus nymph	1	168	69	59	3
	5	110	49	30	8

1/These columns indicate the number of samples with species present whose distributions agree with the expectations of the indicated probability distribution.

Once the sampling distributions have been determined, certain statistical techniques may be utilized to develop sampling systems of maximum efficiency. For example, operating characteristic curves based on both the negative binomial and Poisson distributions have been developed for insects and seeds (Ives and Warren, 1965; Lyons, 1964; Harcourt, 1966; Kuehl and Foster, 1967; Kozak, 1964).

The negative binomial distribution places certain restrictions on the use of sequential sampling systems, operating characteristic curves, and the estimates of confidence limits. This limitation is a result of the constant k which is calculated with the variance estimate; therefore, the estimate of k is variable and may be dependent on the density of the population (Kuehl and Fye, Unpublished). Such a case is demonstrated by the typical variance estimates for the predaceous ladybird beetle, Hippodamia convergens Guèrin-Mèneville, which are associated in Figure 2 with the respective mean populations. Note that as the density of the population increases, so does the variance associated with the mean population estimate. Some success has been achieved in similar situations by pooling the estimates of k and utilizing the resulting estimates for sequential sampling systems though sequential sampling plans based upon the pooled estimates of k are necessarily burdened with broad confidence limits.

Considerations of the broad variance in practical application of field sampling systems led to the use of fields stratified in one-acre blocks (Kuehl and Fye, 1970). One-acre blocks for sampling are selected at random, and plants at several points within each block are inspected. The system may be extended over several fields. The rationale is that with the broad variance, after taking a given number of samples, many more would be necessary to achieve an improved estimate of the variance. Table 2 shows the number of samples that would be required to obtain a standard error of .01 insects per plant on mean populations of several species of insects found in cotton. The .01 limit was utilized to show that if stringent limits are placed on confidence intervals the entomologist is faced with an impossible task. In effect the studies of the variance associated with mean population estimates of insects in cotton have led us to accept broader confidence limits though we are aware that the estimates are burdened with the low confidence.

The research worker sampling insects on large plants would hope that his sampling could be limited to some reduced portion of a plant. For example, preliminary studies of the spruce budworm, Choristoneura fumiferana (Clemens) (Morris, 1955), showed that samples taken from midway in the crown of mature balsam fir gave a relatively good mean population estimate for the tree. Therefore, future samples from the fir trees, which are extremely difficult to obtain, were generally restricted to the midportion of the crown (Morris, ed., 1963). Similar vertical distributions of several insects on cotton were determined on individual cotton plants (Fye, 1972), and the populations in the upper 2 and 3 fifteen-cm increments of the plant were correlated with the populations in the lower portions. Table 3 presents a sample of the correlation coefficients determined by this method. Many are significant, however, the low magnitude precludes the use of subsamples from the top of the plant as an effective means of estimating population in the lower portions of the plant. The research worker in cotton is thus saddled with a large variance associated with the horizontal distributions in the field and within the vertical distributions though large segments of the populations of some pests and predators in cotton are directly associated with the fruiting upper portions (Bonham and Fye, 1971; Fye, 1972).

A research worker approaching the problem of sampling insects frequently utilizes indirect methods such as sweep nets, light traps, and other types of traps to obtain relative estimates of insect populations. However, the indirect methods are limited to certain segments of the population and do not provide an adequate age stratification.

Table 2.--Allocations of plants, blocks, and the man hours for sampling 6 species of cotton insects for a fixed standard error, $\sqrt{v(\bar{X})} = 0.01$, based on variance estimates from 1969 samples of 1 plant units (Kuehl and Fye, Unpublished)

Species	Plants per sample	No. samples per block	No. of blocks	Total No. samples	Estimated man hours
Bollworm larvae	5	21	28	588	56.0
	2	33	19	627	25.7
Cabbage looper larvae	5	14	13	182	18.4
	2	22	9	198	8.9
Lygus adults	5	--	--	130	10.8
	2	--	--	130	4.3
Chrysopa larvae	5	11	37	407	43.2
	2	17	26	442	21.2
Geocoris adults	5	9	100	900	100
	2	14	73	1022	52.3
Ladybird beetle adults	5	16	235	3760	372.1
	2	25	163	4075	176.6

Table 3.--Correlation coefficients for the relationships of fruit and insect populations in the upper portions of a cotton plant with those in the lower portions (Fye, 1972).

	Date	Plant height (ft.)	Correlation coefficients	
			Top 2 increments vs. remainder	Top 3 increments vs. remainder
Squares	Sept. 20, 1967	3-4	0.546[1/]	0.701[1/]
	Aug. 23, 1968	3-4	.620[1/]	.626[1/]
	Sept. 10, 1968	4-5	.550[1/]	.270
Green bolls	Aug. 9, 1967	4-5	.217	.212
	Aug. 23, 1968	3-4	.261[2/]	.286[2/]
	Sept. 10, 1968	4-5	.526[1/]	.479[2/]
Cabbage looper larvae	Sept. 11, 1968	4-5	.301	.479[2/]
Pink bollworm exit holes in dry bolls	Sept. 20, 1967	3-4	.990[1/]	.992[1/]
Chrysopa eggs	Aug. 27, 1968	3-4	.357[1/]	.239
	Sept. 11, 1968	4-5	.101	.110

1/ Significant at 0.01 level.
2/ Significant at 0.05 level.

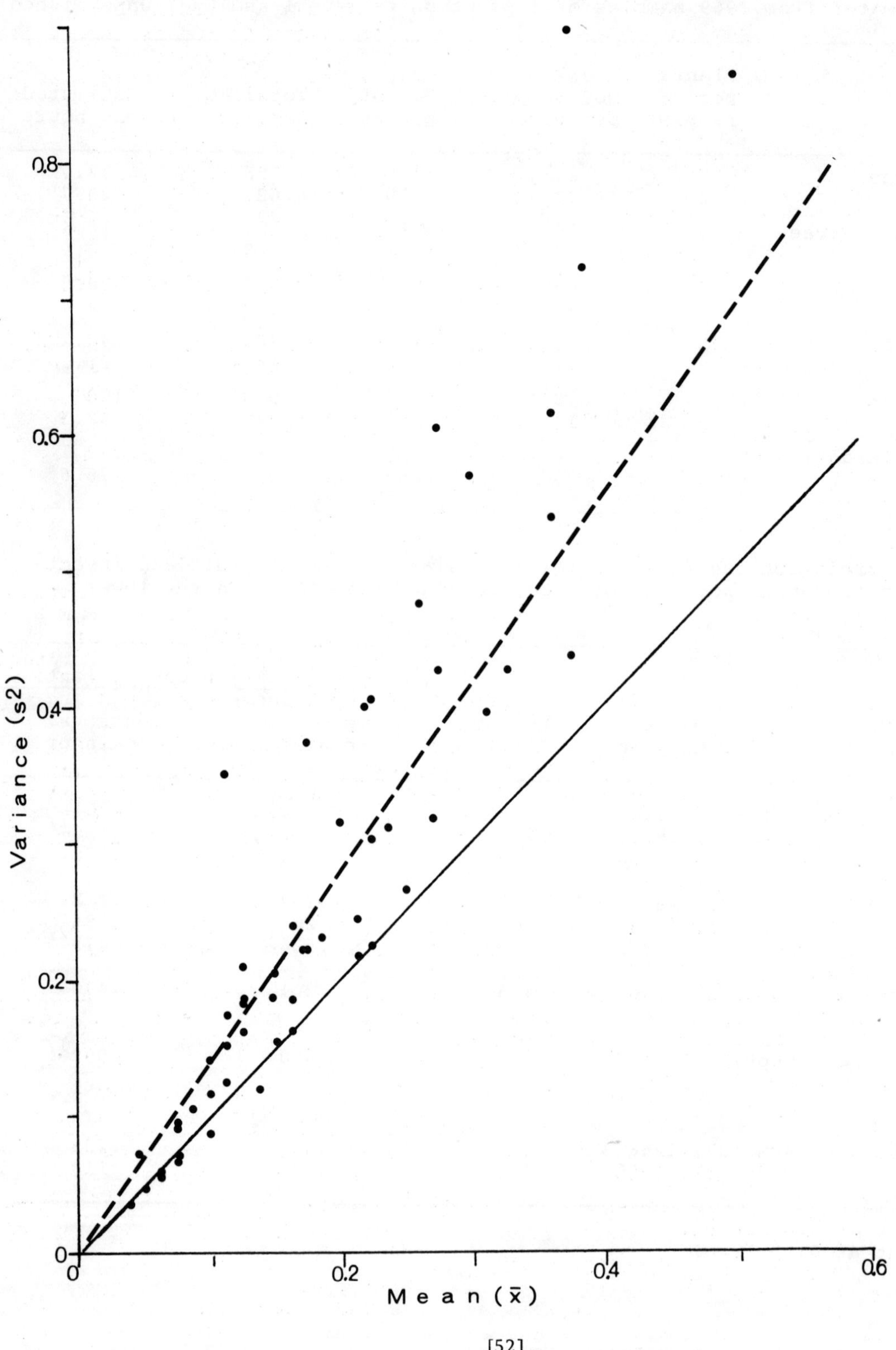

Figure 2.--Plot of sample means vs. sample variances for 1-plant unit samples of adult ladybird beetles in 1969. (Solid line represents equal mean and variance; dashed line represents significant overdispersion by Index of Dispersion test). (Kuehl and Fye, 1972).

To date, few if any correlations of catches by indirect methods with the actual population density have been achieved. The basic difficulty is associated with the behavior of the insects. With nets, many difficulties are involved including the herding of insects ahead of the net, the length of the flight of the insect, the stratification of the insects on plants due to temperature, humidity, or various other physical factors. Therefore, sweepnet samples taken at different times of day, under different climatic conditions, or by different persons are not comparable and provide only very crude estimates of density. Traps of various designs with various baits or attractive factors place maximum importance on the understanding of the behavioral patterns. With traps another pressing question arises: without the attractive components would the insect have been in the trap area? Therefore, many gnawing reservations should pervade the mind of a research worker utilizing sweep or trap data to obtain estimates of population density. From the statistical standpoint such indirect methods that involve the unpredictable behavioral responses of the insect increase the variance and broaden the confidence limits.

Why should we be so concerned about the direct assessment of insect populations? First, and definitely foremost, is the lack of immediately available alternatives. The major alternative that has received much attention is the predictive model, but many entomologists confuse descriptive and predictive models. Perhaps the most elaborate models available are those for the spruce budworm (Morris, ed., 1963) which are descriptive. These models describe about 80 plot years of spruce budworm populations in the Green River area of New Brunswick in the late 1940's and early 1950's. Unfortunately, from the final models, which are mathematically elaborate, little predictive capability can be derived because a predictive model can only be completed from empirical data if all possible major situations are included in the development of the model. The models were based on the life-table concept in which the number of eggs laid by a given generation is placed in the initial cell in the table, the mortality factors through the lifespan of the insect are then elaborated and the mortality subtracted. Finally the number of surviving females in the generation is estimated and by multiplying by the fecundity of the females, the number of eggs available to initiate the next generation is established. The technique is limited because of major sampling problems (Ives, 1964) and can only be used with insects that are univoltine or that have clear-cut generations within a given year. When generations overlap in a multivoltine situation, the technique becomes inapplicable. Therefore, the Canadian research workers (Morris, ed., 1963; LeRoux, et al., 1963) associated with univoltine species have probably developed this technique to its limit. The salient point is that the predictive capability of models derived by descriptive processes is strictly limited.

As we move into detailed efforts to achieve pest management which involves biological control, many entomologists will inevitably fall into the descriptive model trap due to their empirical orientation. Pesticides have been tested successfully by empirical methods since before the advent of the organic insecticides in the mid-1940's because their effects are easily detected. In contrast, biological control, whether it is an attempt to establish a parasite or predator capable of permanent control of a target insect or whether it is an inundative release, can only be premised on effective population dynamics studies; and the effect can only be determined after partitioning the various factors causing mortality in the target population. Fortunately, we may take the key factor approach (Morris, 1959; Morris, ed., 1963) which utilizes certain factors that cause the major part of the mortality ignoring the many minor factors. Even with this restriction to the key factors causing mortality we still have a multivariate situation. If we assume, and it is a valid assumption, that bioclimatic factors are a major cause of mortality and if we consider that the biological factors may or may not be a major cause, we are immediately confronted with determining and predicting the effect of both types of factors. Many researchers are under the impression that the causal effects of population density can be determined by retrospective inspection of the populations of pests and beneficials. They endlessly sift the factors that might cause responses and then attempt to recombine the factors to recreate or predict a situation. More often than not, they fail. In fact, unless they are fully aware of the major factors involved and their quantitative impact, their choice of measurements may actually preclude measurement of the factors that cause major mortality.

Consider the truly predictive model. Basically, such a model involves the determination of the impact of individuals within populations so let us use a predator as an example. To fully understand the predator, we must first understand its basic biology which, in turn, is determined by its genetic background. Concurrently, we must understand its basic behavior patterns which again are determined genetically. The physiological and behavioral potential of the predator therefore combine to give the predator its full bioecological potential. Finally the effect of physical factors on the potential must be established before we can determine the impact of interactions of the physical factors and the physiology and behavioral facets on the individual. Determination of the numbers of individuals in the population of predators will approximate the total impact of the predators upon a target population.

In ideal situations, the impact of individual insects may be determined without investigating the underlying causal factors, and as long as a quantification is possible, the underlying factors may be academic. However, increased precision can usually be obtained by investigation of basic hunger patterns, prey handling time, satiation, feeding patterns, and the other behavioral and physiological functions. In addition, alternate prey outside the target population must be evaluated. If diversion to other species of prey occurs often, the potential predation of the target population by the predator population may be negligible. Conversely the competition for the potential prey with other species of predators must be considered. Finally, the presence of a predator or parasite is of no avail if the organism does not function at an effective level. The researcher therefore needs to understand the underlying phenomena to obtain an adequate approximation of the total potential of an individual. Superimposing the size of the population will obtain an approximation of the impact of the total population. This approach was suggested by Holling (1963) who coined the term "component analysis." Basically his procedure states that the whole is the sum of the parts, as contrasted to the sifting by super-analysis which attempts to determine how many parts the whole can be broken into and then attempts to recombine the factors to recreate or predict a situation.

Figure 3 is a flow chart of the basic systems involved in the consideration of a target population and the various physical and biological factors affecting that population. The upper block presents the two parallel systems that are of interest to researchers in biological control. The oviposition and developmental systems will determine when the population will be in a certain age stratification so introduction of a biological control organism can be made at the most strategic time. The age stratification may be determined by some direct or indirect measurement of temperature. In the case of the spruce budworm, Morris, et al. (1956) associated the development of the larvae with the phenology of the forest trees, but more elaborate systems based on direct temperature input at 2-hr intervals were presented by Fye, et al. (1969b) and Fye and McAda (1972). The latter system is based on the summation of the proportion of development during 2-hr periods with mean temperatures of X_1, X_2, $X_3 \ldots X_n$ estimated by regression models with the log of the mean temperature during the 2-hr period as the independent variable.

Neither the phenological approach nor the temperature input provides a totally accurate estimate. However until such time as the entomologist is able to relate the development of insects to the interaction of the physiological and biochemical facets of the functions of insects and the environmental heat flux by using heat transfer equations, the phenological approach and temperature input will provide reasonably satisfactory answers. In the future, multivariate criteria

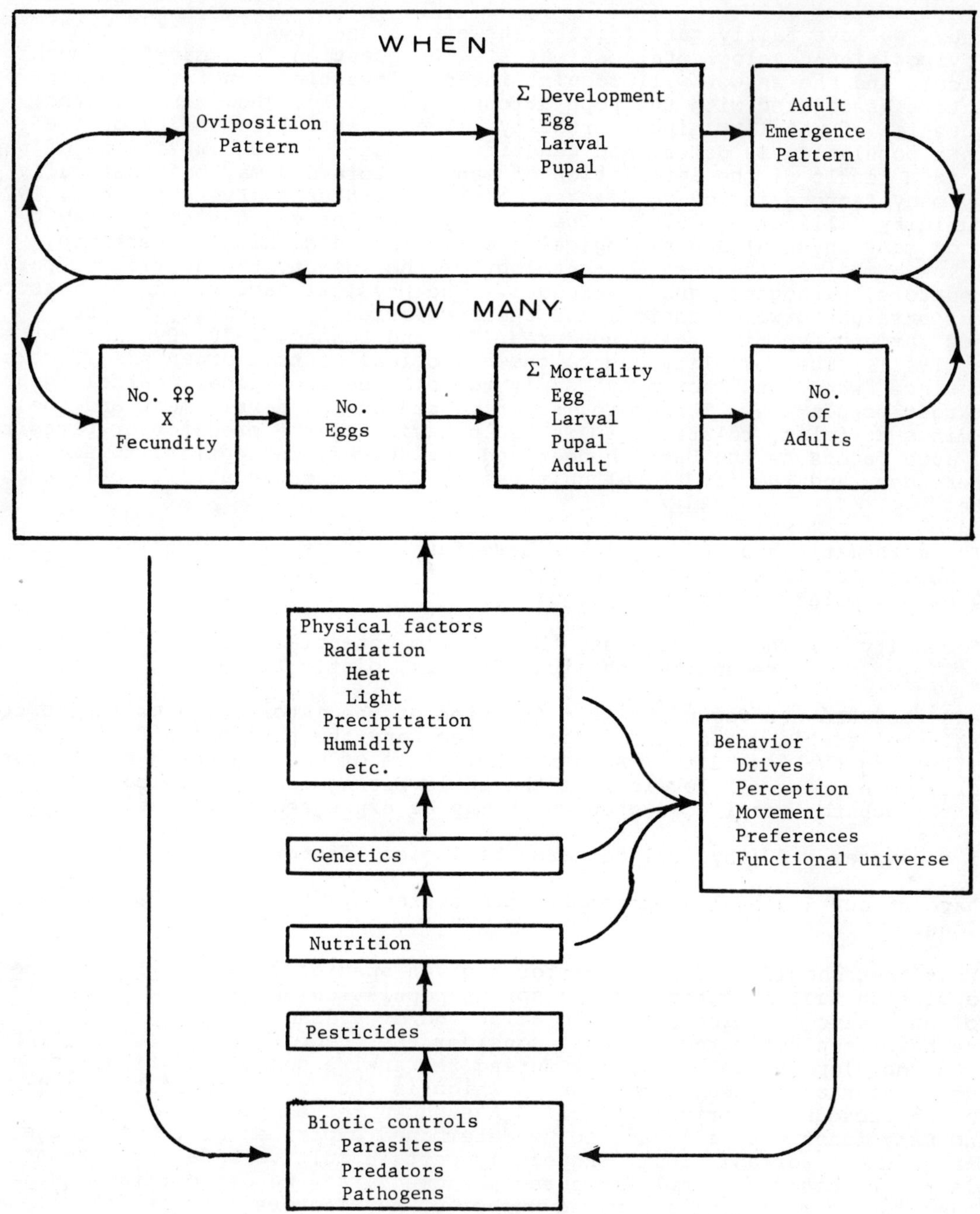

Figure 3.--Flow chart of basic development and mortality systems with associated contributing factors.

measured by simple means will probably be utilized to initiate pest management practices, and phenological observations may provide these simple measurements.

Thus, we have fairly satisfactory answers to the questions of "when" with the oviposition-developmental estimates. The question "how many" is much more difficult and the answer will require intense investigation into the mortality factors associated with the populations. Basically, "how many" is the age-old question of natality minus mortality. The potential fecundity of the females in a population is determined genetically, but the true expression of the fecundity is a result of the interaction of many factors and may be drastically reduced by many factors including disease, high temperatures (Fye and Poole, 1971) and low humidity (Ellington, 1970). The mortality of the various stages may be dependent on many physical and biological factors including high temperature, rainfall, low humidity, the basic gerontology of the insect, the nutrition, parasites, predators, pathogens, and pesticides. The physical factors can be measured in a rather straightforward fashion, and their relation to mortality can be established through experimental procedures (Fye and Bonham, 1970; Fye and Surber, 1971; Fye, 1971). The mortality due to the biological control organisms is more difficult since "when" and "how many" apply to both the biological control agent and the target population. Therefore, superimposed on the development and the natality-minus-mortality relationship is the behavior of the predator or parasite including such facets as the basic hunger and sex drives, perception, movement, food preferences, and the functional universe in which the predator is obligated to operate.

The mathematicians and engineers have furnished us with powerful statements of the interactions noted in the flow chart. For example, the net fecundity of a population may be expressed as:

Net fecundity = f (no. of females, the genetic potential for fecundity, physical factors, biotic factors)

which reads that the net fecundity is a function of the interaction of the number of females, the genetic potential of the females, and the physical and biotic factors affecting the fecundity. Such a simple form of the statement may be used sequentially to express all levels from the basic biochemical physiological reactions through the total ecosystem which may be expressed simply as:

Total ecosystem = f (physical factors, biological factors)

Thus we have at our disposal an uncomplicated statement which has broad implications.

What then should a total model for a given species involve? First, the spring population must be estimated. A spring population may involve either the residue of an active overwintering population and/or a diapausing population. When these have been estimated, we must consider the fecundity, the oviposition pattern, the development and mortality during the entire season and finally the fall triggers of the diapausing population which is the precursor of the population in the subsequent spring. Figure 3 has shown the complexity of the system and many functions will have to be determined experimentally before the models can assume a workable form. Hopefully certain multivariate segments of the models can be based on simply-measured phenomena, and we can obtain a large-scale projection by measuring a limited number of key factors (Morris, 1959). From this brief discussion of the complexity of models, it is obvious that direct accurate assessment systems for field experimentation will be necessary until research and experimental results provide refined models.

The problem of properly assessing populations is complicated by the movement of the members of the populations. Figure 1 was a figurative concept of an insect population with the individual insects pushing outward from a central zone when favorable interactions permitted the extension of the species. This situation is presented more realistically in Figure 4, which shows a group of individuals moved a large distance from the core of the population and its immediate extensions. If these migrant individuals invade an area for which they are ecologically adapted, they will become established as an outlying extension of the population. Quite frequently the barriers that restrict a population to a given central area are physical or geographic in nature, e.g., oceans; if these barriers can be overcome and a suitable ecological niche invaded, the species may establish itself beyond the barriers. Man with his thousands of trans- and intercontinental flights daily in highspeed aircraft has enabled many species to overcome major barriers and our historical records contain numerous examples of populations that have been freed from the natural factors that cause mortality and have become pest species.

However, the insect that moves with normal flight is of more immediate concern. These localized flights are usually relatively short compared with the middle distances achieved by insects that are moved by local thermal currents into higher altitude jet streams or other high altitude wind currents and transported for long distances. During such a movement, mortality is undoubtedly high, but if the conditions are generally favorable, long-distance moves in air currents (and sometimes water currents) may be achieved.

The movement of a mobile population into and out of an area is difficult to assess, and appreciable speculation has necessarily characterized the study of the migration phenomena (Dingle, 1972). Short localized dispersal flight and the extent of such flights have been studied with ingenious flight mills measuring the time an insect can remain in the air. Since most of these tests involve a tethered insect flying in a circle, the distances of the flights can be crudely estimated.

The description of movement in the thermal currents is an even less refined art. The air velocity required to move a passive insect may be fairly readily determined by using wind tunnels (Fye, 1968), and manuals on gliding, e.g. Barringer, 1940, and discussions of local disturbances such as "dust devils" (Sinclair, 1966) provide helpful information concerning the vertical air currents which may be expected in certain situations. As the knowledge of wind currents at various altitudes increases as a result of monitoring modern rocketry, an entomologist can now theorize about where an insect might be transported. In some cases, such theories are purely speculative; in other cases, some documentation is possible (Brown, 1965). Of course, an insect when transported long distances may be deposited in an area that is unfavorable to survival and reproduction; and the introduction is of no consequence. However, the transported insect may survive, reproduce, and become established as a pest or as a beneficial insect.

Undoubtedly, certain techniques for the manipulation of pests in biological control and pest management will involve the movement of insects between crops. The rudiments of such promising systems are apparent in the strip cropping of alfalfa and cotton to retain *Lygus* in the alfalfa which they prefer (Stern, et al., 1967). If alfalfa is maintained in a young vigorously growing condition, the lygus bugs will remain in the alfalfa and will not move into the adjacent cotton. To date, the potential benefits of such a system are overshadowed by the relatively complicated management involved in maintaining the alfalfa stands in the proper condition to retain the lygus bugs. Elsewhere, in localized areas of Arizona, populations of predators that utilize biotype C of the greenbug, *Schizaphis graminum* (Rodani), as a food source reached numbers in excess of one million per acre and as the sorghum matured, they moved into adjacent cotton fields at a time when the cotton had started to square (Fye, 1971a). Such populations of predators were measured by utilizing transect

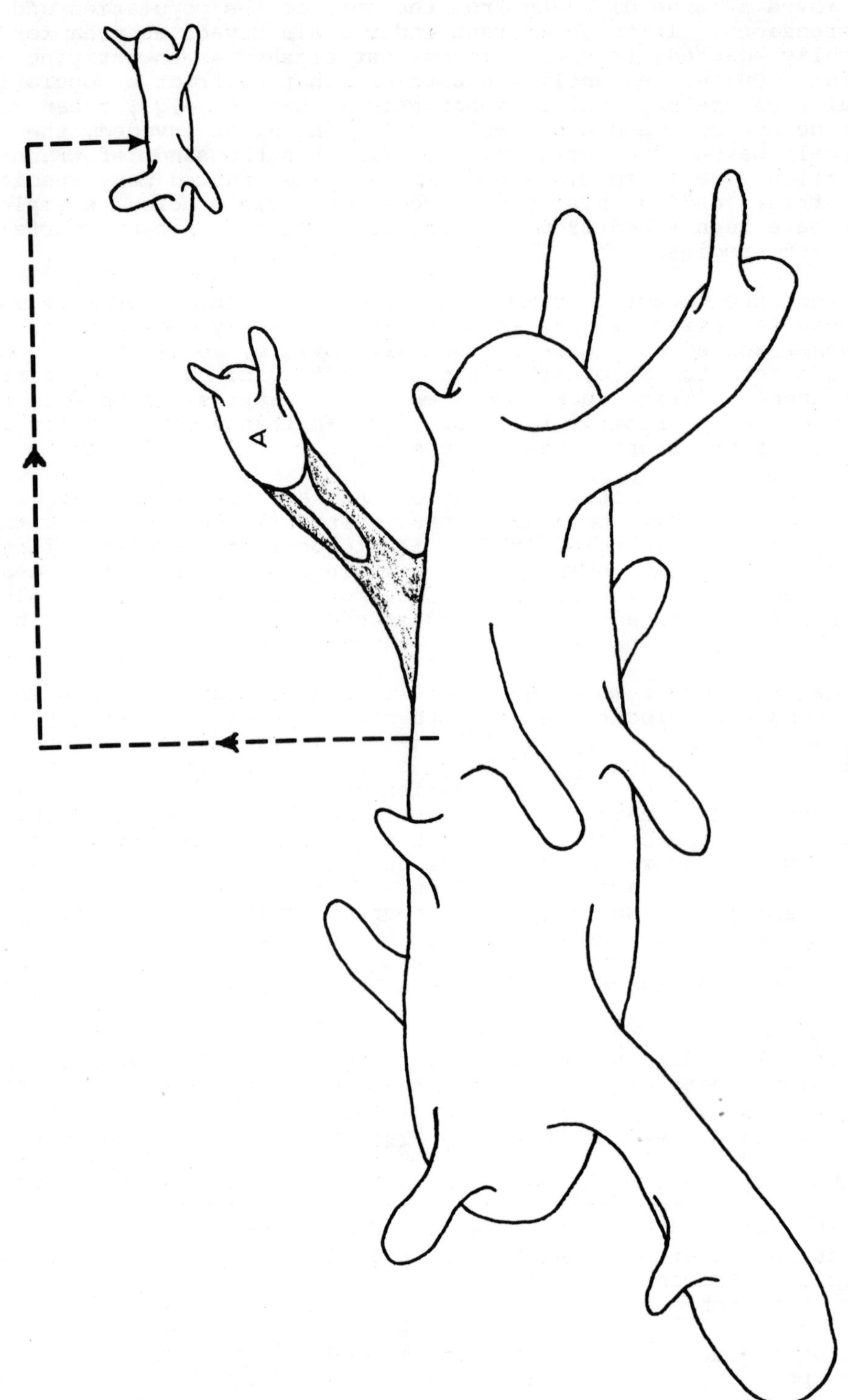

Figure 4.--Schematic drawing of an animal population showing the main central population with an outlying segment established by long distance migration.

methods, and, by inference the movement of the beneficial insects from the sorghum to the cotton was postulated. In the Arizona situation several predators were found to move at least 250 ft (Fye and Carranza, 1972) while Hippodamia convergens and Chrysopa spp. remained near the aphid food source in the sorghum. With such data, the movement of predators can be used to determine the size of cropping blocks that will be best used by migrating predators. If the predators move extended distances, blocks of crops can be used rather than the inconvenient narrow strips which most growers abhor. The size of the block will be of particular interest to the grower if the crops involved require diverse irrigation management.

To recapitulate; regardless of the approach, i.e., the breakdown of the whole into the various parts or component analysis moving from the simple to the complex, somewhere in the procedure it is necessary to assess the populations of various insects directly, therefore, it is essential that we develop the proper sampling and statistical techniques to assess populations adequately. The data required to develop predictive models to substitute for direct assessment are necessarily difficult to obtain because methods associated with component analysis are extremely time consuming and expensive. Still, the component analysis approach is less expensive than the continued costly "shotgun" empirical approach to research which involves consideration of a gross overall situation time after time with little regard for an understanding of underlying factors. Actually the basic information obtained to produce a predictive model gives the research worker a deeper insight into the problem than does the sifting of data in the empirical approach, and these deeper insights should result in more rapid progress in insect control.

The author wishes to express his sincere appreciation to Mr. Harold Brewer, USDA, Agr. Res. Serv., Shafter, California, for his many suggestions incorporated in this discussion. In addition the author expresses his sincere appreciation to the many individuals who have contributed directly or indirectly to the theoretical and experimental work reported.

References Cited

Barringer, L. B. 1940. Flight Without Power. Pitman Publishing Corp., New York, Chicago. 221 pp.

Bonham, C. D. and R. E. Fye. 1971. An empirical model for predicting boll weevil distribution on cotton plants. J. Econ. Entomol. 64(2): 539-540.

Brown, C. E. 1965. Mass transport of forest tent caterpillar moths, Malacosoma disstria Hubner, by a cold front. Canadian Entomol. 97(10); 1073-1075.

Dingle, H. 1972. Migration strategies of insects. Science 175: 1327-1335.

Ellington, J. J. 1970. Approaches to a more meaningful evaluation of host plant resistance to lepidoptera. 1970. Proc. Beltwide Cotton Prod. Res. Conf.

Fye, R. E. 1968. Spread of the boll weevil by drainage water and air currents. J. Econ. Entomol. 61(5): 1418-1424.

Fye, R. E. 1971a. Grain sorghum--a source of insect predators for insects on cotton. Prog. Agr. Arizona. 33(1): 12-13.

Fye, R. E. 1971b. Mortality of mature larvae of the pink bollworm caused by high soil temperatures. J. Econ. Entomol. 64(6): 1568-1569.

Fye, R. E. 1972. Preliminary investigation of vertical distributions of fruiting forms and insects on cotton plants. J. Econ. Entomol. 65: 1410-1414.

Fye, R. E. and C. D. Bonham. 1970. Summer temperatures of the soil surface and their effect on survival of boll weevils in fallen cotton squares. J. Econ. Entomol. 63(5): 1599-1602.

Fye, R. E. and R. L. Carranza. 1972. Movement of insect predators from grain sorghum to cotton. Environ. Entomol. 1(6): 790-791.

Fye, R. E. and W. C. McAda. 1972. Laboratory studies on the development, longevity, and fecundity of six lepidopterous pests of cotton in Arizona. USDA Tech. Bull. 1454. 73 pp.

Fye, R. E. and H. K. Poole. 1971. Effect of high temperatures on the fecundity and fertility of six lepidopterous pests of cotton in Arizona. USDA Prod. Res. Rep. 131. 8 pp.

Fye, R. E. and D. E. Surber. 1971. Effects of several temperature and humidity regimens on eggs of 6 species of lepidopterous pests of cotton in Arizona. J. Econ. Entomol. 64(5): 1138-1142.

Fye, R. E., R. O, Kuehl and C. D. Bonham. 1969a. Distribution of insects pests in cotton fields. USDA Misc. Publ. 1140. 32 pp.

Fye, R. E., R. Patana and W. C. McAda. 1969b. Developmental periods of boll weevils reared at several constant and fluctuating temperatures. J. Econ. Entomol. 62(6): 1402-1405.

Harcourt, D. G. 1966. Sequential sampling for use in control of the cabbage looper in cauliflower. J. Econ. Entomol. 59(5): 1190-1192.

Holling, C. S. 1963. An experimental component analysis of population processes. Entomol. Soc. Canada Mem. 32: 22-32.

Ives, W. G. H. 1964. Problems encountered in the development of life tables for insects. Proc. Entomol. Soc. Manitoba 20: 34-44.

Ives, W. G. H. and G. L. Warren. 1965. Sequential sampling for white grubs. Canadian Entomol. 97(6): 596-604.

Kozak, A. 1964. Sequential sampling for improved cone collection and studying damage by cone and seed insects in Douglas fir. For. Chron. 40(2): 210-218.

Kuehl, R. O. and R. E. Foster. 1967. The use of operating characteristic curves to estimate population proportions. Amer. Soc. Hort. Sci. Proc. 90: 561-566.

Kuehl, R. O. and R. E. Fye. 1970. Efficiency of grid stratification in cotton fields for cotton insect surveys. J. Econ. Entomol. 63(6): 1864-1866.

Kuehl, R. O. and R. E. Fye. 1972. An analysis of the sampling distributions of cotton insects in Arizona. J. Econ. Entomol. 65(3): 855-860.

Lyons, L. A. 1964. The spatial distribution of two pine sawflies and methods of sampling for the study of population dynamics. Canadian Entomol. 96: 1373-1407.

LeRoux, E. J. and others. 1963. Population dynamics of agricultural and forest insect pests. Entomol. Soc. Canada Mem. 32. 103 pp.

Morris, R. F. 1955. The development of sampling techniques for forest insect defoliators with particular reference to the spruce budworm. Canadian J. Zool. 33: 225-295.

Morris. R. F. 1959. Single-factor analysis in population dynamics. Ecology 40(4): 580-588.

Morris, R. F. ed. 1963. The dynamics of epidemic spruce budworm populations. Entomol. Soc. Canada Mem. 31. 332 pp.

Morris. R. F., F. E. Webb and C. W. Bennett. 1956. A method of phenological survey for use in forest insect studies. Canadian J. Zool. 34: 533-540.

Sinclair, P. C. 1966. A qualitative analysis of the dust devil. Ph.D. Dissertation, Institute of Atmospheric Physics, University of Arizona, Tucson.

Stern, V. M., R. van den Bosch, T. F. Leigh, O. D. McCutcheon, W. R. Sallee, C. E. Houston and M. J. Garber. 1967. Lygus control by strip cutting alfalfa. Univ. California Agr. Ext. Serv. Bull. AXT-241.

THE LIFE SYSTEM CONCEPT AS A GUIDE TO UNDERSTANDING POPULATION DYNAMICS

F. M. Davis
Entomologist, USDA, ARS
Boll Weevil Research Laboratory
Mississippi State, Mississippi

Man has been studying insect population dynamics for many years, especially on those insects that he has considered as his competitors for food, fiber and shelter and those insects that affect his health and general well-being. Since the mid-1800's masses of ecological literature on insects has been compiled by scientists all over the world. Some of these scientists have put forward their theories as to why insect populations fluctuate as we observe them to do in nature. By no means were these workers in complete agreement as to which environmental factors were most important in determining the abundance of insect populations. There were those that strongly advocated that climatic conditions were the major factors involved in population fluctuations and one did not need to seriously consider other factors. Others were equally strong in stressing that density dependent factors such as parasites, predators, and diseases or intraspecific competition was sufficient to explain observed fluctuations. Also there were those that considered the genetic constitution of the population as having a strong influence on fluctuations. Even though there is disagreement in the literature, these theories can and do serve as the foundation for us to develop a well balanced approach to understanding population dynamics.

If there was ever a time for us to have a broad understanding of population dynamics, it is now. For those of us who face the task of control of insects by either pest management or integrated control procedures, it is essential that we have a clear picture of the interactions that occur between an insect pest's population and those environmental factors that affect it, since the success of most management programs will be based upon the facts known about the pest's ecology.

Life System

The guide to understanding the insect populations that I am about to discuss was developed by the authors by incorporating "(i) all that is useful in ideas of theorists, and (ii) the latest information on natural populations, much of which was not available to them." (Clark, et al., 1967). It is called the life system.

The life system concept was developed by several Australian entomologists (chiefly L. R. Clark (1964) and P. W. Geier (1964). A full discussion of the life system and its relationship to pest management can be found in the book The Ecology of Insect Populations in Theory and Practice by L. R. Clark, P. W. Geier, R. D. Hughes and R. F. Morris (1967).

The name "life system" is closely related to Tansley's (1935) "ecosystem." It varies from ecosystem in that the life system involves only those elements of the ecosystem that are directly involved in the life of one kind of organism which is called the subject species. Therefore, the life system can be considered as a subsystem of the ecosystem (Geier; 1966).

Thus, the life system of an insect population is that part of the ecosystem that determines its existence, abundance, and evolution. It involves interactions between the subject population and its effective environment which is that part of the ecosystem that provides supplies of all kinds (e.g., food, shelter, oviposition sites, overwintering quarters, etc.) required for the maintenance of the population and those biotic and abiotic factors that oppose the survival and reproduction of the invidivuals within the subject population (e.g., weather, predators, parasites, diseases, interspecific and intraspecific competition).

The interactions that occur between the codeterminants of abundance result in observable events that we can measure. The authors of this concept termed these as ecological events and further divided them into primary and secondary events.

The primary events are observable demographic occurrences such as birth and death rates, dispersal and migration, annual generations, phenology, voltism etc. These events are the expression of the inherited ability of individuals to survive and multiply.

The secondary events are those environmental factors that influence the magnitude, extent, frequency of duration of the primary events by altering the supply of required resources needed by the subject population or by directly affecting the individual in some harmful or helpful manner.

Involved in the secondary events are ecological processes which affect the population by being additive or subtractive processes. The additive processes being those that contribute to the addition of individuals to the subject population, whereas, subtractive processes being those that operate to lower the numbers of individuals in the subject population. Some factors may act as an additive process some of the time and as a subtractive process at other times.

The subtractive forces have been classed by most ecologists and by the authors of this concept into those factors that are density-dependent and those that are density-independent. The density-independent factors are those whose effect are not related to the subject population's density. The density-dependent factors are those that destroy an increasing proportion as the subject population's density increases.

The density-related processes such as parasites, certain predators, diseases, and intraspecific competition can serve as stabilizing mechanisms when other conditions are favorable for indefinite increase in the subject population's density.

In my opinion, the life system concept has great merit for those of us involved with population dynamics of insects. Perhaps, its greatest significance is that it keeps us ever mindful that we are dealing with a dynamic system made up of the subject population and its effective environment. It is unbiased as to what environmental factors (intrinsic or extrinsic) are the most important in any insect population's life system and relies on us through proper experimental procedures (e.g., life tables, key factor studies, in-depth life history studies and mathematical modeling) to determine the elements involved in the functioning of a life system.

European Corn Borer

To illustrate some points about the life system of an insect, I have chosen the European corn borer (Ostrinia nubilalis (Hubner). The ecological research on this insect is massive and it is not my intention to review all of the ecological research findings, but to select a few for illustrative purposes.

This insect is a well-known pest of field crops (especially on corn) in this country and other parts of the world. In the United States the European corn borer's geographical range extends from the New England States down to the southern gulf states and westward across the midwest to the Rocky Mountains (Brindley and Dicke, 1963).

Part of the research I shall be discussing is the results obtained under the North Central Regional Projects NC-20 and NC-87. NC-20 was activated in 1953 under the direction of the North Central Regional Association of Experiment Station Directors to study the factors influencing corn borer populations. State Experiment Station and USDA personnel cooperated in this project. From time to time the objectives of NC-20 were revised and in 1966 the project became NC-87.

This cooperative research effort has great significance because with group planning of objectives and standardization of techniques, many areas over the geographical range of this insect could be compared and resulting conclusions would have greater biological meaning.

Some of the environmental factors that are involved in the life system of the European corn borer are:

(1) Weather - Hill, et al. (1967) summarized the effect of climatic factors on born borer populations for 10 years in the North Central States. They concluded that climatic conditions have a significant influence on corn borer populations and that it may act as an additive or as a subtractive process.

Chiang and Hodson (1972) studied corn borer populations fluctuations in one early planted field at Waseca, Minnesota from 1948 to 1970. They concluded that climatic conditions affected corn borer populations in many ways such as "low temperatures (below 50°F) during the season would reduce the size of fall populations. Winds may cause an influx of moths from the south. Rains may have positive or negative effects depending upon the period of season and perhaps even the hour of the day."

Also, they commented that the long-term cooling trend in Minnesota from 1953 to 1969 was one of the main factors responsible for lower borer populations during this period and that under favorable temperature conditions the borer population could be expected to return to an economically significant level.

C. A. Barlow (1971) reported that 1st brood corn borer populations in southern Canada are primarily controlled by 2 key factors. One being rainfall. He stated that the greatest numbers of eggs are laid and survival is longest at rates of rainfall between 25 and 76 mm. per day. Smaller amounts are sub-optional. The rain serves as a source of drinking water that is essential for survival and reproduction of this insect.

(2) Host - Traditionally ecologists have considered the host plant(s) as a major element in the life system of insects. Abundance: usually emphasis is placed on the abundance of the host and not the effect the host might have on the insect. Scarcity of corn would not normally appear to be a factor in the corn borer's life system but the morphology, physiology and chemistry of the host could be a factor. Stage of growth: for many years researchers have known that the stage of corn growth has an influence on oviposition and survival of the

European corn borer. For example Beard (1943) found that when all stages of growth were present, moths preferred the early silking stage to any other, and that relatively few eggs were deposited on plants younger than the mid-whorl stage. Also he found that few larvae survived until the plants reached the late whorl stage (Beard and Turner, 1942). Turner and Beard (1950) brought out the point that 2nd generation eggs were deposited more in relationship to growth than to height of plants. Most of the eggs being laid on plants in the late tassel or early silking stages. Recently Guthrie and co-workers (1969) have shown that development and survival of 2nd generation corn borer larvae was associated with pollen shed. The pollen was found to be beneficial to larval survival and development. Therefore, stage of growth can be an important factor in the survival of larvae depending upon what stages of corn growth are available during moth flights. Resistant varieties: Host plant resistance in corn has been reported for 1st and 2nd brood corn borers by several researchers (varietal resistance review by Brindley and Dicke (1963). Chiang (1968) recently brought out the importance of the genetic makeup of the food plant in the corn borer's population dynamics. He compared corn borer survival, development, and tunneling activity on two groups of corn varieties, (1) those popular in 1955 and (2) those popular in 1965. His findings showed that the general levels of host-resistance was about 5 to 15% higher in 1965 than in 1955.

(3) Parasites and Predators - Brindley and Dicke (1963) in reviewing European corn borer research stated that some 24 parasite species have been introduced into the United States, 22 have become numerous enough to permit colonization. Most workers agree that the fly, *Lydella grisescens* Robineau-Desvoldy, is the most important exotic parasite of the corn borer in this country.

Sparks and co-workers (1966) evaluated the influence of predation on corn borer populations in the North Central states. They concluded after several years of research that predators play an important part in population fluctuations of the corn borer at some locations during some years. But, predators could not be depended upon year after year, or in any given year, to play a major role in managing a population of corn borers at a specific location.

(4) Diseases - Steinhaus (1949) pointed out that the European corn borer has been reported to be susceptible to a number of disease organisms. One of the most common diseases of the European corn borer in nature is *Perezia pyraustea* Paillot. Zimmack, et al. (1954 and 1958) discussed the distribution of this parasite in corn borer populations over the North Central states and the effect of this parasite on the corn borer. Infected corn borer moths were reported to lay fewer eggs and live shorter lives. Also progeny from infected moths survived at a much lower level than those from uninfested moths.

(5) Agricultural Practices - The practices used in producing corn have certainly changed from time to time, e.g., recommended planting dates, plant densities fertilization, mechanization of harvesting, crop rotation, resistant varieties. These practices may act as additive or subtractive processes in the life system of the corn borer.

Chiang and Hodson (1972) in summarizing agricultural practices lists those that favor increases of corn borer populations as use of soil insecticides and increase of plant densities and those that cause a decrease as picker shellers, the reduction of corn-oats sequence, and the use of resistant varieties. Scott, et al. (1964) reported that the addition of N fertilizer apparently was responsible for increased borer survival.

Many factors appear to be involved in the life system of the European corn borer. However, certain ones may be exerting a dominant influence on corn borer populations. Chiang and Hodson (1972) stated that the major factors responsible for keeping the corn borer population at a low level in Minnesota since 1952 are climatic and agricultural practices. In southern Canada, Barlow (1971) has shown that 2 key factors are responsible for population fluctuations of the first brood. They are rainfall during the growing season, and the number of female corn borers emerging in the spring. These 2 factors have been used to explain fluctuations in the numbers of first brood corn borers in one county in southern Ontario between 1947 and 1961. A partial regression analysis using these 2 factors has accounted for 94% of the variance in estimating size of the first brood.

The elements of the life system of the corn borer vary from one geographical area to another. For instance, one factor may be present in one area and not in another such as some agricultural practice or environmental factors such as temperature, moisture and photoperiod may vary appreciably between areas. Thus, one might visualize that the corn borer may have many life systems in which it evolves and exists. This being true, then one might also expect that different populations of the same species living in different life systems would be different in their ecological responses to certain environmental factors.

The differences between populations of the corn borer was exhibited by Sparks, et al. (1966a, b). They concluded (after finding differences between populations from Minnesota, Iowa and Missouri in numbers of moths emerged, numbers of diapausing larvae, and percentages of surviving forms that diapaused), that naturally occurring biotypes of the corn borer do exist in the mid-west. Chiang, et al. (1968) compared differences in the ecological responses of 3 biotypes of the corn borer from the North Central states. They found differences between the 3 populations for the following ecological characteristics (1) percent survival, (2) percent larvae entering diapause, (3) number of tunnels made by each borer, (4) rate of development, and (5) sensitivity in survival to host-resistance factors. The Minnesota population was distinctly different in all areas from the Missouri population, whereas the Iowa population was intermediate. They also concluded that in regard to the relative sensitivity of the biotypes to changes in ecological conditions that: "(a) the biotype which adapted to warm conditions is more sensitive to temperature changes than is the biotype which is adapted to cooler conditions, and (b) the biotype which is adapted to shorter days is more sensitive to changes in photoperiodism than is the biotype which is adapted to long days."

Concluding Remarks

The life system concept offers guidance to the researcher and to the teacher of population dynamics, mainly by stressing that both the subject population of insects and the environmental elements that affect the population are interacting together in determining the evolution, abundance and existence of the population. Once life systems of insect pests, beneficial insects and insect pathogens are available, the chances of success in managing pest insects below economic levels will increase.

References Cited

Barlow, C. A. 1971. Key factors in the population dynamics of the European corn borer, Ostrinia nubilalis (Hbn.). Proc. 13th Int. Congr. Entomol. Moscow, 1968 1: 472-473.

Beard, R. L. and N. Turner. 1942. Investigations on the control of the European corn borer. Connecticut Agr. Exp. Sta. Bull. 462.

Beard, R. L. and N. Turner. 1943. The significance of growth stages of sweet corn as related to infestation by the European corn borer. Connecticut Agr. Exp. Sta. Bull. 471.

Brindley, T. A. and F. F. Dicke. 1963. Significant developments in European corn borer research. Ann. Rev. Entomol. 8: 155-176.

Chiang, H. C., A. J. Keaster, and G. L. Reed. 1968. Differences in ecological responses of three biotypes of Ostrinia nubilalis from North Central United States. Ann. Entomol. Soc. Amer. 61: 140-146.

Chiang, H. C., A. J. Keaster and G. L. Reed. 1968. Host variety as an ecological factor in the population dynamics of the European corn borer, Ostrinia nubilalis. Ann. Entomol. Soc. Amer. 61: 1521-1523.

Chiang, H. C., A. J. Keaster and G. L. Reed. 1972. Population fluctuations of the European corn borer, Ostrinia nubilalis, at Waseca, Minnesota 1948-70. Environ. Entomol. 1: 7-16.

Clark, L. R. 1964. The population dynamics of Cardiaspina albitextura (Psyllidae). Australian J. Zool. 12: 362-380.

Clark, L. R., P. W. Geier, R. D. Hughes and R. F. Morris. 1967. The Ecology of Insect Populations in Theory and Practice. Metheren and Co., Ltd., London.

Geier, P. W. 1964. Population dynamics of codling moth, Cydia pomonella (L.) (Tortricidae), in the Australian capital territory. Australian J. Zool. 12: 381-416.

Geier, P. W. 1966. Management of insect pests. Ann. Rev. Entomol. 11: 471-490.

Guthrie, W. D., J. C. Huggans and S. M. Chatterji. 1969. Influence of corn pollen on survival and development of second-brood larvae of the European corn borer. Iowa State J. of Sci. 49(2): 185-192.

Hill, R. E., A. N. Sparks, C. C. Burkhardt, H. C. Chiang, M. L. Fairchild and W. D. Guthrie. 1967. European corn borer, Ostrinia nubilalis (Hbn.) populations in field corn, Zea mays (L.) in the North Central United States. Nebraksa Agr. Exp. Sta. Res. Bull. 225 (North Central Regional Publ. 175). 100 pp.

Scott, G. E., F. F. Dicke, and L. H. Penny. 1965. Effects of first brood European corn borers on single crops grown at different nitrogen and plant levels. Crop Sci. 5: 261-263.

Sparks, A. N., H. C. Chiang, C. C. Burkhardt, M. L. Fairchild and G. T. Weekman. 1966. Evaluation of the influence of predation on corn borer populations. J. Econ. Entomol. 59: 104-107.

Sparks, A. N., T. A. Brindley and N. D. Penny. 1966. Laboratory and field studies of F_1 progenies from reciprocal matings of biotypes of the European corn borer. J. Econ. Entomol. 59(1): 915-921.

Sparks, A. N., H. C. Chiang, A. J. Keaster, M. L. Fairchild and T. A. Brindley. 1966. Field studies of European corn borer biotypes in the mid-west. J. Econ. Entomol. 59(1): 922-928.

Steinhaus, E. A. 1949. Principles of Insect Pathology. McGraw-Hill Book Co., Inc., New York. 737 pp.

Tansley, A. G. 1935. The use and abuse of vegetational concepts and terms. Ecology 16: 284-307.

Turner, N. and R. L. Bread. 1950. Effect of stage of growth of field corn inbreds on oviposition and survival of the European corn borer. J. Econ. Entomol. 43(1): 17-22.

Zimmack, H. L., K. D. Arbuthnot and T. A. Brindley. 1954. Distribution of the European corn borer parasite, *Perezia pyraustae*, and its effect on the host. J. Econ. Entomol. 47: 641-645.

Zimmack, H. L., and T. A. Brindley. 1957. The effect of the protozoan parasite, *Perezia pyraustae* Pallot on the European corn borer. J. Sci. 49(2): 185-192.

STATUS OF BIOLOGICAL CONTROL PROCEDURES THAT INVOLVE PARASITES AND PREDATORS

Bryan P. Beirne
Pestology Centre
Simon Fraser University
Burnaby, British Columbia

In this paper I am critical of certain defensive actions of biological control workers that can be harmful to the interests of this subject in the long run. And I am dubious about the future of biological control as a separate subject or discipline. But I should make it clear at the beginning that I am a proponent of biological control as a class of valuable pest damage control procedures and I am confident that its use and importance in that respect will increase substantially in the future.

Biological control is the use by man of living organisms in attempts to reduce the harm caused by pest organisms. Parasites and predators are among the living organisms chiefly used. They are manipulated in three main ways to reduce the general population levels or the amplitude of population fluctuations of pest insects:--(a) the environment is managed to benefit existing parasites or predators, which is biological control by conservation. (b) The existing populations of parasites or predators are augmented temporarily by mass liberation, which is biological control by inundation. And (c) new kinds of parasites or predators are introduced and established as permanent additions to the environment, which is biological control by introduction, or inoculation, or establishment, or colonization.

All these procedures are useful. All will be used increasingly as integrated control programmes and pest management systems are developed. At present each has one or more important obstacles to its widespread or optimum use--obstacles that are in addition to the standard one of considerable gaps in our knowledge of what specific parasites and predators might be manipulated and how and when. It is those obstacles that I wish to discuss, using as far as possible Canadian examples and with emphasis on some that, I believe, retard the development of work in biological control by introduction.

The first of the three procedures, biological control by conservation, is an aspect of cultural control. Like most other aspects of cultural control it has been relatively neglected in the past, it is rapidly becoming increasingly important as integrated control programmes and pest management systems are developed, and it can be relatively ineffective if applied only in a small area because the pests may soon migrate in from and the parasites and predators out to adjacent untreated areas. The effectiveness increases as the size of the treated area increases. The obstacle to the optimum use of this and other cultural controls is the normal lack of means of ensuring that the controls are applied uniformly and simultaneously throughout a region. To overcome it involves regulating human behavior. The chief potential solution is therefore primarily legislative.

The chief obstacle to the widespread application of biological control by inundation is the difficulty in mass-producing or mass-collecting most species, other than some trichogrammids and coccinellids, at economic cost. To overcome this obstacle, effective techniques will have to be developed for mass-producing parasites and predators on artificial diets. The chief potential solution is therefore a technological one.

Biological control by introduction is, when it works, the best and most economical method of pest control next to pest eradication. It is at present one of the two practical methods of wide application that can convert a pest into a lesser pest or a non-pest and then keep it permanently at the lower level without continuing human assistance, and do so at no direct cost to the producers and without causing undesirable side effects to the environment. The other method is of course the development and use of pest-resistant plants.

It is safe to predict that biological control by introduction will continue to be applied in the future at least as frequently as in the past, and quite possibly more so. One reason is that it sometimes can be a permanent alternative to the use of temporarily-effective, environment-polluting chemical pesticides. Another reason is that it is inevitable that sooner or later every pest will be distributed unwittingly by man to every part of the world where it can survive; its parasites and predators will then have to be distributed deliberately through this biological control procedure. A third reason is that the addition of a new pest mortality factor in the form of an introduced parasite or predator may have a major effect in relatively simple environments. With the green revolution such environments are increasing in area and kind.

I do not intend to attempt to evaluate the specific successes and failures that followed the various introductions of parasites or predators and their establishment in regions that they did not inhabit before. This subject has been surveyed and analyzed in detail by the Department of Biological Control of the University of California, Riverside, for the United States Department of Agriculture. The results of this survey were summarized by DeBach (1971) and will be published soon in a USDA Memoir (in press).

Instead I will discuss what I regard as obstacles to the optimum use of this method. I will use mostly examples of occurrences in Canada. In doing so I do not mean to imply that those were in any way uniquely Canadian. They are undoubtedly merely examples of what has happened or will happen to varying degrees in other regions where biological control by introductions is attempted.

This biological control procedure suffers from an enormous handicap. It is that the majority of attempts to control individual pest species by it were not successful. In Canada this procedure has not been tried or has not worked with such major kinds of pests as grasshoppers, root maggots, wireworms, cutworms, leafhoppers, aphids, mosquitoes, or blackflies, or worked significantly with such notorious individual pest species as the spruce budworm, European corn borer, Western wheat stem sawfly, balsam wooly aphid, and European pine shoot moth.

From the administrative viewpoint biological control by introduction consists of a series of gambles. Most of the gambles fail and some succeed only nominally, so there is apparently much loss of money. A few succeed. The returns from them can far exceed the total losses from the remainder, in part because the returns can recur annually indefinitely. In the long run, therefore, the gambles as a whole more than pay off, provided that there are enough of them and that they are big enough. They pay off because the results are not by chance but are biased by us in our favor to the best of our abilities. However, fiscal administrators like to know in advance what is going to happen to money, even if it is going to be wasted. They don't like uncertainties, and biological control attempts are uncertain. They tend to comment on failures and this tends to induce a somewhat defensive attitude among proponents of biological control which in part can lead them to any or all three things that, in my opinion, could be to the disadvantage to biological control in the long run.

The three things are:--I. To be over-charitable in assessing results of biological control attempts and uncritical of claimed successes; that is, not to point out all the gambles that failed. II. To over-react to criticisms of aspects of biological control procedures; that is, fail to accept that the means employed to bias the results of a gamble may not be as effective as generally supposed. And III. To promote the wrong subject. Such actions may win individual battles in defense of biological control but they can increase the possibility of losing the war.

I. The result of biological control attempts in Canada by introducing natural enemies was evaluated by Turnbull and Chant (1961), by DeBach (1964) and by Munroe (1971). Between them these judges considered that a total of 21 pest species of insects were subjects of attempts that were at least substantially successful. But there were wide differences in opinion of the identities of such species. In fact there were only three species that all three evaluators agreed were substantially or completely controlled and only six others that any two agreed upon. This shows the extent to which evaluations can be matters of opinion, which in turn demonstrates that standard criteria for success are not always used.

Criteria are easy to define. Biological control attempts by introducing natural enemies are not merely interesting exercises in establishing organisms in new areas. They are practical attempts to reduce the harm that pests cause by reducing pest populations and keeping them at the new lower levels. The successful establishment of a deliberately introduced natural enemy of an insect (or of any other kind of pest) can therefore be termed a successful attempt if it can be demonstrated, actually or even with a high degree of probability, that: the insect was a pest that needed control; its average level of abundance was substantially reduced by the natural enemy that was deliberately introduced for that purpose and has been subsequently maintained at the new lower level primarily by it; and this causes substantial economic savings annually.

If the nine apparently most successful attempts in Canada--the three that Turnbull and Chant, DeBach and Munroe all agreed were substantial or complete successes and the six which on any two of them agreed--are evaluated in relation to those criteria the conclusion is not that there were nine successes. It is that there were not more than five. There were two that should not be regarded as attempts at persistent control, because they were in greenhouses and had only temporary effects, and two that were failures.

Two of the five were virtually predictable as successes as they resulted from the relocation of established parasites from one part of Canada to another in the same habitat. These were the attempts against the oystershell scale, *Lepidosaphes ulmi* (L.), and the wooly apple aphid, *Erisoma lanigerum* (Haus.). A third was in part fortuitous. This was the attempt against the European spruce sawfly, *Diprion hercyniae* (Htg.), which was ultimately controlled by a combination of parasites introduced deliberately from abroad--a total of some 900 million specimens of a total 27 species were liberated in Canada--and by a virus disease that apparently was introduced accidentally from Europe, perhaps on introduced parasites. Not only did the virus come in by chance rather than by design, but the initial design of the biological control people was to keep it out. Earlier the sawfly larvae containing the virus had been shipped from Europe to the Belleville Laboratory in Canada as a possible new biological control agent for liberation, but by administrative decision the diseased sawfly larvae were quickly incinerated for fear that the disease might escape to harm the mass-production of sawfly parasites that was in progress at Belleville at that time (Finlayson, personal communication).

The two attempts that apparently were successful were against the larch sawfly, *Pristiphora erichsonii* (Htg.), and (at least in the East) the larch casebearer, *Coleophora laricella* (Hbn.).

The two attempts that the evaluators regarded as successes but that actually were failures illustrate how misinterpretations can arise. I will merely summarize the conclusions here because the examples will be reviewed in greater detail elsewhere.

The European wheat-stem sawfly, *Cephus pygmaeus* (L.), has not been a pest of any significance in Ontario since the early 1940's when an European parasite was introduced, became established, increased, spread, and maintained fairly high rates of parasitism at low sawfly population densities. On a basis of those facts this was obviously a successful biological control attempt and was rated as such by Turnbull and Chant and by DeBach (Munroe did not mention it at all). But consider additional facts--the sawfly never was a pest in Ontario--in fact before the biological control attempt was initiated there was only one record, and that about 50 years before; the introduced parasite apparently cannot survive on dense sawfly populations; and it largely replaced an existing parasite that can. I cannot rate as a successful biological control attempt the conversion of a non-pest into a non-pest with a parasite that may not be effective if the non-pest ever becomes a pest and that has partly replaced one that might be effective.

A major outbreak of the European lecanium scale, *Lecanium tiliae* L. (earlier referred to as *Eulecanium coryli* L.), developed in southwestern British Columbia over several years in the 1920's. An imported parasite was liberated. Parasitism of the scale increased rapidly over the next few years and the outbreak collapsed. A success, according to Turnbull and Chant and to DeBach (Munroe queried it), and "an outstandingly-successful small-scale experiment in biological control," to quote Imms (1947). But consider additional facts: the present parasites do not include the species that was allegedly introduced; the original outbreak apparently was controlled by a native parasite; the scale is now at more or less constant outbreak levels and has been for years; and parasitism is normally insufficient to control it. Therefore, no good evidence exists that the introduced parasite became established or, if it did, that it had any effect. The conclusion is that there is no good evidence that this biological control attempt was anything but a failure.

These two examples between them illustrate some of the assumptions that may be implicated in some over-optimistic conclusions that biological control gambles were successful: assumptions that the insect was in fact a pest before the attempt was made; that the mortality caused by the introduced agent was additional to rather than instead of that caused by other factors; that a population reduction that followed the establishment of an introduced agent was caused by it and not, through a coincidence in timing, by something else; that a population reduction apparently caused by an introduced agent was the start of a permenent maintenance by it of the population at a lower level. If assumptions such as these are not taken for granted then some other so-called successes must be rated as, at best, not proven.

In summary, I rate as actual successes in Canada about four and a half of the nine top so-called successes. The proportion of dubious cases is higher when examples that were rated as successes in single evaluations are considered. However, there are conspicuous apparent exceptions: two species for which there are factual life-table data that indicate success, the European winter moth, *Operophtera brumata* (L.) (Embree, 1971), the pistol casebearer, *Coleophora malivorella* Riley (Paradis and LeRoux, 1971). They were not evaluated by Turnbull and Chant or by DeBach merely because the biological control attempts against them were not completed until after those authors had published their evaluations. However, the control of the pistol casebearer was completely fortuitous: the parasite responsible was introduced against the larch casebearer and attacked the pistol casebearer on apple without human design or assistance.

I have tried to show here that some biological control workers in Canada may have been influenced by wishful thinking in evaluating results of some biological control attempts. I do not believe that in this respect they differed greatly from biological control workers elsewhere. I believe that in general biological control proponents have been insufficiently critical of biological control results. From the short-term viewpoint this may have benefited biological control as it increased the number of apparent successes. From the long-term viewpoint it can only be harmful as it can cast unjust doubts on the validity of all attempts.

II. My second main criticism of biological control workers is that they sometimes tend to over-react to criticisms of aspects of biological control practices. One manifestation is emotionally-biased verbiage. A section in the recent book on Biological Control by Huffaker, et al. (1971), pp. 42-55, illustrates this. Though this is not a Canadian example it rehashes controversies in which Canadians were involved. This section is entitled "Challenging concepts relative to the theory and practice of biological control." Incidentally, if you wish to avoid using the phrase "criticisms of" you may use "challenging concepts relative to." In this section criticisms of biological control practices become "attacks" on them. They are "continuing attacks" where the same individual was critical more than once, and when different individuals made the same criticism the more recent "renewed an old attack." A strong statement that the authors agree with is made "forcefully;" one that they disagree with is "overly emphatic." A defender of biological control "poignantly stated;" a critic made a "dour prognostication." Colleagues that the authors agree with have provided "adequate refutation" (incidentally, it is not clear to me why it was necessary to re-refute something that was already refuted adequately) of or have "admirably dealt" with criticisms that included statements or views that are "odd," "strange," "astonishing," "faulty," "erroneous," "damaging," and "irresponsible." Some of the verbal bias is rather more serious as it descends to lower levels of ad hominem argument by imputing less-than-honest behavior to one critic. It was alleged that in one instance he "misleadingly presents a half-truth;" in another he was alleged to "distort."

This kind of abuse is to be deprecated. Moreover, it is contrary to the basic spirit of pest management as it tends to cause work segregation rather than to encourage its integration. To call a criticism an attack and then to discredit the so-called attacker because he has attacked does not bring credit to anybody. I suggest that it would be better for biological control in the long run to encourage criticisms and discuss them with professional objectivity than to try and frighten off potential critics. I suggest that arguments against criticisms are most constructive, most convincing, and most dignified when they are based on objective evaluations of facts. If the facts and arguments do not refute adequately, then perhaps there may be something valid in the criticisms.

A reason why criticisms of biological control procedures can produce emotional reactions is indicated in the same section of the same book. It is that administrators and entomologists in decision-making positions may notice the criticisms and be influenced by them to the disadvantage of biological control activities. This might be related to the fact that some of those who work in biological control are employed specifically as biological control specialists in agencies that have the words "biological control" in their titles. To preserve and promote the interests of their specialties and of their agencies they are in effect forced into a position of having to defend biological control against so-called attacks on it.

III. This leads to my third criticism of biological control workers: That they promote the wrong subject. By this I do not mean to imply that it is wrong to promote biological control; I mean that the best way to promote it is to promote the subject of which it is now becoming an integral part: the management of pests through integrated programs.

In developing an optimum pest management system biological control by conservation, by introduction, and by inundation should be considered, in that sequence, and applied where feasible before resorting to applying non-selective pest-killing procedures such as chemical pesticides (Beirne, 1967). It would seem, therefore, that the best way of promoting the use of biological control procedures to the extent that their capabilities and potentialities warrant is to promote development of the pest management systems of which they are integral parts. They would then be considered automatically when every pest management system that involves a spectrum of applied controls and their integration with one another and with natural regulatory factors is being developed. Instead of promoting biological control gambles individually, promote as a standard practice the testing of biological control possibilities. Or, instead of promoting biological control as an alternative to chemical control, promote chemical control as a supplement to biological control.

The work activities and the titles of most of the agencies that are concerned directly or indirectly with pest control are based on kinds of pests or of pest situations. Biological control agencies are exceptions. They are based on the use of a group of procedures. Agencies based on procedures tend to be inherently and administratively restricted by their titles to those procedures. This does not matter when they exist to provide services that are required for a procedure to be applied and that are unique to that procedure: for example, the biological control overseas exploration agencies that find, procure, and supply biological control agents for colonization, the quarantine laboratories that import and distribute them, and the rearing laboratories that develop and use mass-propagation techniques. But it does matter, insofar as the long-term viability of the procedure-based agencies are concerned, when the main function is pest control and related research. With the increasing development of the concepts of control integration and of pest management the promotion of individual classes of procedures becomes decreasingly realistic and increasingly anachronistic and therefore progressively susceptible to being changed.

Consider what has happened in Canada. In the early 1950's - before biological control was a popular subject: before Silent Spring - there were 35 to 40 entomologists in the Biological Control Investigations Unit of the Federal Department of Agriculture. Twenty years later there apparently will be only one entomologist employed specifically and exclusively by that Department for biological control work--the man in charge of the importation service. This does not mean that there were or will be corresponding reductions in the amount of work or the numbers of workers in biological control. It means that biological control is now being done in Canada as integral parts of broad studies on kinds of pests or of pest situations instead of as a separate subject.

This evolution was in three stages and in three different ways. First, by formal agreement the responsibility for all work on biological control of forest insects in Canada was taken from the Biological Control Investigations Unit by the Forest Biologists. This happened in 1954. It was the beginning of the end for biological control as a separate entity in Canada. The Unit was eliminated soon after. Next, in the mid-1960's, a group of eight entomologists moved themselves voluntarily out of biological control into pestology and pest management. And, finally, the Department of Agriculture made the decision last year to move almost all the remaining biological control workers to pest-based studies: the largest group to Winnipeg, Manitoba, allegedly to develop integrated control programs, especially for cereal insects, and most of the remainder to research centers in various parts of the country to join existing groups that are based on the study and management of particular kinds of pest organisms.

What has happened in Canada can happen elsewhere, and it is likely to because the same kinds of trends and pressures towards holism exist elsewhere to varying degrees. As biological control procedures come to be regarded fully as areas in an integrated control or a pest management spectrum instead of as comprising a separate subject, biological control agencies (other than the service ones) will evolve into or be combined with pest-based agencies or be replaced by them. Attempts to retard this evolution by trying to include integrated pest control under biological control are unrealistic as they are in effect attempts to include the whole within one of its parts. The reverse is happening: biological control is being encompassed and engulfed by pest management.

The chief cause of the expansion of biological control as a distinct subject, anywhere, was the existence of major pest problems for which it provided the best apparent solution. The chief cause of the diminution of biological control as a separate subject is the development of the concept of pest management through integrated controls, because it offers an apparently better solution to pest problems than does biological control alone. How and how fast and to what extent this evolution of biological control occurs is determined by people. If we accept that the evolution is inevitable, as I believe it to be, then biological control people and agencies are faced with a choice of three possible courses of action. One is to remain as biological control specialists, which means concentrating on those aspects that are distinctive to biological control, namely procuring, propagating, and providing biotic agents. A second is to evolve into pest managers in both name and deed, which is the logical choice, and one in which they can excel because of their ecological outlook. The third is to resist change actively or positively, which means waiting to have it imposed upon them.

The crucial decision that most biological control specialists have to make is this: either to be realistic and try to manage their evolution themselves to the advantage of themselves and their specialty; or to be reactionary and resist evolution and thereby leave themselves and their subject open to management by others. Biological control workers still have the choice; but if they wait too long they will lose it. At least that is the theory, though what has happened in Canada does not provide much support for it. There, administrators were faced on three occasions with decisions on what to do with biological control studies: to maintain them or to evolve them. On one of the three occasions, in 1967, biological control people advocated evolution and the administrators decided to maintain the existing situation. On the two other occasions, in 1954 and 1971, the administrators decided to evolve biological control studies though most of the biological control people wanted to maintain the existing situations. Despite all this, if evolution is inevitable biological control people should try to take positive leadership in it.

References Cited

Beirne, B. P. The biological control attempt against the European wheat-stem sawfly, Cephus pygmaeus (Hym.), in Ontario. Canadian Entomol. (In press).

Beirne, B. P. 1967. Pest Management. Internatl. Textbook Co., London.

Clausen, C. P. (ed.) A world review of parasites, predators, and pathogens introduced to new habitats. USDA Publ. (In press).

DeBach, P. (ed.). 1964. Biological Control of Insect Pests and Weeds. Reinhold Publ. Co., New York.

DeBach, P. 1971. The use of imported natural enemies in insect pest management ecology. Proc. Tall Timbers Conf. on Ecol. Animal Control by Habitat Management 3: 211-233.

Embree, D. G. 1971. The biological control of the winter moth in Eastern Canada by introduced parasites. In Huffaker, C. B. (ed.), Biological Control, pp. 217-226. Plenum Press, New York and London.

Finlayson, T. personal communication.

Huffaker, C. B. (ed.). 1971. Biological Control. Plenum Press, New York and London.

Imms, A. D. 1947. Insect Natural History. Collins, London.

Munroe, E. 1971. Status and potential of biological control in Canada. In Biological Control Programmes against Insects and Weeds in Canada, 1959-1968. Commonwealth Inst. Biol. Cont. Tech. Comm. 4: 213-255.

Paradis, R. O. and E. J. LeRoux. 1971. Coleophora malivorella Riley, pistol casebearer (Lepidoptera: Coleophoridae). In Biological Control Programmes in Canada, 1959-1968. Commonwealth Inst. Biol. Cont. Tech. Comm. 4: 15-16.

Rubin, A. Thesis study on the lecanium scale in British Columbia currently in progress. Pestology Centre, Simon Fraser University.

Turnbull, A. L. and D. A. Chant. 1961. The practice and theory of biological control of insects in Canada. Canadian J. Zool. 39: 697-753.

IDENTIFICATION AND CLASSIFICATION IN PEST MANAGEMENT CONTROL

Paul W. Oman
Department of Entomology
Oregon State University
Corvallis, Oregon

I hope what I discuss will be in tune with the spirit of this Summer Institute. It seems to me that our consideration of pest management programs, of the strategies and tactics that we plan to employ, has reached the stage where we should give relatively less attention to theory, and more to the day-to-day activities upon which the success or failure of programs will ultimately depend. Accordingly, my discussion will be directed toward the practical aspects of identification and classification, and will be concerned primarily with two questions--"What are the problems connected with identification of biological organisms?" and "How can we improve the situation?"

Concise but reasonably comprehensive statements of the role of taxonomy, which embodies identification and classification, in pest management and control are available in two recent publications issued, respectively, by the National Academy of Sciences and the National Research Council. These publications are:

NAS, 1969. Principles of Plant and Animal Pest Control, vol. 3, Insect-Pest Management and Control. xxii + 508 pages. Ch. 2, pages 9-22, deals with Identification and Classification.

NRC, 1970. Systematics in Support of Biological Research. 25 pp.

More detailed discussions of the relation of taxonomy to applied biology have been given by numerous authors. Because I shall not otherwise have occasion to cite these numerous papers, a list of a few that seem to me particularly relevant to the matters under consideration is appended (Appendix I). The list also includes several publications of a more general nature that deal with systematic biology, since these give perspective to the field, and contain information about possible future developments of importance to pest management.

The title of my topic, considered together with the theme of this Institute, suggests that there is again concern about the identification and classification of pest species. There was a time, not many years ago, when the prevailing attitude was "Hell, we don't need to know what it is--we can kill it!" But we no longer hear talk of using chemicals to create "biological deserts," and there is a growing realization that to "manage" a pest species entails more than just the ability to kill a small segment of its population. If it is to be effective in its objectives, and economically feasible, pest management requires more subtle and ingenious methods; methods that may often need to be used in combination, either concurrently or sequentially, over considerable periods of time and large areas.

All this presupposes not only knowledge of what the pest is, but a great deal about it and the environment in which it operates--what plants it depends upon for population survival, what enemies it has, its physical and chemical tolerances and requirements, and its biology and behavior. To know these things about a pest and its biological associates places a considerable premium on the basic question of "What is it?" when we start talking about pest population ecology.

We begin with the assumption that identification and classification of organisms contributes constructively to pest management. How is this so? What good are identifications?

In the simplest, most direct way, identifications tell us whether something is important or not, whether it is potentially harmful, whether it is good or bad from a particular point of view in time. In more concrete terms, to identify means to distinguish (for example) between Virginia creeper and poison ivy, between a wasp that stings and its syrphid fly mimic that only acts like it will sting, between the moth that lives in wool clothing in the closet and the one that prefers cereals in the kitchen cabinets.

But "identification," in the sense of contributing to pest management, involves more than just providing a name. To be useful the name supplied must either evoke associated information that is common knowledge, or provide ready access to the desired information. Unfortunately, all too frequently requests for identifications of organisms produce nothing but names, or worse yet, nothing.

Of course there are reasons for this. And as we explore the relations of identification and classification to pest management programs I hope we can at least improve our understanding of the situation, if not find solutions to the shortcomings.

Classification is a means by which we can deal with many objects under a few headings. Its objective is to systematize knowledge or data, and in the process reveal associations that may lead to new knowledge. To classify we commonly arrange information in terms of a heirarchy of nested, mutually exclusive groups for convenience of reference. In biological classification these groups are the species, genera, families, orders, and so on, that we recognize.

Theoretically, at least, classification provides a framework within which all knowledge regarding each species may be recorded. To the extent that a classification reflects genetic relationships, it permits useful generalizations and contains a high degree of predictability regarding pest species and their ultimate control. For example, Rosen (1969) states that "All the known species of *Acerophagus* (members of the Hymenoptera family Encyrtidae) are parasitic in mealybugs." It would be a reasonable assumption, therefore, that the as yet undiscovered species of *Acerophagus* will also be parasitic in mealybugs.

Another example: In Japan and adjacent areas of the Far East, Japanese B encephalitis is transmitted primarily by *Culex tritaeniorhynchus*. During investigations in Japan during the early 1950's this mosquito could not be found during the winter months, although knowledge of its behavior during that period of its developmental cycle might obviously be important from the standpoint of suppressing the disease that it transmits. *C. tritaeniorhynchus* belongs to the typical subgenus *Culex*, all members of which, so far as were then known, overwinter in the adult stage in temperate climates. It was therefore predicted (Oman, 1957) that *C. tritaeniorhynchus* would overwinter in Japan as adults. Subsequent investigations established that this was so (Bullock, et al., 1959).

Predictability, on the basis of taxonomic correlations, becomes more important as pest control tactics become more complex. It should also be apparent that the more we know about species--their structure, behavior, biologies--the better will be our classifications and hence the predictions based upon them.

Our ability to make precise identifications is severely handicapped by the multiplicity of organisms with which we must deal, the fantastic variety of forms in which they occur, and the generally inadequate state of our knowledge of the organic world. Estimates of the number of kinds of living things on earth go as high as 10 million. But these estimates are at best educated guesses; no one really knows the extent of our fauna and flora. The number of kinds of insects that are known--in the sense of having been named and "described"--stands somewhere around 800,000. How many yet remain to be discovered and characterized?

Although estimates of the total kinds of animals, or of selected groups of animals, may vary widely, most informed scientists are in agreement on one point--we are still far short of having accomplished anything like a complete inventory. Some workers (Raven, et al., 1971) express the view that only about 10 to 15 percent of the total kinds of life is now known, and make the gloomy prediction that only another 5% will become known before the remaining 80% becomes extinct under the impact of increasing pollution and environmental deterioration.

But as with the total kinds of organisms, these estimates as to the state of our progress in enumerating them are only guesses; few attempts have been made to analyze the situation critically. One interesting approach has been made by Steyskal (1965, 1967) who has used "trend curves" to gain an understanding of how this enormous job is progressing.

The use of "trend curves" in this manner, as Steystal points out, requires acceptance of the premise that there exists a finite and relatively fixed number of species on the earth, and that we have gradually been approaching a state in which all species will be known. With these 2 assumptions, by plotting the rate at which knowledge of species has been gained from 1758 (1760) to the present, the shape of the curve so produced should give a rough prediction of where we stand in our progress from scant to relatively complete knowledge, at the taxonomic level, about living organisms.

Steyskal's trend curves for several groups of animals were prepared by plotting accumulated numbers of species on ordinates, and years from 1960 to 1965 on abscissas. When the resulting curve was sigmoid he assumed the process of naming and describing species to be relatively complete. From these curves he inferred that the job of describing and naming North American birds to be virtually complete. But the curves plotted for the fleas and mosquitoes, and most other groups of insects considered, suggest that we are still considerably short of the mid-point in the task of enumerating them. An analysis of the state of our knowledge of flies in America North of Mexico, which stood at ca. 16,800 in 1967, predicts that the number will have been increased to at least 20,000 by the year 2100.

Perhaps this is still a rather discouraging outlook, and certainly would be if we assumed that complete knowledge of a fauna is necessary to make valid predictions. Personally, I am inclined to a rather optimistic view of the situation for the simple reason that if we can do as well as we have in the prediction game, knowing considerably less than half of our arthropod fauna in a very sketchy way, my faith in the taxonomic method is confirmed, rather than being shaken.

Knowing what we do about the numbers of kinds of biological organisms, it should be easy to understand why so many "identifications" are incomplete, incorrect, or if correct, often furnish little biological information.

The situation is further aggravated by the fantastic intraspecific diversity and variation that occurs in many organisms, particularly arthropods. Some of the more obvious reasons for this great diversity are:

Life cycle differences, especially in the developmental stages of holometabolous insects.

Sexual differences, extremes of which occur in dimorphic species such as bagworms, Stylopidae and many Hymenoptera.

Polymorphic differences, as found in social insects that have many castes; e.g., ants and termites.

Host-induced differences, as found in armored scale insects, seasonal variations, etc.

But the problems of identification and classification imposed by numbers of species, and kinds and numbers of variants, are not the end of the story by any means. Identifications are customarily made on the basis of individuals (nearly always dead ones), and primarily on the basis of structural evidence. But biological potential is concerned with populations rather than individuals, and changes in the genetic composition of a population may not be accompanied (at least initially) by overt structural changes in the individuals comprising the population. Then there are the biological species, strains, or races--whatever you wish to call them--that are important in pest management because they are adapted to particular environmental conditions or hosts. They may or may not be distinguishable by anatomical differences; usually they are not. Problems of this sort occur with annoying frequency in the parasitic wasps used in biological control, as Paul DeBach (1960, 1969) has so clearly documented. Sometimes, as in the case of _Aphytis holoxanthus_ and _A. coheni_, parasitic in scale insects, correlated structural differences that will permit identification of specimens can be discovered, once the biological differences are recognized. But among the uniparental _Aphytis_, which constitute about 30% of the species for which the sexuality status is known, structural characters that will differentiate these biologically distinct lines are not known. In a situation such as this, conventional taxonomic methods are not helpful; whatever "identifications" are required must be made experimentally.

Considering, then, the complexity of the job, and the limited manpower and facilities that are committed to it, we can only conclude that we are not now equipped to handle the identification and classification problems that will be required to support pest management programs, and I hesitate to predict when we will be. But I consider it a part of my obligation to this Institute to attempt an appraisal of the situation, and suggest possible courses of action that might help improve it.

In recent years there has been frequent (and justified) criticism of the antiquated methods used in taxonomy, much discussion of the feasibility of automated procedures in taxonomy, and of the desirability of using new kinds of evidence that will improve our understanding of phylogenetic relationships of groups of organisms. How practical are these methods?

The automated procedures would be expected to (a) determine the characteristics of a specimen, (b) place those characteristics in computer memory, (c) develop a classification of the specimens, including a diagnostic key to identify additional specimens as either the same species as those previously processed, or new, (d) revise the classification and key as new material is added, and (e) supply on request the stored information about any species included. All these processes except the first are within the capacities of modern electronic computers. Optical scanning devices are being developed that will accomplish at least part of the operation of determining characteristics of a specimen.

The new kinds of evidence that we hope will improve our understanding of organic relationships include primary structure of some proteins such as insulin (which suggests that pigs are more closely related to whales than they are to cattle, horses, or sheep), hemoglobin, dehydrogenases and cytochrome c (which appears to be useful as an evolutionary clock to measure phylogeny). DNA base composition has been very useful in bacterial taxonomy, and offers considerable hope of being helpful in other groups. The chemistry of animal pheromones and defensive secretions promises to be helpful in species determination, as do behavioral characteristics such as courtship songs. Minute details of arthropod structure, as revealed by the scanning electron microscope, will likewise contribute to our ability to discriminate among species once a fund of knowledge sufficient for comparative purposes has been accumulated.

All these lines of study, and others, are highly desirable and essential for the future of a viable taxonomy that in turn will be more useful to pest management. But as for making more than very limited contributions to pest management programs within the immediate future, I think they are largely pie-in-the-sky. In my opinion the development of pest management programs cannot wait for the development and perfection of these more sophisticated, and ultimately much more efficient, methods of classification and identification of organisms. Even the very desirable goal of computer storage and automatic retrieval of information needed in pest management is still not a reality in spite of several years of effort by committees of our scientific biology societies (Foote, 1969, 1970). And even if Bossert's (1969) estimate that the establishment of one or more central computer taxonomic information stores during the 1970's is economically and technically feasible proves to be an accurate one, I am doubtful that we will get much help from that source within the next five years. We are still operating in the Stone Age of automated taxonomic endeavor, and likely to remain there for some time.

So we are stuck with what we have. What do we have?

We have a badly overloaded Federal Systematic Entomology Laboratory, several equally overloaded state organizations that attempt to meet the demands placed upon them, some contributed assistance from specialists located at various institutions, and a potential demand for taxonomic services that far exceeds the capacities of all these together, even if efficiently organized. What can we do about it? I shall discuss this question briefly from 2 points of view: (1) Organizational structure, and (2) individual actions that should improve one's chance of obtaining identifications under our present system.

The NRC publication, "Systematics in Support of Biological Research," outlines a proposal for an "American Institute of Applied Systematics" that would be expected to provide identifications and ancillary information of a taxonomic nature as needed in the U.S. interests. Basically, this proposal calls for a central secretariat to serve as an umbrella agency, and satellite taxonomic centers located in different geographic regions of the country. In addition to the administrative (and clerical) personnel, the taxonomic personnel would consist of technicians, taxonomic specialists (identifiers), and research scientists. A fee system, through which users would in part support the service, is proposed.

In justification of the system of satellite centers that would result in dispersion of identification activities, 3 advantages are cited. These are: Utilization of specialists and collections that already exist--a presumed economic benefit; stimulating taxonomic work through interaction between local systematist and those of the Institute; and facilitating training of new workers in fields where they are most urgently needed.

I accept the validity of the arguments with respect to the last of these cited advantages, but have serious doubts about the alleged economic benefits to be derived from using geographically dispersed facilities and scientific talent for this job. My own conviction is that we should first develop a strong central organization with which satellite centers can later be associated as required. I believe my recommendation for such an organization is not materially at variance with the core plan for an Institute as outlined in the NRC publication.

I believe our immediate needs for taxonomic support of pest management programs can most efficiently be met by a strong, mission-oriented federal organization concerned with taxonomic biology.

The kinds of personnel needed for a service organization of this sort are:
(a) A corps of expert identifiers, also responsible for research on specific, limited problems of identification.
(b) Research taxonomists engaged primarily in improvement of classifications in groups of concern to pest management.
(c) Information and bibliographic specialists, needed to assemble and collate information about taxa under study, including information for transmission along with identifications. At the time identifications are made and reported seems the logical time to "plug in" a summary of information about the organisms, as may be necessary to meet the needs of field biologists.
(d) Support staff of technicians and clerical personnel.
(e) Administrative staff.

So much for meeting _immediate_ needs. To serve the _long-range_ needs of developing pest management programs we will need both better quality in our classifications of biological organisms, and greater precision and efficiency in our identifications. To accomplish these aims I suggest:

(a) Commitment of specialized personnel--geneticists, biochemists, physiologists, molecular biologists, anatomists and others--to production of knowledge of a comparative nature about organisms belonging to taxa of concern in pest management. This sort of knowledge is needed to improve our understanding of biological relationships, and hence our ability to make useful deductions and predictions. It is really immaterial whether or not such personnel are an integral part of a taxonomic organization so long as the desired information, and cooperation with taxonomists, is forthcoming. To date development of knowledge of this sort has been on too much of a hit or miss basis; concentrated, sustained effort is needed.
(b) Precision in identifications, of the sort needed in pest management, may well require discrimination among intraspecific elements as well as among biological species that can usually be recognized only by non-structural evidence. Hopefully most of the cryptic species may eventually be identifiable on the basis of morphological evidence. Recent critical work on some species of the parasitic wasp genus _Trichogramma_ by Nagarkatti and Nagaraja (1971) suggests that this is indeed a reasonable expectation.
(c) Greater efficiency in the identification process is badly needed to cope with what will surely be a considerable increase in demand for this sort of service. Interestingly enough, during the last decade there have been relatively large investments of effort and money in attempts to make classifications more objective, but relatively little attention to the practical problem of getting an answer to the eternal question, "What is it?" Here is an area where pest management programs might profit enormously from adapting modern, highly sophisticated equipment and technologies to their problems. In the medical sciences computer interfaced fast analyzers (Anderson, 1969) have already been adapted to similar procedures. Once the pressure for taxonomic support for pest management is sufficient to finance operation, and the bugs have been worked out (no pun intended), some devices of this sort may be entirely practical. Concentrated effort on improving the speed and efficiency of the identification process is strongly indicated.

There are some things that individuals can do to improve chances of obtaining identifications when needed or requested. A few of these are:
(a) Do-it-yourself identifications. The advice that initial taxonomic identifications of pest species (or other species) should be made by a specialist in taxonomy is generally sound, particularly if there is reason to believe more than one closely related species may be involved. But any alert individual working intensively with a pest

species, or engaged in sequential sampling of a faunal complex, should become sufficiently familiar with most species routinely encountered to make accurate identifications in nearly all cases. Taxonomic specialists should, of course, be consulted in doubtful cases.

(b) Kinds and amounts of material. If possible, get advice on the kinds and amounts of material needed for identification before samples are taken. Many factors--amount of infraspecific variation, occurrence of polymorphic forms, difference in developmental stages, etc.--will influence the decision. If in doubt, take a big sample. It is easier to throw away samples than to augment them.

(c) Association of developmental stages. Field biologists are the ones best able to associate correctly the different developmental stages of polymorphic organisms. Because correct association of stages may be necessary to identify those that have been little studied, attention to this matter may result in acquisition of information not otherwise available. Further, development of means for identification of immature stages depends upon availability of adequate samples known to represent single species.

(d) Preservation and handling of material. Admittedly a difficult subject on which to provide advance information, but if material is important enough to require naming it deserves special care and handling. Consultation with a taxonomist during the planning stage is indicated, if possible. In lieu of that, there are some general guides (Anon., 1965; Oman and Cushman, 1948; Sabrosky, 1971) to the preservation and handling of specimens for identification, and these should be consulted if more information is not readily available.

(e) Preparation of Material. Like preservation of specimens, the exact method of preparation for definitive study by a taxonomic specialist is not easy to predict, for specialists tend to have their own idiosyncracies and preferences. Yet one of the greatest deterrents to prompt and complete identifications in many groups of arthropods is the matter of making dissections or similar preparations, or the slide mounting of minute forms. A taxonomic specialist who is contributing personal time will understandably be reluctant to do technician tasks that someone else can do as well with a little practice.

(f) What needs to be identified, and how completely. Often a partial identification is as useful as a complete one, and the questions "What, and how complete?" can usually be determined by the taxonomic specialist if the reason for requesting the identification is indicated.

(g) Information associated with samples. Anticipating that information that accompanies specimens submitted for identification will eventually be included in information storage centers, it is of interest to the field biologist to provide as much relevent information as possible. Further, such information usually facilitates or expedites identifications.

Of course, there is another way by which taxonomic support of pest management and other phases of applied biology may be considerably improved. That is through better cooperation and coordination of effort. My views in this connection have been stated previously (Oman, 1960) but I venture to reiterate some generalities.

What seems most needed are some adjustments in attitudes and objectives. Traditionally, taxonomists have not been directly involved, and too often not at all concerned about the problems of pest management or control, and consequently disinterested in cooperation. As Sailer (1961) has aptly remarked, "many taxonomists regard utility with suspicion if not outright hostility." Couple that attitude with the too widespread belief among applied biologists that taxonomy has nothing useful to offer, and you don't have a good environment in which to develop cooperation.

It would no doubt be helpful if taxonomists could be induced to reorder their work priorities so as (for example) to direct relatively more attention to the development of sound classifications, and less to the search for and description of taxonomic rarities; to be at least as much concerned about the taxonomy of groups like the corn rootworms and billbugs as they are with the whirligig-beetles; in general terms, to look with greater frequency for their intriguing taxonomic problems among the economically important taxa. I have often argued, usually with little success, that such problems are fully as interesting and challenging as can be found anywhere, and they have the added advantage that more well documented biological and behavioral evidence is readily available.

Perhaps one way of generating meaningful cooperation would be to include taxonomists as partners in pest management programs, particularly in the planning and investigative phases so that an understanding of objectives and methods exists when requests for service identifications arrive. Surely taxonomists cannot be indifferent to the evolutionary significance and taxonomic importance of knowledge of pheromones and other means of infraspecific communication by arthropods that offer so much promise in pest management. By broadening their investigations in these fields, to include studies of siblings of pest species, those concerned with pest management research could produce information of great significance to taxonomy as well as contributing to their own objectives, which include understanding the complex interactions among organisms that occupy a common ecosystem. The evidence from Roelofs and Comeau (1969) on specificity of pheromones in moths of the families Tortricidae and Gelechiidae, that from Moreno, et al. (1972) dealing with pheromones of closely related species of scale insects, and information about the composition of the sting venom constituents of related species of fire ants cited by Buren (1972), are contributions of this nature. Equally intriguing is the mass of information about pheromones of bark beetles and other economically important groups being assembled by many investigators (e.g., Pitman, et al., 1969; Renwick and Vite, 1970; Sanders, 1971). Studies by Rudinsky and Michael (1972) reporting the chemostimulus of sonic signals in a bark beetle should be of special interest to taxonomists interested in evolutionary theory. When our knowledge of the roles of kairomones, allomones and other natural and synthetic chemical messengers becomes more comprehensive, it may provide evidence of comparable importance to taxonomy.

Finally, a brief reference to education of personnel for future pest management programs. Without implying in any way that concern with evolutionary theory is not essential to the education of taxonomists, I believe emphasis on that topic, to the virtual exclusion of training in methods and techniques of identification and classification, is not likely to produce scientists helpful in pest management programs. The process of adjusting the attitudes of future taxonomists to one of more concern with pest management problems must begin during their academic training. And perhaps that is also the time to convince pest management specialists that taxonomy <u>does</u> have utilitarian aspects.

References Cited

Anderson, N. G. 1969. Computer Interfaced Fast Analyzers. Science 166: 317-324.

Anonymous. 1964. The preservation and shipment of dead insects for determination. Survey and Detection Operations, PPC-ARS-USDA. 4 pp.

Bossert, W. 1969. <u>In</u> Systematic Biology, pp. 595-605. Computer techniques in systematics. Publ. 1962. Nat. Acad. Sci. xiii + 632 pp.

Bullock, H. R., W. P. Murdoch, H. W. Fowler and H. R. Brazzel. 1959. Notes on the overwintering of Culex tritaeniorhynchus Giles in Japan. Mosq. News 19(3): 184-188.

Buren, W. F. 1972. Revisionary studies on the taxonomy of the imported fire ants. J. Georgia Entomol. Soc. 7(1): 1-26.

DeBach, P. 1960. The importance of taxonomy to biological control as illustrated by the cryptic history of Aphytis holoxanthus n. sp. (Hymenoptera: Aphelinidae), a parasite of Chrysomphalus aonidium, and Aphytis coheni n. sp., a parasite of Aonidiella aurantii. Ann. Entomol. Soc. Amer. 53(6): 701-705.

DeBach, P. 1969. Uniparental, sibling and semi-species in relation to taxonomy and biological control. Israel J. Entomol. 6: 11-28.

Etzioni, A. 1971. (Editorial) The need for quality filters in information systems. Science 171: 133.

Foote, R. H. 1969. Recent advances in bioscience information--entomology's role. Bull. Entomol. Soc. Amer. 15(3): 233-234.

Foote, R. H. 1970. New directions for commitment. J. Washington Acad. Sci. 60(4): 136-140.

Moreno, D. S., R. E. Rice and G. E. Carman. 1972. Specificity of the sex pheromones of female yellow scales and California red scales. J. Econ. Entomol. 65(3): 698-701.

Nagarkatti, S. and H. Nagaraja. 1971. Redescription of some known species of Trichogramma (Hymenoptera: Trichogrammatidae), showing the importance of the male genitalia as a diagnostic character. Bull. Entomol. Res. 61: 13-31.

Oman, P. W. 1957. The relation of insect taxonomy to mosquito control. Mosquito News 17(3): 149-151.

Oman, P. W. 1960. The relation of insect taxonomy to applied biology. Bull. Entomol. Soc. Amer. 6(1): 3-5.

Oman, P. W. and A. D. Cushman. 1948. Collection and preservation of insects. Misc. Publ. No. 601, USDA. 42 pp. (rev. 1967).

Pitman, G. B., J. P. Vite, G. W. Kinzer and A. F. Fentiman, Jr. 1969. Specificity of population-aggregating pheromones in Dendroctonus. J. Insect Physiol. 15: 363-366.

Raven, P. H., B. Berlin and D. E. Breedlove. 1971. The origins of taxonomy. Science 174: 1210-1213.

Renwick, J. A. A. and J. P. Vite. 1970. Systems of chemical communication in Dendroctonus. Contrib. Boyce Thompson Inst. 24(15): 283-292.

Roelofs, W. L. and A. Comeau. 1969. Sex pheromone specificity: Taxonomic and evolutionary aspects in Lepidoptera. Science 165: 398-400.

Rosen, D. 1969. A systematic study of the genus Acerophagus E. Smith with descriptions of new species (Hymenoptera: Encyrtidae). Hilgardia 40: 4-72.

Rudinsky, J. A. and R. R. Michael. 1972. Sound production in Scolytidae: Chemostimulus of sonic signal by the douglas-fir beetle. Science 175: 1386-1390.

Sabrosky, C. W. 1971. Packing and shipping pinned insects. Bull. Entomol. Soc. Amer. 17(1): 6-8.

Sailer, R. I. 1961. Utilitarian aspects of supergeneric names. Syst. Zool. 10(3): 154-156.

Sanders, C. J. 1971. Sex pheromone specificity and taxonomy of budworm moths (*Choristoneura*). Science 171: 911-913.

Steyskal, G. C. 1965. Trend curves of the rate of species description in zoology. Science 149(3686): 880-882.

Steyskal, G. C. 1967. Another view of the future of taxonomy. Syst. Zool. 16(3): 265-268.

Appendix I

Selected References Dealing with Systematic Biology and Pest Management and Control (Listed by Title)

Entomological Taxonomy: Its Aims and Failures. From a Taxonomic Viewpoint (S. A. Rohwer); From an Economic Viewpoint (A. C. Baker); From an Educational Viewpoint (E. D. Ball), J. Washington Acad. Sci. 16(3): 53-67, 1926.

Modern Taxonomy, Reality and Usefulness. Darlington, P. J., Jr., Syst. Zool. 20(3): 341-365, 1971.

Problems and Trends in Systematics. Munroe, E., Canadian Entomol. 96: 368-377, 1964.

Rainbow's End: The Quest for an Optimal Taxonomy. Johnson, L. A. S., Syst. Zool. 19(3): 203-239.

Systematic Biology. Proceedings of an International Conference. Nat. Acad. Sci. Publ. 1962. xiii + 632 pp., 1969.

The Interrelations of Biological Control and Taxonomy. Sabrosky, C. W., J. Econ. Entomol. 48(6): 710-714, 1955.

The Purpose and Judgments of Biological Classification. Inglis, W. G. Syst. Zool. 19(3): 240-250, 1970.

The Purposes of Classification. Warburton, F. E., Syst. Zool. 16(3): 241-245, 1967.

The Relation of Taxonomy to Biological Control. Clausen, C. P., J. Econ. Entomol. 35(5): 744-748, 1942.

The Significance of Economic Entomology in the Field of Insect Taxonomy. Frison, T. J., J. Econ. Entomol. 35(5): 749-752. 1942.

The Significance of Taxonomy in the General Field of Economic Entomology. Essig, E. O., J. Econ. Entomol. 35(5): 739-742, 1942.

CRITERIA FOR DETERMINATION OF CANDIDATE HOSTS AND FOR SELECTION OF BIOTIC AGENTS

F. D. Bennett
Entomologist in Charge, W.I. Station
Commonwealth Institute of Biological Control
Gordon Street, Curepe, Trinidad

I. Criteria for Determination of Candidate Hosts

The motives for setting classical biological programmes in motion are varied; an analysis of the insect pests on which classical biological control has been attempted would show that few criteria would apply to them all. There may be instances where the insect in question is not a serious pest and hence while it holds true for most investigations even this factor, which would appear to be a logical criterion, breaks down.

In general biological control programmes are initiated as a result of one or more of the following: (1) a recently introduced pest is discovered; (2) other methods of control have been tried and failed; (3) other methods of control are too costly or produce detrimental side effects; (4) a pest is under partial biological control or natural control but an improved level of control is necessary; (5) one species (sometimes more than one) amongst a group of pests is considered a key species which if "taken out" by biological control would permit the relaxation of pesticide treatments and allow established natural enemies to bring the other species of the complex under control.

Likewise the impetus for initiating a biocontrol program may come from various sources; for example: (a) a government agency responsible for biological control activities; (b) a large organization representing farmers, horticulturists etc., with a vested interest in the crop being attacked by the pest under question; (c) an international advisory board or a professional group working on biological control; (d) an individual pest control specialist running into difficulties in a pest management program; or (e) a politician or other important dignitary who has a problem in his own garden, orchard or farm.

Regardless of how the request originated, projects, particularly when actual introductions are contemplated, are brought to the attention of the pertinent government agency for permission to import the biotic agents. Usually their assistance is sought much earlier for advice, finance, professional contacts, etc. There is little doubt that the efforts of an enthusiastic well-informed individual through his persistent prodding and in some instances "salesmanship" can go a long way in getting a project off the ground.

An administrative unit faced with several pest problems requiring biological control and with limited funds and staff may have to produce a list of priorities as to the order in which projects should be undertaken. There is available to assist in this a growing list of references giving the opinions of various authorities in the field of biological control. DeBach (1964) and Huffaker (1971) provide excellent background information. Reviews of classical biological control projects for several particular regions are also available: Wilson (1960, 1963) for Australia and New Guinea; McLeod, McGugan and Coppel (1962); and Corbet, et al. (1971) for Canada; Greathead (1971) for Africa; Rao (1971) for Fiji; Rao, et al. (1972) for Southeast Asia and the Pacific; Clausen (1956) for the USA; Bennett and Hughes (1959) for Bermuda, etc. Among the most outspoken current day proponents for "pressing on" with attempted biological control of insect pests (and weeds) by the classical method are DeBach (1971) and Simmonds (1972). The following statement from Huffaker, Messenger and DeBach (1970: "Native as well as exotic pests are suitable subjects for biological control" sums up the view shared by the author that classical biological control should be among the first means of control to be considered whenever an insect problem arises. This does not mean that other approaches, including eradication when practical, breeding for plant resistance, chemical control when necessary, or the manipulation of natural enemies already present should be ignored, nor does it mean that all programs set in motion will be successful, but as classical biological control still affords the best chance of achieving a permanent solution and also a method which once operative, calls for little or no additional expenditure, it should be given high priority.

Introduced Versus Native Pests

While the statement that pests of both categories are candidates for biological control programs holds true, the statement that introduced pests are better candidates for biological control, with some notable exceptions, particularly where weed problems are considered, is usually valid. Recently introduced pests frequently "explode" because their customary population controls--parasites, predators and pathogens--have usually been left behind. The classical example of biological control of cottony cushion scale, *Icerya purchasi* Mask, i.e., the accidental importation of the pest followed by the introduction of its natural enemies *Rodolia cardinalis* Muls., initially and at a later date *Crypotochaetum iceryae* (Williston), into California and often repeated in other countries (see DeBach, 1964) demonstrates this only too well. Pimental (1963), after pointing out that relatively few attempts had been made to control native pests with introduced parasites and predators, argued that as many introduced species had been brought under control by the introduction of natural enemies obtained from other hosts, the same approach, i.e., the introduction of natural enemies of allied species should also work against native pests. While some of the inferences drawn by Pimental in this article have been criticized (Huffaker, et al., 1972), there is general agreement that answers to native pests may be found by searching elsewhere within the natural distribution of the pest or amongst the natural enemies of allied species. Thus, whereas satisfactory control of the native sugarcane pest *Diatraea saccharalis* (F.) in St. Kitts, W.I., was achieved by the introduction of the Tachinid *Lixophaga diatraeae* Tns. from the same host in the Greater Antilles (Box, 1960), the biological control of several other species of *Diatraea* native to Venezuela was achieved by the introduction of *Metagonistylum minense* Tns., another parasite of *D. saccharalis* originating in Brazil (Box, 1956; Bennett; 1971); and in Barbados *Apanteles flavipes* Cam., an Asian parasite of *Sesamia* spp., and other unrelated borers, has provided excellent control of *D. saccharalis* (Alam, Bennett and Carl, 1971). The frequently quoted examples wherein the Asian fungus *Endothia parasitica* decimated the native chestnuts in North America and the diaspine scales, *Carulaspis minima* (Targ.) and *Lepidosaphes newsteadi* Sulc. played similar havoc to the endemic cedar *Juniperus bermudiana* L., following their introduction into Bermuda, are ample proof that the same approach may be applicable to native weed problems (Wilson, 1970).

Direct Versus Indirect Pests

The statement "Indirect pests are suitable subjects for biological control; direct pests are not" by Turnbull and Chant (1961) has brought forth vigorous rebuttal by Huffaker, et al. (1971). The latter quote the olive scale as an example of a direct pest (it preferentially attacks the fruit although it also is an indirect pest in that it attacks the leaves and twigs), brought under satisfactory control by the introduction of parasites. It is accepted that increasingly higher standards of perfection have meant that additional control measures have been required at times to aid what was previously considered adequate biological control. It does not negate the fact that an appreciable level of control was being exerted, nor does it mean that further attempts to improve the level of control by additional introductions or manipulations of natural enemies would be useless. There is little doubt that the development of the chlorinated hydrocarbons and other more recent "efficient" insecticides was responsible for raising the standard of perfection of fruit and vegetables to the level the housewife now expects. It did, of course, often mean sole reliance for control on the use of indiscriminate pesticides. After initial successes these have all too often "turned sour" when other previously minor pests released from their natural enemies which were killed off by the insecticides, necessitated even further pesticide applications. Thus in South Africa, on citrus where partial permanent biological control was replaced by temporary chemical control following the introduction of initially DDT and later parathion treatments, the frequency and cost of pesticide applications spiralled to the point where in certain areas citrus production became uneconomic. The persistent, painstaking efforts of Eric Bedford and his colleagues in rectifying this situation, wherein classical biological control plus the encouragement of endemic natural enemies and the use of selective pesticides have led to one of the most successful, yet delicately balanced, pest management programs devised to date (Bedford, 1968a, b; 1969; 1971; Broodryk, 1964). This system, apart from being considerably cheaper than the regime of repetitive sprays previously practiced, also meets the demand for high quality blemish-free fruit, indicating that direct pests can be successfully controlled by natural enemies.

In many developing tropical countries where the grading standards for edible produce are low or nonexistent any significant reduction in the level of damage by a direct pest is worthwhile. For example, in Mauritius the lepidopterous pod-borers, *Etiella zinckenella* Tretischke and *Maruca testulalis* Geyer inflicted damage to the point where *Cajanus cajan* (L.) Mill. sp. was little cultivated. Following investigations of the parasites of *Ancylostomia stercorea* (Zell.) the most common pod-borer of *C. cajan* in Trinidad, Bennett (1960) shipped six species of hymenopterous parasites to Mauritius. Two of them, *Eiphosoma annulatum* Cress. and *Bracon cajani* (Mues.) became established. Losses, which were in the order of 60% prior to parasite establishment, dropped to 20 to 30%, an acceptable level to encourage an increase in production of this crop.

Key Pests

All too frequently a complex of pests attacks a single crop. Whereas, many or most of these are under good natural or biological control, this is often upset by applications of pesticides against the remaining pest species. These then become the key factors in pest management programs. They are key pests because they lack adequate natural enemies. For example, following the successful outcome of his investigations on citrus pests in the Rustenberg area, South Africa, Bedford (1968b) considers that the citrus thrips *Scirtothrips aurantii* Faure and the citrus bud-mite *Aceria sheldoni* Ewing, are the two most important annual pests in orchards where biological control is practiced. These two species currently requiring pesticide treatment are the key pests for which biocontrol is most urgently required. In the meantime only the carefully worked out pest management scheme which permits the restricted use of recommended pesticides at the appropriate times allows the biotic agents to check other pest species.

Similarly, in Peru, where studies have indicated (fruit flies excepted) that parasites and predators if undisturbed provide satisfactory control of most citrus pests except the rufous or W.I. red scale, *Selenaspidus articulatus* (Morgan), and the citrus wooly whitefly *Aleurothrixus flocossus* (Maskell) (Beingolea, et al., 1969) emphasis should now be placed on the procurement of suitable natural enemies against these two pests.*

Monoculture Versus Mixed Crops

The value of maintaining some degree of vegetational diversity to encourage or to retain effective populations of natural enemies is well recognized (Southwood, 1971). The general trend toward larger fields and thereby to monocultures has been brought about for reasons of economy and efficiency in agricultural practices other than for insect control. Van den Bosch and Stern (1967) have demonstrated that even in monocultures the value of natural enemies (predators and parasites, introduced as well as native), can be enhanced by altering certain agricultural practices, e.g., strip cutting of alfalfa rather than harvesting the entire field at the same time. In many areas in the Neotropics the biological control of stalk borers, *Diatraea* spp., in monocultures of sugarcane has remained effective despite an increase in the size of the individual fields. In this region the annual reaping season frequently lasts for five to seven months and hence all fields are not reaped simultaneously. The last fields to be cut are often in close proximity to those cut at the beginning of crop and there has been ample time for the migration of natural enemies from field to field in the interim.

Pests of Perennial Versus Annual Crops

The significance of the type of host-plant crop in successful programs of biological control of insect pests analyzed by Lloyd (1960) indicated that most completely successful and partially successful examples were against pests of perennial crops. DeBach (1964) also referred to this, but while he did not pursue it explicitly, there are sufficient at least "partial" successes against pests of annuals that this area should certainly not be overlooked. Also, of the numerous scarce non-economic pests on annuals it is highly probable that many are kept in check by biotic factors. With increasing possibilities to replace chemical pesticides with highly host-specific microbial agents to control some of the major pests the chances of achieving satisfactory control of others by classical biological control are enhanced.

Biological Groupings

DeBach (1964, 1971) has given consideration to the successes achieved among pests of the various categories of insects. Successes have been achieved in nine orders, but principally in the Diptera, Coleoptera, Lepidoptera and Homoptera. More than half of the successes are against homopterous pests, but he had stressed that the high number of successes among this order is more

---

*It appears likely that the answer to *S. articulatus* may be at hand; *Aphytis* sp. shipped to Peru from Kenya by Dr. D. J. Greathead, East African Station, Commonwealth Institute of Biological Control, has been successfully recovered following field releases during the past few months; levels of parasitism are most encouraging (personal communication, O. Beingolea G., May 1972).

probably due to the disproportionate amount of effort directed against scales and mealybugs than that they are necessarily more amenable to biological control.

Experimental Biocontrol Programs

Occasionally the opportunity presents itself wherein a phytophagous insect may be common on a plant or crop of low economic importance or where there is no immediate pressure to attempt control of any type. The occurrence of promising parasites elsewhere may provide the opportunity of testing the two contrasting strategies of biological control, i.e., the release of only one parasite species versus the release of a complex. Dr. H. Pschorn-Walcher of the European Station, CIBC (unpublished 1970 Annual Project statement to the Canadian Department of Agriculture) proposed such an experiment when recommending the trial of European parasites of the birch casebearer *Coleophora fuscedinella* (Zell.) into the Canadian Maritime Provinces where it is usually not an important economic pest. Unfortunately, personnel and adequate funds to carry out such projects properly are usually lacking.

General Conclusions

There are few if any absolute criteria that apply to all problems for selecting candidate hosts for biocontrol other than the contention that every pest is a potential candidate to which some consideration of this method should be given. DeBach (1971) has repeatedly emphasized that the number of successful biocontrol projects in an area bears a direct relationship to the emphasis and resources put into biological control research and importation. This coupled with his statement that "no geographic area or crop or pest should be prejudged as being unsatisfactory for biological control attempts" leaves the field wide open.

II. Criteria for Selection of Biotic Agents

While much has been written about the attributes of a biotic agent there are basically only two major considerations when selecting control agents: (a) that the natural enemy should appear to have the potential to at least partially control the target host species; and (b) that the introduced organism will not cause detrimental effects to the ecosystem.

While with certain projects and with certain groups of predators and parasites detailed investigations may be required to satisfy these criteria--for example where one sex may develop as a hyperparasite, or where species are facultatively hyperparasitic, and in some instances where a large complex of natural enemies already occurs--there are many instances where natural enemies have been shipped and released almost as soon as they were located. The "policy" of the country or organization conducting the investigations may decide the degree of research necessary before a choice is made. Whereas some countries, e.g., Canada attempt to take into account many factors and attempt to find a "fit" among the biotic agents available and those already present, the policy in other countries is often at the other extreme, i.e., "if a natural enemy is available, let's try it."

The procedures adopted during the past decade by the Canadian authorities for the selection of species for introduction against forest pests are described by Reeks and Cameron (1971). As their remarks also sum up several of the factors taken into account when selecting candidate biotic agents and as my views are in agreement with theirs on the previously controversial topic of multiple species releases as opposed to the release of only one or two species, I will quote extensively from their paper--"Appreciation of a pest problem is usually accomplished through a study of the population dynamics of the pest in Canada. As soon as a pest is recognized as one of foreign origin the possibility of biological control is investigated through cooperation with the Commonwealth Institute of Biological Control. The latter organization, chiefly by its staff at the Delemont Station, Switzerland, surveys and appraises the natural enemies in the countries where the pest is indigenous. Two additional features have been added to the C.I.B.C. studies during the present reporting period. More emphasis has been placed on studying competition between species of parasites to determine which ones are intrinsically or extrinsically superior, and in the case of *N. sertifer* studies, the relationship between parasitism and prey density has been examined more critically within limits opposed by availability of staff.

"In the preliminary search for candidates for introduction, the first step was to find host species identical or allied to the pest in Canada. When considering allied host species as a source of suitable material, care was taken to assure that the forest communities in both countries were similar in composition and ecology. Both monophagous and polyphagous biotic agents were considered in the preliminary selection of candidates and invariably there had to be strong evidence that the organism would attack the prey species in its new environment. This consideration is consistent with theories of Turnbull and Chant (1961) but they chose a poor example to demonstrate their point. They suggested that *Dahlbominus fuscipennis* (Zett.) was an unsuitable parasite for use in the biological control of the European spruce sawfly program because it was "never found parasitizing *D. hercyniae* or its close relative, *D. polytomum*, in their native environments. . . ." "Actually, it has been known for over 30 years that the parasite does attack *D. polytomum* in its native environment and the failure of *D. fuscipennis* in Canada must be explained by a combination of other factors, as reviewed by Reeks."

"The selection process continued throughout the European and Canadian programs, and decisions took into account the controversial question as to the number of species or organisms that should be introduced against a single pest. Several candidates were generally selected for importation and rearing with the view to releasing more than one species if circumstances warranted. Multiple species releases are contrary to the views of Turnbull and Chant (1961) who favor release of only one or two promising species against each target species. One must agree with these authors that the selection of 27 species for release against the European spruce sawfly could have been narrowed. However, on biological grounds there has been nothing to indicate that single-species releases have any advantage over multiple-species introduction. The multiple releases against the European spruce sawfly evidently did not have a detrimental effect because, as shown by Neilson, Martineau, and Rose (1971), either parasites, the virus, or combination of both are capable of regulating host densities. Furthermore, there is now excellent evidence from recent experience and inductive population models of Hassell and Varley (1969) that the practice of multiple or successive introduction is sound, the advantages of multiple introductions being: (a) greater chance that at least one species will be established; (b) possibility of better control offered by two species than by one; and (c) dominance of different species by competitive displacement in different climatic zones." Simmonds (1972) has defended the so-called "hit or miss" system of introductions wherein entomophagous species should be tried as they become available rather than attempting to analyze the very complex biological systems to select what appears to be the "best" parasite or predator. As he states "The introduction of a natural enemy into a new area is in fact the final and crucial experiment and the outcome to date never absolutely predictable except insofar as harmful repercussions can be guarded against. For all the careful selection of promising species for introduction and for all the knowledge of the ecological requirements --it may on introduction into a new area with a somewhat different climate, flora, and "ecology" react in a different way. Promising biological control agents have failed to live up to expectations whereas apparently unlikely species have been very successful."

Obviously the phrase of Reeks and Cameron (1971) already quoted "within limits imposed by availability of staff" is frequently the factor determining the intensity of the investigations, prior to, during and following release of organisms. All too frequently the number of problems on which investigations are required preclude studies in depth and under these conditions the "hit and miss" approach of Simmonds (1972) is often the easiest to adopt.

An exploratory entomologist undertakes much of the foreign investigations for Hawaii and sends back material of any potential agent that he encounters. Each species is screened under quarantine and those species found to be primary parasites or predators capable of developing on the pest under study are mass-bred for release.

The attributes of an effective natural enemy are frequently itemized and discussed. Doutt and DeBach (1964) preparatory to listing as attributes (1) a high searching capacity or ability to find its host when scarce; (2) a fairly high level of host specificity; (3) a potential rate of increase; and (4) an ability to occupy all host niches and to survive well; have pointed out that it is not possible to make an entirely reliable prediction on how effective a given exotic entomophagous species will be. Reference has already been made to Simmonds (1972) who expresses the same views and hence it is difficult to suggest absolute criteria for selecting control agents. Whereas some consider that a species should be host-specific, there are times when the best available biotic agent is polyphagous. Thus Hichiki (1971) stated that one of the reasons why *Colpoclypeus florus* (Walker) was selected for trial against the red-banded leaf-roller *Argyrotaenia velutinana* (Walker) because it is polyphagous on tortricids; it is also multi-voltine and increases rapidly. He added that resistance to commonly used insecticides would also be a useful attribute. In instances where an endemic or long-established species occasionally flares up, a "high density" biotic agent may be released to attempt initial control and to lower the density of the pest to the point where another "low density" agent can cope. Recently in St. Kitts, West Indies, populations of the coconut mealybug *Nipaecoccus nipae* (Maskell) increased to the point where control measures were indicated. It was recommended that the high density predator *Cryptolaemus montrouzieri* Muls. be introduced immediately and if permanent establishment did not occur, released periodically until releases of certain parasites credited with suppressing this mealybug in Trinidad (but more difficult to procure in adequate quantities for immediate release) could be arranged. In Canada two introduced parasites of the winter-moth, *Operophtera brumata* (L.) apparently act in this manner. The Tachinid *Cyzenis albicans* Fallen is a more effective control agent at high host densities than is the Ichneumonid *Agrypon flaveolatum* (Gravely) which is most effective at low densities (Embree, 1971).

Concluding Remarks

As stated initially, it is difficult to define criteria for selecting biocontrol agents which are applicable in all instances. It is often equally difficult to explain why certain "promising" species have failed and why "unpromising" ones have succeeded. For some of the former DeBach (1971) has suggested that attempts to obtain possibly slightly different genetic material from areas where the microclimate more closely matches the area where control is desired may be rewarding.

Embree (1971) postulates that if investigations such as those undertaken on the winter-moth are pursued, "ultimately the selection of biological control agents and the manner in which they are manipulated will be based on the results of simulation studies using standard mathematical models encompassing the prospective controlling agent and the ecosystem in which it is to act." It is probable that this method of selecting biotic agents is still far in the future and that the "specific" aspects by which no two problems are exactly the same will continue to require that final experiment, the field release, to determine whether a biotic agent will be effective or not.

References

Alam, M. M., F. D. Bennett, and K. P. Carl. 1971. Biological control of *Diatraea saccharalis* (F.) in Barbados by *Apanteles flavipes* Cam. and *Lixophaga diatraeae* T.T. Entomophaga 16: 151-158.

Bedford, E. C. G. 1968a. The biological control of red scale, *Aonidiella aurantii* (Mask.) on citrus in South Africa. J. Entomol. Soc. South Africa 31(1): March 1968.

Bedford, E. C. G. 1968b. Biological and chemical control of citrus pests in the Western Transvaal; An integrated spray program. South African Citrus J. 417: 9, 11, 13, 15, 17, 19, 21-28.

Bedford, E. C. G. 1969. The selection of insecticides for the integration of biological and chemical control of citrus pests. South African Citrus J. 431: 3, 5, 7, 9, 10.

Bedford, E. C. G. 1971. The effect of Abate and Delnav on populations of red scale, *Aonidiella aurantii* (Mask.), in citrus orchards under integrated control. J. Entomol. Soc. South Africa 34: 159-178.

Beingolea, O., J. Salazar and I. Murat. 1969. La rehabilitacion de un huerto de citricos como ejemplo de la practibilidad de aplican sistemas de control integrado de la plagos de los citricos en el Peru. Rev. Peru Entomol. 12: 3-45.

Bennett, F. D. 1960. Parasites of *Ancylostomia sterocorea* (Zell.) (Pyralidae, Lepidoptera) a pod borer attacking pigeon pea in Trinidad. Bull. Entomol. Res. 50: 737-757.

Bennett, F. D. 1971. Current status of biological control of the small moth borers of sugarcane *Diatraea* spp. (Lepidoptera: Pyralididae). Entomophaga 16(1): 111-124.

Bennett, F. D. and I. W. Hughes. 1959. Biological control of insect pests in Bermuda. Bull. Entomol. Res. 50: 423-436.

Box, H. E. 1960. Status of the moth-borer, *Diatraea saccharalis* (F.) and its parasites in St. Kitts, Antigua and St. Lucia, with observations on Guadeloupe and an account of the situation in Haiti. Proc. Entomol. Soc. Sugarcane Technol. 10: 901-904.

Box, H. E. 1956. The biological control of moth-borers (*Diatraea*) in Venezuela. Battle against Venezuela cane borer. Sugar My 51(5): 25-27, 30, 45 (6): 34-36, 57 (7): 30-34.

Broodryk, S. W. 1964. Biological control of circular purple scale. South African Citrus J. 372: 7, 9, 11, 13.

Clausen, C. P. 1956. Biological control of insect pests in the continental United States. USDA Tech. Bull. 1139. 151 pp.

Corbet, P. S., et al. 1971. Biological control programmes against insects and weeds in Canada. 1959-1968. Tech. Commonwealth Inst. Bio. Control 4. 266 pp.

DeBach, P. (ed.) 1964. Biological control of insect pests and weeds. Chapman and Hall, Ltd., London 1: 844 pp.

DeBach, P. 1971. The use of imported natural enemies in insect pest management ecology. Proc. Tall Timbers Conf. Ecol. Animal Cont. Habitat Management 3: 211-233.

DeBach, P. 1971. Fortuitous biological control from ecesis of natural enemies. Kyushu Univ. Publ. Fukuoka, Japan. 293-307.

Doutt, R. L. and DeBach, P. 1964. Some biological control concepts and questions. pp. 118-124 In Biological Control of Insect Pests and Weeds. P. DeBach ed. Chapman and Hall, London. 844 pp.

Greathead, D. J. 1971. A review of biological control in the European region. Tech. Commun. Commonwealth Inst. Biol. Control 5: 162 pp.

Hassell, M. P. and G. C. Varley. 1969. New inductive population model for insect parasites and its bearing on biological control. Nature 223: 1133-1137.

Hikichi, A. 1971. *Argyrotaenia velutinana* (Walker), red-banded leaf roller (Lepidoptera: Tortricidae). Pp. 10-11 In Biological Control Programmes Against Insects and Weeds in Canada 1959-1968. Tech. Commun. Commonwealth Inst. Biol. Control 4: 266 pp.

Huffaker, C. B. 1970. Summary of a pest management conference--a critique, pp. 227-244 In Concepts of Pest Management, R. L. Rabb and F. E. Guthrie (eds.) North Carolina State University, Raleigh. 242 pp.

Huffaker, C. B. (ed.) 1971. Biological Control. Plenum Press, New York. 511 pp.

Huffaker, C. B., P. S. Messenger and P. DeBach. 1971. The natural enemy component in natural control and the theory of biological control, pp. 16-67 In Biological Control. C. B. Huffaker, ed. Plenum Press, New York. 511 pp.

Lloyd, D. C. 1960. The significance of the type of host plant crop in successful biological control of insect pests. Nature. London 187: 430-431.

McLeod, J. H., B. M. McGugan and H. C. Coppel. 1962. A review of the biological control attempts against insects and weeds in Canada. Tech. Commun. Commonwealth Inst. Biol. Control 2: 216 pp.

Neilson, M. M., P. Martineau and A. H. Rose. 1971. *Diprion thorcyniae* (Hartig.) European spruce sawfly (Hymenoptera: Diprionidae). Tech. Commun. Commonwealth Inst. Biol. Control 4: 137-143.

Pimentel, D. 1963. Introducing parasites and predators to control native pests. Canadian Entomol. 95: 785-792.

Rao, B. P. 1971. Biological control of pests in Fiji. Miscl. Publ. Commonwealth Inst. Biol. Control 2: 38 pp.

Rao, B. P., et al. 1972. A review of the biological control of insects and other pests in Southeast Asia and the Pacific Region. Tech. Commun. Commonwealth Inst. Biol. Control 6: 149 pp.

Reeks, W. A. and J. M. Cameron. 1971. Current approach to biological control of forest insects. Tech. Commun. Commonwealth Inst. Biol. Control 4: 105-127.

Southwood, T. R. E. 1971. Farm management in Britain and its effect on animal populations. Proc. Tall Timbers Conf. Ecol. Animal Control Habitat Mgt. 3: 29-41.

Turnbull, A. L. and D. A. Chant. 1961. The practice and theory of biological control of insects in Canada. Canadian J. Zool. 39: 697-753.

Simmonds, F. J. 1971. Biological control of pests. Trop. Sci 12(3): 191-201.

Simmonds, F. J. 1971. Biocontrol--the alternative pesticide. Your Environ. 11(2): 61-63.

Simmonds, F. J. 1972. Approaches to biological control problems. Entomophaga 17: 251-264.

Van den Bosch, R. and V. M. Stern. 1961. The effect of harvesting practices on insect populations in alfalfa. Proc. Tall Timbers Conf. on Ecol. Animal Control Habitat Mgt. 1: 47-54.

Wilson, C. L. 1969. Use of pathogens in weed control. Ann. Rev. Phytopathology 7: 411-434.

Wilson, F. 1960. A review of the biological control of insects and weeds in Australia and Australian New Guinea. Tech. Commun. Commonwealth Inst. Biol. Control 1: 104 pp.

Wilson, F. 1963. Australia as a source of beneficial insects for biological control. Tech. Commun. Commonwealth Inst. Biol. Control 3: 28 pp.

Wilson, F. 1970. Biotic agents of pest control as an important natural resource. The Fourth Gooding Memorial Lecture 1-12.

FOREIGN EXPLORATION AND IMPORTATION OF EXOTIC ARTHROPOD PARASITES AND PREDATORS

R. I. Sailer
Entomology Research Division
ARS, USDA
Beltsville, Maryland

Presently
Professor of Entomology
University of Florida
Gainesville, Florida

Introduction

For more than 80 years entomologists have been aware that insect pests can, in varying degrees, be controlled through movement of parasites, predators and diseases from one part of the world to another. More recently similar movement of plant feeding insects has been repeatedly shown to be an effective method of suppressing weeds. Where importation of beneficial species has eliminated or greatly alleviated pest problems the economic value of the method is obvious. Usually the total cost of such introductions, including exploration, research on biology, behavior, propagation, colonization, distribution and evaluation, has represented but a small part of the annual loss caused by the pest. The reason for this is clear. Energy stored in the form of the pest's own biomass fuels a self-perpetuating control system.

The fact that such systems pervade all biotic communities has only recently become apparent to applied entomologists. Despite the many successful examples of pest suppression following introduction of what I would like to call counterpests, the full value of parasites, predators and diseases did not become apparent until after their all too frequent annihilation following widespread use of highly effective broad spectrum insecticides. The numerous examples of obscure plant feeding insects and mites that suddenly gained prominence as major pests and the violent "flare backs" of pests following large-scale use of insecticides has now convinced most entomologists that natural enemies are in fact a potent force in the regulation of insect populations. As a result, economic entomology is moving toward the concept of control through management practices that maximize the effect of natural enemies. Resettlement and encouragement of natural enemy populations are basic elements of the pest management concept.

While my topic specifically concerns activities related to finding and importing parasites and predators from foreign countries, the underlying concepts and principles do not differ appreciably from those relating to movement of a useful scale insect parasite from one orchard to another. The differences are simply those of methodology, logistics, and degree of scientific and operational complexity.

Why Import Exotic Parasites and Predators?

Not infrequently I encounter people who express surprise and even incredulity when they learn that entomologists are paid to find and import foreign insects into the United States. This attitude does not surprise me greatly. After all, to the average person, an insect is something that bites, stings, chews up vegetable gardens, spoils prize roses, or is an unwelcome picnic guest. What has surprised me over the years is the large number of entomologists who question the value of parasite introduction programs. Even now, in a time when public interest in biological control has revived and entomologists are eagerly searching for non-chemical methods of pest control, there is no concerted effort to increase research on introduction of beneficial insects. Such interest has translated into a very considerable increase in research on resident parasites, and particularly on pathogens. Important as this research is, it will not add new biological agents to our armory of weapons needed to combat pests. That such agents exist and have completely or partially solved many pest problems in the United States and elsewhere in the world is evident to anyone who wishes to acquaint himself with the facts.

Without enumerating all the examples of pests that have been effectively controlled or partially controlled by introduced natural enemies, it should be sufficient to cite figures given by DeBach (1972). According to his estimates each dollar spent on parasite introduction has returned $30 in benefits to agriculturalists. By comparison, each dollar spent on insecticides returns only $5 in benefits (Pimentel, et al., 1965). However, to place these figures in proper perspective it should be noted that all money spent during the past 80 years in parasite introduction would be but a small part of the $420,000,000 spent annually on insecticides (Agricultural Research Service, 1965).

Many entomologists who do not question the benefits of past introduction programs nonetheless feel that most of the useful species have probably already been found and introduced. After all, they have read in semi-popular and even scientific literature accounts of the parasite explorers who have "combed" the world for useful parasites and predators. It is true that during the past 80 years USDA, University of California and Hawaiian entomologists have logged a lot of miles, and at least 420 beneficial species have been imported and colonized in some numbers. Another 200 have been imported and studied to some extent but not released. Of those colonized, a recent tabulation shows that 128 are now established in continental United States.

Some 60 insect pests and 10 weeds have been targets of this work which has resulted in complete or substantial control of at least 15 insect pests and 2 weeds. However, in the United States there are about 10,000 kinds of insects and mites with some degree of importance as pests. Of approximately 700 that fall in the important pest category 212 are of foreign origin.[1/] The latter include many of our most serious pests and it is against these that parasite introduction work offers the greatest promise.

Experience has shown that few insects, injurious or beneficial, are not at some place and time the host or prey of at least 3, more or less specific entomophagous insects. The number of associated oligophagous and polyphagous parasites and predators is normally many times greater. As another way of looking at the problem, all species of the hymenopterous family Icheneumonidae are parasitic on other insects. Townes (1969) the foremost authority on this family, estimated that it contains 60,000 species, of which less than 15,000 are described.

[1/] Unpublished data, USDA. Import Inspection Task Force.

The Braconidae, another family of Hymenoptera in which all members are parasitic on other insects, is believed to contain an even larger number of species. Add to these, the entomophagous species belonging to other families of Hymenoptera, and of Diptera, Coleoptera, Hemiptera, Neuroptera, together with elements of most other orders, and the total number of beneficial species should easily top 200,000. Now scatter these species over the earth's continents and major islands. Then recall that for most we lack even a name, and some picture of both the potential as well as the magnitude of the problem begins to emerge. Clearly the parasite explorer's "comb" had few teeth.

Now, I would like to return to the point alluded to earlier, namely that alien pests offer the most promising targets for a parasite introduction program. Reasons for this lie deeply rooted in ecology and the evolution of host-parasite relationships within ecosystems. First, the hosts, insofar as we are concerned with plant-feeding insects, are primary consumers. The parasites and predators are secondary consumers and are themselves a food resource for tertiary consumers. The latter fall into the category of secondary parasites or hyperparasites that are also of immediate concern to parasite introduction programs.

Now why is all this important to the question of importing beneficial insects into the United States? For an answer we need only look at the post-Pleistocene history of the North American flora and fauna. Following the withdrawal of the last continental ice sheet about 10,000 years ago the North American biota reoccupied the glaciated areas and species assorted themselves into life zones and biomes according to climate and internal ecological or geographic barriers. Except as he made use of fire, the American Indian had little influence over the history of the ecosystems that evolved in North America prior to the arrival of Europeans. Within a few years after Columbus made landfall at the island of San Salvador in 1492 events were set in motion that were to affect the North American biota more profoundly than did the glaciers of the earlier epoch.

With the arrival of the Spaniards, Frenchmen, and Englishmen, came the livestock, agricultural crops, horticultural crops, and ornamental plants of Europe accompanied by camp following insects, weeds, and diseases. As agriculture and industry spread across North America, commerce reached out to all parts of the world and additional alien species gained entry and became part of the North American biota. A recent tabulation[1/] of immigrant insect species now in the 48 contiguous states includes 1,115 species. Of these only 128 have been purposely introduced. Of the remainder, 616 are of some importance as pests with 212 falling in the important category. It is of more than passing interest that 152 of the accidental immigrants are either known to be or can be regarded as likely to be beneficial. This figure reflects little credit on the efforts of entomologists who after 80 years have succeeded in importing and establishing only 128 beneficial species.

Remembering that each plant-feeding insect in its homeland has at least 3, more or less host-specific natural enemies, why the great disparity between immigrant pest and immigrant counterpest? The answer is obvious. Given equal opportunity to gain entry, the secondary consumer cannot colonize a new area until its primary consumer host is present and has attained a population density level sufficient to sustain the parasite. Add to this the more restrictive environmental requirements of parasites, i.e., alternate hosts, different adult food requirements, and narrower tolerance to varying climatic factors, and it is clear that probabilities of establishment are much more favorable for the primary consumer.

[1/] Unpublished data, USDA. Import Inspection Task Force.

Thus, in breaking down the geographic barrier that so long isolated North America from the Old World, a selective filter was established that admitted plant-feeding insects much more readily than their natural enemies. In the absence of the natural enemies with which they had co-evolved an interacting relationship that contributed to the stability of their ecosystems, the alien plant-feeding species thrived. Having found an otherwise favorable habitat, they were able to increase their numbers rapidly and become a disruptive influence in the invaded ecosystems. We are all too aware of consequences following the establishment of such species as the European corn borer, gypsy moth, smaller European elm bark beetle, Japanese beetle, and many others.

The obvious disparity between the immigrant pests and immigrant counter-pests serves as another measure of the magnitude of effort needed to fully exploit the potential of exotic arthropod parasites and predators. Assuming that there are 212 economically important pests among the 1115 alien species now resident in the United States, a little simple arithmetic suggests that somewhere there are more than 600 beneficial insects that should be useful in this country. Since 128 have been imported and established during 80 years, and assuming no acceleration of the program, it would require another 376 years to finish the job!

Clearly, alien insects have repeatedly posed major threats to American agriculture and today they constitute a majority of our most costly pests. Moreover, despite our foreign quarantine regulations, new alien pests of major importance seem to turn up at the rate of one every 3 years. Short of expecting to rely indefinitely on chemical control or on the hope of developing resistant crop varieties, the only prospect for effective control is through a parasite introduction program. By introducing their primary parasites and excluding associated secondary parasites we can expect a more favorable degree of control than the parasites provided their home country. Here we have an advantage not enjoyed in the home country of the alien pests.

In final analysis, parasite introduction is or at least should be approached as a problem in ecosystem engineering. The entomologist must endeavor to create a system of interacting primary and secondary consumers that will allow the farmer the largest possible share of the products of the agro-ecosystems on which we all depend. However, we have far to go before enough is known about the comparative performance of the vast multitude of parasites and predators available in the world fauna to enable entomologists to confidently select the best parasite or combination of natural enemies required to manage an agro-ecosystem with predictable results.

Methodology of Parasite Introduction Programs

Selection of the Target. Many factors should enter into a determination of when and against what pests exploration for natural enemies should be undertaken. However, projects have generally been initiated in response to urgent pest problems. Funds for parasite introduction have been forthcoming usually as part of a very much larger amount appropriated specifically for research to control a new and alarming pest.

This pattern began with the decision in 1905 to search for parasites of the gypsy moth. Similar work on the alfalfa weevil followed in 1911, the European corn borer in 1919, Japanese beetle in 1920, and Oriental fruit moth in 1929. During the early 1930's substantial additional funds were made available for work on the corn borer. Between 1949 and 1952 following World War II and a period when support for biological control had all but disappeared, funds were allocated for work on the spotted alfalfa aphid, fruit flies, citrus blackfly, and pink bollworm. These injected new life into foreign exploration. Funds for work on the cereal leaf beetle were forthcoming in 1963 and by 1971 the wheel had turned once more to the gypsy moth.

For the most part, this work has been productive and has yielded results useful for control of the pest against which work was undertaken. However, in at least one case, a project that was eventually written off as a failure, yielded, as a by-product, four beneficial species of outstanding value. The project in question was that involving work on pink bollworm in India. Although many thousands of pink bollworm parasites belonging to several species were shipped and released in Texas and Mexico, none became established. However, George W. Angalet (USDA, ARS), who was stationed in India to work on the pink bollworm was alert for opportunities to do useful work on other pest problems. As a result of this secondary activity, we now have in the United States the pea aphid parasite *Aphidius smithi* Sharma and Subba Rao and the parasite of Rhodesgrass scale, *Neodusmetia sangwani* (Rao), as well as the two puncturevine weevils *Microlarinus lareynii* (Jacq. du Val) and *M. lypriformis* (Wollaston). At this point in time, benefits resulting from introduction of these 4 species have many times repaid the cost of keeping Angalet in India for 6 years.

This example points to the need for flexibility in parasite introduction programs, as well as to the obvious advantage of having people in foreign countries who are alert to opportunities aside from their primary objective. This is particularly important in view of the often fortuitous nature of such opportunities. However, such flexibility in the foreign phase of an introduction program will serve little purpose if there is not similar flexibility in the domestic phase. Unquestionably, the uncertainty of a flexible program is unsettling to program planners and even to the working scientists at domestic locations; but, without such flexibility, opportunities will be missed and unnecessary pest losses will continue perhaps indefinitely.

While I am not thoroughly familiar with procedures of the University of California's introduction program, it is my impression that their selection of pest species is also influenced by the factor of economic urgency. However, because of their practice of sending staff scientists on temporary assignments of short duration, the choice of pest species is often influenced by the special interests of the entomologist available for assignment.

Early work by the Hawaiian Sugar Planters' Association was also directed toward pests of immediate economic urgency. In later years, much of the introduction work conducted by the Hawaiian Department of Agriculture has been done by a "roving explorer" who usually had one major pest as his primary objective but who collected and shipped any and all promising parasites and predators as well as weed-feeding insects encountered in his travels. Such material is received at Hawaii's quarantine laboratory and held there until it has been studied and evaluated sufficiently to warrant release and colonization.

Canada has followed a somewhat different approach. As a member of the Commonwealth, most of Canada's parasite introduction work overseas has been handled by the Commonwealth Institute of Biological Control (CIBC). Each year, the Canadian agriculture and forestry agencies responsible for research on control of pests review their situation and determine what information or parasite material they need from Europe during the following year. This information is then sent to the CIBC Delemont, Switzerland laboratory where, after consideration of costs, a plan of work is developed and mutually agreed to. Thus the Canadian introduction program would appear to be better planned, and the foreign and domestic phases of the work better coordinated than is the case with the three U.S. agencies that engage in similar work.

Australia, another country with a major beneficial insect introduction program, again selects pests on which to work largely on the basis of economic urgency. From an organizational standpoint, they use a mixture of arrangements for carrying out foreign work. As another Commonwealth country they sometimes use the services of the Commonwealth Institute of Biological Control. For other problems, they maintain permanent stations where work is conducted over a period of years. In still other instances, they send an entomologist out on temporary assignment.

U.S. Organizational Support for Foreign Work. The problem of getting an entomologist into the part of the world where he might expect to find useful parasites has been with us since the beginning of parasite introduction activities. C. V. Riley, appointed Chief of the Division of Entomology in the Department of Agriculture in 1878, was convinced that great benefits would result from importation of parasites from Europe, where he already had many contacts as a result of French efforts to import natural enemies of the grape phylloxera from the United States. Riley made several official trips to Europe and in 1883 imported *Apanteles glomeratus* (L.), a parasite of the imported cabbageworm. This was the second beneficial insect to be purposely introduced into the United States, the honey bee having preempted first place.

Aware of the great potential of introduced parasites, Riley planned to visit Europe regularly to search for useful species. However, in 1887 no funds were provided to be used to defray foreign trips by employees of the Department. The obvious intent was to put a stop to Dr. Riley's transatlantic "junkets," but the effect was that of disrupting plans then being made to send Albert Koebele to Australia to search for enemies of the cottonycushion scale (Doutt, 1958). However, Dr. Riley was not a man to be easily dissuaded. Through some political connections, an arrangement was made whereby the U.S. Commissioner to the International Exposition in Melbourne set aside $2,000 to pay the expenses of an entomologist who would ostensibly represent the State Department but actually collect enemies of cottonycushion scale. Koebele accomplished his mission. The vedalia beetle was introduced to California where it rescued the citrus industry and Koebele returned $500 of unexpended funds (Doutt, 1958).

Times have changed, but only in degree. Today we have ceilings on the number of employees we can station overseas, and foreign travel ceilings that restrict and effectively prevent federal personnel from undertaking temporary foreign assignments to conduct exploration work. Thus, we now have no greater capability to search for beneficial insects than we had 10 years ago.

The successful control of the cottonycushion scale stimulated much interest and activity in parasite introduction work. This was, for the most part, centered in California and Hawaii. It was soon recognized that in shipping beneficial insects into the United States there was also the danger of secondary parasites and additional insect pests. With passage of the 1912 Plant Quarantine Act it was no longer possible for parasite explorers to travel about the world and ship insects into the United States with little or no quarantine precautions. Growing recognition of the need to handle such material under strict quarantine, together with growth of entomology as an organizational element of the U.S. Department of Agriculture, had the effect of consolidating most parasite introduction work in this federal agency.

Until 1934 all phases of the biological control work of the old Bureau of Entomology were conducted independently by the several Divisions concerned with research on the different insect pests. In that year, the Division of Foreign Parasite Introduction was organized, with responsibility for foreign investigations and for quarantine handling of the imported material. After passing through quarantine, clean stocks of parasites and predators were released to field stations of the respective Divisions or to state cooperators. These agencies were then responsible for rearing, colonization and evaluation programs. This arrangement persisted up to World War II when the Division had entomologists stationed in France and Japan.

Following the war and the advent of the new highly efficient insecticides, interest in biological control declined. The Division of Foreign Parasite Introduction was dismantled in the years between 1952 and in 1954 the remnants were attached to the Division of Insect Identification. This then became the Insect Identification and Parasite Introduction Research Branch (IIPI) of Entomology Research Division, ARS, USDA, and gradually the biological control part of the Branch rebuilt from a low point about 1951 when the staff consisted of 4.5 entomologists located at 5 locations - Beltsville, Maryland; Moorestown, New Jersey; Albany, California; Paris, France; and Mexico City, Mexico. By contrast, within the present organizational framework of IIPI, 26 entomologists are now employed in biological control. There are distributed among 6 domestic and 3 foreign locations. This, of course, includes research on biological control of both insect pests and weeds. Only 16 of these entomologists are directly engaged in introduction work.

The European Parasite Laboratory at Sevres, France, near Paris, is headquarters for research in Europe on parasites and predators of insect pests. The laboratory at Rome, Italy, concentrates on insect enemies of range- and cropland weeds and that at Buenos Aires on insect enemies of aquatic weeds. Quarantine receiving stations are located at Moorestown, New Jersey, where most insect enemies are handled and Albany, California, where weed-feeding insects are studied. Under an agreement with the Animal and Plant Health Inspection Service of the U.S. Department of Agriculture the University of California maintains quarantine stations at Riverside and Albany. Here personnel of the University's Division of Biological Control receive and study material collected by staff members who are on foreign assignments as well as material that may come in from other sources. Hawaii also maintains a quarantine laboratory in Honolulu and the state of Florida is now preparing to build a similar facility at Gainesville.

The current reorganization of Agricultural Research Service promises to greatly alter the organizational structure of the Federal biological control programs. This will be in the direction of regionalization of functions that require a high degree of coordination and rapid exchange of information as well as smooth flow of material across international boundaries and into all parts of the United States.

During the past 12 years ARS has had a second important source of biological control information and material from foreign countries. This has been the Public Law 480 (PL 480), or Special Foreign Currency Program. Currently there are research projects involving many important insect pests and weeds underway in Poland, Yugoslavia, Egypt, Israel, Pakistan, Morocco, and India. These projects have been the source of many parasites, predators, and weed-feeding insects that have been shipped to the United States in recent years. Most noteworthy are those of the cereal leaf beetle from Yugoslavia. In the next year we expect to receive many gypsy moth parasites from projects in Yugoslavia and Morocco.

Planning and Conduct of Foreign Work. Perhaps the best way to present a picture of how federal parasite introduction operations are planned and carried out is to cite a case history. In 1962, the cereal leaf beetle, *Oulema melanopus* (L.), was discovered in southwestern Michigan. By the time it came to the attention of state and federal entomologists, this insect had already demonstrated an alarming ability to destroy oats and other small grains. Men and resources were quickly mobilized to combat the new pest. During the winter of 1962-63, as Director of the European Parasite Laboratory at Nanterre, France, I was informed that funds had been made available to undertake research in Europe on the cereal leaf beetle.

At that time neither I nor any members of my staff had knowingly seen a cereal leaf beetle (CLB). We quickly consulted the "Review of Applied Entomology" and soon acquired as much published information as was available in Europe. This was not a lot but there was one good paper on bionomics of the insect by an Italian entomologist. I assigned Mr. George Angalet responsibility for the initial surveys needed to find populations of the beetle. Wherever possible, he was to collect eggs, larvae, and adults in an attempt to recover parasites. On April 1, 1963, he left for Italy where he established temporary headquarters at our Biological Control of Weeds Laboratory in Rome. He was able to work in southern Italy during early May and then move north as the season progressed. By late May when he returned to Nanterre, he brought with him several hundred cereal leaf beetle pupal cells and extensive notes on the insect's distribution, abundance, and economic importance in Europe.

Soon after the pupal cells arrived at Nanterre, a eulophid parasite began to emerge. We sent specimens posthaste to Dr. W. H. Anderson, Chief, IIPS, who forwarded them to Dr. B. D. Burks, chalcidoid specialist on the staff of the Branch's Systematic Entomology Laboratory. Within two weeks we had a letter from Dr. Anderson advising that the specimens belonged to the species *Tetrastichus julis* (Walker).

After all the adult beetles had emerged from the pupal cells we began examining those that were still intact and discovered that most of the intact cells contained either diapausing larvae of *Tetrastichus julis* or cocoons of an ichneumonid. The latter contained diapausing larvae. Thereupon, the shallow box containing the soil in which the cereal leaf beetle pupal cells were imbedded was placed in an unheated "emergence room." Occasionally during the winter we moistened the soil. In early May, 1964, the first adult ichneumonids emerged. These were promptly dispatched to Beltsville and identified by the ichneumonid specialist of the Systematic Entomology Laboratory as the species we now call *Diaparsis carinifer* (Thomson).

Mr. Angalet, accompanied by Dr. J. D. Paschke of Purdue University, returned to Italy in April 1964, and obtained more detailed information on the parasites. Additional parasite material was collected in Italy and southern France. This was to provide the stock used in the first releases in Indiana and Michigan in 1964-65. Work in Europe during 1965 resulted in the discovery of the egg parasite *Anaphes flavipes* (Foerster), found almost simultaneously in Italy by Dr. J. J. Drea (at the Nanterre station) and in our Nanterre garden where we had planted oats and wheat. Once found, we quickly recorded presence of *Anaphes flavipes* throughout western Europe. Why we missed the species in 1963-64 is a good question. Primarily it was a matter of experience with the problem. Dr. R. C. Anderson, then a graduate student from Purdue University, sent to assist the program and gain experience, was assigned to work on the egg parasite and stock was sent to him as soon as he returned to Purdue. This stock, reinforced by later shipments, provided the source of material mass-reared at the USDA parasite rearing laboratory at Niles, Michigan, for general release in infested areas.

Between 1966 and 1969, populations of the cereal leaf beetle were studied from Spain to Sweden and we had information and a small amount of material from the USSR. Everywhere in Europe there was essentially the same complex of parasites, one of the egg, and 3 of the larvae; however, we have since discovered that one of the larval parasites identified earlier as *Diaparsis carinifer* is actually a complex of two and possibly three species. Also in 1969, we began to receive large numbers of parasites from the PL 480 project in Yugoslavia. Material from this source was hand-carried by Dr. P. Bjegovic (Institute of Plant Protection, Belgrade) from Belgrade to Paris where the non-diapausing parasites were collected and sent to Moorestown for shipment to Michigan and Indiana. Pupal cells containing non-diapausing parasites were sent later to be overwintered at Moorestown, New Jersey. The discovery in 1966 of a *Mesochorus* sp., a secondary parasite, in material collected in France made it essential that all parasites leaving quarantine at Moorestown be subjected to specimen by specimen examination and the sheer quantity of material that had to be handled required a heavy expenditure of time and effort at that station.

By now, nine years after the work on cereal leaf beetle started in Europe, I am highly gratified with reports just received from Michigan. Without describing the work done at Michigan State University and Niles, suffice it to say that the egg parasite Anaphes flavipes is now established over a huge area of southern Michigan and northern Indiana. Both Tetrastichus julis and Diaparsis sp. are established at several places in Michigan and at least one locality in Indiana. Moreover the third larval parasite, Lemophagus curtus Townes, now appears established near Niles, Michigan. The parasitization rate in May 1972, at the Gulf Lake Experimental Farm of Michigan State University was found to average 85% and field days have been held both this year and last to collect and distribute parasites throughout Michigan.

No doubt, to the entomologist accustomed to insecticidal control of insects, the progress that I find gratifying would seem of little or no consequence. For less than $3 per acre he could apply an insecticide and destroy 99% of the larvae--with a certainty that the job would have to be repeated next year and the next, ad infinitum. However, with the state and federal entomologists cooperating to aid natural dispersal of the established parasites, I have every confidence that we have reached the turning point in the battle against CLB. Within a few more years the CLB problem will join the list of insect pest problems that have "disappeared" following introduction programs.

This is but one example of the foreign work carried out by USDA laboratories. During the years 1960-66 while I was at the European Parasite Laboratory, we worked on balsam woolly aphid, European pine shoot moth, face fly, vetch bruchid, grasshoppers, smaller European elm bark beetle, European corn borer, lygus bugs, alfalfa weevil, and as already mentioned, the cereal leaf beetle. One or more species of parasites and/or predators found in association with each of the above pests were shipped to Moorestown. In addition to incidental observations on many other insects, the European Parasite Laboratory also cooperated with the Biological Control of Weeds Laboratory in Rome in work on Scotch broom and tansy ragwort and shipped enemies of both these weeds to the United States.

Obviously, with work underway on such a variety of pests the research on individual species was by no means exhaustive. Here the policy of the USDA introduction program has been more or less intermediate between that of Canada and that of the University of California. Canada, through its arrangements with the CIBC Laboratory at Delemont, Switzerland, has supported much more detailed studies involving host-parasite interactions and population dynamics with the objectives of predetermining which beneficial species should be shipped to Canada.

California, on the other hand, has tended to carefully select the geographic areas which they think would be favorable for finding useful parasites or predators adapted to climatic conditions of that state. Their explorers then ship any primary parasites and predators they find to the California quarantine laboratories to be propagated and studied sufficiently to learn general characteristics of their biology and behavior before they are released.

Operational Aspects of Foreign Work

Personnel and Overseas Assignment. There are many people who like to travel and there is a considerable number of entomologists well-qualified to conduct research on parasites and predators. The problem is to find these characteristics in the same person. For purposes of staffing the American positions at the biological control laboratories overseas, we would look for the following characteristics and qualifications:

1. Excellent training in entomology
2. Sound background in ecology
3. Broad knowledge of American agricultural practices and pest problems.
4. Proficiency in 1 and hopefully 2 major European languages (fluency in English assumed)
5. No family responsibilities

6. Physical stamina
7. Ability to deal effectively with foreign farmers, scientists and bureaucrats of both the homegrown and foreign varieties
8. High frustration tolerance

Any entomologist possessing these qualities should encounter little competition if he applies for one of the USDA biological control positions overseas.

Seriously, it is not easy to find qualified American personnel to staff foreign positions. Many people think such assignments would be interesting, but after reflecting on problems relating to their professional careers and family life they conclude that they are well off to stay home. Employment of foreign nationals also presents problems. First, they work in their own country for a foreign government. Secondly, they work with Americans who may be no more competent but nonetheless have much higher salaries, plus quarters allowances, and other benefits. Finally, conditions of employment are unlikely to offer opportunities for professional advancement equal to that of their own institutions. As a result our best prospects are more often nationals of a third country.

Insofar as possible we attempt to transfer American personnel between our foreign and domestic stations. This affords our scientists the benefit of foreign experience and tends to compensate for loss of career momentum due to "culture shock," disruptions resulting from home leave, and the 6-year limitation on foreign service.

These problems do not affect the overseas programs of Canada, California, and Hawaii to the same degree as they affect ARS operations. By relying on the Commonwealth Institute of Biological Control Laboratory in Switzerland, with its staff of Swiss, German and Austrian entomologists, Canada is little involved in such problems. The temporary or roving assignments characteristic of California and Hawaiian operations also minimize staffing difficulties.

There is one thing the American entomologist who hopes to work effectively in a foreign country must understand. He must speak the language of the country or employ an assistant who can serve as an interpreter. While he may get along well with hotel keepers and scientists of the laboratories he visits, farmers who find him poking around in their fields will want explanations they can understand.

In the early days of parasite introduction work entomologists were given a passport and a few thousand dollars travel advance and wished Godspeed. They often received invaluable help from American Consuls located in foreign cities. There was little red tape involved in undertaking such assignments. This situation continued for many years after the USDA established permanent laboratories in Europe for work on the European corn borer, Oriental fruit moth and gypsy moth. However, the education of dependent children and official association with an American Embassy has gradually drawn the USDA foreign stations into the Embassy orbit. Along with commissary, PX and APO privileges, as well as payroll and personnel services, came Embassy concurrence for staffing plans, security clearance for personnel and status of an attached agency responsible to the Ambassador. One effect of this association has been that of greatly increasing the amount of lead time required to establish a new station in a foreign country and to obtain required clearances needed to send an employee to a foreign duty post.

Facilities and Equipment. The USDA laboratories overseas bear little resemblance to recent vintage USDA and State Experiment Station laboratories at home. For the most part they are rented residences. From 1956 to 1966 the European Parasite Laboratory occupied a renovated stable. Even though there has been a great improvement in recent years, equipment tends toward the do-it-yourself variety. Microscope, camera, and vehicle needs are generally adequate but otherwise sophisticated equipment generally regarded as necessary for experimental work is lacking. In view of the laboratories' present missions, this is not a serious handicap.

For the most part, the essential equipment needs are those required for collecting and transporting insect host material to the laboratory where it is then held in cages for parasite emergence or for use in other studies. Each insect species has its peculiar requirements. This places a high premium on ingenuity and flexibility in design of rearing and holding cages. The shipment of parasites and predators to the Moorestown Quarantine Receiving Station is a critical function. Many styles of shipping boxes and packaging have been used over the years. With the advent of air travel the elaborate provisions to insure survival of a long sea voyage was no longer necessary. With present pressurized and heated compartments in which air freight moves, the problem has been further simplified. The problem now is one of insuring security of containment and protection against thermal shock while packages wait on loading or off-loading docks.

The insects should always be contained in a box adapted to easy introduction and removal of the shipped material. Care should always be exercised to insure that the material of which the box is made contains no toxic substances. This can be a problem with some plastics. It has been my experience that most insects survive best in a wooden box. Besides providing a resting surface on which the insect seems to feel more comfortable, the wood is moisture absorbent. While there is always danger of excessive desiccation, the danger of excessive humidity and condensation of free water is a more frequent problem. The box should also contain some material that will provide additional resting surface. Very loosely packed excelsior is often used. Wedged pipe cleaners have also proved serviceable. The important thing is that such material be fixed. Any movement within the box during transit will have serious consequences. Finally, the box containing the insects should be placed in a larger carton surrounded by packing material that will cushion shock and provide insulation.

Although well known to the veteran parasite explorer the requirements of agricultural quarantine regulations must be observed. Each package containing insects to be shipped into the United States should bear an official importation permit label. Otherwise such packages will probably be held up by port inspectors until proper clearance is forthcoming.

Strategy and Organization of an Optimized Beneficial Insect Introduction Program

With evidence of the substantial benefits of past beneficial insect introduction programs, it is reasonable to ask whether current programs are adequate. If not, what should be done to accelerate the work in order to obtain maximum benefits to American agriculture in the shortest length of time? In my opinion, the current programs are not adequate. Assuming that this is true, what can be done to improve the situation?

Obviously, increased resources in the form of men and money are needed, but even more we need an improved delivery system. Without improved planning and coordination, increased resources would only result in a lower level of overall efficiency and, more important, there would be increased probability of individual project failure. Planning and coordination are of paramount importance in biological control work generally, but become critically important to success of parasite introduction programs. Because of the highly perishable nature of living insects and the sequential nature of the different phases of the work which begins in a foreign country and ends with colonization and evaluation, failure at any intermediate point can be disastrous. This, together with the sheer number of kinds of organisms involved, as well as the number of national and foreign agencies directly or indirectly concerned, suggests need for strong centralized program planning and coordination.

Foreign Phase. The preponderance of European species among the alien pests in the United States gives Europe number 1 priority for location of foreign work. At present the USDA has two stations in Europe. One is located near Paris, France, for work on insect pests, the other at Rome, Italy, for work on weeds. The present laboratories are actually converted dwellings located in congested urban areas. To carry out the work that needs to be done in Europe during the next 20 years, these laboratories should be consolidated into a single, centrally-located, modern, well-equipped laboratory building. A location in Switzerland near the International Airports of either Zurich or Geneva would be ideal. The laboratory should be staffed by an American Director and co-Director, one a specialist in biocontrol of insects and the other in biocontrol of weeds. The non-American professional staff should be expanded to provide specialized coverage of pests associated with different commodities and to include competence in insect and weed pathogens. The importance of wild bees as specialized pollinators suggests that attention should be given to the search for more efficient pollinators of the numerous agricultural and horticultural crops that originated in Europe and the Near East. Each member of the staff should be bilingual with English as one language. It would also greatly facilitate field operations if, in addition to English, language competency within the staff also covered French, Italian, German and Russian.

An important feature of the laboratory would be provision to accommodate visiting scientists who would spend periods of a few days to several weeks at the laboratory or use it as a base of operations. The primary purpose of the provision to accommodate visiting scientists would be that of improving cooperation between the scientists conducting the foreign and domestic phases of introduction research. At the time a new project was to be undertaken the American scientists (either federal or state) most concerned with research on the problem in the United States would be sent to the laboratory. There they would work with the laboratory scientist assigned to handle the research in Europe.

The American would bring with him knowledge of the pest as well as the climate and cultural practices of the region where the pest is a problem. The laboratory scientist assigned to the problem would have knowledge of European agriculture and institutions. With the laboratory scientist available to accompany him the American could be in the field within hours after his arrival. Together they could bring a maximum amount of competence to bear on the problem in the shortest length of time. After acquainting his colleague with any special techniques needed to study the pest and familiarizing himself with the pest's behavior in Europe, the American would return home better able to assess prospects for effective biological control of the pest and better able to cooperate on a co-author basis with his European laboratory counterpart.

A well-equipped and staffed laboratory in Switzerland would have in addition to central location, the advantages of a multilingual country accustomed to serving as host to international agencies. Of all the countries of Europe, Switzerland offers perhaps the most congenial, social, and political climate for an institution that employs people of different national origins. This is of special importance since the laboratory would be staffed in part by personnel transferred from the existing USDA laboratories in France and Italy.

Quarantine Phase. With the increased interest in biological control and an increase in foreign introduction programs, the facilities and staff devoted to quarantine functions would have to be expanded. Otherwise incoming shipments could not be processed effectively and there would be an intolerable hazard that secondary parasites or other unwanted insects would gain entry to the United States. The quarantine laboratory should have staff sufficient to handle propagation of species received in too small numbers to warrant their being forwarded directly to other field stations. There should also be facilities to conduct host specificity and other experimental research where there is need to conduct such studies in quarantine.

Domestic Field Stations. Although there are numerous federal and state laboratories where biological control research is now in progress, there is need for Regional Beneficial Insects Introduction Centers. These centers should be specially staffed and equipped to conduct research on biological control of insect pests and weeds. The scientists located at these centers would cooperate with those of other federal and state laboratories located in their region and serve as liaison between these laboratories and the USDA foreign beneficial insect introduction program. There should be at least 4 such regional laboratories of which one might be located with the Central Quarantine Receiving Laboratory.

One of the functions of the regional biological control laboratories would be to provide staff for roving parasite introduction assignments or for temporary assignments to foreign locations where the USDA does not maintain a permanent station within operating range.

In order for a program such as the one described above to function successfully, and produce maximum results, close coordination of activities would be essential. Such coordination would involve all levels of activity, from long-range planning and budgets to field and laboratory operations. Because of the wide range of commodity and regional problems involved, it would be essential to have these interests represented in developing program plans and fixing priorities for work to be undertaken. Once immediate program objectives were established, coordination of the research and other activities essential to attainment of the objectives should be the responsibility of a single office or National Center. This Center should also maintain records needed to evaluate the results of introducing beneficial insects. It should also serve as a center for dissemination of information useful to biological control workers and provide liaison with the ARS Systematic Entomology Laboratory and other federal and state agencies who have expertises essential to a successful beneficial insect introduction program.

At the present time we are not equipped to effectively exploit the benefits to be gained from introduction of beneficial insects from foreign countries. A program is needed to make these benefits available to American agriculture quickly at a time when restrictions on use of pesticides greatly increases the importance of other methods for controlling pests. While a parasite introduction program cannot promise immediate control of a pest on this year's crop, it does offer one of the few means of eventually reducing a pest to non-economic status.

References Cited

Agricultural Research Service. 1965. Losses in Agriculture. USDA, Agr. Handb. 291, 119 pp.

DeBach, P. 1972. The use of imported natural enemies in pest management ecology. Proc. Tall Timbers Conf. Ecol. Anim. Control Habitat Managemt. 3: 211-233.

Doutt, o. L. 1958. Vice, Virtue and the Vedalia. Bull. Entomol. Soc. Amer. 4: 119-123.

Pimentel, D., D. A. Chant, A. Kelman, R. L. Metcalf, L. D. Newsom and C. N. Smith. 1965. Improved pest control practices, Appendix Y 11, pp. 227-291. In President's Science Advisory Committee Report, "Restoring the Quality of Our Environment." The White House, Washington, D.C., 1965. 317 pp.

Townes, H. 1969. The Genera of Ichneumonidae, Part I. Mem. Amer. Entomol. Inst. 11, 300 pp.

PRODUCTION AND SUPPLEMENTAL RELEASES OF PARASITES AND PREDATORS FOR CONTROL OF INSECT AND SPIDER MITE PESTS OF CROPS

R. L. Ridgway, R. E. Kinzer and R. K. Morrison
Agricultural Research Service
U.S. Department of Agriculture
College Station, Texas

The value of parasites and predators in regulating pest populations is well known. However, adequate numbers are not always available to provide the desired levels of control. This is particularly true where intensified agricultural practices have so altered the environment that it provides particularly suitable conditions for survival and reproduction of certain pests. One method of providing the desired balance between the pest and its natural enemies is the production and release of parasites and predators. Such releases may be referred to as periodic colonizations, programmed releases, strategic releases, or supplemental releases. Information concerning the role of supplemental releases as compared with other methods of biological control can be obtained from several sources (DeBach, 1964; Huffaker, 1971; Ridgway, 1972).

Effective and efficient supplemental releases usually will require (1) an understanding of the major ecological parameters governing the principal interactions between the parasite or predator to be released and the pest to be controlled; (2) an ability to rear predictable quantities of insects of known quality; and (3) an ability to store, transport, and release the parasites and predators in such a manner that they will become a competitive part of the life system in which they are expected to operate. Only production and supplemental releases will be considered here, since considerable insight into the various factors influencing the interactions between parasites or predators and their hosts or prey may be obtained from other sources (Holling, 1961; Clark, et al., 1967; Huffaker, et al., 1971).

Many studies of supplemental releases of parasites and predators have been conducted with only limited knowledge of the many factors that influence the effect on pest populations. Also, in many cases, the releases were not critically evaluated. However, a number of more recent studies provide some definitive results that are useful in examining the potential of this approach to pest control. An attempt will be made here to provide a general review of the production and supplemental releases of parasites and predators.

In cooperation with the Texas Agricultural Experiment Station.

Production

The production of parasites and predators has been of considerable interest to biological control enthusiasts for over half a century. Most of such activities up until 1960 were reviewed previously (DeBach and Hagen, 1964). Some representative examples of instances when significant numbers of parasites and predators were reared or field collected for release in the United States are listed in Table 1. Rearing of large numbers of parasites and predators also occurs in a number of foreign countries. Possibly the largest activity is in the USSR (Dysert, 1973).

In the past, the selection of insects for mass production has been heavily oriented toward those insects that are easiest to rear. More consideration should be given to such factors as searching ability, preference or host specificity, and adaptation to the environment. In this regard, studies of the comparative efficiency of candidate species such as reported by Lingren, et al. (1968) can provide some of the kinds of information needed to make intelligent selection of insects for mass production. In addition, the potential for the development of satisfactory storage, transportation, and release systems should be considered.

Production techniques for a wide range of insects was reviewed by Smith (1965). Many techniques have been developed for rearing insects, including parasites and predators; however, the application of comprehensive production technology, including time and motion studies, has been applied to only a limited extent. Also, detailed analysis of the cost of rearing parasites and predators are almost nonexistent though Scopes (1968) did make an effort to determine the cost of using predators and parasites for control of greenhouse pests. The development of specific production technology will be required to provide large numbers of parasites and predators at minimum cost. Also, the utilization of artificial diets may allow for a considerable cost reduction (Vanderzant, 1969; USDA, 1971). Although producing large numbers of parasites and predators at a low cost is important, an insect of acceptable quality is essential. Genetic deterioration in mass culture can be expected (Mackauer, 1972). Such deterioration can lead to the loss of behavioral traits that are basic to the effectiveness of released parasites and predators (Boller, 1972). Thus, in our studies with *Chrysopa carnea* Stephens and *Trichogramma* sp., the fecundity of the insects is routinely checked. In addition, specific tests are conducted to evaluate searching efficiency. Reductions in searching efficiency of *C. carnea* have been detected after 1 year in mass culture. Also, the searching efficiency of *Trichogramma* sp. has been reduced by length of time in mass culture and kind of rearing host. As a result, cultures are now reestablished from field-collected material annually to compensate for this deterioration.

Successful continuous mass rearing of parasites and predators for use in supplemental releases will therefore undoubtedly require routine testing for those behavioral traits required if the insects are to be effective when they are released. Also, some genetic manipulation will probably be required to maintain the desired traits.

Supplemental Releases

Since many of the earlier studies of supplemental releases of parasites and predators were reviewed previously (DeBach and Hagen, 1964), only some of the earlier more significant studies and some of the more recent studies will be considered here.

Doutt and Hagen (1949) placed eggs of *Chrysopa* sp. in pear trees and controlled mealybugs, *Pseudococcus* sp. Huffaker and Kennett (1956) demonstrated significant control of predaceous mites on strawberries when they released both phytophagous and predaceous mites. Chant (1961) controlled phytophagous mites on greenhouse crops by releasing predaceous mites, and Hussey, et al. (1965) and Parr and Hussey (1967), in related studies, also obtained effective control of spider mites in greenhouses. Other successful releases in greenhouses for control of whiteflies and aphids were reviewed by Hussey and Bravenboer (1971). In our studies, we have successfully controlled the bollworm, *Heliothis zea* (Boddie), and tobacco budworm, *Heliothis virescens* (F.), on cotton in field cages and in small plots by releases of larvae of the green lacewing, *C. carnea* (Ridgway and Jones, 1968, 1969). Also, Lingren (1970) successfully controlled bollworms and tobacco budworms on cotton by releases of *Trichogramma* sp., and we were subsequently able, at our laboratory, to obtain high rates of parasitism with similar releases. For example, in 1971 field-releases of *Trichogramma* sp. reared on eggs of the Angoumois grain moth, *Sitotroga cerealella* (Olivier) increased parasitism of *Heliothis* eggs in both corn and cotton; rates of parasitism of *Heliothis* eggs as high as 95% were obtained. Parker, et al. (1971) successfully controlled the imported cabbageworm, *Pieris rapae* (L.), on cabbage in Missouri by releasing fertile cabbageworms, an egg parasite, *Trichogramma* sp., and a larval parasite, *Apanteles* sp.

Although a number of examples of successful supplemental releases can be cited, controlled studies designed to determine the numbers of released parasites and predators required to obtain various levels of pest control and to develop techniques for large-scale releases are more limited. Parr and Hussey (1967) introduced different numbers of predaceous mites on cucumbers in greenhouses (1 to 10 mites per 10 plants) and concluded that 5 mites per 10 plants were needed to prevent economic damage. Likewise, at our laboratory, an effort has been made to determine the number of larvae of *Chrysopa carnea* that might be needed to control *Heliothis* spp. on cotton; the results of these field studies indicated that about 50,000 larvae per acre might be required (Figure 1). Similar studies with *Trichogramma* sp. for control of the same pests indicated that releases of 100,000 per acre at 2-3-day intervals may be required to insure substantial parasitism of eggs (Figure 1). Thus, considerably larger numbers are required than were expected from earlier estimates (Boyd, 1970; Knipling and McGuire, 1968). However, there probably is a great deal lacking in our ability to rear and release either *C. carnea* or *Trichogramma* sp. in such a manner as to provide insects that are competitive with naturally occurring insects.

Although *C. carnea* and *Trichogramma* sp. are currently being used commercially, our studies indicate that improved methods of rearing and distribution will be required before they can provide practical control of *Heliothis* spp. on cotton. Also, studies of supplemental releases of *C. carnea* and *Trichogramma* sp. in the presence of a full complement of naturally occurring parasites and predators must be conducted before the optimum numbers for such use can be determined.

Considerations for Practical Application

The advantages (such as selectivity) and the possible limitations (such as the need for large area management and problems with insecticide drift) of utilizing biological agents (including supplemental releases) for practical control of pests were discussed previously (Ridgway, 1972). However, reiteration of some of the characteristics of supplemental releases of parasites and predators as a means of biological pest control seems appropriate as follows:

(1) The establishment of a predictable number of parasites and predators in a given area is possible when proper procedures are used.
(2) Intensive management and improved methods of mass production and distribution will be required.
(3) When supplemental releases are integrated with natural control over a large area, only 10-20% of a crop may need releases.
(4) Supplemental releases may provide a useful tool for making a transition from a pesticidal system to an ecological system of control.

The extent to which the supplemental release of predators and parasites may be used in the future for practical pest control is difficult to predict. However, the increasing number of cases in which this method is being demonstrated as effective would indicate that the probability of reliable practical use is increasing. Certainly, the extent to which supplemental releases are used in the future will depend on whether effective and acceptable insecticides are available, since insecticides are easier for producers to manage. However, regardless of future events, the research and development of the production and supplemental release of predators and parasites are providing a better understanding of the behavior and population dynamics of predators and parasites. This improved understanding will be extremely valuable in designing improved pest control programs.

Table 1.--Some examples of parasites and predators produced or collected in sizable numbers in the United States.[a]

Parasite or predator	Estimated annual production in millions	Location	Year
Metaphycus sp.	>5	Filmore Citrus Protective District Insectary, California	1958-59
Aphytis sp.	>50	Filmore Citrus Protective District Insectary, California	1971
Trichogramma sp.	1,500	Agr. Res. Serv., USDA College Station, Texas	1971
	>2,000	Rincon-Vitova Insectaries Rialto, California	1970-71
Leptomastix sp.	56	Associated Insectaries California	1958-59
Cryptolaemus sp.	40	14 insectaries, California	1926-57
	31	Associated Insectaries California	1958-59
Chrysopa sp.	>50 (eggs)	Rincon-Vitova Insectaries Rialto, California	1965-71
	40 (eggs)	Agr. Res. Serv., USDA College Station, Texas	1971
Hippodamia sp.	>700[b]	3-5 dealers, California	1965-71

[a] Source: DeBach and Hagen (1964) and personal communications received by the authors. The list is not intended to be complete. The mention of a commercial firm does not imply endorsement by the USDA.

[b] Field collected

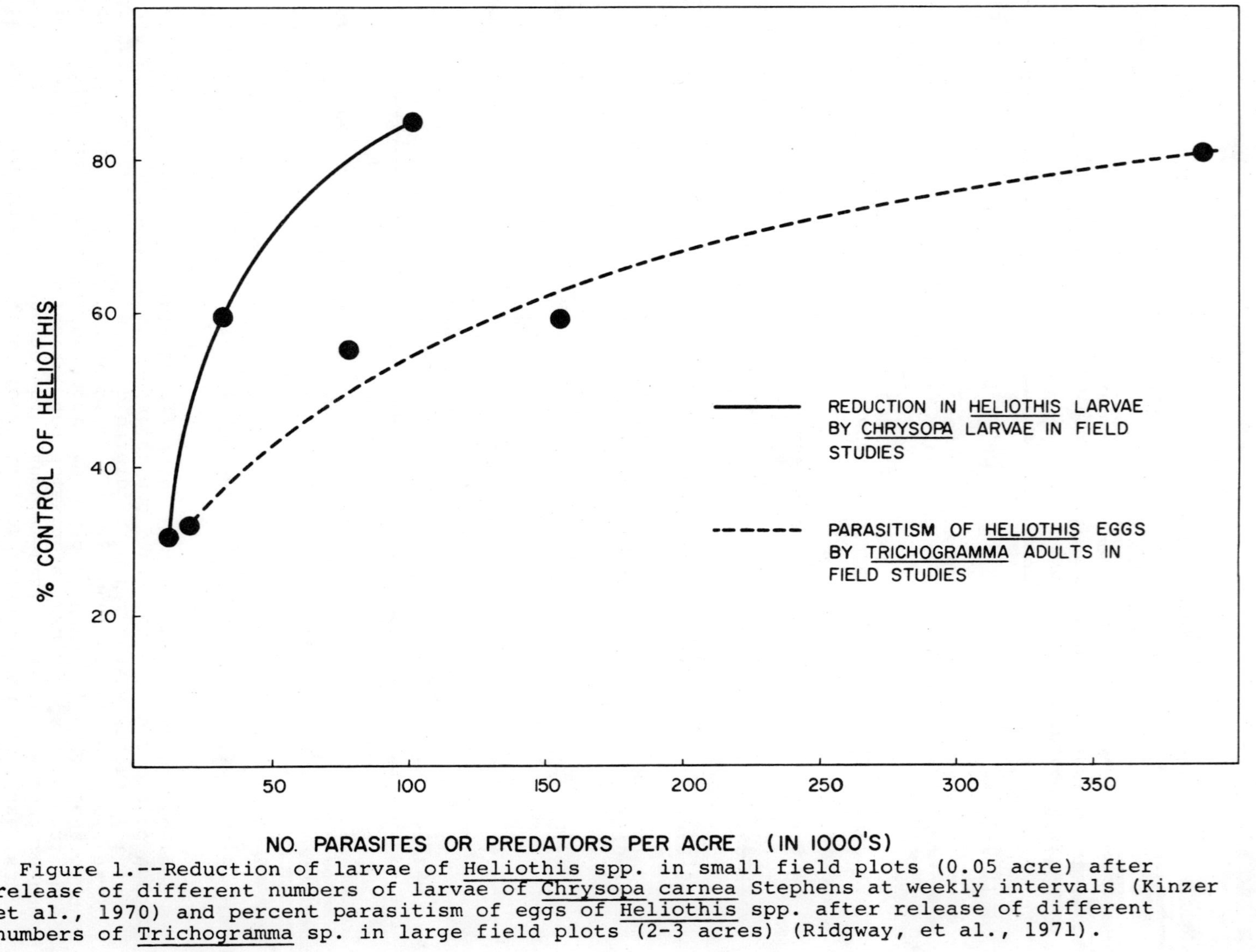

Figure 1.--Reduction of larvae of *Heliothis* spp. in small field plots (0.05 acre) after release of different numbers of larvae of *Chrysopa carnea* Stephens at weekly intervals (Kinzer et al., 1970) and percent parasitism of eggs of *Heliothis* spp. after release of different numbers of *Trichogramma* sp. in large field plots (2-3 acres) (Ridgway, et al., 1971).

References Cited

Boller, E. 1972. Behavioral aspects of mass-rearing of insects. Entomophaga 17(1): 9-25.

Boyd, J. P. and R. L. Ridgway. 1970. Feeding and searching behavior of *Chrysopa carnea* Stephens. Ph.D. Dissertation, Texas A&M University. College Station. 108 pp.

Chant, D. A. 1961. An experiment in biological control of *Tetranychus telaris* (L.) in a greenhouse using the predaceous mite, *Phytoseiulus persimilis* A.-H. Canadian Entomol. 93: 437-443.

Clark, L. R., P. W. Geier, R. D. Hughes, and R. F. Morris. 1967. The Ecology of Insect Populations in Theory and Practice. Methuen & Co., Ltd., London. 232 pp.

DeBach, P. (ed.) 1964. Biological Control of Insect Pests and Weeds. Reinhold Publ. Corp., New York. 844 pp.

DeBach, P. and K. S. Hagen. 1964. Manipulation of entomophagous species, Ch. 15, pp. 429-458. *In* P. DeBach (ed.) Biological Control of Insect Pests and Weeds. Reinhold Publ. Corp., New York. 844 pp.

Doutt, R. L. and K. S. Hagen. 1949. Periodic colonization of *Chrysopa californica* as a possible control of mealybugs. J. Econ. Entomol. 42(3): 560.

Dysart, R. J. 1973. The use of *Trichogramma* in the USSR. Proc. Tall Timbers Conf. Econ. Animal Control Habitat Mgmt. 4: 165-173.

Holling, C. S. 1961. Principles of insect predation. Ann. Rev. Entomol. 6: 163-182.

Huffaker, C. B. (ed.) 1971. Biological Control. Plenum Press, New York. 511 pp.

Huffaker, C. B. and C. E. Kennett. 1956. Experimental studies on predation: Predation and cyclamen-mite populations on strawberries in California. Hilgardia 26(4): 191-222.

Huffaker, C. B., P. S. Messenger and P. DeBach. 1971. The natural enemy component in natural control and the theory of biological control, Chp. 2, pp. 16-67. *In* C. B. Huffaker (ed.) Biological Control. Plenum Press, New York. 511 pp.

Hussey, N. W. and L. Bravenboer. 1971. Control of pests in glasshouse culture by the introduction of natural enemies, Ch. 8, pp. 195-215. *In* C. B. Huffaker (ed.) Biological Control. Plenum Press, New York. 511 pp.

Hussey, N. W., W. J. Parr and H. J. Gould. 1965. Observations on the control of *Tetranychus urticae* Koch on cucumbers by the predatory mite *Pytoseiulus riegeli* Dosse. Entomologia Exp. Appl. 8: 271-281.

Kinzer, R. E., S. L. Jones and R. L. Ridgway. Unpublished report, Entomol. Res. Div., Agr. Res. Serv., USDA. College Station, Texas.

Knipling, E. F., and J. U. McGuire, Jr. 1968. Population models to appraise the limitations on potentialities of *Trichogramma* in managing host insect populations. USDA Tech. Bull. 1386. 44 pp.

Lingren, P. D., R. L. Ridgway and S. L. Jones. 1968. Consumption by several common arthropod predators of eggs and larvae of two *Heliothis* species that attack cotton. Ann. Entomol. Soc. Amer. 61(3): 613-618.

Lingren, P. D. 1970. Biological control--can it be effectively used in cotton production today? Proc. 2nd Ann. Texas Conf. on Insect, Plant Disease, Weed and Brush Control. Texas A&M University, College Station, pp. 236-240.

Mackauer, M. 1972. Genetic aspects of insect production. Entomophaga 17(1): 27-48.

Parker, F. D., F. R. Lawson and R. E. Pinnell. 1971. Suppression of *Pieris rapae* using a new control system: Mass releases of both the pest and its parasites. J. Econ. Entomol. 64(3): 721-735.

Parr, W. J. and N. W. Hussey. 1967. Rep. Glasshouse Crops Res. Inst. 1966: 135-139.

Ridgway, R. L. 1972. Use of parasites, predators and microbial agents in management of insect pests of crops, pp. 51-62. *In* Implementing Practical Pest Management Strategies, Proc. Natl. Exten. Insect-Pest Management Workshop, Purdue University. 206 pp.

Ridgway, R. L. and S. L. Jones. 1968. Field-cage releases of *Chrysopa carnea* for suppression of populations of the bollworm and the tobacco budworm on cotton. J. Econ. Entomol. 61(4): 892-898.

Ridgway, R. L. and S. L. Jones. 1969. Inundative releases of *Chrysopa carnea* for control of *Heliothis* on cotton. J. Econ. Entomol. 62(1): 177-180.

Ridgway, R. L., R. E. Stinner and R. K. Morrison. 1971. Unpublished rept. Dept. Entomol., Texas Agr. Exp. Sta., and Agr. Res. Serv., USDA, College Station, Texas.

Scopes, N. E. A. 1968. Mass-rearing of *Phytoseiulus riegeli* Dosse for use in commercial horticulture. Plant Pathol. 17: 130-132.

Smith, C. N. (ed.). 1965. Insect Colonization and Mass Production. Academic Press, New York. 618 pp.

USDA. 1971. Packaged meals for insects. Agr. Res. 19(11): 3-4.

Vanderzant, E. S. 1969. An artificial diet for larvae and adults of *Chrysopa carnea*, an insect predator of crop pests. J. Econ. Entomol. 62(1): 256-257.

REARING AND QUARANTINE OF IMPORTED PARASITES AND PREDATORS

J. S. Kelleher
Importation Officer
Research Program Services Section
Canada Department of Agriculture
Ottawa, Ontario

The word "quarantine" is derived from the Italian "quarantina" meaning forty days. By a system of trial and error and without knowing the cause of disease, port authorities of the 14th century found that if ships were isolated after arrival, disease among passengers and crew would become manifest within forty days. We use this term to denote the restrictions imposed to prevent the introduction of undesirable species of phytophagous insects, hyperparasites or other harmful organisms. Theoretically, if only the desired entomophagous insects are shipped then there would be no need for quarantine but in practice there are several reasons why potentially dangerous insects are imported as well: (1) parasites survive the rigors of shipping if they are sent as immatures within their hosts, rather than as adults, and field-collected material will consist of unparasitized hosts as well as primary and secondary parasites, (2) parasites can be synchronized with susceptible host stages in the new environment when received as immatures, by delaying or accelerating development with appropriate temperatures, (3) predators may be held in quarantine while dissections and rearings are made which will preclude the possibility of introducing its parasites, (4) parasites and predators may have to be imported before proper identifications are made--this applies in particular where time for foreign collecting is limited. As a matter of principle, the University of California propagates each newly-introduced species for at least one generation in quarantine before it is released in the field (Fisher, 1964). This enables biological characteristics to be checked and provides a permanent taxonomic record of the total introduction. On the other hand, the USDA operates a laboratory near Paris and we in Canada sponsor laboratories of the Commonwealth Institute of Biological Control where the same information is obtained with a lesser need for quarantine rearing.

One of the objectives of this series of lectures is to explain the various techniques which may be employed in Biological Control and Pest Management. It is within this context that I will try to show the type of facilities, equipment and techniques used in operating a quarantine laboratory. In doing so I will draw heavily from a chapter by T. W. Fisher in the book Biological Control of Insect Pests and Weeds edited by P. DeBach.

Importations of insect collections must be sanctioned by Plant Quarantine officials and before a permit is issued, certain safeguards must be met. For approved quarantine facilities Plant Quarantine (USDA, 1971) will require such items as:

(1) an anteroom entryway with doors of insect proof design

(2) provision for a shower room and change of clothes
(3) insect and rodent proof floors, walls, ceilings and windows
(4) sealed electrical system including floor plugs, switches and lights
(5) heating and exhaust system, preferably a closed-air system, fitted with adequate filters
(6) plumbing system with screens in floor drains, and other drainlines
(7) pressurized air system - positive pressure in non-containment areas, negative pressure in containment areas.
(8) autoclave or incineration system within the containment area
(9) access to quarantine area limited to workers assigned to program
(10) traps effective for pest species placed in anteroom
(11) insects confined to cages within the quarantine facility and overcrowding of cages avoided.

A year ago we at Belleville were presented with these criteria plus the usual budgetary restrictions, and told to move our Importation Service to Ottawa. Ottawa had been chosen over Belleville for three main reasons: (1) an international airport is nearby for receiving and dispatching shipments, (2) taxonomic experts are located there and can be reached more readily, (3) agriculture and forestry headquarters staff are also in closer proximity for consultation.

A separate building could not be provided, only space within an existing structure where every square foot had to be justified. Three rooms, each 16 x 22 feet were available, exclusive of office space. These rooms had several desirable features: they were on the top floor so incinerator and fumigator exhausts were easier to install, air-conditioning compressors could be located in a roof penthouse rather than in the quarantine area, the outside windows faced north providing more constant natural lighting, and the rooms were at the end of a corridor where traffic is relatively light.

The main quarantine area can be entered through only one door although one-way fire doors are provided. Shipments are opened in the first room and the insects removed for processing. Entry is gained to other rearing and emergence rooms through an anteroom and door switches are provided to darken this anteroom whenever these rooms are entered. Cold storage rooms (45°F and 35°F) lead from the main rearing room and the temperature in the rearing rooms can be individually adjusted from 55° to 80°F. Humidity is maintained at 50%; higher humidities can be obtained in enclosed cabinets. One room is used for both disposal and packaging. Two disposal systems are used: fumigation and incineration. Fumigation with methyl bromide is a hazardous operation and incinerators are often forbidden by anti-pollution laws. Large autoclaves are therefore employed by most establishments.

Rather than seal the existing windows and heating radiators our architect found it more feasible to build a new glazed wall with sealed panels and plastic windows which could be removed for cleaning or emergency repairs. I should also add that a false ceiling 7 feet high is to be constructed in the rearing room so that any escaped insects can be readily captured and to house permanent ductwork. Vents are covered by 50-mesh wire screening which must be frequently cleaned and insect escapes are further prevented by filters in the exhaust duct. The only significant departure from USDA Plant Quarantine requirements is the substitution of a "water closet" for a shower room.

The preoccupation with facilities suggests that these are the main feature of quarantine whereas in fact, the personnel who operate the facility are far more important. First, they must have conscientious attitudes and an awareness of the dangers involved. Secondly, they should possess an appreciation for the value of each imported entomophagous species being handled. For this reason some feedback on the results of attempts at biological control are always solicited. It is valuable to have a general taxonomic knowledge of the major entomophagous groups and an ability to distinguish closely related species with the guidance of expert systematists. At Belleville, our taxonomy is primarily on a host basis, that is identifications are based on reference specimens reared from the same host collected in the same locality. But very frequently some entomophagous species are obtained which cannot be categorized in this manner and the worker must recognize the need for further taxonomic guidance. Lastly, the quarantine worker should possess those difficult-to-define characteristics, akin to the "green thumb" gardener, which enable him to sense differences and changes in the culture he is holding and which alert him to the presence of disease or other catastrophic conditions so that remedial action can be taken.

The importation of biological control material involves four groups of workers: (1) those who request the entomophagous species, (2) those who make the collections, (3) quarantine workers who receive, process and distribute the imported material, and (4) those who make the liberations and subsequent surveys and who are usually the same group who made the initial request. The importance of adequate communication among these groups should be apparent and because of his position in the stream of events, the quarantine worker often provides the necessary liaison. In any event quarantine personnel are concerned first with the foreign collector and then with the receiving field laboratory.

In preparation for importing insect collections survey reports from the foreign collector will be carefully studied for information on the presence of hyperparasites, and the identity of primaries and pertinent details of their biologies which will aid in rearing the insects in the laboratory. Sufficient advance notice must be obtained to enable the provision of any plant and host insect cultures that may be required.

Shipments of living insects imported into this country must bear a Plant Quarantine import permit label supplied to the foreign collector in advance. Shipping containers must be of approved construction. Basically they must remain insect-tight despite any rough handling that may be endured in transit; sometimes an outer container of metal or an outer box enclosed by heavy cloth with sewn seams is required. At one time a ventilated wooden box was commonly used but the frequency and speed of modern transport and the possibility of insecticide sprays being used to decontaminate cargo has led to the adoption of insulated and refrigerated containers with an outer wrapping that can be removed and incinerated.

Both air mail and air cargo are used. Mail is so much cheaper that it is tempting to use this service and in many cases the difference in transit time is negligible. Air cargo for valuable or difficult-to-replace insect collections is justified, as transit is usually quicker and shipments can be readily traced if delayed. However, we find mail service from our nearest airport faster than rail service. The same principles apply to domestic air shipments, but in this case we always use cooled, insulated containers. The container is insulated with 2 inches of commercial styrofoam within a sturdy cardboard carton. Space is provided for refreezable ice packets and the insects are carried in appropriate cages in an inner cardboard or wooden carton. Such containers will hold living insects at 45°F for 24 to 48 hours depending on outside conditions*. But if it is necessary to send colonies to dozens of places, as is done by the USDA Parasite Laboratory at Moorestown, New Jersey, then mail service must be used.

Prior contact with officials of customs and carriers invariably fosters cooperation from them. We find it helpful to renew these contacts at the beginning of each major shipping season to acquaint new personnel with our

---

*Such low temperatures can seriously affect the reproductive potential of some parasites--some *Aphytis* species are even affected after 24 hours at sub-optimal temperatures of about 60°F (DeBach and Argyriou, 1966).

operations and to obtain any new information on routings. We provide telephone numbers where we can be reached at all times and insure that these are placed on package labels and express waybills, to insure quick contact on arrival. When shipments are dispatched the sender alerts the consignee by cable, indicating the number, species and routing and if applicable, the air waybill number. Abbreviated names and cable addresses reduce the cost. Similar alerting telegrams are sent when dispatching domestic shipments although long distance telephone can also be used. Such communication keeps members of the team fully informed and enables last-minute details to be checked.

Each package contains a shipment form giving details of the collection and particularly the species and stages shipped, secondary species possibly present, and date and place of the collection. The consignee adds the date the package was received and its condition on arrival. One copy is returned to the sender. When a series of shipments is expected, the sender will want to know by cable, if adjustments in packaging or routing can be made.

Packages containing active insects must be opened in a special examination room or in a large sleeve cage where any insects that escape can be readily recovered. Personnel engaged in this and other recovery operations are required to wear white laboratory coats which are left in the quarantine area and which are decontaminated in an autoclave or fumigation chamber before laundering. Because of the possibility of different strains or races of parasites and predators, collections from different areas are separated and reared independently. This also indicates the best collection areas for future consideration. Depending on the amount of information the collector has provided, it may be advantageous to dissect a sample to ascertain the incidence of parasitism or predators and their stage of development. If the number is too small to sacrifice some in dissection, it may still be possible to detect parasites by microscopic examination of hosts over light transmitted through flashed opal glass or by X-ray photographs (Holling 1958). In some cases, e.g., that of _Olesicampe benefactor_ attacking larch sawfly, the size of the host is markedly reduced in comparison to non-parasitized ones (Muldrew, 1967).

The daily record of emergence is placed on a form which is kept with the collection. At Belleville we assign an accession number to each collection; this is used on the rearing form, and is again referred to when shipment is made or insects placed into culture. The rearing form is used to record the number and sex of each host and parasite adult removed daily. If a species cannot be readily identified it is assigned a number and notes on its taxonomic features made on the reverse side of the form. Also on the reverse side we note the various temperatures at which a collection may be held.

Individuals must be reared separately if disease is apt to spread to the other insects or if parasites are present which will emerge early and parasitize and kill other parasites in the hosts. This can happen, for example, with a pupal parasite of European pine sawfly, _Dahlbominus fuscipennis_, which in nature must search the soil diligently for a single cocoon but when presented with a concentrated mass of cocoons will readily parasitize and kill other parasites in them. To prevent this, the cocoons are placed individually in 1 1/2 inch vials stoppered with absorbent cotton. The vials can be spread one layer deep on shallow trays where emergence can be readily checked. There are other advantages to this method of rearing, but it is laborious and is difficult to maintain a steady level of contact moisture.

In the absence of more precise data, collections are usually reared at constant conditions of 72°F and 50% RH in the Belleville Institute, whereas at California 76°F is more common. This probably reflects the acclimitization of the workers usually handling them. Sometimes optimal conditions for rearing the hosts are known in which case these conditions are used. Particular attention must be given to humidity. If it is too high, mold problems inevitably will occur;

if too low then of course insects will fail to survive or be unable to emerge or spread their wings properly. We have found it preferable to add moisture periodically and sparingly spraying a fine mist or sprinkling the substrate. Mold inhibitors such as sodium proprionate can also be used. It is difficult to generalize with regard to humidity requirements but we have found that tachinids often require contact moisture for survival. On one occasion almost an entire collection of *Myxexoristops blondeli* larvae failed to pupate for this reason. The cocoons in which they were contained had been individually vialed so that another parasite, *Mesoleius tenthredinis* could be obtained as virgin females for genetic studies. On the other hand ichneumonids often survive under much drier conditions; on another occasion most of the larvae of a eulophid, *Tetrastichus incertus* died when normal moistening procedures were carried out.

When emergence does occur the various species must be segregated. In mass rearings of field-collected material the hosts can often be readily distinguished with the naked eye and large numbers removed with power-suction apparatus. The same apparatus can be used to transfer large numbers of a desired entomophagous species but great care must be taken to prevent damaging the insects by using too much suction. When there are a mixture of species, each specimen must be identified under the microscope. Carbon dioxide is frequently used to immobilize the insects for such an examination but care must be taken to administer a safe level, i.e., a mixture of 20-40 percent carbon dioxide with air (Patton, et al., 1968) and to keep the time under anesthesia at a minimum. Low temperatures can also be used for immobilizing insects (Harris and Frazar, 1968). And other anesthetics such as Freon have been successful. But if the insects are confined in the 1 1/2 inch glass vials mentioned previously, the particular diagnostic features can usually be seen during a momentary pause in movement and if necessary carbon dioxide can be introduced through the cotton plug. Some features are easier to see on living specimens than on pinned ones, colors are sharper and activity can be characteristic.

If the entomophagous species is a primary and is sufficiently numerous for release, it can be shipped to the field directly. In practice emergence will occur over a certain period and males will often emerge before females so the insects must be held for several days at a time. This is usually done in cages of various sizes depending on the size of the insects and their number. We offer them a 50 percent mixture of honey and water by dipping a short length of dental cotton in the mixture and shaking off the excess liquid. The dental cotton is placed on a piece of waxed paper on the floor of the cage and is changed daily. The insects are often shipped in a smaller container, e.g., a one-pint waxed paper carton, with strands of wooden excelsior to which they can cling and which have been dipped in honey solution to provide food. A hole is cut in the top of the carton for introducing the insects and it and the top are then sealed with masking tape. We have also used a copper screen container of similar size as insects can be loaded with power-suction apparatus and a moistened cloth can be wrapped around the outside to provide a high humidity. A form is enclosed with each shipment showing the species, and number of each sex. A photocopy of the original shipment form received with the overseas collection is also attached. The recipient is expected to record the date and condition when the shipment is received as well as the number liberated, the locality and the date, and to return one copy of the form. This forms a permanent record which can be referred to when required.

There are occasions when the quarantine laboratory would be expected to culture the parasites. This may be because further screening is necessary to establish the parasitic nature of the species, because further tests of host specificity are warranted or because the numbers obtained are too small for field release. For these purposes it is desirable to have ready access to common laboratory insect colonies such as aphids or greater waxmoth, which might serve as factitious hosts (Simmonds, 1944). Often it is a more cumbersome procedure to rear the natural host especially if plants must be provided for it. However, great advances have been made in the use of artificial diets although many workers prefer to make releases with parasites reared from the natural host.

Mating is normally essential to laboratory propagation and some species require specific physical conditions to mate successfully. Various techniques have been employed when difficulties are encountered. Some females will mate only shortly after emergence with males one or two days old. When this is not possible, females can be stunned with carbon dioxide or chilled with cold temperatures to make them more receptive. Another common technique is to confine males and females in a small vial.

In screening for host preferences, the procedures are analagous to those used in screening insects feeding on noxious weeds, that is, hosts closely-related taxonomically to the target host are exposed to the parasites. For example in the proposed introduction of parasites of the Mexican bean beetle it was necessary to determine if they would attack native beneficial coccinellids. Usually the parasites are confined with a single host species to force them to attack and to determine if they can develop successfully on these hosts. But in other tests they are allowed a multiple choice of host species; this more closely approximates natural conditions. Fortunately most of our importations are against pests whose close relatives are also considered undesirable.

A much commoner danger is the possibility of introducing hyperparasites which will attack the primary being colonized and others already present. Hyperparasites can be broadly divided as obligatory or facultative. Obviously those which can be either primary or secondary are the most difficult to detect. Obligatory hyperparasites are regarded as highly undesirable but facultative hyperparasites can be studied for their possible impact on primaries. For example Californian workers (Flanders, 1959) found that the females of some Aphelinidae will act as normal primaries but that males of the same species must develop as obligatory secondaries. It is only through careful studies that such relationships can be shown. The quarantine worker should be alert to parasites that spend an unusual amount of time searching among parasitized hosts or species whose numbers increase with time as further collections are received from the field.

Certain taxonomic groups contain many members which are known to be obligatory hyperparasites so others in that group will be immediately suspect. The habit is common in such chalcidoid families as Elasmidae, Pteromalidae, Eupelmidae and Thysanidae and in the Mesochorini and Cryptini of the Ichneumonoidea but it is rare among the braconids and absent in the tachinids.

If individuals in a collection are reared separately it is possible to check host remnants for two different types of meconia or exuviae. Should these be found then hyperparasitism is obviously indicated.

Dissections also provide a positive indication of hyperparasitism if larvae are found feeding in or on other parasites.

Because of its unique and costly facilities quarantine centres usually service a large area and are involved in many projects. During the current year at Belleville we will be importing entomophagous species for use against seven insect pests at different regional establishments from four different countries. The variety and challenges of these exercises and the contact with many research workers are, in my estimation, a very satisfying occupation.

References Cited

DeBach, P. and L. C. Argyriou. 1966. Effects of short-duration suboptimum preoviposition temperatures on progeny production and sex ratio in species of Aphytis (Hymenoptera: Apheliniade). Res. Popul. Ecol. 8: 69-77.

Fisher, T. W. 1964. Quarantine handling of entomophagous insects. In DeBach, P. (ed.). Biological Control of Insect Pests and Weeds. Chapman and Hall, London.

Harris, R. L. and Frazar. 1968. A device for immobilizing insects with cooled air. J. Econ. Entomol. 61: 1755-1757.

Holling, C. S. 1958. A radiographic technique to identify healthy, parasitized and diseased sawfly prepupae within cocoons. Canadian Entomol. 90: 59-61.

Muldrew, J. A. 1967. Biology and initial dispersal of Olesicampe (Holocremnus) sp. nr. Nematorum (Hymenoptera: Ichneumonidae), a parasite of the larch sawfly recently established in Manitoba. Canadian Entomol. 99: 312-321.

Patton, R. L., L. J. Edwards and S. K. Gilmore. 1968. Delivering safe levels of CO_2 for insect anesthesia. Ann. Entomol. Soc. Amer. 61: 1046-1047.

Simmonds, F. J. 1944. The propagation of insect parasites on unnatural hosts. Bull. Entomol. Res. 35: 219-226.

USDA. 1971. Issuing permits for the movement of plant pests, pathogens, and vectors. PA-967. Agricultural Research Service.

RELEASE, ESTABLISHMENT AND EVALUATION OF PARASITES AND PREDATORS

Frederick W. Stehr
Department of Entomology
Michigan State University
East Lansing, Michigan

Introduction

In the history of biological control, many natural enemies have been released, maybe 20 to 25% have been established (DeBach, 1964; Clausen, 1956; Turnbull, 1967), and a number have been extremely effective (DeBach *in* DeBach, 1964). Experimental documentation has been accomplished for some biological control successes (DeBach, *in* Huffaker, 1971), but a complete understanding of the mechanisms involved in pest regulation by natural enemies has yet to be achieved (Varley and Gradwell, *in* Huffaker, 1971). Numerous papers and several chapters in two recent books on biological control (DeBach, 1964 and Huffaker, 1971)[1/] have been devoted to release, establishment, and evaluation of insect natural enemies. It is immediately obvious that releases are relatively easy and establishment is more difficult. Experimental evaluation of the effect of introductions is possible in some cases but difficult in others, and achieving an understanding of the mechanisms of population regulation is a difficult task. Two types of methods are used in evaluation. One can be termed a "quantitative approach" using key factor, life table and modeling techniques, and the other is termed the "experimental check method," wherein various techniques are used to demonstrate the response of the pest in the presence and absence of natural enemies. Some workers (DeBach and Huffaker, *in* Huffaker, 1971) contend that the quantitative methods do not provide the rigorous proof of control or regulation by enemies that is needed and provided by the experimental check methods.

Biological control by parasites and predators is not as spectacular as the massive and immediate results obtained with insecticides (although the action of microbial diseases can be striking at times, but is not considered here). Therefore, if we expect to maintain and increase the necessary financial support for biological control, we must make a major effort to adequately demonstrate its effectiveness by experimental and/or quantitative techniques.

[1/] Citations to chapters in both of these books have been used frequently instead of giving complete citations of many original papers which can be found there.

Releases and Establishment (they are inseparable)

The simplest way to release natural enemies is to remove the lid and throw the carton out the window as you drive by an area where you think there are hosts. This has been done, and I know of at least one case where establishment was successful, but that is not surprising since it is analagous to what must happen when many of our exotic pests are established by the accidental importation of a few individuals.

Obviously, there are better methods, but before I discuss them I would like to cover a related point concerning the collection of enemies for release. The chances for establishment and the degree of pest control achieved are often directly related to the quality of the natural enemies obtained. In other words, if an efficient natural enemy (or an efficient strain or subpopulation of the species), is not established, control may not be obtained. The topic of foreign exploration and importation of exotic parasites and predators has been covered by Dr. Sailer in this book but, as Flanders (1959) pointed out, the scarcity of the host in the area to be searched is often a primary requirement for the collection of an efficient natural enemy (assuming the area is otherwise suitable for the host). Too much emphasis has been put on collecting large numbers of natural enemies at times, and we would be ahead to concentrate on the quality of the specimens collected, rather than simply trying to collect as many as possible.

Releases of natural enemies are easy enough to make, but what should we consider if we are to have the best chance of establishment? There are two different kinds of releases (1) original establishment releases and (2) subcolonization releases. I will concentrate on original establishment releases but there are procedures that can be used for subcolonization which are inappropriate for original establishment releases because of the necessity to screen out hyperparasites that may be present, and the relatively small number of natural enemies usually available for original establishment releases.

Factors to Consider in Establishment Releases

1) Is the climate suitable? All species (pest and non-pest) are adapted to some climates better than others, and subpopulations have evolved to be more effective in one climate than in others. This means releases should be made in climatically similar areas to those from which the natural enemy was collected if at all possible. Many factors must be considered, including latitude (=photoperiod), altitude, temperature and rainfall. Extreme climatic conditions should be examined (not the average), since survival may be affected by the extremes when there appear to be only superficial differences in the averages. Messenger (*in* Rabb and Guthrie, 1970) thoroughly reviews the bioclimatic input to biological control, and cites examples of climatically selected parasites from California where climatic conditions are very diverse.

Messenger and Van den Bosch (*in* Huffaker, 1971) cite several samples where climate has affected the success of introductions. A most interesting case is the release of *Trioxys pallidus* (Halliday) to control the walnut aphid, *Chronaphis juglandicola* (Kaltenbach). In 1959 this parasite was introduced from France throughout southern California and in the next two years it was colonized throughout central and northern California. It rapidly became an important control agent in the coastal and intermediate zones of southern California, but failed to become effective in the northern two-thirds of the state after 5 or 6 years of intensive colonization. It was concluded that the more severe summer and winter climates in the north were limiting the effectiveness, so the "same" parasite was imported from the central plateau of Iran which is climatically similar to the Central Valley of California, but with somewhat more severe winters.

In less than 2 years it was apparent that the Iranian stock was well adapted to the climate of northern California, and excellent biological control was obtained at some release sites. The possibility exists that two species of parasites may be involved, but the benefits obtained by using populations which are climatically adapted are obvious.

2) What is the detailed life history of the natural enemy? Synchronization of the parasite with the host is obviously essential, and this is most often affected by the climate, but regardless of the climate, it is absolutely essential that the proper stage of the host be available at the proper time. This may be the primary host, an alternate host necessary to get through the season, or even female larvae of the same species of natural enemy in cases where the male larvae are obligate hyperparasites of their own species (Flander, 1936).

A knowledge of the overwintering site of the natural enemy may be important in non-tropical areas, and a knowledge of aestivation sites may be essential in wet-dry climates, but establishment is possible in some cases even if such gaps exist. The example I am most familiar with is that of *Anaphes flavipes* (Foerster), a mymarid egg parasite of the cereal leaf beetle, *Oulema melanopus* (L.), which has been established (Maltby, et al., 1971) without knowledge of its overwintering site, although this knowledge might be useful in trying to manage it.

Knowledge of the overwintering or aestivating site is particularly important when the site is going to be disturbed by some management practice which will destroy a large number of the inactive parasites. In the Midwest, we believe the successful establishment and buildup of the larval parasites of the cereal leaf beetle, *Tetrastichus julis* (Walker), *Diaparsis carinifer* (Thomson), and *Lemophagus curtus* Townes to be due in large part to (1) the complete protection from plowing or disking of the overwintering sites in the soil until after the parasites have emerged in late spring, and (2) to the planting of several ages of grain which provides cereal leaf beetle larvae for a longer period of time than is normally the case, and tends to negate any synchronization problems which might exist. (Stehr, 1970; Stehr and Haynes, 1972).

The diapause requirements of the host and natural enemy can also be important. Clausen (1956) outlines the history of the release of a Mexican tachinid, *Paradexodes epilachnae* Aldrich, against the Mexican bean beetle, *Epilachna varivestris* Mulsant. Field parasitization of up to 90% was obtained after release, but for 5 years the parasite was never recovered in the season after colonization. A close study of the biology showed that the parasite could not survive the winter without host larvae, and since the Mexican bean beetle overwintered as an adult in the United States, the parasite was doomed. In Mexico there were always a few larvae available in the winter to maintain the parasite.

A knowledge of both climate and biology were essential for the establishment in Michigan of *Microctonus aethiops* (Nees), a parasite of adult alfalfa weevils, *Hypera postica* (Gyllenhall) (Stehr and Casagrande, 1971). This braconid has two generations per year, the overwintering generation spending most of the summer and the winter as a diapausing first instar larva inside the overwintering adult weevils. It resumes development when the weevils become active in the spring, killing them and parasitizing remaining unparasitized overwintered weevils. The spring generation does not diapause, and emerges to parasitize the newly emerged alfalfa weevils in which the 1st instar larvae go through the winter, thus completing the yearly cycle. The season of central Michigan is 10-14 days later than the Philadelphia area, thus it was possible to sweep adult weevils containing non-diapausing spring generation *M. aethiops* larvae near Philadelphia in mid-May, hold them for adult parasite emergence, and release the parasites in Michigan in late May and early June where they parasitized overwintered weevils and produced a non-diapausing generation which emerged and parasitized the newly emerged weevils. By having a detailed knowledge of the parasite's biology and by taking advantage of the 10-14-day earlier season in the Philadelphia area we obtained the non-diapausing generation in Michigan the first year. That would not have been possible if establishment had been attempted in an area with a climate similar to Philadelphia's unless a very large effort had been put into collecting the overwintered weevils containing diapausing M. aethiops early in the spring (when they are much more difficult to collect).

3) What is the host density and population trend? The number of hosts available to a natural enemy and the condition of these hosts are both related to the time when the release should be made. It is obvious that if the host density is too low, the parasites[2/] (1) will spend most of their time looking for hosts, (2) may not be able to realize all of the fecundity, (3) may end up superparasitizing the few hosts available (which will either exhaust the food available in the host before the parasites can complete development or result in lost production if competition is such that only 1 parasite survives in each host), or (4) dispersal in search of hosts may be so extensive that the problem of finding mates in the next generation will be considerable. However, low populations of hosts are not normally a problem since work on biological control is often not even started until the pest becomes abundant enough to cause economic losses.

A more serious problem which is often not considered is the release of parasites when host populations are at or near their peak density, or even when they have started to decline. It is undesirable (but sometimes unavoidable) to release at these peak or declining densities because of the great loss of parasite reproduction which is bound to occur when the host population crashes because of starvation, disease, or other density dependent mortality often associated with peak population densities. Competition with established parasites may also be more severe when populations are declining.

A similar pest "population crash" can result when pest populations reach levels which require extensive insecticide treatments. Although economic levels requiring insecticide treatment are usually far below natural peak populations, the effect is the same--a drastic reduction in available hosts and decimation of parasitized hosts, in addition to direct adult parasite kill. Insecticide treatments can of course be avoided by full control of release areas, but there is no way to avoid a natural population crash.

This means the ideal host population level for parasite releases is a vigorous, expanding population which is high enough to provide sufficient hosts so excessive searching, dispersal or superparasitism do not occur, yet low enough to avoid all the mortality factors associated with peak or declining populations, and low enough to provide a period of insecticide-free growth. In short, this ideal host population level must be determined for each target insect so as to insure that the reproduction of the parasite will be maximal for several generations. With obvious exceptions (such as excluding superparasitism) similar considerations apply to the release of predators, too.

Embree and Underwood (1972) have described the establishment of Olesicampe benefactor Hinz against the larch sawfly, Pristiphora erichsonii (Htg.) in Maine, Nova Scotia and New Brunswick, and have discussed the optimum range of host densities to have the greatest chance for establishment of this parasite. Stehr and Haynes (1972) have suggested optimum densities for colonization of cereal leaf beetle larval parasites based on the results of establishment releases made in Kalamazoo Co., Mich. Similar studies should be carried out for other pests and parasites whenever possible.

4) How many parasites to release? It is impossible to state how many of a given natural enemy will be needed to effect establishment, but several considerations should be examined. Perhaps the most elementary and important is how

---

2/The term "parasite" is used in a general sense to include predators.

many are available. If only a few are available, they should all go to the best available site where they may even be caged (although some parasites don't do well in cages), or possibly they should be cultured in the lab. If the supply is moderate or unlimited (as may be the case in subcolonizations) more options are open. Major considerations should be the diversity of the release material and the diversity of the areas where colonization is desired. In areas like the Midwest, climatic and physiographic diversity are low, so material of diverse origins can often be released at the same site with little loss of potential for establishment, and perhaps even an increase due to genetic recombination. The only loss may be in knowing which material was established if populations of different origins are not morphologically distinctive.

In areas where climatic and physiographic diversity are great and the number of parasites for release is limited, some hard decisions will have to be made, but the primary objective of original establishment releases is to get establishment. In these instances the best strategy is "large" numbers (500-1000 or more) at the best possible site, but this does not mean smaller numbers will not work.

Clausen (1939) indicates 10-20 females is enough for many lady beetles, and a single fertile female could be sufficient for the establishment of a species if everything was perfect.

In contrast, there are times when the release of thousands does not result in establishment, but if establishment is not obtained following a release of 500-1000, factors other than the number released should be carefully examined before a major effort is undertaken to obtain larger numbers.

5) Are there any competitors? In most cases an inherently superior parasite would not be affected by other parasite species since it will kill the inferior parasite if in the same host, or displace it if superior in other ways. However, it may be impossible to establish an inferior species in the presence of a superior one. DeBach *in* Baker and Stebbins (1965) gives the example of *Aphytis fisheri* imported from Burma and its closest relative *Aphytis melinus* imported from India and West Pakistan at the same time against the California red scale, *Aonidiella aurantii* (Maskell). They are sibling species and do not hybridize. Despite large and frequent releases of both species in the same orchards, *A. fisheri* never became permanently established while *A. melinus* became the dominant parasite and even replaced another species, *Aphytis lingnanensis* Compere, which had been established earlier and was widespread and reasonably effective, and had previously displaced *Aphytis chrysomphali* from certain areas.

It is difficult to predict with any degree of certainty which species will become established and which ones will be the most effective. However, it is obvious that if a sequence of species is to be tried as advocated by Turnbull and Chant (1961) and others, the best approach is to initially introduce one which is killed by all others in a multiple parasitism situation, thus enabling it to demonstrate its potential in the absence of competitors. If it is ineffective then the next one up the scale of competitive superiority has a good chance for establishment. If the establishments are reversed, the poorer competitor has a considerably reduced chance of becoming established in the presence of the superior competitor, and even less chance to become effective.

Many workers do not agree with the philosophy of single species introductions. They believe control will never be reduced by the establishment of additional natural enemies, and the establishment of a superior natural enemy will improve it. See Huffaker, et al. *in* Huffaker (1971) for a discussion of the subject from the multiple species introduction viewpoint.

The history of biological control to date certainly supports multiple species introductions, but carefully designed programs to test the single species introduction philosophy are lacking. Even if there is no pressure to obtain satisfactory biological control quickly, the single species introduction philosophy suffers from the almost insurmountable obstacle of knowing how to rank the known natural enemies from worst to best before introductions are made to a new area. And even if a ranking of natural enemies is established in the source area, it does not necessarily follow that the same rank will be maintained in the new area.

6) Direct release or laboratory production? Both techniques have their place and both have advantages and disadvantages. Laboratory production is desirable when there is any possibility of hyperparasites being present, when the biology is imperfectly known, and when very limited numbers are available. The major disadvantage of laboratory production is the risk of selecting out a strain which is adapted to laboratory conditions and does poorly under field conditions, with the related possibility of eliminating genes that may be adaptive in the field.

For direct field releases each individual must be examined to eliminate hyperparasites (time-consuming), there is no chance to investigate the biology more thoroughly, and the chances of establishment are minimal if small numbers are involved. Direct field release is required if the host or the parasite cannot be reared satisfactorily in the lab, but the possibility of eliminating desirable genes in the lab is avoided, so the maximum genetic variability is retained. A situation where laboratory production for a few generations might increase the genetic variability and/or adaptability of a population could arise when populations of natural enemies from different geographic areas are sufficiently different so that little or no mating takes place between these populations under field conditions. If certain desirable traits from one population need to be incorporated in the other population(s) the laboratory may be the best (or quickest) place to do it since some isolating mechanisms which operate in the field can often be negated in the laboratory. Nevertheless, in most situations we don't know which traits are best, so I am inclined to agree with Simmonds (1963) who suggested that releases of collections from various sites of the native home without propagation would provide maximum genetic variability, and field selection would provide the best genotypes for the target area. DeBach and Hagen (*in* DeBach, 1964), White, et al. (1970), and Wilson (*in* Baker and Stebbins, 1965) have suggested that strains could be produced in the laboratory that would be better able to cope with some limiting factors and hence colonize previously unsuitable areas. However, to date, this has not been demonstrated to be effective, and it certainly is not a necessary approach toward an original establishment unless all other methods have failed. Obviously, the optimum way to make original establishment releases is to colonize in the field and to retain a culture in the lab if at all possible.

7) Should adults or immature stages be released? It goes without saying that an original establishment release should be composed of adults to avoid the possibility of introducing hyperparasites that may be present in immature parasites. To my knowledge parasites (= hyperparasites) are unknown from adult parasites, so the only hyperparasites to be sorted out of adult releases are those adult hyperparasites that resemble the primary parasites. In addition, adults are by far the most effective searchers, which can be critical at low pest populations. Predators are a different matter, of course, since immatures and adults can be parasitized. Subcolonization from an original establishment site is another matter, and the shipment of hosts containing immature parasites is often the most efficient method since the hosts are usually much easier to find and collect than the adult parasites, and may be easier to handle. We use it for the cereal leaf beetle larval parasites (Stehr and Haynes, 1972), it is commonly used for alfalfa weevil larval parasites, and numerous other examples could be cited. Subcolonizations are critical for parasites which have limited powers of dispersal. Perhaps the best recent example of an efficient subcolonization system is that used for the dispersal of the parasite *Neodusmetia sangwani* (Rao) against rhodesgrass scale, *Antonina graminis* (Maskell), using aerial drops of parasitized infested grass

culms in southern Texas (Schuster and Boling, 1971). This was essential because natural spread of the parasite was no more than 1/2 mile per year under optimum conditions.

8) Is there food, water and cover for adult parasites? The parasite must be attracted to the habitat where the host is found, and the lack of suitable food, water and cover in many areas has probably been responsible for the failure of releases, since many parasites require sugars, protein and moisture for egg production and survival. Some obtain this by host feeding, but most do not. Clausen (1932) notes that a parasite of the Japanese beetle, *Tiphia vernalis* Roh., requires food consisting largely of honeydew produced by aphids. A host-feeding parasite is ideal, but even host-feeders may require other sources of carbohydrate that they obtain from preferred species of wild flowers (Leius, 1967).

Modern agricultural systems (especially those for annual crops) lack diversity and are essentially devoid of all plants except the crop. Therefore, original establishments of parasites of pests of annual crops should be made on fully controlled land which is managed for a diversity of crops, weeds, etc., especially if the nutritional requirements of the parasites are not known. Crops such as alfalfa are inherently favorable except for the cutting, abandoned orchards are good for tree fruit pests, and of course rangelands and forests ordinarily are not as biologically sterile as annual crops. Lands withdrawn from cultivation through federal acreage restrictions and soil conservation programs, as well as wildlife areas and natural areas are ideal, since such areas are undisturbed and can sometimes even be managed to favor natural enemies.

9) Handling before release? In most cases, parasites should be released as soon as possible after receipt if they are to realize their maximum fecundity. Some are very short-lived, such as *Neodusmetia sangwani* (Rao), the parasite of Rhodesgrass scale, *Antonina graminis* (Maskell), which lives less than 2 days under field conditions in Texas (Schuster and Boling, 1971). They should be mated before release, and if they must be temporarily stored, temperatures of 7° to 15°C are good, provided they are warmed up periodically and given food and water. However, some species have narrow temperature tolerances, and things such as sterilization can result at non-lethal temperatures, so the optimum temperatures must be determined for each species or closely related group of insects. Containers and methods of release too numerous to mention here have been developed and described for many species.

10) Spacing of releases in the field? Spacing depends on the number available, the density of hosts available, mobility of parasites, etc., but in general, they should be spaced so superparasitism is avoided, but not so distant that mates will be hard to find.

11) Time of day for release? Hot, bright, windy or rainy conditions are to be avoided since these can result in excessive dispersal or mortality, or both. Late afternoon or early evening when cool and calm conditions prevail are probably best. This gives the parasites a chance to find suitable resting spots before dark, and when they resume activity the following morning, they will experience a completely normal day from sunrise to sunset. Parasites released in the morning are rarely released at dawn and thus are subjected to rapidly increasing temperatures and wind which may result in excessive activity and dispersal, if not mortality. Good comparative studies between early morning and evening releases are needed. Of course, if subcolonizations of parasitized hosts are being made, the parasite is going to emerge at its normal emergence time (usually early morning) under the best possible conditions, no matter when the parasitized hosts are placed in the field.

12) Preservation of specimens! This would seem to be self evident, but some specimens must be preserved for a reference standard in determining if the released species has become established. Ordinarily, there will be a few dead

ones in the containers, but if not, some living ones must be preserved (both sexes, but females are required for identification in some cases.) Specimens from laboratory cultures should be periodically preserved, too, as vouchers for any changes that might occur in the lab. If sibling species are involved, it is essential that living cultures be maintained in the lab if specimens recovered from the field are to be identified (and it must be remembered that lab cultures can and do change through time.)

Determination of Establishment

Recovery is usually made by rearing the host for parasite emergence or by dissection of the host. Collecting of adult parasites is ordinarily unproductive, but a collection is required for predators since there are no parasitized hosts. For positive identification of parasites it is necessary to rear adults out, since the immature stages can rarely be positively identified to species in the present state of our knowledge.

In some cases, it is quite obvious that a parasite has become established, but we are often faced with the question of what do we call establishment? I think the minimum requirement for establishment is that the parasite has survived for at least one full year and has been recovered in "adequate numbers" (depends on the biology of the parasite and the host) in relation to the number released. This ensures that it has survived all conditions (climatic and otherwise) that it encountered in a full year, but does not mean it is permanently established, since it has not survived all possible extremes and could very well die out in future years.

One of the reasons for "lack of establishment" in some cases is lack of effort in trying to recover the parasite. If you don't look for it, you aren't likely to find it. And even if you do look for it the first year (or generation) following a relatively small release in a large host population the chances for recovery are not great. Another reason for failure is sampling at the wrong time or the wrong life stage. This is especially so when the biology is imperfectly known. Nevertheless, if a natural enemy is at all effective, there should be no difficulty in recovering it after a few generations.

Evaluation

It only takes one recovery to give an indication of establishment, but what this means in terms of future host regulation is a difficult question. The effectiveness of natural enemies varies from negligible to giving complete control. Complete control is the easiest to "evaluate," since without setting up experiments or taking samples, it is quite obvious something has reduced the host to non-economic levels. Often we assume it was the natural enemy, but this is not necessarily true. It is also common to assume a natural enemy is having a negligible effect if the pest population remains at economic levels. The partial successes appear to be the most difficult to document, but in fact the same techniques can be used to evaluate all levels of success from negligible to complete.

As DeBach and Huffaker (in Huffaker, 1971) have pointed out, a clear distinction must be made between (1) whether or not prey regulation by natural enemies actually occurs? and (2) how it occurs? Regulation has been defined by them as the maintenance of an organism's density over an extended period of time between characteristic upper and lower limits. They also defined a regulatory factor as one which is wholly or partially responsible for the observed regulation under the given environmental conditions, and whose removal or adverse change in efficiency or degree will result in an increase in the average pest population density. This definition can be tested experimentally, using techniques which exclude the possible effects of factors other than the natural enemies in question.

The selection of appropriate test areas is critical; areas should be large enough to exclude insecticide drift and abnormal movements of either prey or predators into or out of the plots. Cultural practices (especially cultivation) must be selected so as to minimize the effects on natural enemies or their habitats. The effects of dust and other "non-toxic" materials such as fertilizer must be considered. Insecticide residues in the soil of heavily treated perennial crops such as tree fruits are often so great as to virtually eliminate the study of natural enemies which spend part of their life cycle in or on the soil. Such areas have a better chance of being used to study the lack of natural enemies (the insecticidal check method).

We are primarily concerned here with the evaluation of newly established natural enemies, but the importance of native ones in the regulation of native insects has been repeatedly demonstrated by the use of insecticides which have eliminated the natural enemies and permitted formerly innocuous species to attain pest status. This is, of course, an example of the unintentional use of the insecticidal check method to demonstrate the regulatory ability of natural enemies. An extensive review of naturally occurring biological control has been given _in_ Huffaker (1971) by Hagen, et al., for the western United States, by Rabb for the eastern United States, and by MacPhee and MacLellan for Canada.

DeBach and Huffaker (_in_ Huffaker, 1971) have described the experimental check methods in detail, and have classified them into 3 groups (1) Addition, (2) Exclusion (or subtraction), and (3) Interference.

I prefer to subdivide the "Addition Method" into "before and after" and "present and absent" forms.

1) The "_before and after_" form of the _Addition method_ compares an area before the natural enemy is established with the situation after establishment. This is relatively easy to do, can be repeated in many different areas which are not necessarily the same climatically or biologically and it makes no difference how rapidly the natural enemy disperses.

Another form of the Addition method can be termed the "_present and absent_" method where "identical" plots are set up, natural enemies released in half of them, and the results are compared. This works best for natural enemies which are slow to disperse, since separation of the "present" and "absent" plots by any great distance to negate rapid dispersal tends to decrease the similarity of the plots, and local variations in factors such as rainfall can affect the results.

2) _The Exclusion or Subtraction method_ requires the elimination of natural enemies, followed by the use of techniques which keep them excluded. Cages of various kinds are most commonly used, but similar cages which permit entry of natural enemies must be used to eliminate any possible effects of the cage.

3) _The Interference or Neutralization method_ employs anything which selectively reduces the natural enemies while having a minimal effect on the pest. Complete removal of natural enemies is not obtained, but they are reduced or disturbed enough so the pest population escapes to a higher level when compared with untreated plots (if the natural enemies were responsible for the lower pest population). DeBach and Bartlett (_in_ DeBach, 1964) and DeBach and Huffaker (_in_ Huffaker, 1971) have described and discussed these methods in detail, including the "insecticidal check method," the "biological check method," the "hand removal method" and the "trap method."

a) _The insecticidal (or chemical) check method_ uses any material which selectively kills the natural enemies through differences in toxicity, dosage, formulations, timing, etc., and has a minimal effect on the pest. DeBach (1946) first used DDT to reduce natural enemies of the long-tailed mealybug, *Pseudococcus longispinus* (Targoni-Tozzetti) on citrus, and the method has been used in various ways since.

b) *The biological check method* takes advantage of the aggressive reactions of honeydew-seeking ants against natural enemies, and has essentially the same effect as selective insecticides, but without any of the complications associated with them or with other methods such as exclusion cages. It is worth noting that the interference against natural enemies by ants is not restricted to honeydew-producing species, so increases in pests which are of no interest to ants have been observed (DeBach and Huffaker, *in* Huffaker, 1971).

c) *The hand-removal method* is ideal from the standpoint of avoiding any influence on anything except the natural enemies. The obvious disadvantage is the excessive work required to find and remove most of the natural enemies, and the necessity of working with pests which have many generations per year or which continually invade the protected area if populations are to build up and results are to be obtained in a reasonable period time. Fleschner, et al. (1955) provide an excellent example of the use of hand removal of natural enemies in the evaluation of the effectiveness of natural enemies in avocado groves in California.

d) *The trap method* is a modification of the insecticidal check method where an untreated central area is surrounded by a treated area that kills the natural enemies which try to leave or enter the untreated area. It is obviously restricted in use to those situations where the pest moves very little such as mealybugs and scales, and where the natural enemy is mobile.

Discussion

The experimental check method has been successfully used in many cases to demonstrate the regulatory ability of natural enemies. However, with the exception of the "before and after" method, the experimental check methods have been most effective against pests which have limited mobility and many generations per year.

Trying to experimentally test the regulating ability of a natural enemy against an insect such as the cereal leaf beetle which has one generation per year, leaves the fields to overwinter, and returns to different fields in different locations the following year is difficult. It is further complicated by the fact that the beetle density per field is affected by the species, age, quality, and acreage of the crop planted. In addition, the larval parasites exhibit delayed mortality since they don't kill the host until after the cereal leaf beetle larva pupates and the damage is done. The alfalfa weevil is a similar pest, but somewhat less complicated to evaluate, because the crop and the acreage don't move or vary as much from year to year. We need experimental techniques which can be used on such pests, but these inherent problems are going to be difficult to overcome.

The use of experimental check methods to demonstrate the effectiveness of natural enemies is believed by DeBach and Huffaker (*in* Huffaker, 1971) and others to be the only satisfactory proof of their regulating power, but in the present state of our knowledge, experimental techniques simply cannot be satisfactorily applied to every pest or crop.

So what do we do? Use "before and after" comparisons, or use census, life table, key factor, modeling and systems techniques to document the changes that appear to take place and then project what will take place? All have been used with varying degrees of success. The "before and after" technique is by far the least expensive, but it only suggests that regulation occurs without explaining how it occurs. The theoretical arguments over population regulation, density dependence vs. density independence, etc., have gone on for years (see Huffaker, et al. chap. 2 in Huffaker, 1971; Huffaker and Messenger, Chap. 3 & 4 in DeBach, 1964) for review and discussion), but progress has been slow in resolving the arguments. The subject is complex to say the least, and methods of analysis have evolved from census data through key factors and life tables to relatively simple models to complex systems models. Numerous examples of census, life table and key factor analysis and the use of relatively simple models are cited and discussed by Varley and Gradwell 1970; by Varley and Gradwell (in Huffaker, 1971); by Varley in Rabb, 1970; and to some extent by DeBach and Bartlett (in DeBach, 1964). It is evident that these methods have not been eminently successful to date. More recently the advent of computers which can swiftly handle tremendous amounts of data has led to the development of systems modeling. This, combined with adequate online input (from the field) of biological and environmental parameters, may shortly make it possible to evaluate the current situation, predict what is likely to happen given certain conditions, and recommend procedures to obtain a desired result.

In order to do this effectively, correct models and subsystems must be incorporated in the system. Being able to prove that pest regulation by natural enemies occurs is one thing, but explaining how and why it occurs is another, and this may be essential for the correct design and operational success of the biological control component of a pest management system model.

Huffaker and Kennett (1969) have discussed some of the problems encountered in using the experimental check methods as compared with the analytical key factor, life table, and modeling approaches. They have used a combination of applicable methods to convincingly document the complete biological control of Klamath weed (St. John's wort), Hyperium perforatum L., olive scale, Parlatorea oleae (Colvee), and cyclamen mite, Stenotarsonemus pallidus (Banks) in California. This is a reasonable approach--using whatever methods are applicable to a given problem to prove that biological control works, and if possible, to explain how it works.

However, some long-held fundamental assumptions may not be entirely correct, and Varley and Gradwell (in Huffaker, 1971) have questioned the common assumption that any form of density dependent factor will regulate a population, and that parasites and predators necessarily act as density dependent factors. Related to this, Hassel and Varley (1969) have proposed a parasite "quest" theory in which mutal interference between parasites increases as parasite density increases, resulting in a smaller area of discovery, and permitting the formulation of a model with stable parasite-host interaction that was not possible in earlier models. This quest theory also permits the coexistence of 2 or more specific parasites on one host species, and predicts that the introduction of additional parasites for biological control of a pest is much more likely to have a beneficial effect, and only rarely might there be an adverse effect (which has never been observed in nature to date, according to Varley and Gradwell (in Huffaker, 1971).

I am not a biomathematician or systems scientist, but it seems evident that mathematical models of pest management systems have the potential to make predictions about future population levels based on past experience and a current input of on line biological and climatological data. Perhaps these systems models will even eventually be able to explain how and why a regulatory natural enemy operates, but even if they are not completely successful at that, the one thing they will do is force us to obtain quantitative data for everything, including some things that we always took for granted before. The prospects for better evaluation and understanding of biological control have never been better, and they should rapidly improve in the next few years.

Literature Cited

Baker, H. G. and G. L. Stebbins. (ed.) 1965. The genetics of colonizing species. Academic Press, New York. 588 pp.

Clausen, C. P., T. R. Gardner and K. Sato. 1932. Biology of some Japanese and Chosenese grub parasites (Scoliidae). USDA Tech. Bull. 308. 26 pp.

Clausen, C. P. 1939. Some factors relating to colonization, recovery and establishment of insect parasites. Proc. 6th Pacific Sci. Congr. (1939) 4: 421-428.

Clausen, C. P. 1956. Biological control of insect pests in the continental United States. USDA Tech. Bull. No. 1139. 151 pp.

DeBach, P. 1946. An insecticidal check method for measuring the efficacy of entomophagous insects. J. Econ. Entomol. 39: 695-697.

DeBach, P. (ed.) 1964. Biological control of insect pests and weeds. Reinhold Publ. Corp., New York. 844 pp.

Embree, D. G. and G. R. Underwood. 1972. Establishment in Maine, Nova Scotia and New Brunswick of *Olesicampe benefactor* (Hymenoptera: Ichneumonidae), an introduced ichneumonid parasite of the larch sawfly, *Pristiphora erichsonii* (Hymenoptera: Tenthredinidae). Canadian Entomol. 104(1): 89-96.

Flanders, S. E. 1936. A biological phenomenon affecting the establishment of Aphelinidae as parasites. Ann. Entomol. Soc. Amer. 29: 251-255.

Flanders, S. E. 1959. The employment of exotic entomophagous insects in pest control. J. Econ. Entomol. 52: 71-75.

Fleschner, C. A., D. J. Hall and D. W. Ricker. 1955. Natural balance of mite pests in an avocado grove. Yearbook, California Avocado Society 39: 155-162.

Hassell, M. P. and G. C. Varley. 1969. New inductive population model for insect parasites and its bearing on biological control. Nature 223: 1133-1137.

Huffaker, C. B. (ed.) 1971. Biological control. Plenum Press, New York. 511 pp.

Huffaker, C. B. and C. E. Kennett. 1969. Some aspects of assessing efficiency of natural enemies. Canadian Entomol. 101: 425-447.

Leius, K. 1967. Food sources and preferences of adults of a parasite, *Scambus buolianae* (Hymn: Ich.), and their consequences. Canadian Entomol. 99: 865-871.

Maltby, H. L., F. W. Stehr, R. C. Anderson, G. E. Moorehead, L. C. Barton and J. D. Paschke. 1971. Establishment in the United States of *Anaphes flavipes*, an egg parasite of the cereal leaf beetle. J. Econ. Entomol. 64(3): 693-697.

Schuster, M. F. and J. C. Boling. 1971. Biological control of rhodesgrass scale in Texas by *Neodusmetia sangwani* (Rao); effectiveness and colonization studies. Texas A&M Univ., B-1104, 15 pp.

Rabb, R. L. and F. E. Guthrie. 1970. Concepts of pest management. North Carolina State University. 242 pp.

Simmonds, F. J. 1963. Genetics and biological control. Canadian Entomol. 95: 561-567.

Stehr, F. W. 1970. Establishment in the United States of *Tetrastichus julis*, a larval parasite of the cereal leaf beetle. J. Econ. Entomol. 63(6): 1968-1969.

Stehr, F. W. and R. A. Casagrande. 1971. Establishment of *Microctonus aethiops*, a parasite of adult alfalfa weevils, in Michigan. J. Econ. Entomol. 64(1): 340-341.

Stehr, F. W. and D. L. Haynes. 1972. Establishment in the United States of *Diaparsis carinifer*, a larval parasite of the cereal leaf beetle. J. Econ. Entomol. 65(2): 405-407.

Turnbull, A. L. 1967. Population dynamics of exotic insects. Bull. Entomol. Soc. Amer. 13: 333-337.

Turnbull, A. L. and D. A. Chant. 1961. The practice and theory of biological control of insects in Canada. Canadian J. Zool. 39: 697-753.

Varley, G. C. and G. R. Gradwell. 1970. Recent advances in insect population dynamics. Ann. Rev. Entomol. 15: 1-24.

White, E. B., P. DeBach and M. J. Garber. 1970. Artificial selection for genetic adaptation to temperature extremes in *Aphytis lingnanensis* Compere. Hilgardia 40: 161-192.

INCREASING NATURAL ENEMY RESOURCES THROUGH CROP ROTATION AND STRIP CROPPING

Marion L. Laster
Entomologist
Delta Branch of the Mississippi Agricultural
and Forestry Experiment Station
Stoneville, Mississippi

Cropping practices vary from location to location and farming practices vary within locations. Because of the broad scope of farming practices it is evident that one management program is not suitable for all situations. The cropping system within which natural enemy resources are to be increased must be defined in order to discuss the topic of increasing natural enemy resources through crop rotation and strip cropping. The cropping systems for this purpose will be defined as an optimally organized or modified 1000-acre farm in the Mississippi Delta area of Mississippi (Figure 1) with average soil resource distribution. Within the structure of the linear programming model used to determine the optimum solutions, one set of farm organizations was determined assuming that beef feeding was possible. Another set of solutions was derived under the assumption that no beef feeding was possible. Crop and insect management practices will be directed toward bollworm, _Heliothis_ spp., control in cotton on the specified farm assuming that boll weevils, _Anthonomus grandis_ Boh., are not a problem.

Selected crops grown in the Mississippi Delta that affect beneficial insect populations are shown in Table 1. Changes in cropping practices are illustrated for the years 1949-69 (Miss. Crop Reporting Service). Cotton and soybeans occupy most of the cropland in the Delta and much of the area is concentrated in large fields with inadequate reservoirs for beneficial insects. Consequently, if beneficial insects are to be used in a pest control program, they must be provided with a suitable habitat and managed in such a manner that they will be effective in pest control.

Crop rotation, as defined by Isely (1957), involves a sequence of crops grown in a given field during a cycle of successive years and rotated again in the same sequence. Isely indicated how rotations have been used in controlling the northern corn rootworm, _Diabrotica longicornis_ (Say), wheat straw-worm, _Harmolita grandis_ (Riley), and the sweet potato weevil, _Cylas formicarius elegantulus_ (Summers). Control in these cases required removal of the host crop from the infested area. Crop rotation for increasing natural enemy resources can be defined as a sequence of crops grown in a given field during a cycle of successive years and repeated again in the same sequence for the purpose of enhancing beneficial insect populations. The concern here is for maintaining beneficial insect populations in a given area and utilizing their assistance in controlling a particular pest species. Since cotton and soybeans occupy most of the cropland in the Delta area of Mississippi, it is essentially a monoculture system. Cotton, the primary cash crop, occupies the best soils and a crop of lesser value cannot compete for these soils at the present time. Therefore, crop rotation specifically for an insect management program in this area has little or no chance of being accepted and the remainder of the discussion will be devoted to strip cropping.

Kell (1937), defined field strip cropping for soil erosion control as the production of regular farm crops in more or less uniform parallel strips laid out crosswise of the general slope but not parallel to the true contour. Strip cropping for increasing natural enemy resources can be defined as interplanting primary crops with uniform parallel strips of secondary crops in sufficient density to harbor ample beneficial insect populations to combat insect pests of the primary crop. Strip cropping for erosion control illustrates how a management system has been developed and employed to effectively serve a specific purpose. It further indicates that for a pest management system to be effective, it too must be developed to serve a specific purpose.

Increase of natural enemy resources through strip cropping requires a complex management system that must be well planned if it is to serve its intended purpose. Each crop and pest-enemy relationship constitutes a separate management system. Unfortunately, none of these systems have been adequately developed at the present time. Knipling (1970), pointed out that entomologists have not yet solved the important problem of determining the actual number of the host insect that is present at a given time and place, and even less information is available concerning the actual number of parasites and predators that prey on a given host species. A better understanding of the interrelationships between host insects and the complex of parasites and predators that prey on them is needed before a pest management system can be developed and employed for maximum pest control efficiency.

The various situations encountered in a strip cropping program to increase natural enemy resources for insect pest control make it nesessary that certain factors be considered: (1) the management situation must be defined; (2) the pest insect must be identified; (3) crops for stripping must be selected; and (4) a management plan must be formulated.

Management Situation

A program for increasing natural enemy resources must be designed to fit into the production system where it is to be used. Acceptance of the program depends upon its effectiveness, cost, and ease of operation. If the program is not competitive with a total chemical control program in these areas, then chemicals will continue to be the primary means of insect control.

Beef finishing on high energy silage has recently been introduced in the Mississippi Delta (Heagler, et al., 1967 and Pund, 1970). This enterprise has the possibility of blending with cotton production and permitting production of good reservoir crops for beneficial insects.

Since a specific management situation must be defined, an example of a 1000-acre Mississippi Delta farm with optimum farm organization and labor supply, average soil resource distribution, and a beef feeding program will be used in the first example. Crop allocations for the land resources are presented in Table 2 (Anderson, 1972) with cotton as the main cash crop. It will be necessary to impose hypothetical strip cropping situations on this farm since real programs of this nature do not exist at the present time.

Pest Insect

A program directed toward bollworm (*Heliothis* spp.) control in cotton offers the greatest potential in the specified management situation. Bollworms, *H. zea*, and tobacco budworms, *H. virescens*, are common pests of cotton and rank second only to the boll weevil in importance. Larvae of both species are alike in appearance to the unaided eye and they damage cotton in the same way. They will be referred to collectively as bollworms in the remainder of this discussion except where specified.

There are approximately 4 generations of bollworms per year on cotton in the Delta area of Mississippi. The seasonal occurrence in east-central Mississippi are presented in Figure 2 for both species (Snow and Brazzel, 1965). An approximate seasonal average is imposed on the graph as indicated by the broken line, to illustrate the normal occurrence of bollworm populations in the Delta area. The first generation on cotton normally occurs in mid-June and seldom causes economic damage. However, insecticide applications are frequently applied to control these populations. When these applications are made over large areas, most of the natural enemies of bollworms are killed.

The second bollworm generation in cotton occurs during the latter part of July. These populations are usually much larger than those of the first generation. Populations during this generation often cause considerable damage to cotton if control measures are not applied. These populations are the ones that natural enemies can be most helpful in controlling. If previous insecticide applications were made, the initial migration of beneficial insects were killed. With inadequate beneficial insect reservoirs there are not sufficient natural enemies to control the second generation of bollworms and the only alternative is chemical control. For this reason, reservoir areas for natural enemies must be provided and managed in such a way to effect control of the target pest.

Bollworm generations in cotton overlap after the second generation. Consequently, subsequent generations are not as distinguishable as the first two. Economic infestations usually occur during late August and early September. These infestations must also be considered in a control program. Larvae of the last generation that mature on cotton develop on late vegetative growth and vegetative regrowth after the cotton has been defoliated. These larvae do not reduce the yield of cotton but they do contribute to the overwintering population (Laster and Furr, 1971).

Strip Crops

All insects, pest or beneficial, must grow and reproduce to survive. These life functions require energy which is derived from food. Therefore, all populations will be ultimately limited in size by the amount of food in their particular environment. Knipling (1970) stated that the growth of the predaceous population must follow the growth of the population of its host(s) and collapse of the prey population leads to collapse of the predaceous population.

A crop for strip planting in cotton to increase natural enemies of the bollworm must be one that will maintain an ample food supply for the natural enemy populations. Wene and Sheets (1962) stated that lady beetles migrated to cotton from alfalfa fields infested with aphids. Since aphids are important hosts for most bollworm predators, a crop which harbors high populations of aphids should be considered.

Strip crops may influence the total farm economy in two ways other than serving as natural enemy reservoirs. Certain of these crops may be harvested and used for livestock feed or sold for cash income. Crops disposed in this manner would be complementary and supplementary to the farm economy and are most desirable if they can serve the intended purpose. Other crops used for strip planting may have no economic value other than natural enemy reservoirs. These crops must be destroyed and are considered sacrificial to the total farm economy and the costs must be absorbed by the primary crop. Therefore, strip planting of sacrificial crops should be at the minimum density required to serve the intended purpose.

Unfortunately, total programs for strip planting both types of crops to aid in pest control have not been satisfactorily developed.

Some strip crops attract pest insects and also serve as natural enemy reservoirs. Corn is the preferred host for the bollworm, *H. zea*, and may serve satisfactorily as a trap crop for this species. Lincoln and Isely (1947) found that corn in the silking stage appeared to be effective in attracting moths away from cotton, but if scattered stalks or single rows of corn were planted in cotton fields, moths were attracted and deposited eggs not only on the corn but also on nearby cotton plants. The tobacco budworm, *H. virescens*, is not attracted to corn. It was collected in Mississippi for the first time from whorl stage corn at the Delta Branch Experiment Station in 1971. Corn leaf aphids, *Rhopalosiphum maidis* (Fitch), are usually present in corn in the whorl stage. Other host species during the whorl and silking stages of corn make it a good reservoir for beneficial insects.

Grain sorghum is grown in the Delta for silage and grain. It grows well on most Delta soil types and fits well in many farming programs. Aphid species, particularly corn leaf aphids, are usually abundant in grain sorghum. Since many bollworm predators are attracted to the aphid populations in the grain sorghum, this crop is an excellent reservoir for natural enemies of the bollworm. Robinson, et al. (1972), from strip cropping studies in Oklahoma, reported that lady beetles may have been attracted to aphid populations in sorghum and migrated to adjacent cotton. Consequently, bollworm square damage in cotton strip-planted with sorghum was less than that in cotton strip-planted with other crops. Fye (1971) gave an excellent example of the potential of grain sorghum interplanted in cotton for increasing natural enemies of cotton pests. *Heliothis zea* is a pest of grain sorghum and the larvae of this species destroy the grain by feeding on the heads. *H. virescens* is not attracted to this plant.

Alfalfa serves as a host for many pest species and is an excellent reservoir for many beneficial insects. It is a host of both bollworm species, an excellent food for livestock, but it has some disadvantages for strip cropping in cotton. The alfalfa weevil, *Hypera postica* (Gyllenhal), is a serious pest of this crop in Mississippi and usually requires insecticidal control in early spring (Laster and Davis, 1967). Insecticides applied to control this pest also eliminate the beneficial insects and lower the value of alfalfa as a natural enemy reservoir. Persistent pesticide residues on alfalfa strip planted in cotton under present cropping practices would practically eliminate its use for livestock feed. Perhaps with more intensive research, programs can be developed to fit alfalfa into a cropping system as a supplemental strip crop.

Sesame, *Sesamum indicum* L., a crop that has only been grown experimentally in Mississippi, offers some promise for increasing natural enemies of the bollworm when strip planted in cotton. Sesame was grown experimentally at the Delta Branch Experiment Station from 1952 through 1964. Snow and Brazzel (1965) found that 100% of a larval collection from these experimental plots were tobacco budworms, *H. virescens*. Rivers, et al. (1965) reported that a collection of *Heliothis* larvae from sesame at College Station, Texas contained approximately equal numbers of *H. zea* and *H. virescens*. Laster and Furr (1972) reported that collections taken from sesame indicated that sesame was more attractive to *H. virescens* than *H. zea*. Although high larval populations were found in the sesame parasitic and predaceous insects prevented these larvae from reaching maturity. The insecticide resistance problem that has been encountered in many cotton producing areas (Harris, 1970; Lukefahr, 1970; Nemec, 1970; Phillips, 1971) serves to emphasize the importance of natural enemies in a control program. If resistance levels continue to increase in tobacco budworm populations, sesame is the most logical crop to fit into a strip cropping program to control this species. It has good potential as a supplemental crop but lack of available markets limit this crop to sacrificial status at the present time. Perhaps other crops may prove more desirable as further research is conducted.

Management Plan

After the management situation has been defined, the insect pest problem identified and strip crops considered, an overall management plan must be formulated. Optimum land allocations as determined by linear programming procedures for crops and soil types for the example in Table 2 show that 180 acres of sandy loam and 60 acres of mixed soils are alloted to corn. Cotton is alloted 215 acres of mixed soils. Corn yields best on the sandy loam soils and consequently has taken these soils from cotton in this example. Competition of corn for the best soils is one disadvantage of a beef feeding program on a cotton farm.

There has not been an urgent need for a complete management program including strip cropping to control bollworms in cotton. Since there has not been an immediate need for a program of this nature, there has been very little effort toward developing a complete program. During the past several years insecticides, principally DDT, have been used for bollworm control. Recent action on the ban of DDT and the organophosphate resistance problems in our bollworm populations indicate the immediate need for alternate control programs.

Although a complete management program has not been developed, some excellent work has been reported that should be helpful in developing such a program. Stern, et al. (1964) reported control of Lygus hesperus Knight in alfalfa by strip cutting. Stern (1969) reported Lygus hesperus control in cotton by interplanting alfalfa. Robinson, et al. (1972) reported that cotton strip planted with sorghum suffered less square damage from bollworms than cotton strip planted with corn, soybeans, alfalfa, and peanuts. These examples show that certain strip cropping programs are effective in reducing insect damage and that insect populations can be manipulated by proper harvest management.

The corn and cotton acreage on mixed soils (Table 2) offer an excellent opportunity for a strip cropping program. These 215 acres of cotton and 60 acres of corn permit alternate 36-acre blocks of cotton with 10-acre strips of corn. In order to follow through with a complete program, a planting date and variety of corn should be selected that will be ready for harvest about July 20, when Heliothis populations begin increasing (Figure 2). Cutting the corn for silage at this time forces the beneficial insects into adjacent blocks of cotton to prey on the bollworm populations (Figure 3). If these beneficial populations are sufficient to overwhelm the bollworm population, many of them will be moving out of the cotton due to lack of prey. Some means for holding these beneficials in the field must be provided. Since fence rows and border areas are insufficient to hold these insects, an additional strip crop is needed in the cotton for a reservoir area. This is better suited for the management program than fence rows and ditch banks because the natural enemies cannot be forced out of these areas when needed. The natural enemies will migrate to areas where food is available. If these are areas other than cotton, they can be expected to remain in these areas as long as they are more suitable than cotton or until they are forced to move out. A well planned strip cropping program with a well-timed harvest of these crops (Figure 3) permits organized manipulation of the population.

Crops possibly effective for the additional reservoir strips in cotton are alfalfa, mustard, or other crops revealed through research. Assuming this crop to be alfalfa, the density should be kept within the economical limits of a normal insect control program. A good harvest method for the alfalfa during the season appears to be that reported by Stern (1969). Approximately one-half of each alfalfa strip in this report was harvested and the remainder was harvested 2 to 3 weeks later. Vegetative alfalfa growth present at all times when harvested in this manner permits movement of insects from the freshly cut areas to the vegetative growth. The alfalfa can be managed in this manner until bollworm populations begin building up in August. At this time the alfalfa should all be cut to force the natural enemies into the cotton.

In the meantime, the strips that had been planted in corn should be prepared and planted in a winter cover crop. A mixture of small grain (oats or wheat) and winter peas would seem to fit well here. This cover crop when planted in early September would serve as a winter reservoir for the natural enemies. The cover crop would be plowed under in early spring and the area planted back in corn. At this time, vegetative growth on the alfalfa would provide a place for the natural enemies to go when the cover crop was destroyed.

This is a hypothetical example of how one situation might be managed to increase natural enemy resources through a strip cropping program. This is not presented as a total solution to the bollworm problem but it could serve as a part of a program which would contribute to a solution. Other practices that would lessen the insect problem should be included in the overall program where practical One such practice would be the selection of a suitable cotton variety. A cotton variety with characteristics such as smooth leaf and nectariless that would give a 10% reduction in bollworm numbers could well make the difference between success and failure of the entire program. Therefore all proven practices should be fitted into an overall program where feasible.

Land and crop allocations for a 1000-acre farm with 5 man-years labor and excluding the possibility of a beef feeding program are presented in Table 3. In order to apply a strip cropping program in this situation to increase natural enemy resources for bollworm control, certain modifications in land and crop allocations are necessary. Using milo in 5-acre strips alternated with 25-acre blocks of cotton, 30 acres of sandy soils are needed for the milo. These 30 acres of cotton would be moved into the mixed soils and 10 acres of milo would be needed for strip planting. These changes would move 40 acres of milo from the clay soils which would be replaced by 40 acres of soybeans. Land and crop allocations would then appear as indicated in Table 4.

A variety of milo should be selected which would mature in late July to correspond with the bollworm increase. The milo would be harvested at this time or mature and force the natural enemies to the cotton. A strip of alfalfa would be needed in each block of cotton as in the previous example. These strip crops would be handled in the same manner as previously indicated to manipulate the natural enemy populations. A strip cropping program in this situation could possibly be used to increase natural enemy resources for bollworm control in cotton.

It is evident from these two examples that a strip cropping program to increase natural enemy resources for pest control must be designed for each management situation. Many questions remain to be answered and many problems remain to be solved in this regard. In the final analysis, proper management will remain as the key to success or failure of any program to increase natural enemy resources for pest control.

Acknowledgments

Grateful acknowledgment is expressed to Dr. J. M. Anderson, Agricultural Economist, Delta Branch Experiment Station, and Mr. F. T. Cooke, Agricultural Economist, FPED-ERS, USDA, for their advice and assistance on economic aspects of the manuscript.

References Cited

Anderson, J. M. 1972. Personal communication on file in the author's office.

Fye, R. E. 1971. Grain sorghum--a source of insect predators for insects on cotton. Prog. Agr. in Arizona 23(1): 12-13.

Harris, F. A. 1970. Monitor of insecticide resistance. Mississippi Farm Res. 33(6): 3.

Heagler, A. M., B. Bolton and P. G. Hogg. 1967. Costs of beef gains with high-energy corn silage - Mississippi River Delta Area. DAE Res. Rept. No. 365, Louisiana Agr. Exp. Sta.

Isely, D. 1957. Methods of insect control. Braun-Brumfield, Inc., Ann Arbor, Michigan. 208 pp.

Kell, W. V. 1937. Strip cropping for soil conservation. USDA Farmers' Bull. 1776. 37 pp.

Knipling, E. F. 1970. Influence of host density on the ability of selective parasites to manage insect populations. Proc. Tall Timbers Conf. on Ecol. Animal Control by Habitat Mgmt. 2: 3-21.

Laster, M. L. and L. B. Davis. 1967. Effects of insecticides on the alfalfa weevil. Mississippi Farm Res. 30(6): 1, 4.

Laster, M. L. and R. E. Furr. 1971. Relationship of regrowth cotton to over-wintering populations of the bollworm complex. J. Econ. Entomol. 64(4): 974-975.

Laster, M. L. and R. E. Furr. 1972. *Heliothis* populations in cotton-sesame interplantings. J. Econ. Entomol. (in press).

Lincoln, C. and D. Isely. 1947. Corn as a trap crop for the cotton bollworm. J. Econ. Entomol. 40: 437-438.

Lukefahr, M. J. 1970. The tobacco budworm situation in the lower Rio Grande Valley and northern Mexico. Proc. 2nd Ann. Texas Conf. on Insect, Plant Disease, Weed and Bush Control. Texas A&M Univ., College Station, Texas.

Nemec, S. J. 1970. Present status of tobacco budworm and bollworm resistance. Proc. 3rd Ann. Texas Conf. on Insect, Plant Disease, Weed and Brush Control. Texas A&M Univ., College Station, Texas.

Phillips, J. R. 1971. The bollworm complex--a scourge spawned by modern farm technology. The Cotton Ginners Journal and Yearbook, March 1971, pp. 36-37.

Pund, W. A. 1970. Finishing yearling steers with high energy grain sorghum silage. Mississippi Agr. Exp. Sta. Bull. 780.

Rivers, G. W., M. V. Meisch, and P. J. Hamman. 1965. Sesame: A new host for tobacco budworm and bollworm. J. Econ. Entomol. 58(5): 1003-1004.

Robinson, R. R., J. H. Young and R. D. Morrison. 1972. Strip-cropping effects on abundance of *Heliothis*-damaged cotton squares, boll placement, total bolls, and yields in Oklahoma. Environ. Entomol. 1(2): 140-145.

Snow, J. W. and J. R. Brazzel. 1965. Seasonal host activity of the bollworm and tobacco budworm during 1963 in Northeast Mississippi. Mississippi Agr. Exp. Sta. Bull. 712.

Stern, V. M. 1969. Interplanting alfalfa in cotton to control *Lygus* bugs and other insect pests. Proc. Tall Timbers Conf. on Ecological Animal Control by Habitat Mgmt. 1: 55-69.

Stern, V. M., R. van den Bosch, and T. F. Leigh. 1964. Strip cutting alfalfa for *Lygus* bug control. California Agr. 18(4): 4-6.

Wene, G. P. and L. W. Sheets. 1962. Relationships of predatory and injurious insects in cotton fields in the Salt River Falley area of Arizona. J. Econ. Entomol. 55(3): 395-398.

Table 1. Production of selected crops in the Delta area of Mississippi.[1/]

	Acres			
	1949	1954	1964	1969
Approximate land area	----	----	4,728,320[2/]	----
Crop:				
Cotton	1,376,904	925,473	796,428	725,700
Soybeans	141,289	440,597	895,940	1,365,000
Hay	84,969	42,540	20,271	----
Pasture[3/]	104,350	258,838	387,200	
Corn	319,634	212,234	42,877	8,900[4/]
Sorghum	4,614	11,363	10,259	
Total	2,031,760	1,891,045	2,152,975	2,090,700

1/ Includes the 10 pure Delta counties plus Tallahatchie and Yazoo counties.

2/ Total Delta area.

3/ Does not include woodland.

4/ 1970 acreage.

Table 2. Optimum farm organization assuming 7 man years labor supply for a 1000-acre[1/] Mississippi Delta farm with average soil resource distribution and beef feeding as a potential activity.

Crop	Soil type			Total
	Sandy	Mixed	Clay	
	Acres			
Corn silage	180	60		240
Oat-wheat silage-soybeans (double cropped)			145	145
Soybeans		85		85
Small grain-soybeans (double cropped)			215	215
Cotton		215		215
Total	180	360	360	900

1/ There are actually only 900 acres of land available for crops on a typical 1000-acre farm. The remaining 100 acres are accounted for in buildings, roads, turnrows, ditches, and land agronomically unsuitable for tillage.

Table 3. Optimum farm organization assuming 5 man years labor supply for a 1000-acre[1/] Mississippi Delta farm with average soil resource distribution without beef feeding as a potential activity.

Crop	Soil type			Total
	Sandy	Mixed	Clay	
	Acres			
Soybeans		345	238	583
Cotton	180	15		195
Milo			122	122
Total	180	360	360	900

1/ There are actually only 900 acres of land available for crops on a typical 1000-acre farm. The remaining 100 acres are accounted for in buildings, roads, turnrows, ditches, and land agronomically unsuitable for tillage.

Table 4. Organization of a 1000-acre[1/] Mississippi Delta farm without a beef feeding activity and modified for a strip-cropping program.

Crop	Soil type			Total
	Sandy	Mixed	Clay	
	Acres			
Soybeans		305	278	583
Cotton	150	45		195
Milo	30	10	82	122
Total	180	360	360	900

1/ There are actually only 900 acres of land available for crops on a typical 1000-acre farm. The remaining 100 acres are accounted for in buildings, roads, turnrows, ditches, and land agronomically unsuitable for tillage.

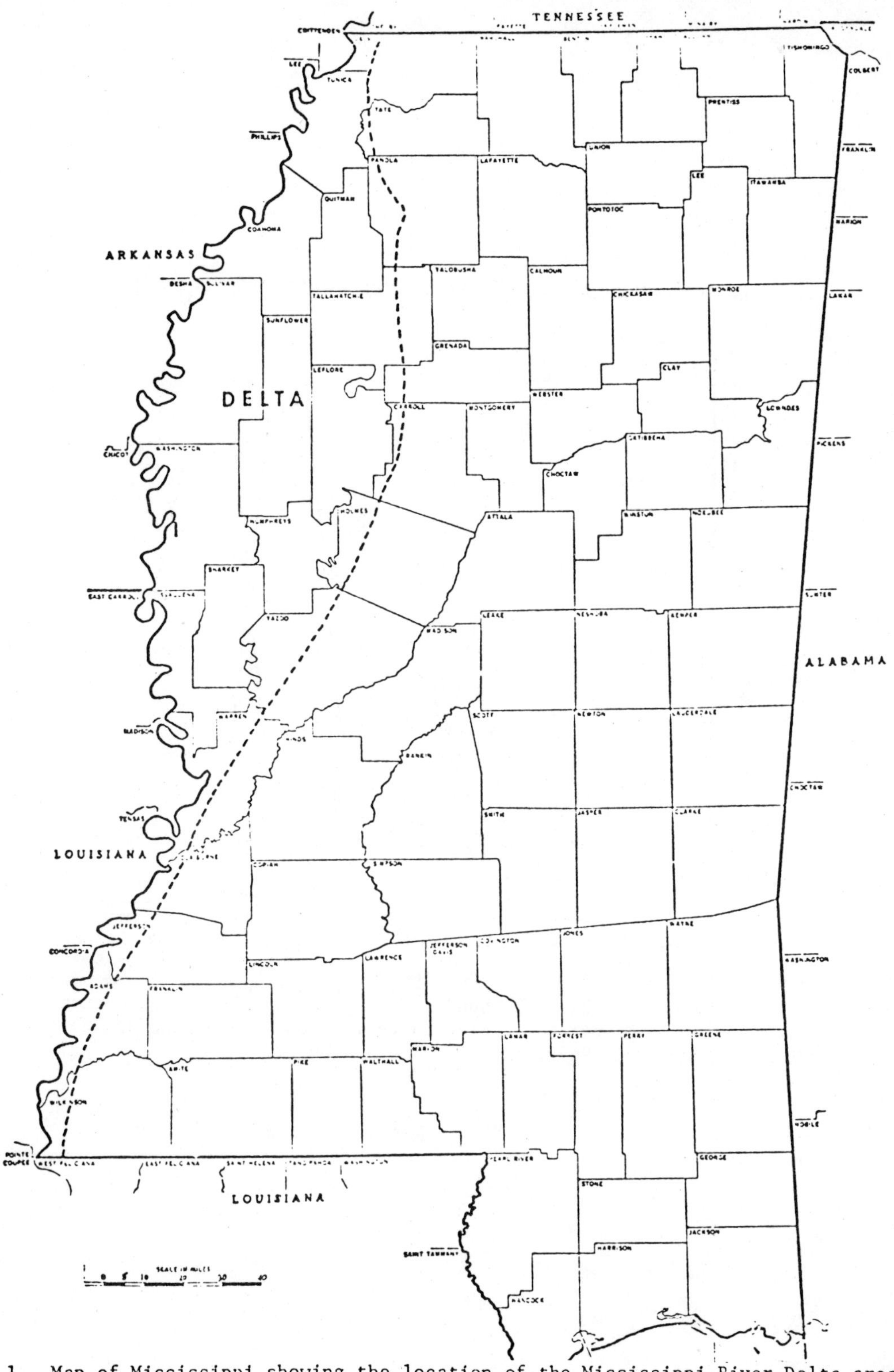

Figure 1. Map of Mississippi showing the location of the Mississippi River Delta area.

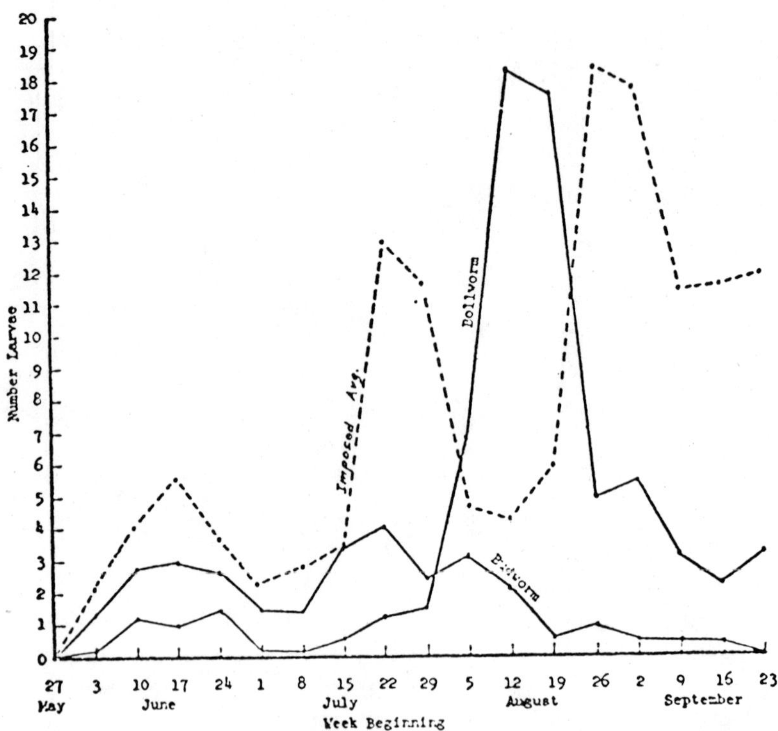

Figure 2. Seasonal abundance of Heliothis larvae on cotton per 100 whole plants in 1963. (Snow and Brazzel 1965).

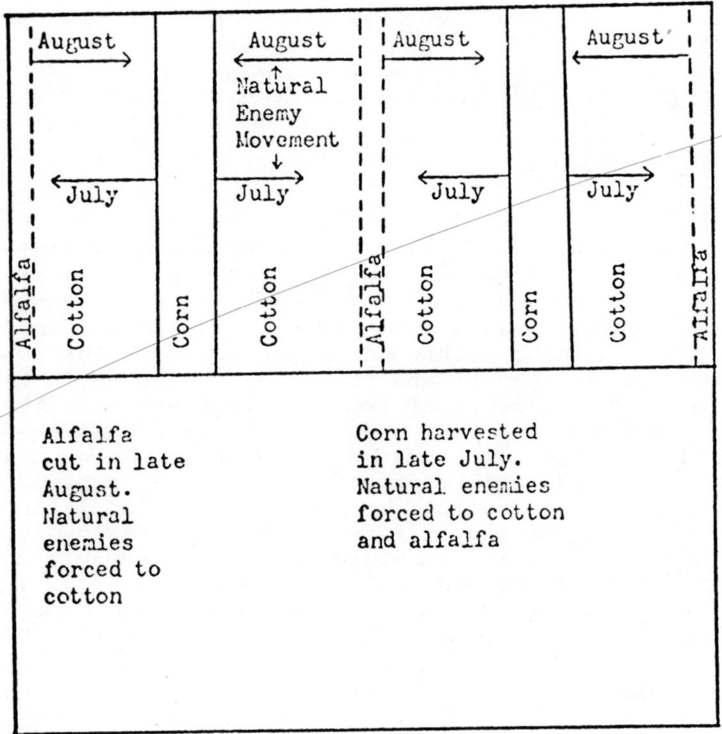

Figure 3. Strip-crop and harvest practice proposed to manage beneficial insects for cotton bollworm control on a 1000-acre Delta Farm with a beef feeding program.

NATURAL POPULATIONS OF ENTOMOPHAGOUS ARTHROPODS AND THEIR EFFECT ON THE AGROECOSYSTEM

W. H. Whitcomb
University of Florida
Department of Entomology
Gainesville, Florida

Abstract

The author has reviewed 35 years of experience with field populations of predators and parasitoids. He has stressed certain of the problems involved, particularly how to identify and work with the tremendous numbers of entomophagous species. The steps leading to actual population management are described. The various groups of entomophagous arthropods are discussed, but not in phylogenetic order. Several families of spiders and their various roles in the ecosystem are mentioned. Many of the more important families of predaceous mites are mentioned and feeding habits discussed. Much space is given to Formicidae and their role in the ecosystem. Dragonflies as predators receive attention. The true bugs are found to be less spectacular, but nonetheless effective as predators. Among the beetles, Carabidae, including arboreal forms, are studied. Staphylinidae are among the most important predators, but research has been stymied because of taxonomic difficulties. Lady beetles are discussed. Neuroptera, especially Chrysopidae and Hemerobiidae, proved to be especially important. Among the Hymenoptera, both parasitic and predaceous forms are discussed. The tremendous variety of Diptera allows for only a cursory coverage. Some species of birds are mentioned as effective predators under certain circumstances. The prospects of manipulating the environment as an integral part of pest management systems are considered.

The modeling and mathematical analyses referred to several times by previous speakers are becoming increasingly important in helping us understand the complex web of multi-variate phenomena involved in insect population management. We must not forget, however, that such models must be built on carefully gathered field data and accurate identifications. I would like to present to you some observations and conclusions of 35 years of field experience on the part of my students, co-workers, and myself. My earliest experience was in the Mid-West then in New England, and was continued in the forests of Germany under such ecological pioneers as Karl Friedrichs and Paul Schulze. As most of you know, our greatest experience has been in such row crops as cotton, corn and soybeans. However, this discussion will move rapidly from Venezuela to Arkansas, from Brazil to Florida, from tobacco to peaches, from rice to sugarcane, and from spiders to woodpeckers.

In studying predation in any agroecosystem, the insect field ecologist finds himself with an almost impossible problem. There are so many species of arthropods involved that even their identification is a serious problem. Moreover, these species interact with each other and with other components of the ecosystem. To determine what is actually happening to a pest, especially in the southeastern United States or in the tropics, at first glance appears almost impossible because of the complexity of the situation. Whitcomb and Bell (1964) reported the presence of over 600 predators in the Arkansas cotton field. In Florida, we are finding well over 1,000 species of predators in soybean fields. Also, the numbers of species of parasitoids in soybeans is high.

Two approaches to predation studies have often been taken, both of which have limited value at best. The first is to confine one's study to the 4 or 5 most important enemies of the target pest. This is a fallacious oversimplification that usually creates a misleading picture. In my experience, it takes a long time and much work before one can hazard a guess as to what are the important predators. They are rarely the ones one would expect. Moreover, predators and parasitoids differ in importance from field to field and from year to year. Most overlooked of all is the indirect importance of supposedly uninvolved predators and parasitoids. They may affect the entomophagous forms by depriving them of food. The target pests and secondary prey may be affected in an unexpected manner as well.

The second approach is even less tenable; that of lumping related species together and counting as single items green lacewings, lady beetles, spiders, or nabids. This is of little value, since each species has its own niche. There is no ecological significance to be derived from research based on mixed species counts. In the past, much of the data on predators in cotton, corn, and sorghum have been gathered in this manner. Unfortunately, thousands of acres of row crops are still being scouted for arthropods in terms of grouped species.

I find that the first step should be a general survey of predators and parasitoids present and a determination of their populations in actual numbers per acre. It is true that ecological studies of this type put a serious strain on the entire identification system in this country. Synoptic collections are only a partial answer. If individual field ecologists begin doing much of their own species identification, it will set the profession of entomology back a hundred years. In many cases, cryptic species show crucial ecological differences. The staff of the United States identification services will have to be increased and regional centers will need to be developed. Possibly a type of technologist, working under the direction of a taxonomist, could be trained for routine identification and to facilitate the rapid and efficient handling of material. This would relieve the research taxonomist from much drudgery and encourage the taxonomic revision of groups crucial to ecological studies.

The second step is to find which predators directly affect the target prey. This is usually accomplished either by field observations or by the use of radioactive tracers. Where field observations are employed, both natural and artificial populations are used. Vigils should last at least 12 hours at a time and must be both diurnal and nocturnal. Various techniques have been developed to determine principal predators by marking prey with radioactive isotopes.

The third step is to find the niche and life cycle of these key insectivores. Before extensive field and laboratory research is begun on this phase, a thorough perusal of the literature is mandatory. In some cases, one can find 17 or 18 life histories of a given predator, not one adding anything essential to the others, while other predators go unstudied. Studies of the niche of a given predator again demand long hours of actual field observation.

Moreover, it is not until after one has found the factors affecting both the numbers and efficiency of crucial predators that one should consider attempting manipulation. One must study the effect of both abiotic and biotic factors; many distinct techniques are required. Two full years of field research are often necessary to collect these data.

I will not discuss predators and parasitoids order by order, in the normal phylogenetic sequence, rather I will deal with the important groups as you would come to them in most field or orchard sampling.

Araneida

Spiders are particularly intriguing. Even though they are found everywhere in almost all cultivated fields, little is known of their ecology and effect on the agroecosystem. Years of study have emphasized, to us, 4 important roles of spiders.

First, they prey on destructive insects. Because of their high populations and the fact that the various species of Araneida have complementary niches in most situations, they cannot be overrated. For example, one pest may be preyed on in 8 different ways by 8 different species of spiders.

Secondly, spiders serve as food for predators. Hundreds of spider eggs are laid for every individual that matures. Ballooning deters cannibalism, but is only partially successful against other arthropods. In the author's experience, 1 *Lycosa rabida* Walckenaer female usually carries 1000 or more spiderlings of which 4 to 5 reach maturity. In Florida, second and third instar larvae of *Chrysopa rufilabris* Burmeister gorge themselves on half-grown green lynx spiders, *Peucetia viridans* (Hentz), in the field. In the laboratory, Hydorn (1971) reared this lacewing repeatedly on green lynx spiderlings. *Coleomegilla maculata* (DeGeer), a coccinellid, has been observed to consume 180 eggs of the spider *Chiracanthium inclusum* (Hentz) at a single site under the bracts of a cotton square.

Thirdly, since spiders tend to be general feeders, they are natural enemies of most beneficial insects. The orb weavers, theirediids, and other spiders destroy large numbers of parasitoids and predators. An orb weaver, *Neoscona* sp., has been observed feeding on the arboreal ground beetle, *Calleida decora* (Fab.) in kudzu.

Fourthly, spiders compete with insect predators for prey. When secondary prey is scarce this can be important. Stewart, et al. (1967) correlated lack of prey and temporary estivation in *Coleomegilla maculata*. Often, in Arkansas cotton fields, large populations of striped lynx spiders appeared to be contributing factors to low prey populations.

Among spiders, general patterns of predation vary from group to group. The field entomologist should recognize most temperate zone spiders to family and many to genus. He should also know their habits. For instance, wolf spiders, recognized by the 4 large eyes in 2 rows above the 4 small eyes on the front edge of the carapace, are abundant, patrolling the soil surface of both temperate and tropical cultivated fields. Their impact on the insect population is grossly underrated because above 59°F most species tend to be nocturnal. When a headlight, worn on the forehead, is used at night the spider's eyes shine back like blue diamonds (Wallace, 1937). On April 22 of this year, in Mato Grosso of Brazil at 11:00 p.m. we found one species of wolf spider lined along the cutter ant trail like hogs feeding from a trough; the workers of an *Atta* sp. were being slaughtered wholesale. On summer nights, hayfields in the United States are alive with species of *Lycosa*, *Schizocosa*, and *Pardosa* (Figure 1) preying on crickets, lepidopterous larvae, and other quarry. The seasonal shift in adult spider populations is fascinating to watch month by month. It appears to be attuned to abundance of preferred prey. *Lycosa punctulata* Hentz, mating in the late fall and bearing eggs in the spring in Arkansas, tend to feed on early season noctuid larvae on the ground. *Lycosa rabida*, similar in appearance, mates and carries its young in July. They are most often taken from plants while feeding on adult noctuids.

Lynx spiders, family Oxyopidae, tend to have all 8 eyes grouped together on the carapace. Especially characteristic are the large spines on the legs. The striped lynx spider, *Oxyopes salticus* Hentz, searches for lepidopterous larvae and other prey on foliage. If one species of spider could be said to be most common in North American row crops, *O. salticus* is it. The immature green lynx spider, *Peucetia viridans*, also searches foliage, but as an adult female it awaits its prey in ambush, which may be a bollworm moth or a wasp of the genus *Polistes* in search of bollworm larvae.

Jumping spiders, family Salticidae, are characterized by the 4 large eyes on the front of their cephalothorax. Some of you have often heard me say that the jumping spider is the only "critter" which would stare back through the microscope so that it seemed as if I were the one being observed. Because of their excellent eyesight, jumping spiders are efficient at patrolling plant foliage. Prey does not have to move to be attractive to many of them. I know of no other spiders which feed on insect eggs. The number of second and third instar bollworm larvae consumed by them is out of proportion to their numbers.

The first two pairs of legs on crab spiders, family Thomisidae, are unusually long, with the exception of the genus *Philodromus* and its relatives. These spiders lie in wait for their prey. They are capable of overcoming a moth many times their size. In Florida and Brazil, I have been surprised by their numbers and diversity.

Dwarf spiders, family Erigonidae, often construct road blocks at the junction of the main stem of a plant and a branch, to intercept larvae moving up and down. The tangled webs of the black widow spider, family Theridiidae, often entrap even the largest beetles.

The most efficient traps constructed by arthropods are the orb webs of the family Araneidae, which are spun between the branches. The silk of *Nephila* is so strong that South Sea islanders use it for fish nets. The spinning of webs by spiders is instinctive, but it may be triggered by hunger. In a sorghum head or in the web of the fall webworm, *Hyphantria cunea* (Drury), prey can be taken as fast as it can be consumed. Under such circumstances *Araneus* and *Neoscona* tend not to spin webs. On the other hand, *Geolycosa missouriensis* (Banks), which in normal situations never spins a web, when gradually starved has been observed to make a beautiful agelenid type web.

Unfortunately, there is not sufficient time to deal with the role of all families of spiders, nor to involve oneself in the many details of their biology. Few things in nature can be as interesting as the mating of spiders.

Acarina

The full significance of mites as predators has been recognized only in the last 10 years. For some time predation studies have been underway on members of a few families such as Phytoseiidae on tree crops. However, the tremendous scope of the field escaped many earlier workers. Usually, mite studies cannot be done in conjunction with other predator observations since exacting techniques and special precision are required. Because so little is known of the feeding habits and biology of predatory mites, only a few of the families can be dealt with in this discussion; again they will be mentioned as they most frequently occur in the field.

The Phytoseiidae is one of the best known and most predaceous of the families. Muma and Denmark (1970) reported 86 species from Florida alone. They tend to be predators on other mites, but Muma and Denmark (1967) reported *Macroseius biscutatus* Chant, Denmark, and Baker, as feeding exclusively on nematodes. Many species are well known as predators on crawlers of armored scales. Muma and Denmark (1970) stated that at least 1 genus is believed to feed and reproduce on pollen alone. More typical are the genera that feed on spider mites. Muma (1965) showed that the most abundant predatory mite on Florida citrus was the yellow mite, Typhlodromalus peregrinus (Muma). J. R. Reid (personal communication, 1972) found this was also one of the commonest predaceous mites on soybeans. He also found 6 other phytoseiids on soybeans, the most abundant of which was Proprioseiopsis asetus (Chant).

The Cheyletidae are of interest to Florida citrus growers for the big headed mite, Cheyletia wellsi Baker. This species, while predaceous, is certainly not beneficial. Muma (1961) reported that it feeds largely on mites of the beneficial family Phytoseiidae. The family Hemisarcoptidae is best known in Florida for Hemisarcoptes malus (Shimer), which, according to Muma (1961), is an important enemy of Florida red scale, Chrysomphalus aonidum (Linn.), and especially its eggs and crawlers. This species has also been reported to feed widely in North America on oystershell scale, Lepidosaphes ulmi (Linn.).

The family Anystidae includes the whirligig mite reported by Muma (1961) as the largest mite on citrus. Its predaceous habits are well known, but its primary prey in Florida citrus was still in question at the time of Muma's report. The families Teneriffiidae and Pseudocheylidae, in the same superfamily Anystoidea, have also been collected in Florida and are predaceous, but nothing of their biology is known.

The long-nosed mites of the family Cunaxidae are also predaceous. We found 4 specimens of an unidentified species in soybeans in 1971. Muma (1960) reported the bull mite, Cunaxa taurus (Kramer), as an active predator in citrus. He also found Andre's long-nosed mite, Cunaxoides andrei Baker and Hoffman, in citrus. This is a bright red, tiny mite with short palpi. Several other genera and species are present in citrus.

The family Erythraeidae was common in cotton fields in Arkansas. Two species, Balaustium dowelli Smiley and Erythremis whitcombi Smiley, pierce and suck many bollworm eggs, particularly under hot dry conditions. In the field, one individual was observed to destroy 25 bollworm eggs in 5 1/2 hours.

Ascidae is another family taken in Florida, both in soybeans and citrus. Blattisocus keegani Fox and Lasioseius sp. were taken on citrus (Muma, 1961). Blattisocus keegani, a species of Lasioseius, and 1 of Asca were taken on soybeans by Reid (personal communication, 1972). Chant (1963) reported that most blattioscines are probably predaceous. He also stated that a number of species of Ascidae are specific parasites of certain insects. Muma (1961) reported that Proctolaelaps hypudaei (Oudemans) has been maintained through several generations in the laboratory on fungus cultures.

The snout mites of the family Bdellidae were reported by Atyeo (1960) as active, fast-running mites, predaceous on small arthropods and arthropod eggs. They took Collembola readily in the laboratory. Bdella distincta Baker and Balock was reported by Muma (1971) as common in citrus. Bdellidae has been found in litter under row crops such as corn and soybeans.

The family Tydaeidae, which also contains predaceous forms, was found in Florida soybeans. Brickhill (1958) found that Tydeus backeri Brickhill and Lorryia ferulus Baker could be reared freely on twospotted spider mite, Tetranychus urticae Koch, eggs.

Although members of the family Trombidiidae are fairly abundant in row crops throughout the Southeast, few have as yet been identified to species. A dense coat of setae and their bright color gives them the appearance of bits of red velvet. Many trombidiid larvae parasitize arthropods. Some species are apparently capable of transmitting microorganisms from one arthropod to another. Both nymphal and adult stages attack eggs and early larval stages of many insects. Mites taken while feeding on bollworm eggs in cotton in 1962 were identified as members of this family. Mites taken in corn in Florida also belonged to this family, but have not yet been identified. The family Johnstonianidae is very similar to the Trombidiidae in habits, but very little is known about the group in Florida.

There are other families of predaceous mites such as Stigmaeidae, Penthalodidae, Ereynetidae, and Podapolipodidae, collected in Florida. However, to date few observations on their feeding habits and biologies are available.

Formicidae

Among the insects, ants form an outstanding group of dominant arthropod predators whose importance has often been underestimated. True, they themselves can be serious pests; however, eliminate them from the soybean or sugarcane field and outbreaks of other serious pests tend to result.

If the objective is to control an ant pest, then William Morton Wheeler's words (1910) should be remembered, "An ant's worst enemy is another ant." We would be in serious trouble if it were not for the destruction of red imported fire ant, *Solenopsis invicta* Buren, queens by other ants such as species of the genera *Hypoponera* (Figure 2F), *Pogonomyrmex* (Figure 2D), *Aphaenogaster* (Figure 2D), *Pheidole* (Figure 2D), *Conomyrma* (Figure 2B), *Iridomyrmex* (Figure 2B), *Formica* (Figure 2A), *Camponotus* (Figure 2A), *Lasius* (Figure 2A) and *Paratrechina* (Figure 2A). In Florida, the most effective natural enemy of the red imported fire ant queens, other than the red imported fire ant itself, is *Conomyrma insana* (Buckley). In the "Pantanal" of Brazil, species of *Pheidole* appear to be more important.

The difference in habits of the subfamilies of ants are so great and they are so easily separated that every field ecologist should be able to do this at sight. Two important subfamilies of ants, Dolichoderinae and Formicinae, can be distinguished from the other subfamilies by a one-segmented petiole and no sting (Figure 2). The Formicinae stand out as, in place of the sting, they have a highly developed organ known as the acidopore, from which toxic substances are expelled (Hung and Brown, 1966). Many species are among the finest predators; the effectiveness of the *Formica rufa* group in European forests is well known (Pavan, 1959). Many formicines are effective killers of other ants. When a *Lasius neoniger* Emery worker confronts a *S. invicta* worker (Bhatkar, et al., 1972), the *L. neoniger* worker first takes one of the red imported fire ant antennae between its mandibles, secreting on it what appears to be a toxic substance, then turns and applies a killing poison from its acidopore to the facial region. The so-called "Cuiabana" ants, *Nylanderia* spp., have been used for 20 years in Brazil against *Atta* species. It is also an effective enemy of *S. invicta*. Since it builds no big mound, has no sting, and attacks no crops, why don't we import it into the United States? The consequences could be serious. The destruction of native ants would almost certainly bring on pest outbreaks of which we have never dreamed.

A second subfamily of ants with 1 segment in the petiole and no sting (Figure 2) is the Dolichoderinae. They also depend on their mandibles and secretions from the tip of the gaster for both defense and attack, but do not have an acidopore. The "eleven o'clock ant of Arkansas," *Iridomyrmex pruinosum* (Roger) is an unusually effective predator of bollworm eggs. *Conomyrma insana* is one of northern Florida's most effective predators. On the other hand, *Iridomyrmex humilis* (Mayr), the Argentine ant, is one of the world's most serious ant pests, partially because of its association with aphids and scale insects even though it too can be a predator.

The driver ant, subfamily Dorylinae (Figure 2) is recognized by the absence or vestigal nature of the eyes. They tend to have an all too effective sting. In the Mato Grosso of Brazil, the genus Eciton is so destructive to S. invicta colonies that at times, in entire areas, colonies of red imported fire ants can be found only in such protected situations as in termite nests and under stones and logs. In general, the Dorylinae tend to attack other ants. In northern Florida, Neivamyrmex opacithorax Emery was found attacking C. insana.

The subfamily Myrmicinae (Figure 2) in the southern United States is the most varied and most abundant of all. It is characterized by two segments in the petiole, a sting, and normal eyes, neither vestigal as in the Dorylinae or over-developed as they are in the long, slender, hollow twig-dwelling Pseudomyrmecinae (Figure 2). Pogonomyrmex badius (Latreille), the grain and grass seed feeder, can at times be an effective predator, such as when it is attacking red imported fire ant queens. Members of the tribe Dacetini are highly specialized predators feeding exclusively on Collembola and related groups. Most notorious of this subfamily are the fire ants, tribe Solenopsini. What we used to think of as one species, Solenopsis saevissima F. Smith (Buren, 1972), is now known to be at least 40 species with their homelands stretching from the Rio Negro in Argentina to the Guianas. Several species are sympatric, but most are allopatric, constantly battling each other as well as other even more successful ant species. No wonder the red imported fire ant is such a problem in the United States. No wonder the populations have exploded. Fortunately, the 2 species of the Solenopsis saevissima complex, S. invicta and S. richteri Forel, established in the southern United States are 2 of the least aggressive. We might be able to eliminate the red imported fire ant in the United States by bringing in a new species of Solenopsis from Corumba, Brazil, but the human population would certainly suffer. Solenopsis invicta does on occasion attack crops and it definitely cultivates scale insects, but it is also one of our most successful predators. My biggest objection to the red imported fire ant is that it is too successful as a predator. It attacks many other predators, thereby simplifying the ecosystem. A simplified ecosystem is often an unstable ecosystem.

Pheidole, another myrmicine genus, provides us with several predators of cultivated crops in Florida. Pheidole morrisi Forel tends to be everywhere in soybean fields of northern Florida where infestations of the red imported fire ant are not too heavy. Aphaenogaster spp. are effective predators, attacking a wide array of prey from termites to the red imported fire ant queen; they tend to prefer woodlands, but some species range widely in cultivated fields. As if to compensate for providing so many fine predators, the Myrmicinae does have one tribe, the Attini, whose members not only are almost never predaceous, but include some of the world's most destructive pests. They cultivate fungi in their underground nests. To provide the substrate for these gardens, they defoliate thousands of acres of cotton, sesame, citrus, peanuts, pine seedlings, and other crops. They have taken corn seed out of the ground as fast as I could plant it.

Odonata

To many an entomologist, dragonflies appear inconsequential, both in numbers and activities. To the sharp-eyed ecologist, dragonflies are important predators, each species flying differently, most species occupying a particular part of the aerospace. Ecologists also see that dragonflies are more numerous than one would think. In Florida, Neal and Whitcomb (1972) found that they were missing some crucial species as these dragonflies flew at dusk.

Dragonflies are obligate carnivores which feed on almost any animal small enough to be captured. However, it is a matter of record that, as an apparent result of feeding habits and size of prey taken, the same species of dragonfly has been repeatedly captured with the same species of prey between its mandibles. Similar prey records are taken often thousands of miles apart. The number per acre and flight habits of dragonflies largely determine the impact of any single species on the agroecosystems.

Gomphus ivae Williamson, family Gomphidae, rested more than it flew. In the soybean field, it appeared to be in the air only long enough to take low-flying prey, capturing it less than a foot above the plants.

Perithemis tenera seminole Calvert, family Libellulidae, is a small low-flying dragonfly. It flies continually between the soybean plants throughout the field from early morning until dusk, taking small prey such as dolichopodid flies. The green jacket dragonfly, *Erythemis simplicicollis* (Say), is the commonest dragonfly in north Florida (Neal and Whitcomb, 1972). It flies very close to the top of the plants, often only a few inches from the leaves. Bell and Whitcomb (1961), Whitcomb and Bell (1964) and Lincoln, et al. (1967) reported it as an effective predator of adult bollworm moths, *Heliothis zea* (Boddie). *Pachydiplax longipennis* (Burmeister) captures alate red imported fire ants as they leave the mounds on the nuptial flight. *Anax junius* (Drury), family Aeshnidae, are abundant high above cotton and corn fields throughout the Southeast. These strong fliers are often observed feeding on small Diptera or alate ants, but will take larger prey. Members of the family Cordulidae, such as *Somatochlora filosa* (Hagen), are high fliers which often feed on swarming alate ants.

Hemiptera

Predators of the order Hemiptera are less spectacular than those of some other orders. They tend to be ubiquitous in distribution and often are the only predators of importance present. The family Anthocoridae is well represented. The insidious flower bug, *Orius insidiosus* (Say) is especially effective destroying mites as pointed out by Iglinsky and Rainwater (1950) and others. It is an important predator of thrips, aphids, and noctuid eggs. In northern Florida, *Orius insidiosus* also has been observed by us destroying the eggs of microlepidopterous pests of pecans. Numerous authors have stressed its effectiveness against *Heliothis zea* eggs in corn silks and on cotton plants. The resulting collapsed egg cannot be distinguished easily from those destroyed by other sucking insects. In the west, the minute pirate bug, *Orius tristicolor* (White), fills a similar niche.

Another anthocorid, *Cardiastethus assimilis* (Reuter), has been observed late at night destroying noctuid eggs in Florida soybeans and has been reared through two generations on soybean looper eggs in the laboratory. Few row crops in the United States do not benefit from the activities of anthocorids.

Members of the family Nabidae present a very special problem. Individual species of the genus *Nabis* are almost impossible to distinguish in the field, yet each species has a distinct niche. To further complicate the problem they are among the most effective predators of noctuid eggs, fleahoppers, and other pests. The best solution to date is to bring representative samples into the laboratory and prorate results to the proportion of each species. Members of the family Lygaeidae are mostly plant feeders, many are destructive pests. However, the genus *Geocoris* is an important exception (Mead, 1972). *Geocoris punctipes* (Say), *G. uliginosus* (Say), and *G. bullatus* (Say) routinely destroy large numbers of *Empoasca* spp., *Heliothis zea* eggs, and other pests.

A few Miridae are beneficial; *Spanogonicus albofasciatus* (Reuter) has been reported by Neal, et al. (1972) as destructive to noctuid pests of soybeans in Florida. The genus *Deraeocoris* feeds consistently on aphids; *Deraeocoris nebulosus* (Uhler) is reported by Whitcomb and Bell (1964) feeding on *Aphis gossypii* Glover.

The family Reduviidae is especially important in tropical and subtropical regions. The wheel bug, *Arilus cristatus* (Linn.), is known to most cotton scouts.

It is largely destructive to fourth instar bollworm larvae, but seldom abundant. It is a far more effective predator of the fall webworm, Hyphantria cunea (Drury). Members of the genus Zelus are largely predators on lepidopterous larvae; Zelus bilobus Say is an important enemy of third and fourth instar noctuid larvae, including the velvetbean caterpillar Anticarsia gemmatalis Hubner in central and southern Florida. The preferred prey of Sinea spinipes (Herrich - Schaeffer) appears to be coccinellids and chrysomelids; it often attacks the boll weevil in the cotton field.

Pentatomids of the subfamily Asopinae are excellent predators. They are distinguished from other stink bugs because the first segment of the beak is not covered by the gula. Stiretrus anchorago (Fab.) feeds freely on Mexican bean beetle larvae. Perillus bioculatus (Fab.) attacks potato beetle larvae. Podisus maculiventris (Say) attacks large bollworm larvae; it is common in the cotton fields to see the beak of one inserted near a bollworm head and the beak of another inserted in the posterior end. Podisus placidus Uhler destroys fall webworm larvae (Tadic, 1964). Alcaeorrhynchus grandis (Dallas) and Euthyrhynchus floridanus (L.) are important predators of soybean pest insects in Florida.

Coleoptera

It is no surprise that the tremendously large order Coleoptera should be of utmost importance among the entomophagous arthropods. What is surprising is that many of the important predators belong to small families, like the Anthicidae and the Melyridae. Unexpected also is the fact that many are parasitoids. Bombardier beetles, in their larval stage, parasitize pupae of the family Hydrophilidae. Larval Lebia spp. attack individual pupae of chrysomelids. Important also is the fact that larvae of some families, such as Meloidae, are predators, but the adults are plant-feeding pests.

Important as the ground beetle group is, almost nothing is known of the larvae other than most are either predators or parasitoids. The larvae of Calosoma sayi DeJean destroy large numbers of Heliothis spp. and Spodoptera spp. pupae in the soil. The spotty distribution of many soil pests can be explained by the presence or absence of carabid immatures. Even when the larvae are observed feeding on a given pest, it still does not tell us much because the identification to species of most carabid larvae is still virtually impossible and rearing methods are inadequate despite the work of Vernon Kirk and others.

Foliage dwelling groundbeetles, especially of the subfamily Lebiinae, have a greater importance and potential than previously suspected. We have 10 species and 4 genera in Florida soybeans. Calleida decora (Fab.) can reach a population of 1400 per acre and will destroy fourth instar soybean looper larvae. The larvae of C. decora are as active on the foliage as adults, tending to avoid last instar velvetbean caterpillar larvae, but destroying almost every other noctuid larvae on the soybean foliage. Lebia analis Dejean and L. viridis Say are present throughout the South and as adults can be general predators. However, their populations depend on the presence of certain weeds since their larvae feed on the pupae of chrysomelids that attack these weeds.

The list of species of ground beetles that, like wolf spiders, police the surface of the ground in cultivated crops is almost endless. Calosoma sayi builds up to high numbers in Florida by late August and September, early enough to do some good against the velvetbean caterpillar, but too late for many other pests. The genera Pasimachus, Progaleritina, Anisodactylus, Selenophorus, Agonum, Harpalus, Agonoderus are only a few of the many other ground beetles.

The family Staphylinidae is one of the most important of all families of arthropod predators. Research on them has been almost paralyzed because of the complexity of the taxonomy. Thousands of observations and hundreds of life histories mean nothing as the beetles could not be identified. No one questions their effectiveness; I have seen a *Staphylinus* sp. take a muscoid fly out of the air a quarter of an inch above itself. Something must be done as soon as possible to make identification feasible so that the role of each species can finally be established.

The lady beetles, "Marien Kafer" of Germany or "Joaninchen" of Brazil who are in the family Coccinellidae tend to be the "work horses" of hundreds of natural control situations. Two lady beetles, 1 large, *Chilocorus* sp. and 1 small, *Lindorus* sp., are surprisingly effective against white peach scale when not killed off by insecticides. In the author's experience, the population level of three lady beetles, *Hippodamia convergens* Guerin-Meneville, *Coleomegilla maculata*, and *Cycloneda sanguinea* (Linn.), determine whether there is going to be a *Heliothis zea* outbreak in thousands of cotton and corn fields in the United States. The genus *Stethorus* is vital to the natural control of dozens of species of mites. Often, tiny lady beetles of the genus *Scymnus*, along with their naked or wax-covered larvae, are most important of all. The specificity or near specificity of many lady beetles has been underestimated; in our research *Delphastus pusillus* (LeConte) laid its eggs only in the pupal skins of whiteflies. On the other hand, many species, such as *Coleomegilla* and *Hippodamia* spp., are general feeders. There is no excuse, except plain laziness, for any entomologist not to recognize to species the larvae of all the larger and some of the smaller lady beetles. According to Dr. R. E. Waites (personal communication, 1972), the pattern of the dorsum of the abdomen is a major part of the secret. Whether black extends all the way across the segment, or whether the center third is yellow helps identification.

Dozens of other families, including Cicindelidae, Cleridae, Anthicidae, Melyridae, and Cantharidae, should be dealt with, but time will not allow.

Neuroptera

Neuroptera, although a small order, is among the more important groups of entomophagous arthropods since so many species are efficient predators and consume unusually large numbers of prey. Even though they are mandibulate, their mouthparts are modified for piercing and sucking. A bollworm egg attacked by Neuroptera is completely collapsed and can seldom be distinguished from one destroyed by a hemipterous predator. Green lacewings, family Chrysopidae, are possibly the best known; *Chrysopa carnea* Stephens is being mass produced for field release in California, Texas, Missouri, and other points. It is not present in Florida, nor have field releases been successful in this state. *Chrysopa rufilabris* Burmeister (Figure 3) is the commonest species in citrus, pecans and row crops in Florida at certain times of the year. Mass production is being developed at Gainesville. Entomologists are just now beginning to understand the importance of the phenology of species of *Chrysopa* and *Nodita* in Florida and across the nation.

Brown lacewings, family Hemerobiidae, are a largely neglected, but important group. We have thought of them as more abundant in more northern regions, but in Florida they can be very numerous in pecans, peaches, and citrus. *Hemerobius* and *Micromus* are among our more important genera; they deserve a far more careful study. Antlions, family Myrmeleontidae, their doodlebug larvae and their funnel ant-traps have been written off as interesting curiosities. This may be a serious mistake. A careful study of population dynamics may disclose that certain species are indeed serious biotic factors affecting some ants.

Hymenoptera

Time will not allow us to discuss the hymenoptera parasites. Even to deal lightly with a few of the most important ones would take many lectures. The family Ichneumonidae is one of the largest of all insect families. Its place in biological control has been ably portrayed by H. Townes (1971). One simple generalization is in order. In natural biological control, few families of hymenopterous parasites have anywhere near the importance of the Ichneumonidae; however, in man's attempts to introduce parasites for the control of pests, the Braconidae and the various families of chalcidoids have proven much more useful. Even in natural biological control, few genera can equal in importance the genus Aphidius of Braconidae in aphid control or the chalcidoid genus Aphytis in control of scale insects. The chalcidoid genus Trichogramma is almost synonymous with mass rearing and release.

Three distinct groups of predaceous wasps attack spiders exclusively. The pompilids distinguish themselves not in the taxonomic group of spiders they capture but that they typically take their prey as it moves about on the soil surface. Certain species prey upon tarantulas, and others upon wolf spiders. The common mud daubers in the southeastern United States belong to the genera Sceliphron and Chalybion of the subfamily Sphecinae. The organ-pipe mud daubers of the genus Trypoxylon are also sphecids. Both groups take a great many orb weavers, along with theridiids and other prey. Mud nests of Chalybion californicum (Saussure) often are packed with black widow spiders.

Both the social and solitary Vespidae supply their nests with lepidopterous larvae. Lawson (1959), Rabb (1953), Rabb and Lawson (1957) and Kirkton (1970) have written major works on the genus Polistes as predators of tobacco and cotton pests. By the judicious use of wasp boxes, populations of these insects can be increased under some circumstances. In large areas of the Southeast, the only effective predator of fourth instar Heliothis zea and Heliothis virescens (Fab.) in cotton are species of Polistes. The yellow jacket, Vespula sp. can be an effective predator of several noctuids, yet to increase populations of this wasp would not be in the public interest.

Sphecids are almost exclusively solitary and store their nests with a wide variety of prey. The confusion over the name Sphex has been unfortunate, since 2 distinct genera are involved and both are excellent predators, one genus attacking longhorn grasshoppers and the other lepidopterous larvae. Species of the genus Cerceris attack small beetles, including both buprestids and weevils. The horse guard, Sticia carolina (Fab.) is a well known predator of Tabanidae near horses. I was astounded while digging up the nest of a Bombicini type wasp to find large adult skippers in cells at the bottom of a tube the size of a pencil.

Diptera

The huge order Diptera is a very strange and intriguing one in relation to entomophagous forms. There are 6 or more families that stand among the most beneficial families of arthropods and 60 or more other predaceous or parasitoid families that are dismissed as unimportant. When the scientific facts are in, the unimportant entomophagous dipterous families may not only be more significant than the important ones, but they may also answer many of the unsolved problems of population dynamics of the last 30 years. The first and most common mistake which the field entomologist, uninterested in aquatic arthropods, is likely to make, is to assume that because a fly is nemocerate it is of no importance. Hardly a month goes by without new parasitoid or predaceous midges being reported. I do not mean to underrate the Syrphidae. Although the larvae of many species are not predaceous, there is no questioning that a wide variety of species exert tremendous biotic pressure on most species of aphids. The percentage of syrphid flies parasitized by Diptera and Hymenoptera may be the factor determining the aphid population in many situations. Mesograpta, Allograpta, Metasyrphus, and Syrphus are only a few genera of particular significance.

The Asilidae in a few cases are definitely beneficial, such as Diogmites symmachus Loew feeding on the threecornered alfalfa hopper, Spissistilus festinus (Say), but many more species attack wasps of the genus Polistes and honey bees.

Careful personal monitoring, plus Malaise trapping indicate that in general, asilid populations in Florida are much lower than in the Arkansas-Missouri area. Few people realize that the larvae of Tabanidae are predators of invertebrates. Sciomyzidae have far greater importance as predators of snails than formerly realized. Dolichopodidae is a large family of metallic-colored flies found almost everywhere in Florida, a state rich in species. They were often observed taking aphids from plants such as goldenrod, *Solidago* spp. Empididae, or dance flies, take a heavy toll of prey early in the spring when this loss is significant. The list of small families exerting biotic pressure on various groups is long, but would be much longer if more information were at hand.

Tachinidae is the giant of the parasitic families, as important among the Diptera as Ichneumonidae is among the Hymenoptera. In the tropics, tachinids seem to come in all shapes, sizes, and colors. The postscutellum and large abdominal bristles are always present. In Florida, few lepidopterous larvae and not many other insect pests escape attack by these insects. Not many entomologists realize that Sarcophagidae, another muscoid family, is largely parasitic and also exerts an important influence on arthropod populations. For the field entomologist a good rule of thumb is that once he is certain that the fly is not Tachinidae and finds the parallel stripes on the pronotum, he is probably dealing with Sarcophagidae. Anthomyiidae, a family full of destructive pests, has many important predators. In this family, the fact that a voracious predator and a destructive pest may belong to the same genus and even look much alike leads to confusion.

Dermaptera

Of all the small orders, possibly the most underrated are the Dermaptera, or earwigs. In the southern states, *Labidura riparia* Pallas appears to be on and in the ground everywhere, especially in disturbed soil habitats. It attacks noctuids as larvae, pupae, and adults, feeding voraciously on a wide range of prey. Another earwig, *Euborellia annulipes* (Lucas), has also proven to be an effective ground-feeding predator, mostly of smaller prey. Not much is known about *Doru lineare* (Escholtz) despite the tremendous populations in late summer in several southern states. It is a flier and climber that is seldom found on the ground. At times it appears to be a scavenger, at times a plant feeder, and at times a predator. It fed on both aphids and bollworm eggs in the laboratory.

Aves

Although this is a discussion of entomophagous arthropods, it is hardly complete without mentioning the vertebrates, especially certain birds. Two stand out in my memory. The effect of their feeding had definite clear-cut economic implications. In some years they were a bit more effective than in other years. It was, however, an annual phenomenon, and they could be counted on to be of important benefit at the proper season.

The first was the flicker, *Colaptes auratus* (Linn.) reported first by Wall and Whitcomb (1964) as a predator of the Southwestern corn borer. In the winter and spring in Mississippi, Louisiana, and Arkansas, they work near the base of the stalks, making irregular rectangular shaped holes, and removing the larvae with their tongues. This is of special significance, since recommendation of early planting of corn to reduce damage from the southwestern corn borer depends on low borer populations in the spring.

The second, the tufted titmouse, *Parus biocolor* (Linn.), is an effective predator of the pecan nut case bearer, *Acrobasis caryae* Grote. This insect overwinters in a tough, saucer-shaped hibernaculum which is attached to the bud and lies between the bud and twig. The tufted titmouse rips apart the twigs and removes the larvae. Whitcomb (1970) dissected 26 nut case bearer larvae from the crop of single tufted titmouse.

In my experience, effective bird predators have three characteristics in common: 1) they strike when the pest population is low, 2) at the time of their importance they specialize on one species of insect, and 3) they have a specialized behavioral pattern involved in their feeding on particular pests.

Discussion

Factors affecting both numbers and efficiency of beneficial insects are proving to be more complex than the most profound scientist could have predicted ten years ago. Now, we know that an almost endless chain of reciprocating and non-reciprocating density-dependent factors are involved. Hydorn and Whitcomb (1972) showed that even parental age at oviposition affects survival of progeny of *Chrysopa rufilabris* (Table 1). Efficiency of predators, as well as reproductivity, is certainly affected by quantity and kind of parental food. Type of larval food also affects efficiency. The plant on which the predator is active has both obvious and unexpected effects. Natural enemies of the beneficial arthropods, including pathogens, are especially important under semitropical and tropical conditions, but they also have been underrated in more northern conditions.

The size of the predator population is determined to an unusual degree by the presence or absence of prey. The prey may be the target pest or secondary prey and is equally important in cultivated or noncultivated areas. The destruction of composite aphid populations, *Dactynotus* spp., on goldenrod by a fungus disease may affect the predator population in 7 or 8 counties in northern Florida. Late July droughts in Arkansas often affect secondary prey populations outside the cotton field, and contribute to bollworm outbreaks in both cotton and soybeans. Systemic insecticides, under some circumstances, can seriously reduce prey population such as thrips and aphids and virtually eliminate *Orius insidiosus* over wide areas.

I am convinced that much of the future of entomology lies in habitat manipulation both inside and outside of the cultivated field. In Figure 4, the effects of changing the date of plowing is shown. The species of plants on highway rights-of-way or on canal banks may decide the number of insecticide applications that must be used in cultivated crops a mile away. The opportunities for manipulating the populations of ants, ground beetles, spiders, and many other predators have not yet been fully explored.

Population manipulation will have to be based on a sound and thorough understanding of the agroecosystem. The predators and parasitoids present must be known. Knowledge of what they are feeding on and how they capture their prey is essential. Both the direct and indirect effect of these insectivores on the target pests will require careful study. The life history of the important predators and parasitoids will need investigation. The source of these beneficial insects and the cause of population fluctuation is almost a science in itself. One must build toward management of the environment step by step, adding one bit of information to another while keeping an eye on the ultimate objective so that even the computer is not overwhelmed. Empirical and trial and error information combined with ample practical agricultural experience can then be meshed into this theoretical understanding to produce a workable part of a pest management program.

References Cited

Atyeo, W. T. 1960. A revision of the mite family Bdellidae in North and Central America (Acarina: Prostigmata). University of Kansas Sci. Bull. 40(8): 345-499.

Bhatkar, A. P., W. H. Whitcomb, W. F. Buren, P. S. Callahan and T. Carlysle. 1972. Confrontation behavior between Lasius neoniger (Hymenoptera: Formicidae) and the imported fire ant. Environ. Entomol. 1(3): 274-279.

Bell, R. and W. H. Whitcomb. 1961. Erythemis simplicicollis (Say), a dragonfly predator of the bollworm moth. Florida Entomol. 44(2): 95-97.

Brickhill, C. D. 1958. Biological studies of two species of tydeid mites from California. Hilgardia 27(2): 601-620.

Buren, W. F. 1972. Revisionary studies on the taxonomy of the imported fire ant. J. Georgia Entomol. Soc. 7: 1-26.

Chant, D. A. 1963. The subfamily Blattisocinae Garmen (=Aceosejinae Evans) (Acarina: Blattisocidae Garmen) (=Aceosejidae Baker and Wharton) in North America, with descriptions of new species. Canadian J. Zool. 41: 243-305.

Hung, A. C. F. and W. L. Brown, Jr. 1966. Structure of gastric apex as a subfamily character of the Formicinae. J. New York Entomol. Soc. 74: 198-200.

Hydorn, S. B. 1971. Food preferences of Chrysopa rufilabris Burmeister in North Central Florida. Unpublished master's thesis. University of Florida. 79 pp.

Hydorn, S. B. and W. H. Whitcomb. 1972. Effect of parent age at oviposition on progeny of Chrysopa rufilabris (Neuroptera: Chrysopidae). Florida Entomol. 55(2): 79-85.

Iglinsky, W. and C. F. Rainwater. 1950. Orius insidiosus, an enemy of a spider mite on cotton. J. Econ. Entomol. 43(4): 567-568.

Kirkton, R. 1970. Habitat management and its effect on populations of Polistes and Iridomyrmex. Proc. Tall Timbers Conf. on Ecol. Animal Control Habitat Mgmt. 2: 243-246.

Lawson, F. R. 1957. The natural enemies of the hornworms on tobacco (Lepidoptera: Sphingidae). Ann. Entomol. Soc. Amer. 52(6): 741-755.

Lincoln, C., J. R. Phillips, W. H. Whitcomb, G. C. Dowell, W. P. Boyer, K. O. Bell, G. L. Dean, C. J. Matthews, J. B. Graves, L. D. Newsom, D. F. Clower, J. R. Bradley and J. L. Bagent. 1967. The bollworm-tobacco budworm problem in Arkansas and Louisiana. Arkansas Agr. Exp. Sta. Bull. 720: 1-66.

Mead, F. W. 1972. Key to the species of big-eyed bug, Geocoris spp., in Florida. Florida Dept. Agr. Consumer Serv., DPI, Entomol. Circul. 121.

Muma, M. H. 1960. Predatory mites of the family Cunaxidae associated with citrus in Florida. Ann. Entomol. Soc. Amer. 53(3): 321-326.

Muma, M. H. 1961. Mites associated with citrus in Florida. Univ. of Florida. Agr. Exp. Sta. Bull. 640: 1-39.

Muma, M. H. 1965. Populations of common mites in citrus groves. Florida Entomol. 48(1): 35-46.

Muma, M. H. and H. H. Denmark. 1967. Biological studies on Macroseius biscutatus Chant, Denmark and Baker (Aranina: Phytoseiidae). Florida Entomol. 50: 249-255.

Muma, M. H. and H. Denmark. 1970. Phytoseiidae of Florida. Florida Dept. Agr. Consumer Serv. Arthrop. of Florida 6: 1-150.

Neal, T. M. and W. H. Whitcomb. 1972. Odonata in the Florida soybean agroecosystem. Florida Entomol. 55(2): 107-114.

Neal, T. M., G. L. Greene, F. W. Mean and W. H. Whitcomb. 1972. Spanagonicus albofasciatus (Hemiptera: Meridae): a predator in Florida soybeans. Florida Entomol. (in press).

Pavan, M. 1959. Attivita Italiana per la lotta Giologica con formiche del gruppo Formica rufa contro gli Insetti dannosi alla forests. Ministero dell Agricoltura e Foreste, Collana Verde. 4. 79 pp.

Rabb, R. L. 1953. Observations on the predator activities of Polistes annularis (Linnaeus) P.f. fascatus (Fabricius), P.c. exclamans Viereck (Hymenoptera: Vespidae). Unpublished Ph.D. dissertation. North Carolina State College, Raleigh. 99 pp.

Rabb, R. L. and F. R. Lawson. 1957. Some factors influencing the predation of the Polistes wasp on the tobacco hornworm. J. Econ. Entomol. 50(6): 778-784.

Stewart, J. W., W. H. Whitcomb and K. O. Bell. 1967. Estivation studies of the convergent lady beetle in Arkansas. J. Econ. Entomol. 60(6): 1730-1735.

Tadic, M. 1964. Possibilities for introducing into Yugoslavia Podisus placidus Uhl. an American predator of the fall webworm. J. Sci. Agr. Res. 16(54): 42-52.

Townes, H. 1971. Ichneumonidae as biological control agents. Proc. Tall Timbers Conf. on Ecol. Animal Control Habitat Mgmt. 3: 235-248.

Wall, L. and W. H. Whitcomb. 1964. The effect of bird predators on winter survival of the Southwestern and European corn borers in Arkansas. J. Kansas Entomol. Soc. 37(3): 187-192.

Wallace, H. K. 1937. The use of the headlight in collecting nocturnal spiders. Entomol. News 48: 107-111.

Wheeler, H. K. 1910. Ants. Columbia Univ. Press, New York. 633 pp.

Whitcomb, W. H. 1970. The tufted titmouse, Parus biocolor as a predator of the pecan nut caseborer, Acnobasis caryae. Proc. Tall Timbers Conf. on Ecol. Animal Control Habitat Mgmt. 2: 305-308.

Whitcomb, W. H. and K. O. Bell. 1964. Predaceous insects, spiders, and mites of Arkansas cotton fields. Arkansas Agr. Exp. Sta. Bull. 690: 1-84.

TABLE 1. RELATIONSHIP BETWEEN AGE OF C. *rufilabris* ADULTS AT OVIPOSITION AND TOTAL LARVAL AND PUPAL DEVELOPMENT TIME OF PROGENY REARED ON *T. castaneum* at 26±°C [1]

Progeny from:	Av. larval and pupal development time (days)*, **	Range (days)	No. lacewings tested
First 10 days of oviposition	23.3±0.3	21-26	27
Second 10 days of oviposition	24.4±0.2	22-26	28
Third 10 days of oviposition	26.0±0.3	25-28	23

* Mean ± Standard Error of mean

** All values are significantly different from each other at the 5% level.

[1] From p. 80 of Hydorn, S. B. and W. H. Whitcomb. 1972.

Figure 1.--Dorsal view of a wolf spider, *Pardosa melvina* Hentz.

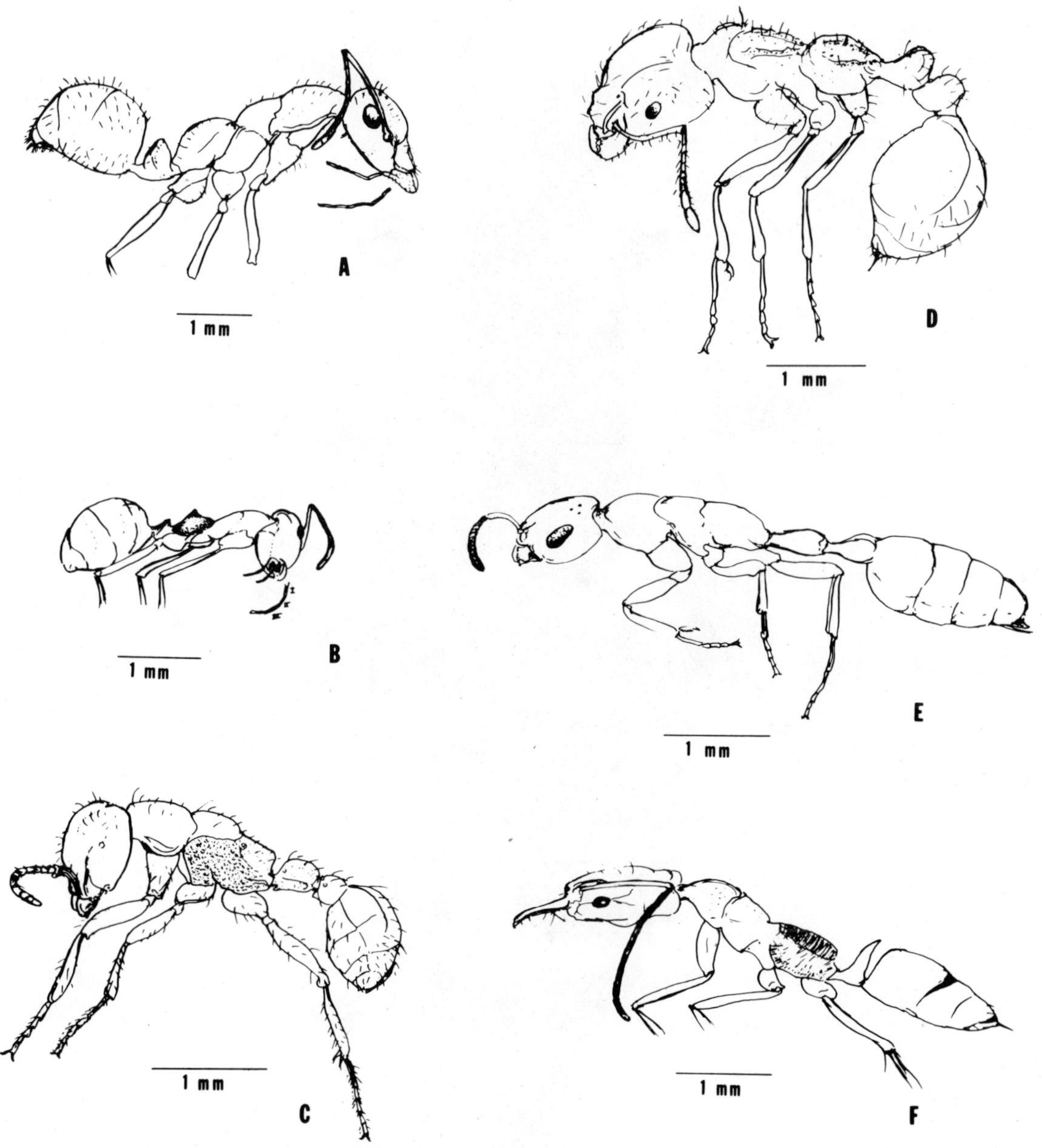

Figure 2.--Examples of ant subfamilies:

A. Formicinae
B. Dolichoderinae
C. Dorylinae
D. Myrmicinae
E. Pseudomyrmecinae
F. Ponerinae

Figure 3.--Larva of Chrysopa rufilabris Burmeister, a green lacewing common in Florida.

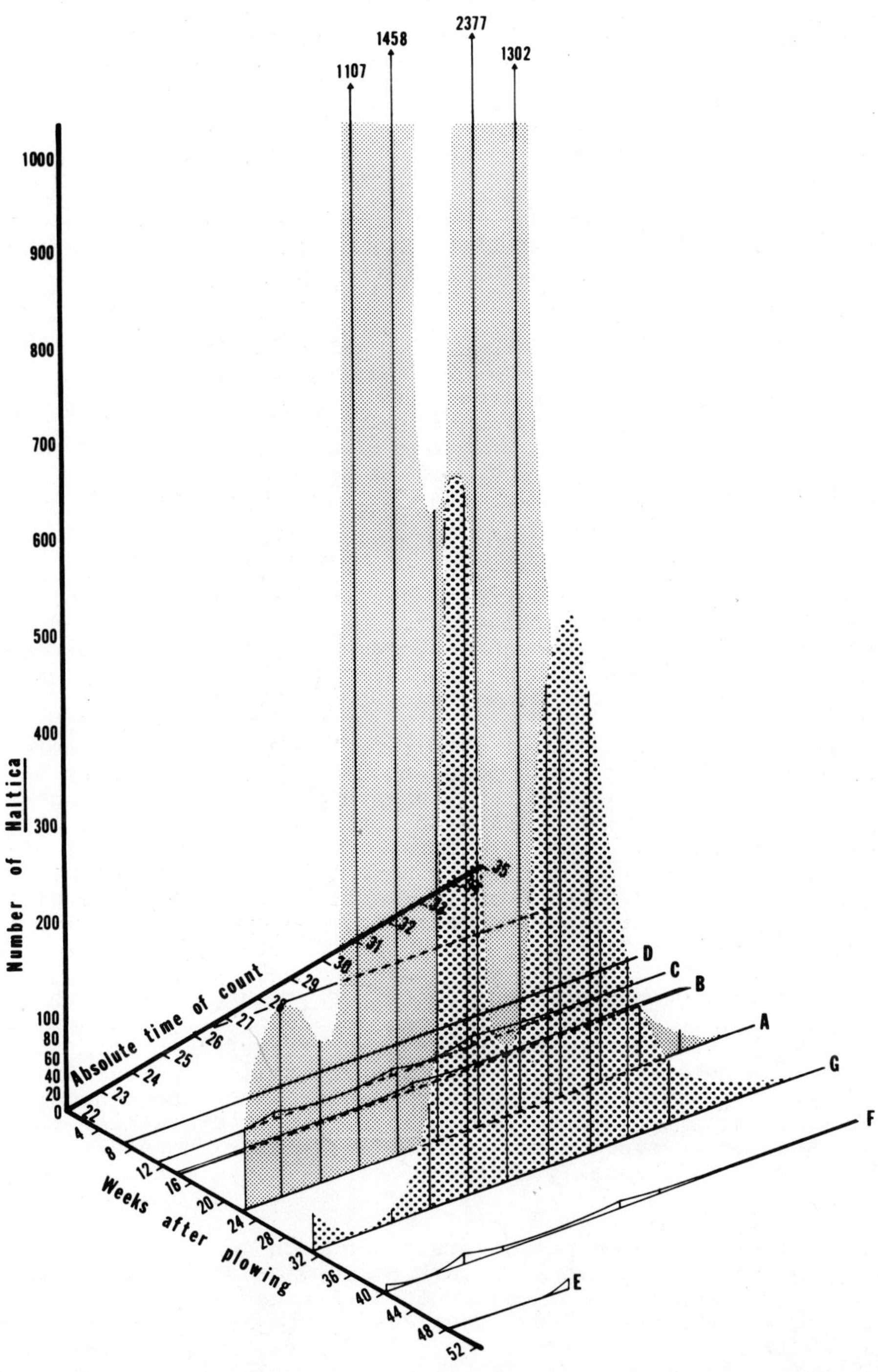

Figure 4.--Graph showing effect of time of plowing on populations of <u>Haltica</u> sp.

INCREASING NATURAL ENEMIES THROUGH USE OF SUPPLEMENTARY FEEDING AND NON-TARGET PREY

K. S. Hagen and Roy Hale
Department of Entomological Sciences,
Division of Biological Control
University of California, Berkeley
Department of Entomology
Division of Entomology
University of California, Riverside

Monoculture in modern agriculture particularly annual crops strongly precludes classical biological control and often limits natural biological control. Introduction of imported natural enemies of arthropod pests and subsequent establishment is difficult enough in perennial crops but to expect effective parasite or predator populations to persist in non-diverse ephemeral agroecosystems is indeed calling for extraordinary responses.

Monocultures actually discriminate against natural enemies and favor development of "exploding" pest populations. Insect parasites and predators usually have more complex food requirements than most phytophagous insects. The hemimetabolous insect plant feeders in both nymphal and adult stages need only one plant species as food to develop and reproduce, and the adults of holometabolous pest phytophagous insects can usually mate and oviposit without any feeding. On the other hand, most predaceous and parasitic insects require different sources of food in larval and adult stages to develop, reproduce and to survive throughout the year. Areas where literally thousands of acres of a single crop are present only during the summer, leaving vast areas of only fallow land during the fall and winter, invite and encourage plant pests. Phytophagous insects by either long distance flights, or emerging from pupae in the once fallow soil readily find the crops, and with built-in metabolite reserves transferred from their larval food are loaded with hundres of eggs ready for deposition. These insects find little opposition to depositing their eggs and having a high percentage of their progeny survive. The parasites and predators that neither overwinter in soil, nor have nearby vegetation to overwinter, nor pollen, nectar, honeydew nor alternate insect hosts or prey as food usually arrive too late in the monoculture crops to be of much value.

The lack of ecological diversity in both surrounding crops or within crops favors the arthropod pests. Also vast harvesting of a crop further disrupts the predator-prey relationships in favor of pest organisms. The role of ecological diversity in agroecosystems have been dealt with in recent years by DeBach, 1964; DeLoach, 1971; Hodek, 1966: Huffaker, 1971; Muma, 1971; Pickett and Patterson, 1953; Pimentel, 1971; Pollard, 1971; Southwood, 1972; Southwood and Way, 1970; Stern, 1969; Turnbull, 1969; Uvarov, 1964; van den Bosch and Stern, 1969; van Emden, 1965a, b; 1974; Way, 1966.

Because of a lack of knowledge, an entomologist can rarely recommend the planting of non-crop plants that would provide the requisites certain parasites and predators require, for such plants could also be reservoirs of plant pathogens and pest insects (Lewis, 1965; van Emden, 1965a). Attempts are being made to provide artificial foods and artificial shelters to make up for the environmental deficiencies. There are, however, other methods that are also being practiced in the integrated control approach that can shift the predator-prey or parasite-host ratios in favor of natural enemies (van den Bosch and Telford, 1964; Wilson, 1966). The following is a brief description of some various ways of modifying the environment and manipulating entomophagous species to increase the effectiveness of natural enemies.

Methods of Modifying Predator-Prey Ratios in Favor of Natural Enemies

The approach to modifying parasite-host or predator-prey ratios in order to achieve the biological control of an insect pest or spider mite populations differs depending upon the host specificity of the natural enemies involved and upon the population densities of both the natural enemies and the pests. Whatever strategy is chosen to manipulate natural enemy populations, it is important to act when the pest populations are low. One of the few recourses available when pest populations are approaching economic damaging levels is the use of selective pesticides. More specifically, the following strategies have been used to modify predator-prey or parasite-host ratios.

Resistant Plant Varieties

The use of fairly tolerant plant varieties or varieties that need not be completely unacceptable to pests but slow the development and reproduction of the phytophagous insects can be utilized. Natural enemies then have an opportunity to find prey and reproduce before the pest population gains enough momentum to surpass economic injury levels, but some natural enemies that have the potential of controlling the pest must exist in the community (Horber, 1972; Starks, et al., 1972; van Emden, 1966).

Non-Crop Plant Manipulation

Two studies in California grape vineyards may serve as examples of the influence of non-crop plant manipulation on increasing the effectiveness of natural enemies by providing an environment for alternate hosts. In one case a parasite survives the winter in an alternate leafhopper on a planted non-crop, and in the other case, weeds left in the crop serve as a plant host for alternate prey which a spider mite predator feeds upon during periods of low pest mite abundance.

Doutt, et al. (1966) planted wild blackberry, *Rubus*, in small areas in vineyard districts far removed from wild areas to provide overwintering alternate leafhopper host eggs for mymarid egg parasite. These plantings of *Rubus* led to a distinct increase in parasitization of grape leafhoppers on grape vines up to 4 miles away. Since the grape leafhopper overwinters in the adult stage, the egg parasite becomes completely disengaged from grape leafhopper populations in monoculture areas remote from wild refuges which harbor non-pest leafhoppers that overwinter in the egg stage.

Flaherty (1969) found that weedy vineyards have more stable populations of Willamette mite pests, *Eptetranychus willamettei* Ewing, than in weed-free vineyards. The small numbers of twospotted mites, *Tetranychus urticae* Koch, which moved from the weeds onto the grapevines served as an alternate prey species to maintain effective populations of the predaceous phytoseiid mite, *Metaseiulus occidentalis* (Nesbitt) during low densities of Willamette mites.

Release of Natural Enemies

Mass culture and periodic colonization of natural enemies against specific pests has been successful in controlling a few pest species (DeBach and Hagen, 1964; Monastero and Delanoue, 1966; Ridgway, 1969; Shands and Simpson, 1972). Inoculative releases of both pest and natural enemies has also worked in a few crops (Huffaker and Kennett, 1956; Parker, 1971), and inoculating crops with pests, adults or eggs alone was found to increase parasitization of pests (Shutte and Franz, 1961; Thewke and Puttler, 1970). Lack of technology has limited the use of periodic natural enemy releases, but this will become a common approach once inexpensive mass culture techniques have been developed for a variety of different species of natural enemies. Ecological knowledge of timing, quantities of natural enemies that should be released as well as the temperature thresholds of candidate predators must also be known. Hukusima (1971) used a combination of releasing coccinellids with spray programs in apple orchards in Japan.

Selective Pesticides

The use of selective pesticides that permits enough target pests and non-target phytophagous arthropods to survive to provide the naturally occurring parasites and predators the quantities of hosts or prey necessary for retention and reproduction can shift the predator-prey ratios in favor of natural enemies. Here again it is necessary to have some natural enemies present in the treated fields, otherwise the pest population will resurge faster than if depressed to extremely low levels by a highly toxic pesticide (Hukusima, 1963; Ripper, 1956; Smith and van den Bosch, 1967; Stern, et al., 1959). Preserving non-target mites has recently been demonstrated to be important for increasing the effectiveness of predaceous spider mites during periods when pest spider mites are low in abundance. Timing pesticide applications or again using selective pesticides so that the relatively inocuous rust mites, blister mites or tydeiid mites are spared has been found to provide suitable prey for predaceous phytoseiid mites to prevent disengagement of the predators from the target pests (Flaherty, et al., 1971, 1972; Hoyt and Caltagirone, 1971).

Supplementary Feeding

Application of supplementary foods in crops can also be used to retain, to arrest, to attract and to sustain natural enemies when natural prey populations are low or where non-prey food is lacking such as pollen or honeydew. It has been only a few predaceous insects and mites that have been manipulated by applying supplementary foods to crops. Simple diets have been used to arrest and retain predators, and complex diets have been employed to attract and induce oogenesis in certain predators. Pollen has also been applied in the field to increase the effectiveness of certain predaceous mites indirectly and sunflower seeds have been scattered in a crop to increase egg deposition of certain predaceous bugs.

By simply spraying sucrose solutions in corn fields, adult coccinellid populations were increased on the plants (Ewert and Chiang, 1966a). A higher adult predator coccinellid and chrysopid ratio to aphids was achieved by sucrose sprays resulting in a lower aphid population even though there were less predator immatures present in corn field treated with sucrose (Schiefelbein and Chiang, 1966).

Coccinellid adults, mainly *Hippodamia* spp. and *Chrysopa carnea* were not found to be attracted to sucrose sprayed alfalfa plots but they were arrested and the coccinellids aggregated where they found the sugar during their random searchings (Hagen, et al., 1971). A diet of sucrose alone does not induce oogenesis in common coccinellids, and only a few eggs developed from reserves are deposited by C. carnea when fed only sugar in the laboratory (Hagen, 1962; Hagen and Tassan, 1966).

Applications of complex diets which induce oogenesis and oviposition in certain predators have been made in alfalfa, cotton and peppers with desirable results. It was found that Chrysopa carnea Stephens could be attracted and induced to oviposit in alfalfa fields sprayed with a mixture of enzymatic protein hydrolysate of yeast plus sugar and water even though there were scarcely any aphids present. This diet also induced oogenesis. The six-fold increase of Chrysopa oviposition plus the increased activity of the coccinellids in the food sprayed plots prevented aphid populations from attaining damaging levels which occurred in the control plots (Hagen, et al., 1971).

The ecological basis accounting for the above reaction of C. carnea is that the adults are attracted naturally to various homopterous honeydews upon which they feed; oogenesis is stimulated and the Chrysopa oviposits in the general area of feeding. The yeast hydrolysate plus sugar is essentially like honeydew in composition, free amino acids preserved in high sugar concentrations (Hagen, 1950). Therefore applying the artificial honeydew simulates the presence of a high homopterous population density to which C. carnea is attracted, feeds and oviposits even though no homoptera may be present (Hagen, et al., 1970, 1971).

The yeast hydrolysate mixture was found to be phytotoxic to cotton, but by providing the artificial honeydew on feeding stations in cotton fields C. carnea was attracted and deposited over twice as many eggs in the general area of the stations as compared to the number of eggs counted in the control plots (Hagen, et al., 1971). The expense of the yeast hydrolysate, $1.50 per lb., and its phytotoxicity to certain plants led to a search for a cheaper yeast protein product.

Fortunately an inexpensive dairy product, (about .15¢ per lb.) composed of the yeast Saccharomyces fragilis and its whey substrate commercially known as Wheast® (Knudsen Creamery Co., Los Angeles, California) could be substituted for the yeast hydrolysate in the food spray. Wheast plus sugar contain all the essential nutritional components for oogenesis, and the particulate nature is similar to pollen. Wheast also possesses the attractant that yeast hydrolysate contains and is not phytotoxic to a variety of crops (Hagen and Tassan, 1970; Hagen, et al., 1971). A diet of Wheast plus sugar and water sprayed on alfalfa and cotton increased the effectiveness of C. carnea and certain coccinellids against aphids in alfalfa and C. carnea against Heliothis zea eggs and larvae in cotton (Hagen, et al., 1970, 1971). (See addendum).

Recently an experiment using predator food sprays on bell peppers was conducted that increased the effectiveness of C. carnea against the green peach aphid, Myzus persicae. This experiment will be described here.

Influence of high protein food sprays on Chrysopa in Bell peppers.--Three alternate rows in two plots (10 rows wide, 50 yards long) of Bell peppers, Capsicum frutescens var. grossum at Santa Maria, California were sprayed 4 times beginning July 15 with two different predator food sprays. One plot was sprayed with an enzymatic protein hydrolysate of yeast (Type pH yeast hydrolysate, Yeast Products Co., Patterson, New Jersey), plus sugar (sucrose) and water (400g, 700g, 2000ml, respectively) and another plot was sprayed with Wheast plus sugar and water (480g, 580g, 2000ml, respectively). The food spray experiment was started on July 15 when the target pest, Myzus persicae, was just beginning to alight on pepper plants.

The data in Figure 1 indicate that more *Chrysopa* eggs were deposited and less aphids were found in both the yeast hydrolysate and Wheast plots than sampled in the untreated control plot. The July 15 application of the predator food sprays was the most critical application date, for by two weeks later there was an average of twice the number of *Chrysopa* eggs on treated plants compared to control plants. The *Chrysopa* larvae hatching from the previous week and those larvae from eggs deposited on July 29 turned the aphid population trend downward while the aphid population of the control plot were still increasing. It is interesting to note the greater number of *Chrysopa* eggs sampled in the control plot on September 3 than were obtained in the treated plots. It appears that the greater amounts of honeydew on the control plants out-competed the applied artificial honeydew.

The aphid population crash which occurred in the control plot after August 26 was largely the result of fungal epizootic. The fungus involved was *Entomopthora planchoniana* identified by Dr. I. Hall, University of California, Riverside. Similar epizootics occurred in the predator food sprayed plots on August 26. However, the aphid population had already been depressed two weeks earlier by the *Chrysopa* predation. Fungal epizootics of *M. persicae* in peppers were observed to occur during the two years pepper plantings were under surveillance. Prevailing morning fogs coupled with hot afternoon temperatures, an *Entomopthora* epizootic seems to be triggered when the aphid population attains about 20 per leaf. However, the epizootics occurred too late in control plots to prevent damage to the plants.

Populations of other aphid natural enemies such as parasitic aphidiines and predaceous coccinellids and syrphids were too low in all three pepper plots to have an important influence on the reduction of aphid populations.

Influence of pollen dusting on predaceous mites in grapes.--Pollen is an important food for certain predaceous mites. Different species of the predaceous phytoseiid mites show different degrees of dependency on pollen feeding from continuous development on pollen alone, *Amblyseius hibisci* (Chant), only one generation developing on pollen *Typhlodromus pyri* Scheuten, to not feeding on pollen at all, *Metaseiulus occidentalis* (Huffaker, et al., 1970). McMurtry and Scriven (1966) demonstrated that the avocado brown mite was controlled by *A. hibisci* when pollen was added but not where pollen was omitted.

The general predaceous phytoseiid mite, *Metaseiulus occidentalis* (Nesbitt) does not feed upon pollen, but feeds readily upon pollen feeding tydeid mites. Thus, the nontarget tydeiid mites can act as an effective alternate prey when the target pest, the Willamette mite, in vineyards is scarce. Where tydeids are key alternate prey species, pollen producing flora in and around vineyards may influence more effective predator control of primary spider mite pests (Flaherty, 1969; Flaherty and Hoy, 1971).

To determine if tydeid mite populations could be artificially increased by applying pollen to the vines and thus increase the predaceous mites, cattail pollen (*Typha* sp.) was dusted on vineyard test plots. The hypothesis being that the tydeids increase in numbers by feeding on the pollen, the non-pollen feeding predaceous mite, *M. occidentalis* would feed on the tydeids and thus be maintained at high enough numbers to control the Willamette mite. The numbers of tydeid mites and *M. occidentalis* obtained in the pollen treated plot and control plots are shown in Table 1. A significantly greater number of tydeid mites and predaceous mites occurred in pollen treated plots. Also the target pest mite, the Willamette mites, remained low the rest of the season (Flaherty and Hoy, 1971; Flaherty, et al., 1971).

Pollen is also an important food for various predaceous insects. For example, adults of most predaceous syrphid species require pollen for egg production (Barlow, 1961; Schneider, 1969). In the absence of aphids some coccinellid larvae can develop on pollen alone and pollen can serve as a sustaining food for many different coccinellid adults (Hagen, 1962; Smith, 1961). Certain Chrysopa species also can feed upon pollen and produce eggs (Sheldon and MacLeod, 1971). Some parasitic Hymenoptera are also sustained and lay more eggs if certain pollens are available and less pest insects are present in certain crops where flowers are available (Leius, 1963, 1967).

Dusting pollen on crops may increase the activities of these various entomophagous species, and since Wheast plus sugar simulates pollen as well as honeydew, perhaps some of the pollen feeding entomophagous insects may be influenced beneficially by spraying crops with Wheast plus sugar. However, the presence of the flower itself with its color and odor is also required to attract certain natural enemies to the pollen; thus dusting pollen or applying Wheast and sugar would be of no value to strictly flower visiting adult parasites or predators.

Influence of sunflower seeds scattered on sugarbeet plants on Geocoris spp.--Tamaki and Weeks (1972) found in the laboratory that Geocoris bullatus (Say) and G. pallens Stal reared on a combination diet of green plant, insect prey and sunflower seeds had the shortest developmental period and highest egg production than obtained from feeding any of the single diet components. Four field plots with chopped sunflower seeds scattered on plants (1/4 lb. to 180 sq. ft.) twice per plot on a sugarbeet field had over twice as many Geocoris eggs as plots without sunflower seeds. Since Geocoris are rather general predators feeding upon aphids, lepidopterous eggs and spider mites, an increase of these bugs will certainly have a beneficial effect toward controlling pests.

Discussion

The few trials using supplementary foods in crops to increase the effectiveness of predators appears to be one promising approach to rectify poor predator-prey ratios. The use of sugar sprays alone can be used to arrest and sustain searching coccinellids and Chrysopa adults, but sugar is neither an attractant nor nutritious enough to induce oogenesis in these predators. With complex diets composed of sugar and enzymatic protein hydrolysate of yeast or Wheast plus sugar not only are certain Chrysopa spp. attracted but oogenesis and oviposition is induced, and the resulting increase in predaceous larvae can prevent aphids in alfalfa and peppers, and bollworms in cotton from attaining their potential high population levels in the absence of these predators.

The application of predator food sprays in crops does not necessarily insure pest reduction. Before considering the use of predator food supplements, it must first be determined that candidate adult predators are in the crop, in the surrounding crops or in the nearby wild vegetation. Secondly, the density of natural enemies has to be determined. Our present knowledge tells who these predators are, but we have little information on how many there have to be. By no means a definitive number, it seems that in cotton if there are two Chrysopa carnea adults in 50 sweeps, it may be feasible to attempt a food spray test with a Wheast plus sugar diet.

In conducting experiments with attractant predator food sprays for Chrysopa, the unsprayed control plots should be situated in line so that the prevailing evening winds do not blow across the treated food spray plots first. Chrysopa carnea are attracted to the Wheast while in flight, but the adults frequently land on plants before reaching the attractant, if small amounts of pollen or honeydew are present they may be arrested and delayed before reaching the predator food sprayed plots. The food sprays used today apparently will not attract C. carnea from plants having large amounts of honeydew present.

There is little value in applying yeast hydrolysates or Wheast alone without sugar, for although C. carnea may be attracted to the protein source, little feeding will occur on the attractant material. Shands, et al. (1972) sprayed only yeast hydrolysate on potatoes and obtained no appreciable effect upon aphids, coccinellids or *Chrysopa* larvae on treated plants. Since no sugar was available, their results are not surprising, for it is suspected that little or no feeding actually occurred on just the yeast hydrolysate; thus little egg deposition could be expected.

Sufficient quantities of predator food must be applied to a crop if egg production is to be increased. Butler and Ritchie (1971) applied Wheast plus honey and glycerin at 3 gallons per acre (the actual amount of Wheast applied was estimated to be about 5 pounds per acre by aircraft). A 3-fold increase of *Chrysopa* adults was obtained in their treated plots but egg deposition was not increased consistently. It appears from our work that a minimum of 10 pounds of Wheast plus 10 pounds of sugar in 7 to 10 gallons of water applied to an acre in alternate airplane swaths (1/2 acre actually receiving the 10 pounds Wheast) is enough needed to attract and induce oviposition from *C. carnea*. The food spray seems to be effective for at least one week and at most two weeks.

Predator food sprays may arrest or attract certain harmful insects. Sucrose alone apparently will arrest *Lygus* bug adults in alfalfa (Linquist & Sorensen, 1970). The sucrose Wheast mixture has not been observed to aggregate *Lygus* in our studies, but in alfalfa hay there have been greater numbers of *Lygus* adults at times in yeast hydrolysate plus sugar plots. The inconsistent appearance of *Lygus* adults in treated alfalfa appears to be an arrestment of *Lygus* on the food when *Lygus* are migrating. Since yeast hydrolysate and other protein hydrolysates attract tephritid flies (Steiner 1952), the use of predator food sprays should not, at this time, be applied in crops where tephritid flies are potential pests, for their fecundity will be increased by the supplementary protein diets (Hagen, 1958). It is possible in the future by including specific antibiotics in attractant food sprays that a differential effect could be induced between "pest" population and a predatory population. The tephritid adults attracted along with *Chrysopa carnea* adults feeding on the food spray could result in the fruit fly populations being reduced by destroying their bacterial symbiotes, and simultaneously increasing the *C. carnea* populations since the *Chrysopa* symbiotes are yeasts and unaffected by the antibiotics (Hagen, et al., 1971; Hagen and Tassan, 1973).

Although the pest control approach of applying supplementary foods on crops is in its infancy, the results indicate that such a strategy well may be worth pursuing in monoculture crops. However, the degree of understanding the ecology of the particular agroecosystem under consideration will determine the success achieved.

References Cited

Barlow, C. A. 1961. On the biology and reproductive capacity of *Syrphus corollae* Fab. (Syrphidae) in the laboratory. Entomol. exp. & appl. 4: 91-100.

Butler, G. D., Jr. and P. L. Ritchie, Jr. 1971. Feed Wheast and the abundance and fecundity of *Chrysopa carnea*. J. Econ. Entomol. 64(4): 933-934.

DeBach, P., ed. 1964. Biological control of insect pests and weeds. Reinhold Publ. Corp. 844 pp.

DeBach, P. and K. S. Hagen. 1964. Manipulation of entomophagous species. Ch. 15 *in* DeBach, P., ed. Biological control of insect pests and weeds. Reinhold Publ. Corp. pp. 429-458.

DeLoach, C. J. 1971. The effect of habitat diversity on predators. Proc. Tall Timbers Conf. Ecol. Animal Control Habitat Mgmt. 2: 223-241.

Doutt, R. L., J. Nakata and F. E. Skinner. 1966. Dispersal of grape leafhopper parasites from a blackberry refuge. California Agr. 20(10): 14-15.

Ewert, M. A. and H. C. Chiang. 1966. Dispersal of three species of Coccinellids in corn fields. Canadian Entomol. 98(9): 999-1003.

Flaherty, D. L. 1969. Ecosystem trophic complexity and Willamette mite, *Eotetranychus willamettei* Ewing (Acarina: Tetranychidae) densities. Ecol. 50: 911-915.

Flaherty, D. L. and M. A. Hoy. 1971. Biological control of Pacific mites in San Joaquin Valley vineyards: Part III. Role of tydeid mites. Res. Popul. Ecol. 13: 80-96.

Flaherty, D., C. Lynn, F. Jensen and M. Hoy. 1971. Influence of environment and cultural practices on spider mite abundance in southern San Joaquin Thompson seedless vineyards. California Agr. 25(11): 6-8.

Flaherty, D., C. Lynn, F. Jensen and M. Hoy. 1972. Correcting imbalances--spider mite populations in southern San Joaquin vineyards. California Agr. 26(4): 10-12.

Hagen, K. S. 1950. Fecundity of *Chrysopa californica* as affected by synthetic foods. J. Econ. Entomol. 43(1): 101-104.

Hagen, K. S. 1958. Honeydew as an fruitfly diet affecting reproduction. Proc. Internatl. Congr. Entomol. 10: 25-350.

Hagen, K. S. 1962. Biology and ecology of predaceous Coccinellidae. Ann. Rev. Entomol. 7: 289-326.

Hagen, K. S., E. F. Sawall and R. L. Tassan. 1971. The use of food sprays to increase effectiveness of entomophagous insects. Proc. Tall Timbers Conf. Ecol. Animal Control Habitat Mgmt. 2: 59-81.

Hagen, K. S. and R. L. Tassan. 1966. The influence of protein hydrolysates of yeasts and chemically defined diets upon the fecundity of *Chrysopa carnea* Stephens. Vestnik Cs. spol. Zool. 30(3): 219-227.

Hagen, K. S. and R. L. Tassan. 1970. The influence of food Wheast and related *Saccaromyces fragilis* yeast products on the fecundity of *Chrysopa carnea*. Canadian Entomol. 102(7): 806-811.

Hagen, K. S. and R. L. Tassan. 1973. Exploring nutritional roles of extracellular symbiotes on the reproduction of honeydew feeding adult chrysopids and tephritids. *In* Rodriguez, J. G., ed., Insect and mite nutrition. North-Holland Publ. Co. pp. 323-352.

Hagen, K. S., R. L. Tassan and J. R. Sawall. 1970. Some ecophysiological relationships between certain *Chrysopa*, honeydews and yeasts. Boll. Lab. Entomol. Agr. "Filippo Silvestri" Portici, 28: 113-134.

Hodek, I., ed. 1966. Ecology of aphidaphagous insects. Academia, Prague, 360 pp.

Horber, E. 1972. Plant resistance to insects. Agr. Sci. Rev. 10(2): 1-11.

Hoyt, S. C. and L. E. Caltagirone. 1971. The developing programs of integrated control of pests of apples in Washington and peaches in California. Ch. 18 in Huffaker, C. B., ed. Biological control. Plenum Press, New York, pp. 395-421.

Huffaker, C. B., ed. 1971. Biological Control. Plenum Press, New York. 511 pp.

Huffaker, C. B. and C. E. Kennett. 1956. Experimental studies on predation: Predation and cyclamen-mite populations on strawberries in California. Hilgardia 26: 191-222.

Huffaker, C. B., M. van de Vrie and J. A. McMurtry. 1970. Tetranychid populations and their possible control by predators: An evaluation. Hilgardia 49(11): 391-458.

Hukusima, S. 1963. Comparative acaricidal activity on the pest-predator complex in young apple orchards. Res. Bull. Facul. Agr., Gifu Univ. 18: 76-87.

Hukusima, S. 1971. Simultaneous suppression of major phytophagous arthropods in apple orchards by combination spray programs with release of Harmonia axyridis Pallos. Res. Bull. Facul. Agr., Gifu Univ. 31: 113-135.

Leius, K. 1963. Effects of pollens on fecundity and longevity of adult Scambus builianae (Htg.) Hymenoptera: Ichneumonidae). Canadian Entomol. 95(2): 202-207.

Leius, K. 1967. Influence of wild flowers on parasitism of tent caterpillar and codling moth. Canadian Entomol. 99(4): 444-446.

Lewis, T. 1965. The effects of shelter on the distribution of insect pests. Scientif. Hort. 17: 74-84.

Lindquist, R. K. and E. L. Sorensen. 1970. Interrelationships among aphids, tarnished plant bugs and alfalfas. J. Econ. Entomol. 63(1): 192-195.

McMurtry, J. A. and G. T. Scriven. 1966. The influence of pollen and prey density on the number of prey consumed by Amblyseius hibisci (Acarina: Phytoseiidae) Ann. Entomol. Soc. Amer. 58(1): 147-419.

Monastero, S. and P. Delanoue. 1966. Un grande esperimento di lotta biologia artificiale contro la mosca delle olive (Dacus oleae G.) a mezzo dell Opius C. sz Siculus Mon. in Sicilia Boll. Instit. Entomol. Agr. Osserv. Fitopath. Palermo 6(50): 1-53.

Muma, M. H. 1971. Preliminary studies on environmental manipulation to control injurious insects and mites in Florida citrus groves. Proc. Tall Timbers Conf. Ecol. Animal Control Habitat Mgmt. 2: 23-40.

Parker, F. D. 1971. Management of pest populations by manipulating densities of both hosts and parasites through periodic releases. Ch. 16 in Huffaker, C. B., ed. Biological control. Plenum Press, pp. 365-376.

Pickett, A. D. and N. A. Patterson. 1953. The influence of spray programs on the fauna of apple orchards in Nova Scotia. IV. A review. Canadian Entomol. 85(12): 472-478.

Pimentel, D. 1971. Population control in crop systems: monocultures and plant spatial patterns. Proc. Tall Timbers Conf. Ecol. Animal Control Habitat Mgmt. 2: 209-221.

Pollard, E. 1971. Hedges VI. Habitat diversity and crop pests: A study of Brevicoryne brassicae and its syrphid predators. J. Appl. Ecol. 8: 751-780.

Ridgway, R. L. 1969. Control of the bollworm and tobacco budworm through conservation and augmentation of predaceous insects. Proc. Tall Timbers Conf. Ecol. Animal Control Habitat Mgmt. 1: 127-144.

Ripper, W. E. 1956. Effect of pesticides on balance of arthropod populations. Ann. Rev. Entomol. 1: 403-438.

Schiefelbein, J. W. and H. C. Chiang. 1966. Effects of spray of sucrose solution in a corn field on the populations of predatory insects and their prey. Entomophage 11(4): 333-339.

Schneider, F. 1969. Bionomics and physiology of aphidophagous Syrphidae. Ann. Rev. 14: 103-124.

Schutte, F. and J. M. Franz. 1961. Untersuchungen zur Apfelwickler-bekampfung (Carpocapsa pomonella (L.) mit hilfe von Trichogramma embryophagum Hartig. Entomophaga 6(4): 237-47.

Shands, W. A., G. W. Simpson and M. H. Brunson. 1972. Insect predators for controlling aphids on potatoes. I. In small plots. J. Econ. Entomol. 65(2): 511-518.

Shands, W. A. and G. W. Simpson. 1972. Insect predators controlling aphids on potatoes. 7. A pilot test of spraying eggs of predators on potatoes in plots separated by bare fallow land. J. Econ. Entomol. 65(5): 1383-87.

Sheldon, J. K. and E. G. MacLeod. 1971. Studies on the biology of the Chrysopidae II. The feeding behavior of adult Chrysopa carnea. Psyche 78: 107-121.

Smith, B. C. 1961. Results of rearing some coccinellid larvae on various pollens. Proc. Entomol. Soc. Ontario 91: 270-271.

Smith, R. F. and R. van den Bosch. 1967. Integrated control in Kilgore, W. W. and R. L. Doutt, eds. Pest control. Biological, physical and selected chemical methods. Academic Press, New York, pp. 295-340.

Southwood, T. R. E. 1972. Farm management in Britain and its effect on animal populations. Proc. Tall Timbers Conf. Ecol. Animal Control Habitat Mgmt. 3: 29-51.

Southwood, T. R. E. and M. J. Way. 1970. Ecological background to pest management. In Rabb, R. L. and F. E. Guthrie, eds. Concepts of pest Management. Proc. Conf. North Carolina State Univ. Raleigh, pp. 6-29.

Starks, K. J. R. Muniappan and R. D. Eikenbary. 1972. Interaction between plant resistance and parasitism against the greenbug on barley and sorghum. Ann. Entomol. Soc. Amer. 65(3): 650-655.

Steiner, L. F. 1952. Fruit fly control in Hawaii with poison bait sprays containing protein hydrolysates. J. Econ. Entomol. 45(5): 838-843.

Stern, V. M., R. F. Smith, R. van den Bosch and K. S. Hagen. 1959. The integrated control concept. Hilgardia 29(2): 81-101.

Tamaki, G. and R. E. Weeks. 1972. Biology and ecology of two predators, *Geocoris pallens* Stal and *G. bullatus* (Say). USDA Tech. Bull. 1446, 46 pp.

Thewke, S. E. and B. Puttler. 1970. Aerosol application of lepidopterous eggs and their susceptibility to parasitism by *Trichogramma*. J. Econ. Entomol. 63(3): 1033-1034.

Turnbull, A. L. 1969. The ecological role of pest populations. Proc. Tall Timbers Conf. Ecol. Animal Control Habitat Mgmt. 1: 219-232.

Uvarov, B. P. 1964. Problems of insect ecology in developing countries. J. Appl. Ecol. 1: 159-168.

Van den Bosch, R. and V. M. Stern. 1969. The effect of harvesting practices on insect populations in alfalfa. Proc. Tall Timbers Conf. Ecol. Animal Control Habitat Mgmt. 1: 47-69.

Van den Bosch, R. and A. D. Telford. 1964. Environmental modification and biological control. Ch. 16 *in* DeBach, P., ed. Biological control of Insect pests and weeds. Reinhold Publ. Corp., New York, pp. 459-488.

Van Emden, H. F. 1965a. The role of uncultivated land in the biology of crop pests and beneficial insects. Scientif. Hort. 17: 121-136.

Van Emden. H. F. 1965b. The effect of uncultivated land on the distribution of cabbage aphid (*Brevicoryne brassicae*) on an adjacent crop. J. Appl. Ecol. 2: 171-196.

Van Emden, H. F. 1966. Plant insect relationships and pest control. World Rev. Pest Contr. 5(3): 115-123.

Van Emden, H. F. and G. C. Williams. 1974. The insect stability and diversity in agroecosystems. Ann. Rev. 19 (in press).

Way, M. J. 1966. The natural environment and integrated methods of pest control. J. Appl. Ecol. 3 (suppl.) 29-32.

Wilson, F. 1966. The conservation and augmentation of natural enemies. Proc. FAO Symp. Integrated Pest Control 3: 21-26. Rome.

Addendum

The Wheast® products used in all experiments contained 10% ash. New Wheast products (1973) contain 20% ash, and these Wheasts with higher ash content give poor responses from *Chrysopa*.

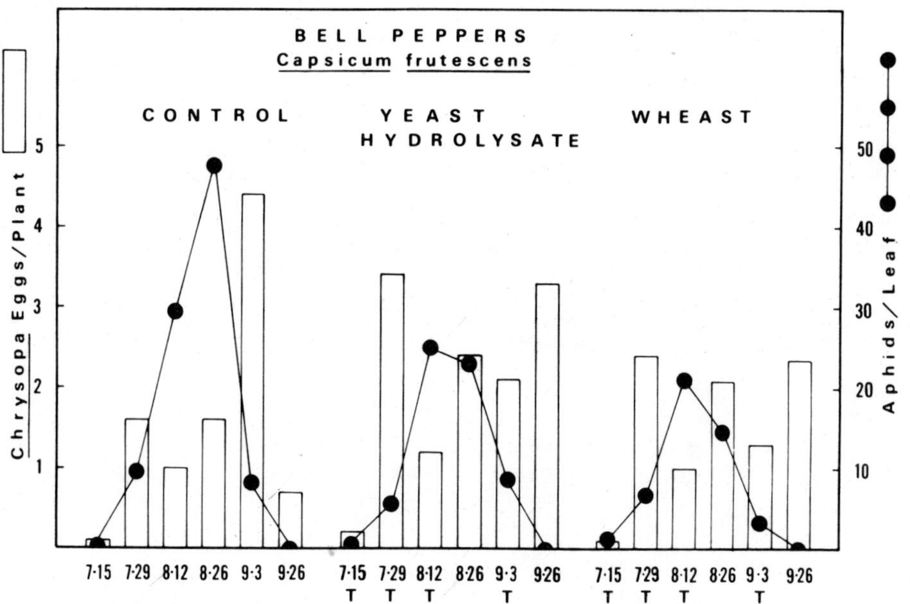

Figure 1.--Influence of two types of predator food sprays applied to bell peppers on *Chrysopa carnea* egg deposition and abundance of green peach aphids. (The T's beneath the dates are when the treatments of food sprays were made.)

Table 1.--Influence of cattail pollen on tydeid mite and the predaceous mite, *Metaseiulus occidentalis* populations on grapevines (Flaherty, et al., 1971).

Treatment	Tydeid mites	*M. occidentalis*
No pollen	1,199 [1,2]	107 [1]
Pollen [3]	2,065	171

[1] Significantly different at the 5% level (t test).

[2] Number of prey and predators on 480 leaves. Seven samples of 60 leaves (6 leaves from each of 10 vines) were taken from each treatment from August 19 through November 10.

[3] Cattail pollen (*Typha* sp.) freely dusted on 10 grapevines August 6 and October 7 with Hudson garden duster.

USE OF ECONOMIC THRESHOLDS AND SCOUTING AS THE BASIS FOR USING PARASITES AND PREDATORS IN INTEGRATED CONTROL PROGRAMS

Charles Lincoln
Entomologist
University of Arkansas
Fayetteville, Arkansas

Application of insecticides should be on the basis of need. Treatment at sub-economic levels of infestation adds unnecessarily to the cost of production, increases selection pressure for resistance to insecticides, decimates populations of parasites, predators and other nontarget organisms, and pollutes the environment.

A great deal of research has been done on economic thresholds. Much of it has been incidental to testing of insecticides. The most clear-cut economic thresholds have been obtained by simulating insect damage by such methods as removal of fruit or foliage. Maintaining a pre-determined population level, usually by use of cages, is frequently attempted, usually with erratic results. Research on economic thresholds frequently goes unpublished. It is reported at various conferences or spread by word of mouth. After considerable soul-searching, I have decided not to attempt a thorough literature review, but will cite only a few key references.

The economic threshold depends on the crop and end use of the product as much as on the level of the pest population.

The threshold of *Heliothis zea* is about 110,000/A on grain sorghum, about 25,000-30,000/A on soybeans, about 2,500-3,000/A on cotton, and only one per acre, if found, on strawberries. Economic thresholds on cotton are high enough to give parasites and predators a working margin before insecticidal controls must be invoked, and a wealth of information is available. My presentation will largely concern itself with cotton insects. Hopefully, the principles expounded will have more general application.

A system for treating as needed requires determination of thresholds for each pest in relation to the stage of growth and fruiting of cotton. Desirable supporting information includes the potential yield, optimum fruiting period, prediction of population trends of pests, and the status of biocontrol agents. All of these factors are brought to bear on the situation in each cotton field and the process repeated weekly or more often. The system is usually used by a person of limited entomological training who spends from one-half to one hour in each cotton field on each visit. Such systems have been in use for almost half a century. Many farmers scout their own cotton, while others hire the service.

Unlike most field crops, cotton sets fruit over a long period of time. For instance, heavy blooming commonly lasts for seven or eight weeks in eastern Arkansas. However, the effective blooming period is only four or five weeks. As the boll load accumulates, subsequent flowers are shed as small bolls. Normally, about one-half of the season's production of flowers is shed.

Excessive loss of early fruit can subsequently be compensated for, but this late fruit may be subject to greater hazards from insects and weather. From the above, it is obvious that economic thresholds cannot be fixed. Rather, we must think in terms of treatment thresholds developed for limited areas similar in climate, production capabilities and pest problems.

To give biological control a chance to work, treatment thresholds should be set at the maximum tolerable levels. Where possible, selective insecticides should be used or applications made at a time that will cause minimal disruption of biological control. Let us now consider the problem pest by pest.

Thrips may invade seedling cotton in massive numbers. The resulting stunting, curling of leaves, and loss of terminals may be spectacular, but the real damage directly attributable to thrips is small and difficult to measure. Given healthy seedlings and good growing conditions, thrips damage, at most, delays early fruit set a few days and this is later compensated for with no delay in harvest date. After cotton has two fully expanded true leaves, it is almost immune to damage although heavy thrips populations may persist. However, in cold or dry weather, seedling plants are under stress and their slow growth keeps them susceptible to thrips, which, in turn, slows their growth still more.

The expression, "The hoe must go," is obsolete. The hoe has gone. Weeds in the drill are controlled by herbicides or flame or not controlled at all. If early growth is slowed, cotton does not stand enough taller than the weeds for herbicides or flame to work. The contribution that thrips make to this situation, as opposed to the contributions of weather, plant diseases, and chemical change, defies analysis.

Automatic applications for thrips control are widely practiced. In eastern Arkansas, most fields are not heavily infested in most years. Automatic treatment of lightly infested fields is a needless expense, selects for resistance of spider mites, thrips and aphids, and contributes to environmental pollution. The impact of such treatments on parasite and predator populations has not been defined. Most beneficial insects move into cotton after treatment for thrips is terminated. However, thrips are an excellent food source for some predators. Aphids are also excellent food for certain predators and treatment for thrips ordinarily reduces aphid numbers.

Timing of applications as needed for thrips control poses certain problems. Use of damage symptoms as criteria is of no value. By the time symptoms are evident the damage has been done. Furthermore, thrips damage symptoms are mimicked by damage from cool weather, diseases, and chemicals. The best scouting approach is by use of a beater-box, made from hardware cloth and a cigar box. Cotton plants in the cotyledon stage are pulled gently and beaten against the hardware cloth. Thrips in the cigar box are then counted. One adult per plant represents a potentially damaging infestation. Plants with true leaves do not need to be pulled. The box is tipped at an angle greater than 45° and the plants slapped against the hardware cloth. In the two-leaf stage, five thrips per plant is considered a treatable infestation. This system has been recommended in Arkansas for some five years but has not been widely accepted. Part of the problem is that our scouting program is based on the use of students and they are not in the field this early in the season.

Cutworms cut off seedling plants and may reduce the stand along field borders or in fields where vegetation has not been thoroughly destroyed prior to planting. Firm thresholds have not been developed. The Arkansas recommendations, which are typical, use such terminology as, "If cutworms threaten the stand," and "problem fields."

Fleahoppers and Plant Bugs may cause pinhead squares to be aborted. Later in the season anthers are fed on in larger squares and blooms, increasing the tendency to shed. Feeding on small bolls may result in shedding or in poorly developed locks.

In the rainbelt the accepted scouting method has been to make terminal counts, usually examining the top six inches of 100 dominant plants per field. In the Far West sweepnets are more commonly used. Both methods are inaccurate and results are influenced by time of day, temperature, and wind. Given accurate counts, setting economic thresholds is fraught with error. Infestations are often transient and early season loss of squares is readily compensated for by later fruit set. For example, the late Al Hamner showed that complete removal of all young squares through the third week of July caused no reduction in yield, (Hamner, 1941).

Caging plant bugs to maintain pre-determined infestation levels has produced variable results. The shading by the cage tends to promote vegetative growth, thus introducing an unwanted variable. Introduced bugs often fail to survive or to feed.

Evaluation of a field problem should include both insect counts and damage symptoms. The first squares may be set by the sixth node. If squares are not set by the ninth node, trouble is indicated. Insect damage is only one of several factors that may delay the initiation of squaring. In such problem fields, very careful plant bug counts should be made and treatments applied for light infestations, as low as 10 to 15 per 100 terminals. If cotton is setting squares by the ninth node, higher infestations should be tolerated, 25 to 40 per 100 terminals.

Applications of insecticides at the initiation of fruiting comes at a critical time from the standpoint of biological control. Decimation of populations of predators and parasites at this time may induce outbreaks of other pests. The blow can be softened by using a partially selective insecticide, such as dimethoate or trichlorfon at low dosages. By treating promptly as soon as the threshold is reached, more time is allowed for recovery of populations of beneficial insects before the critical bollworm season.

Later in the season the decision to treat or not treat should consider feeding signs on anthers of large squares and blooms, feeding punctures on small bolls, and bug populations in terminals and squares. This is presently a murky area and good treatment thresholds have not been developed. Variations in varietal susceptibility to damage and in attraction to plant bugs apparently are adding greatly to the confusion. In any case, treatment is disruptive of biological control.

Lack of sound economic thresholds and reliable scouting methods for plant bugs and fleahoppers is a big problem in developing pest management programs in many areas.

Boll Weevil is the key pest over most of the rainbelt. It is not subject to effective biological control from parasites and predators. Varieties and cultural practices that make for high yields favor boll weevil buildup.

Full yields of cotton may be made despite continuously heavy weevil infestations with more than 40% punctured squares (Young, 1935; Fife, et al., 1949; Isely and Barnes, 1951). This is not to disparage the potential of boll weevil as a pest of cotton. In one of the tests just referred to the check plot yielded 16 pounds per acre, a 98% reduction in yield. A friend recently told me that he and his brother once picked 14 acres of cotton in one day.

Most of the work on damage thresholds of boll weevil has dealt in terms of percentage of punctured squares. This is as much a function of the rate of squaring as it is of the boll weevil population. Most professional cotton entomologists are cognizant of this trap and make some allowance for it. Few farmers or scouts have the necessary experience to do so.

Traditionally, weevil infestations have been determined by examining squares "at random" while covering a field in an "X" or "S" pattern. This is another trap. Weevil punctures are concentrated in the upper part of the plant. A tall scout will find a higher percentage of punctured squares than will a short scout. True random sampling of boll weevil punctures is not possible under field conditions.

The point sample system of scouting (Lincoln, et al., 1963) solves these problems. Fifty consecutive squares, 1/4 inch or larger in diameter, are examined for weevil punctures. The number of row feet is measured. This process is repeated four times within a field. The counts can readily be converted to squares per acre and punctured squares per acre. An estimate of weevils per acre can be obtained by dividing punctured squares per acre by a factor of 20. Since all squares are examined, selection for or against vertical concentration of punctured squares is eliminated.

Unfortunately, the point sample system introduces problems of its own. Boll weevil punctured squares are neither randomly nor evenly distributed throughout a field. Four sub-samples per field are obviously inadequate. Sampling error is great and hot spots may be missed.

Some compromise systems are being developed. Boll weevil punctures may be counted "at random" and number of squares per acre estimated in a separate operation.

I am not aware of any work on economic or treatment thresholds for boll weevil based on number of punctures per acre as opposed to percentage of punctures. In eastern Arkansas the average maximum number of 1/4 inch and larger squares per acre at the seasonal peak is almost 200,000. By extrapolation, cotton can tolerate 80,000 punctured squares per acre at the peak of fruiting.

Given favorable weather and abundant food, the weevil population in a cotton field doubles or triples each week. For approximately five weeks the square load also increases, masking much of the increase in weevil population and yielding a slow rise in percentage of punctured squares. Then the squaring rate levels off for some three weeks. The boll weevil increase does not level off and an apparent population explosion results. As an example, let us assume a constant rate of squaring for a three-week period and 10% punctured squares the first week. There would be 20 to 30% punctured squares the next week and 40 to 90% punctured squares the third week. In this hypothetical situation beginning applications at 25% punctured squares, a long-time recommendation, is barely adequate. One ineffective application will result in economic loss. Starting at 10% punctured squares would allow adequate lead time in this situation. However, starting at 10% punctured squares earlier in the season would be unnecessary and undesirable. We have struggled with this problem for years in Arkansas and it is the basic reason that we went to point sample scouting. Using this method, we recommended initiation of treatment at 60% punctured squares the first week of squaring and dropped the percentage each week until we were recommending treatment for 6 to 10% punctured squares after five to eight weeks of squaring. However, the treatment threshold was increasing in terms of punctured squares per acre. Although we were still using percentage of punctured squares, we were thinking in terms of punctured squares per acre. To legitimize what we were doing, we changed the recommendations to start treating for 20,000 punctured squares per acre from the second week of squaring until cutout was approaching. This required use of a conversion table and a change in thinking from percentages. It created a lot of confusion. This year we have adopted a very simple method of doing the same thing. Instead of 20,000 per acre, the treatment threshold is set at 1 1/2 punctured squares per row foot.

There are three approaches to insecticidal control of boll weevil for which punctured squares do not serve as a useful criterion. One of these approaches is to suppress the overwintered weevil population at the initiation of squaring. Counting weevils at this stage is very time-consuming. Applications at this stage drastically reduce populations of beneficial insects at a sensitive period. This approach has largely been abandoned except for its use in the boll weevil eradication experiment.

Another approach is treatment for boll protection during the cutout period. Usually a farmer has already started treatment based on punctured squares and he continues on schedule until all bolls to be protected are 16 to 20 days old. Occasionally, weevil populations first reach an economic level after squaring has dropped to a low level. Arkansas recommendations are to inspect small bolls instead of squares and to treat for one punctured boll per row foot, but this is based more on logic than research. Determination of economic thresholds for boll protection is a murky area. By this time biological control has served its purpose or else failed. It is better to err on the side of over-treatment because losses at this stage are irrecoverable.

A third approach is diapause control. Applications are made late in the season to prevent weevils from entering diapause. This reduces the overwintering population. Hopefully, a diapause program delays the start of insecticide application the following year, permitting parasites and predators to work in bollworm, tobacco budworm, spider mites, whiteflies and other pests.

Timing of applications for diapause control is based on condition of the plants with little attention to counts of weevils or weevil damage. It is assumed that there are enough weevils to treat if (a) in-season applications have been required; and (b) there are squares or small bolls to serve as food.

The <u>Bollworm-Tobacco Budworm</u> complex is susceptible to biological control and is difficult to control with insecticides. Innate susceptibility of the two species differs and varying patterns of insecticide resistance have developed between populations within each species.

The damage threshold is a seasonal average of two larvae to 10 row feet, between 2,500 and 3,000 per acre, or 3% damaged squares or 3% damaged bolls (Adkisson, et al., 1964). This research was done in walk-in cages. Different infestation levels were obtained by liberation of moths and small larvae. Infestation levels fluctuated and were highest late in the season.

Simulating bollworm damage by removing squares and using a cork borer on bolls showed no reduction in yield from the equivalent of four larvae per 10 row feet as one time infestations in early, mid, or late season in the Mississippi Delta (Kincade, et al., 1970). Damage equivalent to four larvae per 10 row feet three times during the season reduced yields by 285 pounds lint per acre in one of two years but the reduction was not statistically significant. These workers considered eight squares and three bolls to be equivalent to the damage caused by one larva.

Use of these research results in setting treatment thresholds poses certain problems. Heavy infestations prior to bloom can be dismissed as unimportant. The resultant square loss is readily compensated for and may even increase yields slightly.

During the period of heavy square production, bollworms feed more on squares than on bolls and the damage threshold appears to be greater than four larvae to 10 row feet or 5,000 per acre.

Late in the season boll feeding increases and newly-hatched larvae from eggs laid on dried blooms or bracts may attack bolls directly, probably lowering the economic threshold.

Two methods of scouting are commonly used for the bollworm-tobacco budworm complex. One is to examine dominant terminals for eggs and small larvae. Applying insecticides on this basis sacrifices biological control at the time when biological control may be very effective. Work by Whitcomb and associates over a three-year period showed more than 30% reduction in bollworm eggs from predations in 12 daylight hours. Somewhat lower but significant predation occurred at night. Predation of small larvae is also important, 10-20% of second instar larvae in six hours of exposure.

We attempted to solve this problem by having the scouts make counts of key predators, the number found while examining 200 squares, 100 terminals, and 100 leaves. Results were discouraging. Apparently, terminal counts for bollworm eggs and small larvae and the predator counts were subject to too much sampling error.

The other common approach to scouting for bollworm is to count damaged squares. Five damaged squares is equivalent to one bollworm (Sterling, 1972). Putting all the pieces together, a reasonable treatment threshold would be 20 damaged squares to 10 row feet or 25,000 damaged squares per acre when cotton is squaring heavily. Based on average peak squaring rates in Arkansas, this would be 12-13% damaged squares.

No one that I know has the nerve to recommend such a high treatment threshold for general use. Mississippi recommends starting application during the heavy squaring period for 5% damaged squares with continued pressure indicated by eggs and small larvae in terminals. Texas recommends first treatment for 5 to 8% damaged squares. Arkansas recommends first treatment for one damaged square per row foot, equivalent to 6-7% at the average peak. In all three states certain adjustments are made after first treatment or as cutout develops.

Question: Why are we all recommending initiation of treatment at subeconomic levels?

Answer: Because we have a healthy and well-founded respect for bollworm. If we wait until bio-control has had full opportunity to express itself, worms are in the third instar. We have no more than three days in which to kill them, because tolerance to insecticides increases rapidly with worm size. Control is difficult at best because of the secretive habits of the worms and insecticide resistance patterns are variable and unpredictable.

Having effective and cheap insecticides, we have erred on the side of overtreatment for control of all cotton insects. Readjustment toward minimal use of insecticides poses more problems with the *Heliothis* complex than with any other insect.

Before abandoning the bollworm-tobacco budworm problem, we should consider a new insecticide and a biocide. The insecticide (Galecron or Fundal) is primarily an ovicide but is also effective if taken by the moth. The biocide is the *Heliothis* nuclear polyhedrosis virus and it is most effective against newly-hatched larvae. The virus is completely host specific and the insecticide at low dosages is relatively harmless to predators. For these materials to have a chance, we must be able to do a more accurate job of scouting for eggs, i.e., terminal counts, or learn to make adult counts, or accept high levels of damage for brief periods, i.e., permit the larger worms to mature unmolested before the treatment becomes effective. This last course presents the fewest problems to the scout but calls for drastic reappraisal of values by farmers and professional agricultural workers, not excluding entomologists.

Boll rot is a major problem in rank cotton with high humidity. Estimates of crop loss from boll rot generally exceed 10% in Louisiana. Several pathogens are involved; most are weak parasites. Entry into bolls is often associated with mechanical or insect damage. Little work has been done in the area of insect population or damage thresholds in relation to incidence of boll rot.

Pink Bollworm populations must exceed one worm per boll before direct economic damage results. However, in the Imperial Valley of California, the economic threshold is considered to be only about 5% boll infestation because of the interaction of pink bollworm and boll rot.

Faced with a serious new pest, many farmers in the Imperial Valley went onto automatic schedules of application in an area where insecticides had been little used in recent years. This upset the balance between cotton leaf perforator and a host specific parasite. Induced outbreaks of perforator proved difficult to control.

The pink bollworm does not appear to be readily controlled by parasites and predators. The cultural methods so effective in Texas were either ineffective or unacceptable. A partial solution is to set treatment thresholds as high as is reasonable and to base insecticidal treatment on scouting.

The accepted method of scouting is to slice hard bolls and look for worm entries. The method is tedious. The worms found have developed beyond the stage where they can be controlled with insecticides. To use this method as a guide to insecticide application, certain assumptions must be made about the rate of population buildup, maturity of the crop and the boll rot hazard.

A sex pheromone is a practical trapping tool for detection and for overall population monitoring of pink bollworm. Recent work suggests that sex pheromone traps might also be used to set treatment thresholds for pink bollworm on a field by field basis. This would be a marked advance in scouting for pink bollworm. Interestingly enough, sex pheromones are very useful tools in boll weevil and bollworm research and survey but presently offer little hope of being useful in timing field applications of insecticides.

Discussion

My presentation has largely dealt with use of economic thresholds and scouting to reduce or eliminate unnecessary application of insecticides. I have also discussed selective insecticides and diapause control of boll weevil. These measures tend to conserve populations of parasites and predators.

Except for one slip, I have avoided the subject of scouting and setting effective population thresholds for parasites and predators. Despite our fiasco alluded to earlier, I want to be realistic, not pessimistic. With better training and supervision, scouting accuracy can be improved, and with research, better scouting methods can be developed.

Much of the problem lies in the number and diversity of predators. Of the hundreds of species found in a cotton field, only a few species have the potential for controlling a serious pest like the bollworm. Effectiveness depends on numbers and stadia of the host and the predator. Competing prey and other factors further confuse the situation. We need a great deal of quantitative research on each of several species of predators and parasites before their contributions can be evaluated. Inauguration of the Extension-APHIS pest management program will make available a lot of scouting information that includes predator counts. Computerization of this information will make possible an evaluation of the importance of each species or group of predator in relation to pest problems. Crude as the data are, its sheer mass should yield useful information that can be extracted by computers.

Summary

Use of parasites and predators in control of pest insects is approached in a negative manner. By setting realistically high treatment thresholds and measuring them by scouting methods that can be used by a non-professional or at least by a sub-professional, natural populations of parasites and predators are not destroyed unnecessarily.

There is a great need to develop methods of scouting for parasites and predators and to set effectiveness thresholds.

References Cited

Adkisson, P. L., C. F. Bailey and R. L. Hanna. 1964. Effect of the bollworm, *Heliothis zea*, on yield and quality of cotton. J. Econ. Entomol. 57: 448-450.

Fife, L. C., R. L. Walker and F. F. Bondy. 1949. Boll weevil control with several organic insecticides during 1948. J. Econ. Entomol. 42: 682-683.

Hamner, A. L. 1941. Fruiting of cotton in relation to cotton fleahopper and other insects that do similar damage to squares. Mississippi Agr. Exp. Sta. Bul. 360.

Isely, D. and G. Barnes. 1951. Boll weevil control in late summer. Arkansas Agr. Exp. Sta. Bul. 510.

Kincade, R. T., M. L. Laster and J. R. Brazzel. 1970. Effect on cotton yield of various levels of simulated *Heliothis* damage to squares and bolls. J. Econ. Entomol. 63: 613-615.

Lincoln, C., G. C. Dowell, W. P. Boyer and R. C. Hunter. 1963. The point sample method of scouting for boll weevil. Arkansas Agr. Exp. Sta. Bul. 666.

Lincoln, C., W. P. Boyer, G. C. Dowell, G. Barnes and G. Dean. 1970. Six years experience with point-sample cotton insect scouting. Arkansas Agr. Exp. Sta. Bul. 754.

Sterling, W. 1972. Personal communication.

Whitcomb, W. H. and K. O. Bell, Jr. 1967. In "The bollworm-tobacco budworm problem in Arkansas and Louisiana." Arkansas Agr. Exp. Sta. Bul. 720: 34-41.

Whitcomb, W. H. 1967. Field studies of predators of the second instar bollworm, *Heliothis zea* (Boddie). J. Georgia Entomol. Soc. 2: 113-118.

Young, M. T. 1935. Boll weevil control with calcium arsenate on field plots in Madison Parish, Louisiana, from 1920 to 1934. USDA Tech. Bul. 487.

SELECTIVE USE OF INSECTICIDES IN PEST MANAGEMENT

Robert L. Metcalf
Department of Entomology
University of Illinois, Urbana-Champaign

The past 25 years has witnessed the steadily increasing use of insecticides both in number of chemical compounds, increasing in the U.S. from about 30 to more than 300, and in volume of use, increasing from about 200 million pounds to 500 million pounds annually. This use of insecticides has resulted in outstanding examples of effective insect control and in dramatic increases in crop yields, and in improved human health. However, there has also been rapidly increasing public concern about the side effects of pesticides on nontarget organisms and upon environmental quality. Moreover, entomologists themselves have become uneasy about the steadily increasing problems of insect resistance to insecticide applications and the mounting indications that regular and repetitive insecticide treatments of crops such as cotton and deciduous fruits can result in spiraling treatment costs and in ecological changes in the nature of the pest complexes involved. All of these danger signals indicate that substantial changes must be made in both the nature and usage of pesticides. It must be remembered always that insecticides are applied to the environment as purposeful contaminants and consequently that benefits from their use must substantially exceed any damages to environmental quality.

In this discussion, selectivity is defined in terms of maximum effect of the insecticide in the target organism with minimal effects on humans, domestic animals, wildlife, beneficial invertebrates, and to the quality of the environment. It is important to determine how insecticides may be used most effectively and harmoniously in integrated pest management programs.

The practice of pest management has been defined (Geier, 1965) as consisting of three phases:

1. Determine how the life system of the pest needs to be modified to reduce its numbers to tolerable levels, i.e., <u>below the economic threshold</u>.
2. Apply biological knowledge and current technology to achieve the desirable modification, i.e., <u>applied ecology</u>.
3. Devise procedures for pest control both suited to current technology and compatible with economic and environmental quality aspects, i.e., <u>economic and social acceptance</u>.

This pest management approach to the use of insecticides may be likened to the use of the surgical scalpel and offers marked contrast to the employment of the broadsword. A more appropriate analogy is to compare the use of insecticides in the practice of pest management to the use of pharmaceuticals in the practice of medicine. Thus we need to vastly increase our knowledge of the environmental properties of pesticides and learn how to apply them specifically and precisely to weak points in the pest's developmental pattern. To quote from the National Academy of Science Publication 1695 (1969), p. 456, Insect Pest Management and Control ". . . each control technique has a potential role to play in concert and harmony with the others, and a major technique such as the use of pesticides can be the very heart and core of integrated systems. Chemical pesticides will continue to be one of the most dependable weapons of the entomologist for the foreseeable future. . . . There are many pest problems for which the use of chemicals provides the only acceptable solutions. Contrary to the thinking of some people, the use of pesticides for pest control is not an ecological sin. When their use is approached from the sound base of ecological principles, chemical pesticides provide dependable and valuable tools for the biologist. This use is indespensible to modern society." Other useful reviews of the general subject of insect pest management include Vanden Bosch and Stern (1962), Geier (1966) and the FAO Symposium on Integrated Control (1966).

Selective Use of Insecticides

As the United States moves into an era of pest management and away from 25 years of indiscriminant usage of pesticides, wholesale changes must be made in our strategies for the employment of chemicals. We are leaving a period when the initial reaction to the presence of insect enemies was to apply a broad spectrum insecticide and where entomological techniques were all too often typified as "squirt and count." In the pest management era the use of insecticides can be categorized in 3 ways: (1) carefully timed and gauged suppressive applications aimed at a weak point in the insect's life cycle, (2) emergency applications reserved for epidemic situations where all other control measures are inadequate and the insect population exceeds the economic threshold, and (3) preventive treatments of highly selective impact made with the least volume of insecticide and calculated to provide the least disturbance of environmental quality. To implement pest management practices on a broad scale will demand more sophisticated entomological knowledge and techniques including a detailed knowledge of the properties of the insecticides themselves in relation to effects on target and nontarget organisms, on human and public health, and on the total quality of the environment.

These premises then suggest that in the future we may well use substantially smaller quantities of insecticides for specific programs of pest control, that the chemicals chosen may be quite different from those heavily relied upon today, and that our standards of efficiency in control, efficacy of crop protection, and economics of pesticide employment are likely to be significantly different from those of the past several decades. Two essential steps in the transition period into the pest management era are:

1. replacement of wasteful "routine-treatment" schedules by treat-when-necessary schedules
2. recognition that 100% control of pests is <u>not</u> required to prevent economic losses (PSAC Report, 1965).

<u>A. Selectivity of Insecticides</u>

Achievement of the aims of pest management programs will depend in substantial measure upon the properties of insecticides available for pest population regulation. At least 6 types of insecticide use are recognized which can be used to exert direct effect on the target organism together with minimal undesirable effect on parasites and predators.

<u>1. Intrinsically selective insecticides</u> are few in number. However, <u>Bacillus thuringiensis</u> toxin "BT" is effective only against a few species of lepidopterous larvae such as the cabbage looper <u>Trichoplusia ni</u> and may be considered a truly selective insecticide. The new mosquito larvicide 2,6-di-tert-butyl-4-(α,α-dimethylbenzyl)-phenol (MON-0585) seems to be active only against a

variety of mosquito larvae and virtually non-toxic to anything else (Sacher, 1971). The acaricides dicofol, ovex, tetrasul, tetradifon, and omite are almost specifically toxic to the Acarina and essentially non-toxic to insects, wildlife, and humans and higher animals.

2. Systemic insecticides show pronounced selectivity especially against sucking plant pests such as aphids, mites, thrips, leafhoppers, and sometimes to chewing insects. Insecticides such as schradan, demeton, and oxydemeton-methyl applied to plant foliage rapidly penetrate the leaf cuticle and are translocated throughout the xylem tissue where they serve as stomach poisons to sucking insects with little or no harm to parasites, predators, and pollinators. The more persistent systemics such as phorate, disulfoton, aldicarb, and carbofuran are best applied as granulars to the soil about roots at time of planting or as seed dressings. They are translocated to the above-ground portions of the plant, concentrating in the most rapidly growing areas such as new leaves and fruits. Aldicarb in particular has pronounced systemic effects for the adult cotton boll weevil, Anthonomus grandis, and when applied to soil at 2 lb/acre as a side dressing (Bariola, et al., 1971), reduced the population of adult weevils from 94-96%.

3. Fugitive broad spectrum insecticides such as tetraethyl pyrophosphate, mevinphos, and nicotine achieve a degree of selectivity when properly timed and applied, as beneficial species can survive an initial onslaught through resistant stages, e.g., pupal, protected locations, or reservoirs in hedge rows or in strip crops sprayed alternately. Thus Trichogramma semifumatum in host eggs survived treatment with mevinphos and the residue dissipated rapidly enough to be innocuous to newly emerging adult parasites, while persistent insecticides such as carbaryl killed the adult parasites after they emerge (Vanden Bosch and Stern, 1962).

4. Ecologically selective insecticides can be used by specific timing or placement of the material to decrease its effects on nontarget organisms. The simplest example is the application of compounds such as methyl parathion, parathion, and carbaryl which are highly toxic to honey bees, at best after bloom is completed, or at least in the evening when bees are not visiting blossoms. The eradication of the Tsetse fly, Glossina swynnertoni, over 35 square miles of Africa, by selectivily treating preferred resting places with 3% endosulfan or dieldrin applied on the underside of tree branches 1 to 4 inches in diameter, 4 to 9 feet above-ground, and inclined less than 35° from the horizontal, is an astonishing example of ecological selectivity (Chadwick, et al., 1965). The gravid female melon fly, Dacus cucurbitae enters tomato fields to oviposit but leaves these fields at dusk and spends the evening on adjacent vegetation. Applications of DDT or parathion to the vegetation adjacent to crop areas, between sunset and 7 a.m., reduced the average infestation of tomato fruits from 65% to 3% (Nishida and Bess, 1950; Nishida, 1954).

The use of DDT as an interior house spray at 1 g per m^2 for malaria eradication is an outstanding example of ecological selectivity which is effective because the female Anopheles mosquito enters human habitation in search of a blood meal and prefers to rest on wall or ceiling during her digestive processes, where she receives a lethal dose of DDT.

Other well-known examples of ecological selectivity include the common practice of seed treatments, especially of corn to control Diabrotica undecimpunctata howardi, Hylemya cilicrura, and species of wireworms. Newsom (1966) reports the successful control of these species with dieldrin at 2 oz/100 lb seed or about 5 g/acre. This contrasts with the 2 lb of aldrin or heptachlor per acre commonly applied broadcast or sidedressed which requires 180 times as much insecticide. Other successful seed treatment practices such as corn with diazinon and oats with propoxur are discussed under Seed Treatment. Treatment of sweet corn silks with an injection of 0.5 ml of 1% DDT in mineral oil to prevent infestation by the corn earworm, Heliothis zea, is a classic example of ecological selectivity.

5. Selectivity through restricted treatment has been known for years but rarely practiced on a substantial scale. The treatment of alternate strips of alfalfa together with alternate strip cutting practices provides a haven for natural enemies of the alfalfa weevil and spotted alfalfa aphid.

The purple scale *Lepidosaphes beckii* attacking citrus has been controlled through successful alternation of biological and chemical control achieved by spraying alternative pairs of tree rows at 6-month intervals with oil spray (Vanden Bosch and Stern, 1962). Such methods need to be explored on a much wider basis.

6. Selectivity through use of baits or attractants is an obvious method which is discussed under Ecological Approach VII.

Choice of Insecticide for Pest Management

The criteria for using insecticides in pest management programs are complex, yet the ultimate success of the program may be determined by the insecticide selected as well as by the method of application employed and by the timing of the treatment. It is this area that the practice of pest management becomes both critical to the health of the total environment and challenging and satisfying to the pest management specialist.

To a considerable extent, the choice of insecticide is biased by federal pesticide registrations and residue tolerances. Yet these permit wide latitude on most important crops. For example for the 4 important crops, alfalfa, apples, corn, and cotton, the following federal tolerances are in effect:

	alfalfa	apples	corn	cotton
Total number of insecticides registered	39	72	66	50
Insecticides registered on all 4 crops	--	20	--	--
Insecticides registered on 3 of 4 crops	--	21	--	--

Who among us can honestly say we are familiar with all their common and tradenames, chemical structures, and individual properties in insect pest control? If we add to these essentials a requisite for knowledge of their toxicity to humans and domestic animals, effects on a variety of nontarget organisms--parasites, predators, pollinators, and wildlife; and their environmental fate in air, soil, water, and food, we have an enormous responsibility when we choose and prescribe or recommend an individual insecticide for a specific pest management program. Yet all the criteria mentioned are important facets of environmental quality.

To provide some rationale for intelligent choice, the data in Table 1 may be helpful. This table attempts to provide a numerical rating for most of the common insecticides, widely used on the important crops above, in regard to their safety and effects on environmental quality, and hence in regard to their preferred suitability for pest management programs. Ratings were made on the basis of average performance in (a) acute toxicity to humans and domestic animals, (b) toxicity to nontarget organisms, and (c) environmental persistence. Each category was assigned a rating of 1 to 5 with increasing hazard, as shown below. Nontarget effects were rated as the average of toxicity to honey bee, trout, and pheasant. These animals were chosen because they represent three typical nontarget groups of animals and because of a substantial amount of data which is available (Pimentel, 1971).

I Mammalian Toxicity - rat oral LD_{50} mg/kg

1 = >1000
2 = 200-1000
3 = 50- 200
4 = 10- 50
5 = <10

II Nontarget Toxicity

Trout - LC_{50} ppm	Pheasant - oral LD_{50}	Bee -topical LD_{50} mg/kg
1 = >1.0	1 = >1000	1 = >100
2 = 0.1 -1.0	2 = 200-1000	2 = 20-100
3 = 0.01 -0.1	3 = 50- 200	3 = 5- 20
4 = 0.001-0.01	4 = 10- 50	4 = 1- 5
5 = <0.001	5 = <10	5 = <1

average value used to rate insecticide

III Environmental persistence - soil half-life

1 = <1 month
2 = 1-4 months
3 = 4-12 months
4 = 1-3 years
5 = 3-10 years

To summarize the data in Table 2, the compounds have been segregated in 4 classes:

1. suitable for use in pest management (rating 4-7), carbaryl, chlordane, Gardona, malathion, methoxychlor, naled, trichlorfon.
2. caution for use in pest management (rating 8-10), azinphos-methyl, demeton, diazinon, dicofol, dimethoate, Dursban, lindane, mevinphos, methyl parathion, phosphamidon, Dursban, oxydemeton-methyl, tetraethyl pyrophosphate, toxaphene, Zectran, endosulfan.
3. to be used for pest management only under restricted conditions (rating 11-13), such as seed or soil treatment with aldicarb, carbofuran, disulfoton, phorate, parathion, EPN, or indoor treatment with DDT.
4. little if any place in pest management (rating 13-15), aldrin, dieldrin, endrin, heptachlor.

Examples of Ecological Approach to Pest Management

The categorization of pest management practices in Table 2 have been adapted from Geier (1966). Table 2 relates the basic procedures of intervention against insect pests in terms of (a) the relative frequency of employment and (b) their intrinsic value, i.e., unsatisfactory, acceptable, and good. The Roman numerals reflect an increasing efficiency and desirability in pest management from I to IX. Examples of the use of insecticides in a variety of pest management practices follow:

(1) Illustrates the customary way in which insecticides have been used in a curative fashion to reduce pest populations after they have surpassed the economic threshold. Examples include the repeated spring and summer applications of acaricides such as omite, chlorphenamidine, dicofol, etc. in deciduous fruit orchards to control the European red mite, *Panonychus ulmi*, or in citrus orchards for the

citrus red mite, *Panonychus citri*. Another example is the repeated *ad lib* applications of calcium arsenate, toxaphene, or methyl parathion to control the cotton boll weevil, *Anthonomus grandis*. These practices are clearly unsatisfactory in that they lead to the development of insecticide resistance, to steadily increasing frequency of applications reaching as high as 20 to 50 per year, e.g., cotton insect control in southern USA or in Central and South America; and to increasing numbers of undesirable effects on nontarget organisms and to environmental quality. In California for example, the indiscriminant application of insecticides for the control of flood water *Aedes* mosquitoes has produced resistance successively to DDT, the cyclodienes, and to organophosphorus insecticides of all descriptions--so that presently there are very few insecticides to which these pests are susceptible.

(II) In this procedure, the insecticide is applied to a weak point in the life cycle where the pest is most susceptible, and it is kept from exceeding the economic threshold. This method is essentially the essence of the proper use of pesticides in pest management and is largely preventive. It requires strict timing of application and is not, of course, free from the disadvantages of the development of resistance or of undesirable effects on nontarget organisms. Examples include the soil application of heptachlor, aldrin, and more recently diazinon, phorate, carbofuran, or Bux for the control of the larvae of the corn rootworms *Diabrotica* spp.

(III) This procedure centers around the methodical and repeated withdrawal of an important requisite of the pest in an otherwise favorable environment. The use of pasture rotations for the control of the cattle tick *Boophilus annulatus* in the U.S. or *B. microplus* in Australia is an appropriate example. Through pasture rotation, the young seed ticks hatching from eggs dropped by the adult female tick will starve to death before host cattle are returned to the area. This method of pest management is most effective when used in conjunction with routine dipping of the cattle in arsenic trioxide 0.175%, toxaphene 0.5%, or lindane 0.025% to free them from ticks before they are returned to pasture (Metcalf, et al., 1962).

Another example of this principle is crop rotation to control the northern corn rootworm, *Diabrotica longicornis*, which overwinters as the egg is laid around corn roots and attacks only corn. Rotation of soybeans, alfalfa, or clover in place of corn causes the newly hatched spring larvae to die of starvation as they have no corn roots on which to feed (Metcalf, et al., 1962). This procedure could be used routinely together with corn seed treatment with diazinon or carbofuran in a pest management program. Type III procedures are simplistic, highly effective and generally involve no major side effects. However, they complicate routine agronomic practices especially in intensively farmed areas and thus are generally neglected.

(IV) This procedure involves the supplanting of natural populations of parasites or predators with artificially reared species. The mass release of the egg parasite *Trichogramma pretiosum* in tomato fields at 465,000 per acre to reduce populations of tomato fruitworm (*Heliothis zea*), cabbage looper, *Trichoplusia ni*, and tobacco hornworm, *Manduca sexta*, with rates of parasitism from 39 to 64% (Oatman and Platner, 1971). The mass release of *Chrysopa carnea* eggs or larvae in cotton fields at about 290,000 per acre reduced populations of the bollworm *Heliothis zea* and the tobacco budworm, *H. virescens*, as much as 96% (Ridgway and Jones, 1969).

(V) This procedure involves the use of precisely timed and applied insecticides to control numbers of target organisms below the economic threshold without completely suppressing the host and its parasites and predators or adversely affecting other injurious organisms. Such use of pesticides are true pest management operations and require precise knowledge of the life table of the target organism and full appreciation of the properties of the pesticide to be used.

Examples: (1) The employment of a spring spray of methoxychlor, by hydraulic spraying or mist blowing, in an area or community program along with pruning and burning of dead branches, and removal and burning of dead trees to control the elm bark beetles, Scolytus multistriatus and Hylurgopinus rufipes, which are the vectors of Dutch elm disease, Ceratocystis ulmi. Years of experience have shown that this program can be effective in limiting the progress of the disease to 2% or less trees dying per year without unacceptable damage to robins and other important elements of environmental quality (Wallner and Hart, 1971).

(2) The use of Zectran spray, (4-dimethylamino-3,5-xylenyl N-methylcarbamate), at 0.12 lb/acre as a highly selective forest treatment to control the spruce budworm Choristoneura fumiferana, considered the most widely distributed and destructive forest tree defoliator in North America (Tucker and Crabtree, 1969). Zectran although highly acutely toxic to mammals and birds apparently can be used specifically and safely for spruce budworm control because of the low dosage required, rapid photo- and biological degradation, low degree of cumulative toxic action, and low toxicity to fish. The development of this use of Zectran and the demonstration of its safety are the result of a 10-year search for a suitable replacement for DDT.

(3) The control of the Culex mosquito vectors of the equine encephalitis viruses by ULV spraying of 98% malathion at 1 lb/acre is included as a special application of Type V involving a carefully studied response to an emergency outbreak. Such applications gave a very high degree of control of an outbreak of St. Louis encephalitis in Dallas in 1969 (WHO, 1971) and were employed over more than 30 million acres of southern United States in the summer of 1971 to control an outbreak of Venezuelan encephalitis in horses. There is no reason to believe that such applications had appreciable adverse effects on environmental quality because of the intensive study given malathion for general environmental effects (Pimental, 1971).

Dependence on Economic Threshold. Where the economic threshold levels are very low in terms of pest infestation as in protecting apples from codling moth (Carpocapsa pomonella), apple maggot (Rhagoletis pomonella), apple curculios (Tachypterellus) and plum curculio (Conotrachelus nenuphar); sweet corn from corn earworm (Heliothis zea); or Christmas trees from European pine shoot moth (Rhyaciona buoliana) (Wallner and Butcher, 1970); reliance on pesticides must be relatively greater in order to produce a marketable product. The problems of the precise dosage, timing, and choice of pesticide in pest management programs for such pests become critical. To avoid unnecessary treatments the reliable sampling of insect population is essential. The future will see increased reliance on pheromone traps for determining population densities and consequent timing of pesticide applications (Roelofs, et al., 1970). Generally, for crops of high market value and very low economic thresholds, the pest management program should be built around the proper use of insecticides, considering such items as planting to provide proper spacing and access for application of insecticides, proper pruning or growing practices to increase spray coverage, strip cultivation to harbor and retain natural enemies, and selection of varieties which are especially compatible to other agronomic practices.

(V) This procedure denies the pest species its customary access to requisites such as food, shelter, or mates. The basic methodology consists in providing a lethal substitute which is more attractive than the natural requisite. This methodology can provide highly effective control over long periods without hazards to nontarget organisms or to the environment. The methods, however, generally need to be coordinated over large areas by an authority such as a control district or suppression or abatement program. Examples include: (1) the eradication of the oriental fruit fly, Dacus dorsalis from the island of Rota by dropping fiberboard squares treated with the male attractant methyl eugenol containing 3% naled (DiBrom) using a total dosage of 3.4 g of toxicant per acre (Steiner, et al., 1965); (2) the use of protein-hydrolysate-malathion bait sprays to eradicate the Mediterranean fruit fly, (Ceratitis capitata) from Florida at a dosage of 1.2 lb of toxicant per acre (Steiner, et al., 1961); (3) the use of sugar containing 1% dichlorvos, trichlorfon, dimethoate, or other toxicant as a bait for the control of the house fly is a more familiar example of this method of control.

(VII) This procedure applies most typically to the introduction of insect enemies of the target organism which reduce the pest to levels of abundance that are generally below the economic threshold but that occasionally exceed it when pest outbreaks occur. This is the situation under which most examples of the biological control of insects exist and is generally the result of the impact of density independent factors such as extremes of weather which are unfavorable to the parasite. Examples include the control of black scale of citrus Saissetia oleae by the encyrted parasite Aphycus helvolus which failed dramatically after unusually cold winters, and the control of the olive scale, Parlatoria oleae by the parasite Aphytis maculicornis which is poorly adapted to summer conditions (Huffaker, 1962).

(VIII) This procedure implies the establishment of a complex of natural enemies and diseases so as to contain the pest at numbers below the economic threshold over long periods of time despite fluctuations in density independent factors affecting the individual components of the complementary repressive system. (1) The complex procedures for the repression of the spotted alfalfa aphid Therioaphis maculata, include (Stern, et al., 1959; Stern and Vanden Bosch, 1959; (a) planting of resistant varieties of alfalfa, (b) establishing and maintaining a vigorous stand by watering, fertilizing, and weed control, (c) establishment of imported hymenopterous parasites Praon pollitans, Trioxys utilis, and Aphelinus semiflavus, (d) encouragement of natural enemies such as lady beetles, syrphids, lacewings (Chrysopa), and hemipterans (Nabis and Geocoris) by strip cutting, (e) encouragement of entomogenous fungi Entomophthora, and (f) timing of alfalfa cuttings when there is little green leaf residue to kill aphids by exposure. When, in spite of the practices, the aphid population exceeds the economic threshold of 20 to 40 per stem, (g) spraying with demeton or mevinphos at 1-2 oz/acre will repaidly reduce the aphid population without severely damaging beneficial insects. (h) Seed treatment with phorate or disulfoton at 1.5 lb/100 lb seed or comparable application of granular will effectively prevent young seedlings from being killed by the aphid.

An example (2) in the making is the repression of the cereal leaf beetle (Oulema melanopus) by the establishment of (a) the eulophid parasite, Tetrastichus julus, and the mymarid parasite, Anaphes flavipes, (b) planting of resistant varieties of cereals, (c) spray programs where necessary with carbaryl, malathion or endosulfan, and (d) treatment of cereal seeds with propoxur or carbofuran at 4 oz/100 lb of seed to control adults early in the season (Ruppel, et al., 1970; Stehr, 1970; Maltby, et al., 1971).

(IX) This procedure represents perhaps the most satisfactory and at the same time the most difficultly obtainable pest control. Under this heading are included measures which more or less permanently solve the pest problem, e.g., the development of a wholly satisfactory resistant variety of host, e.g., wheat varieties resistant to the Hessian fly, Mayetiola destructor. Some chemical control measures approach this goal, as in the use of timbers impregnated with pentachlorophenol for protection against termites and powderpost beetles, or the use of aldrin, chlordane, dieldrin, or lindane as a soil treatment for subterranean termites, especially under concrete slab construction, where protection is given for 5 or more years.

Returning again to Table 2, it is apparent that Type A procedures are basically unsatisfactory because of uncertain and often unpredictable efficacy. Entomological practices based on their use largely represent emergency responses to pest outbreaks. These responses often produce second and third order consequences outside the effects on the target organism which in the long run may outweigh the immediate benefits of the pest control procedure. As an example the area-wide spray campaign with carbaryl against the pink bollworm, Pectinophora gossypiella in southern California in 1967 resulted in destruction of >30,000 colonies of bees vitally involved in pollination.

In moving into an era of pest management, emphasis will be directed at Type B procedures which plan to contain the population densities of pests at levels substantially below those which environmental requisites normally permit, i.e., the economic threshold. These Type B procedures are most immediately available and applicable and involve the least drastic modifications of present agriecosystem strategies. They also contain the most options which will produce the most flexible and manipulatable pest management programs. This is particularly necessary during the transition period from Type A to Type B programs which may require a decade or more.

Of the Type B procedures, II offers again the most logical transition from Type A and may be expected to be the area of choice for the majority of pest management procedures. Procedure II has again the greatest flexibility and the most options and is also best suited to maintaining the very low population densities of target organisms that are demanded in commercial production of high value agricultural commodities. Procedure II will almost always involve the logical, precise, and careful use of insecticides for intervention to destroy large fractions of target pests. Procedure V will also be substantially based on interventions of insecticides as occasional low level suppressants for the target pest population. However, the sophistication of technology necessary is substantially greater than for Procedure II and the developmental time will be longer and the skilled supervision necessary will be of higher order. Procedure VIII requires the utmost sophistication in pest management and long-term study and observation. Intervention here with insecticides will be done under the most carefully controlled circumstances, with very specifically chosen selective insecticides or ecologically selective methods, only when the pest population exceeds economic threshold, as for example after extreme weather variations which totally upset biological balances. This method is less well suited for complexes of pests attacking high value crops grown under highly specialized agronomic practices.

Type C procedures have in common the need to restrict the target organism from available requisites. They offer the most constant effects on population control but also afford decreased flexibility in operation and a necessarily wider scale of practice on a community, regional or commodity basis. Procedure VI is certain to be increasingly studied as we gain greater knowledge of the behavioral stimuli, light, temperature, tactile, pheromones, etc. that are responsible for selection of sites for feeding, oviposition, mating, and shelter of the target pest. Procedure IX represents truly successful host modification or manipulation and essentially solves the pest problem for long periods of time. Chemical treatment of modification of requisites by genetic change or by human intervention will become increasingly useful here. However, the Type C procedures have limited flexibility and are poorly suited to control complexes of pests attacking intensively produced crops. In these situations, Type B methods offer the most feasible solutions to pest management.

Where there are complexes of pests attacking a single crop such as the various deciduous fruits or corn, cotton, or alfalfa, the ultimate aim must be to manage several pests simultaneously and thus to integrate a variety of control procedures into a single manipulated ecosystem. In such endeavors which are just now beginning to appear on the drawing boards, the flexibility of specific insecticidal use appears indispensible. The urgency is to develop technological refinements for the selection and use of presentday insecticides and to develop new compounds and procedures that are specifically designed for pest management programs.

Pesticide Application

The broadscale application of sprays and dusts for insect control is a very inefficient process and estimates suggest that only about 10 to 20% of the insecticides applied as dusts and 25 to 50% of those applied as sprays is deposited on plant surfaces for effective insect control and less than 1% is applied to the insect pests themselves. These data suggest that even under the most optimum conditions, 50 to 75% of sprayed or dusted insecticide is useless for pest control and falls to the ground or drifts away from the treatment area where it becomes an undesirable environmental contaminant (PSAC 1965). This wastage also causes a substantial and unnecessary economic loss.

It seems clear that pest management programs must minimize these wasteful environmental overtreatments which are generally inimical to the pest management concept. This view strongly supports a new look at the role of pesticides in pest management.

A. Seed treatments:

Application of insecticides to seeds before or at time of planting offer the most efficient and concentrated means of protecting the germinating seed and the seedling plant. Such applications are minimal in dosage and least disturbing to the environment. The savings in application costs and in total amount of pesticide are striking. (1) Treatment of seeds of a wide variety of field and vegetable crops with lindane, aldrin, heptachlor, dieldrin, or endrin at about 0.25 oz. (7 g) per acre has given 70 to 95% mortality of wireworms and ensured the production of satisfactory stands at a reduction of over 99% of the conventional dosage of 2 to 3 lb/acre. (2) Use of planter box treatment of corn seeds with diazinon at 1.3 oz (37 g) per 100 lb of seed provides adequate wireworm protection at 0.4 oz (11 g) per acre, a reduction of 98% of the conventional dosage of 24 oz/acre as a 7 in band at planting. (3) Treatment of oats seed with propoxure (Baygon) at 4 oz per 100 lb of seed or 2.8 oz (80 g) per acre provided high kill of the cereal leaf beetle (*Oulema melanopaus*) over 40 to 50 days after planting and gave a reduction of 83% of the conventional application of 1 lb of carbaryl per acre (Ruppel, et al., 1970). (4) Application of the systemic insecticide phorate or disulfoton to alfalfa, sugarbeet, or cottonseed at 4 to 8 oz per acre gave control of aphids, thrips, leafhoppers attacking the seedling plants for several weeks (Reynolds, et al., 1957).

B. Ecologically Selective Applications:

The highly precise placement of insecticide treatment to effect the pest at the most vulnerable ecological niche or in a place most suitable for control has been mentioned previously under *ecologically selective insecticides* with examples such as corn silk or ear treatment for corn earworm larvae control, precise spraying of tree branches for tsetse fly control, or residual house spraying for malaria eradication. For many agricultural pests, sufficient knowledge about microhabitats and ecological behavior is not available to take full advantage of the many possibilities. Meyer and Luckman (1970) have described a pest management program for the European red mite, *Panonychus ulmi*, which uses a single and precise spray of acaricide such as omite, chlorphenamidine or Plictran applied to the periphery of the tree around the tips of the branches to suppress the overwintering mites without affecting materially the predaceous fallacis mite, *Neoseiulus fallacis* which overwinters on the trunks. The fallacis mite increases in numbers by preying on the two- and four-spotted mites, *Tetranychus* spp., which also overwinter on the trunks. The increasing population of predator mite then moves up the tree to control all 3 species of phytophagous mites for the remainder of the season. The predator is very tolerant of the codling moth spray schedule of such insecticides as azinphos-methyl and Gardona. This precisely timed and applied pest management spray program enables a single acaricidal spray to replace 6 or 7 acaricides formerly applied each season.

Summary and Conclusions

1. Concern for the quality of the environment and for balanced ecology of host-pest-natural enemy complexes will dictate major changes in the use of pesticides in the immediate future.

2. Pest management practices will generally involve the logical, precise, and careful use of insecticides.

3. Methods must be developed to use insecticides specifically and selectively and with particular concern for the economic thresholds of pests attacked, so that unnecessary treatments are avoided and the insecticidal intervention at the target site involves maximum ecological selectivity.

4. Insecticides must be selected for use in specific pest management schemes on the basis of overall safety to humans and domestic animals, to nontarget organisms, and to the overall quality of the environment, as well as for specific effectiveness against the target species.

5. The requisites of successful pest management programs will provide an enormous challenge to entomological practitioners who must be equipped with knowledge about host-pest-natural enemy complexes, about the toxicological and environmental properties of insecticides, about the practical economics of risks versus benefits, and with a code of ethics to match the responsibilities which they will assume.

Table 1.--Pest management rating of widely used insecticides.

Insecticide	Mammalian toxicity	Nontarget toxicity				Environmental persistence	Overall rating
		Fish	Pheasant	Bee	Average		
aldicarb	5	3	5	5	4.3	3	12.3*
aldrin	4	4	4	4	4	5	13
azinphos-methyl	4	3	2	4	3	3	10
carbaryl	2	1	1	4	3	2	7
carbofuran	5	2	5	5	4	3	12*
carbophenothion	4	2	4	4	3.3	2	9.3
chlordane	2	3	2	2	2.3	3	7.3
DDT	3	4	2	2	2.7	5	10.7
demeton	5	2	5	2	3	2	10
diazinon	3	2	5	4	3.7	3	9.7
dicofol	2	1	2	1	1.3	4	7.3
dieldrin	4	4	3	4	3.7	5	12.7
dimethoate	3	1	4	5	3.3	2	8.3
disulfoton	5	3	5	2	3.3	3	11.3
Dursban	3	3	3	5	3.7	3	9.7*
endosulfan	4	4	2	1	2.3	3	9.3
endrin	5	5	5	2	4	5	14
ethion	3	2	3			2	7*
EPN	4	2	3	4	3	4	11
Gardona	1	4	1	4	3	1	5*
heptachlor	4	3	4	4	3.7	5	12.7
lindane	3	3	2	4	3	4	10
malathion	2	2	1	4	2.3	1	5.3
methoxychlor	1	3	1	1	2.3	2	5.3
methyl parathion	4	1	5	5	3.7	1	9.7
mevinphos	5	3	5	4	4	1	10
naled	2	2	3	4	3	1	6
oxydemeton-methyl	3	2	4	2	2.7	2	7.7*
parathion	5	2	4	4	4	2	11
phorate	5	4	5	2	3.7	3	11.7

Table 1.--Continued.

Insecticide	Mammalian toxicity	Nontarget toxicity Fish	Pheasant	Bee	Average	Environmental persistence	Overall rating
phosphamidon	4	1	5	3	3	2	9
tetraethyl-pyrophosphate	5	2	5	5	4	1	10
toxaphene	3	4	4	1	3	4	10
trichlorfon	2	1	2	1	1.3	1	4.3
Zectran	4	1	5	5	3.7	2	9.7

*Insufficient data for accurate rating.

Table 2.--Ecological synopsis of pest management.[1/]

INTRINSIC VALUE OF PEST CONTROL

FREQUENCY OF INTERVENTION	A. Unsatisfactory	B. Acceptable	C. Good
Continuing	I. Arbitrary intervention to reduce population curative sprays in cotton, orchards, mosquito control, etc	II. Systematic intervention to reduce population preventive seed, soil, or fruit or ear treatment	III. Manipulation of environment crop or pasture rotation
Periodic	IV. Intermittent intervention to supplement undependable density mechanism mass release of parasites	V. Intermittent intervention to restore population balance ecologically planned selective or spot treatment	VI. Diversion from requisites bait sprays, poison lures
Nil	VII. Permanent unreliable intervention producing undependable control establishment of parasite	VIII. Permanent collective intervention producing dependable control establishment of complex of natural enemies	IX. Permanent intervention to eliminate pest mothproofing, timber treatment

[1/]Modified after Geier (1965).

References Cited

Bariola, L. A., R. L. Ridgway, J. R. Coppege. J. Econ. Entomol. 64: 1280. 1971.

Chadwick, P. R., J. S. S. Breesley, P. J. White, H. T. Matechi. 1965. Bull. Entomol. Res. 55: 411.

FAO Proc. Symposium Integrated Pest Control. Rome. 1966.

Geier, P. W. 1965. Ann. Rev. Entomol. 11: 471.

Huffaker, C. 1962. Hilgardia 32: 541.

Maltby, H. L., F. W. Stehr, R. C. Anderson, G. E. Moorehead, L. C. Barton, J. D. Pashke. 1971. J. Econ. Entomol. 64: 693.

Metcalf, C. L., W. P. Flint and R. L. Metcalf. 1962. Destructive and Useful Insects, 4th ed. McGraw-Hill Book Co., New York.

Meyer, R., W. H. Luckmann. 1970. Proc. 1st Allerton Conf. on Environmental Quality. University of Illinois, College of Agriculture.

National Academy of Sciences. 1969. Insect Pest Management and Control. Publ. 1695. Washington, D.C.

Nishida, T. 1954. J. Econ. Entomol. 47: 226.

Nishida, T., H. A. Bess. 1950. J. Econ. Entomol. 47: 877.

Oatman, E. R. and G. R. Platner. 1971. J. Econ. Entomol. 64: 510.

Pimentel, D. 1971. Ecological Effects of Pesticides on Nontarget Species. President's Office Science Technology.

President's Science Advisory Committee (PSAC) Rept. 1965. Restoring the Quality of our Environment. Washington, D.C.

Reynolds, H. T., T. R. Fukuto, R. L. Metcalf and R. B. March. 1959. J. Econ. Entomol. 50: 527.

Ridgway, R. L., and S. L. Jones. 1969. J. Econ. Entomol. 62: 177.

Roelots, W. H., E. H. Glass, J. Tette, and A. Comeau. 1970. J. Econ. Entomol. 63: 1162.

Ruppel, R. F., J. Valarde and S. L. Taylor. 1970. Michigan State Univ. Res. Rept. 122.

Sacher, R. 1971. Chem. Eng. News Nov. 29, p. 9.

Stehr, F. W. 1970. J. Econ. Entomol. 63: 1968.

Steiner, L. F., W. C. Mitchell, E. J. Harris, T. T. Kozama, M. S. Fujimoto. 1965. J. Econ. Entomol. 58: 961.

Steiner, L. F., G. G. Rhower, E. L. Ayers and L. D. Christenson. 1961. J. Econ. Entomol. 54: 30.

Stern, V. M. and R. Van den Bosch. 1959. Hilgardia 29: 103.

Stern, V. M., R. F. Smith, R. Van den Bosch and K. S. Hagen. 1959. Hilgardia 29: 81.

Tucker, R. K., and D. G. Crabtree. 1969. J. Econ. Entomol. 62: 1307.

Van den Bosch, R. and V. M. Stern. 1962. Ann. Rev. Entomol. 7: 367.

Wallner, W. E. and J. W. Butcher. 1970. Michigan State Univ. Ext. Bull. 353.

Wallner, W. E. and J. H. Hart. 1971. Michigan State Univ. Ext. Bull. 506 rev.

WHO Chronicle. 1971. 25(6): 209.

BIOLOGICAL CONTROL OF WEEDS: INTRODUCTION, HISTORY, THEORETICAL AND PRACTICAL APPLICATIONS

Kenneth E. Frick
Research Entomologist
Southern Weed Science Laboratory, ARS, USDA
Stoneville, Mississippi

Introduction

The biological control of weeds may be defined as the use of plant-feeding organisms or diseases to reduce the population of a plant species that has become weedy. The objective is simply to reduce a weed to a status of little or no economic significance (Huffaker, 1958). The employment of biological means does not lead to eradication. For biological control to be continuously successful, small numbers of the host must always be present to assure the survival of its natural enemies and the natural enemies must always be present to attack those hosts that survive and to prevent an increase in their numbers. When suitable and effective phytophagous agents are available, this method is inexpensive and permanent, involving no repetitious treatments or corrective measures year after year (National Academy of Sciences, 1968).

Weeds are plants that have become overly abundant in the wrong places. Fortunately, like most plants including crops, weeds generally have natural enemies. However, unlike in the case of crops, forest trees, and ornamentals, that man wishes to protect from their natural enemies, it is the purpose of biological control of weeds programs to increase, manipulate, and otherwise encourage the phytophagous natural enemies of certain weedy plants.

Insects constitute the largest group of natural enemies of weedy plants. Among insects that have preved effective in the biological control of weeds are species of Hemiptera, Homoptera, Thysanoptera, Coleoptera, Lepidoptera, Diptera, and Hymenoptera. Also, mites have been occasionally utilized in biological control, e.g., a spider mite was introduced into Australia to feed on prickly-pear cactus (Mann, 1969). Fish, snails, ducks, and the manatee have been used in the control of aquatic weeds. Weeds may be damaged by a wide variety of pathogens (fungi, viruses, bacteria) or by parasitic plants (dodder, witchweed). Many of these are obligate parasites that develop only on specific live host plants, but the degre of host plant specificity is unknown for many insects, pathogenic and other biological organisms. Several pathogens are presently under study and their host plant ranges are being determined.

Because insects have been the group of phytophages most studied and used on weedy plants, this paper will be confined to a discussion of them. Insects are of importance because of their great variety and numbers and frequent (1) high degree of host specialization, (2) intimate adaptation to their host plants, and (3) availability of a range of natural enemies suited to particular ecological situations (National Academy of Sciences, 1968). When "insect" is used herein in a generic sense, i.e., in referring to natural biological control agents in general, "phytophagous organisms" is always implied.

Established weeds are difficult to eradicate. Thus, methods must be developed to provide for coexistence with them. All methods should be considered, such as preventive, cultural, chemical, and biological, and no method should be thought of as the only solution. Insects can often reduce weed numbers to subeconomic levels or make the weed amenable to the more artificial control methods, such as chemical or cultural. Biological control is a natural method of control and should be considered early in a weed control program. Even if only partially effective, it can become a part of an integrated control program. All too often biological methods are considered as a last resort, such as after a weed has escaped over large acreages, or has spread into inaccessible areas, for example range and forest lands, where chemical or cultural methods are out of the question due to the terrain and/or the cost in relation to the value of the land.

Biological control has been applied most often against alien plants that have become weedy (Zwolfer, 1968). Indeed, the weediness of plants introduced from other geographic regions is frequently due to the escape of these plants from their phytophagous enemies of their native environment. The deliberate introduction of such enemies, each cultured in the laboratory so that their own natural enemies (parasites, predators, diseases) are eliminated, has, in a number of instances, resulted in great reductions in the abundance of certain alien weeds.

It was only a few years ago, in 1954 to be exact, that the biological control of weeds was considered to be very limited in scope and as being applicable only to alien plant species (Williams, 1954). It was generally considered as (1) a relatively minor adjunct of better established forms of control, (2) being seldom applicable, and (3) almost a last resort when all other means had failed (Wilson, 1964). But the growing number of successes in biological control and the ever-increasing numbers of phytophagous insects showing a high degree of host specificity, are promoting awareness of the possibilities of deliberately extending the geographic ranges of those insects whose food habits conform to our desires (Wilson, 1964).

Although biological control is continually increasing in importance as a means for control of alien weeds, its scope is steadily expanding, and the full range of possible applications of the use of insects for control of weedy plants has yet to be determined.

Biological control is by its very nature selective. That is, the encouragement of introduced or native insects having feeding and/or ovipositional habits narrowly restricted to a target weed makes biological control ineffective against a complex of weeds, unless host-specific insects are provided for each species in the complex. Biological control achieves its greatest utility where a single species of weed occurs in dense stands, a condition that permits a rapid increase in its natural enemies with a resulting quick decline in the weed's abundance. But, after the phytophages have reduced their host to non-economic levels, these same natural enemies must then be able to seek, find, and attack the small numbers of their host plants that continue to appear. The employment of such effective natural enemies has occurred several times throughout the 70-year history of the biological control of weeds. These cases have been thoroughly documented and some of them will now be discussed as part of the history of the biological control of weeds.

History

The use of insects to reduce the abundance of a weedy plant is not new. The first attempt appears to have been prior to 1902 by planters in Hawaii who were greatly troubled by the rapid spread of lantana, *Lantana camara* L., which had been planted as an ornamental beginning in 1860. A scale insect, *Orthezia insignis* Browne, had been accidentally introduced from Mexico with the plants and it was observed that the insect injured and even killed plants in some localities; so the planters spread scale-infested branches through lantana-infested areas. However, the extent of lantana in Hawaii was described in 1902 as follows: "When one looked over miles of country covered with almost continuous growth of the plant, every bush in its season with masses of flowers or fruit, one might well doubt whether anything could be done to check a growth that had already acquired so strong a hold on the land" (Perkins and Swezey, 1924).

However, an entomologist, Albert Koebele of the Hawaiian Sugar Planters' Association, while collecting in Mexico in 1898, noted that lantana seeds were damaged by the larvae of a small agromyzid fly, *Ophiomyia lantanae* (Froggatt), and it was felt that the profuse seeding of lantana might be prevented if the fly could be established in Hawaii. By 1902 his idea was officially approved and he returned to Mexico to study lantana insects. He recorded about 75 insect species feeding on lantana and an attempt was made to introduce 18 into Hawaii. Primarily because of the slow rail and boat transportation of that period few of the insects reached Hawaii alive and only 8 species became established. No tests were conducted with these insects in order to determine their host plant ranges before they were released. It is fortunate that of the 8 species, only 2 attacked other plants: one of them fed on 2 shrubs related to lantana and the other fed to some extent on eggplant, in pepper pods, and on an ornamental borage (Perkins and Swezey, 1924).

Control was moderately successful in the drier areas of the Islands, primarily because of the destruction of buds, flowers, and seeds by 6 of the 8 species of insects. However, greater control was desired and, with the development of air transportation, the importation of lantana insects was resumed in Hawaii by the Hawaiian Department of Agriculture after a lapse of 50 years.

By 1966, 15 introduced species of insects had been established in Hawaii. One, *Hypena strigata* F., a noctuid moth, was found on a related plant in East Africa while the others were from the Western Hemisphere. In many of the drier areas of Hawaii (less than 30 inches rainfall), lantana has been suppressed or eliminated completely as a result of defoliation by *Hypena* in the cooler months and, during the hotter months, defoliation by the lantana lace bug, *Teleonemia scrupulosa* Stal (Waterhouse, 1967). The latter is one of the original 8 species that became established in Hawaii. In the wetter areas (30-80 inches rainfall), 3 introduced beetles look promising and hopes are high that the complete biological control of lantana in Hawaii finally will be successful (Waterhouse, 1967). The larvae of one beetle, the cerambycid, *Plagiohammus spinipennis* (Thomson), bore into lantana stems while the larvae of the other 2, *Octotoma scabripennis* (Guerin-Meneville) and *Uroplata girardi* Pic, both chrysomelids, mine in the leaves.

Lantana is a tropical New World plant distributed from Mexico to northern Argentina that has become a pest in several parts of the world that have mild climates. Eastern Australia, for example, has some 10 million infested acres in the coastal areas where lantana is essentially a weed of misused or unused land. As of 1966, Australia had introduced 9 insects species for lantana control, 7 of which became established. However, only the lantana lace bug has caused significant damage and Australian investigators are actively studying other promising insects in Hawaii (Waterhouse, 1967) and in Mexico and South America (Harley and Kassulke, 1971).

One of the greatest successes in the biological control of weeds is that of the pricklypear cacti (*Opuntia* spp.) in Australia. The acreage covered by *Opuntia* spp. in 1925 had reached a maximum of 60 million acres, of which half was dense growth completely covering the ground to the exclusion of all grass and herbage. The search for insects got underway in late 1920 under the auspices of the Commonwealth Prickly Pear Board organized by the governments of the

Commonwealth of Australia and the states of Queensland and New South Wales (Mann, 1969). Studies were conducted primarily in the United States, Mexico, and Argentina but the West Indies, Central America, and 8 other South American countries were included.

A total of 150 species of insects were discovered whose life cycles were restricted to species of *Opuntia*. Fifty of these were shipped to Australia, where 12 were released and became established. The spider mite, *Tetranychus opuntiae* Banks, was accidentally introduced about 1922 from Texas and had given promising results by 1927 on the 2 most common species of pricklypear, *Opuntia inermis* DC. and *O. stricta* Haworth (Dodd, 1940). However, in that year, the tunneling caterpillar, *Cactoblastis cactorum* (Berg), from Argentina became established (Wilson, 1960) and liberations of additional species were stopped when it became apparent that larval feeding of this pyralid moth would result in control of both *O. inermis* and *O. stricta*.

Such a vast project involved many specialists, including 9 that studied in various countries of the New World, and 20 years of time. But by 1940, 95% of the land was reclaimed and the pricklypears had been reduced to incidental status and the strong flying *Cactoblastis* moths and their eggs were found everywhere. Except for higher elevations, the 2 primary species of pricklypear have remained under control by insects in Australia (Dodd, 1940). It has not been necessary for many years to rear *Cactoblastis* in the laboratory or to distribute moths or their larvae. However, in the cooler regions, *Cactoblastis* is ineffective, and furthermore some minor species of *Opuntia* are not attacked by *Cactoblastis*. For these situations, species of *Dactylopius*, cochineal scale insects, are extensively used for control and are systematically distributed each season in the infested areas, giving varying degrees of control (Wilson, 1960).

Cactoblastis and species of dactylopiid scales have since been introduced into South Africa, India, and Hawaii, where partial or completely successful control of several species of *Opuntia* has been achieved. The most recent success has been in the Leeward Islands, where a number of *Opuntia* spp., primarily *O. dillenii* (Ker-Gawler) Haw., *O. lindheimeri* Engelmann, and *O. triacantha* (Willd.) Sweet, have been brought under control (Simmonds and Bennett, 1966).

St. Johnswort, *Hypericum perforatum* L., commonly called Klamath weed on the west coast of the United States, is the first weedy plant for which insects were imported into the United States. In 1952 about 2.3 million acres were infested in northern California (Murphy, et al., 1954), about 1.25 million acres in Oregon, and 1.4 million acres in Washington, Idaho, and Montana (Huffaker, 1959) and *Hypericum* was also a problem in southern British Columbia.

Through the cooperative efforts of the U.S. Department of Agriculture and the University of California, a joint research team began to study the possibilities of biological control for St. Johnswort in 1944. The first insect liberated in the field was the northern European leaf beetle, *Chrysolina hyperici* (Forster), released in the spring of 1945 in northern California. This was followed in February, 1946 with the release of the southern European *C. quadrigemina* (Suffrian) (Holloway, 1958). Because World War II had been in progress, these insects had been obtained from Australia, where a biological control program for St. Johnswort had begun in 1920. These 2 beetles, plus 35 other insects specific to *Hypericum*, had been found and studied in France as candidates for importation into Australia (Wilson, 1943).

In California, success was immediate and rapid and it soon became evident that *C. quadrigemina* had a much greater rate of increase than *C. hyperici* (Holloway, 1958). In the fifth year after the initial 1946 release of 5,000 beetles in Humboldt County, 3 million beetles were collected for redistribution. Generally in the third year after release in a locality, *C. quadrigemina* had become so numerous that larval and adult feeding resulted in the death of all host plants.

Three more insects were later introduced directly from France, rather than from Australia. These were (1) the flatheaded borer, *Agrilus hyperici* (Creutzer), whose larvae bore into the rootcrown and tap root, (2) a gall midge, *Zeuxidiplosis giardi* (Kieffer), both introduced and established in 1950, and (3) a third species of *Chrysolina*, *C. varians* (Schaller), released in 1952; the latter did not become established in California (Holloway, 1958).

The biological control of St. Johnswort in California has been a complete and continuing success, largely because of the foliage-feeding larvae and adults of *C. quadrigemina* (Huffaker, 1959). However, the root- and crown-boring *Agrilus* also shows excellent ability to destroy dense stands of *Hypericum*. These 2 species complement each other in that the females of *C. quadrigemina* prefer not to lay eggs in shady areas, whereas the females of *Agrilus* freely lay their eggs in shaded locations (Holloway, 1958). Because of these insects, St. Johnswort has been reduced to less than 1% of its former abundance and no longer constitutes any problem whatever in California (Huffaker, 1959).

Such completely successful control of *Hypericum* as achieved in California has not been duplicated elsewhere. Both species of *Chrysolina*, and in some cases *Agrilus* too, have been introduced into Australia, the Pacific Northwest, British Columbia, and Chile, but the degrees of success have been quite variable (Huffaker, 1967). In Australia, control of infestations of St. Johnswort in cooler and shadier locations and those on rather extensive gravelly soils has been poor, whereas excellent control has been obtained on infestations in pastures in the open sun, particularly in West Australia (Huffaker, 1967). In British Columbia insects have been sought by investigators of the Canada Department of Agriculture, particularly from northern Europe, and the geometrid moth, *Anaitis plagiata* (L.), was liberated in southern British Columbia in 1967 (Harris, 1967).

Although the lantana program started in 1902 and the prickly-pear and St. Johnswort programs in 1920, satisfactory control has not been obtained in all geographic areas for any of the 3 weeds. However, spectacular and continuing control has been achieved in certain areas for each weed: in parts of Hawaii for lantana, an estimated 95% of 60 million acres in eastern Australia for prickly-pear, and an estimated 99% of 2.3 million acres in California for St. Johnswort.

These and several smaller, but definite and self-perpetuating, successes with other weedy plant species, have resulted in an overall increase in interest in and employment of biological methods to control certain weedy plants, including aquatic weeds. There are too many weed projects to enumerate them here but a number of reviews exist that include some or all of the weeds that are currently under study (Holloway, 1964; Huffaker, 1959; National Academy of Sciences 1968; Wilson, 1960; 1964; Zwolfer, 1968).

The continuing expansion in the use of biological agents for control of weeds has resulted in an increase in the numbers of countries and agencies involved. The largest involved agency is the Commonwealth Institute of Biological Control (CIBC), which has research stations in South America, Europe, Africa, and Asia. In addition to studies on the biological control of insect pests, this Institute conducts studies on the insects associated with species of weedy plants; such studies are requested and paid for by member nations of the Commonwealth and others, e.g., the University of California. The U.S. Department of Agriculture also is actively engaged in this field of study through its biological control of weeds laboratories at Albany, California, Stoneville, Mississippi, Gainesville and Fort Lauderdale, Florida, Buenos Aires, Argentina, and Rome, Italy. Weed projects are also supported by funds provided under Public Law 480, The Agricultural Trade Development and Assistance Act, with studies being conducted at present in Israel, Yugoslavia, Poland, Pakistan, India, and Egypt.

Other agencies active in the biological control of weeds include the Hawaiian Department of Agriculture and the University of California. The latter has 3 investigators studying the biological control of weeds, one at Albany, one at Riverside, and one at Rome, Italy. The Canada Department of Agriculture, through its Research Institute, Belleville, Ontario, works cooperatively with the CIBC in studying and obtaining weed insects for release throughout Canada. The Australian Commonwealth Scientific and Industrial Research Organization at present has a Biological Control Unit at Montpellier, France. The Republic of South Africa also has an active biological control of weeds program. The USSR has recently become interested in the biological control of several alien weeds; biological control of weeds investigations are part of the All-Union Plant Protection Institute at Leningrad.

The successes, including partial ones, in the biological control of weeds since the first organized importation of insects into Hawaii in 1902 have recently been listed (National Academy of Sciences, 1968). These successes, which involve 10 genera of plants, have occurred on 8 islands and in 6 countries located in 5 continents. The insects that were employed in these projects originated from 11 countries located in 4 continents. Thus, the biological control of weeds can now be considered as worldwide.

Theoretical and Practical Applications

"The climatic, edaphic [soil], and biotic factors that characterize an environment, provide a variable though essentially stable matrix within which the responses of all plants, including weeds, determine their occurrence, abundance, range, and distribution." (National Academy of Sciences, 1968). Important climatic factors include light, temperature, water, wind, humidity, and the seasonal changes in these. Edaphic factors include structure, soil water, aeration, temperature, pH, fertility, and cropping systems as well as slope and exposure. Both plants and animals are among the biotic factors that affect plant growth. Plants, including weeds and crops, compete with each other for available resources including space. Animals of primary importance to plant growth are grazing animals, insects, the soil fauna, and man himself.

Of these 3 factors (climatic, edaphic, and biotic), the easiest to manipulate is the biotic (Andres, 1966b). This has been done inadvertently in the past in the spread of plants from one region to another. Many of our worst weeds are of foreign origin and their aggressiveness in their new country is often, but not necessarily, caused by the absence of their phytophagous enemies (Huffaker, 1959). The aggressiveness of an alien weed in a new area may also be due to more favorable climatic and/or edaphic conditions or to relative freedom from competition with other plants. However, the fact that many alien weeds have no or only a few insects specifically feeding on them in their new homes as compared to their country of origin, is the basis upon which the study of the biological control of weeds rests. Theoretically, the introduction of host-specific phytophagous insects, minus their own complements of natural enemies (parasites, predators, diseases) should reduce the abundance of an alien weed in a typical host-parasite interaction. In practice, this has happened in a number of instances, including the 3 examples previously described.

<u>Methods Used in the Biological Control of Weeds</u>.--Biological control by introduction of host-specific insect enemies of alien weeds has received the greatest attention from investigators and will probably continue to do so because (1) past successes have been encouraging and (2) alien weeds are numerous on most continents (for example, Zwolfer (1968) noted that of 107 noxious Canadian weeds, 78 had been introduced from Europe or Asia). However, there are 2 other methods that can be used to control a weedy plant by biological means. These are biological control by (1) conservation and (2) periodic release and/or distribution of phytophagous natural enemies (Andres, 1966b).

Biological control by conservation is applicable primarily to native weedy plant species. Although alien weeds are frequently fed upon by indigenous or cosmopolitan insects, these insects are oligophagous or polyphagous in their feeding habits and most are casual feeders and generally have no impact on the target weed. For example, along the Pacific Coast, 42 insect species have been found to feed on the alien tansy ragwort, *Senecio jacobaea* L. (Frick, 1964; Frick and Hawkes, 1970; Frick, 1972). The impact that 3 *Eupithecia* spp., native geometrid moths, and one native tephritid fly, *Paroxyna genalis* (Thomson), have on seed production has been determined: these 4 species destroyed 4.8% and 5.8% of the seeds in 1961 and 1962, respectively, at Fort Bragg, California (Frick, 1964). The damage that any of the remaining species inflicts on *S. jacobaea* has yet to be determined.

The conservation of native phytophagous insects that feed on native weedy plant species will primarily involve a reduction in the numbers of native parasites, predators, and diseases that attack them and keep their numbers from remaining high generation after generation. This technique would have the same effect as importing alien insects free of the natural enemies that hold each species in check in its native region. Conservation of phytophagous insects is one aspect of weed control that has received little attention. Their conservation and increase could prove to be useful in the control of certain weedy plants.

The second method of biological control, that of the periodic release and/or redistribution of natural enemies, appears to have been applied only in the case of *Opuntia* and the dactylopiid scales in Queensland and New South Wales, Australia. Each year these cochineal insects are systematically and artificially distributed to *Opuntia* species that are not heavily attacked by *Cactoblastis* or to areas that are too cool for *Cactoblastis* to be fully effective (Wilson, 1960).

What appears to be the first attempt to rear large numbers of a phytophagous insect on an artificial diet for periodic release in the field against its host weeds is now underway at the USDA Southern Weed Science Laboratory, Stoneville, Mississippi. It was found last winter that a native tortrid moth, *Bactra verutana* Zeller, specific to plants in the family Cyperaceae, could be artificially cultured in the laboratory. This insect occurs locally on the native yellow nutsedge, *Cyperus esculentus* L., and the introduced purple nutsedge, *C. rotundus* L. In the numbers at which this insect is found in nature at Stoneville, it is quite ineffective in adequately controlling the nutsedges. Not only are *Bactra* populations slow to increase early in the growing season, but they are attacked by several species of parasitic insects.

Should the method of mass culture and what might be termed inundative releases of this phytophagous insect prove to be successful in suppressing its weedy host plants, this technique could well be given increased importance in biological control of weed programs.

There are many theoretical and practical problems associated with the biological control of weeds. These cannot be dealt with thoroughly in this paper and, for further information, the reader is referred to a number of papers which deal with the fundamentals and ecological bases for the biological control of weeds (Huffaker, 1957, 1958, 1962, 1964; National Academy of Sciences, 1968; Wilson, 1964).

Procedures Used in the Biological Control of Weeds.--The procedures to be described are those used in the third method of biological control, introduction of natural enemies, by far the most common approach so far used in solving a weed problem with biological methods. There are several descriptions of these procedures that may be consulted for additional information (Huffaker, 1957; National Academy of Sciences, 1968; Zwolfer, 1968; Zwolfer and Harris, 1971).

Preliminary Considerations.--There are many factors to consider before deciding that biological means should be used to control a certain weed. Some of these are:

1. Eradication may be desired, for example because a weed is toxic to livestock. If so, biological control may have to be ruled out. (However, reduction of stand is adequate protection of livestock for some poisonous weed species).

2. Biological control is selective and aimed at one species only. If the target weed is widely scattered throughout a complex of other weeds, biological control at best would suppress only one species. However, if the weed is devastatingly abundant and aggressively spreading in dense stands, biological control may be strikingly successful.

3. Biological control is a relatively slow method. In the past an average of 5 years has elapsed between initiation of a project and the importation of the first enemy insect. In addition the insect enemies have not been imported and released in large populations, but in what have been termed inoculative numbers (Andres, 1966b) and therefore, it may require several more years following release before significant damage occurs. For example, Holloway (1958) reported that death of St. Johnswort stands occurred in the third year after beetle introduction and Hawkes (1968) reported essentially complete defoliation of tansy ragwort infestations by the larvae of the arctiid moth, *Tyria jacobaeae* (L.) in the fifth year following release. Thus, if there is an urgent need for quick control, biological methods should not be given first priority.

4. It is commonly considered that there are risks involved in importing plant-feeding organisms and that biological control should be resorted to only if other methods have failed or the weed covers large areas of land with (1) such low value, i.e., range land, that chemical treatments are too expensive or (2) such rough terrain that cultural or chemical treatments are precluded. However, regarding the risks involved in the introduction of phytophagous organisms, Wilson (1964) has succinctly said: "No errors have yet been made, and none should be made." He feels that excessive anxiety has been felt about the matter of risks. As scientific understanding of the factors involved in host plant specificity continues to increase, the danger of introducing insects that will pose a threat to desirable plants is continually lessened.

5. Biological control cannot be limited in area, like chemical and cultural treatments, because living organisms will disperse from the areas onto which they are liberated. Because of this it must be agreed by all agencies and groups involved that the target plant is a "weed" (defined as any useless, troublesome, or noxious plant, especially one that grows profusely). Our weedy plants extend into Canada quite frequently and less so in Mexico. If the weed does have such distribution, appropriate agencies in these countries, as well as each state in which it occurs, should be notified of the intent to introduce alien biological control agents. In some cases not all parties will agree that a weedy plant should be controlled by natural organisms and a conflict of interest arises. For example, our native pricklypears are used as emergency forage during times of drought in the southwest. Also, the fruit and cladodes of several species are used as food in Mexico and, to a small extent, in the United States. Because of these uses of *Opuntia* spp., *C. cactorum* has not been introduced into the United States. However, repeated requests for *Cactoblastis* larvae were made up to 1958, because the insect was desired for Santa Cruz Island, 20-25 miles off the coast of southern California (Goeden, et al., 1967). Instead, because it was feared that the moths would reach the mainland, a scale insect, *Dactylopius* sp., native to southern California but absent in Santa Cruz, was introduced from Hawaii in 1951 without its own natural enemies. By 1966, partial to substantial control of the 2 *Opuntia* species has been achieved (Goeden, 1970; Goeden, et al., 1967).

Domestic Surveys.--Once it has been decided to attempt biological control of a weed, two surveys should be undertaken: (1) to glean from literature and specialists as much information pertaining to the target weed and its natural enemies as possible and (2) to determine the organisms feeding on the target weed in the area for which control is desired. The first has been briefly outlined: "Prior to beginning the actual search for biological control agents, the following basic information on a weed should be established, if possible: (1) taxonomic position, biology, ecology, and economic importance; (2) native geographic distribution; (3) total present distribution; (4) probable center of its origin and that of its close relatives (section, genus, tribe); (5) coextensive occurrence of related species; (6) occurrence of related and ecologically similar species in regions where the weed does not occur, but where exploration for any enemy agent seems desirable; and (7) the literature record of the weed's natural enemies." (National Academy of Sciences, 1968).

The other survey is concerned with determining what native insects are already present on the target weed and the extent of their damage. Such a study is a necessary preliminary and can save time and funds because effective local insects may be present or some of the natural enemies from the native home of the target weed may have been already imported with their host. Scotch broom presents a good example because 4 European phytophagous insects were found naturalized on this shrub in California (Waloff, 1966). A fifth European insect, an oecophorid moth, *Agonopterix nervosa* (Haworth) [= *A. costosa* (Haworth)], was not included (Frick, unpublished data). In British Columbia, Waloff (1966) reported the presence of 10 European species on Scotch broom, and along the east coast of the United States, a European seed beetle, *Bruchidius ater* (Marsham) is naturalized on this host (Bottimer, 1968). In the future, should additional insects be desired for the biological control of Scotch broom, these species could be neglected during foreign exploration. However, their transfer from one region in North America to another could prove useful, for example, the release of *B. ater* in California.

Foreign Exploration.--Based upon the information gained from the literature and from correspondence or discussions with specialists, one or more geographical areas are chosen for a survey of the target plant, plants related to it, and the insects and other natural enemies associated with these plants. The records in the literature are usually sparse, because, aside from European countries that have a long history of scientific endeavor, records of insects from non-crop plants can be assumed to be very incomplete. And even in Europe, biological control of weeds investigators frequently discover new undescribed insect species. These discoveries are most encouraging because it indicates that some of the insects are so intimately associated with their host plants that they have remained unnoticed by economic entomologists and even by systematists since they are associated only with non-economic plants. For example, recent studies of the insects attacking Mediterranean sage, *Salvia aethiopis* L., broadened naturally to include several closely related *Salvia* spp. in 5 countries. Weevils in the genus *Phrydiuchus* were always found associated with species of *Salvia* in France, Italy, Greece, Turkey, and Iran (Andres, 1966a; Andres and Rizza, 1965). When the weevils were identified, a taxonomic revision of the genus *Phrydiuchus* became necessary because 2 new *Phrydiuchus* species were found (Warner, 1969). [These were *P. spilmani* Warner, reared from *S. verbenaca* L. in France and Italy and *P. tau* Warner, reared from *S. aethiopis* in Turkey but also known from eastern Europe and Iran. Two other species of *Phrydiuchus* had previously been known: *P. topiarius* (Germar), reared from *S. pratensis* L., but associated with 3 other *Salvia* spp. in central and eastern Europe, and *P. speiseri* (Schultze) from *Salvia* in central Europe (Andres, 1966a)].

During the foreign surveys, as much attention as possible should be given to plants related to the target weed. The value of such a broad approach is exemplified by the thistle tribe Cynareae in the family Compositae. Cooperative studies were conducted by the European Station of the Commonwealth Institute of Biological Control, Delemont, Switzerland, and the U.S. Department of Agriculture

Biological Control of Weeds Laboratory at Rome, Italy, on insects attacking Canada thistle, Cirsium arvense (L.) Scop., and musk thistle, Carduus nutans L. (Zwolfer, Frick, and Andres, 1971). A total of 70 species of thistles were surveyed in western Europe, including the crops safflower, Carthamus tinctorius L., and artichoke, Cynara scolymus L. A total of 17 species of weevils in the genus Larinus were found to be associated with these 70 thistles. Some Larinus species were restricted to a single plant species while others had quite broad host plant range attacking thistles in 3 or 4 genera. Expanding the surveys from insects attacking the 2 target thistles to include 70 species in 13 genera in 4 subtribes, helped to prevent possible errors that could have been made in estimating the potential value or specificity of some species.

Areas where the target plant is in low numbers, rather than where the species is abundant, should be concentrated upon. This is based on the assumption that effective and/or efficient natural enemies are present where the plant is relatively scarce. Although there may also be only small numbers of insects present, their numbers may be low because of the few host plants available as food. However, as many geographical and ecological situations as possible should be surveyed since not all of the natural enemies will be found in every possible situation that will support the target plant. But, when choices are available, natural enemies should be sought from climatic zones similar to that in which the plant is a weed in its new home.

In outlining the early obstacles to be overcome by a newly introduced insect species, Remington (1968) considered that the severest threat to successful establishment would theoretically come from oligophagous parasitic, predaceous or phytophagous insects. In the case of insects introduced for control of a weed, any potential danger of massive destruction of the target weed by a competing native phytophagous insect should be known as a result of prerelease surveys. However, subsequent attack of the introduced insect by native parasites and predators of native phytophages has been a serious threat to establishment or effective control in a number of instances. The next most dangerous threat to an introduced species would be the climate of the new environment into which the insects are to be released. Biological control workers have given little attention to the possibility of postliberation adaptation; instead they usually try to ensure that the individuals to be released are from a population preadapted to the new environment. Much emphasis has been placed on the climatological and ecological similarities between the native and projected environments. These have been termed "ecological analogs" (Wilson, 1965).

An example of this is given by Harris, et al. (1969), who correlated the effectiveness of the introduced insect species attacking St. Johnswort in British Columbia with aridity and moisture indices. Chrysolina hyperici was generally effective in the moist subhumid region and C. quadrigemina in the dry subhumid region. The different moisture requirements for these 2 species compared favorably with those reported earlier in Rumania. However, neither beetle was effective in the semiarid region of steppe-like grasslands. Thus, an insect adapted to attack St. Johnswort in dry situations, such as rocky ground and open sandy places, was sought in Europe. The moth, Anaitis plagiata, appeared to be suitable and was released in dry, open, grassy sites in 1967. It has not been liberated long enough to prove its value and ability to survive (Harris, et al., 1969).

The choice of which organisms to select for further study as potential biotic control agents is not an easy one to make. However, safety to desirable plants is the first consideration or, as Williams (1954) put it: "The critical phase of biological control work against weeds is the selection of species that will not harm other plants, or at least useful plants. All other considerations are subordinate, and a suitable species for introduction into a country against a weed is one that is safe to introduce, irrespective of its other characteristics."

Which kind of insect would eventually prove to be a significant agent of control is difficult to determine a priori. Insects that have been highly effective in the biological control of weeds include a stem-borer (*Cactoblastis cactorum*), plant suckers (*Dactylopius* spp.), a leaf-feeder (*Chrysolina quadrigemina*), a gall insect (*Procecidochares utilis* Stone), and a seed-feeder (*Eurytoma* sp.) (Wilson, 1964). To these should be added the root- and rootcrown-borer, *Agrilus hyperici*, which can be very effective in shaded situations (Holloway, 1958; D. M. Maddox and K. E. Frick, unpublished observations). It is generally agreed that the ability of an insect to control a plant is not readily evident, and may require much study for its determination. The fact that no one anticipated the collapse of vast areas of pricklypear within a few years after the establishment of *Cactoblastis* in Australia (Wilson, 1960), is an example of the difficulty in making accurate prejudgments; until *Cactoblastis* began to destroy the 2 primary pest pricklypears, plans to establish an insect complex of many species were continuing.

However, there are 2 criteria of primary importance in choosing candidate insects for further study: (1) a narrow range of host plants, none of which can be crop plants, and (2) an attack by the plant-feeding stages (larvae, nymphs, adults) of growing tissues vital to the plant rather than tissues of little importance to plant growth, such as senescent foliage or pith.

An additional consideration of importance is the rate of population growth of the candidate species. This statistic, *r*, is called the intrinsic rate of increase and is usually expressed as growth per unit of time (Dingle, 1972). Factors included in the rate of population growth are survivorship, fecundity, and developmental rate, which would include the average number of generations per year. An attempt to effectively measure *r* should be undertaken for each species appearing to have potential value in a weed control program.

Another decision to be made is whether to concentrate upon a single species or to select 2 or more species that might work well together and not directly compete with each other. For example, the defoliation of lantana by *Teleonemia* during the warmer months and by *Hypena* during the cooler months has brought about significant suppression of lantana in parts of Hawaii. In another instance, *Chrysolina* is primarily responsible for maintaining the low populations of St. Johnswort in California, but *Agrilus* also contributes by attacking plants in the shaded locations generally shunned by *Chrysolina*. In a third instance, the flea beetle, *Longitarsus jacobaea* (Waterhouse) was chosen to complement *Tyria*, whose larvae defoliate tansy ragwort plants during the summer (Hawkes, 1968). *Longitarsus* larvae bore in the rootcrown of ragwort in the fall, winter, and spring (Frick, 1970). *Longitarsus* was released in 1969 and has not had time to show its potential.

Even though a decision is made to intensively study one or more kinds of natural enemy, continued survey is desirable. Additional species may be found and the population trends and significance of damage of the selected insects should be followed in nature. Also, damage caused by other organisms not selected for further study should be observed. It is entirely possible that species previously evaluated as insignificant may be more important than originally believed.

Biological and Host Plant Specificity Studies.--Before a phytophagous insect is acceptable for importation, a great deal must be known about its annual life history. In addition, the plants that it selects in nature to feed and/or oviposit on, and the plants that it will feed and/or oviposit on under stress in the absence of its host plants, are very important and practical pieces of information that must be obtained. This is because, if successful, an "explosive" stage is reached in biological control in which hoards of the insect enemy must either starve or disperse as their feeding causes dramatic collapse or defoliation of their host plants, especially if these plants occur initially in large "monoculture" situations. This phase has been documented for a number of weeds including St. Johnswort in California where it occurred in the fifth year following initial release of 5000 beetles. During that year (1950), 3 million adults were collected for distribution (Holloway, 1958). Davies and Greathead (1967) reported that the lantana lace bug reached the "explosive" stage in the sixth year following release in Uganda. In the case of tansy ragwort, it was in the seventh year that the population "explosion" of *Tyria* caterpillars took place; as many as 100 *Tyria* larvae per tansy ragwort plant were found near Fort Bragg, California. When their food supply was depleted, the *Tyria* larvae crawled away searching for their host plant, ignoring other plants in their path except for occasional plants of ornamental florists cineraria, *Senecio cruentus* DC., in home gardens, which they consumed, according to reliable observers. Many larvae undoubtedly died of starvation (Hawkes, 1968).

The crucial aspect in host plant specificity lies in the capacity of a candidate insect to breed on economically important plants. This depends on the behavioral, including ovipositional, characteristics of the adults, the yearly life cycle of the insect as it is synchronized to the phenology of its host plants vs. that of economic plants, as well as the food requirements of the adults and larvae.

The food plants acceptable to mobile larvae or adults, particularly under stress, must be determined in order to prevent introduction of a species which will assume a pest status by damaging useful plants on which it will not reproduce (Wilson, 1964). It is the function of the testing program to guard against such possibilities. Probably the only important damage caused by an imported insect is that of *Teleonemia* on sesame, *Sesamum indicum* L., in Uganda (Davies and Greathead, 1967). *Teleonemia* was released in Uganda in 1960 and by 1965 had reached the "explosive" stage where there was a massive population buildup; defoliation of lantana was complete within a radius of 2-3 miles and the lace bugs were spreading naturally over a wide area. It was then that large numbers, up to 5-10/leaf, appeared on sesame. The plants were defoliated and fewer capsules were produced than in blocks of sesame that had been sprayed for control of other pests. Eggs were laid on sesame but nymphal mortality was high. Although some adults were produced on sesame, no eggs were laid by these individuals (Davies and Greathead, 1967). Harley and Kassulke (1971) noted that sesame was not among the 50 plants previously tested with *Teleonemia* in Australia, Fiji, and India. However, of interest here is the fact that lantana (family Verbenaceae) and sesame (family Pedaliaceae) are in the same order Tubiflorae (Fernald, 1970) and therefore are somewhat related systematically.

Behavior under stress in artificial laboratory conditions should not be the sole basis for rejection of a candidate species. For example, in testing the lantana stem-boring beetle, *Plagiohammus spinipennis* (Thomson), larvae received from Mexico were transferred to holes bored in sugarcane stems, in which the larvae completed their development. This species might well have been excluded on the basis of the larvae completing their development on sugarcane. However, the female beetles failed to oviposit on sugarcane, which provided a safety factor that allowed liberation in Hawaii and continued consideration for importation into Australia (Waterhouse, 1967). Andres and Goeden (1971) reported that in Hawaii the feeding of several larvae per plant has frequently resulted in branches breaking off so that only stumps remain and that prospects for additional control of lantana by *P. spinipennis* are excellent.

The techniques used in host specificity studies have been thoroughly discussed in recent years (Harris and Mohyuddin, 1965; Harris and Zwolfer, 1968; National Academy of Science, 1968; Waterhouse, 1967; Zwolfer and Harris, 1971). Such studies require laboratories in 2 locations: one abroad as near the target plant and its selected natural enemies as possible and the other a quarantine laboratory in the United States. The studies conducted at each will here be considered separately.

Studies Abroad: The overseas laboratory should be near the target weed and its natural enemies, not only to avoid the problem of crossing political borders with plants and insects, but to have a ready supply of material for study and to observe the host-parasite relationships under natural conditions throughout the year.

There are 2 general types of laboratory tests that can be conducted at the overseas laboratory: the starvation and the multiple choice tests. In the first, only a single kind of plant is provided per cage and the insects can either starve or feed and/or withhold egg laying or oviposit. Should a promising species feed and/or oviposit on some of the test plants other than the target plant, thus indicating that a narrow host specificity is in doubt, multiple choice tests should then be conducted in order to determine the insect's preferences for those test plants. In the multiple choice type of test, the candidate insects are presented 2 or more kinds of plants to feed and/or oviposit on in the same test cage. Usually one of the plants is the target plant, and therefore presumably the natural host of the candidate insect.

Four classes of plants may be included in a testing program: (1) representative economic and desirable plants to which the insects would be exposed in the target weed area so that questions may be answered as to the danger posed to such plants by the suggested import; (2) a systematically arranged spectrum of plants related to the target plant; (3) plants that are known to contain chemical constituents that are similar to those of the target plant; and (4) plants that the candidate insect has been reported collected from or in association with in the literature. In testing the host plant specificity of *Longitarsus jacobaeae*, Frick (1970) selected 56 plant species in 31 genera, which are in 12 tribes of the family Compositae. Within the genus *Senecio*, 18 species were tested, with 5 of the species being in the section Jacobaea, which includes *S. jacobaea*. A number of plants are known to contain alkaloids similar to those in *Senecio* species. Of these, 4 species of 4 genera in 2 families, were included in the tests. One plant in the family Cistaceae was tested because it was stated in the literature that adult *Longitarsus* fed on the foliage.

Although testing is seldom done outdoors, Waterhouse (1967) has reported 2 techniques that might be used where applicable in order to test candidate organisms under more natural conditions than can be obtained in the laboratory. One method utilizes large outdoor cages placed over infested target plants. Then potted test plants are introduced to the cage so that the insects have a multiple-choice type of test that approaches field conditions. This technique can only be used in the country of origin of the candidate insect. The other method involves the testing of potted economic and ornamental plants outdoors, not in the geographic area of origin, but in a country where the candidate had previously been introduced for weed control. Thus, the weed might still be plentiful and its introduced natural enemies in high densities. This method has been successful in Hawaii, where Australian investigators placed economic plants among dense stands of lantana to test several tropical American insects that had been introduced into Hawaii, where they became abundant. These insects were desired for Australia. This method is most useful where only a few other pest insects capable of attacking the test plants are present. Otherwise, various polyphagous plant-feeding insect species might also damage the test plants, making it difficult to assess which species was responsible.

Domestic Studies: After an insect has been studied sufficiently in its native location so that its annual life cycle is known and understood, its feeding habits have been established, and the damage that it inflicts on its host plant is found to be significant, permission is requested to import the candidate species into quarantine in the United States. Quarantine laboratories are designed to prevent the escape of insects. The laboratory construction and procedures necessary to assure that imported phytophagous species may be safely studied are discussed by Fisher (1964).

In addition to life history studies, much of which can also be made at the overseas laboratory, the completion of host plant specificity testing is a major phase of work at the domestic laboratory. Additional test plants are included that are designated by interested agencies in the country in which the releases are planned. These test plants invariably include (1) ornamental, crop, and forage plants that would have been difficult to obtain or grow abroad, (2) native North American plants that afford browse for domestic and wild animals, and (3) other native plants that are deemed necessary as food and shelter for wildlife (Martin, et al., 1951). These plants should be tested in the country where releases are intended, rather than abroad, so that there is no danger of establishing North American plants in foreign countries, where they might become weedy. As an example of native plants included in a quarantine testing program, 3 species of *Senecio* considered to have forage value on western ranges were tested with the European fly, *Hylemya seneciella* (Meade), whose larvae feed on the seeds of tansy ragwort (Frick and Andres, 1967), and 2 were also included in the *Longitarsus* testing program (Frick, 1970). In the case of purple nutsedge, *Cyperus rotundus*, Australian investigators tested 2 Asiatic insects, the weevil *Athesapeuta cyperi* Marshall, and the tortricid moth *Bactra venosana* (Zeller) (= *B. truculenta* Meyrick; Diakonoff, 1964), that had been introduced into Hawaii for control of nutsedge. These insects were not introduced into Australia because of the risk of attack on 2 nutsedges considered to be useful; *C. retzii* Nees, as an excellent fodder plant, and *C. victoriensis* Clarke, in preventing erosion of stream banks. In addition, the insects were considered as unlikely to have a significant effect on the abundance of purple nutsedge because vegetative propagation below ground is very rapid and takes place even under adverse conditions (Wilson, 1960).

To date the methods for testing host specificities of insects to be released against weeds are considered adequate and there has been no example of a basic change in diet of any kind that would constitute a real challenge to the use of insects for weed control.

Domestic Release.--When it is believed that sufficient data have been gathered to be certain that the candidate species is safe to release in the field, a report, containing all relevant information, is submitted to the "Working Group on Biological Control of Weeds," part of the Joint Weed Committees of the U.S. Departments of Agriculture and of Interior, and the concurrence of Canadian authorities is solicited on a reciprocal basis (Coulson, 1971). If importation is approved by this Working Group, permits are sought from the Animal and Plant Health Inspection Service of the USDA and approval from the Departments of Agriculture of the states in which introductions are planned. If permission is granted, a population, usually limited in numbers, is prepared for liberation in the field.

It is seldom advisable to release specimens received directly from abroad because unwanted parasites and diseases may well be harbored by them. Rather, the specimens to be taken to the field should be laboratory-reared for one generation in quarantine, taking all precautions to eliminate their natural enemies so that as clean a culture as possible is actually released. However, laboratory rearing should not be relied upon to make mass releases from an initial stock of one or a few individuals (Remington, 1968). Such a technique could result in a population closely adapted to one narrow set of environmental parameters, i.e., laboratory conditions. Wilson (1965) stated that the primary need is to establish in the field the maximum genetic variability in a species, so that gene combinations appropriate to each environment may be produced by natural selection. To accomplish this Remington (1968) suggested the introduction of "a large, wild sample from a large, central source population which has an environment most similar to that of intended establishment." Then Remington suggested a second method that appears to lend itself to a short period of laboratory rearing before release, a precaution usually deemed necessary to initially free a colony of its native natural enemies. Remington would "introduce a closely spaced succession of wild samples from several source populations from various environments moderately like the area of intended colonization; this maximizes the relevant genetic variability on which selection can then act to produce an optimal genotype in the new environment." A succession of samples of moderate size could be handled and reared in a quarantine laboratory, thus assuring reasonably healthy individuals having, in the aggregate, a broad genetic variability.

When liberated, the candidate species will be involved in a "colonizing episode," which has been defined by Lewontin (1965) as the establishment of a population of a species in a geographical or ecological space not already occupied by that species. A high potential for population increase (r) is a distinct advantage to a colonizing species, because the species usually, especially in the case of insects introduced on alien weeds, enters a new and quite unlimited environmental niche that was previously empty. There is, therefore, an absence of both intra- and interspecific population pressures, and until the available host plants have been consumed, the population is essentially density independent. In spite of these opportunities, colonies are not always successful, perhaps because many involve expansion of area and ecological tolerance (Lewontin, 1965) or because of biotic resistance, e.g., polyphagous parasites and predators. Because of this, the selection of optimal genetic characteristics becomes important and this is discussed in a theoretical analysis of genetic strategies of colonization by Lewontin (1965). On the other hand, if the colony is successful, the population can be assumed to be in the exponential (logarithmic) growth phase with unlimited resources, at least for a short time (Lewontin, 1965). The exponential growth phase terminates in an "explosive" stage when great numbers of the insect have consumed their host plant and must either migrate or starve.

Prior to actual release, several sites should be selected that have an abundance of the target weed. The sites should provide protection of the plants and insects from such hazards as grazing animals, vehicles, farm implements, pesticides, and, where the insects are large and showy, from man himself. It is frequently difficult to establish a species from small numbers and every protection is necessary until the insect is well-established and can safely be distributed to new areas.

The chosen sites should not all be of the same ecological type but some should closely conform to the requirements of the candidate species (Huffaker, 1957). If possible, several sites rather distant from each other should be selected if the weed is widespread over a range of geographic and climatic conditions. The value of using more than one site was shown in the case of *Tyria* introduced against tansy ragwort. Because of the abnormally dry conditions prevailing in 1959, 4400 of 4800 larvae were placed in a damp creek bottom at Fort Bragg, California. However, as a precaution against losing the entire shipment, 400 larvae were also released on a few stunted plants in an open field (Frick and Holloway, 1964). The creek bottom population disappeared after the 1962 season because of flooding, trampling by cattle, removal by persons desiring to infest their own properties, and possibly because of a lack of sunlight. In contrast, the population in the open field grew steadily until a population "explosion" occurred in 1965 (Hawkes, 1968). Fortunately, the release in the open field was not known to the public and it was not pastured, so that site was not disturbed by man or animals.

Decisions as to the best time of day to make a release, the season in which to make it, and whether to use field cages will be up to the investigators who are most familiar with the candidate insect. The conditions to be used will vary with the behavioral characteristics of each species.

Following release, a population should be observed as often as possible. Preferably, the introduction should not be publicized, at least until a population "explosion" has occurred and a "field day" can be held during which interested persons and agencies may collect larvae for their respective locations. This was done in August, 1965, at Fort Bragg, California, so that surplus *Tyria* larvae, many that would otherwise have starved, could be distributed to new areas infested with tansy ragwort. An estimated 50,000 larvae were collected and released in 19 new areas in northwestern California and 28 in southwestern Oregon (Harkes, 1968).

Evaluation of Results.--Of first consideration in evaluating the results of successful colonizations is a statistical record of the plant cover and status of the target weed before the natural enemy is liberated. This record can be started by selecting release sites several years before release, i.e., as soon as anticipated that a promising candidate insect will be released in the future. Following release, detailed description of the numbers of the introduced species that are present, the amounts of destructive feeding, and the patterns of damage as the insects disperse, are essential if the results of a successful introduction are to be evaluated accurately. Meanwhile, the statistical record of any change in the plant cover and the status of the target weed must be continued.

Direct evidence of the effect phytophagous natural enemies may be having on the abundance of their host plants, even during the "explosive" phase, is generally lacking because, unlike herbicide tests that can be confined to certain plots with provisions for untreated checks, insects and other organisms are mobile and actively seek out their host plants. In order to provide areas for comparison, Huffaker (1958) suggested that insect exclosures be established, using selective insecticides and acaracides. These would exclude, not only introduced insects from certain plots, but also any indigenous species that may be grazing on plants in the release area, including the target weed. These may be quite numerous (Frick, 1964; Frick and Hawkes, 1970; Frick, 1972), and each will have some effect on the competition between plant species and therefore on the resulting plant composition. Thus, the results obtained in these exclosures may not be comparable with real situations.

Introduced natural enemies have a continuing vital role in the determination of plant cover once they are well established. For example, St. Johnswort was reduced in 12 years from the status of a primary noxious weed that was an exceedingly important pest on California range lands to a casual roadside weed by the action of imported beetles (Holloway, 1958). At the present overall very low density of *Hypericum*, the action of the few remaining beetles appears trivial. But the beetles still hold the key role because their feeding "action alone is so geared to increase of the weed as to preclude return to the former states" (Huffaker, 1962), while other factors in the environment influencing local changes in the abundance of St. Johnswort (rainfall, temperature, fire, pesticide usage, soil disturbance, soil water-logging) are transitory. A 10-year study has shown that the beetles continue in a weed-insect mutually density-dependent host-parasite relationship that keeps *Hypericum* reduced to a level less than 1% of its former abundance (Huffaker and Kennett, 1959).

In evaluating results, the degree of weed reduction is of primary importance. However, the replacement vegetation is likewise of great importance. [Although it is not the responsibility of the biological control of weeds investigators to reseed an area as it becomes cleared of the target weed, it is gratifying to be able to record that plants of greater forage value replace the weedy species in forage-type habitats (Huffaker, 1951).] Determination of replacement vegetation can only be made as a result of pre- and post-release statistical records of plant cover. These may have to be continued for many years, as was done in the case of St. Johnswort in northern California, the results of which were published after 4 years (Huffaker, 1951) and 10 years (Huffaker and Kennett, 1959). An additional statistical record of plant cover was made again in 1966, 21 years after the first introduction of *Chrysolina quadrigemina*, and as a result it was found that the degree of control of St. Johnswort has been maintained at more than 99% for 17 years (Huffaker, 1967).

Conclusions

Considering the biological control of St. Johnswort as a model, it is evident that, even if highly successful, biological control is a slow-starting form of weed control that is self-maintaining and therefore continuing. DeBach (1964) has estimated the savings (over previous losses plus pest control costs) of $20,960,000 in the 6 years from 1953 through 1958. This amount includes 8 million dollars in increased land values, 12 million dollars realized from the weight gain in cattle, and $960,000 saved in the costs of chemical control. In the 13 years since 1958, the additional amounts realized from cattle weight gains ($2 million/year) and savings from a lack of need for annual chemical controls ($160,000/year), total an additional $28,080,000. Unfortunately, DeBach does not provide an estimate of the costs of the biological control program; thus a complete cost-benefit statement cannot be provided. However, these monetary gains have been without additional cost (i.e., other than for the initial program), since the beetles have continued to control St. Johnswort satisfactorily throughout California without periodic recolonization or other expenditure of effort by man.

References Cited

Andres, L. A. 1966a. Host specificity studies of *Phrydiuchus topiarius* and *Phrydiuchus* sp. J. Econ. Entomol. 59: 69-76.

Andres, L. A. 1966b. The role of biological agents in the control of weeds, pp. 75-82. *In* E. F. Knipling (Chairman), Pest Control by Chemical, Biological, Genetic, and Physical Means. A Symposium. ARS 33-110, 214 p. U.S. Department of Agriculture.

Andres, L. A. and R. D. Goeden. 1971. Biological control of weeds by introduced natural enemies, pp. 143-164. *In* C. B. Huffaker (ed.), Biological Control Plenum Publ. Corp., New York. 511 p.

Andres, L. A. and A. Rizza. 1965. Life history of *Phrydiuchus topiarius* on *Salvia verbenaca*. Ann. Entomol. Soc. Amer. 58: 314-319.

Bottimer, L. J. 1968. On the two species of *Bruchidius* established in North America. Canadian Entomol. 100(2): 139-145.

Coulson, J. R. 1971. Prognosis for control of water hyacinth by arthropods. Hyacinth Control J. 9(1): 31-33.

Davies, J. C. and D. J. Greathead. 1967. Occurrence of *Teleonemia scrupulosa* on *Sesamum indicum* Linn. in Uganda. Nature 213(5071): 102-103.

DeBach, P. 1964. The scope of biological control, p. 3-20. *In* P. DeBach (ed.), Biological Control of Insect Pests and Weeds. Reinhold Publ. Corp., New York, 844 p.

Diakonoff, A. 1964. Further records and descriptions of the species of *Bactra* Stephens. Zool. Verh. Rijksmus. Nat. Hist. Leiden 70: 1-81.

Dingle, H. 1972. Migration strategies of insects. Science 175(4028): 1327-1335.

Dodd, A. P. 1940. The biological campaign against prickly pear. Commonwealth Prickly-Pear Board, Brisbane, Australia. 177 p.

Fernald, M. L. 1970. Gray's Manual of Botany. 8th ed. Van Nostrand Reinhold Co., New York, 632 p.

Fisher, T. W. 1964. Quarantine handling of entomophagous insects, pp. 305-327. In P. DeBach (ed.), Biological Control of Insect Pests and Weeds. Reinhold Publ. Corp., New York, 844 p.

Frick, K. E. 1964. Some endemic insects that feed upon introduced tansy ragwort in western United States. Ann. Entomol. Soc. Amer. 57: 707-710.

Frick, K. E. 1970. *Longitarsus jacobaeae*, a flea beetle for the biological control of tansy ragwort. I. Host plant specificity studies. Ann. Entomol. Soc. Amer. 63: 284-296.

Frick, K. E. 1972. Third list of insects that feed upon tansy ragwort, *Senecio jacobaea*, in the western United States. Ann. Entomol. Soc. Amer. 65: 629-631.

Frick, K. E. and L. A. Andres. 1967. Host specificity of the ragwort seed fly. J. Econ. Entomol. 60: 457-463.

Frick, K. E. and R. B. Hawkes. 1970. Additional insects that feed upon tansy ragwort, *Senecio jacobaea*, an introduced weedy plant, in western United States. Ann. Entomol. Soc. Amer. 63: 1085-1090.

Frick, K. E. and J. K. Holloway. 1964. Establishment of the cinnabar moth, *Tyria jacobaeae*, on tansy ragwort in the western United States. J. Econ.

Goeden, R. E. 1970. Current research on biological weed control in southern California, pp. 25-28. In F. J. Simmonds (ed.), Proceedings of the First International Symposium on Biological Control of Weeds, March, 1969. Commonw. Inst. Biol. Control Misc. Publ. 1, 110 p.

Goeden, R. E., C. A. Fleschner and D. W. Ricker. 1967. Biological control of prickly pear cacti on Santa Cruz Island, California. Hilgardia 38(16): 579-606.

Harley, K. L. S. and R. C. Kassulke. 1971. Tingidae for biological control of *Lantana camara*. Entomophaga 16(4): 389-410.

Harris, P. 1967. Suitability of *Anaitis plagiata* for biocontrol of *Hypericum perforatum* in dry grassland of British Columbia. Canadian Entomol. 99(12): 1304-1310.

Harris, P. and A. I. Mohyuddin. 1965. The bioassay of insect feeding tokens. Canadian Entomol. 97(8): 830-833.

Harris, P. and H. Zwolfer. 1968. Screening of phytophagous insects for biological control of weeds. Canadian Entomol. 100(3): 295-303.

Harris, P., D. Peschken and J. Milroy. 1969. The status of biological control of the weed *Hypericum perforatum* in British Columbia. Canadian Entomol. 101(1): 1-15.

Hawkes, R. B. 1968. The cinnabar moth, *Tyria jacobaeae*, for control of tansy ragwort. J. Econ. Entomol. 61: 499-501.

Holloway, J. K. 1958. The biological control of Klamath weed in California. Proc. Int. Congr. Entomol., 10th, Montreal, 1956, 4: 557-560.

Holloway, J. K. 1964. Projects in biological control of weeds, pp. 650-670. In DeBach, P. (ed.), Biological Control of Insect Pests and Weeds. Reinhold Publ. Corp., New York, 844 pp.

Huffaker, C. B. 1951. The return of native perennial bunchgrass following the removal of Klamath weed by imported beetles. Ecology 32(3): 443-458.

Huffaker, C. B. 1957. Fundamentals of biological control of weeds. Hilgardia 27(3): 101-157.

Huffaker, C. B. 1958. Principles of biological control of weeds. Proc. Int. Congr. Entomol., 10th Montreal, 1956, 4: 533-542.

Huffaker, C. B. 1959. Biological control of weeds with insects. Ann. Rev. Entomol. 4: 251-276.

Huffaker, C. B. 1962. Some concepts on the ecological basis of biological control of weeds. Canadian Entomol. 94(5): 507-514.

Huffaker, C. B. 1964. Fundamentals of biological weed control, pp. 631-649. In P. DeBach (ed.), Biological Control of Insect Pests and Weeds. Reinhold Publ. Corp., New York, 844 p.

Huffaker, C. B. 1967. A comparison of the status of biological control of St. Johnswort in California and Australia. Mushi 39, suppl.: 51-73.

Huffaker, C. B. and C. E. Kennett. 1959. A ten-year study of vegetational change associated with biological control of Klamath weed. J. Range Manage. 12: 69-82.

Lewontin, R. C. 1965. Selection for colonizing ability, pp. 77-91. In H. G. Baker and G. L. Stebbins (eds.), The Genetics of Colonizing Species. Academic Press, New York and London, 588 pp.

Mann, J. 1969. Cactus-feeding insects and mites. U.S. Natl. Mus. Bull. 256, 158 p.

Martin, A. C., H. S. Zim and A. L. Nelson. 1951. American wildlife and plants. Dover Publ., Inc., New York, 500 p.

Murphy, A. H., R. M. Love and L. J. Berry. 1954. Improving Klamath weed ranges. California Agr. Exp. Sta. Circ. 437, 16 pp.

National Academy of Sciences - National Research Council. 1968. Principles of Plant and Animal Pest Control. Vol. 2, Weed Control. Natl. Acad. Sci. Publ. 1597, 471 p.

Perkins, R. C. L. and O. H. Swezey. 1924. The introduction into Hawaii of insects that attack Lantana. Hawaii Sugar Plan. Assoc. Exp. Sta. Entomol. Serv. Bull. 16, 83 pp.

Remington, C. L. 1968. The population genetics of insect introduction. Ann. Rev. Entomol. 13: 415-426.

Simmonds, F. J. and F. D. Bennett. 1966. Biological control of *Opuntia* spp. by *Cactoblastis cactorum* in the Leeward Islands (West Indies). Entomophaga 11(2): 183-189.

Waloff, N. 1966. Scotch broom and its insect fauna introduced into the Pacific Northwest of America. J. Appl. Ecol. 3: 293-311.

Warner, R. E. 1969. The genus *Phrydiuchus*, with the description of two new species. Ann. Entomol. Soc. Amer. 62: 1293-1302.

Waterhouse, D. F. 1967. The entomological control of weeds in Australia. Mushi 39, Suppl.: 109-118.

Williams, J. R. 1954. The biological control of weeds. Rep. Commonw. Entomol. Conf., 6th, London, pp. 95-98.

Wilson, F. 1943. The entomological control of St. Johnswort with particular reference to the insect enemies of the weed in southern France. Counc. Sci. Ind. Res. Australia, Bull. 169, 87 p.

Wilson, F. 1960. A review of the biological control of insects and weeds in Australia and Australian New Guinea. Commonw. Inst. Biol. Control Tech. Commun. 1, 102 pp.

Wilson, F. 1964. The biological control of weeds. Ann. Rev. Entomol. 9: 225-244.

Wilson, F. 1965. Biological control and the genetics of colonizing species, pp. 307-325. In H. G. Baker and G. L. Stebbins (eds.), The Genetics of Colonizing Species. Academic Press, New York and London, 588 p.

Zwolfer, H. 1968. Some aspects of biological weed control in Europe and North America. Brit. Weed Control. Conf., 9th, Proc.: 1147-1156.

Zwolfer, H., K. E. Frick and L. A. Andres. 1971. A study of the host plant relationships of European members of the genus *Larinus*. Commonw. Inst. Biol. Control Tech. Bull. 14: 97-143.

Zwolfer, H. and P. Harris. 1971. Host specificity determination of insects for biological control of weeds. Ann. Rev. Entomol. 16: 159-178.

BIOLOGICAL CONTROL OF AQUATIC WEEDS

F. D. Bennett
Entomologist in Charge
West Indies Station
Commonwealth Institute of Biological Control
Gordon Street, Curepe, Trinidad

Although the practice of biological control of aquatic weeds is relatively new and progress on only a few projects has reached the stage where a meaningful appraisal of the value of introductions can be undertaken, the first attempt to control an aquatic weed, *Alternanthera philoxeroides* (Mart.) Griseb., by the introduction of insects shows excellent promise of providing effective control and has created optimism that this method will provide a satisfactory solution to numerous other aquatic weed problems. Similarly, the experimental use of higher animals, particularly fish, also shows promise. Increasing attention is also being paid to the possibilities of plant pathogens as control agents either for classical control (the introduction of self-perpetuating organisms into a new geographic area) or, if necessary, by routine inoculative releases as an alternative to the use of chemical herbicides.

Because this field was largely neglected by entomologists until the last decade overall reviews on biological control of weeds (Huffaker, 1959; Wilson, 1964) did not discuss the topic of biological control of aquatic weeds or were able to deal with it in one or two paragraphs. This does not hold true today and as great progress is now being made in this particular field it will certainly be less true after the next decade.

Reasons for Controlling Aquatic Weeds

Aquatic weeds affect man and his well-being in several ways. The following will suffice to indicate their importance. (1) Agriculture demands that irrigation and drainage canals flow unimpeded by weed growth, and that water must be readily available in predictable amounts and of acceptable quality. (2) Dense weed mats affect the oxygen level of the water, thereby altering food chains, affecting fish and fishing and the quality of water for domestic and commercial use. (3) They impede or completely prevent navigation on river systems and pose a threat to hydro-electric plants. (4) They may be responsible for flooding or at times water shortages. (5) They may encourage the buildup of those snails which are the essential intermediate hosts for bilharzia organisms and liver-flukes, as well as providing breeding places for certain disease-carrying mosquitoes. (6) Aquatic weeds may be detrimental to rice cultivation, and in a few instances may serve as alternative hosts and as a means of spread of pathogens affecting cultivated crops. (7) They also affect man's recreation, i.e., swimming, boating, fishing, etc.

Factors Leading to the Increased Interest in Biocontrol of Aquatic Weeds

It is obvious that with the growing list of successes in controlling terrestrial weeds by the introduction of natural enemies, attention would eventually be turned to aquatic weeds. Just as the first attempts at biocontrol of terrestrial weeds were triggered by serious repercussions following the introduction and rapid spread of an exotic species, Lantana camara Linn., in Hawaii and subsequently in many other areas, attention to aquatic weeds was focused on such introduced aggressive invaders as Alternanthera philoxeroides (Mart.) Griseb. and Eichhornia crassipes (Mart.) Solms. in the USA and Salvinia auriculata (auct. nec. Aubl.) in Africa, Ceylon, etc. The construction and development of man-made lakes and the rapid explosive colonization of them by aquatic weeds, e.g. S. auriculata on Lake Kariba (Rhodesia-Zambia) and E. crassipes on the Brocopondo Lake in Surinam, have pointed to the need for urgent and permanent solutions. In developed countries the increased time for leisure and sport has placed a heavy demand on the construction and maintenance of recreation lakes, canals, etc., in a condition suitable for boating, swimming and fishing. In developed countries the increased use of fertilizers in farming, and of detergents in household use, as well as many other materials rich in plant nutrients which have ended up in ponds, lakes, rivers, etc., have led to eutrophication often producing almost optimum conditions for some of the most aggressive aquatic weeds.

Factors Which May Render Aquatic Weed Problems More Difficult to Solve than Terrestrial Ones

Considering the vast array of aquatic weed problems throughout the world and of the organisms being considered as potential control agents it is difficult to make general statements on this topic. Andres (1968), after mentioning that there were more than 52 aquatic weeds in the USA belonging to four categories, i.e., floating, emersed, submersed and algae, raised doubts as to whether host-specific insects to control many submerged species would be found.

An even more important factor (Bennett, in press) may be the absence of the effect of competition from other plants. Most terrestrial weeds have to some extent to compete with crop or other plants for growing space, nutrients, water, etc.; relatively minor injury inflicted to the weed by a phytophagous species or a disease may tip the balance in favor of the other, possibly allowing them to outgrow and suppress the competing weeds. By contrast, the only competing plants in many aquatic situations are frequently other weeds which may or may not be less desirable. If there are no other desirable species then the important element of crop competition is lacking and accordingly the extent of damage or the feeding pressure required to suppress the weed may have to be very high and continuous. This is particularly true for floating weeds such as Salvinia spp. and water-hyacinth, Eichhornia crassipes, which rapidly proliferate new plants by vegetative means. One might even postulate that the end results of minor damage by an insect could be an increase in the number of plants if not in their actual weight. Vogel and Oliver (1969) reported that damage to an apical bud of E. crassipes by the Noctuid Arzama densa Walker, induced a significant increase in the number of lateral buds, each being capable of producing a new plant.

While in some terrestrial situations, e.g., poorly managed pastureland, weeds may increase as fertility levels drop (although selective grazing on the desirable plants may also be a factor), it is ironic that most of the aquatic weed problems are aggravated by an overabundance of nutrients.

Factors Which May Render Aquatic Weed Problems More Amenable to Biological Control

A larger range of candidate control organisms may be available for aquatic than for terrestrial weed control. Apart from pasture management programs and where domesticated animals, cattle, pigs, goats, ducks and geese, etc., are occasionally manipulated to suppress weeds, serious consideration is seldom given to the introduction of vertebrates for the biological control of terrestrial weeds. On the other hand, several vertebrates are being evaluated for use against aquatic weeds. Similarly certain groups of invertebrates, e.g., snails, are under scrutiny as potential agents against undesirable aquatic vegetation, such as would not be considered for terrestrial weed control because of their omnivorous feeding habits. As in aquatic situations certain of these can be manipulated and controlled or eradicated if they prove undesirable at high population levels, and as there are relatively few aquatic crop plants we can at times recommend species with somewhat polyphagous feeding habits. While increasing concern is being evinced in the possible detrimental effect on native weeds and a correlated chain reaction on wildlife systems, the qualification that an introduced control agent will not feed on the half-dozen or so economic aquatic crops, and is tied by some stage of its biology to an aquatic environment, is sometimes considered sufficient insurance that certain organisms are safe for introduction. On this basis, the release of the grasshopper, *Paulinia acuminata* DeG. in several African countries was recommended and approved (Bennett, 1966). Obviously, in instances such as these, each candidate control agent must be evaluated and, on the basis of accumulated data, approved or rejected pending further tests. While *P. acuminata* has been cleared for release against Salvinia, another grasshopper, *Cornops longicorne* Bruner, attacking *E. crassipes* which is somewhat less tied to an aquatic environment has not been recommended for release. This species will also develop from small nymph to adult on *Commelina* spp., but suitable oviposition sites are present only in the spongy tissue of the stems of its host. While ootheca have never been found in the stems of plants other than the Pontederiaceae either in the field or in laboratory tests, adults have been taken some distance from *E. crassipes* and have been observed to feed on *Equisetum* sp. under field conditions. While investigators are fairly certain that the species is unlikely to cause significant damage to any economic crop, it is unlikely that its release will be recommended, at least until the several other oligophagous species known to attack *E. crassipes* have been tried (see below).

It has also been postulated that relatively minor damage by a primary feeder may render aquatic plants more prone to attack by secondary organisms than does a comparable level of attack on terrestrial plants other than succulents. While this generalization is difficult to prove, plants of *Pistia stratiotes* frequently collapse rapidly following minor insect damage.

Biocontrol of Introduced Versus Native Weeds

While the concensus of opinion expressed in most review papers is that introduced weeds are the best candidates for classical biocontrol, Harris and Piper (1970) argue that the species of *Ambrosia* native to Canada and the northern United States may be amenable to biological control. Their argument is based on the related paucity of oligophagous or host-specific native natural enemies, to its resemblance to a well-established introduced pest and to the possibility that elsewhere there may be natural enemies of related species suitable for introduction. The oft-quoted examples of the decimation of the American chestnut by the introduced chestnut blight, a pathogen of a related species, and the drastic mortality of the Bermuda cedar *Juniperus bermudiana* L., caused by the introduced diaspine *Carulaspis minima* (Targ.) are ample testimony that plant populations can be held in spectacular check by an organism previously unrecorded from the plant in question (Doutt, 1964).

While major concern over aquatic weeds deals with introduced species, e.g., *Eichhornia crassipes*, listed as one of the ten most serious weed pests and according to Holm (1969) "the most terrible and frightening weed problem I have ever known" and *Salvinia auriculata* which caused grave concern when the Kariba Lake was formed on the Zambezi water-system, and is a major problem in the Chobe River and Lake Naivasha in Africa and several water systems in Asia, problems with native weeds are also legion. While many of these problems are the end-product or perhaps an intermediate succession stage resulting from eutrophication, some form of biological control is desired. *E. crassipes* is also causing increasing concern in South America where it is endemic. It has been suggested (Bennett, 1970) that biotic agents which have transferred from native Pontederiaceae in North America and Asia, if sufficiently oligophagous, might be introduced into areas of South America where *E. crassipes* is a problem, e.g., on Lake Brokopondo, Surinam (Bennett, and Zwolfer, 1968) or into Guyana on transport and drainage canals in the sugar plantations. Sankaran, et al. (1966) have reported that the native *Gesonula punctifrons* Stal. which attacks the introduced *E. crassipes* in India might be suitable for trial in other areas. While it is not host-specific, its ecological requirements restrict it to an aquatic or semi-aquatic habitat. Hence, while introduced weeds are often more logical targets for biological control, native species should also be considered.

Plant Pathogens as Control Agents

The use of pathogens for control of aquatic weeds still remains a relatively unexplored field. Wilson (1969) has summarized the current state of knowledge. Progress to date indicates that certain groups of lower plants, for which little or no hope may be held for control by insects, may be amenable to control by plant pathogens. Blue-green algae have been successfully controlled by a virus in large-scale experiments in sewage disposal pools (See Wilson, 1969). Similarly, it has been suggested that a virus may be responsible for the reduction of water-milfoil, *Myriophyllum spicatum* L., in Chesapeake Bay (Bayley, et al., 1968). The reduction of phytoplankton blooms by the attack of aquatic fungi also indicates the promise that may lie in the field of plant pathology for aquatic weed control. A virus transmitted by the aphid *Rhopalosiphum nymphae* L., has been suggested as an important cause in the annual die-back of *Pistia stratiotes* L. in Nigeria (Pettet and Pettet, 1970). It is probable that other factors including fluctuations in nutrient levels are also involved, but if the virus is the main factor it is essential that additional investigations, including a study of the host range of the virus as well as of *R. nymphae* and other potential vectors, be carried out before introductions are warranted. *R. nymphae* is, of course, known to attack many other aquatic plants. Diseases of several aquatic weeds have been under study in India (Nag Raj, 1965, 1966; Nag Raj and Ponnappa, 1969a, b; Ponnappa, 1970). Recently, a laboratory aimed to exploit this approach has been established in Gainesville, Florida (Zettler, et al., 1971).

Vertebrates as Control Agents

Fish: Hickling (1965) has summarized the role of fish as well as other vertebrates in the biological control of aquatic vegetation. While several fish, chiefly tropical species such as *Tilapia* spp., have been utilized, frequently with the aim of fulfilling the dual function of weed control and providing human food, the species currently attracting most attention is the amur or grass-carp, *Cteropharyngodon idella* Val., native to east Asia and one of the few vegetarian fish in temperate climates of commercial interest. Michewicz, et al. (1972a), provide an up-to-date evaluation of investigations on this species. While not cleared for unrestricted distribution in the USA it is under study in certain ponds and lakes (Bailey and Boyd, 1972). It is being stocked widely in Europe (Krupauer, 1971; Robson, 1971), Asia (Meta and Sharma, 1972), and Fiji (Hughes, 1971), and appears to acclimatize to tropical as well as temperate waters including those with considerable salinity levels. This species has spawned naturally in only a few areas outside its native home. This may offer an advantage initially as it affords an opportunity to study the effect of the fish on the native fauna and flora in the field with little risk of permanent establishment. It may also permit correct stocking rates to achieve the desired level of control of aquatic vegetation in water bodies of varying sizes (Michewicz, et al., 1972a). Once it reaches a length of 2.5 cm., the amur feeds almost entirely on higher plants and to achieve control of such weeds in some situations simultaneous stocking with other species, e.g., _Tilapia_, may be necessary (Blackburn, et al., 1971). The use of sterile F_1 crosses between two species of _Tilapia_ should permit considerable latitude in the level of weed control desired because again given numbers of a non-breeding organism can be stocked. Studies to determine the effect of the amur on native fish are underway in Europe and at least certain species continue to exist with it (Stott, et al., 1971).

In view of the changes in chemical analysis of the water in pools stocked with the amur (Michewicz, et al., 1972b), the overall effects on water quality resulting from the activities of the fish need to be determined before recommending its introduction into water systems where these effects would be detrimental. Probably the greatest stumbling block to permitting a wide distribution of the amur in the USA is the concern as to the overall effect that it may have on other desirable components of the aquatic ecosystem.

Other Vertebrates: Domesticated animals including cows, goats, pigs, ducks and geese are often cited as control agents of aquatic weeds, but while they may be useful in restricted situations they cannot be satisfactorily manipulated to cope with most aquatic weed problems.

Similarly many wild animals, e.g., the hippopotamus, play an important role in weed control in their native habitats, but are not likely to be candidates for release for weed control elsewhere! The manatee, _Trichechus manatus_ L., has been under investigation since Allsopp (1960) drew attention to its potential for weed control. More recently he (Allsopp, 1969) pointed out some of its limitations. While under certain conditions it may be possible to "manage" the manatee, its widespread use as a weed control agent is unlikely to develop. While investigations in Florida have also indicated that this animal consumes impressive amounts of aquatic weed (Sguros, et al., 1965), the difficulties in securing adequate numbers, due to its scarcity and its low reproductive rate, coupled with the high costs of moving it from canal to canal militate against its selection as a control agent. If allowed to roam in large reservoirs it grazes selectively, often ignoring or avoiding those species which to man are the most serious weeds.

The South American coypu or nutria, has also been suggested as a useful animal for control of aquatic weeds. It has been widely disseminated as a fur-bearing animal in North America and Europe and released in parts of Africa. The relatively low value of its pelt, the damage caused by its burrowing activities and its omnivorous feeding habits should raise second thoughts as to its further deliberate dissemination either for weed control or for fur-bearing.

Invertebrates as Control Agents

Among the invertebrates only insects and mites have been introduced into new areas for the biological control of terrestrial weeds but these have been used extensively. As already noted, the choice is wider for aquatic weed control. To date snails as well as insects and mites have been exploited.

Snails.--An early attempt at biological control of aquatic weeds involved the use of the snail _Pomacea caniculata_ Lamer to combat a submersed weed, reported as _Anachares densa_ but probably _Elodea_ sp., in ponds in Brazil (Silva, 1960). Recent investigations on snails as weed control agents have been centered in the continental United States although intensive investigations have also been carried out in Puerto Rico, where the value of the Marisa snail, _Marisa cornuarietis_ L., as a predator of and a habitat competitor of _Biomphalaria glabrata_ (Say), the snail vector of schistomiasis, has been of greater importance than its potential as an agent of weed control (Seaman and Porterfield, 1964; Ferguson and Butler, 1966). In Florida, pilot trials have indicated that _M. cornuarietis_ can be manipulated to provide control in certain situations and Blackburn, Taylor and Sutton (1971) have worked out stocking rates necessary to control _Hydrilla verticillata_ Presl. and the southern naiad, _Naias guadalupensis_ Morong. Attempts have also been made to develop mass-production techniques to enable re-stocking of _Marisa_ in areas where its populations are killed annually by low winter temperatures (Rich and Rouse, 1970). Several aspects of the biology and of the behavior of this snail have also been investigated in Sweden (Hubendick, 1966) and Egypt (Demain and Ibrahim, 1969a, b; Demain and Lutfy, 1965a, b) to determine whether it can be safely released in Africa for the control of schistomiasis and liver-flukes (by predation on their snail hosts), as well as certain aquatic weeds.

Insects and Mites.--When Wilson (1964) reviewed the topic of biological control of weeds he noted that at that time no insects had been used for the control of aquatic weeds and raised the question as to whether in the aquatic environment phytophagous insects exhibited a sufficient degree of monophagy to permit introduction from one country to another. Andres (1968) has also suggested the possibility that water may constitute a sufficient barrier to deter the exploitation by insects of submersed weeds or submersed portions of emergent plants. These two factors, if they hold true, would greatly restrict the species of weeds against which insects might play a useful role. However, in his concluding remarks, Wilson (1964) stated "that research on the biological control of weeds is in its infancy," and that "extension of research in the general field is very desirable, particularly in relation to entomological control."

Bennett (in press) has considered that a greater degree of polyphagy is permitted for aquatic than for terrestrial weed insects as biocontrol agents provided that they are tied to an aquatic environment by their biology and do not regularly attack economic or otherwise desirable aquatic plants.

On this basis the semiaquatic grasshopper, _Paulinia acuminata_, which must deposit its eggs in ootheca underwater if they are to develop, has been introduced into Africa despite the knowledge that it develops successfully and feeds in the field on botanically unrelated plants, e.g., the ferns _Salvinia_ spp. and aroid _Pistia stratiotes_ L. Many of the phytophagous insects which have given spectacular control of terrestrial weeds seldom achieve the required population levels in their country of origin because they are held in check by their specific natural enemies. Bennett (in press) has suggested that certain groups of insects attacking submersed weeds may be less prone to attack by parasitoids than are terrestrial insects belonging to the same groups. If this is so we can assume that some other factor holds the aquatic phytophagous insects in check in instances where they do not inflict serious injury to their hosts in the country of origin. We should perhaps not anticipate the spectacular population buildup that occurs when terrestrial insects are freed of parasite attack following their introduction into other areas unless of course the other factors limiting the population of aquatic species are absent in the country where introductions are contemplated. The absence of attack by specific parasitoids may be compensated for by the activities of more generalized predators but they may be just as abundant in the area of introduction as they are in the native habitat of the phytophage. Difficulties in the laboratory rearing of certain of the Lepidoptera attacking aquatic weeds have been attributed to a high incidence of attack by insect pathogens and we may perhaps infer that the microhabitat in which these develop may be very favorable for certain groups of these pathogens. Accordingly every possible effort should be made to ensure that disease-free stocks are available for introduction into new areas.

While the techniques for rearing and testing insects of aquatic weeds differ somewhat from those used for terrestrial insects, Bennett (1966) has suggested that these differences may be no greater than those for testing two terrestrial insects belonging to different orders. Harris (1971); Harris and Zwolfer (1968) and Zwolfer and Harris (1971) have critically reviewed the procedures for determining the host-specificity of insects for the biological control of weeds.

Presently, active projects in which insects and mites are under investigations will now be dealt with individually.

Present Projects Involving Insects and Mites

1. Alligator weed, Alternaria philoxeroides.--Of South American origin Alternaria philoxeroides was recorded in Florida over 75 years ago and subsequently from all of the southern states. Surveys were undertaken in South America by Vogt (1960, 1961) and based on his findings, studies on promising organisms were commenced in Argentina and Uruguay. Following field-screening, three species were cleared for release (see Maddox, et al., 1971, for details). The first of these, a Chrysomelid, Agasicles hygrophila Selman & Vogt, was first released in 1964 in South Carolina and in 1965 in Florida and subsequently in most states. One site, a small inlet on the Ortega River near Jacksonville, Florida, where a release was made in early 1965, was practically clear of the weed by July 1966. Similar outstanding control has resulted in many other areas in both Florida and in other states and natural dispersal of the beetles over distances of several miles has occurred (Maddox and Hanbric, 1970, 1971; Maddox, et al., 1971). Amynothrips andersoni O'Neill, the second insect, became established following its introduction in 1966 but has not caused spectacular damage (Maddox, et al., 1971). The third species, the Phycitid stem-borer, Vogtia malloi Pastrana, was cleared for introduction in 1970 and releases effected in 1971. Detailed studies to document its progress are currently underway at Gainesville and it is anticipated that this species, either by itself or in combination with A. hygrophila, will provide adequate permanent control of alligator weed under a wide range of conditions (Mr. N. B. Spencer, personal communication, May 1972).

2. Salvinia auriculata.--A floating aquatic fern indigenous to South America, S. auriculata, has been introduced into several countries in Africa and Asia, as well as into Australia, and has become a major problem in rice paddies in Ceylon and on several other water-systems in both Africa and Asia. Following its invasion and rapid spread on the Kariba Lake, a search for potential controlling agents was instigated in 1961. Surveys undertaken in Trinidad and northern South America were followed by host-specificity and evaluation tests of three species, the semi-aquatic Acridid Paulinia acuminata, the Curculionid Cyrtobagous singularis Hulst., and the Pyralid Samea multiplicalis Guenee. The Kariba Lake Coordinating Committee provided a list of only five economic crop-plants on which host specificity tests with insects known to be restricted by their biology to an aquatic environment were required. While C. singularis is specific to Salvinia, P. acuminata is known to feed on several other botanically unrelated aquatic weeds, and S. multiplicalis also develops readily on Pistia stratiotes and has also been reared from E. crassipes. These three insect species have been shipped for release in Africa and for further study to India. P. acuminata was released initially in a cage in the Rhodesian side of Lake Kariba during 1969, and on the Zambian side in 1970. In August 1971, 2000 nymphs and adults from a tropical climate (Trinidad) and 1500 from a cooler climate (Uruguay) were released in separate areas on the Rhodesian side (D. J. W. Rose, personal communication, October 1971). Establishment and minor damage were noted during a survey in October (author's unpublished notes) and in January 1972 rapid distribution and population densities ranging from 1.6 to 9.4 adults per square metre were recorded (J. T. Kabayadondo, unpublished data, 1972). P. acuminata was released on Lake Naivasha, Kenya, in 1970 but permanent establishment has apparently not occurred. The low night temperatures are considered to militate

against establishment. Releases of *P. acuminata* and *C. singularis* on the Chobe River, Botswana, initiated in December 1971 are continuing. It is anticipated that the release of one or more of these species in India will commence in the near future.

3. *Eichhornia crassipes*.--The distribution and pest status of waterhyacinth, *E. crassipes*, are too well-known to require elaboration (see Little, 1965). As the plant and other members of the genus are native to South America, initial basic investigations on their natural enemies have been undertaken (Silveira-Guide, 1965; Bennett, 1968; Bennett and Zwolfer, 1968; Coulson, 1971). Surveys in India (Rao, 1965) and the USA (Vogel and Oliver, 1968, 1969; Coulson, 1971; Bennett, 1969), areas where the weed has long been a pest, have also been undertaken to determine what biotic agents already occur in the respective areas. Detailed investigations on the South American species *Acigona infuscatella* (Walk.) (= *A. ignitalis* Hmps.), Pyralidae, *Epipagis albiguttalis* Hmps., Pyralidae, *Cornops* spp. (*longicorne* Bruner and *acquaticum* DeGeer) Acrididae, *Neochetina bruchi* Hulst. and *N. eichhorniae* Warner, Curculionidae, as well as the mite *Orthogalumna terebrantis* Wallwork, Galumnidae, have been carried out or are in progress, and certain of these organisms are in various stages of being cleared for release in the United States; the first species to be released will probably be *N. eichhorniae* (N. R. Spencer, personal communication, 1972). *O. terebrantis*, which is already present in the USA, has been shipped to India for laboratory trials and subsequently to Zambia where it has been released and temporarily established on the Kafue River. *N. eichhorniae* and *E. albiguttalis* were also released in Zambia in 1971 and *N. eichhorniae* has now been released in Rhodesia.

4. Other aquatic weeds.--The insect fauna of several other weeds is under investigation in several parts of the world. *Myriophyllum spicatum* L., in Yugoslavia where the Pyraustid *Parapoynx strationata* L., and the Curculionid *Litodactylus leucogaster* Marsh are considered to be important biotic agents (Lekic, 1970; Lekic and Mihajlovic, 1970). In India studies on a broad range of aquatic weeds, e.g., *Myriophyllum intermedium*, *Nymphoides indicum* (L.) O. Kuntz, *Ludwigia adscendeus* (L.) Hara, *Potomogeton nodosus* Poir., etc., have been undertaken (Sankaran, et al., 1970). The fauna of *Myriophyllum* spp. has also been studied in Pakistan where the weevils *Bagous geniculatus* Huchuth, *B. vicinus* Hust. and *Phylobus* sp., as well as the Gelechiid *Aristotelia* sp., were considered to be specific to these hosts and hence merit consideration as control agents (Baloch, et al., 1972). Spencer (1971) has also suggested *Parapoynx* sp. as a potential control agent for *M. spicatum*.

Concluding Remarks

The author (Bennett, in press) has recently summarized certain of his views on the biological control of aquatic weeds as follows:

1. Workers in the field of biological control of aquatic weeds may have recourse to a broader spectrum of potential control agents than those working with terrestrial weeds.

2. Groups of organisms other than insects, which traditionally have been used for the control of terrestrial weeds, may play a greater role in the field of aquatic weed control. Conversely, insects, particularly for submersed weeds and algae, may perhaps be only of relatively minor importance.

3. Insects with a wider potential host range may be acceptable for aquatic weed problems than is permissible for species selected for terrestrial weeds when by their biology they are tied to an aquatic environment.

4. Parasitic insects are less likely to be a limiting factor to population growth of their phytophagous host in aquatic situations than they are to be terrestrial ones, although this may not hold true for general predators and insect pathogens.

5. The field of plant pathology may hold solutions to problems of the more serious aquatic weeds not amenable to control by invertebrates.

6. Finally, the element of competition existing between cultivated terrestrial plants and weeds, which may be important in tipping the balance in favor of a crop, is frequently lacking in aquatic situations. Hence the intensity of damage required may have to be greater, particularly in instances where eutrophication of the aquatic environment is a continuing process.

It should be pointed out that these conclusions are tentative and some of them based on little factual information. In view of the rapid acceleration of investigations currently occurring these may require considerable modification or expansion in the near future.

References Cited

Allsopp, W. H. L. 1960. The manatee: Ecology and use for weed control. Nature London 188: 762.

Allsopp, W. H. L. 1969. Aquatic weeds control by manatees--its prospects and problems. Man-made Lakes, The Accra Symp., Ghana, University Press.

Andres, L. A. 1968. Insects and the control of aquatic weeds in the United States. Weed Soc. of Amer. meeting, New Orleans. 17 pp.

Avault, J. W., Jr. 1965. Biological weed control with herbivorous fish. Proc. Southern Weed Conf. 18: 590-591.

Bailey, W. M. and R. L. Boyd. 1972. Some observations on the white amur in Arkansas. Hyacinth Control J. 10: 20-22.

Baloch, G. M., A. G. Khan and M. A. Ghani. 1972. Phenology, biology and host-specificity of some stenophagous insects attacking *Myriophyllum* spp. in Pakistan. Hyacinth Control J. 10: 13-16.

Bayley, S., H. Rahin and C. H. Southwick. 1968. Recent decline in the distribution and abundance of Eurasian millfoil in Chesapeake Bay. Chesapeake Sci. 9: 177-181.

Bennett, F. D. 1966. Investigations on the insects attacking the aquatic ferns *Salvinia* spp. in Trinidad and northern South America. Proc. Southern Weed Conf. 19: 497-504.

Bennett, F. D. 1968. Insects and mites as potential controlling agents of water hyacinth (*Eichhornia crassipes* (Mart.) Solms.). Proc. 9th British Weed Control. Conf. 832-835.

Bennett, F. D. 1970. Insects attacking water hyacinth in the West Indies, British Honduras and the USA. Hyacinth Control J. 8(2): 10-13.

Bennett, F. D. 1972. Some aspects of the biological control of aquatic weeds. Proc. 2nd Int. Symp. on Biological Control of Weeds, Rome (in press).

Bennett, F. D. and H. Zwolfer. 1968. Exploration for natural enemies of the water hyacinth in northern South America and Trinidad. Hyacinth Control J. 7: 44-52.

Bennett, F. D. and H. Zwolfer. 1968. Preliminary report on the investigations on the fauna associated with water hyacinth, Eichhornia crassipes in Surinam. CIBC unpubl. rept. 3 pp.

Blackburn, R. D. and L. A. Andres. 1968. The snail, the mermaid, and the flea beetle. Yearbook of Agriculture, Washington, D.C. 229-234.

Blackburn, R. D. and D. L. Sutton. 1971. Growth of the white amur (Cteropharyngodon idella Val) on selected species of aquatic plants. Proc. European Weed Res. Cour. 3rd Symp. Aquatic Weeds: 87-93.

Blackburn, R. D., D. L. Sutton and T. M. Taylor. 1971. Biological control of aquatic weeds. J. Irrigat. and Drainage Div. 97: 421-432.

Blackburn, R. D., T. M. Taylor and D. L. Sutton. 1971. Temperature tolerance and necessary stocking rates of Marisa cornuarietis L. for aquatic weed control. Proc. European Weed Res. Coun. 3rd Sym. Aquatic Weeds: 79-85.

Butler, J. M., F. F. Ferguson and L. A. Berrios. 1968. Significance of animal control of aquatic weeds. Proc. Southern Weed Conf. 21: 304-308.

Coulson, J. R. 1971. Prognosis for control of water hyacinth by arthropods. Hyacinth Control J. 9: 31-34.

Cross, D. G 1969. Aquatic weed control using grass carp. J. Fish Biology 1: London, England, 27-30.

Demian, E. S. and R. G. Lutfy. 1964. Prospects of the use of Marisa cornuarietis in the biological control of Limnaea caillaudi in the UAR. Proc. Egypt. Acad. Sci. 18: 46-50, Pls. 1-111.

Demain, E. S. and R. G. Lutfy. 1965a. Predatory activity of Marisa cornuarietis against Bullinus (Bullinus) truncatus, the transmitter of urinary schistosomiasis. Ann. Trop. Med. Parasit. 59: 331-336.

Demain, E. S. and R. G. Lutfy. 1965b. Predatory activity of Marisa cornuarietis against Biomphalaria alexandrina under laboratory conditions. Ann. Trop. Med. Parasit. 59: 337-339.

Demain, E. S. and A. M. Ibrahim. 1969a. Feeding activities of the snail Marisa cornuarietis (L.) under laboratory conditions. Sixth Arab Sci. Cong., Damascus, November 1-7, 1969, Pt. 1: 145-165.

Demain, E. S. and A. M. Ibrahim. 1969b. Tolerance of the snail Marisa cornuarietis (L.) to desiccation and to continuious submersion, under laboratory conditions. Bull. Zool. Soc. Egypt. 22: 73-88.

Doutt, R. L. 1964. The historical development of biological control. Ch. 2 in Biological Control of Insect Pests and Weeds. P. Debach (ed.) Reinhold Publ. Co., New York. 844 pp.

Ferguson, F. F. and J. M. Butler. 1966. Ecology of Marisa and its potential as an agent for the elimination of aquatic weeds in Puerto Rico. Proc. Southern Weed Conf. 19: 468-476.

Grizzell, R. A., Jr. and W. W. Neely. 1962. Biological control for waterweeds. Transatlantic North American Wildlife Conf. 27: 107-113.

Harris, P. and Zwolfer, H. 1968. Screening of phytophagous insects for biological control of weeds. Canadian Entomol. 100: 295-303.

Harris, P. 1971. Biological control of weeds. Environ. Letters 2(2): 75-88.

Harris, P. and G. L. Piper. 1970. Ragweed (Ambrosia spp.: Compositae): Its North American insects and the possibilities for its biological control. Tech. Bull.: Commonw. Inst. Biol. Control 13: 117-140.

Hickling, C. F. 1965. Biological control of aquatic vegetation. PANS 11: 237-244.

Holm, L. 1969. Weed problems in developing countries. Weed Sci. 17: 113-118.

Hubendick, B. 1966. Some aspects of vector snail control. Malacologia 5(1): 31-32.

Huffaker, C. B. 1957. Fundamentals of biological control of weeds. Hilgardia 27(3): 101-157.

Huffaker, C. B. 1958. Principles of biological control of weeds. Proc. 10th Internatl. Congr. Entomol. 4(1956): 1958.

Huffaker, C. B. 1959. Biological control of weeds with insects. Ann. Rev. Entomol. 4: 251-276.

Huffaker, C. B. and L. A. Andres. 1970. Biological weed control using insects. Proc. FAO Internatl. Conf. on Weed Control 2: 222-230.

Hughes, H. R. 1971. Control of the waterweed problem in the Rerva River. Fiji Agr. J. N.S. 33: 67-72.

Hunt, B. P. 1958. Introduction of Marisa into Florida. Nautilus 52: 53-55.

Krupnauer, V. 1971. The use of herbivorous fishes for ameliorative purposes in Central and Eastern Europe. 3rd Symp. Aquatic Weeds, EWRC, Oxford, England 95-103.

Lekic, M. 1970. Ecology of the aquatic insect species Parapoynx stratiotata L. (Pyraustidae, Lepidoptera). Arh. poljopr. Nauke 23: 49-62.

Lekic, M. and Lj. Mihajlovic. 1970. Entofauna of Myriophyllum spicatum L. (Halorrhagidaceae) and aquatic weed on Yugoslav Territory. Arh. poljopr. Nauke 23: 63-76.

Little, E. C. S. 1968. The control of water weeds. Weed Res. 8: 75-105.

Maddox, D. M. 1968. Bionomics of the alligatorweed flea beetle, Agasicles sp. in Argentina. Ann. Entomol. Soc. Amer. 61(5): 1299-1305.

Maddox, D. M. 1969. Sex determination of pupae of Vogtia malloi (Lepidoptera: Phycitidae). Ann. Entomol. Soc. Amer. 62(5): 1212-1213.

Maddox, D. M., L. A. Andres, R. D. Hennessey, R. D. Blackburn and N. R. Spencer. 1971. Insects to control alligatorweed: an invader of aquatic ecosystems in the United States. Bioscience 21(19): 985-991.

Maddox, D. M. and R. N. Hambric. 1970. Use of alligatorweed flea beetle in Texas: an exercise in environmental biology. Proc. Southern Weed Conf. 23: 283-286.

Maddox, D. M. and R. N. Hambric. 1971. A current examination of the alligatorweed flea beetle in Texas. Proc. Southern Weed Conf. 24: 343-348.

Mehta, I. and R. K. Sharma. 1972. Control of aquatic weeds by the white amur in Rajasthan, India. Hyacinth Control J. 10: 16-19.

Michewicz, J. E., D. L. Sutton and R. D. Blackburn. 1972a. The white amur for aquatic weed control. Weed Sci 20: 106-110.

Michewicz, J. E., D. L. Sutton and R. D. Blackburn. 1972b. Water quality of small enclosures stocked with white amur. Hyacinth Control J. 10: 22-25.

Nag Raj, T. R. 1965. Thread blight of water hyacinth. Curr. Sci. 34(21): 618-19.

Nag Raj, T. R. 1966. Fungi occurring on witchweed in India. Tech. Bull. Commonwealth Inst. Bio. Control 7: 75-79.

Nag, Raj, T. R. and K. M. Ponnappa. 1969a. Some interesting fungi occurring on aquatic weeds and Striga spp. in India. J. Indian Bot. Sco. XLIX(1-4). 1970.

Nag, Raj, T. R. and K. M. Ponnappa. 1969b. Blight of water hyacinth caused by Alternaria eichhorniae sp. nov. Trans. British Mycol. Soc. 55(1): 123-130.

Penfound, W. T. and T. T. Earle. 1948. The biology of the water hyacinth. Ecol. Monogr. 18(4): 447-472.

Pettet, A. and S. J. Pettet. 1970. Biological control of Pistia stratiotes L. in Western State, Nigeria. Nature 226: 282.

Ponnappa, K. M. 1970. On the pathogenicity of Myrothecium roridum - Eichhornia crassipes isolate. Hyacinth Control J. 8: 18-20.

Randall, J. E. 1965. Grazing effect on sea grasses by herbivorous fish in the West Indies. Ecology 46: 255-260.

Rao, V. P. 1965. Survey for natural enemies of witchweed and water hyacinth and other aquatic weeds affecting waterways in India: Report for period January to December 1964. Commonw. Inst. Biol. Con., unpubl.

Rao, V. P. 1969. Possibilities of biological control of aquatic weeds in India. Water Resources J. C/82: 40-50.

Rich, E. R. and Rouse, W. 1970. Mass producing a tropical snail for biological control. Proc. Southern Weed Conf. 23: 288-298.

Robson, T. O. 1971. The European Weed Research Council's 3rd Symposium on Aquatic Weeds. Conf. Rpt. PANS 17: 515-516.

Sankaran, T., D. Srinath and K. Krishna. 1966. Studies on Gesonula punctifrons Stal. (Orthoptera: Acrididae: Cyrtacanthacridinae) attacking water hyacinth in India. Entomophaga 11(5): 433-440.

Sankaran, T., Pk. B. Menon, E. Narayanan, K. Krishna and Y. Ranganath Bhat. 1970. Studies on natural enemies of witchweed, nutsedge and several aquatic weeds (for USA) in Commonw. Inst. Bio. Control Ann. Rpt. for 1970.

Schuster, W. H. 1962. Fish culture as a means of controlling aquatic weeds in inland waters. FAO Fish. Bull. 5(1): 15-24.

Seaman, D. E. and W. A. Porterfield. 1964. Control of aquatic weeds by the snail, *Marisa cornuarietis*. Weeds 12: 87-92.

Sguros, P. L., T. Monkus and C. Phillips. 1965. Observations and techniques in the study of the Florida Manatee-reticent, but superb weed control agent. Proc. Southern Weed Conf. 18: 588.

Silva, L. de Olivieri de. 1960. Control biologicao de *Anacharis densa* (Planch.) Vict. nos lagos de Universidad Rural de Rio de *Janeiro. Agron.* 18: 117-27.

Silveira-Guido, A. 1965. Natural enemies of weed plants. Final Rpt. (Unpubl. rpt., Department Sanidad Vegetal, Univ. de La Republic, Montevideo, Uruguay).

Spencer, N. R. 1971. The potential usefulness of an aquatic Lepidoptera as a control agent for *Myriophyllum spicatum*. Southern Weed Conf. 24: 348.

Stott, B., D. G. Cross, R. E. Iszard and T. O. Robson. 1971. Recent work on grass carp in the United Kingdom from the standpoint of its economics in controlling submerged aquatic plants. 3rd Symp. Aquatic Weeds, EWRC, Oxford, England: 105-116.

Swingle, H. S. 1957. Control of pond weeds by the use of herbivorous fishes. Proc. Southern Weed Conf. 10: 11-17.

Vogel, E. and A. D. Oliver, Jr. 1969. Evaluation of *Arzama densa* as an aid in the control of water hyacinth in Louisiana. *J. Econ. Entomol.* 62: 142-145.

Vogt, G. B. 1960. Exploration for natural enemies of alligatorweed and related plants in South America. USDA Special Rpt. Pl-4, 58 pp.

Vogt, G. B. 1961. Exploration for natural enemies of alligatorweed and related plants in South America. USDA Special Rpt. Pl-5, 50 pp.

Weldon, L. W. and W. C. Durden. 1970. Integrated biological and chemical control of aquatic weeds. Proc. Southern Weed Conf. 23: 282.

Wilson, F. 1964. The biological control of weeds. Ann. Rev. Entomol. 9: 225-244.

Wilson, C. L. 1969. Use of pathogens in weed control. Ann. Rev. Phytopathology 7: 411-434.

Yeo, R. R. 1967. Silver dollar fish for biological control of submersed aquatic weeds. Weeds 15: 27-31.

Yeo, R. R. and T. E. Fisher. 1970. Progress and potential for biological weed control with fish, pathogens, competitive plants, and snails. FAO Internatl. Conf. on Weed Control, Rome, Italy.

Zettler, F. W., T. E. Freeman, R. E. Retz and H. R. Hill. 1971. Plant pathogens with potential for biological control of water hyacinth and alligatorweed (Presented at 11th Ann. Meet. of Hyacinth Cont. Soc., Tampa, Florida, July 1971).

Zwolfer, H. 1968. Some aspects of biological weed control in Europe and North America. Proc. 9th British Weed Control Conf. 1147-1156.

Zwolfer, H. and P. Harris. 1971. Host specificity determination of insects 6006
for biological control of weeds. Ann. Rev. Entomol. 16: 159-178.

HOST PLANT RESISTANCE TO PLANT PATHOGENS AND INSECTS: HISTORY, CURRENT STATUS, AND FUTURE OUTLOOK

John F. Schafer
Department of Plant Pathology
Washington State University

Abstract

Host plant resistance is a significant aspect of biological control of insects and pathogens. It may be considered as any character that causes a plant to have less disease, insect attack, or overall loss than another. Its use was first seriously begun late in the 19th century. Possible classification is into antibiosis, nonpreference, and tolerance for insects, and escape, exclusion, inhibitory host-parasite interactions, and tolerance for diseases. Resistance may also be distinguished as specific or general, and by the mode of inheritance. Resistance appears best utilized as a rational part of a broad scheme of integrated control, set in an epidemiological context. Some resistance has been long-lasting; other has been effective but ephemeral; whereas in other cases it has been difficult to identify or to dissociate from adverse characteristics. Several strategies are proposed to lengthen the effective lifetime of resistance in its evolutionary setting. A major factor in these strategies is the avoidance of host uniformity.

Introduction

Use of inherent resistance of certain plants is a significant segment of biological control of insects and pathogens as viewed in a broad sense (Dicke, 1972). Resistance may be examined in several contexts. Its use should be considered in an ecological and dynamic evolutionary system. We propose to manipulate resistance through breeding, to our advantage in plant production, but with the realization of its place in this dynamic system with interorganismal and environmental constraints which provide evolutionary pressure on the pathogen or pest. This introductory presentation is to provide general background for the subsequent considerations of specific crops.

History

Resistance to disease and insects was recognized and reported early in the 19th century. Obviously, it must have been observed earlier. Painter (1951) and Walker (1966) have briefly reviewed this history. Painter notes Lindley's report on an apple variety resistant to the woolly aphid in 1831. Darwin subsequently proposed searching for late blight resistance in potato at the time of the infamous disaster in that crop. Conscious use of resistance as an economic factor in disease and insect control was seriously begun later in the 19th century. The grape phylloxora became an important pest in Europe. Resistant American grape rootstalks were utilized in protection from attack (Painter, 1951). Stem rust was an important problem of wheat in Australia. Cobb (1890, 1892, 1894) began a study of stem rust resistance, and Farrer (1898) included resistance as a breeding objective in his program about 1890. Orton (1909) in the U.S. began field selection for cotton plants resistant to wilt about 1900, and this was soon followed by similar procedures for watermelon, flax, and cabbage.

Inheritance of resistance to yellow rust of wheat was reported by Biffen (1905) in England in 1905. He happened to work with a simply inherited character which easily fit the mathematical model of Mendel, publicized shortly before, and thus fitted disease resistance into genetic theory. Biffen's work was controversial at first, for the now obvious reasons of differential pathogenicities and environments. However, it began a long history of determining the genetics of resistance, to provide background for manipulating it as a breeding character in plant improvement.

From this early history began the look at resistance as an entity, as a "something" that could be considered as a unit, that was useful as such an item, and which could be recombined as a character with other desired characters in plant breeding. In this sense, resistance was considered a "thing," as we do in titles today. Obviously, in total it is not, but represents several phenomena, is comprised of various components, conditioned by various mechanisms, effective in varying degrees, and affected by the environment. Thus, some definitions, categorizations, and analyses are called for to further our understanding.

Definitions and Categories

Resistance, in the broad sense, may be considered a character of the host plant, causing it to have less disease, less insect attack, or less overall loss than another plant, cultivar, or species subjected to the same attack or epidemic. In order to study it or adequately utilize it, we must consider resistance in greater detail. Various categorizations may be made, depending on the particular purpose. Painter (1951) classified resistance to insects as being based on three possible types of mechanisms: antibiosis, nonpreference, and tolerance. He described these as three interrelated components, one or more of which is frequently present in resistant cultivars. Antibiosis provides an adverse effect by the plant on the insect, thus restricting its development. Nonpreference implies some characteristic that inhibits use of the plant for oviposition, food, or shelter. Tolerance represents the capacity to withstand the presence of the insect with lesser damage than on a susceptible host. This may be due to repair, recovery, or a less severely damaging response to the presence of insect infestation.

In turn, disease resistance has been classified various ways. One breakdown is into escape, exclusion, inhibitory host-parasite interactions, and tolerance (Schafer, 1971). Walker (1941) in 1941 listed the first three and in 1966 (Walker, 1966) the last three. A classical presentation was that of Cobb (1890, 1892, 1894), in 1894, including rust-proof, rust-resistant, rust-escaping, and rust-enduring in regard to his work on wheat. By far the most important category is that resulting from host-parasite interactions. This appears somewhat analogous to Painter's antibiosis. Exclusion and host-parasite interactions have been distinguished from each other on the basis of a pre-existing barrier (physical or chemical) on the one hand in contrast to no such initial barrier, but only a response after infection occurs. Earlier workers expected some barrier or at least an inhibitory chemical. However, presence of a genetic basis for triggering of resistance responses upon infection has been more common (Allen, 1959). Walker (1941) proposed several mechanisms of exclusion: 1) repellance of a vector, 2) mechanical (physical), 3) chemical, and 4) physiological. Examples include the Lloyd George and other raspberry cultivars, indicated by Huber and Schwartze (1938) to escape mosaic infection because of resistance to the aphid vector. Physical exclusion by corn of the smut fungus was found to be effected by the degree of husk coverage of young ears in studies by Kyle (1929). The relation of pigmentation of the outer scales of onion to resistance to smudge and neck rot was shown to be a chemical exclusion by Walker and associates (1930). A physiological or functional exclusion may be harder to identify, but one example was reported by Pool and McKay (1916) in which the resistance of sugar-beet leaves to Cercospora beticola was an age-related phenomenon based on amount of stomatal movement.

The major grouping of resistance to plant disease is that which involves the host-parasite interaction upon infection. This may occur with evident histological responses, or there may be no such obvious expression. Conant (1927) showed that in tobacco roots infected by Thielaviopsis basicola resistance was brought about by a proliferation of cork cells around the invading hyphae.

As the study of the biochemistry and physiology of plant disease has expanded in recent years, studies of the nature of host-parasite interactions bringing about resistance have greatly increased. This is now a major field of endeavor, in a relatively early stage of development, which I shall not attempt to review. Agrios (1969) includes an extended consideration of how plants defend themselves against pathogens.

Disease tolerance, like the analogous insect tolerance, is also a useful category of resistance. Its presence implies less damage to a tolerant cultivar than to an equally diseased susceptible cultivar. Its conceptualization in regard to disease, in contrast to relating to the pest or pathogen, depends on differentiating disease from damage or loss (Schafer, 1971). Some classical measures of disease include only certain obvious signs and symptoms. If, on the other hand, expression of disease is considered to include all aspects of damage, there is no distinction left by which to measure disease tolerance. However, tolerance to a given level of the pathogen may be identified. Tolerance has the advantage of not affecting the reproduction of the pest or pathogen, and thus does not apply evolutionary pressure for the development of a strain for which tolerance is no longer effective.

It may be argued whether escape is a bona fide inclusion in the broad concept of resistance. Nevertheless, when due to a character of the host plant, and useful in reducing disease, it fits the general scheme. Early maturing cultivars which replaced older, later maturing types have, in fact, been effective in lessening losses from the rusts (Caldwell, et al., 1954; Heald, 1937).

It should be noted that the term disease resistance is used in differing ways. Often, it is used in the broad sense just discussed, encompassing all of the various phenomena which bring it about, including even escape and tolerance. On the other hand, resistance may be viewed in a narrower sense, specifically to describe direct restrictive host-parasite interactions, as they are contrasted to the more compatible interactions, designated as susceptibility. Such variable use of a word is not unusual, but may confuse and require extra explanation. In this narrow sense, escape and tolerance would not be included, and might be contrasted to resistance as distinct possible mechanisms. The glossary of the recent plant-disease treatise (National Academy of Sciences, 1968) lists "resistance--the inherent capacity of a plant to prevent or restrict the entry or subsequent activities of a pathogenic agent when the plant is exposed to inoculum under environmental conditions suitable for infection."

I would further note that any of these various phenomena, described generally as resistance, are comparative to something else which is less resistant. We may consider resistance as a phenomenon or entity and manipulate it as a character, but as defined, resistance represents a relationship that is comparative to a second host-pathogen relation, which by the comparison is less resistant, or susceptible. When viewing disease or insects in a crop production situation, we start from an existing condition of disease or insect attack. Any host characteristic that lessens this is a form of resistance (whatever its degree or mechanism) and is comparative to the initially found condition. Whether we exploit it depends on a judgment as to whether it is economically useful, compared to costs of development. We may thus be interested in high types (possibly immunity) or much more limited degrees if they show value and can be identified.

There are other useful ways to classify resistance. Van der Plank (1963) has developed a valuable classification, which he calls vertical and horizontal. I prefer more descriptive terms, such as specific and general (Caldwell, 1968; Thurston, 1971). Specific resistance is effective against some races of a pathogen or pest but not others. This is found in the case of those organisms which also have a high degree of specificity, and frequently are obligate parasites. Examples are the resistances to the rust, powdery mildew, and apple scab fungi, and to the hessian fly of wheat. Occurrence of specific resistance implies races of the pathogen, and vice versa. General resistance, in contrast, is that which is equally useful against all collections or isolates of the pathogen. This total distinction of specific and general may break down as a biological possibility when examined closely (Aslam, 1972), as do almost all categorizations when pressed to their limits. However, it is a useful, practical distinction for disease and insect resistance breeding applications. It is important to know if resistance is broadly effective or will break down to race development. General resistance may often be found to be of a lower degree. Thus, it is sometimes identified and searched for on this basis, and intermediate resistance is initially assumed to be general (Poehlman, 1952). This may be a useful procedure, but is not necessarily true. Van der Plank (1968) and Caldwell (1968) have provided examples of specific and general resistance as can also be done in the specific crop areas to follow.

Resistance may also be separated into monogenic and polygenic (Shay, 1960; Van der Plank, 1968). The separation may again not necessarily be sharp, but the distinction is important for exploitation through plant breeding, in the handling of progenies. Identification and reassembling of resistance requires smaller breeding populations for monogenic or oligogenic (Van der Plank, 1968) resistance and usually a less precise test. The association of monogenic and polygenic resistance with specific and general resistance, respectively, is frequent enough that the numerical situation may be used to search for or identify the general resistance (Nelson, 1972; Shay, 1960). One may even describe a resistance as polygenic, and fully infer that it is thus general rather than specific, or vice versa. Although such a particular conclusion may be valid, I believe this usage is to be avoided, as the relation is not obligatory (Simons, 1972).

Utilization

I would like to turn to the utilization of resistance and some of the factors that bear on this. No matter how intricate and interesting the details of an individual program become, we must focus on what I would choose to call the "big picture." Entomologists have used the phrases pest management and integrated control. Previous presentations bear on these. I believe the use of resistance is one of the most effective and economical means of control, possibly the least well exploited in relation to its potential. Nevertheless, it should be fitted into the best, net mix of practices. The ecological and epidemiological relations of these should be understood. Resistance is a means to an end, not a cure-all or objective of its own. I would cite some examples from previous work in our programs at Purdue. Late fall planting of wheat is a cultural practice used to avoid severe infestation by hessian fly. With Dual wheat in 1955 (Caldwell, et al., 1957), we had fly resistance and could now plant earlier, violating the previous cultural practice, but achieving advantages of reduced potential winter-kill and increased economical potential from cattle grazing. We, of course, found that planting after the "fly-free" date had not only assisted in avoiding attack by hessian fly, but also avoided seedling infection by leaf rust, stem rust, and powdery mildew fungi, and yellow dwarf virus. Thus, resistance was obviously narrower in its effect than late planting. The broad ecological picture and its dynamic nature must be understood. Progress occurred with the exploitation of hessian fly resistance, but numerous problems remained.

Apple scab has been limiting to apple production in Indiana. As many as 13 sprays may be needed for adequate control. An aggressive scab resistance program was generated (Williams and Kuć, 1969). This will be valuable, but it will not totally eliminate spraying of apples in Indiana, as is evident to those familiar with the scope of production problems of apples.

These examples point out a broad picture of pest management of which resistance is an important, but not necessarily the only, part. Progress is real, but total or permanent success from use of resistance is not always the result at a given point in time.

A second, different, broad picture is that of overall plant improvement through breeding. Again, resistance is important, but not the only part. A disease or insect resistant cultivar is of little use unless it is also productive, possesses proper quality, and other essential characteristics. Priorities must be determined, and set, in a plant breeding program. The place of resistance, and which resistance, should be ascertained, and a sensible program implemented, without either undue emphasis, nor ignoring of real needs. The efforts at attaining resistance should be integrated with those aimed at other improvement goals. Suitable organization or cooperation is vitally needed to achieve these ends (Painter, 1951).

Results from Resistance

Use of resistance for different crops and for different insects or diseases may vary greatly. Following early skepticism, yet when resistance exploitation was still a relatively new endeavor, it was thought by some to be a final solution to end many problems. With the development of differentially pathogenic races of some pathogens and insects, which attacked previously resistant cultivars, considerable pessimism set in (Walker, 1959). Van der Plank (1968) suggests that this in turn should pass. We now know that some resistance has been long lasting and very valuable. It may be general in the sense of Van der Plant (1968) and Caldwell (1966), or it may merely relate to an epidemiological pattern that has allowed it to remain functional over a long period. For example, resistance to certain soilborne pathogens has persisted many decades. Although races of some of the Fusaria causing the wilts have been identified (Walker, 1941), many of these have not spread, and the resistances found early by Orton (1909) and Jones (1915), and others have remained valuable for 60 years or so. The example I know best is the soilborne mosaic of wheat, apparently spread through soils of much of the Eastern soft wheat area, but not now of economic significance because of the continued effectiveness of resistance for many decades. Year after year resistant cultivars are effective, whereas, in areas where susceptible cultivars are grown, they may be severely damaged (Sill, et al., 1960).

On the other hand, the airborne, highly specialized, variable, and relatively wide ranging agents such as the rust, mildew, and late blight fungi and the hessian fly remain as important threats. I would note that resistance to these, over the years, has been extremely valuable, and surely paid its way many times over, but nevertheless has often been ephemeral.

There would appear to be a third situation in addition to those two in which resistance has been effective and long lasting on the one hand or effective but ephemeral on the other. This is illustrated by the stalk rot diseases of corn and sorghum. There are differences between hybrids or lines of these crops. Yet they are not easily measurable. Cultivar differences cannot be readily separated from the effects of normal maturation of the crop, nor can they be readily separated from adverse physiological aspects related to resistance such as the lack of translocation of metabolites from stalks to grain. These problems appear to be a common situation of the "low level" parasite in contrast to the highly specialized one. More research is needed on resistance to this nature of disease.

Plant Rust Example

Let me briefly trace the rust picture to summarize this example of history, status, and outlook of the ephemeral type of resistance. In 1894 Eriksson (1894) showed that the stem rust fungi on various cereals did not cross-infect but were independent formae specialis. From an epidemiological point of view this was good, as the source of inoculum would be more limited. With this optimistic outlook, exploitation of resistance was initiated. In 1917 Stakman and Piemeisel (1917) extended Eriksson's finding to show specialized races not only on the different grain species, but on cultivars within these crop species. This was not so good as these were now the entities that overcame resistance within a crop. A race attacking one cultivar and not another would be resisted by the second. Another race attacking that cultivar would be the source of the transient quality of such resistance. In the years following Stakman and Piemeisel's work, effectiveness of many resistant cultivars has, in fact, sooner or later been negated by selection and increase of the specifically pathogenic races. Ceres wheat with stem rust resistance from Kota was removed in 1935 by race 56 (Waldron, 1935). Important durum wheats were decimated in 1953 and 1954 by race 15B (Stakman and Harrar, 1957). Clinton and related oats with Bond resistance were severely damaged in the crown rust epidemic of 1957 (Endo and Boewe, 1958).

Flor in writing since 1942 (1971) has best demonstrated this host:parasite relationship in what has come to be called the gene-for-gene theory. Person (1959) has arranged Flor's proposals into a highly organized, logical system. Interestingly, this showed that to some extent the "race" concept as widely envisioned had gotten in the way in our thinking. What we are concerned with are specific virulences overcoming specific resistances, rather than the logarithmically based numbers of races. Races are merely packages of the characters of virulence and avirulence as measured by cultivars possessing arbitrarily chosen resistance genes. The decimation of a resistance is not brought about necessarily by the presence of a single new race, but by a new virulence, which may be packaged in several possible races. Interestingly, certain ones of these may predominate. The biological basis for such a particular packaging rather than for other combinations of virulence is currently of great interest. Such possible occurrence was reviewed in 1972 at a Discussion Session of the American Phytopathological Society.

Basically, what we have is a step-wise evolutionary sequence in which we start with disease. The host plant provides usually rare resistance to this, which we select and multiply, thus speeding evolution. The common population of the pathogen can not now propagate on this host. Thus, there is pressure on the pathogen to provide a variant which can survive, if it has the biological capacity to provide such a variant (Person, 1959). If this does occur, the new strain has little competition and spreads quickly and wide, forming a new population which can attack the previously resistant cultivar. Obviously, such parasites need this capacity of variation to survive such a situation. In that they have survived many centuries, it is equally obvious that they are adapted to face such problems placed before them by the host plant. The major difference we have provided in manipulation of resistance is one of timing.

Our big challenge, it seems to me, is how to increase the length of time of the functioning of resistance in this sort of system that we have now identified. The answer would appear to be based at least partly in diversity. There are several possibilities (Nelson, 1972). Careful hoarding of resistance genes and slow use over time would be one. This would need regulation and still does not appear too promising. It would be a slight improvement presumably over our present random usage. Alternatively, we could systematically arrange for interregional diversity (Browning, et al., 1969; Caldwell, et al., 1957; Knott, 1972). This will take more organization than we have been able to generate so far, but appears promising. There could also be local or intraregional diversity. A little of this has been achieved with use of several differing resistant cultivars (Caldwell, et al., 1957), but mostly we haven't been this far ahead of the diseases or insects when races are a problem to develop a systematic plan. A current example is the soft winter wheat, Redcoat, with leaf rust resistance from Surpresa, as a variant resistance, when most cultivars of the area possess that from Chinese. We could have intravarietal diversity, the multiline concept of Jensen (1952); Borlaug (1959); and Browning and Frey (1969). Finally, there could be combined resistance, more than one gene per plant, to pile up the genetic barriers to selection of virulence (Nelson, 1972; Schafer, et al., 1963; Watson and Singh, 1952). These latter two types of approach are in preliminary developmental stages or in trial runs.

The main point I would leave is the plea to avoid wide areas of uniformity. This is the lesson of the oat Victoria blight of 25 years ago (Wallin, 1948) or the southern corn leaf blight of 1970 (Hooker, 1972). Be sure there is variability. Even when we can't foresee our problems, as in those two cases, when variation is present, those problems which do occur should result in small failures rather than in vast, widespread ones. Communication and transfer of materials, as now practiced, could actually present a worse situation than that faced with more limited, smaller, random efforts if the communication leads to uniformity. Thus, once we move to our current level of sophistication of activity and communication, we should also take the next step of sophistication, that of planning and implementing to avoid those pitfalls, which have now become obvious, and which may have been made worse by our first steps in utilizing resistance.

Summary

In summary, for the future, I would state: 1) Use resistance because it is effective, cheap, and non-contaminating, 2) Fit use of resistance into the overall epidemiology of disease and pest attack, 3) Integrate resistance with other means of control, 4) Integrate resistance with other phases of plant improvement, 5) Clap your hands in glee when resistance appears permanent, and 6) Continue the effort to develop more effective systems, which should be possible, where the biology makes the problem more difficult, and we don't get it all solved before lunch on the first day out.

Literature

Agrios, G. N. 1969. Plant pathology. New York: Academic Press. 629 pp.

Allen, P. J. 1959. Physiology and biochemistry of defense. In Horsfall, J. G. and A. E. Dimond, ed., Plant Pathology. New York: Academic Press. V. 1: 435-467.

Angell, H. R., J. C. Walker and K. P. Link. 1930. The relation of protocatechuic acid to disease resistance in the onion. Phytopathology 20: 431-438.

Aslam, M. 1972. Aggressiveness in Puccinia recondita f. sp. tritici: I. Concepts and terminology. II. Predominance of one culture over another in composites. III. Components of aggressiveness. Diss. Abstr. Inter. 33: 1883B.

Biffen, R. H. 1905. Mendel's law of inheritance and wheat breeding. J. Agr. Sci. 1: 4-48.

Borlaug, N. E. 1959. The use of multilineal or composite varieties to control airborne epidemic diseases of self-pollinated crop plants. Inter. Wheat Genet. Symp. 1: 12-27, Univ. Manitoba, Winnipeg.

Browning, J. A. and K. J. Frey. 1969. Multiline cultivars as a means of disease control. Ann. Rev. Phytopathology 7: 355-382.

Browning, J. A., M. D. Simons, K. J. Frey and H. C. Murphy. 1969. Regional deployment for conservation of oat crown-rust resistance genes. In Browning, J. A., ed., Disease consequences of intensive and extensive culture of field crops. Iowa Agr. and Home Econ. Exp. Sta. Spl. Rpt. 64: 49-56.

Caldwell, R. M. 1966. Advances and challenges in the control of plant diseases through breeding. USDA-ARS 33-110: 117-126.

Caldwell, R. M. 1968. Breeding for general and/or specific plant disease resistance. Inter. Wheat. Genet. Symp. 3: 263-272, Australian Acad. Sci. Canberra.

Caldwell, R. M., L. E. Compton, W. B. Cartwright, J. F. Schafer and F. L. Patterson. 1957. Utilization of the W38 resistance of wheat to hessian fly. Agron. J. 49: 520.

Caldwell, R. M., L. E. Compton, J. F. Schafer and F. L. Patterson. 1954. Knox, a new leaf rust resistant and stem rust escaping soft winter wheat. (Abstr.) Phytopathology 44: 483-484.

Caldwell, R. M., J. F. Schafer, L. E. Compton and F. L. Patterson. 1957. A mature-plant type of wheat leaf rust resistance of composite origin. Phytopathology 47: 690-692.

Cobb, N. A. 1890, 1892, 1894. Contributions to an economic knowledge of Australian rusts (Uredineae). Agr. Gaz. N.S.W. 1: 185-214, 3: 44-48, 181-212, 5: 239-250.

Conant, G. H. 1927. Histological studies of resistance in tobacco to Thielavia basicola. Amer. J. Bot. 14: 457-481.

Dicke, F. F. 1972. Philosophy on the biological control of insect pests. J. Environ. Qual. 1: 249-253.

Endo, R. M. and G. H. Boewe. 1958. Losses caused by crown rust of oats in 1956 and 1957. Plant Dis. Rptr. 42: 1126-1128.

Eriksson, J. 1894. Ueber die Specialisierung des Parasitismus bei den Getreiderostpilzen. Ber. Deut. Bot. Ges. 12: 292-331.

Farrer, W. 1898. Making and improvement of wheats for Australian conditions. Agr. Gaz. N.S.W. 9: 131-168, 241-250.

Flor, H. H. 1971. Current status of the gene-for-gene concept. Ann. Rev. Phytopathology 9: 275-296.

Heald, F. D. 1937. Introduction to plant pathology. New York: McGraw-Hill, 579 p.

Hooker, A. L. 1972. Southern leaf blight of corn--present status and future prospects. J. Environ. Qual. 1: 244-245.

Huber, G. A. and C. D. Schwartze. 1938. Resistance in the red raspberry to the mosaic vector *Amphorophora rubi* Kalt. J. Agr. Res. 57: 623-633.

Jensen, N. F. 1952. Intra-varietal diversification in oat breeding. Agron. J. 44: 30-34.

Jones, L. R. and J. C. Gilman. 1915. The control of cabbage yellows through disease resistance. Wisconsin Agr. Exp. Sta. Res. Bul. 38. 70 pp.

Knott, D. R. 1972. Using race-specific resistance to manage the evolution of plant pathogens. J. Environ. Qual. 1: 227-231.

Kyle, C. H. 1929. Relation of husk covering to smut of corn ears. USDA Tech. Bul. 120. 7 pp.

National Academy of Sciences. 1968. Plant-disease development and control. Nat. Acad. Sci. Publ. 1596. 205 pp.

Nelson, R. R. 1972. Stabilizing racial populations of plant pathogens by use of resistance genes. J. Environ. Qual. 1: 220-227.

Orton, W. A. 1909. The development of farm crops resistant to disease. USDA Yearbook 1908: 453-464.

Painter, R. H. 1951. Insect resistance in crop plants. New York: Macmillan. 520 pp.

Person, C. 1959. Gene-for-gene relationships in host:parasite systems. Canadian J. Bot. 37: 1101-1130.

Poehlman, J. M. 1952. Are our varieties of oats too resistant to disease? Amer. Soc. Agron. Abstr. 1952: 15-16.

Pool, V. W. and M. B. McKay. 1916. Relation of stomatal movement to infection by *Cercospora beticola*. J. Agr. Res. 5: 1011-1038.

Schafer, J. F. 1971. Tolerance to plant disease. Ann. Rev. Phytopathology 9: 235-252.

Schafer, J. F., R. M. Caldwell, F. L. Patterson and L. E. Compton. 1963. Wheat leaf rust resistance combinations. Phytopathology 53: 569-573.

Shay, J. R. 1960. Breeding vegetable and fruit crops for resistance to disease. *In* Reitz, L. P., ed., Biological and chemical control of plant and animal pests. AAAS Publ. 61: 229-244. Washington, D.C.

Sill, W. H., Jr., H. Fellows and E. G. Heyne. 1960. Reactions of winter wheats to soil-borne wheat mosaic virus in Kansas. Kansas Agr. Exp. Sta. Tech. Bul. 112. 7 pp.

Simons, M. D. 1972. Polygenic resistance to plant disease and its use in breeding resistant cultivars. J. Environ. Qual. 1: 232-240.

Stakman, E. C. and J. G. Harrar. 1957. Principles of plant pathology. New York: Ronald. 581 pp.

Stakman, E. C. and F. J. Piemeisel. 1917. A new strain of *Puccinia graminis*. Phytopathology 7: 73.

Thurston, H. D. 1971. Relationship of general resistance: Late blight of potato. Phytopathology 61: 620-626.

Van der Plank, J. E. 1963. Plant diseases: Epidemics and control. New York: Academic Press. 349 pp.

Van der Plank, J. E. 1968. Disease resistance in plants. New York: Academic Press. 206 pp.

Waldron, L. R. 1935. Stem rust epidemics and wheat breeding. North Dakota Agr. Exp. Sta. Cir. 57. 12 pp.

Walker, J. C. 1941. Disease resistance in the vegetable crops. Bot. Rev. 7: 458-506.

Walker, J. C. 1959. Progress and problems in controlling plant diseases by host resistance. *In* Holton, C. S., et al., eds., Plant Pathology problems and progress. Madison: University Wisconsin Press, pp. 32-41.

Walker, J. C. 1966. The role of pest resistance in new varieties. *In* Frey, K. J., ed., Plant Breeding. Ames: Iowa State University Press, pp. 219-42.

Wallin, J. R. 1948. The *Helminthosporium* blight situation in the upper Mississippi Valley region as of December 1947. Plant Dis. Rptr. 32: 94-96.

Watson, I. A. and D. Singh. 1952. The future of rust resistant wheat in Australia. J. Australian Inst. Agr. Sci. 18: 190-197.

Williams, E. B. and J. Kuć. 1969. Resistance in *Malus* to *Venturia inaequalis*. Ann. Rev. Phytopathology 7: 223-246.

UTILIZATION OF BIOLOGICAL AGENTS OTHER THAN HOST RESISTANCE FOR CONTROL OF PLANT PATHOGENS

Thor Kommedahl
Department of Plant Pathology
University of Minnesota, St. Paul

Interest in the biological control of plant pathogens has deepened considerably in the past decade or so. Papers have been published, symposia held, and review articles written on this topic, suggesting a wealth of examples for controlling plant diseases. Yet clear-cut, commercially applicable instances where biological agents are in use are hard to find. Many writers and speakers allude to promising leads for biological control but the investigations reported stress the biology but only suggest hope for control.

As investigators pursue research to control pathogens by biological agents they become aware of the fact that this is a control measure based on certain ecological relationships and that little is known about the relationships among microorganisms that inhabit surfaces of plants. It has become clear that there are resident microorganisms on leaves, buds, and other aerial parts of plants as well as in the better-known rhizospheres of roots. There are a number of general works that treat the ecology of leaf surface organisms (Preece and Dickinson, 1971), and root-infecting fungi (Baker and Snyder, 1965; Griffin, 1972). Thus my presentation will not be an attempt to review the literature on this topic, since this has been done in many review articles (Baker, 1968; Curl, 1963;Starkey, 1958). Instead, I will present kinds of research attempted with selected examples of each kind, and suggest directions that seem likely to be explored to successfully attain a measure of biological control. Sufficient research has been done to show that many pathogens are held in check biologically in nature and that the plant pathologist's aim should be to augment the biological control that already exists.

Investigations on pathogen control can be grouped into topics where microorganisms antagonistic to pathogens are applied to (i) seeds or fruits, (ii) transplants, (iii) foliage, or (iv) where growing plants are used as catch or trap crops, and their residues as substrates for antagonists, to reduce or eliminate inoculum of pathogens in soil.

Organisms Applied to Seeds

Seedling blight of corn.--Work done at Minnesota will be used to illustrate this method of biological control (Chang and Kommedahl, 1968; Mew and Kommedahl, 1972). Fungi are frequently found on kernels of corn (*Zea mays*); some of these are storage and some are field fungi. Field fungi can cause seedling blight under adverse weather and soil conditions. Some of the storage fungi may serve as antagonists to field fungi (Mew and Kommedahl, 1972).

After some fungi and bacteria isolated from both root and leaf surfaces were tested in the laboratory for antagonism to one of the most common root pathogens of corn (Fusarium roseum), two of the best were selected for further study. These were the fungus Chaetomium globosum and the bacterium Bacillus subtilis. Results with these organisms have been published in part (Chang and Kommedahl, 1968; Kommedahl and Chang, 1970).

The application of C. globosum to corn kernels suppressed Fusarium spp. on germinating corn in culture as well as the storage fungi, except Mucor. Bacillus subtilis was less effective. When Aspergillus flavus, A. niger, and four isolates of Penicillium (present as storage fungi on kernels) were used to coat corn kernels, A. niger and three of four isolates of Penicillium suppressed growth of Fusarium spp. on the corn kernels.

When either of these two organisms were used in greenhouse or field trials they gave generally as good results in improvement of stand and yield of corn as the fungicide captan did, as reported previously (Chang and Kommedahl, 1968).

To treat kernels, they were either dipped into a cell suspension of Bacillus subtilis and stored in a refrigerator 1 month before planting, or, dropped into a vessel containing a culture of Chaetomium globosum; by agitating kernels they became coated with spores of this fungus. These too can be stored for a month or so before planting.

In one seed lot of corn, in which kernel infection was high (mainly Fusarium moniliforme), the application of either of the two coating organisms suppressed growth of Fusarium spp. on culture media, and in other lots heavily infested with storage fungi; they too were inhibited by the coating organisms. The action of the coating organisms may not be antibiotic since Chaetomium globosum does not form zones of inhibition on agar although Bacillus subtilis does, when tested against Fusarium roseum.

The purpose of coating kernels is to protect the kernels during germination, and therefore the seminal root system before the secondary (adventitious) root system emerges. Hopefully the organisms may grow further and become established in the rhizosphere before penetrating root epidermis.

In the laboratory, Chaetomium effectively suppressed mycelial development of F. roseum in soil containing growing corn seedlings. The experiment was done also in the greenhouse with similar results. When plants were dug from soil, perithecia of Chaetomium were firmly established on kernel surfaces and gave better root systems of corn than nontreated kernels. Seedling stands were improved from a 50% stand in soil containing the pathogen to 75% when kernels were coated with C. globosum, in the greenhouse at 18 C.

Bacillus subtilis also improved stands in the greenhouse in both sterile and nonsterile soil from 50 to 82% in autoclaved soil and from 40 to 60% in non-autoclaved soil. The improvement in root systems was apparent when plants were dug from soil. Results were similar to those obtained by treating kernels with captan, a common seed treating fungicide.

The trials were made also in the field. Captan gave earlier stands of either organisms or a nontreated control. However, at the end of the season, the results were that there was no difference in stand or yield between the organism-coated and the fungicide-coated kernels; however this depended on year and variety. In general, when captan was effective, the organisms were also. Field trials were made in 1968-70. In 1970, with three varieties and nine replicates per variety, and 100 plants/replicate, grain yield was significantly higher for Chaetomium and captan over the control or Bacillus for the one of the varieties. Whether benefits occur depends on weather conditions for the critical period after planting, between the times that seminal and adventitious roots form. When corn was planted earlier than normal, the period of time of plants in cooler soil was longer, and response to treatment was better than when planted later. This corresponded with earlier greenhouse experiments where control was obtained at 18C but not at 30 C.

When either B. subtilis or C. globosum was applied to kernels, subsequent investigation showed that these organisms multiplied on the root surface of germinating seedlings and could be reisolated from them. Moreover, it was found that in addition to the coating organisms, more organisms antagonistic to the Fusarium pathogen were present in the rhizosphere than where chemical seed treatment was used. Thus there is a shift in the ecological balance of root surface organisms that tends to inhibit pathogens by encouraging antagonists.

Other examples could be cited such as the one by Liu and Vaughan (1965) in which they coated beet seeds with spores of Trichoderma viride or Penicillium frequentans to protect them from seedling preemergence damping off. Also, Mitchell and Hurwitz (1965) reported that tomato seeds inoculated with rhizosphere bacteria (Arthrobacter sp.) suppressed damping off caused by Pythium debaryanum. Other similar reports have been published by Tveit and Moore (1954) and Kommedahl and Brock (1954).

Inoculum Applied to Transplants and Soil

Sclerotial blight of tobacco.--Corticium rolfsii causes a sclerotial blight of tobacco in Japan and biological control measures for this disease have been described by Oshima (1966) working at the Okayama Tobacco Experiment Station. At first, he applied spores of Trichoderma lignorum to tobacco seedlings of transplanting age at the dosage of 3 g spores/plant and planted them in the greenhouse at 28 C at high humidity for 7 days. The success with this method led to field trials. Spores of T. lignorum were collected from a medium, air-dried at 35 C and stored for 1 month in paper bags, 2 kg spores/bag (about 2 million conidia/gram). Experiments were planned so that spores were applied at one or more of the following times: (i) at the time of transplanting, (ii) at the time of the last cultivation (about 1 month later), and (iii) at the time of pruning of leaves from plants. Spores were mixed with soil, and with ground rice bran to make a powder which was applied to plants by hand at the times indicated. This powder was sometimes mixed with soil around tobacco plants, and when this was done the application rate was 2 kg/are. Disease ratings were made 3 months after the time of transplanting.

In four trials in the field in which T. lignorum spores were applied twice. once at transplanting and again 1 month later, the average percentage of infected plants was 13.8 in nontreated plants and 0.5 for treated plants. In trials in which spores were applied to plants only once in the field 1 month after transplanting (May), the average percentage of infected plants for seven trials was 7.1 for nontreated plants and 0.3 for treated ones. Where spores were applied only at leaf cut (June), the average of six trials was 11.1% for nontreated plants and 4.7% for treated ones. Thus it appears that application of spores could reduce incidence of disease in tobacco transplants in Japan. The effect of T. lignorum appears not to be an antibiotic one since no zones of inhibition were seen in culture. It was established by Oshima that T. lignorum digests and destroys sclerotia of C. rolfsii, and causes malformations of hyphae.

Tomato blight.--A similar study was made in Georgia on tomato transplants by Wells, Bell and Jaworski (1972). They used spores of Trichoderma harzianum, probably a synonym of T. lignorum, to control Sclerotium rolfsii, the imperfect stage of Corticium rolfsii (Aycock, 1966) on blue lupines, peanuts and tomatoes in the greenhouse, and on transplant tomatoes in the field.

Wells, et al. (1972) planted tomatoes in a field where S. rolfsii had been a serious problem during the 1970 tomato transplant production season. In a trial of five replicates, spores of T. harzianum (mixed with soil and ground ryegrass grains) were applied by hand at the rate of 1.5 g/cm of row over about a 10 cm band on different dates. Plants were about 4 cm tall at the first application.

Nontreated plots had the highest disease ratings and the lowest yield of healthy tomato plants, and the differences were statistically significant. One application of *T. harzianum* increased the percentage of healthy plants from 22 to 90. Plots receiving three applications in May, produced 99.5% plants that were healthy.

It is likely that Oshima (1966) and Wells, et al. (1972) were using the same fungus species. Both of them applied, to plants and soil, spores of *Trichoderma* in a packaged mixture of spores, soil, and freshly ground plant material. The addition of plant material would ensure a supply of readily available substrate for a food base and thereby enhance the biological activity of *Trichoderma*. Both workers found that zones of inhibition were not produced in culture media and that *Trichoderma* can kill *Sclerotium rolfsii* by destroying sclerotia. Antagonism does not require that antibiosis be observed in culture.

It is necessary now to establish whether either of these diseases can be controlled biologically and economically on a commercial scale in the field. It may be more feasible to control diseases of high-income crops than low-income ones in the way described by these workers.

Bacterial wilt of alfalfa.--Alfalfa is not an example of a plant that is transplanted as a commercial practice. However work by Carroll and Lukezic (1972) illustrates a possible method of control by inoculated roots of seedlings and mature plants. Six-week-old seedlings were inoculated first with avirulent cells of *Corynebacterium insidiosum*, by wounding roots with a needle, then dispensing 2 ml of bacterial cell inoculum into soil around roots. After 24 hr., virulent cells of the same bacterium were introduced to roots. A similar procedure was used for mature plants except that 10 ml of inoculum were applied. The presence of avirulent cells protected plants from infection by virulent ones, in plants grown in a gnotobiotic environment.

Organisms Applied to Foliage

Northern Corn Leaf Blight.--Microorganisms were washed from the leaf surface of corn and tested for their effects on spore germination, appressorium formation, and lesion development of the Northern leaf blight pathogen, *Helminthosporium turcicum*, in Minnesota (Asare-Nyako, 1965).

The application of these organisms, tested singly, to detached corn leaves where *H. turcicum* conidia were present resulted in lower percentages of germination, shorter germ tubes, and fewer appressoria of *H. turcicum* on leaf surfaces. Where spores of *H. turcicum* germinated 56% in the control (only water added to leaves), they germinated 20 and 51% for each of two bacterial species, 1 and 74% for each of two yeast species, and 26% for *Penicillium* sp. Germ tube length of *H. turcicum* averaged 17.8 μ for the control and ranged from 5.3 to 10.2 μ for the five organisms added singly to leaf surfaces.

Where all 63 spores of *H. turcicum* in the control produced appressoria on corn leaves, only 15% of 103 germinated spores produced appressoria when one bacterial isolate was sprayed onto leaves; all germinated spores produced appressoria when another bacterial isolate was applied. Of three yeast isolates applied singly, none of the 100 to 180 germinated spores of *H. turcicum* examined produced appressoria. For an application of *Penicillium* sp. to leaves, 8% of the germinated *H. turcicum* spores produced appressoria.

When one of the most antagonistic bacterial isolates was sprayed onto leaves from three corn hybrids, inoculated previously with H. turcicum, the area of leaf surface infected with H. turcicum ranged from 0.1 to 5% among the three hybrids; however the area of leaf infected ranged from 63 to 65% for the pathogen alone. The addition of some nutrient broth with the inoculum enhanced the disease control.

In another trial, a mixture of leaf-surface resident microorganisms (9 spp. bacteria and fungi) was sprayed simultaneously with inoculum of H. turicicum onto leaf surfaces of corn. The area of the leaf surface infected with H. turcicum ranged from 29 to 46% among the three hybrids of the control, but 2 to 16% among the hybrids where mixed residents were applied. Thus some reduction in area infected was achieved by the presence of leaf-surface organisms. Application of resident organisms prior to application of inoculum of H. turcicum gave complete control of leaf blight in one trial but about 50% control in another trial.

This example represents one in which antagonistic organisms are sprayed directly to leaves of a plant prior to or after inoculation by a leaf pathogen. The next example represents control by application of organisms to leaves after leaf fall which can control a disease that overwinters and completes its life cycle during the winter and spring, namely apple scab.

Apple scab, Venturia inaequalis.--the causal organism of apple scab, must survive on leaves for 4-5 months after leaf fall. During this time, the fungus produces a perithecium which, at the end of the overwintering period, contains mature ascospores ready to infect emerging leaves of apple.

In the first 20-30 days after leaf fall, numbers of bacteria and yeasts increase rapidly, according to work done by Burchill and Cook (1971) at East Malling, England. Numbers of these organisms increase up to 40-50 days and then hold relatively constant. Some of these bacterial isolates caused a partial reduction in numbers of perithecia when leaf discs were inoculated with these bacteria within a month prior to inoculation with V. inaequalis. One isolate, a yellow peritrichous rod, inhibited ascospore production on leaves naturally infected with the apple scab fungus. Also, Fusarium tricinctum (F. sporotrichioides) was found to significantly (P=.001) reduce perithecial formation when applied to leaf discs taken from naturally infected leaves. It was not ascertained whether the microorganisms known to inhibit perithecial development on leaves were present on the leaf surface at leaf fall, or whether they colonized leaf surfaces from soil after leaf fall. They also reported that the addition of 5% urea to leaves stimulated microorganisms antagonistic to perithecial development by V. inaequalis. This is an example of biological control that would prevent infection instead of curing or arresting it in infected leaves.

A third example is that of bacterial wilt of alfalfa in which avirulent cells of the pathogen can protect plants from virulent cells of the same pathogen.

Bacterial wilt of alfalfa.--Carroll and Lukezic (1972) have reported that it was possible to apply avirulent cells of Corynebacterium insidiosum to alfalfa leaves of four susceptible varieties and thereby protect them from the virulent isolates. This was done by infiltrating leaflets of intact plants first with the avirulent cells then with virulent ones. If infiltration with avirulent cells occurs 12 hr before inoculation with the pathogen, protection is induced, and remains unchanged for 36 hr. Unfortunately, the protection induced in one leaf does not give protection to another, i.e., the factor providing protection is not translocated.

Other examples.--Examples similar to the one on bacterial wilt are described by Matta (1971). Another source of examples where nonpathogenic organisms can invade leaves and block infection by foliage pathogens is the volume by Preece and Dickinson (1971).

Interactions between bacteria in leaf diseases are fragmentary and more work is needed to explore possibilities of biological control (1971). It will be necessary to maintain an antagonistic organism at high populations on a leaf surface for extended periods of time before control is possible.

Catch and Trap Crops to Reduce Inoculum

Biological control might be achieved by using catch or trap (decoy) crops. A catch crop is one with plant roots that exude one or more substances that stimulates the propagule of a parasite to germinate and infect that plant in lieu of the wanted host plant. This reduces inoculum available to the wanted host since the infected catch crop could be plowed as green manure and the parasite reduced in amount or destroyed.

A trap or decoy crop is one with roots that release one or more substances to stimulate germination of the parasite propagule but the parasite is not capable of parasitizing or infecting the host or if so only to a limited degree.

White root disease of rubber plants.--An example of a trap or decoy crop might be the creeping legumes used in control of the white root disease of rubber plants (*Hevea brasiliensis*) in Malaysia. As explained by Fox (1965), it is necessary to plant rubber after rubber without benefit of crop rotation for disease control. Thus creeping legumes are sown in drills between young rubber trees so that these cover plants occupy all the ground except for a 6-foot clean-weeded planting row of rubber trees. *Fomes lignosus* is one of the most important pathogens causing white root disease of rubber and the legumes provide conditions that encourage epiphytic growth of *F. lignosus*. The effect of the trap or decoy crop is to (i) hasten the wastage effect of inoculum by the formation of fruiting bodies and rhizomorphs, (ii) speed up the succession of saprophytes which denies food potential to the pathogen, and (iii) enhance microbial antagonisms in soil. This use of legume cover is not as effective for *F. noxius* or *Ganoderma pseudoferreum*. Since the effect of the legume as a trap crop is not a direct one, it could be argued that this is not biological control; however a biological agent has been used to obtain control even though indirectly.

Powdery scab of potatoes.--A more direct example of the use of trap crops for disease control is cited by White (1954) in Australia. He found that resting spores of *Spongospora subterranea*, the cause of powdery scab of potatoes, was stimulated to germinate by roots of the decoy crop, Jimson weed. With no galls formed, resting spores were not produced. This led to field trials in which Jimson weed was sown in a field heavily infested with resting spores of the powdery scab fungus. When Jimson weed came into flower, it was plowed down for green manure and to prevent seed set. Next year potatoes were planted and tubers were subsequently examined for incidence of scab. Where 37% of the control tubers were infected with *S. subterranea*, only 7% of the tubers were infected in plots following Jimson weed. Where the control tubers showed a mean disease rating of 4, tubers from Jimson weed plots showed a mean disease rating of 1. Thus the use of Jimson weed as a decoy or trap crop reduced inoculum of the scab fungus in soil and thereby yielded fewer infected potato tubers. Perhaps with further tests, crops instead of a weed could be found to control scab since Jimson weed is not a desirable crop plant.

Nematodes affected by marigolds.--Suatmadji (1969) summarized work on this topic as well as present data to confirm or substantiate the nematicidal effect of marigolds (*Tagetes* spp.) on certain nematodes, principally *Pratylenchus* spp., *Meloidogyne* spp., and *Tylenchorhynchus dubius*. Other nematodes such as *Ditylenchus dipsaci* and *Aphelenchoides ritzemabosi* were not affected.

Root extracts were shown to contain nematicidal thiophenes, but their production varied with species of *Tagetes*. Use of *Tagetes* spp. can increase yields of crops that follow, but the lack of value of a crop of *Tagetes* is its main drawback. Fall planted marigolds are more effective in nematode control than spring planted marigolds. If marigolds could be more extensively grown and valued as an ornamental crop, it might be useful in agriculture although its prospects may be more promising in tropical and subtropical than in temperate zone agriculture.

Witchweed.--Sunderland (1960) in England suggested that use could be made of nonhost plants to rid soil of parasite seeds by stimulating their germination but not their parasitism. Robinson and Dowler (24) in the USA describe a field study from 1958 to 1964 in which both catch and trap crops were used to reduce seeds in soil of the parasitic seed plant, witchweed (*Striga asiatica*). Soybeans and field peas were the trap crops and corn, millet and sorghum were the catch crops for witchweed. Over a 5-year period, there was an increase in yield of corn, the indicator crop, attributed primarily to the decrease in population of witchweed. Catch or trap crops were about equal in value for witchweed eradication; however, complete eradication was not achieved during the 5-year period.

Soybeans, field peas, and sorghum were grown for 3 years before acceptable corn yields were obtained. Two years of millet and 4 years of corn were needed to get acceptable corn yields. On a practical basis, trap crops can be grown to maturity provided grassy weeds are eliminated but catch crops must be destroyed before witchweed reseeds, either by hand weeding, hoeing or herbicides.

Crop Residues in Relation to Inoculum

Whether the addition or removal of crop residues constitutes biological control depends on the nature of the relationship. Residues may improve the physical conditions of the soil and promote plant growth without there being a direct relationship between microorganisms, although there may be secondary effects that aid in disease control (Cook and Watson, 1970; Kelly and Curl, 1972). Work on the influence of crop residues on fungus-induced diseases has been summarized for western states by Cook and Watson (1970). Many papers also appear in a symposium volume edited by Baker and Snyder (1965). A few examples will be given, however.

Grape replant problem.--Deal, et al. (1972) have recently shown that amending vineyards with fragments of old grape roots suppressed populations of *Penicillium* spp., increased populations of *Trichoderma viride*, and improved growth of grape seedlings. Such amendments gave better development of endophytic phycomycetous mycorrhizae than soil lacking these amendments, However this is only circumstantial evidence that the change in populations or organisms in soil altered the disease problem.

Seedling blight of wheat.--In work done at Minnesota (Warren and Kommedahl 1973), in plots in which wheat has been grown for a decade, the removal of residue but addition of fertilizer resulted in a drop in percentage infected roots from 56% to 20% in plots where residues were removed. The removal of residues probably reduced the population of antagonists which would have suppressed the organisms that infect roots; the two pathogens present in roots were *Helminthosporium sativum* and *Fusarium roseum* (mainly 'Graminearum'). Further studies showed that even from fall to spring, *F. roseum* made up about half the population of *Fusarium* spp. in spring than the previous fall. Our experience shows that much of the inoculum was removed with the residue and that the residue that escaped being collected deteriorated rapidly to increase the proportion of antagonist to pathogen. A search among rhizosphere microflora showed *Fusarium roseum* to be less abundant in the rhizosphere where residue was retained than where removed and that addition of fertilizer following removal of residue meant that *Fusarium* was more abundant in the rhizosphere than in nonrhizosphere soil.

In another experiment where _Fusarium roseum_ and _Rhizoctonia solani_ were compared on wheat (Kommedahl and Young, 1956), the addition of cornstalks to soil increased the incidence of _Fusarium_-infected wheat roots, whereas it reduced the incidence of _Rhizoctonia_-infected roots. Plant roots frequently contact roots of previous crops, and we have frequently seen corn roots, for example, growing into and through old cornstalks.

Crop Rotation in Relation to Inoculum

Wheat after oats.--Wheat was planted in plots following wheat, corn, and oats, in work done at Minnesota (unpublished data). Only the results with wheat after oats will be reported, although results were similar for the other two crops. The difference in these experiments from others tried was that resistant and susceptible oat varieties were compared for their effects on wheat. A root rot resistant wheat variety following a root rot resistant oat variety gave a low infection rating of wheat and a full stand. However, a susceptible wheat variety after a susceptible oat variety had double the infection rating (from 3 to 6 on a scale of 10) and reduced the stand by one-third. Similar results were obtained for wheat after resistant and susceptible varieties of either wheat or corn. Varieties vary in the antagonistic and pathogenic organisms present in the rhizospheres, e.g., one corn hybrid harbored mainly _Fusarium roseum_ isolates while a resistant variety hybrid had mainly antagonistic _Penicillium_ spp.

Organisms were more abundant in the rhizosphere of oats of a root rot susceptible variety than a resistant one, as one would expect, but it was also possible that antagonists to the common root rot fungi (_Fusarium roseum_ and _Helminthosporium sativum_) were most numerous in oat soil rhizospheres than in other cereal crop rhizospheres.

Conclusions

Although only a small sample of papers on biological control have been reviewed, there are some generalizations that might be made regarding this kind of control.

1. There is no one method that is effective in disease control by using biological agents. The method should be investigated that gives the best combination of pathogen and host for a given locality.

2. Application of antagonists to seeds or transplants has proved to be more successful and economically feasible than application of antagonists to foliage or soil.

3. Organisms selected for application to plant parts must be those that are normally present on those parts; otherwise they will not become established.

4. In screening organisms for possible disease control agents, one should not rely on the appearance of a zone of inhibition around the test antagonist as a criterion for further experimentation. Some organisms are antagonistic because they physically occupy the substrate and thereby exclude the pathogen.

5. Addition of nutrients with the antagonistic organisms applied usually enhances its effect whether applied to seeds or leaves, mainly because it builds up the inoculum potential of the antagonist.

6. Application of antagonistic organisms to soil usually ends in failure, mainly because the nutrients available in soil are so low and homeostasis

keeps a microbiological balance to nullify the effects of a sudden increase in populations of a given organism.

7. Trap crops, catch crops, or decoy and cover crops might be used with more success than at present. There are not many instances where this has been tried.

8. Where rotation of crops is not always a way for disease control, rotation of varieties can be used and may be effective in keeping the incidence of disease to minimum levels.

Literature Cited

Asare-Nyako, A. 1965. The role of leaf microflora on epidemiology of the Northern leaf blight of corn. University of Minnesota Ph.D. thesis. 92 pp.

Aycock, R. 1966. Stem rot and other diseases caused by *Sclerotium rolfsii*. North Carolina Agr. Exp. Sta. Tech. Bul. 174. 202 pp.

Baker, K. F. and W. C. Snyder (eds.). 1965. Ecology of soil-borne plant pathogens. University of California Press, Berkeley. 571 pp.

Baker, R. 1968. Mechanisms of biological control of soil-borne pathogens. Ann. Rev. Phytopathology 6: 263-294.

Burchill, R. T. and R. T. A. Cook. 1971. The interaction of urea and microorganisms in suppressing the development of perithecia of *Venturia inaequalis* (Cke.) Wint. pp. 471-483. *In* T. F. Preece and C. H. Dickinson (eds.). Ecology of leaf-surface microorganisms. Academic Press, New York.

Carroll, R. B. and F. L. Lukezic. 1972. Induced resistance in alfalfa to *Corynebacterium insidiosum* by prior treatment with avirulent cells. Phytopathology 62: 555-564.

Chang, I. and T. Kommedahl. 1968. Biological control of seedling blight of corn by coating kernels with antagonistic microorganisms. Phytopathology 58: 1395-1401.

Cook, R. J. and R. D. Watson (eds.) 1970. Nature of the influence of crop residues on the fungus-induced root diseases: a summary of contributions of the Western Regional Project W-38. Washington Agr. Exp. Sta. Bull. 716. 32 pp.

Crosse, J. E. 1971. Interactions between saprophytic and pathogenic bacteria in plant disease. pp. 283-290. *In* T. F. Preece and C. H. Dickinson (eds.) Ecology of leaf-surface microorganisms. Academic Press, New York.

Curl, E. A. 1963. Control of plant diseases by crop rotation. Bot. Rev. 29: 413-479.

Deal, D. R., W. F. Mai and C. W. Boothroyd. 1972. A survey of biotic relationships in grape replant situations. Phytopathology 62: 503-507.

Fox, R. A. 1965. The role of biological eradication in root-disease control in replantings of *Hevea brasiliensis*, pp. 348-362. *In* K. F. Baker and W. C. Snyder (eds.). Ecology of soil-borne plant pathogens. University of California Press, Berkeley.

Griffin, D. M. 1972. Ecology of soil fungi. Syracuse University Press, New York. 193 pp.

Kelly, W. D. and E. A. Curl. 1972. Effects of cultural practices and biotic soil factors in *Fomes annosus*. Phytopathology 62: 422-427.

Kommedahl, T. and T. D. Brock. 1954. Studies on the relationship of soil mycoflora to disease incidence. Phytopathology 44: 57-61.

Kommedahl, T. and I. Chang. 1970. Biological seed coating for control of seedling diseases--a principle for the future? Proc. Ann. Corn and Sorghum Conf. Publ. 25: 84-89. American Seed Trade Assn., Washington, D.C.

Kommedahl, T. and H. C. Young. 1956. Effect of host and soil substrate on the persistence of *Fusarium* and *Rhizoctonia* in soil. Plant Dis. Reptr. 40: 28-29.

Liu, S. and E. K. Vaughan. 1965. Control of *Pythium* infection in table beet seedlings by antagonistic microorganisms. Phytopathology 55: 986-989.

Matta, A. 1971. Microbial penetration and immunization of uncongenial host plants. Ann. Rev. Phytopathology 9: 387-410.

Mew, I. C. and T. Kommedahl. 1972. Interaction among microorganisms occurring naturally and applied to pericarps of corn kernels. Plant Dis Reptr. 56: 861-863.

Mitchell, R. and E. Hurwitz. 1965. Suppression of *Pythium debaryanum* by lytic rhizosphere bacteria. Phytopathology 65: 156-158.

Oshima, S. 1966. Antagonisms of *Trichoderma lignorum* (Tode) Harz. to *Corticium rolfsii* Curzi and their applications to control the fungus. Okayama Tobacco Exp. Sta. Bull. 27. 56 pp. (English summary).

Preece, T. F. and C. H. Dickinson (eds.). 1971. Ecology of leaf-surface microorganisms. Academic Press, New York.

Robinson, E. L. and C. C. Dowler. 1966. Investigations of catch and trap crops to eradicate witchweed (*Striga asiatica*). Weeds 14: 275-276.

Starkey, R. L. 1958. Interrelationships between microorganisms and plant roots in the rhizosphere. Bacteriol. Rev. 22: 154-172.

Suatmadji, R. W. 1969. Studies on the effect of *Tagetes* species on plant parasitic nematodes. H. Veenman and Zonen, N. V., Wageningen. 132 pp.

Sunderland, N. 1960. Germination of the seeds of angiospermous root parasites. pp. 83-93. *In* J. L. Harper (eds.). The biology of weeds. Blackwell Sci. Publ., Oxford, England.

Tviet, M. and M. B. Moore. 1954. Isolates of *Chaetomium* that protect oats from *Helminthosporium victoriae*. Phytopathology 44: 686-689.

Warren, H. L. and T. Kommedahl. 1973. Fertilization and wheat refuse effects on *Fusarium* species associated with wheat roots in Minnesota. Phytopathology 63: 103-108.

Wells, H. D., D. K. Bell and C. A. Jaworski. 1972. Efficacy of *Trichoderma harzianum* as a biocontrol for *Sclerotium rolfsii*. Phytopathology 62: 442-447.

White, N. H. 1954. Decoy Crops. Australian J. Sci. 17: 18-19.

BREEDING FOR DISEASE RESISTANCE IN CEREALS

E. G. Heyne
Professor of Plant Breeding
Agronomy Department
Kansas State University, Manhattan

Effective cereal breeding programs generally are the result of cooperative efforts of breeders with geneticists, pathologists, entomolgists, chemists, statisticians, and others. Every problem dealt with in breeding is a special one often requiring new or modified techniques to carry out the work.

Control of plant diseases by breeding long has been used as an example of how effectively diseases can be controlled by host resistance. However, many pathogens can be controlled economically and effectively by use of fungicides and cultural practices. An effective procedure is to utilize all those approaches--host resistance, fungicides and cultural practices--to control or reduce losses due to plant diseases.

Thus, we need to recognize that all problems related to host-pathogen relationships cannot be solved by plant breeding. Source of resistance, heritability of the trait and adequate facilities should be available before the breeding program is started. Research in the area of host-pathogen problems to learn as much basic information may lead to control measures other than breeding for resistance.

For convenience of discussion, I will consider breeding for protection against pathogens under pathogen escaping; tolerance to pathogen attack; and host resistance (Heyne and Smith, 1967). All are real but appear different to various individuals.

Pathogen escaping. The pathogen responds to the environment as does the host. Management can aid in escaping or retarding disease development. Late seeding of winter wheat increases the possibility of bunt infection because the pathogen develops better under cooler temperatures and increased available moisture. On the other hand, hard winter wheat in the Southern Great Plains seeded too early (soil temperature over 65°F) encourages certain root rot organisms. Timely seeding of winter wheat helps avoid these optimum infection periods of the pathogen. This is a cultural approach but is an important aspect of disease control. Breeding earlier maturing wheats also aids in avoiding excess damage from leaf and stem rust in the Southern Great Plains and is a character that can be manipulated by breeding. Triumph wheat is susceptible to both stem and leaf rust but often escapes damage because it matures before sufficient inoculum is built up to cause damage. This is especially true with stem rust which requires warmer temperatures to develop than does leaf rust. The breeding of early-maturing cultivars is of some assistance in rust control but should be combined with other control measures, when possible.

Tolerance. Use of this term has been controversial but I use it to mean that the pathogen develops normally on the host, but the plant suffers less damage than other cultivars that have similar development of the pathogen. Actually, these plants are susceptible, but because of "vigor" the plant produces more grain, forage, etc. than a nontolerant cultivar in spite of the pathogen being present. Another explanation of tolerance is that the response of the host is such that the plant continues reasonable growth and development without retarding the development of the pathogen. Blackhull-type wheats are fully susceptible to leaf and stem rust but suffer less damage than other cultivars growing under similar infections.

Host-resistance. This category also has several different definitions. I use it as a relative term referring to any retardation in the development of the attacking pathogen, whether it is due to a metabolic deficiency of the pathogen or some character attributed to the host. This will vary from only slight retardation of the pathogen to complete limitation of growth and reproduction.

The broad definition of host-resistance covers any of the responses that retard pathogen development that have a host-parasite genetic relationship. Terms as specific or vertical and general or horizontal resistance are used (Caldwell, 1968; Van der Plant, 1968). As long as any resistance can be transferred by breeding, it really is immaterial as to what the genes are called. Resistance of Hussar wheat to leaf rust often appears as general resistance in the field but under specific environmental conditions in the greenhouse it is clearly race-specific and monogenic in inheritance.

Plants vary in their response to pathogen attack and in some cases the plant is so sensitive to infection that the surrounding area becomes necrotic very soon and the pathogen is isolated in an island of dead tissue (Luke, et al., 1960). This sensitivity can be readily transferred from one parent to another and becomes a useful means of plant protection against certain pathogens.

The basic principles for breeding pathogen resistant cultivars are similar to other plant characters except that we must be aware we are working with two genetic systems and their interrelationships. Disease resistant breeding programs, to be most effective, need to be cooperative efforts between the plant pathologist and the plant breeder. The following items should receive careful consideration in planning and conducting a disease breeding program.

1. Obtain a thorough knowledge of the pathogen on etiology, distribution, ecological relationships, epidemiology, range of the host plants, and genetic variation. This is primarily the responsibility of the plant pathologist and he should be the leader of this phase of the project.

2. Survey of germ plasm (Creech and Reitz, 1971) for sources of resistance is a joint effort. At the present time there are good collections of various cereals throughout the world, for example, rice at International Rice Research Institute in the Philippines, maize in Mexico, and the small grains in the United States. Domesticated cultivars, as well as related species of the economic crop, should be screened for resistance. Occasionally, resurveys of wild germ plasm is valuable. The recent collections of *Avena sterilis* in the Mediterranean area yielded new genes for resistance to crown rust and virus (Murphy, et al., 1967).

3. Inoculation techniques and creation of artificial epidemics are necessary so large populations of breeding material can be studied. In some cases this is relatively easy, such as soil-borne mosaic virus of wheat or dwarf bunt of wheat where the organisms live in the soil. Relatively simple techniques of inoculating large numbers of oat kernels as developed by Wheeler and Luke (1955) for Victoria blight in oats illustrate how large populations can be handled efficiently. They isolated the toxin (victoxin) from the fungus and applied this to germinating seed. They treated 100 bushels of oat seed (about 45 million oat seedlings) of two susceptible cultivars, Victograin 48-93 and Fulgrain. There were 973 seedlings

which survived the test and after further screening for crown rust resistance and plant characters, 72 selections were obtained that were essentially the same as one or the other parent. Such effective screening tests are not common but many have been developed that are useful and efficient. Frey and Caldwell (1961) discuss in detail techniques for use in breeding disease resistant oats which are applicable to other cereals.

The development of equipment for the inoculation of numerous breeding lines and different cultures of the pathogen are described in considerable detail by Browder (1971) for studies in leaf rust of wheat. The sophisticated equipment and procedures he developed for studies of pathogenic specialization in cereal rust fungi are readily adapted to screening and testing of breeding lines.

4. Sources of inoculum should be representative of the pathogen in the area or region where the new cultivars will be used. Most studies of the pathogens include the variation within the organism and should provide this information. Be sure to have a representative sample of the parasite preferably made up of a number of collections. However, in a number of instances it is desirable to study race-specific situations. This is generally conducted under controlled conditions in a greenhouse or growth chamber. Creating "epidemics" in the field allows agronomic as well as disease response evaluations of the breeding material. Thus, both field and laboratory techniques are needed for a complete program.

5. Plan a breeding program using the sources of resistance found and the techniques developed or borrowed. A practical breeding program does not lend itself to accumulation of data satisfactory for publication. Random samples of breeding populations are not available, the size of the population generally limits note taking and often subjective instead of objective notes are recorded. However, plant breeding material can be used for detailed study, for example, inheritance of disease resistance, but such studies must be planned at the time the crosses are made.

Notes taken should describe the disease response readily and quickly. Subjective or response categories can be set up such as resistant (R), intermediate response or moderately resistant (MR) and a susceptible class (S). Three such categories generally are sufficient but easily can be expanded into more classes such that the R class could include some better than usual (1, or HR), R becomes (2) and slightly below R becomes (3). This would give nine classes, 1-9, easily divided into three major groups.

Group	Class	Response	Description
	(1	(HR))	This group of plant responses is the most desirable and often
R	(2	(R))	monogenic and race-specific.
	(3	)	
	(4	)	This group of responses is effective in reducing damage of the
MR	(5	(MR))	pathogen and may serve as an example of horizontal resistance.
	(6	)	
	(7	(MS))	Generally such responses are of no value but slight retardation
S	(8	(S))	of disease development may be useful if no other sources of
	(9	(VS))	resistance are available.

In some cases more objective notes may be taken. Percentage of bunted spikes of wheat in a plot can be arrived at quite accurately by estimation and accomplished rapidly in the field. Whatever scheme is used, keep it simple.

Breeding Procedures

Every time a cross is made the parental combinations are broken up and many new recombinations result. The breeding objective is to find new recombinations that are superior to the parents not only for disease resistance but all other characters that make a cultivar accepted for production. A cultivar that resists a disease, has good quality, or the proper maturity does not assure that its performance will be satisfactory. For disease resistance to be useful as a means of control it has to be combined with the present good cultivars and transferred to new improved cultivars (Walker, 1959).

Breeding procedures are outlined in numerous plant breeding texts and will not be discussed in detail. The pedigree and bulk approaches are the two major procedures, both with many modifications. In some cases the backcross approach is the fastest and may be the method to use. On the other hand, the bulk approach may be the most efficient and practical procedure. Improved cultivars have been developed by many procedures and the choice often is decided by the whim of the breeder but generally dictated by funds and facilities available.

Mutation breeding for disease resistance has received considerable emphasis the past two decades (Frey, 1968; Knott, 1971). A number of instances have been reported where new genetic sources of disease resistance have been obtained. Mutation should be used as a supplement to other procedures of breeding and in some cases may be a last resort in trying to locate genetic resistance.

Early workers suggested that breeding for resistance would be easy as they experienced success with single gene transfers and were not cognizant of the variability of the pathogen. The terms monogenic and polygenic are in common use. Van der Plant (1968) suggests the use of oligenic (meaning few) resistance to contrast with polygenic (many) resistance. Whether it is one or more has some relation to breeding procedures used. Monogenic resistance for some host-parasite interactions has been short-lived as demonstrated by the changes in wheat rust plant and pathogen genotypes. However, in bunt in wheat the Oro gene for resistance has been effective for control of this disease in Kansas since 1930.

In leaf rust studies of wheat, twenty different genes for resistance to leaf rust have been described (Heyne, 1972). Many of these have lost their value because they were used one at a time. Knott (1971) stresses the point that such resistant genes should never be used singly. Those that have lost their present value may be removed from the present host population and if the virulence of the leaf rust fungus is lost, in time these resistant genes could be used again, preferably two or more resistant genes in one cultivar. This has been demonstrated even though virulence to LR 1, LR 2 and LR 3 are still present in the leaf rust populations in Kansas. Neither gene by itself shows resistance in the field in Wichita lines but with a combination of any two there is some reduction in amount of infection and when all three are present the amount of infection is much reduced. Judicious use of such genes are required. There appears to be enough genes to go around if their use is deployed intelligently (Browning, et al., 1969); Knott, 1971).

Widespread use of a source of resistance or genotype has resulted in a rapid turnover in cultivars, especially in oats (Browning and Frey, 1969), and wheat. The recent experience (1969-1971) of the extensive use of Texas cytoplasm in maize illustrates how serious a problem the use of one genotype can be (Plant Dis. Reptr. 1970). The ineffectiveness of leaf and stem rust genes, when used singly, has resulted in similar losses.

Singly used genes in homozygous lines in mixtures have been suggested as a means of using monogenic sources of resistance. The Iowa research workers (Browning and Frey, 1969) are doing this in their oat disease breeding program. The common term for this approach is "multiline cultivars." A multiline cultivar is a mixture of lines of the same basic genotypes but each line carries a specific gene for resistance not found in the other lines. A mixture of these lines includes as many sources of resistance as can be found or adequately handled. This mixture does not serve as a screening cultivar for virulent genes in the pathogen.

Browning and Frey (1969) refer to multilines as instant synthetic horizontally resistant cultivars.

We do not seem to learn too well from past experiences for at present (1972) the Agent gene (Plant Dis. Reptr. 1970) for leaf rust resistance is being used in most wheat breeding programs in North America. The cultivar Agent was first released in Oklahoma in 1967. The source of resistance is an alien transfer of genetic material from Agropyron to common wheat. No leaf rust cultures attacked this source until several years ago. Fox (Gilmore, et al., 1971) has been released in Texas and Blueboy II (Murphy, 1972) in North Carolina and a number of hard red winter wheats are being increased for distribution in the Southern Great Plains. Virulent collections were common on Agent-derived wheats in 1972 (Browder, L. E., personal communication) and it could be possible that the usefulness of this source of resistance could be lost before the cultivars carrying the resistance are extensively grown. Browning, et al. (1969) urged that a "planned gene deployment, as part of an integrated continental program of use and conservation of resistance genes against continental pathogens, should undergird other programs, whether those utilize generalized resistance, specific resistance, tolerance, or combinations of them possibly with fungicides." This will require local and wide-spread cooperation among plant research workers and needs the active participation of all involved in breeding disease resistant cereals that involve pathogens that can spread readily in an area or region.

I have tried to point out and to emphasize the following points:

1) Breeding for disease resistance is a cooperative team effort.

2) Check the literature for details on techniques used in creating artificial epidemics using the procedures directly or as guides to developing your own.

3) Basic information of host and pathogen are essential for successful disease breeding programs.

4) There is no <u>one</u> way to conduct a disease breeding program; the techniques and procedures used need to be adjusted to the local problems and facilities.

5) Consider and use more than one means of control whenever possible.

6) Judicious use and deployment of resistant genes must be done on a regional or continental basis, to conserve the resistant resources now available.

References

Browder, L. E. 1971. Pathogenic specialization in cereal rust fungi, especially <u>Puccinia recondita</u> f. sp. <u>tritici</u>: Concepts, methods of study, and application. ARS-USDA Tech. Bul. No. 1432, 1971. 51 pp.

Browning, J. A. and K. J. Frey. 1969. Multiline cultivars as a means of disease control. Ann. Rev. Phytopathology 7: 355-382.

Browning, J. A., M. D. Simons, K. J. Frey and H. C. Murphy. 1969. Regional deployment of conservation of oat crown-rust resistance genes. pp. 49-56. <u>In</u> Disease consequences of intensive and extensive culture of field crops. J. A. Browning (ed.). Iowa Agr. and Home Econ. Exp. Sta. Spl. Rept. 64.

Caldwell, R. M. 1968. Breeding for general and/or specific plant disease resistance. <u>In</u> Proc. 3rd Int. Wheat Genet. Symp. Canberra, Australia Acad. Sci. W. K. Finlay and W. K. Shepherd (eds.) pp. 263-272.

Creech, J. L. and L. P. Reitz. 1971. Plant germ plasm now and for tomorrow. Adv. in Agron. 23: 1-49.

Frey, K. J. 1968. Induced variability in diploid and polyploid *Avena* sp. *In* Gamma Field Sym. No. 7, The Present State of Mutation Breeding, 1968. pp. 41-56.

Frey, K. J. and R. M. Caldwell. 1961. Oat breeding and pathologic techniques. *In* Oats and Oat Improvement. F. A. Coffman (ed.) Amer. Soc. Agron. Mono. 8: 227-262.

Gilmore, E. C., K. B. Porter and K. A. Lahr. 1971. Cultivar improvement. Ann. Wheat Newsletter 17: 127.

Heyne, E. G. 1972. Leaf rust genes in wheat. Ann. Wheat Newsletter 18: 97-99.

Heyne, E. G. and G. S. Smith. 1967. Wheat breeding, pp. 269-306. *In* Wheat and Wheat Improvement. Amer. Soc. Agron. Mono. 13. K. S. Quisenberry and L. P. Reitz (eds.). Madison, Wisconsin.

Knott, D. R. 1971. The transfer of genes for disease resistance from alien species to wheat by induced translocations. *In* Mutation Breeding for Resistance. Inter. Atomic Energy Agency, Vienna. pp. 67-77.

Knott, D. R. 1971. Can losses from wheat stem rust be eliminated in North America? Crop Sci. 11: 97-99.

Luke, H. H., H. E. Wheeler and A. T. Wallace. 1960. Victoria-type resistance to crown rust separated from susceptibility to *Helminthosporium* blight in oats. Phytopathology 50: 205-209.

Murphy, C. F. 1972. Registration of Blueboy II wheat (Reg. No. 514). Crop Sci. 12: 398.

Murphy, H. C., I. Wahl, A. Dinoor, J. D. Miller, D. D. Morey, H. H. Luke, D. Sechler and L. Reyes. 1967. Resistance to crown rust and soil-borne mosaic virus in *Avena sterilis*. Plt. Dis. Reptr. 51: 120-124.

Plant Disease Reporter - Special Issue. 1970. Vol. 54, No. 12 (Part II). Thirteen papers on Southern Corn Leaf Blight.

Smith, E. L., A. M. Schlehuber, H. C. Young, Jr. and L. H. Edwards. 1968. Registration of Agent wheat (Reg. No. 471). Crop Sci. 8: 511-512.

Van der Plank, J. E. 1968. Disease resistance in plants. Academic Press: New York. 206 pp.

Walker, J. C. 1959. Progress and problems in controlling plant diseases by host resistance. *In* Plant Pathology, Problems and Progress 1908-1958. C. S. Holton, ed. pp. 32-41.

Wheeler, H. and H. H. Luke. 1955. Mass screening for disease resistant mutants in oats. Sci. 122: 1229.

TECHNIQUES, ACCOMPLISHMENTS, AND FUTURE OUTLOOK IN BREEDING VEGETABLES FOR RESISTANCE TO DISEASE

Raymon E. Webb
Plant Pathologist
Plant Genetics and Germplasm Institute, ARS, USDA
Beltsville, Maryland

The first efforts to develop disease-resistant cultivars of vegetables were undertaken to protect crops against loss in yield, or destruction by pathogens that could not be effectively controlled by chemicals or by other practical means. At the turn of the century, methods of vegetable production and market requirements were such that the objective of the plant pathologist-breeder was essentially that simple. As the science and art of plant breeding has developed, and the economics and technology have changed radically in the production, handling, and processing of vegetables, the vegetable breeders' objectives and methods have become vastly more complex. This paper briefly sketches the salient features of these developments and changes.

Techniques

Evaluations: Breeding of vegetable crops specifically for resistance to disease was begun in the United States in the late 1800's by Orton (1911) of the U.S. Department of Agriculture. He crossed a wilt-resistant (*Fusarium oxysporium* f. *Niveum* (E. P. Sm.) Snyder and Hansen), nonedible citron melon with the wilt-susceptible watermelon cultivar 'Eden.' By subsequent selection of wilt-resistant plants growing in heavily infested soil and backcrossing these to edible types, he developed the wilt-resistant cultivar 'Conqueror.' Orton also found cowpea cultivar 'Iron' to be highly resistant to both the Fusarium (*F. oxysporum* f. *tracheiphilum* (E. F. Sm.) Snyder and Hansen) wilt disease and root-knot nematodes (*Meloidogyne* spp.). Although the resistant cultivars of watermelon and cowpea developed by Orton did not make a major direct impact on the vegetable industry, their genes for resistance have been invaluable in subsequent breeding programs. Moreover, his successes in developing disease-resistant vegetable cultivars demonstrated certain principles which were basic to future progress by others. These principles are: 1. useful resistance to a major disease may be found in advanced cultivars as well as among their wild relatives; 2. progeny evaluation for resistance to disease must be done under conditions which will minimize the chance of susceptible escapes; and 3. resistance factors transferred by crossing can be incorporated into types with superior horticultural properties.

The next major advance in breeding vegetable crops for disease resistance occurred at the Agricultural Experiment Station, University of Wisconsin. During the development there of cultivars of cabbage with resistance to yellows caused by *F. oxysporum* f. *conglutinans* (Wr.) Snyder and Hansen, an attempt was made to move Orton's method of evaluation for resistance to disease from the field to controlled conditions in the greenhouse. Tisdale (1923), using young plants of the 'Wisconsin Hollander' cabbage cultivar, found that plants which were resistant to yellows in the field, except during warm growing seasons, showed little resistance when young plants were grown in infested soil in the greenhouse. Later Walker and Smith (1930) developed cabbage lines which were completely free of symptoms, under similar greenhouse conditions, that were highly favorable for disease development in susceptible plants. They found that young susceptible and "field"-resistant plants of 'Wisconsin Hollander,' dipped in washed, macerated mycelium and grown in quartz sand at 24°C, became diseased while plants with the higher type of resistance did not. Above 24°C, the highly resistant plants succumbed to a cortical root decay. Below 24°C, some of the plants of 'Wisconsin Hollander' survived. This illustrated the critical effect of temperature of the growing medium on the development of this disease.

The basic principles used for assaying progenies for resistance to yellows in cabbage have been adapted to the development of resistance to other major vegetable diseases. Inoculum levels, age of plant, manner of timing of inoculation, moisture and nutritional levels of the growing medium, temperatures of the growing medium, and of the air and the relative humidity of the air for optimum disease development of foliar pathogens have all been manipulated and controlled to meet the requirements for efficient and rapid assays in the greenhouse (see Walker, 1965; 1966). For example, when seeds of the garden pea (*Pisum sativum* L.) are planted directly into sand infested with *F. oxysporum* f. *pisi* (Lindford) Snyder and Hansen (race 1), and maintained at 22°C, susceptible seedlings are wilted or killed within 21 days. Resistance to pea near-wilt (*F. oxysporum* f. *pisi* (race 2) requires root wounding for optimum infection and disease development. Here the seedlings are removed 10 days after seeding; the roots are clipped while immersed in inoculum; and the seedlings are replanted in sand where the temperature is maintained at 21°C. The susceptible and resistant seedlings are readily discernible 20 days after inoculation.

Because field evaluations proved unreliable, Wellman (1939) introduced a greenhouse method for evaluating young tomato plants for resistance to Fusarium wilt caused by *F. oxysporum* f. *lycopersici* (Sacc.) Snyder and Hansen. Young tomato plants were pulled from the seed flat, roots washed in running water, dipped in the inoculum, and transplanted in the steam-sterilized soil maintained between 25-28°C with electrically controlled heating elements. With this method he could distinguish susceptible, moderately resistant, and highly resistant types. This evaluation method was largely responsible for the discovery of the monogenic dominant resistance in an accession of the 'Red Current' tomato cultivar by Bohn and Tucker (1940). Presently breeders routinely use a slightly modified steam sterilization, soil temperature-controlled method for assaying tomato seedling progenies for resistance to Fusarium wilt. Essentially the same procedure is used to assay large populations of tomatoes for resistance to Verticillium wilt (*Verticillium albo-atrum* Rinke and Berth. and *V. dahliae* Kleb.), Fusarium wilt in watermelon, Fusarium wilt or stem rot (*F. oxysporum f. battatas* (Wr.) Snyder and Hansen) in sweetpotato, and the root-knot nematode complex in some crops (Harrison, 1960; Steinbauer, 1956; Walker, 1965).

Resistance evaluations of several vegetable crops to foliar, stem, and fruit pathogens, including the viruses, are now done in the seedling or young plant stage under controlled conditions in the greenhouse. Resistance to gray leaf spot (*Stemphylium solani* G. F. Weber) of tomato can be readily detected in segregating populations by adjusting relative humidity to 100 percent for 12 to 14 hours after inoculation with a spore suspension. Relative humidity is unimportant in disease development after the infection period. Similar methods are used to assay for resistance to downy mildew (*Phytophthora phaseoli* Taxt.) of lima bean (Thomas, et al., 1952) and downy mildew (*Peronospora effusa* (Grw. and Desm.) Ces.) of spinach (Webb, 1955) except that intermittent high relative humidity is necessary after the initial infection period for maximum disease development on susceptible types.

Resistance to spinach blight caused by the cucumber mosaic virus is conditioned by a single dominant gene, and resistance is complete at and below 24°C (Pound and Cheo, 1952). Webb (1955), using seedlings grown in thumb pots in the greenhouse, reported a method of inoculating spinach plants in the fully expanded cotyledonary stage. Because mosaic symptoms developed in the first true leaves, these young blight-resistant plants, without repotting, can be subjected to an assay for resistance to the downy mildew organism (*P. effusa*).

Resistance to bacterial canker, caused by *Corynebacterium michiganense* (E. S. Smith) Jensen, is determined by cutting through the petioles of young greenhouse-grown plants with a sharp knife which has been dipped in an inoculum suspension (Thyr, 1968). Infection of young susceptible plants has been successful with as few as 9 bacterial cells/ml of inoculum (Strider, 1970).

Some greenhouse evaluations for resistance to diseases have been adapted to the laboratory to shorten further the test period, avoid possible contamination with other organisms, and to improve reproducibility of tests. Among such techniques are: evaluation of tomato seedlings growing in test tubes (Dropkin, et al., 1967) and tomato and cantaloupe seedlings growing in clear plastic seed growth pouches (Fassuliotis and Corley, Jr., 1967) for resistance to root-knot nematodes; resistance to anthracnose fruit rot (*Colletotrichum phomoides* (Sacch.) Chester) in tomato (Robbins and Angell, 1970); and studying the host pathogen relationship of pea wilt (Roberts and Kraft, 1971).

Assays to detect "field" or polygene resistance and to separate susceptible plants from those showing slight resistance under greenhouse conditions have been developed. "Field" resistance to late blight in potato, caused by *Phytophthora infestans* (Mont.) DBy, is conditioned by such a polygene system. Whole plants, leaves, or leaf discs inoculated with a standardized spore suspension have been used to find a greenhouse assay that correlates well with plant reaction to infection under field conditions. Resistant and susceptible ratings are based on susceptibility of plant part to infection, rate of lesion development after infection, and extent and intensity of sporulation (Hodgson, 1961, 1962).

Root rot, caused by *Aphanomyces euteiches* Drechs., is a serious disease of the garden pea, and despite numerous intensive studies no significant resistance has been found. However, Lockwood and Ballard (1959) developed a sensitive technique which can detect low levels of resistance among cultivars and breeding lines. They used a sand substrate and maintained the temperature at 24°C. They found that by adhering precisely to the following assay procedure variability in host reaction could be minimized. Their procedure is: apply 4-6 pounds of water to each 36 pounds of sand; plant seed 2 cm deep; apply inoculum when plants are 2-5 cm high; use 4- to 5-day-old cultures as inoculum; standardize zoospore suspension at 1.5×10^5 per ml; apply zoospores 2-14 hours old in a suspension at a rate of 10 ml to each 10-inch row as close to the plants as possible; and saturate the sand once after inoculation. Differences in susceptibility among stocks could be evaluated after 12 days.

Greenhouse techniques have not yet been developed that will yield highly valid disease evaluations of breeding lines and progenies for polygenic resistance to some major diseases such as Alternaria leaf spot (*Alternaria cucumerina* (Ell. and Ew.) J. A. Elliott) of cantaloup; white rust (*Albugo occidentalis* G. W. Wilson) of spinach; downy mildew (*Pseudosperonospora cubenis* (Buk. and Curt.) Rottow) of cucumber; early blight (*Alternaria solani* (Ell. and Mart.) L. R. Jones and Grout.) of tomato; and to other diseases. Such evaluations must be obtained by growing the crop where the disease occurs naturally in epidemic proportions or where artificial epidemics can be produced. Spreader rows of susceptible types, interplanted among test lines and inoculated at the proper time, have effectively spread and intensifield disease incidence in assaying for resistance to early and late blight of potato, early blight of tomato, and to other diseases.

How resistance is inherited: Van der Plank (1968) has proposed the term "oligogenic" resistance for host resistances which are conditioned by one to several genes, often called major genes, whose individual effects can be easily detected. Effects of such genes may be dominant, incompletely so, or recessive. Linked and allelic genes for resistance may differ only in their reaction to specific races of an organism and may be difficult to identify. Where genes are not allelic or closely linked, a breeding test will show they are inherited independently (Day, 1972).

Most "vertical" resistance (genes which are effective only against certain specialized races of the pathogen) is generally also oligogenic although this may not be entirely so (see Van der Plank, 1963, 1968). And not all oligogenic resistance is "vertical." Several vegetable cultivars have major genes for resistance which have remained stable for many years; among them are resistance to Fusarium yellows in cabbage, to scab and mosaic in cucumber, and to gray leaf spot in tomato.

Cultivars of potatoes which possess a generalized (polygenic) resistance to late blight respond similarly when infected with any or all races that may arise in a given season. Van der Plank (1963) labels this type of resistance "horizontal" resistance in contrast to "vertical" resistance. The degree of expression of resistance in those cultivars possessing polygenic (horizontal) resistance is dependent upon environmental conditions during the growing season.

Hare (1965) has recently reviewed the work on inheritance of resistance to parasitic nematodes among crop plants. Resistance is inherited in ways similar to the inheritance of resistance to other pathogens that attack crop plants.

Variability in the pathogen: Some plant pathogens possess almost infinite variability. Sexual reproduction, mutation, and heterocaryosis are means by which variability may be enhanced among plant pathogens. The presence of different host genotypes providing for selection and survival of pathogenic variants among large populations is a major force that influences the degree to which physiologic races of an organism may become predominant on a crop. *P. infestans*, the cause of late blight on potato, is an example of such a highly variable organism. Shortly after the hypersensitive resistance (due to R genes) to late blight found in *Solanum demissum* Lindl. was introduced into the *S. tuberosum* L. types, biotypes of the pathogen were found which would break the dominantly inherited resistance. A series of 9 differentials have since been synthesized which differ only by a single gene based on their reaction to specific races of the late blight organism. With these it is possible to identify 512 races of late blight (see Van der Plank, 1968) Not all of the possible races have been identified in the laboratory, and a much lesser number have been found in the field, probably because host genotypes are not present in cultivated types. Those cultivars with 1, 2, and 3 genes for resistance, that are widely grown, are frequently attacked by the appropriate race(s) specific for those genotypes. Monogenic resistance to leaf mold (*Cladosporium fulvum* Cke.) in tomato has suffered a fate similar to that of resistance to late blight in potato. Almost as soon as a new leaf mold-resistant cultivar of tomato is developed, a pathogenic race attacks it. To date, 6 resistant genes have been postulated, and 11 pathogenic races of leaf mold have been identified (see Walter, 1967).

Resistance breaking races or biotypes have been found in other organisms that cause serious diseases in vegetables. Original resistance in each case was controlled by a major gene system. Resistances in vegetable crops which have been nullified by an appropriate gene(s) change for virulence in the pathogen include resistance to anthracnose (*Colletotrichum lindemutheanum* (Sacc. & Magn.) Briosi & Cav.) of beans, anthracnose (*C. lagenarium* (Pass.) Ell. & Halst.) of watermelon, downy mildew of lima bean, downy mildew (*Bremia lactucae* Reg.) of lettuce, rust (*Uromyces phaseoli typica* Arth.) of bean, and club root (*Plasmodiophora brassicae* Wor.) of cabbage. Resistance breaking strains in both tobacco etch virus and potato virus Y have delayed the development of multiple virus-resistant cultivars in pepper.

Natural race formation among the nematodes affecting vegetable crops occurs in the golden nematode, Heterodera rostochiensis Wollenweber. Resistance in potato was initially derived from S. tuberosum spp. andigena (Juz. & Buk.) Hawkes and is inherited as a monogenic dominant character. Nine years after golden nematode resistance was found, a resistance breaking physiologic race appeared in England (see Kehr, 1966). Subsequently additional such races have since been found and potato breeders have turned to a source of polygenic resistance found in S. vernei Bitter & Wittm. ex Engl. In general, natural physiologic race formation has been sparse in nematode species controlled among crops through host resistance. Race formation and survival in a competitive environment have been reported under experimental conditions (Winstead and Riggs, 1963). Herein lies an imminent danger as good vegetable land becomes more intensively cropped.

Accomplishments

Prior to the discovery of monogenic resistance to Fusarium wilt (race 1) of tomato, serious losses to this disease occurred in all production areas each year. Beginning in the 1930's, when resistant cultivars became available that were highly acceptable commercially, losses from Fusarium wilt have practically disappeared. The same is true with respect to the Verticillium wilt. This latter disease seriously limited productivity in the Western United States until cultivars with monogenic resistance became available in the 1950's. Both Verticillium and Fusarium wilt resistances were incorporated into cultivars adapted to the West. Now, neither wilt is a limiting factor in producing tomatoes commercially in that area. Shortly after Fusarium wilt was brought under control through resistance in the Midwest and East, Verticillium wilt began to appear in these areas as well as in a localized area in southern Florida. Breeders rapidly transferred resistance to Verticillium wilt into Fusarium wilt-resistant breeding stocks and developed acceptable cultivars with resistance to both diseases for those regions.

Gray leaf spot was a very serious foliar disease of the tomato in the humid areas east of the Rocky Mountains until resistant cultivars became available during the 1950's. Resistance is easily transferred to commercially acceptable types possessing resistance to both Verticillium and Fusarium wilts. Consequently, tomato cultivars grown east of the Rocky Mountains are resistant to at least two serious diseases and most are resistant to three.

Resistance to cabbage yellows is of inestimable value to the cabbage industry. Since the first cultivars with the high types of monogenic resistance were released from Wisconsin in the 1920's, numerous resistant cultivars and hybrids have been developed which have rendered yellows insignificant in cabbage production. Monogenic resistances to pea wilt (races 1 and 2) have been incorporated into numerous cultivars adapted to freezing, canning, and the fresh market. Pea wilt is no longer a threat to the industry except in a localized area in northwest Washington (Haglund and Kraft, 1970). Blue mold and mosaic resistances in spinach have brought this crop up from a seriously threatened status to a highly satisfactory one.

Scab causes necrotic lesions on cucumbers, and mosaic infection results in malformed and discolored fruits. Resistance to each disease is now available in a large number of cultivars and hybrids adapted to areas where one or both diseases are production factors. Pink root of onion has been brought under control in most areas of production through host resistance.

Lettuce breeders have managed to keep ahead of changes in physiologic races in the downy mildew organism (B. lactucae). Currently used resistance is inherited as a monogenic dominant, and cultivars with resistance to a new race have been rapidly developed through the backcross method. There is, however, the danger that in the future breeders may not have available major genes for resistance to new physiologic races as they arise (Sleeth and Leeper, 1966).

Host resistances, controlled by polygene systems, have been generally most significant in alleviating losses to diseases. Their effectiveness is, however, conditioned by prevailing weather conditions. Occasionally, in an adverse season, resistance must be supplemented with some applications of a fungicide or the crop must be grown in rotation with nonsusceptible crops.

Resistance to Fusarium wilt of watermelon, in combination with resistance to races 1 and 3 of anthracnose, has been of untold value to the watermelon industry. 'Charleston Gray', a cultivar with such resistances, accounts for almost 95 percent of the watermelons produced east of the Rocky Mountains. Resistant cultivars adequately control Fusarium wilt in the West. Resistance to downy mildew of both cantaloup and cucumber in the East and South, anthracnose of cucumber, and early blight of tomato gives growers a real measure of assurance of profitable crops, particularly if fungicides are applied during prolonged periods of weather favorable for disease development.

Many of the host resistances mentioned above are combined in single cultivars of the respective crops. Some widely grown cultivars of tomatoes are resistant to Fusarium and Verticillium wilts, gray leaf spot, nailhead spot (*Macrosporium tomato* Cke.), and early blight. Some cucumber cultivars are resistant to mosaic, downy and powdery mildews, anthracnose, and angular leaf spot (*Pseudomonas lachrymans* (E. F. Sm. & Bryan) Carsner).

Future Outlook

Breeders and pathologists have done a creditable job in breeding vegetable cultivars for resistance to diseases since the turn of the century and particularly so during the past 3 decades. During this latter period many vegetable crops had to be rather markedly redesigned to meet the requirements for harvesting by machine and handling the produce in bulk, or to meet requirements of changes in processing, or marketing practices. The productivity and the fresh-market and processing qualities of the old varieties had to be retained insofar as possible or, better yet, improved. And the spectrum of resistance to both parasitic and physiological diseases had to be broadened. The challenge of developing these new cultivars has been met without sacrificing resistance to disease. The job was not an easy one. I will use the tomato for processing as an example: the vine had to be made compact with a profuse fruit bearing habit. Flowering and fruit setting had to be completed in about 3 weeks; maturity had to be medium to medium-late. Fruits were desired that average 4 ounces in size, round to slightly oblong in shape, and highly resistant to circular and radial cracking and to bursting following unseasonable rains. Sixty percent of the fruits had to ripen to full ripeness within 3 weeks with the remaining fruit ripening at the rate of about 3 percent per day. Ripe fruit must "store" well on the vine for weeks, remain firm, and detach readily at the calyx but not so easily as to shatter onto the ground when vines are cut by the harvester. The fruits must also be highly resistant to breakage during passage through the harvester and when placed to a depth of 30 inches in bulk and finally transported up to 150 miles to a factory. Qualities to meet the requirements (pH, solids, flavor, texture) of whole-pack tomatoes, juice, soup, paste, and various other products had to be retained. Resistance to Fusarium and Verticillium wilts, gray leaf spot, nailhead spot, and tolerance to early blight, some forms of the blotchy ripening complex, certain fruit rots, and seedling diseases had to be added. Before release of a variety, adaptability and acceptability for commercial production had to be confirmed. This was accomplished over a 2-year period. This new combination of properties was achieved all within 12 plant generations, granting, of course, that most of the basic work on inheritance of certain specific properties had been done over a much longer time. Similar progress can be cited for cucumber, cabbage, cowpea, lima beans, spinach, and other crops (Barnes, 1966; Crill, et al., 1971; Hare, 1967; Wester, 1967; Williams, 1968; Walker, 1966). I don't mean to imply that all the disease problems of these reengineered vegetable crops have been solved.

Nor have we solved the "quality" problem. Actually, we have just made a good beginning.

Because of the frequency in the failure of resistances controlled by some major genes, there is a strong upsurge of feeling that we need a change in our current approach to breeding vegetables for resistance to disease. The rapid appearance of physiologic races of late blight on potato cultivars, golden nematode on potato, leaf mold on tomato, mildew on lettuce and lima bean, and others is a strong signal that in some instances we should strongly consider an alternate approach to control of these diseases through breeding. Currently used resistance to many vegetable and other crop diseases is of the monogenic vertical type (Person and Sidhu, 1971; Robinson, 1971), and resistant cultivars have been developed largely by the backcross method. Stringent evaluation procedures for this high type of resistance have tended to eliminate the "field" or horizontal type of resistance from breeding stocks and cultivars. Consequently, when a new resistance breaking race appears, often the resistant cultivars are more susceptible to the disease than currently grown susceptible ones. This has been the case in potato breeding for the hypersensitive type of resistance to late blight. Breeders now are trying to eliminate the "R" genes from their stocks in favor of horizontal resistance to late blight. Horizontal resistance to the highly variable golden nematode of potato appears to be the best approach to control of that organism in some areas. Some breeders struggling to control the leaf mold organism through vertical resistance are giving some serious thought to the horizontal route.

Resistances to a number of diseases of vegetable crops, each of which is controlled by a single major gene, have remained remarkably stable over a long time. These include resistances to Fusarium and Verticillium wilts of tomato, yellows in cabbage, scab and mosaic of cucumber, pink root of onion, Fusarium wilt of garden pea, blue mold of spinach, gray leaf spot of tomato, and other organisms. One or more of the above-named pathogens was a limiting factor in production of the respective host crop over wide areas before acceptable resistant cultivars became widely grown. New races affecting most of the stable monogenic host resistances have rarely appeared. Single gene resistances developed against two such races (race 2 of pea wilt and race 2 of spinach blue mold) have remained stable for a long time under commercial conditions. Race 2 of *F. oxysporum* f. *lycopersici* has been reported from three widely separated locations (Alexander and Tucker, 1945; Jones, 1966; Miller and Kanenen, 1968) and may become a real menace over a broad area unless the monogenic resistance to it is rapidly incorporated into commercially acceptable tomato cultivars. Race 5 of pea wilt is known to occur in northwest Washington (Haglund and Kraft, 1970), and major gene resistance against the pathogen is being sought.

I do not expect breeders to change their major gene, including vertical, resistance approach to disease control in vegetables except where circumstances dictate a change to the horizontal type. Major gene resistance is relatively easy to manipulate in a breeding program, and the results are usually dramatic. Breeding for field or horizontal resistance can be frustrating, costly, and time consuming. No doubt, this is the unavoidable long-range approach to most successful control of some diseases through breeding. Perhaps, in some instances, both major and polygene resistances can be incorporated into commercially accepted cultivars (Abdalla and Hermsen, 1971; Day, 1972), to delay the onset of disease and reduce inoculum potential once the disease has appeared.

We have a wealth of materials in vegetable crops at many points in this country, collected by the Plant Introduction Investigations of the U.S. Department of Agriculture, having potential as sources of disease resistance. This material is being screened for resistance to numerous diseases on a regular basis, and new resistances, both major and polygene, are reported frequently in leading agricultural journals (see Walker, 1965; Cook, 1968; James and Leeper, 1971; Kraft and Roberts, 1970; Natti, et al., 1967; Strider and Konsler, 1967; Sowell, et al., 1965; Wester, 1972). Also, many breeders develop and maintain germplasm with major and polygenes for resistance to many of the diseases of concern in their respective areas. As new breeding stocks with new or increased resistances are developed, it has become customary to share such materials with plant breeders domestically as well as internationally. Such resistances enter the germplasm pool of the breeder, and he uses the characters as his circumstances dictate.

Much time can be spent in speculating on the probability of the appearance and subsequent effect of resistance breaking races which will nullify vertical resistances now used to control major diseases of vegetables. No doubt, there will be additional such instances, particularly with those pathogens which are distributed through airborne inoculum. I believe both breeders and pathologists have such breeding stocks available, adequate technical expertise, and communication with coworkers to be able to respond adequately to a potentially dangerous disease situation which might develop among vegetable crops now protected by major genes. In most instances, newer fungicides, nematicides, and adjustments in cultural practices will provide the necessary lead time while new resistant, acceptable cultivars can be developed.

Literature Cited

Abdella, M. M. F. and J. G. Th. Hermsen. 1971. The concept of breeding for uniform and differential resistance and their integration. Euphytica 20: 351-361.

Alexander, L. J. and C. M. Tucker. 1945. Physiological specialization in the wilt fungus *Fusarium oxysporum* f. *lycopersici*. J. Agr. Res. 70: 303-313.

Barnes, W. C. 1966. Development of multiple disease resistant hybrid cucumbers. Proc. Amer. Soc. Hort. Sci. 89: 390-393.

Bohn, G. W. and C. M. Tucker. 1940. Studies on Fusarium wilt of tomato. I. Missouri Agr. Exp. Sta. Res. Bul. 311. 82 pp.

Cook, A. A. 1968. Virus disease resistance in some *Capsicum* species from South America. Plant Dis. Reptr. 52: 381-383.

Crill, J. P., D. S. Burgis and J. W. Strobel. 1971. Development of multiple disease-resistant fresh market tomato varieties adapted for machine-harvest. Phytopathology 61: 888 (Abstr.).

Day, P. R. 1972. Crop resistance to pests and pathogens, pp. 257-271. *In* Pest Control, Strategies for the Future. National Academy of Sciences, Washington, D.C.

Dropkin, V. H., D. W. Davis and R. E. Webb. 1967. Resistance of tomato to *Meloidogyne incognita acrita* and to *M. hapla* (root nematodes) as determined by a new technique. Proc. Amer. Soc. Hort. Sci. 90: 316-323.

Fassuliotis, G. and E. L. Corley, Jr. 1967. Use of seed growth pouches for root-knot nematode resistance tests. Plant Dis. Reptr. 51: 482-486.

Haglund, W. A. and J. M. Kraft. 1970. *Fusarium oxysporum* f. *pisi*, race 5. Phytopathology 60: 1861-1862.

Hare, W. W. 1965. The inheritance of resistance of plants to nematodes. Phytopathology 55: 1162-1167.

Hare, W. W. 1967. A combination of disease resistance in a new cowpea, Mississippi Silver. Phytopathology 57: 460 (Abstr.).

Harrison, A. L. 1960. Breeding disease-resistant tomatoes, with special emphasis on resistance to nematodes. Proc. Plant Sci. Seminar, Campbell Soup Co., pp. 57-78.

Hodgson, W. A. 1961. Laboratory testing of the potato for partial resistance to *Phytophthora infestans*. Amer. Potato J. 38: 259-264.

Hodgson, W. A. 1962. Studies on the nature of partial resistance to *Phytophthora infestans*. Amer. Potato J. 39: 8-13.

James, B. L. and P. W. Leeper. 1971. Sources of immunity from Races 5 and 6 of the lettuce downy mildew fungus (*Bremia lactucae*). Plant Dis. Reptr. 55: 794-796.

Jones, J. P. 1966. Distribution of Race 2 of *Fusarium oxysporum* f. *lycopersici* in Florida. Plant Dis. Reptr. 50: 707-708.

Kehr, A. E. 1966. Current status and opportunities for the control of nematodes by plant breeding, 33-100; 126-138. *In* E. F. Knipling (ed.) Pest Control by Chemical, Biological, Genetic, and Physical Means. USDA.

Kraft, J. M. and D. D. Roberts. 1970. Resistance in peas to Fusarium and Phythium root rot. Phytopathology 60: 1814-1817.

Lockwood, J. L. and J. C. Ballard. 1959. Factors affecting a seedling test for evaluating resistance of pea to Aphanomyces root rot. Phytopathology 49: 406-410.

Miller, R. E. and D. L. Kananen. 1968. Occurrence of *Fusarium oxysporum* f. sp. *lycopersici* Race 2 causing wilt of tomato in New Jersey. Plant Dis. Reptr. 52: 553-554.

Natti, J. J., M. H. Dickson and J. D. Atkin. 1967. Resistance of *Brassica oleracea* varieties to downy mildew. Phytopathology 57: 144-147.

Orton, W. A. 1911. The development of disease resistant varieties of plants. IV. Conf. Int. Genet. Paris, C. R. et Rapp. 247-265.

Person, C. and G. Sidhu. 1971. Genetics of host-parasite interrelationships. *In* Mutation Breeding for Disease Resistance. Proc. of a Panel, Vienna, Oct. 12-16, 1970, jointly organized by the IAEA and FAO: 31-38.

Pound, G. S. and P. C. Cheo. 1952. Studies on resistance to cucumber virus 1 in spinach. Phytopathology 42: 301-306.

Robbins, M. L. and F. F. Angell. 1970. Tomato anthracnose: A hypodermic inoculation technique for determining genetic reaction. J. Amer. Soc. Hort. Sci. 95: 118-119.

Roberts, D. D. and J. M. Kraft. 1971. A rapid technique for studying Fusarium wilt of peas. Phytopathology 61: 342-343.

Robinson, R. A. 1971. Vertical resistance. Rev. Plant Pathol. 50: 233-239.

Sleeth, B. and P. W. Leeper. 1966. Mildew resistant lettuce susceptible to a new physiologic race of *Bremia lactucae* in South Texas. Plant Dis. Reptr. 50: 460.

Steinbauer, C. E. 1956. Types of sweetpotato cuttings for most precise evaluation of Fusarium wilt susceptibility. Proc. Amer. Soc. Hort. Sci. 68: 394-399.

Strider, D. L. 1970. Tomato seedling inoculations with Corynebacterium michiganense. Plant Dis. Reptr. 54: 36-39.

Strider, D. L. and T. R. Konsler. 1965. An evaluation of the cucurbita for scab resistance. Plant Dis. Reptr. 49: 388-391.

Sowell, G., J., K. Prasad and J. D. Norton. 1966. Resitance of Cucumis melo introductions to Mycosphaerella citrullina. Plant Dis. Reptr. 50: 661-663.

Tisdale, W. B. 1923. Influence of soil temperature and soil moisture upon the Fusarium disease in cabbage seedlings. J. Agr. Res. 24: 55-86.

Thomas, H. R., H. Jorgensen, and R. E. Wester. 1952. Resistance to downy mildew in lima bean and its inheritance. Phytopathology 42: 43-45.

Thyr, B. D. 1968. Resistance to bacterial canker in tomato, and its evaluation. Phytopathology 58: 279-281.

Van der Plank, J. E. 1963. Plant diseases: epidemics and control. Academic Press: New York and London. 349 pp.

Van der Plank, J. E. 1968. Disease resistance in plants. Academic Press: New York and London. 206 pp.

Walker, J. C. and R. Smith. 1930. Effect of environmental factors upon the resistance of cabbage to yellows. J. Agr. Res. 41: 1-15.

Walker, J. C. 1965. Use of environmental factors in screening for disease resistance. Ann. Rev. Phytopathology 3: 197-208.

Walker, J. C. 1965. Disease resistance in the vegetable crops. III. Bot. Rev. 31: 331-380.

Walker, J. C. 1966. The role of pest resistance in new varieties, 219-242. In K. J. Frey (ed.). Plant Breeding. The Iowa State University Press, Ames, Iowa.

Walter, J. M. 1967. Hereditary resistance to disease in tomato. Ann. Rev. Phytopathology 5: 131-162.

Webb, R. E. 1955. Cotyledonary inoculation, a method for screening spinach for blight resistance. Phytopathology 45: 635.

Wellman, F. L. 1939. A technique for studying host resistance and pathogenicity in tomato Fusarium wilt. Phytopathology 51: 614-616.

Wester, R. E. 1967. Two new green-seeded baby lima beans resistant to two strains of Phytophthora phaseoli. Phytopathology 57: 648-649 (Abstr.).

Wester, R. E., V. J. Fisher and V. L. Blount. 1972. Multiple resistance in lima beans to downy mildew (Phytophthora phaseoli). Plant Dis. Reptr. 56: 65-66.

Williams, P. H., J. C. Walker and G. S. Pound. 1968. Hybelle and Sanibel, multiple disease-resistant F_1 hybrid cabbages. Phytopathology 58: 791-796.

Winstead, N. N. and R. D. Riggs. 1963. Stability of pathogenicity of B biotypes of the root-knot nematode Meloidogyne incognita on tomatoes. Plant Dis. Reptr. 47: 870-871.

SOME ASPECTS OF DISEASE RESISTANCE IN COTTON

A. B. Wiles
Research Plant Pathologist
Plant Science Research Division, ARS, USDA
Plant Pathologist
Mississippi Agricultural and Forestry Experiment Station

For many years world production of Upland cotton (*Gossypium hirsutum* L.) has relied on host resistance as a primary means of disease control. Resistance to the Fusarium wilt - root knot nematode complex (*Fusarium oxysporum* f. sp. *vasinfectum* (Atk.) Sny. & Hans. - *Meloidogyne incognita* (Kafoid & White) Chitwood); resistance to bacterial blight (*Xanthomonas malvacearum* (E. F. Sm.) Dows.); and greatly increased tolerance to Verticillium wilt (*Verticillium albo atrum* Reinke & Berth.) are notable examples of host reactions that have been successfully utilized in developing practical disease control measures. Indeed, disease resistance combined with various cultural practices such as sanitation and rotation have been our chief weapons against pathogens that attack cotton. The uses of chemical measures for controlling cotton diseases have been almost entirely limited to seed and soil treatment materials for seed- and soil-borne pathogens and soil fumigants for nematodes.

Because of our rapidly expanding knowledge of the various aspects of this subject, future advances in plant disease control by means of host resistance can be expected to come at a more rapid rate than it has in the past. Time does not permit an examination of the many inoculation techniques and evaluation methods that have been and are being used in disease resistance studies concerning cotton. Therefore, I shall first make a few general observations concerning plant inoculations and then use Verticillium wilt of cotton to illustrate some of the techniques of inoculation, methods of evaluation, and problems that arise in disease resistance studies. Later I will review some past accomplishments in disease resistance in cotton and then discuss some resistances that may be utilized in the future.

Plant Inoculation

While several concepts have been advanced concerning the term inoculation, the following practical definition is proposed: the process of introducing the inoculum (the pathogen) within or onto the surface of host plants. Inoculation of host plants is needed to obtain proof of pathogenicity; this is required by Koch's postulates. Inoculation is also needed for making plant selections in breeding programs, for making host range determinations, for studying life cycles of pathogens, and for such purposes as the development of supplementary control measures.

The method of plant inoculation that is employed will depend on a number of factors and considerations. Some of these are: the nature of the pathogen, the mode of host entry, the part of the plant involved, the nature and site of host resistance, and the effects of various environmental factors on disease development. Additional factors that must be considered are type of available equipment, the size of the facilities, and the amount of labor required. In screening for disease resistance in plants, seedling inoculation techniques are often used because large plant populations can be inoculated and rapidly evaluated using a minimum amount of space. Whether the disease reactions of seedlings and those of mature plants are sufficiently comparable to obtain the desired information must be determined by the researcher.

When screening for disease resistance, the techniques employed and the pre- and post-inoculation environments should simulate natural conditions as closely as possible. All types of host responses--susceptible, tolerant, resistant, and even immune--are subject to the effects of the environment on the pathogen, the host, and their ultimate interaction in disease development. Therefore, failure to obtain visible symptoms cannot be regarded as proof of resistance or of non-pathogenicity, since some influencing factor might have been overlooked. Other items that require consideration will be mentioned later. General descriptions of inoculation techniques for plant pathogens have been presented by Riker and Riker (1936); Altman (1966); Waterston (1968); and Tuite (1969).

Verticillium Wilt of Cotton

Now I shall examine some specific methods of plant inoculation that have been used in studying Verticillium wilt of cotton. It should be remembered that this destructive disease of cotton is incited by a soil-borne fungus that in nature enters the plant by direct penetration of the roots. Thus, it was a rather normal development that field nurseries were the earliest method of evaluating the reaction of cotton plants to this pathogen. Field nurseries are normally located where Verticillium wilt has regularly occurred, where susceptible plants are rotated with the nursery strains, or where fungal inoculum can be added to the soil. Such disease nurseries are still extensively used in both the U.S. and the U.S.S.R. In the U.S. field selection in wilt nurseries has played an important role in the development of our present commercial varieties with higher tolerance to Verticillium wilt.

Beginning about 1950, attempts were initiated to develop seedling inoculation techniques for breeding programs that would give critical differentiation of resistance and susceptibility. A "solution-culture method" was utilized where cotton seedlings were grown on excelsior in shallow tanks of nutrient solution and subsequently inoculated by placing measured amounts of a spore-mycelium suspension in each tank (Presley, 1950). Later a seedling inoculation method that had been found satisfactory for testing tomato seedlings for resistance to Verticillium wilt was adapted for use on cotton (Virgin and Maloit, 1947; Wiles, 1952). This was termed the "root-dip method" of inoculation. The seedlings were inoculated in the 4-leaf stage by lifting them with a spatula from flats of soil, washing the roots in tap water, and then dipping the roots in a liquid spore-mycelium inoculum. The plants were then reset in the soil and disease symptoms occurred 7-8 days after inoculation. Another seedling inoculation technique termed the "root-ball method" utilized 6-week-old potted plants that were gently removed from their pots and sprayed with a conidial suspension using an aerosol spray. The plants were then returned to the pots and disease readings were later made (Garber and Houston, 1966, 1967; Schnathorst and Mathre, 1966a, b).

During this period still another method of plant inoculation was being employed (Erwin, et al., 1965). This was termed the "stem-puncture method" of inoculation and utilized a hypodermic syringe to inject measured amounts of liquid inoculum into the hypocotyl of field-grown seedlings. This method produces very high percentages of infection. Later this same technique was used except that the center of the cotyledonary node was punctured to a controlled depth with a hypodermic needle which had been dipped in a spore suspension of *V. albo-atrum*

(Bugbee and Presley, 1967). Others (Schnathorst and Mathre, 1966a, b) have used slight modifications of this method.

Inoculation of the roots and hypocotyls of cotton plants have also been made by placing blocks of potato dextrose agar (PDA) containing conidia of V. albo-atrum on selected sites of the plant (Garber and Houston, 1966). Other methods employ the placing of fungal microsclerotia in soil prior to seeding (Schnathorst and Mathre, 1966a, b), or using liquid inoculum applied to the growth medium after root pruning (Ranney, 1962).

Now let us briefly examine some factors that may influence the results of these inoculation studies. Plants for such studies may be grown in fields, greenhouses, or growth chambers but as already noted, the environment should closely simulate the conditions favoring natural infection. It should be remembered that inoculation procedures themselves may exert an influence on the host plants. Lifting of young plants for inoculation causes some root injury that may facilitate fungal entry. The same situation is true to some extent where the root-ball is sprayed. Hypodermic injections tend to produce more severe infections than natural infections. Another factor to consider is that the fungus causing Verticillium wilt of cotton varies in virulence due to numerous factors, thus different isolates may give different experimental results. The amount of potentially infective material in a given environment (inoculum potential) largely determines the probability of successful infection. Thus a standardization of inoculum density (spores/ml) must be made, using some type of instrument such as a hemacytometer or a colorimeter. Temperature is an important factor that influences the host, the pathogen, their interactions, and disease development. It is critical in the development of Verticillium wilt of cotton, and a recent study showed that accurate classification of genetically tolerant and susceptible cotton plants was dependent upon carefully controlled temperatures (Barrow, 1970).

Difficulties often arise in assessing the results of plant inoculations. In field nurseries plant breeders and pathologists have often encountered difficulties in relating disease severity and yield. Significant yield differences have been recorded between strains of cotton that appeared nearly equal in regard to disease symptoms. Various field grading schemes have been proposed. In controlled environments, assessments of host reaction have been made on the basis of wilting plus internal discoloration, plant death, and by measuring the downward bending of petioles of healthy and inoculated plants (Presley, 1950; Wiles, 1953, 1960, 1963). Another difficulty arises when it is noted that resistance or tolerance may be related to root structure or function (Garber and Houston, 1967). If this be the case, stem puncture inoculations might preclude or bypass and prevent expressions of resistance to Verticillium wilt by plants inoculated in this manner. Finally, if we assume that resistance is a function of the total plant then this alone gives sufficient reason for some type of field evaluation. A plant that shows severe disease symptoms but maintains a high yield is far more desirable than an immune, non-productive plant.

Most of the early research objectives were directly related to finding sources of resistance to Verticillium wilt in cotton and then attempting to utilize them in breeding programs. In more recent years many of the objectives have been more basic in nature. Thus, the objectives of the researcher and his available facilities still have a bearing on the type of inoculation technique he will use and the ultimate choice of techniques is a question that he must decide.

Past Accomplishments

There are notable past accomplishments in developing disease resistance in cotton and some of these have already been cited. Fusarium wilt is among the oldest recognized cotton diseases in the United States, and in former years caused heavy losses. It is closely interrelated with the activities of certain plant parasitic nematodes, most commonly the root knot nematode. Therefore, field studies on the inheritance of resistance to Fusarium wilt have been complicated by the presence or absence of nematodes. Early workers (Smith and Dick, 1960) reported that a single dominant gene controlled Fusarium wilt resistance in Upland cotton and additional genes determined root knot nematode resistance. A later study that involved resistant and susceptible parents as well as F_1, F_2 and F_3 populations, indicated that resistance may be a quantitative character (Jone, et al., 1967). Kappelman (1971) recently reported that additive, dominance, and epistatic effects were involved in the inheritance of resistance of cotton to Fusarium wilt. At any rate, by means of recurrent selection and pure line breeding techniques plant breeders and pathologists have successfully incorporated the Fusarium wilt resistance that occurs in Upland cotton into some highly resistant commercial cotton varieties (Kappelman, 1971). Since Fusarium wilt is caused by a soil-borne fungus, artificial inoculation techniques that are used in resistance studies for this disease are similar to those employed by Verticillium wilt.

Resistance to the root knot nematode has been reported in such Upland cottons as Clevewilt-6; in a primitive strain of _G. hirsutum_, Louisiana Mexico Wild; and in _Gossypium barbadense_ var. _darwinii_ (Watt.) Hutch. (Jones, et al., 1958; Smith, 1941). In Clevewilt-6, resistance was found to be inherited as a quantitative character and probably controlled by two or more genes (Jones, et al., 1958). Resistance in _G. barbadense_ var. _darwinii_ was reported to be controlled by a pair of recessive genes (Turcotte, et al., 1963). The resistance that is currently being utilized in commercial cotton varieties has been derived from Upland types such as Clevewilt-6. A greenhouse method of determining the reaction of cotton plants to the root knot nematode would consist of planting seed in containers of soil previously infested with a known population of root knot larvae. After about 3 months, a count of the nematode population in each container would be made and the roots of the plants rated for nematode injury and egg mass production (Jones, Birchfield, 1967).

Resistance to bacterial blight in cotton has been the subject of intensive study and investigation. In 1950 Knight and Hutchinson (1950) listed 5 major genes for resistance to blight: B_1, B_2, B_3, B_4, and B_5. He further noted that minor genes occurred but their value was quite limited when used alone. In 1963 Knight (1963) reported the occurrence of 10 major blight resistance genes. The influence of minor genes and the influence of genetic background on major genes appear to be of considerable importance in blight resistance. The occurrence of biotypes or races of the causal organism has been demonstrated and has further complicated the development of resistant varieties. Many of our present cotton varieties contain some of the major or minor genes for blight resistance. In blight breeding programs, plants are usually inoculated by spraying a standardized bacterial suspension onto the lower surface of the leaves and forcing the inoculum into the substomatal cavities. Disease evaluations are then based on the extent of disease development on the inoculated leaves.

As already noted, tolerance to Verticillium wilt has been substantially increased in our commercial cotton varieties, primarily through the use of recurrent selection in field nurseries. Sappenfield (1963) reported that genetic interrelationships may exist among factors which condition resistance to the Fusarium wilt - root knot nematode disease complex and verticillium wilt. Varieties highly resistant to the Fusarium wilt - root knot disease showed some degree of tolerance to Verticillium wilt, but some strains tolerant to Verticillium wilt were not resistant to the Fusarium wilt - root know complex. While resistance to Verticillium wilt has not been found in Upland cottons, high levels of tolerance most likely of multigenic nature, do occur and this has been incorporated in a number of commercial varieties. Resistance to Verticillium wilt does occur in _Gossypium barbadense_ L. and appears to be inherited as an incompletely dominant factor (Wilhelm, et al., 1969). This type of resistance is not present in commercial varieties of Upland cotton.

Future Outlook

The future outlook for the continued use of resistance as a major disease control measure in cotton appears favorable. This seems particularly true regarding the development of strains with multiple disease resistances that are adapted for use in specific areas of the cotton belt, but we must remember that disease resistance in crops must be combined with high levels of production. Another favorable factor is that sources of resistance to other diseases of cotton are still being reported. In some cases this resistance has already been incorporated into usable breeding stocks.

Recently, it was announced that 3 strains of cotton resistant to Southwestern cotton rust (*Puccinia cacabata* Arth. & Holw.) were developed by isolating resistance from 2 seed lots of interspecific origin, and breeding stocks have been released. Plants from these 2 sources which were resistant were crossed with Acala lines to produce the 3 resistant strains. Nanking (*Gossypium arborum* L.) has been reported to be a source of resistance to the cotton anthracnose organism (*Glomerella gossypii* (South.) Edg. (Bollenbacher and Fulton, 1967).

Certain boll characteristics such as the absence of nectaries, tightly closed sutures, and tightly closed apices have been shown to lessen boll rot losses by altering the mode of entry of boll rot pathogens (Bagga, 1970). These plant characters, together with the okra leaf character which tends to reduce boll rot by reducing leaf canopy, might be termed disease escape mechanisms. They could, however, contribute to the overall disease control picture in cotton.

In short, there have been some notable achievements and progress in the control of cotton diseases by host resistance, but, obviously, there is room for additional progress.

Literature Cited

Altman, J. 1966. Laboratory Manual: Phytopathological Techniques. Pruitt Press, Inc., Boulder, Colorado.

Bagga, J. S. 1970. Mode of entry of boll rot pathogens in selected cotton varieties and strains. Plant Dis. Reptr. 54: 719-721.

Barrow, J. R. 1970. Critical requirements for genetic expression of Verticillium wilt tolerance in Acala cotton. Phytopathology 60: 559-560.

Bollenbacher, K. and N. D. Fulton. 1967. Susceptibility of *Gossypium arboreum* and *Gossypium hirsutum* to seedling anthracnose. Plant Dis. Reptr. 51: 632-636.

Bugbee, W. M. and J. T. Presley. 1967. A rapid inoculation technique to evaluate the resistance of cotton to *Verticillium albo-atrum*. Phytopathology 57: 1264.

Erwin, D. C., W. Moje and I. Malca. 1965. An assay of the severity of Verticillium wilt on cotton plants inoculated by stem puncture. Phytopathology 55: 663-665.

Garber, R. H. and B. R. Houston. 1966. Penetration and development of *Verticillium albo-atrum* in the cotton plant. Phytopathology 56: 1121-1126.

Garber, R. H. and B. R. Houston. 1967. Nature of Verticillium wilt resistance in cotton. Phytopathology 57: 885-888.

Jones, J. E. and W. Birchfield. 1967. Resistance of the experimental cotton variety, Bayou, and related strains to root knot nematode and Fusarium wilt. Phytopathology 57: 1327-1331.

Jones, J. E., S. L. Wright and L. D. Newsom. 1958. Sources of tolerance and inheritance of resistance to root knot nematode in cotton. Cotton Impr. Conf. Proc. 11: 34-39.

Kappelman, A. J. 1971. Fusarium wilt resistance in commercial cotton varieties. Plant Dis. Reptr. 55: 896-897.

Kappelman, A. J. 1971. Inheritance of resistance to Fusarium wilt in cotton. Crop Sci. 11: 672-674.

Knight, R. L. 1963. The genetics of blackarm resistance. XII. Transferrence of resistance from Gossypium herbacearum to G. barbadense. J. Genet. 58: 328-346.

Knight, R. L. and J. B. Hutchinson. 1950. The evolution of blackarm resistance in cotton. J. Genet. 50: 36-58.

Presley, J. T. 1950. Verticillium wilt of cotton with particular emphasis on variation of the causal organism. Phytopathology 40: 497-511.

Ranney, C. D. 1962. Effects of nitrogen source and rate on the development of Verticillium wilt of cotton. Phytopathology 52: 38-41.

Riker, A. J. and R. S. Riker. 1936. Introduction to research on plant diseases. John S. Swift Co., St. Louis.

Sappenfield, W. O. 1963. Fusarium wilt - root knot nematode and Verticillium wilt resistance in cotton: Possible relationships and influence on cotton breeding methods. Crop Sci. 3: 133-135.

Schnathorst, W. C. and D. E. Mathre. 1966a. Host range and differentiation of a severe form of Verticillium albo-atrum in cotton. Phytopathology 56: 1155-1161.

Schnathorst, W. C. and D. E. Mathre. 1966b. Cross-protection in cotton with strains of Verticillium albo-atrum. Phytopathology 56: 1204-1209.

Smith, A. L. 1941. The reaction of cotton varieties to Fusarium wilt and root knot nematodes. Plant Dis. Reptr. Suppl. 227: 90-91.

Smith, A. L. and J. B. Dick. 1960. Inheritance of resistance to Fusarium wilt in Upland and Sea Island cottons as complicated by nematodes under field conditions. Phytopathology 50: 44-48.

Tuite, J. 1969. Plant pathological methods: Fungi and bacteria. Burgess Publ. Co., Minneapolis, Minn.

Turcotte, E. L., H. W. Reynolds, J. H. O'Bannon and C. V. Feaster. 1963. Evaluation of cotton root knot nematode resistance of a strain of G. barbadense var. darwinii. Cotton Impr. Conf. Proc. 15: 36-44.

Virgin, W. J. and J. C. Maloit. 1947. The use of the seedling inoculation technique for testing tomatoes for resistance to Verticillium wilt. Phytopathology 37: 22 (Abstr.).

Waterston, J. M. 1968. Inoculation. Rev. Appl. Mycology 47: 217-222.

Wiles, A. B. 1952. A seedling inoculation technique for testing cotton varieties for resistance to Verticillium wilt. Phytopathology 43: 288. (Abstr.).

Wiles, A. B. 1953. Reaction of cotton varieties to Verticillium wilt. Phytopathology 43: 489 (Abstr.).

Wiles, A. B. 1960. Evaluation of cotton strains and progenies for resistance to Verticillium wilt. Plant Dis. Reptr. 44: 419-422.

Wiles, A. B. 1963. Comparative reactions of certain cottons to Fusarium and Verticillium wilts. Phytopathology 53: 586-588.

Wilhelm, S., J. E. Sagen and H. Tietz. 1969. Dominance of Sea Island resistance to Verticillium wilt in F_1 progenies of Upland x Sea Island cotton crosses. Proc. Beltwide Cotton Prod. Res. Conf. 1969, p 31 (Abstr.).

CORN BREEDING AND THE ASSOCIATION OF MALE STERILITY WITH INFECTION BY *Helminthosporium maydis* AND POSSIBLE ACTION OF *Fusarium moniliforme*

M. C. Futrell and Nusrat Z. Naqvi
Research Plant Pathologist, Plant Science Research Division, ARS, USDA
Post Doctoral Fellow
Plant Pathology and Weed Science Department
Mississippi Agricultural and Forestry Experiment Station
Mississippi State, Mississippi

Abstract

Texas male sterile corn seed infected with race T of *H. maydis* were soaked in benomyl, embryos excised and studied histologically. Mycoplasmic bodies of *H. maydis* were found in the scutellar cells immediately adjacent to the endosperm. As the corn seed began to germinate these bodies increased in length in the scutellar cells giving rise to intercellular mycelia. The scutellar cells that gave rise to the *H. maydis* mycoplasma appeared to be female in origin which could account for the relationship between male sterility in corn and susceptibility in race T of *H. maydis*. No mycoplasmic bodies were found in seed of normal cytoplasm corn. The theory is proposed that male sterility is due to mycoplasmic bodies functioning as genes in the cytoplasm producing estrogenic compounds as reported for *F. moniliforme*.

Introduction

The epiphytotic of southern corn leaf blight caused by race T of *Helminthosporium maydis* Nisikado & Miyake gave us the possibilities of new approaches in studying disease resistance in corn, *Zea mays* L. Male sterility is female inherited and the gene that controls this is located in the cytoplasm. In 1896 Eriksson (1922) postulated that plant diseases could be female inherited. This was part of his mycoplasm theory. Most plant pathologists of that era were too busy criticizing Eriksson to really take a good look at what he was theorizing. His original theory (1897) was set forth with stem rust of wheat in 1897, and by 1921 he had shown that twenty-one other species of fungi could live a symbiotic life in the cytoplasm of host plant cells. A very minute quantity of fungal life lives in the cytoplasm of the host cells and can be transmitted from one generation to the next by female inheritance and then, at the opportune time and under the proper stimulation, it becomes parasitic on the host cells. The fungal mycoplasma in the cytoplasm are probably in the range size of genes.

What stimulation could cause a submicroscopic gene to become parasitic? One possibility is a similar gene outside the host in a spore, bacterium, virus, etc. This type of infection process supports Flor's gene for gene (Eriksson, 1922; Flor, 1942) theory of disease development. Person, et al. (1962) further applied the gene for gene theory to include two organisms living together as symbiotic partners. These workers pointed out that there is no distinct dividing line between mutualistic and antagonistic symbiosis. The gene for gene concept comes into operation only when one organism becomes parasitic on the other. This appears to be what is happening between the corn plant and *Fusarium moniliforme* Sheldon. The fungus appears to be present in most corn plants, but only under certain conditions does it become parasitic. Lucado (1970) and Naqvi (1971) found mycoplasmic stages of *F. moniliforme* in the corn plant or in the corn seed at varying stages of development. Futrell (1972) showed that benomyl and thiabendazole applied to corn seed will prevent the fungal mycoplasma of *F. moniliforme* from developing into mycelium.

New data are presented in this paper showing that *H. maydis* has a mycoplasmic stage in corn seed.

Materials and Methods

Texas male sterile corn seed known to be infected with *H. maydis* were soaked in 200 ppm of benomyl for 18 hours. This was done to prevent *F. moniliforme* from developing in the corn seed. The embryos were then excised and fixed in Randolph-Navashin fixative (Johansen, 1940), embedded, sectioned and stained with a safranin-fast green double stain. Photomicrographs were made with a 35 mm camera.

Normal cytoplasm corn seed were treated in the same manner and used as checks.

Results

An example of corn seed infected with race T of *H. maydis* typical of that used in this experiment is shown in Figure 1. The internal infection of *H. maydis* was first found as mycoplasmic life inside the scutellar cells adjacent to the endosperm (Figures 2 and 3). These cells were either female in origin or were affected by the 3N female endosperm cells. These mycoplasmic bodies inside the cells begin to elongate at both ends inside the cells as shown in Figure 3. The cell walls at the end of the cell then break down and the mycelium moves out of the cell into the intercellular spaces (Figure 4). At this stage of development and not before can the mycelium be found in the intercellular spaces (Figure 5). When the intracellular mycoplasmic life becomes active inside the cell and gives rise to intercellular mycelium a toxin is produced and embryonic cells in the root tips begin to die (Figure 6). This causes seedling blight and death of corn plants.

No mycoplasmic bodies or internal *H. maydis* infection were found in normal cytoplasm corn seed.

Discussion

Eriksson (1897, 1922) described mycoplasm (also mycoplasma) as a form of a fungus body which may be merged with the protoplasm of the host in the seed or other dormant structures and later gives rise to a typical mycelium. This work was published in 1897. The presentday authoritative classification (Breed, et al., 1957), used for the order Mycoplasmatales completely ignores Eriksson's work. The presentday classification of the Mycoplasmatales includes a group of organisms between the viruses and bacteria. They use the pleuro-pneumonia organism (*Mycoplasma mycoides*) as the form species of the order Mycoplasmatales. In 1935 Turner (1935) described a mycelial stage of the pleuro-pneumonia organism which would make this organism fit into the description given by Erikkson in 1897 for mycoplasma. Turner's description is shown in Figure 9. Instead of "mycoplasma" being a group of organisms, it appears the word describes a stage in the development of a pathogen probably fungal in nature.

Let's compare the internal infection of H. maydis and F. moniliforme in corn seed. Both fungi are carried internally in the seed as mycoplasmic life inside the cells of the scutellum lying near the endosperm. The mode of development of the mycoplasmic body inside the cells to form a mycelium differs with the two species of fungi. In the case of F. moniliforme, Naqvi (1971) showed that from one to three mycoplasmic bodies can develop in a cell (Figures 7 and 8). These may develop in the center of the cell (Figure 7) as a proliferation around the primary nucleus, or they may develop throughout the cell and proliferate out as a mycelium as shown in Figure 8. When the mycoplasmic bodies reach the size shown in Figure 8 the cell wall disintegrates and the mycelium grows out into the intercellular spaces producing toxin and destroying other cells. By use of a crude extract of toxin produced by F. moniliforme Lucado (1970) was able to get the same damage to the embryonic root cells as that produced by the fungus. The toxin action caused by F. moniliforme appears to be similar to that shown for H. maydis given in Figure 6.

What is the relationship between H. maydis race T infection and male sterility? Race T of H. maydis is highly pathogenic on Texas male sterile corn and less pathogenic on normal cytoplasm corn. This gives strong evidence that the male sterile gene in the cytoplasm has more control over susceptibility to race T than does nuclear genes. Further evidence of this close association is given by the fact that the mycoplasmic bodies in the seed are located in the outer cells of the scutellum. These cells may be female in origin, but if they are not they could be influenced by the 3N female endosperm cells adjacent to the scutellum. In this way, both male sterility and susceptibility to race T of H. maydis is female inherited. Eriksson (1922) proposed in his mycoplasma theory that susceptibility to some diseases in plants was female inherited. The 1970 epiphytotic of Southern corn leaf blight on Texas male sterile corn in the United States suggested that Eriksson was right because susceptibility to this disease was female inherited.

F. moniliforme has been reported to affect estrogen activity in animals. Mirocha, et al. (1969) described the estrogenic syndrome of female dominance in swine. They point out that the biosynthesis of the fungal estogrens F-2 and F-3 by F. moniliforme is the primary cause of the estrogenic syndrome. They further point out that these estrogens are important sex regulating hormones in many fungi.

Mycoplasmic material of F. moniliforme in the corn cell cytoplasm could bring about the synthesis of estrogenic substances which could inhibit pollen formation.

Virus infection has been associated with male sterility in pepper, Capsicum annuum L. (Ohta, 1970). Duggar and Armstrong (1923) postulated that a virus was a gene that had gone wild in the cytoplasm. When corn plants are infected with maize dwarf mosaic virus the incidence of F. moniliforme increased in virus infected plants (Futrell, 1971; Futrell and Scott, 1969).

Literature Cited

Breed, R. S., E. G. D. Murray and N. R. Smith. 1957. Bergey's Manual of Determinative Bacteriology. The Williams and Wilkins Co., Baltimore, 7th ed. 1094 pp.

Duggar, B. M. and J. K. Armstrong. 1923. Indications respecting the nature of the infective particle of the mosaic disease of tobacco. Ann. Missouri Bot. Garden 10: 191-212.

Eriksson, J. 1897. Vie lantente et plasmatique de certaines Uredinees. Comptes Rendus 124(9): 475-477.

Eriksson, J. 1922. La theorie du mycoplasma. Sa portee scientifique et sa perspective pratique. Bull. Des Renseignements Agricoles et des maladies des Plantes 13: 283-294.

Flor, H. H. 1942. Inheritance of pathogenicity in Melampsora lini. Phytopathology 32: 653-669.

Flor, H. H. 1955. Host-parasite interaction of the flax rust--Its genetic and other implications. Phytopathology 45: 680-685.

Futrell, M. C. 1971. Association of maize dwarf mosaic virus and Fusarium moniliforme with yellows and stunting of corn and mosaic of grain sorghum. Proc. 8th BiAnn. Meet. Grain Sorghum Prod. Assoc. Lubbock, Texas 8: 10-12.

Futrell, M. C. 1972. New concepts in chemical seed treatment of agronomic crops. J. Environ. Qual. 1: 240-243.

Futrell, M. C. and G. E. Scott. 1969. Effect of maize dwarf mosaic virus infection on invasion of corn plants by Fusarium moniliforme. Plant Dis. Reptr. 53: 600-602.

Johansen, D. A. 1940. Plant Microtechnique. McGraw-Hill Book Co., Inc., New York. 523 pp.

Lucado, J. S., Jr. 1970. Pathogenesis of corn seedlings infected with Fusarium moniliforme Sheldon. M.S. Thesis. Mississippi State University. 29 pp.

Mirocha, C. J., C. M. Christensen and G. H. Nelson. 1969. Biosynthesis of the fungal estrogens F-2 and a naturally occurring derivative (F-3) by Fusarium moniliforme. Appl. Microbiology 17: 482-483.

Naqvi, N. Z. 1971. The mycoplasmic stage of Sclerospora sorghi Weston and Uppal and Fusarium moniliforme Sheldon. Ph.D. Dissertation. Mississippi State University. 53 pp.

Ohta, Yasuo. 1970. Cytoplasmic male sterility and virus infection in Capsicum annuum. Japanese J. of Genet. 45: 277-283.

Person, C., D. J. Samborski and R. Rohringer. 1962. The gene-for-gene concept. Nature 194: 561-562.

Turner, A. W. 1935. Study of the morphology and life cycles of the organism of Pleuropneumonia contagiosa Baum (Borrelomyces peripneumoniae Nov. Gen.) by observation in the living state under dark ground illumination. J. Pathol. & Bacteriol. 41: 1-32.

Figure 1. Corn ears infected with H. maydis; L. to R. healthy ear, apical tip infection, base infection, severe base infection and spot infection.

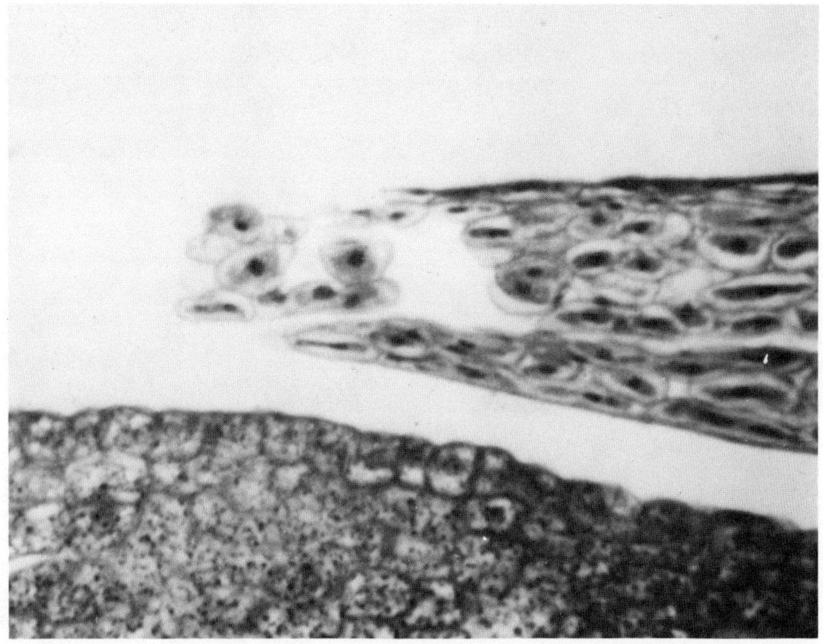

Figure 2. Scutellar cells lying next to the endosperm.

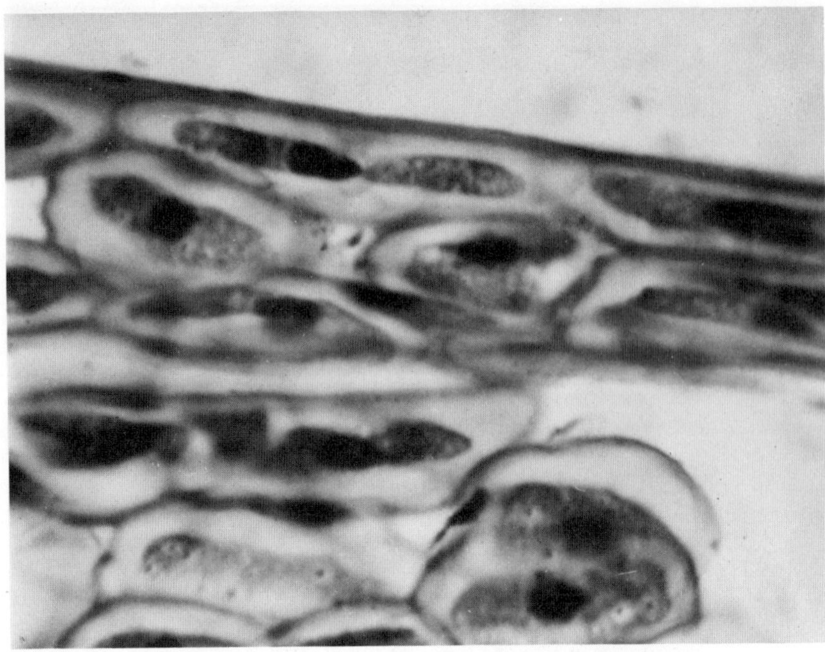

Figure 3. Elongation of mycoplasmic bodies and breakdown of cell walls in scutellum cells next to 3N endosperm tissue.

Figure 4. The cell walls have broken down (especially at the ends of the cells) and the nycoplasmic bodies have given rise to intercellular mycelium.

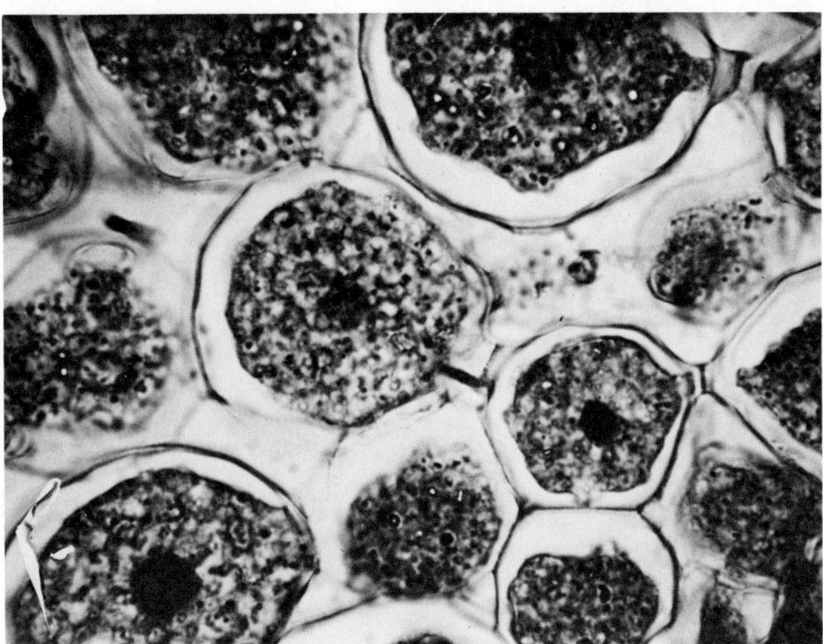

Figure 5. Intercellular mycelium in scutellar cells.

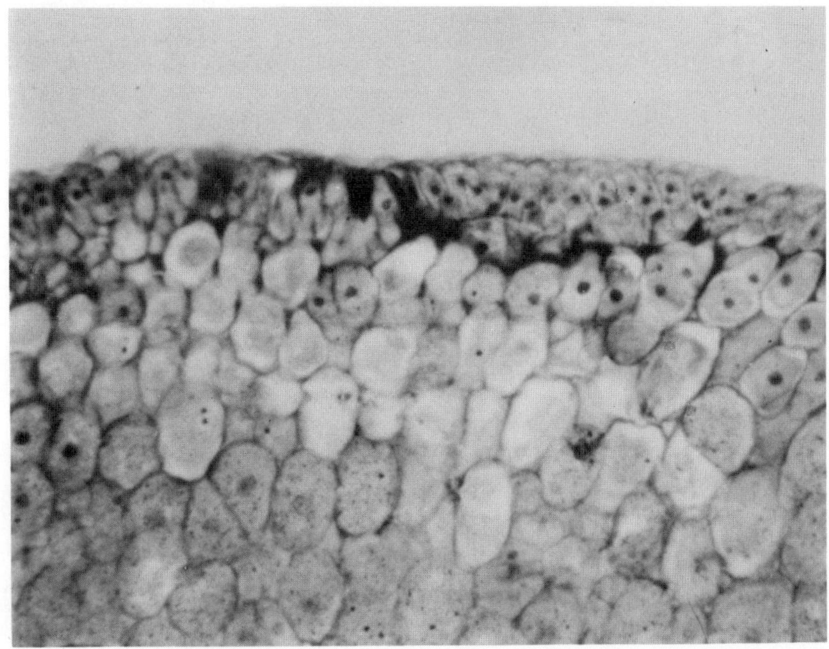

Figure 6. Breakdown of meristematic cells (probably caused by *H. maidis* toxin) in the root tip.

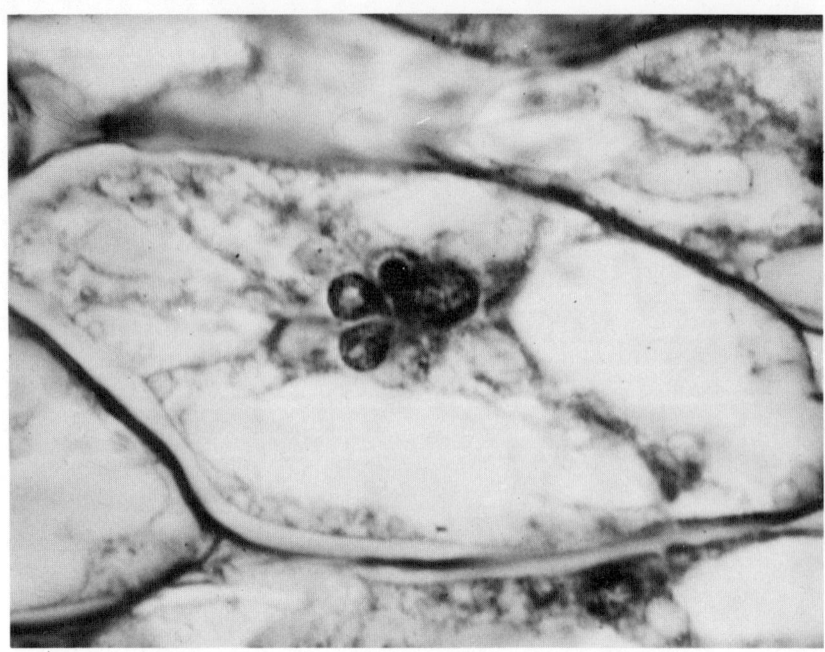

Figure 7. Mycoplasmic bodies of F. moniliforme near the nucleus of scutellar cell of a corn embryo--after Naqvi (1971).

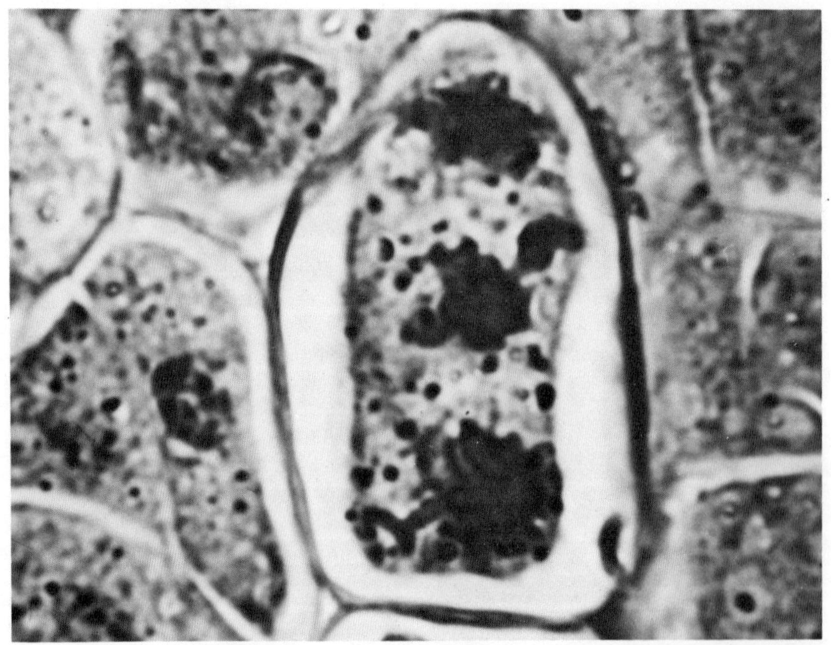

Figure 8. Three mycoplasmic bodies of F. moniliforme in a scutellar cell. Note the cell wall breakdown and the formation of a mycelium from the top mycoplasmic body--after Naqvi (1971).

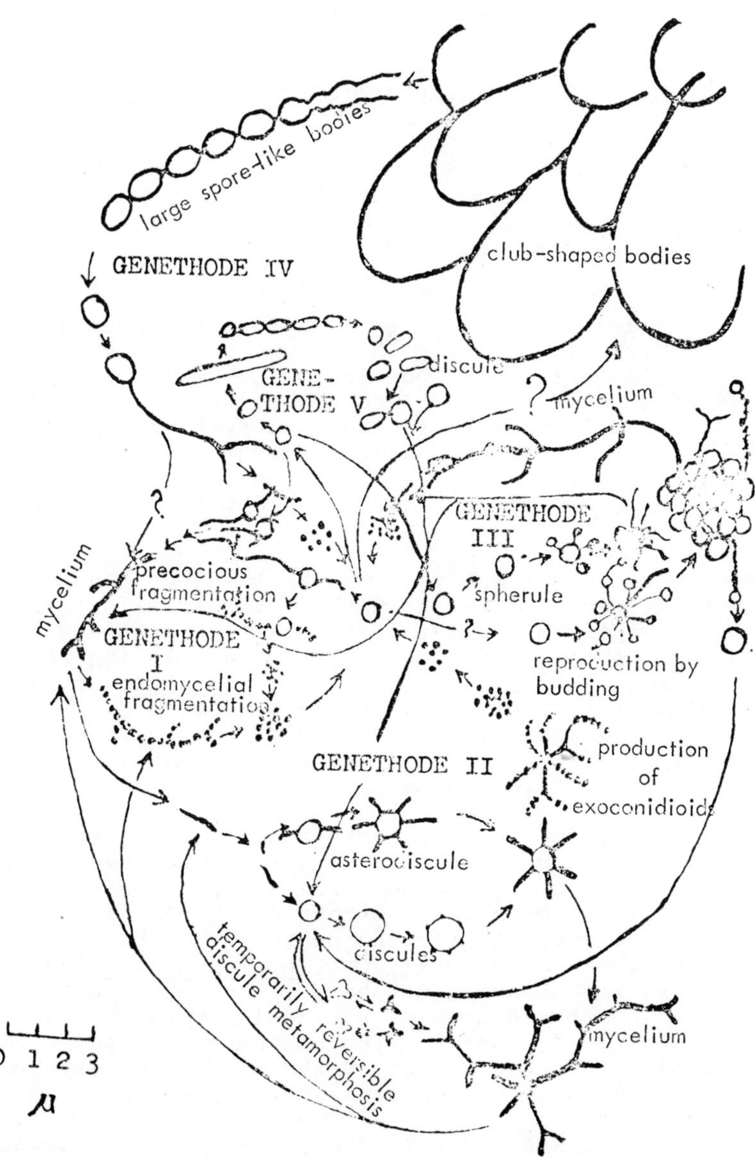

Figure 9. Five genothode stages of the pleuro-pneumonia organism. Data from Turner (1935).

THEORETICAL ASPECTS OF HOST PLANT SPECIFICITY IN INSECTS

Stanley D. Beck
Department of Entomology
University of Wisconsin
Madison, Wisconsin

Why is it that larvae of the northern corn rootworm, *Diabrotica longicornis* Say, will feed on the roots of only corn, but the corn earworm, *Heliothis zea* (Boddie), will inflict serious damage on cotton, tobacco, tomato, and a number of other plants as well as corn? And why can the pea aphid, *Acyrthosiphon pisum* (Harris) live successfully only on legumes; whereas the green peach aphid, *Myzus persicae* (Sulzer), does very well on a wide range of hosts, including deciduous trees, shrubs, and garden plants? The question of host specificity might be posed for each of the hundreds of thousands of plant-feeding insect species and their multitude of hosts.

On the basis of what is currently known about insect-plant interactions, it should be possible to develop a reasonable theoretical basis for dealing with the factors determining insect-plant relationships. A good understanding of this subject is of great importance to the whole effort of developing resistant crop plants for use in pest management systems. When entomologists and plant breeders develop insect-resistant plant varieties, they are directing an accelerated plant evolution of adaptations to minimize insect damage and maximize plant survival and reproduction. For it must be appreciated that the plant is an active evolving participant in every insect-plant relationship.

It is a common observation that phytophagous insects are usually specialized as to the parts of the host utilized. Some insects consume flower structures, or seeds, or leaves, or roots, or they bore in the stem. Thus the range of insect specificities possible is much larger than the number of plant species available. As an insect develops, its feeding behavior may be different from one stage to the next. The European corn borer, *Ostrinia nubilalis* (Hubner), as just one example, is a leaf-feeder when newly hatched; it becomes a stalk borer only in the last two instars. The food habits of the adult of a given species may bear little resemblance to those of the larvae. The northern corn rootworm was mentioned above as a virtually monophagous form in the larval stages; but the adult of the species will feed on a large number of different plants.

The insect's requirements and behavior may change during development, but the plant, too, is a dynamic biological system. The plant undergoes developmental changes that exert considerable influence on its suitability as a host for many insects. Very young corn plants are highly resistant to European corn borer and the corn leaf aphid, *Rhopalosiphum maidis* (Fitch), but become more susceptible as the plant reaches the tasseling stage (Beck, 1956, 1957). With concurrent developmental changes in both insect and plant, the synchronization of the two becomes an important factor in the insect-plant relationship. The developing insect lives in a microenvironment that is unstable, being determined by the developmental and phenotypic state of the plant, which in turn is an expression not only of genotype but also of the effect of the environment in which the plant is growing. It is not surprising, therefore, that genetic varieties showing insect resistance in one agricultural area may be more susceptible when grown in another area--soil type, water relations, daylength, temperature patterns, etc., may all influence the phenotypic expression of resistance (Isaak, et al., 1965). Similarly, biological variability has been frequently observed among insect populations; biotypes of aphids and European corn borers, for example, have been shown to differ in their abilities to utilize resistant and susceptible genetic lines of their preferred host plants (Cartier, et al., 1965; Sparks, et al., 1966; Chiang, et al., 1968). Both of these effects--geographical variability of both plant and insect--post important problems to those who would develop and employ resistant crop plants.

If one were to ask the naive question, "Why do corn borers eat corn?", an appropriate answer would be, "Borers eat corn because they like it and it's good for them." Although undeniably anthropocentric, the answer is not facetious, because the host plant must meet the insect's requirements in respect to both behavior and general dietetics. A rather simplistic depiction of this idea is shown in Figure 1. The relative effect of the plant on the insect's growth and well-being is represented on the ordinate, in which the arbitrary range is from very good (+) to quite deleterious (-). The relative scale on the abscissa expresses the insect's behavioral responses to the plant, ranging from avid acceptance (+) to complete avoidance (-). Plants that would fall in the (++) quadrant for a particular insect would be those that are very attractive, offer good feeding and orientation stimuli, and are both nutritious and nontoxic; these would be acceptable or preferred hosts. Conversely, plant species that are unattractive, repellent, physically unmanageable, and toxic would be nonhosts under all circumstances, and such plants would fall in the (--) quadrant.

The (+-) and (-+) quadrants are of more interest to our subject, because these quadrants describe at least roughly, the two principal categories of plant resistance as formulated by Painter (1951). A plant which would be suitable for growth and survival of the insect, but to which the insect responds negatively in some aspect of behavior--either orientation, oviposition, or feeding--would fall in the (+-) quadrant and would be an example of resistance of the <u>nonpreference</u> type. On the other hand, a plant that met the behavioral requirements of the insect, but was inadequate for normal larval growth and survival would also be characterized as resistant. This resistance would be of the <u>antibiosis</u> type, and the plant would be of the (-+) characteristic.

In respect to the evolution of defenses against insect depredation, plant evolution tends to be in the general direction of (--). Insect evolution toward successful utilization of its host plants is in the form of adaptations that enable the incorporation of its host plants into the (++) quadrant. Because every plant has individual characteristics that tend to fall in different quadrants, the assignment of a given plant to a set of coordinates is at best a rough approximation of the resultant of positive and negative characteristics. Although both axes--behavior and growth--are involved in an insect's host plant specificity and in insect-plant interactions, they are experimentally separable but usually interacting. There is a voluminous body of literature covering experimental investigation of these two aspects, with the greatest emphasis on the behavioral axis, and seldom any detailed consideration of interdependence between the two. The comprehensive reviews of Dethier (1954), Thorsteinson (1960), Schoonhoven (1968), Fraenkel (1969), and Dethier (1970) should be consulted for detailed studies; only a generalized overview will be attempted here.

Insect Behavior and Host Plant Specificity

<u>Recognition and Orientation</u>.--The first requirement of a host plant is that

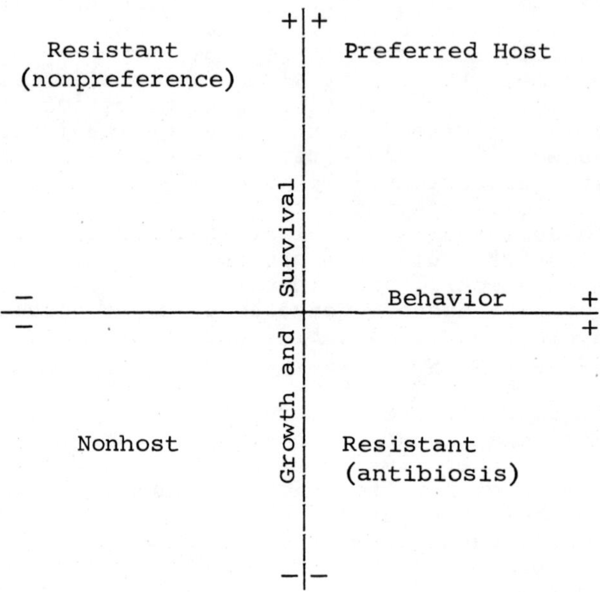

Figure 1. Postulated relationships of plant characteristics with degree of success of host utilization by plant-feeding insects.

it be locatable by the insect. It must not only be present in the insect's environment, but it must also evoke an appropriate orientation response. Most phytophagous insects deposit their eggs on or very near the host plant that will be utilized by their offspring. Except for the more mobile insects, such as cutworms and grasshoppers, the progeny are committed to survive or succumb on the plant selected by their female parents. At the behavioral level, the greatest proportion of host plant specificity is a function of the ovipositing female; the larvae may be much more polyphagous than the adult. The European corn borer is a good example, as the larvae are quite polyphagous and will feed on any plant tissue that is not too physically resistant. We have found that the larvae were attracted to extracts of widely diverse plants, such as apple, gingko, and corn leaves (Beck and Smissman, 1960), and we have reared the larvae on a wide variety of plant tissues. The fact that the adult moths prefer to deposit their eggs on corn leaves accounts for the host plant preference of the species. A similar dependence of host preference on adult behavior has been observed in many species (e.g., Force, 1966a, b).

Orientation of a gravid insect to a prospective host plant may involve visual as well as chemical stimuli. Two phases of moth orientation to the plant were observed in the tobacco hornworm, *Manduca sexta* (Johan.)--the approach and the landing (Yamamoto, et al., 1969; Sparks and Cheatham, 1970). The approach appeared to be based on visual clues, and the landing was in response to olfactory stimuli. Visual responses have also been demonstrated to be involved in the orientation of aphids (Kennedy, et al., 1961), corn earworms (Callahan, 1957), and others (see Beck, 1965).

The importance of plant-borne attractants in insect orientation is unquestioned, having been demonstrated in a large number of species (see reviews by Thorsteinson, 1960; Schoonhoven, 1968). What effects might be exerted by nonhost plants occurring in the insect's environment? Odors from nonhosts may simply fail to evoke any orientation response, or they may elicit a negative response in which the insect moves away from the odor source (deWilde, et al., 1960; Maxwell, et al., 1969)., suggesting that the insect is repelled by volatile chemicals emanating from nonhosts.

Host plants may also be encountered during the course of random, undirected locomotor activity. In such cases the insects tend to resume locomotion after encountering a nonhost, but to remain or display a behavioral change upon contacting an acceptable host. Young winged aphids are evoked into flight by visual stimuli from blue sky and are not attracted to foliage. After a period of flight, however, plant foliage provides visual stimuli that cause the aphids to land on plants (Kennedy, 1965). Having landed on a plant at random, chemosensory responses determine whether the aphid will remain or resume flight (Muller, 1958; Wensler, 1962). Goeden and Norris (1965a) reported a similar host finding behavior in the bark beetle, *Scolytus quadrispinosus* Say.

Feeding and Oviposition.--Arriving at the prospective host plant, the insect is responsive to stimuli releasing subsequent components of the ovipositional or feeding behavioral sequence. Oviposition is not usually indiscriminant, but is typically confined to selected plant parts. Specific oviposition sites are selected in response to physical and chemical stimuli, which may vary with leaf maturity and the physiological state of the plant (Goeden and Norris, 1965b; Miller and Hibbs, 1963; Gara, et al., 1971). Specific stimulants may be involved, as has been demonstrated in the carrot rust fly, *Psila rosae* F. (Beruter and Stadler, 1971) and onion maggot, *Hylemya antiqua* (Meigen) (Matsumoto and Thorsteinson, 1968; Matsumoto, 1970).

Tactile and proprioceptive stimuli also influence oviposition. Leaf pubescence may affect egg deposition in several species; the cereal leaf beetle, *Oulema melanopus* (L.) is deterred from oviposition by dense pubescence (Schillinger and Gallun, 1968), but both the soybean pod borer, *Grapholitha glicinivorella*

Matsumura, and the green cloverworm, Plathypena scabra (F.) oviposit more readily on pubescent than glabrous leaves of their host plants (Nishijima, 1960; Pedigo, 1971). Relatively pubescent wheat varieties tend to be more resistant to the cereal leaf beetle than do glabrous varieties, and because a behavioral response is involved, the resistance is considered to be of the nonpreference type. However, of the eggs deposited on highly pubescent wheat, fewer than 10% hatched, and 80% of the hatching larvae succumbed within three days (Schillinger and Gallun, 1968). This effect is clearly one of antibiosis, and illustrates the necessity of considering both behavioral and survival parameters of the effects of plant characteristics on host specificity. Proprioceptive stimuli associated with specific oviposition sites, such as stalk, leaf, petiole, etc., are important in releasing egg deposition behavior in many species.

The feeding behavior of both relatively sessile and highly mobile insects involves a stereotyped behavioral sequence: (1) host recognition and orientation to whole plant or plant part; (2) initiation of feeding (exploratory biting or piercing); (3) maintenance of feeding; and (4) cessation of feeding, followed by dispersal of more mobile forms (Dethier, 1954; Thorsteinson, 1960). Each of these steps is manifested in response to appropriate stimuli.

Feeding specificities have been associated with specific chemical stimuli in a number of cases, and the list of such chemicals grows steadily as more research is reported. Factors that induce the initiation of feeding (feeding incitants) include beta-sitosterol in the case of the silkworm, Bombyx mori L. (Hamamura, 1965); sinigrin and related mustard oil glycosides for the diamondback moth, Plutella maculipennis (Curtis), and a number of other crucifer-feeding insects (Thorsteinson, 1953; Nayar and Thorsteinson, 1963); and the alkaloids phaseolunatin and lotaustrin for the Mexican bean beetle, Epilachna varivestris Mulsant (Nayer and Fraenkel, 1963a).

The distinction between a feeding incitant and a feeding stimulant (evokes continued feeding after the exploratory bite) is not always clear, and many substances may have both effects. Some feeding stimulants have been shown to be host-specific substances, but others have proved to be common chemicals of universal botanical distribution, such as glucose, sucrose, ascorbic acid, a number of amino acids, and some phospholipids (Beck and Hanec, 1958; Thorsteinson, 1960; Gothilf and Beck, 1967). Examples of the more specific feeding stimulants include factors in cotton that stimulate the feeding of the cotton boll weevil, Anthonomus grandis Boheman (Keller, et al., 1962; Maxwell, et al., 1963); catalposides stimulating larvae of the catalpa moth, Ceratomiae catalpae (Boisduval) (Nayer and Fraenkel, 1963b); a series of related essential oils from Umbelliferae that attracted larvae of Papilio ajax and stimulated their feeding (Dethier, 1941).

A series of related plant species may all contain similar arrays of attractants and stimulants. Host specificities of insects utilizing members of the plant group may then be determined by repellents and deterrents present in some of the plants. The Colorado potato beetle, Leptinotarsa decemlineata Say, is confined almost exclusively to plants of the genus Solanum. Solanum species differ in their suitability as hosts, however, with the common potato, S. tubersosum being the most efficiently utilized. S. luteum foliage contains a powerful feeding deterrent (inhibits feeding after the exploratory bite), and larvae that are confined to its foliage die of starvation. S. lycopersicum also contains feeding deterrents, but the larvae will feed if no other foliage is available (Bongers, 1970).

Adult Scolytus multistriatus (Marsham) feed almost exclusively on bark of two-to-four year-old twigs of American elm, Ulmus americana (Loschiavo, et al., 1963; Baker and Norris, 1968). Extracts of the bark of nonhosts, such as poplar, oak, and hickory were found to contain feeding deterrents. Hickory bark was shown to contain juglone (5-hydroxy-1,4-naphthoquinone), which strongly deterred feeding by S. multistriatus, but not that of the hickory bark beetle, S. quadrispinosus.

When juglone was chemically removed from bark extracts, both species fed readily (Gilbert and Norris, 1968). These workers postulated that deterrents may be an important basis for bark beetle distinction of host from nonhost trees.

Feeding deterrents may play a part in the feeding behavior of even the most polyphagous insects. The desert locust, *Schistocerca gregaria* Forsk, will not feed on seeds of neem, *Azadirachta indica*. This behavior was shown to be caused by the presence of a terpenoid, azadirachtin, which acted as a potent feeding deterrent (Butterworth and Morgan, 1971).

Gossypol, a volatile chemical produced by gland cells in the leaves of cotton, is apparently a minor attractant but otherwise has no effect on the host plant relationships of the boll weevil (Maxwell, et al., 1966; Parrott, et al., 1969). The development and use of glandless varieties of cotton has resulted in damaging attacks by blister beetles, *Epicauta* spp., that will not feed on glanded varieties (Maxwell, et al., 1965). Apparently gossypol is a repellent or deterrent to blister beetles, preventing their feeding on glanded cotton.

It is apparent from these several examples that host plant specificity is determined by both positive (acceptance-evoking) and negative (rejection-evoking) stimuli (see also Ishikawa, et al., 1969). The problem of host plant specificity becomes even more complex, however, as it is now known that the plant produces a multiplicity of stimuli to which the insect may be sensitive. And some of these stimuli tend to evoke conflicting responses. For example, cotton buds contain a boll weevil attractant (Keller, et al., 1963), but also two volatile substances that act as repellents (Maxwell, et al., 1963), The anthers of the cotton plant produce an oviposition stimulant (Everett, 1964), but an oviposition deterrent has also been identified (Buford, et al., 1968). Similarly, both feeding stimulants (Keller, et al., 1962) and feeding deterrents have been demonstrated (Maxwell, et al., 1965). The insect, then, is buffeted by a complex of physical and chemical signals, and its resultant behavior must represent a central nervous system integration of the stimuli received. The relative strength of each of the several stimuli is variable, depending on the anatomical and physiological state of the plant, and the effect of the complex of signals on the insect's behavior will also be variable, depending on the growth stage and physiological condition of the insect.

Experimental alterations of either the plant or the insect might be expected to alter the specificity of the insect's behavior. Maxillectomy is a case in point. Removal of the maxillary sense organs from feeding larvae has been shown to cause changes in the feeding behavior, the usual effect being a reduction in food plant specificity. Maxillectomized larvae of the tobacco hornworm, *Manduca sexta*, would feed on dandelions, burdock, and a number of other plants; whereas the normal larvae would feed only on Solanaceae (Waldbauer and Fraenkel, 1961; Waldbauer, 1962). Similar loss of some feeding specificity following maxillectomy has been reported in *Bombyx mori* and *Leptinotarsa decemlineata* (Torii and Morii, 1954; Chin, 1950).

Chemoreceptors of the maxillae include both olfactory and gustatory receptors, with the latter consisting of two sensilla each containing four chemosensory cells (Schoonhoven and Detheir, 1966). The response spectra of the several cells differ, with some responding to general feeding stimulants (sugars, amino acids, inositol, etc.); others are sensitive to feeding deterrents; and still others to host specific stimulants (Schoonhoven, 1969). Information from these receptors as well as from other sense organs (eyes, tarsi, antennae, labium, etc.) feed into the central nervous system, from which motor control of behavior emerges. Removal of the maxillae changes the input information and results in a change of behavior.

The observation that maxillectomized larvae may feed on some plants not normally accepted has led some workers to postulate that host plant specificity is principally determined by repellents and feeding deterrents, so that the insect utilizes only those few plants that it does not find to be repellent or distasteful (Jermy, 1966). But such an interpretation is both unwieldy and unnecessary, if it is remembered that the insect is constantly receiving much more information about the plant than merely the presence of absence of one or more inhibitors (Schoonhoven, 1969).

In considering the input of sensory information, it must be remembered that the anatomical characteristics of the plant contribute to this information. The Frego mutant of cotton displays a rolled bract, and such a flower structure disrupts boll weevil oviposition (Jenkins and Parrott, 1971). The gravid weevils are exposed, and they do not remain on the plant long enough to complete oviposition. Similarly, physical characteristics of different varieties of rice strongly influence their utilization by the striped rice borer, *Chilo suppressalis* Walker (Pathak, 1971).

The role of the host's anatomical development in determining insect feeding behavior may be illustrated with the seasonal development of the European corn borer on corn. The larvae are relatively polyphagous, but corn is the preferred host of the ovipositing moth. Despite their lack of specificity, the larvae display a well-defined pattern of feeding behavior on the growing corn plant. Borer eggs are deposited on the underside of the leaves of the young corn plant; upon hatching, the larvae move immediately into the whorl where they feed on the basal leaf tissue. As the tassel forms and pushes up through the whorl, the larvae transfer their attack to the tassel, invading that structure while it is small and tightly enclosed. As the stem elongates and the tassel emerges and opens, the larvae move down the plant to invade leaf sheaths. As these structures mature, the borers invade the plant stem and become stalk borers. Bioassays of aqueous extracts of corn plant tissues for larval attractants and feeding stimulants demonstrated that all were equally attractive and stimulating. The pattern of feeding behavior could not be accounted for on the basis of tissue-specific factors. Experiments on the behavior of borer larvae showed that they were negatively phototactic and positively thigmotactic. Feeding behavior experiments demonstrated a high sensitivity to sugars, especially fructose and sucrose, both of which greatly stimulated feeding. The larvae were capable of discriminating relatively small differences in sugar concentrations. Analyses of plant tissues revealed that sugar contents of leaves, tassel, sheaths, and stem were different and changed as the plant developed. It was then shown that the seasonal pattern of borer feeding was the result of the insect's phototactic, thigmotactic, and saccharotrophic characteristics in response to the anatomical and chemical (sugar) developmental pattern of the host plant (Beck, 1956a, b, c, 1957; 1961). On the other hand, plant-borne feeding stimulants may vary in different parts of the plant and at different developmental stages, and such effects were found to be important in the feeding pattern of corn earworm, *Heliothis zea* on corn (McMillian, et al., 1970).

Host Specificity and Insect Growth and Survival

In turning our attention from the *behavior* axis to the *growth and survival* axis (Figure 1), it should again be pointed out that these are not completely independent aspects of host plant specificity. A feeding deterrent, for example, may have a measurably important effect on the insect's behavior, but its action may also jeopardize growth and survival, in that food intake may be reduced, growth slowed, and exposure to other deleterious factors prolonged. The same considerations apply to physical characteristics of the plant insofar as they may limit or alter the insect's behavior.

An acceptable host plant must provide the insect a suitable physical environment and a nutritional substrate that is adequate, nontoxic, and utilizable from the standpoint of digestion, assimilation, and conversion into insect tissues. Even the best host plants are suboptimal in some of these characteristics, and mortality among young larvae tends to be very high. We have observed that our best laboratory methods for rearing the European corn borer and cabbage looper, _Trichoplusia ni_ (Hubner), using meridic diets and standardized conditions, promote faster and more uniform larval growth, better survival, greater body weight, and better adult longevity and reproduction than do the insects' normal host plants. Plant hosts to which the insect is adapted contain phenols, flavonoids, alkaloids, and various glycosides that are basically deleterious, and which must be degraded metabolically. The insect's ability to metabolize such plant components constitutes an important aspect of its adaptation to a particular plant as a host (Self, et al., 1964; Krieger, et al., 1971). Inasmuch as different plant species or populations may differ either qualitatively or quantitatively in respect to these substances, an insect that is metabolically adapted to one type of plant may not be able to meet the metabolic demands posed by another plant, and consequently will not be able to survive on it.

Physical Characteristics.--The role of physical characteristics has been studied mainly in relation to host plant resistance, including characteristics such as pubescence, spacing of vascular bundles, tissue silica content, etc. Little emphasis has been put on clarifying the role of physical characteristics in insect utilization of hosts to which they are well-adapted. The previous discussion of the seasonal feeding behavior of European corn borer larvae illustrated the importance of physical form of the plant to successful host plant utilization. Growth and survival as well as behavioral patterns are involved, as developmental changes in the borer larva's nutritional requirements were also found to be coordinated with the feeding pattern (Beck, 1956b).

Tissue proliferation in response to injury is a physical defense against invading insects, and is known to be a reaction of cotton against boll weevil (Hinds, 1907) and of some plants against leaf miners. Gall midges, however, have taken advantage of such a wound reaction in their host plant adaptations, thereby overcoming and exploiting the plant's defense mechanism (Dieleman, 1969; 1970; Miles, 1969). Glandular hairs on the leaves of several _Solanum_ species discharge a sticky substance that entraps aphids and prevents their establishment on the plant (Gibson, 1971). Hook-like trichomes on the leaves of _Passiflora adenopoda_ tend to entrap small caterpillars, effectively preventing their survival and feeding (Gilbert, 1971).

Biochemical Characteristics.--Many of the so-called 'secondary' plant chemicals exert metabolic effects on plant-feeding insects; as discussed in a previous section, a number of these substances are important attractants, deterrents, etc. In the long-run, however, plant chemicals affecting metabolism and thereby the insect's survival and development are at least as important as those affecting behavior in determining host plant specificity. For example, the senita cactus, _Lophocereus schotti_, contains a number of toxic alkaloids. Of eight species of desert _Drosophila_, the alkaloids killed both adults and progeny of all except _D. pachea_, and only _D. pachea_ breeds in the tissues of the senita cactus (Kircher, et al., 1967). The insect's ability to detoxify the alkaloids is critical to its utilization of senita cactus tissues.

Solanaceous plants are well-known for their rich and varied content of alkaloids, and the host specificity of insect parasites of Solanaceae appears to be determined largely by the insects' ability to metabolize the alkaloids of their host species. The host ranges of the Colorado potato beetle, _Leptinotarsa decemlineata_ Say, the three-lined potato beetle, _Lema trilineata daturaphila_ K & G, and the tobacco hornworm, _Manduca sexta_ (Johan.) are apparently determined on this basis (Kogan and Goeden, 1971; Self, et al., 1964; Bongers, 1970). Host plant tissues may vary in concentration of secondary chemicals, depending on the developmental stage of the tissue as well as on the plant organ. Such variability may influence the feeding activity and developmental success of the insect. Colorado potato beetle larvae were observed to feed more readily and to grow more satisfactorily when fed young potato foliage than when given older tissues; senescent foliage of either potato or tomato tended to retard growth. These effects may be partly nutritional and partly the reflection of developmental differences in tissue concentrations of deleterious substances (Bongers, 1970; Cibula, et al., 1967). Similarly, larvae of the winter moth, _Operphtera brumata_ (L.), consume oak leaves early in the growing season, but not later, apparently because of a high concentration of tannins in the older leaves (Feeny, 1968).

Todd, et al. (1971) tested the toxicity of flavonoids and phenolic compounds found in the tissues of the host plants of the greenbug, _Schizaphis graminum_ (Rondani). Many of the normal constituents of the plant (barley) were found to reduce the insect's growth, survival, and reproduction when incorporated into a standard synthetic diet. These workers concluded that small varietal differences in the tissue concentrations of such chemicals might be responsible for varietal resistance to the greenbug.

Tissues of young corn plants have been shown to contain benzoxazolinone compounds that are deleterious to growth and survival of European corn borer larvae, and also to exert a mild feeding deterrent action. The compounds were identified as 6-methoxybenzoxazolinone and 2,4-dihydroxy-7-methoxy-2H-1,4-benzoazin-3-one; the tissue concentrations of these compounds decline as the plants mature, as well as vary among different plant parts (Beck, 1957, 1960; Beck, et al., 1957; Beck and Smissman, 1960; Klun and Brindley, 1966; Klun, et al., 1967). Reed, et al. (1972) showed that borer larvae grew poorly and few completed development when reared only on leaf tissue containing high concentrations of the benzoxazolinone analogs; the insect's normal feeding pattern on corn plants is such that the larvae feed on such tissue for only a short time early in larval life, as was discussed above.

Some plants have been found to contain analogs of the insect hormones ecdysone and juvenile hormone. The role of plant-borne hormone analogs in insect-plant interactions has not been determined, but the subject is of current research interest. Juvenile hormone (JH) is the methyl ester of epoxyfarnesoate (Roller, et al., 1967). A few closely related molecules have been shown to occur in insects and to have comparable hormone activity. Slama and Williams (1966) found that some commercial papers contained a JH-mimic that would prevent the metamorphosis of the plant bug _Pyrrhocoris apterus_ L. The source of the factor was shown to be paper-pulp conifers, especially balsam fir, _Abies balsamea_. The hormone analog was identified as methyl todomatuate and given the trivial name 'juvabione' (Bowers, et al., 1966). A second, closely related JH-mimic was also found in conifer tissues; it proved to be the dehydro analog of juvabione (Cerny, et al., 1967). Bowers (1968, 1969) reported that compounds containing a methyl-enedioxyphenyl grouping are common plant constituents, and that many display JH activity. Compounds with JH activity were found in bark and wood samples of red cedar, spruce, hemlock, and pine as well as balsam fir (Mansingh, et al., 1970).

The insect molting hormone, ecdysone, is a complex steroid, of which a number of active isomers are known to occur in both arthropods and plants. They have been found in high concentrations in many species of ferns of the family Polypodiaceae, and in nearly twenty families of gymnosperms and angiosperms (Slama 1969). Ecdysone analogs from plant tissues were shown to be feeding deterrents for the larvae of _Pieris brassicae_ L. (Wei-Chun, 1969). Working with the southern armyworm, _Prodenia eridania_ (Cramer), Soo Hoo and Fraenkel (1964) reported that fern tissues contained a potent feeding deterrent. Ehrlich and Raven (1964) pointed out that few insects feed on ferns. In view of the high ecdysone concentrations known to occur in ferns, the feeding deterrency may have been related to tissue concentrations of the hormone.

Digestion and Conversion.--The feeding insect must not only ingest the tissues of its host, but the material ingested must also be suitable for conversion into the energy and structural substances required for insect development.

Digestibility is one factor determining the utilizability of plant tissues.

Protease inhibitors occur in some plant tissues as part of the plant's defense mechanism against herbivores. Such inhibitors have long been known to occur in legumes and grains (Borchers, et al., 1947; Vogel, et al., 1968), and have recently been reported in solanaceous plants, at least tomato and potato (Green and Ryan, 1972). Green and Ryan found that feeding activity of larval and adult Colorado potato beetles induced the formation of a protease inhibitor; within a few hours the inhibitor that was formed at the wound site had been translocated to other parts of the plant. The factor was found to be a powerful inhibitor of the major intestinal proteases of animals, including insects. Applebaum and Konijn (1966) demonstrated the presence of a specific *Tribolium*-protease inhibitor in wheat seeds. Applebaum (1964) studied the effects of soybean potease inhibitor on the digestive processes of two bruchid beetles, *Callosobruchus chinensis* L. and *Acanthoscelides obtectus* Say. Proteolytic activity of the digestive tracts was very low in both species, and the protease inhibitor did not inhibit digestion in *Callosobruchus*. Applebaum postulated that the antiproteases of legumes are defense mechanisms, and that the bruchid beetles have overcome this barrier to host specificity.

Assimilation and conversion into insect tissues must follow digestion and plant tissues differ in the degree to which they meet this requirement. Because maxillectomized larvae will feed on a wider than normal host range, this technique has been used to determine differences in digestibility and conversion of host and nonhost plant tissues. Using maxillectomized fourth-instar larvae of the tobacco hornworm, *Manduca sexta*, Waldbauer (1962, 1964) investigated the effects of solanaceous and nonsolanaceous plant foliage on feeding rates, growth rates, digestion, and efficiency of conversion. Tomato, *Lycopersicon esculentum*, was used as the normal solanaceous host; two other Solanaceae, *Solanum tuberosum* (potato) and *S. dulcamara*; one Scrophulariaceae, *Verbascum thapsus*; one Bignoniaceae, *Catalpa speciosa*; two Compositae, *Arctium minus* (burdock) and *Tarazacum officinale* (dandelion) were tested. In terms of Consumption Index (g eaten/g body weight), dandelion was as acceptable as the solanaceous plants; *Arctium minus* was slightly inferior, but *Verbascum thapsus* and *Catalpa speciosa* were distinctly inferior. The latter two plants supported poor larval growth. Coefficients of Digestibility ((g ingested - g feces)/g ingested) x 100) indicated that the solanaceous foliage was most digestible, with dandelion and burdock nearly as good, but *Verbascum* was poorly digested (*Catalpa* was not tested). Conversion of ingested food into body weight was carried out most efficiently with potato foliage, and nearly as efficiently with the other two Solanaceae. Dandelion and burdock were only slightly inferior, but again *Verbascum* was poor. With the exception of *Verbascum*, which appeared to be both poorly digestible and nutritionally inferior, the lower conversion efficiencies of the nonsolanaceous plants were attributed to lower digestibility.

Digestibility and convertibility varies among plants fed upon by polyphagous insects. The southern armyworm, *Prodenia eridania*, was studied on 18 different plant species representing 13 plant families (Soo Hoo and Fraenkel, 1966a, b). The larvae utilized 10 of the plants quite efficiently, but the others supported suboptimal or poor larval growth. Poor host utilization was caused by low feeding rate, poor digestibility, or inefficient conversion. Digestibility ranged from 76% down to a low of 36% in the poorest host tested. Efficiency of conversion ranged from 16 to 56%. Beenakkers, et al. (1971) also reported significant differences among food plants consumed by the highly polyphagous migratory locust. For a comprehensive review of methods and data on the subject of digestion and conversion, the reader is referred to Waldbauer (1968).

Nutritional Requirements.--The role of the insect's nutritional requirements in determining its host plant specificity is still an open question. The idea that differences in food habits reflected differences in nutritional requirements was an attractive theory to most early workers. In recent decades, however, the theory has fallen into disrepute, perhaps in part because of definitional changes as to what is meant by the term "nutrition." The classical definition is that nutrition is the sum of the processes by which an organism takes in and utilizes food substances. By such a definition, host plant specificity differences do indeed represent differences in the insects' nutritional requirements. But by current usage, nutritional requirements is defined as being that set of chemicals obtained from the ingested diet that are indispensible to the metabolic and developmental processes of the organism. Adoption of this narrow definition leads us to the conclusion that the great bulk of evidence indicates that host plant specificity cannot be explained on the basis of specificity of nutritional requirements. There are both qualitative and quantitative differences among the nutritional requirements of plant feeding insects, and some of these differences almost certainly represent adaptations to particular families or species of host plants. But the differences in nutritional requirements are, in all instances so far reported, of relatively small magnitude; they are far too small to account for the very great differences observed in host plant specificities.

Plants vary in their content of nutrients required by insects, with such variation being dependent on the plant part, developmental stage, physiological condition, and plant genotype. These variations have been shown to influence both the behavior and the development of plant-feeding insects. In studies of the pea aphid, *Acyrthosiphon pisum* (Harris), Auclair, et al. (1957) found that the amino acid content of an aphid-resistant variety of peas was quantitatively different from that of an aphid-susceptible variety; they postulated that resistance was caused by a relatively low content of free amino acids. In the absence of experimental data on the aphid's amino acid requirements, the interpretation was speculative at best. Colorado potato beetle larvae have been shown to grow faster with higher survival rates on young foliage than on old foliage of a number of Solanaceæ presumably because of a more favorable amino acid content in the younger tissue (Cibula, et al., 1967). Grison (1958) observed that adult Colorado potato beetles showed egg production rates that were positively correlated with the phospholipid content of the foliage fed; senescent foliage was deficient in the phospholipids required for egg production. A number of other investigators have also reported differences in nutrient content of plant structures of different ages, and their apparent effects on insect growth, survival, and reproduction (Blais, 1952; deWilde, et al., 1969). The importance of the dietary proportions of required nutrients may be of greater importance than their absolute quantities (House, 1969, 1970, 1971; Maltais and Auclair, 1962; McGinnis and Kasting, 1961). Such nutrient ratios may also influence feeding behavior, and they are almost certain to influence the efficiency of conversion of ingested food.

Nutrient compounds in the host plant are involved not only in meeting the insect's requirement for energy and development, but also in other metabolic processes essential to survival. For example, European corn borer larvae are adversely affected by benzoxazolinones contained in the tissues of young corn plants. The toxic effect of these chemicals is diminished or destroyed by relatively high levels of sugar and/or niacin in the larval diet (Beck, 1957; Beck and Lilly, 1949; Klun, et al., 1967). Plant-feeding insects are usually relatively soft-bodied and easily desiccated; water content of the host plant may, therefore, be an important factor in the insect's survival. Water is not usually classified as a nutrient, and its role in insect nutrition has been little studied. The possible involvement of plant tissue water content was considered by Waldbauer (1964) in his study of plant utilization by *Manduca sexta*, and some of the differences between hosts in regard to digestibility and conversion were attributed to differences in moisture content. Mellanby and French (1958) observed that larvae of *Diataraxia oleracea* L. grew normally on turgid cabbage leaves (85% water), but became seriously dehydrated if fed wilted cabbage containing only 70% moisture. Beenakkers, et al. (1971) postulated that the need for water influenced the feeding rate and plant selection of *Locusta*.

Evolutionary Considerations

Questions frequently arise as to the evolutionary history of insect host plant specificities; that is, how insects become adapted to new hosts, the permanence of specificity, and the general trend of evolutionary changes.

Earlier workers generally considered it most likely that the original plant-feeding insects were polyphagous, and that evolution of host specificity was from a primitive polyphagy to the more restricted food habits of oligophagy and monophagy (Brues, 1920; Dethier, 1954; Fraenkel, 1959; Waldbauer, 1963). According to this interpretation, general feeders such as grasshoppers and locusts are considered to be relatively primitive; whereas the more specialized aphids and sawflies are more highly evolved in their host plant specificities. There is, however, very little evidence in support of such a hypothesis. There are more compelling reasons to consider the evolutionary trend to have been much more diversified. The bulk of the evidence supports the hypothesis that there has been a coevolution of plants and insects, involving a wide spectrum of host plant specificities (Ehrlich and Raven, 1964, 1969; Dethier, 1970; Singer, 1971). According to this hypothesis, as new plant populations become available to insects that have been utilizing related plant forms, the new potential host may be incorporated into the insect's host range, or excluded from it, or populations of the insect may become adapted to the new plant form to the exclusion of the original host. On an oversimplified basis of either acceptance or rejection of a host plant, there are 2^n possible combinations and permutations of host plant specificities, where n represents the number of plant forms available. A surprisingly small number of different plant forms can be associated with a very large variety of specificities in insect host range. There is no reason to limit the insect's adaptive capability and evolutionary potential to a progression from polyphagy toward monophagy.

The probable role of secondary plant chemicals in the evolution of insect-plant interactions has been emphasized by a number of investigators. These substances are not known to play a part in the plant's principal metabolic pathways, and are considered to be components of plant defense against pathogens and herbivores, including insects (Fraenkel, 1959, 1969), and as allelopathic agents directed toward other plants (Whittaker, 1970, Whittaker and Feeny, 1971). Having arisen by chance mutation or genetic recombination, such characteristics might tend to confer a selective advantage, enabling the plant to enter a new adaptive zone. Ensuing evolutionary radiation might result in what started as a chance genetic characteristic to become a characteristic typical of an entire plant population, species, or family. On the other hand, a genetic mutation or recombination appearing in a population of insects might enable some individuals to feed on the previously protected plant group, with the mutant insects responding positively in behavior (feeding, oviposition) to the previously prohibitive secondary plant chemicals and having the metabolic capability of detoxifying protective chemicals. Selective advantage thereby conferred on the mutant insects would allow them to enter a new adaptive zone, where diversification in the absence of competition from other herbivores could occur (Ehrlich and Raven, 1964).

Although of unquestioned importance, secondary plant chemicals are not the only factors guiding the evolution of host plant specificity. This point is well-illustrated by a recent study (Singer, 1971) of the evolution of host plant specificity in a nymphaline butterfly, *Euphydryas editha*. Singer investigated the ovipositional preferences of six geographical populations of the species, with field observations and laboratory experimentation. Distinct population differences were demonstrated, but they were in large part attributable to the insects' adaptations to ecological differences between the areas occupied by the several populations. Phenological, geographical, plant population density, and insect genetic differences were all shown to play a part in the evolution of the separated populations, each of which was well-adapted to its geographical location.

Orientation responses to a plant, whether for feeding or oviposition, are genetically determined. The effects of mutation or genetic recombination might be expected to alter the population-typical phenotypic response patterns. The genetics of host plant specificity has been little studied, however, and more investigations of the subject are badly needed. Sibling species and geographical populations of a single species should provide excellent material for such research.

In his very fine review and analysis of insect-plant relationships, Dethier (1954) cited some early genetic studies. In crosses of two sphingid moths, _Pergesa elpenor_ males (feeds on _Epilobium_) were mated with females of _Celerio hippophaes_ (feeds on _Hippophae rhammoides_). The mated females laid their eggs on their natural host, _Hippophae_, but the progeny would feed only on _Epilobium_. Other early hybridization experiments also produced F_1 progeny that displayed the feeding preferences of only the male parent. Hybrid sterility prevented further genetic analysis in most cases. Hybrids of _Poecilopsis pomonaria_ and _P. isabellae_ were found to be cross-fertile. The F_1 progeny of reciprocal crosses fed only on larch, the preferred host of _P. isabellae_, but other crosses and backcrosses would feed only on hawthorn, the preferred host of _P. pomonaria_. Both _Celerio euphorbiae_ and _C. mauretanica_ normally feed on _Euphorbia_ and reject _Salix_; hybrid progeny fed readily on _Salix_. Selected strains of _Pieris rapae_ (L.) were studied by Hovanitz (1969). Crossing experiments between strains that had been conditioned to different host plants demonstrated that multiple genetic factors control both feeding and oviposition preferences.

The feeding experience of a larva may condition its subsequent feeding preferences, but a carry-over of larval conditioning into the adult stage has not been convincingly demonstrated. Jermy, et al. (1968) found that last-instar larvae of _Manduca sexta_ and _Heliothis zea_, which were offered a choice of different host plants, showed a clear preference for the plant on which they had been previously fed. They could not be conditioned to a plant species outside of their normal host range. Tobacco hornworm larvae that had been reared from the 2nd to the 5th instar on artificial diet, fed rather indiscriminantly when transferred to plant material, indicating at least a temporary loss of feeding specificity (Schoonhoven, 1967, 1969). Although larval conditioning could be demonstrated in the Colorado potato beetle, the conditioning did not influence the choice of plant for either feeding or oviposition by the adult beetles (Bongers, 1965). Where a predilection occurs for oviposition on a plant not usually preferred, it has been observed under conditions that probably involved a selective survival of the larval stages (e.g., Hovanitz, 1969). This effect would result in the establishment of a genotype differing from the original population and showing altered phenotypic orientation behavior.

In order to account for the evolution of different host plant specificities on the basis of genetic mutation or recombination, it is necessary to assume the existence of isolation mechanisms. In the absence of genetic isolation, the new genotype would continue to be part of the population gene pool, and this would have the effect of broadening the population's host range. At the present time, we do not have good information concerning isolation mechanisms, other than geographical. The population rather than the species forms the gene pool that is applicable (Ehrlich and Raven, 1969). Isolation mechanisms that might be operative in the case of mutant forms feeding on a new host plant include such factors as markedly different developmental rates, resulting in the adults being available for mating at times differing from that typical of the parent population. Behavioral isolation might also occur, providing that the mutants display behavioral differences in flight behavior, mating behavior, pheromone response, or circadian activity rhythms, in addition to the altered host plant specificity. There is also a possible isolation of the mutant from the parent population as the result of differences in the local distribution and population density of their respective host plants. This aspect of host plant specificity merits much further research.

The use of insect-resistant agricultural plants is an attempt to disrupt existing host plant specificities by removing the protected plant population from the target insect's host range. If it cannot be removed entirely, the aim is at least to make the plant less well utilized. The question of the permanence of such resistance is frequently raised. Might not the insects overcome the resistance, and become well-adapted to the once-resistant crop plant? Unless this possibility is guarded against, eventual adaptation seems highly likely. A few of the requirements for maintaining reasonably permanent resistance are known. Within a geographical area where large plantings of one crop occur, more than one type of resistant plants should be used. A genetic novelty that might enable a particular insect to survive on a resistant variety might be of no selective survival value on another resistant variety, provided that the mechanisms underlying the resistance were different in the two cases. For an example of this type of permanent resistance to the Hessian fly, *Mayetiola destructor* (Say), see Hatchett and and Gallun (1970). If the agricultural area concerned is large with a randomly distributed, fully interbreeding insect population, the chances of an adapted insect population might also be greatly reduced. This aspect of insect-plant interaction is another area that is in need of much further research.

References Cited

Applebaum, S. W. 1964. Physiological aspects of host specificity in the Bruchidae. I. General considerations of developmental compatability. J. Insect Physiol. 10: 783-788.

Applebaum, S. W. and A. M. Konijn. 1967. Factors affecting the development of *Tribolium castaneum* (Herbst) on wheat. J. Stored Prod. Res. 2: 323-329.

Auclair, J. L., J. B. Maltais and J. J. Cartier. 1957. Factors in resistance of peas to the pae aphid, *Acyrthosiphon pisum* (Harr) (Homoptera: Aphididae). II. Amino acids. Canadian Entomol. 89: 457-464.

Baker, J. E. and D. M. Norris. 1968. Behavioral responses of the smaller European elm bark beetle, *Scolytus multistriatus*, to extracts of nonhost tree tissues. Entomol. Exp. & Appl. 11: 464-469.

Beck, S. D. 1956a. The European corn borer, *Pyrausta nubilalis* (Hbn.) and its principal host plant. I. Orientation and feeding behavior of the larva on the corn plant. Ann. Entomol. Soc. Amer. 49: 552-558.

Beck, S. D. 1956b. The European corn borer, *Pyrausta nubilalis* (Hbn.), and its principal host plant. II. The influence of nutritional factors on larval establishment and development on the corn plant. Ann. Entomol. Soc. Amer. 49: 582-588.

Beck, S. D. 1956c. Nutrition of the European corn borer, *Pyrausta nubilalis* (Hbn) IV. Feeding reactions of first instar larvae. Ann. Entomol. Soc. Amer. 49: 399-405, 510.

Beck, S. D. 1957a. The European corn borer, *Pyrausta nubilalis* (Hbn.) and its principal host plant. IV. Larval sacchorotrophism and host plant resistance. Ann. Entomol. Soc. Amer. 50: 247-250.

Beck, S. D. 1957a. The European corn borer, *Pyrausta nubilalis* (Hbn.) and its principal host plant. VI. Host Plant Resistance to larval establishment. J. Insect Physiol. 1: 158-177.

Beck, S. D. 1960. The European corn borer, *Pyrausta nubilalis* (Hbn.), and its principal host plant. VII. Larval feeding behavior and host plant resistance. Ann. Entomol. Soc. Amer. 53: 206-212.

Beck, S. D. 1961. Fundamental studies concerning resistance to the European corn borer. Proc. 15th Ann. Hybr. Corn Res. Conf. pp. 48-53.

Beck, S. D. 1965. Resistance of plants to insects. Ann. Rev. Entomol. 10: 207-32.

Beck, S. D. and W. Hanec. 1958. Effect of amino acids on feeding behavior of the European corn borer, *Pyrausta nubilalis* (Hbn.). J. Insect Physiol. 2: 85-96.

Beck, S. D., E. T. Kaska and E. E. Smissman. 1957. Quantitative estimation of the resistance factor 6-methoxybenzoxazolinone, in corn plant tissue. J. Agr. Food Chem. 5(12): 933.

Beck, S. D. and J. H. Lilly. 1949. Report on European corn borer resistance investigations. Iowa State Coll. J. Sci. 23: 249-259.

Beck, S. D. and E. E. Smissman. 1960. The European corn borer, *Pyrausta nubilalis* and its principal host plant. VIII Laboratory evaluation of host resistance to larval growth and survival. Ann. Entomol. Soc. Amer. 53: 755-762.

Beck, S. D. and E. E. Smissman. 1961. The European corn borer, *Pyrausta nubilalis* and its principal host plant. IX. Biological activity of chemical analogs of corn resistance factor A (6-methoxybenzoxazolinone). Ann. Entomol. Soc. Amer. 54: 53-61.

Beenakkers, A. M. T., M. A. H. Q. Meisen, and J. M. J. C. Scheres. 1971. Influence of temperature and food on growth and digestion in fifth instar larvae and adults of *Locusta*. J. Insect Physiol. 17: 871-880.

Beruter, J. and E. Stadler. 1971. An oviposition stimulant for the carrot rust fly from carrot leaves. Z. Naturforsch. 26: 339-340.

Blais, J. R. 1952. The relationship of the spruce budworm (*Choristoneura fumiferano* Clem.) to the flowering condition of balsam fir (*Abies balsamea* L. M. II.). Canadian J. Zool. 30: 1-29.

Bongers, W. 1965. External factors in the host plant selection of the Colorado beetle, *Leptinotarsa decemlineata* Say. Meded. Landbou. Opzoek. Staat Gent. 30: 1516-1523.

Bongers, W. 1970. Aspects of host-plant relationship of the Colorado beetle. Meded. Landbouwhogeschool Wageningen, Nederland 70-10: 1-77.

Borchers, R., C. W. Ackerson and L. Kimmet. 1947. Trysin inhibitor. IV. Occurrence in seeds of the Leguminosae and other seeds. Arch. Biochem. 13: 291-293.

Bowers, W. S. 1968. Juvenile hormone: Activity of natural and synthetic synergists. Science 161: 895.

Bowers, W. S. 1969. Juvenile hormone: Activity of aromatic terpenoid ethers. Science 164: 323.

Bowers, W. S., H. M. Fales, M. J. Thompson and E. C. Uebel. 1966. Juvenile hormone: Identification of an active compound from balsam fir. Science: 154: 1020-1021.

Brues, C. T. 1920. The selection of food plants by insects, with special reference to lepidopterous larvae. Amer. Nat. 54: 313-332.

Buford, W. T., J. N. Jenkins and F. G. Maxwell. 1968. A boll weevil oviposition suppression factor in cotton. Crop Sci. 8: 647-649.

Butterworth, J. H. and E. D. Morgan. 1971. Investigation of the locust feeding inhibition of the seeds of the neem tree, Azadirachta indica. J. Insect Physiol. 17: 969-977.

Callahan, P. S. 1957. Oviposition response of the imago of the corn earworm Heliothis zea (Boddie), to various wavelengths of light. Ann. Entomol. Soc. Amer. 50: 444-452.

Cartier, J. J., A. Isaak, R. H. Painter and E. L. Sorenson. 1965. Biotypes of pea aphid Acyrthosiphon pisum (Harris) in relation to alfalfa clones. Canadian Entomol. 97: 754-760.

Cerny, V., L. Dolejs, L. Labler and K. Slama. 1967. Dehydrojuvabione--a new compound with juvenile hormone activity from balsam fir. Tetrahedron Letters March 1967, 12: 1053-1057.

Chiang, H. C., A. J. Keaster and G. L. Reed. 1968. Differences in ecological responses of three biotypes of Ostrinia nubilalis from the north central United States. Ann. Entomol. Soc. Amer. 61: 140-146.

Chin, C. T. 1950. Studies on the physiological relations between larvae of Leptinotarsa decemlineata Say and some Solanaceous plants. Tijdschr. Plantenziekten 56: 1-88.

Cibula, A. B., R. H. Davidson, F. w. Fisk and J. B. LaPidus. 1967. Relationship of free amino acids of some solanaceous plants to growth and development of Leptinotarsa decemlineata (Coleoptera: Chrysomlidae). Ann. Entomol. Soc. Amer. 60: 626-631.

Dethier, V. G. 1941. Chemical factors determining the choice of food plants by papilio larvae. Amer. Nat. 75: 61-73.

Dethier, V. G. 1954. Evolution of feeding preferences in phytophagous insects. Evol. 8: 33-54.

Dethier, V. G. 1970. Chemical interactions between plants and insects. In Chemical Ecology, ed. E. Sondheimer and J. B. Simeone, Academic Press, pp. 83-102.

Dielman, F. L. 1969. Effects of gall midge infestation on plant growth and growth regulating substances. Entomol. Exp. Appl. 12: 745-749.

Dielman, F. L. 1970. Gall midges: Feeding behavior and host-plant relationship. EPPO Public. Ser. A. 54: 87-91.

Ehrlich, P. R. and P. H. Raven. 1964. Butterflies and plants: A study in co-evolution. Evol. 18: 586-608.

Ehrlich, P. R. and P. H. Raven. 1969. Differentiation of populations. Science 165: 1228-1231.

Everett, T. R. 1964. Feeding and oviposition reaction of boll weevils to cotton, althea, and okra flower buds. J. Econ. Entomol. 57: 165-166.

Feeny, P. P. 1968. Effects of oak leaf tannins on larval growth of the winter moth Operophtera brumata. J. Insect Physiol. 14: 805-817.

Force, D. C. 1966a. Reactions of the three-lined potato beetle, Lema trilineata (Coleoptera: Chrysomelidae), to its host and certain nonhost plants. Ann. Entomol. Soc. Amer. 59: 1112-1119.

Force, D. C. 1966b. Reactions of the greed dock beetle, *Gastrophysa cyanea* (Coleoptera: Chrysomelidae), to its host and certain nonhost plants. Ann. Entomol. Soc. Amer. 59: 1119-1125.

Fraenkel, G. S. 1957. The raison d'etre of secondary plant substances. Science 129: 1466-1470.

Fraenkel, G. 1969. Evaluation of our thoughts on secondary plant substances. Entomol. Exp. & Appl. 12: 473-486.

Gara, R. I., R. L. Carlson and B. F. Hrutfiord. 1971. Influence of some physical and host factors on the behavior of the Sitka spruce weevil, *Pissodes sitchensis*, in southwestern Washington. Ann. Entomol. Soc. Amer. 64: 467-471.

Gibson, R. W. 1971. Glandular hairs providing resistance to aphids in certain wild potato species. Ann. Appl. Biol. 68: 113-119.

Gilbert, B. L. and D. M. Norris. 1968. A chemical basis for bark beetle (*Scolytus*) distinction between host and non-host trees. J. Insect Physiol. 14: 1063-1068.

Gilbert, L. E. 1971. Butterfly-plant coevolution: Has *Passiflora adenopoda* won the selection race with Heliconiine butterflies? Science 172: 585-586.

Goeden, R. D. and D. M. Norris. 1965a. The behavior of *Scolytus quadrispinosus* (Coleoptera: Scolytidae) during the dispersal flight as related to its host specificities. Ann. Entomol. Soc. Amer. 58: 249-252.

Goeden, R. D. and D. M. Norris. 1965b. Some biological and ecological aspects of ovipositional attack in *Carya* spp. by *Scolytus quadrispinosus*. Ann. Entomol. Soc. Amer. 58: 771-777.

Gothilf, S. and S. D. Beck. 1967. Larval feeding behavior of the cabbage looper, *Trichoplusia ni*. J. Insect Physiol. 13: 1039-1053.

Green, T. R. and C. A. Ryan. 1972. Wound-induced proteinase inhibitor in plant leaves: A possible defense mechanism against insects. Science 175: 776-777.

Grison, P. 1958. L'Influence de la Plant-hote sur la Fecondite de l'insecter Phytophage. Entomol. Exp. & Appl. 1: 73-93.

Hamamura, Y. 1965. On the feeding mechanism and artificial food of silkworm, *Bombyx mori*. Mem. Konan Univ. Science Series 8(38): 17-22.

Hatchett, J. H. and R. L. Gallun. 1970. Genetics of the ability of the Hessian fly, *Mayetiola destructor*, to survive on wheats having different genes for resistance. Ann. Entomol. Soc. Amer. 63: 1400-1407.

Heinrich, G. and H. Hoffmeister. 1967. Ecdyson als Begleitsubstanz des Ecdysteron in *Polypodium vulgare* L. Experientia 23: 995.

Hinds, W. E. 1907. Proliferation as a factor in the natural control of the Mexican cotton boll weevil. U.S. Dept. Agr. Bur. Entomol. Bull. 59: 45 pp.

House, H. L. 1969. Effects of different proportions of nutrients on insects. Entomol. Exp. & Appl. 12: 651-669.

House, H. L. 1970. Choice of food by larvae of the fly, *Agria affinis*, related to dietary proportions of nutrients. J. Insect Physiol. 16: 2041-2050.

House, H. L. 1971. Relations between dietary proportions of nutrients, growth rate, and choice of food in the fly larva Agria affinis. J. Insect Physiol. 17: 1225-1238.

Hovanitz, W. 1969. Inherited and/or conditioned changes in host-plant preference in Pieris. Entomol. Exp. & Appl. 12: 729-735.

Isaak, A., E. L. Sorenson and R. H. Painter. 1965. Stability of resistance to pea aphid and spotted alfalfa aphid in several alfalfa clones under various temperature regimes. J. Econ. Entomol. 58: 140-143.

Jenkins, J. N. and W. L. Parrott. 1971. Crop Sci. 11:

Jermy, T., F. E. Hanson and V. G. Dethier. 1968. Induction of specific food preferences in lepidopterous larvae. Entomol. Exp. & Appl. 11: 211-230.

Keller, J. C., F. G. Maxwell and J. N. Jenkins. 1962. Cotton extracts as arrestants and feeding stimulants for the boll weevil. J. Econ. Entomol. 55: 800-801.

Keller, J. C., F. G. Maxwell, J. N. Jenkins and T. B. Davich. 1963. A boll weevil attractant from cotton. J. Econ. Entomol. 56: 110-111.

Kennedy, J. S. 1965. Coordination of successive activities in an aphid. Reciprocal effects of settling on flight. J. Exp. Biol. 43: 489-509.

Kennedy, J. S., C. O. Booth and W. J. S. Kershaw. 1961. Host finding by aphids in the field. III. Visual attraction. Ann. Appl. Biol. 49: 1.

Kircher, H. W., W. B. Heed, J. S. Russell and J. Grove. 1967. Senita cactus alkaloids: Their significance to Sonoran desert Drosophila ecology. J. Insect Physiol. 13: 1869-1874.

Klun, J. A. and T. A. Brindley. 1966. Role of 6-methoxybenzoxazolinone in inbred resistance of host plant (maize) to first-brood larvae of European corn borer. J. Econ. Entomol. 59: 711-718.

Klun, J. A., C. L. Tipton and T. A. Brindley. 1967. 2,4-Dihydroxy-7-methoxy-1,4-benzoxazin-3-one (DIMBOA), an active agent in the resistance of maize to the European corn borer. J. Econ. Entomol. 60: 1529-1533.

Kogan, M. and R. D. Goeden. 1971. Feeding and host-selection behavior of Lema trilineata daturaphila larvae (Coleptera: Chrysomilidae). Ann. Entomol. Soc. Amer. 64: 1435-1448.

Krieger, R. I., P. P. Feeny and C. F. Wilkinson. 1971. Detoxication enzymes in the guts of caterpillars: An evolutionary answer to plant defense? Science 172: 579-581.

Loschiavo, S. R., S. D. Beck and D. M. Norris. 1963. Behavioral responses of the smaller European elm bark beetle, Scolytus multistriatus (Coleoptera: Scolytidae) to extracts of elm bark. Ann. Entomol. Soc. Amer. 56: 764-768.

Maltais, J. B. and J. L. Auclair. 1962. Free amino acid and amide composition on pea leaf juice, pea aphid haemolymph, and honeydew, following the rearing of aphids on single pea leaves treated with amino compounds. J. Insect Physiol. 8: 391-400.

Mansingh, A., T. S. Sahota and D. A. Shaw. 1970. Juvenile hormone activity in the wood and bark extracts of some forest trees. Canadian Entomol. 102: 49-53.

Matsumoto, Y. 1970. Volatile organic sulfur compounds as insect attractants with special reference to host selection. In Control of Insect Behavior by Natural Products. Academic Press, pp. 133-160.

Matsumoto, Y. and A. J. Thorsteinson. 1968. Olfactory response of larvae of the onion maggot, *Hylemya antiqua* Meigen (Diptera: Anthomyiidae) to organic sulfur compounds. App. Entomol. Zool. 3: 107-111.

Maxwell, F. G., J. N. Jenkins and J. C. Keller. 1963. A boll weevil repellent from the volatile substance of cotton. J. Econ. Entomol. 56: 894-895.

Maxwell, F. G., J. N. Jenkins, J. C. Keller and W. L. Parrott. 1963. An arrestant and feeding stimulant for the boll weevil in water extracts of cotton plant parts. J. Econ. Entomol. 56: 449-454.

Maxwell, F. G., H. N. Lafever and J. N. Jenkins. 1965. Blister beetles on glandless cotton. J. Econ. Entomol. 58: 792-793.

Maxwell, F. G., H. N. Lafever and J. N. Jenkins. 1966. Influence of the glandless genes in cotton on feeding, oviposition, and development of the boll weevil in the laboratory. J. Econ. Entomol. 59: 585.

Maxwell, F. G., W. L. Parrott, J. N. Jenkins and H. N. Lafever. 1965. A boll weevil feeding deterrent from the calyx of an alternate host, *Hibiscus syriacus*. J. Econ. Entomol. 58: 985-988.

McGinnis, A. J. and R. Kasting. 1961. Comparison of tissues from solid- and hollow-stemmed spring wheats during growth. I. Dry matter and nitrogen contents of pith and wall and their relation to sawfly resistance. Canadian J. Plant Sci. 41: 469-478.

McMillian, W. W., B. R. Wiseman and A. A. Sekul. 1970. Further studies on the responses of corn earworm larvae to extracts of corn silks and kernels. Ann. Entomol. Soc. Amer. 63: 371-378.

Mellanby, K. and R. A. French. 1958. The importance of drinking water to larval insects. Entomol. Exp. & Appl. 1: 116-124.

Miles, P. W. 1969. Interaction of plant phenols and salivary phenolases in the relationship between plants and hemiptera. Entomol. Exp. & Appl. 12: 736-744.

Miller, R. L. and E. T. Hibbs. 1963. Distribution of eggs of the potato leafhopper, *Empoasca fabae* on *Solanum* plants. Ann. Entomol. Soc. Amer. 56: 737-740.

Muller, H. J. 1958. The behavior of *Aphis fabae* in selecting its host plants, especially different varieties of *Vicia faba*. Entomol. Exp. & Appl. 1: 66-72.

Nayar, J. K. and G. Fraenkel. 1963a. The chemical basis of host selection in the Mexican bean beetle, *Epilachna varivestis* (Coleoptera: Coccinellidae). Ann. Entomol. Soc. Amer. 56: 174-178.

Nayar, J. K. and G. Fraenkel. 1963b. The chemical basis of the host selection in the Catalpa sphinx, *Ceratomia catalpae* (Lepidoptera: Sphingidae). Ann. Entomol. Soc. Amer. 56: 119-122.

Nayar, J. K. and A. J. Thorsteinson. 1963. Further investigations into the chemical basis of insect-host plant relationships in an oligophagous insect

Plutella maculipennis (Curtis) (Lepidoptera: Plutellidae). Canadian J. Zool. 41: 923-929.

Nishijima, Y. 1960. Host plant preference of the soybean pod borer, Grapholitha glicinivorella Matsumura (Lep., Eucosmidae). 1. Oviposition site. Entomol. Exp. & Appl. 3: 38-47.

Painter, R. H. 1951. Insect Resistance in Crop Plants. Macmillan Publ. Co., New York. 520 pp.

Parrott, W. L., F. G. Maxwell, J. N. Jenkins and D. D. Hardee. 1969. Preference studies with hosts and nonhosts of the boll weevil, Anthonomus grandis. Ann. Entomol. Soc. Amer. 62: 261-264.

Pathak, M. D. 1971. Resistance of rice varieties to striped rice borers. Int. Rice Res. Instit. Tech. Bull. 11: 69 pp.

Pedigo, L. P. 1971. Ovipositional response of Plathypena scabra (Lepidoptera: Noctuidae) to selected surfaces. Ann. Entomol. Soc. Amer. 64: 647-651.

Reed, G. L., T. A. Brindley and W. B. Showers. 1972. Influence of resistant corn leaf tissue on the biology of the European corn borer. Ann. Entomol. Soc. Amer. 65: 658-662.

Roller, H., K. H. Dahm, C. C. Sweely, and B. M. Trost. 1967. The structure of juvenile hormone. Angew. Chem. (Int. Edit.) 6: 179-180.

Schillinger, J. A. and R. L. Gallun. 1968. Leaf pubescence of wheat as a deterrent to the cereal leaf beetle, Oulema melanoplus. Ann. Entomol. Soc. Amer. 61: 900-903.

Schoonhoven, L. M. 1967. Loss of host plant specificity by Manduca sexta after rearing on an artificial diet. Entomol. Exp. & Appl. 10: 270-272.

Schoonhoven, L. M. 1968. Chemosensory bases of host plant selection. Ann. Rev. Entomol. 13: 115-136.

Schoonhoven, L. M. 1969a. Gustation and food plant selection in some Lepidopterous larvae. Entomol. Exp. & Appl. 12: 555-564.

Schoonhoven, L. M. 1969b. Sensitivity changes in some insect chemoreceptors and their effect on food selection behavior. Proc. Koninkl. Nederl. Akad. Wetensch. C72: 491-498.

Schoonhoven, L. M. and V. G. Dethier. 1966. Sensory aspects of host-plant discrimination of lepidopterous larvae. Arch. Neerl. Zool. 16: 497-530.

Self, L. S., F. E. Guthrie and E. Hodgson. 1964. Adaptation of tobacco hornworm to the ingestion of nicotine. J. Insect Physiol. 10: 907-914.

Singer, M. C. 1971. Evolution of food-plant preference in the butterfly Euphydryas editha. Evolution 25: 383-389.

Slama, K. 1969. Plants as a source of materials with insect hormone activity. Entomol. Exp. & Appl. 12: 721-728.

Slama, K. and C. M. Williams. 1966. "Paper factor" as an inhibitor of the embryonic development of the European bug, Pyrrhocoris apterus. Nature 210: 329.

Soo Hoo, C. F. and G. Fraenkel. 1964. The resistance of ferns to the feeding of Prodenia eridania larvae. Ann. Entomol. Soc. Amer. 57: 788-790.

Soo Hoo, C. F. and G. Fraenkel. 1966a. The selection of food plants in a polyphagous insect, Prodenia eridania. J. Insect Physiol. 12: 693-709.

Soo Hoo, C. F. and G. Fraenkel. 1966b. The consumption, digestion, and utilization of food plants by a polyphagous insect, Prodenia eridania (Craner). J. Insect Physiol. 12: 711-730.

Sparks, A. N., T. A. Brindley and N. D. Penny. 1966. Laboratory and field studies of F_1 progenies from reciprocal matings of biotypes of the European corn borer. J. Econ. Entomol. 59: 915-921.

Sparks, M. R. and J. S. Cheatham. 1970. Responses of a laboratory strain of the tobacco hornworm, Manduca sexta, to artificial oviposition sites. Ann. Entomol. Soc. Amer. 63: 428-431.

Thorsteinson, A. J. 1953. The chemotactic responses that determine host specificity in an oligophagous insect (Plutella maculipennis Curt.) (Lepidoptera). Canadian J. Zool. 31: 52-72.

Thorsteinson, A. J. 1960. Host selection in Phytophagous insects. Ann. Rev. Entomol. 5: 193-218.

Todd, G. W., A. Getahun, and D. C. Cress. 1971. Resistance in barley to the greenbug, Schizaphis graminum. 1. Toxicity of phenolic and flavonoid compounds and related substances. Ann. Entomol. Soc. Amer. 64: 718-722.

Torii, K. and K. Morii. 1948. Studies on the feeding behavior in the silkworm. II. Location of the sense organ selecting mulberry leaf by silkworm larva. Bull. Res. Inst. Seric. 2: 6-12.

Vogel, R., I. Trautschold, and E. Werle. 1968. Natural Proteinase Inhibitors. Academic Press, New York.

Waldbauer, G. P. 1962. The growth and reproduction of maxillectomized tobacco hornworms feeding on normally rejected non-solanaceous plants. Entomol. Exp. & Appl. 5: 147-158.

Waldbauer, G. P. 1963. Maxillectomized insects in the study of host plant specificity. Proc. North Central Bran. Entomol. Soc. Amer. 18: 28.

Waldbauer, G. P. 1964. The consumption, digestion and utilization of Solanaceous and non-solanaceous plants by larvae of the tobacco hornworm, Protoparce sexta (Johan.) (Lepidoptera: Sphingidae). Entomol. Exp. & Appl. 7: 253-69.

Waldbauer, G. P. 1968. The consumption and utilization of food by insects. Adv. Insect Physiol. 5: 229-288.

Waldbauer, G. P. and G. Fraenkel. 1961. Feeding on normally rejected plants by maxillectomized larvae of the tobacco hornworm, Protoparce sexta (Lepidoptera: Sphingidae). Ann. Entomol. Soc. Amer. 54: 477-485.

Wei-Chun, M. 1969. Some properties of gustation in the larva of Pieris brassicae. Entomol. Exp. & Appl. 12: 584-590.

Wensler, R. J. D. 1962. Mode of host selection by an aphid. Nature 195: 830-831.

Whittaker, R. H. 1970. The biochemical ecology of higher plants. In Chemical Ecology, E. Sondheimer and J. B. Simeone, eds. Academic Press. pp. 43-70.

Whittaker, R. H. and P. P. Feeny. 1971. Allelochemics: Chemical interactions between species. Science 171: 757-770.

Wilde, J. de, W. Bongers and H. Schooneveld. 1969. Effects of host plant age on phytophagous insects. Entomol. Exp. & Appl. 12: 714-720.

Wilde, J. de, K. Hille Ris Lambers-Suver-Kropp, and A. Van Tol. 1969. Responses to air flow and airborne plant odor in the Colorado beetle. Neth. J. Pl. Pathol. 75: 53-57.

Yamamoto, R. T., R. Y. Jenkins and R. K. McClusky. 1969. Factors determining the selection of plants for oviposition by the tobacco hornworm, Manduca sexta. Entomol. Exp. & Appl. 12: 504-508.

TECHNIQUES, ACCOMPLISHMENTS AND POTENTIAL OF INSECT RESISTANCE IN FORAGE LEGUMES

Ernst Horber
Entomologist
Department of Entomology
Kansas State University
Manhattan, Kansas

Introduction

Economically important plant resistance to insects was recognized to grape Phylloxera in Europe in the 1870's and early in this century in wheat against Hessian fly, *Mayetiola destructor*, and against wheat stem sawfly, *Cephus cinctus*. These were insect problems, which at that time could not have been solved without resistance.

In forage legumes expenditures for pest control have to be minimal to be economical. But, fortunately, the tolerable economic injury level of pest populations in forage crops is comparatively high to obviate complete eradication of insect populations.

All classes of livestock depend on high quality forage production. Increasing apprehension concerning pollution with insecticide residues in animal products such as milk, meat and eggs and elsewhere emphasizes the importance of plant resistance in pest management, because man is recognized as the last link in the food chain.

Most forage crops support with their lush growth an intense insect life and harbor not only pests but many indifferent or beneficial species as pollinators, predators and parasites. Perennial forage crops like alfalfa also are overwinter habitats for insects.

Forage crops therefore play a complex and important role in balancing the agroecosystem beyond field boundaries. Resistant plants only interfere minimally with such delicate balances, as they do not completely eliminate insect populations but react defensively against attacking pests. Certain modes of resistance like tolerance may even favor a population buildup with much less damage than to a susceptible variety. Even neighboring crops may benefit from tolerant forage varieties, once the predator-parasite complex has been built up and is spilling over.

Crop improvement has come a long way from distinguishing between proveniences, selecting of local varieties, creation of cultivars, to assiduous breeding of nearly isogenic lines, differing in a single or a few characteristics only. Alfalfa is outstanding among forage crops in insect resistance. Hopefully, the fast progress achieved and experience gained in alfalfa breeding will reflect on other forage crops in which development of resistance is lagging behind or completely in the dark.

Blanchard (1943), surveying the literature, found more evidence of resistance in forage crops to sucking insects such as the pea aphid and potato leafhopper than to the chewing insects, which is still true, and there is still a pressing need for forage crops resistant to a host of insect pests.

Entomological Techniques and Procedures

Field Evaluation of Forage Legumes Under Natural Infestations.--Natural outbreaks of many insects occur at intervals of several years and may be utilized to select plants showing varying degrees of resistance. Outbreaks have been particularly useful in selecting alfalfa plants with resistance to the pea aphid (Blanchard and Dudley, 1943; Painter and Grandfield, 1935; Dahms and Painter, 1940; Ortman, et al., 1960), and the spotted alfalfa aphid (Smith, et al., 1958; Harvey, et al., 1960). Field populations also have been successfully utilized in recurrent selection programs to develop resistance to the potato leafhopper (Dudley, et al., 1963) and the alfalfa weevil (Barnes, et al., 1970). Areas where infestation and damage are more frequent than elsewhere are valuable sites for resistance-evaluation nurseries. Permanent or semipermanent testing areas help maintain an insect population.

Resistance is usually measured by counting the surviving insect population or estimating damage to plants. Estimates by weight or volume replace exact counts when insects are too numerous or too active. Counts of eggs may suggest whether nonpreference is a part of the resistance phenomena. In addition various indirect methods such as shed skins, parasite exit holes (Painter, 1951) and parasitized insects (Harvey and Hackerott, 1967), may be used to estimate insect populations.

Damage is usually scored on a 1 to 9 scale with 1 = least and 9 = most. When evaluating clones, experimental lines or cultivars in the field for resistance to insects, conditions that may cause deviations in the behavior of the insect and of the plants under study should also be considered. For example, orientation and host selection of tarnished plant bugs are apparently strongly influenced by the presence of aphids or honeydew (Lindquist and Sorensen, 1970).

Controlled Infestation in Laboratory and Greenhouse.--In the laboratory or greenhouse thousands of seedlings can be evaluated in a short time. Mass screening is especially valuable because resistant individuals often exist as a small fraction of a generally susceptible population. Environmental conditions can be adjusted to favor the plants or insects according to the selection pressure required in addition to maintaining the appropriate infestation level and homogeneous distribution of the insects. Knowing how to culture the insect species is necessary for laboratory or greenhouse screening. Rearing techniques are usually intricate and require knowledge of physical and biotic factors that influence fecundity and development of the species.

Measuring Resistance.--Resistance is a relative state obtained by comparing with a known variety of predictable reaction under the same conditions. Any method that measures plant growth and/or insect damage helps. Examples are plant height, number of leaves, chlorotic spots, and area of leaves consumed. For convenience of recording, biometric evaluation and summarizing of large bodies of data, the various degrees of resistance are assigned numerical values. Many preliminary studies may be required to arrive at a satisfactory method of managing the insect population and handling the screening.

Special Studies on Host Plant-Insect Relationship in Forage Legumes.--The need to distinguish the different modes of resistance, e.g., antibiosis, nonpreference, and tolerance, in order to succeed in a breeding program for resistance to leafhoppers was stressed by Moore (1968). The method of phenotypic recurrent selection used to increase frequency of resistant genes without separating tolerant and antibiotic factors seemed to stabilize the ratio of both types of resistance mechanism. The tolerant reaction was assumed to continually dilute the more effective antibiotic mechanism and, thus, retard progress toward more complete leafhopper resistance.

Cooperation with efficient biochemical laboratories seems to be of great advantage to establish the relation between biochemical compound and host plant resistance or to measure the relation between insect injury and level of nutrients and feeding value. Moore (1968) established the relation between carotene content of alfalfa, flowering data, and the potato leafhopper feeding injury. Attention has been drawn to saponins as possible factors for resistance in alfalfa to white grubs (Horber, 1965) and in legume seed to attack by stored-product insects (Applebaum, et al., 1969). Hsiao (1969), who investigated the chemical basis of acceptance or rejection of various leguminous plants by the alfalfa weevil used several saponin fractions isolated from alfalfa leaves that inhibited microbial growth. Incorporated into artificial diets at a 1% concentration, none of the fractions inhibited larval growth of the alfalfa weevil. In fact, some improved larval feeding growth, and tests on agar-cellulose medium confirmed their stimulative effects on feeding. Unfortunately Hsiao did not name the alfalfa variety he used. Probably different saponins or different components of the saponin molecule are responsible for the different effects, as several different saponins may occur in the same plant. It has been difficult to isolate saponins in a pure state so relatively little work has been done on the chemistry of saponins.

Roof, et al. (1972) reported a bioassay technique to differentiate various concentrations of saponins of different plant origins, e.g., *Yucca* sp. and alfalfa, using nymphs of potato leafhopper and pea aphid, respectively. Increasing saponin concentrations from 0.01% to 5% increased mortality in those two insects. Biological activity apparently depends also on the variety and origin of alfalfa. Biological activities of 3 of 10 fractions from Dupuits alfalfa and purified with repeated preparatory thin-layer chromatography exceeded those of any of 10 fraction from Lahontan. Specific differences between varieties susceptible or resistant to leafhopper, pea aphid, and possibly other alfalfa insects may be caused by various degrees of antibiosis or aversion to differences in content and composition of saponins (Horber, et al., 1974). The influence of physical factors and host plant odor on the induction and termination of dispersal flight by adult sweetclover weevil, *Sitona cylindricollis* Fåhraeus, was studied by Hans and Thorsteinson (1961). They found that colors were ineffective to attract weevils to the host but that coumarin was very effective to terminate flight activity.

Appreciable resistance to feeding by the adult sweetclover weevil has been found in only one sweetclover species, *Melilotus infesta* Guss (Gross and Stevenson, 1964; Manglitz and Gorz, 1964; Radcliffe and Holdaway, 1964). Three water-soluble factors from hot water extracts of *Melilotus* leaves (Akeson, et al., 1968), which influenced feeding by the adult sweetclover weevil were detected with a root-disk bioassay (Akeson, et al., 1967). One feeding stimulant (Stimulant A) and one feeding deterrent (Deterrent A) were found in leaves of both the resistant *M. infesta* and susceptible *M. officinalis* (L.) Lam. plants. A second feeding deterrent (Deterrent B) was detected only in leaves of the resistant species. Available evidence suggested that Deterrent B was the factor primarily responsible for resistance of *M. infesta* to the sweetclover weevil.

Deterrent B was isolated from hot-water extracts of *M. infesta* leaves and identified as ammonium nitrate (Akeson, et al., 1969d). Isolated Deterrent B and ammonium nitrate had identical feeding deterrent activities when tested with the root-disk bioassay. Other studies (Akeson, et al., 1969a, 1969c) gave substantial evidence that the nitrate content of leaves of young *M. infesta* plants was associated with resistance to adult sweetclover weevils. Beland, et al. (1970)

studied the accumulation of nitrate in various parts and at different stages of plant maturity. All parts of 6-week-old M. infesta plants were analyzed to determine nitrate distribution within a plant of a given age. Nitrate content was low in newly developed pinched leaves but became progressively higher with each successive stage of leaf development. The nitrate content of the oldest leaves was 2, 3, 9 and 80 times higher than that of the first fully expanded, unfolded, loosely pinched, and tightly pinched leaves, respectively. Nitrate levels in petioles and stems were about equal to those of the older fully expanded leaves and nearly twice as much as in the roots. Susceptibility of the leaves of young M. infesta plants to feeding of the sweetclover weevil was associated with their respective nitrate contents. Pinched leaves were fed on extensively, unfolded leaves were fed upon less extensively than pinched leaves, and fully expanded leaves were not fed upon.

The nitrate content of leaves of a particular stage of development was affected by the age of the plant. With the exception of the pinched leaves, which at all plant ages had a low nitrate content, the leaves of each developmental stage had a high nitrate content in young plants but a low level in the mature plants. The low-nitrate leaves of the mature M. infesta plants were resistant to sweetclover weevil feeding, which fact indicated that some factor other than nitrate was responsible for the feeding resistance in leaves of these mature plants.

The sugars, sucrose, fructose and glucose have been identified by Akeson, et al. (1969b) as the active components of Stimulant A, a water-soluble fraction of Melilotus officinalis (L.) Lam. leaves which stimulates feeding by the sweetclover weevil. Of 21 sugars and selected compounds tested by Akeson, et al. (1970) sucrose was the most effective feeding stimulant, whereas fructose, glucose, galactose, mannose, myo-inositol, and maltose exhibited only moderate stimulant activity. Four compounds, arabinose, ascorbic acid, glucuronic acid and mannitol displayed feeding deterrent activity.

Manglitz, et al. (1971) reported that the sweetclover (Melilotus alba Desv.) variety Denta, which was low in coumarin, was less susceptible to rootborer (Walshia miscecolorella Chambers) than varieties high in coumarin content.

Before a certain attractant, repellent, or insecticidal property is unequivocably attributed to chemicals they should be isolated in pure form and thoroughly investigated as to their composition, structure, activity in both resistant and nonresistant cultivars, preferably in isogenic lines. When adverse properties to insects have been proved, they should be tested with domesticated animals. Foaming properties of alfalfa have been studied in relation to saponin content and bloat-promoting potential in cattle and sheep (Hanson, 1963; Lindahl, et al., 1957). Saponins apparently do not have the same importance in bloat promotion as do proteins, pectins, and poliuronides, favorable media for rapid and intensive CO_2 production, foam promotion, and suppression of eructation (Maymone, 1963).

Mechanical Causes of Resistance.--An insect's choice of plants for feeding and oviposition often may be determined by some external protective feature such as thickened epidermis or cuticle, fibrous or spiny surface, small cavities or crevices, pubescence and hairiness, etc. Taylor (1956) observed that pubescent alfalfa plants were more resistant to infestation by the potato leafhoppers than more glabrous plants were, although differences in resistance were observed among progenies of glabrous plants. He concluded that the leafhopper preferred glabrous to pubescent plants for oviposition. Differences were also observed in stem anatomy, thickness, and localization of sclerenchymatic tissue and formation of several layers of cortex in lupines resistant to pea aphids (Wegorek and Dunajska, 1964).

Role of Host Plant in Wing Polymorphisms in Aphids.--Influence of host plant on wing polymorphisms was studied by several authors to prove or disprove the theory that escape from an existing or impending depletion of nutrient supplies is attempted by production of winged forms. Apparent support for the importance of diminished food supply on the production of alatae was provided by Gregory (1917). She removed adult pea aphids from the plant for daily periods of limited starvation and observed an increased percentage of alate offspring when compared to unstarved parents. Because of the importance of crowding on production of alatae, the validity of most of the early research on the role of nutrition via host plant in which crowding was not controlled has become doubtful (Johnson and Birks, 1960). Crowding effect was considered acting through over-exploitation of the host plant or from aphid interactions. Crowding effects were described by Paschke (1959) in the spotted alfalfa aphid, by Johnson and Birks (1960), in the cowpea aphid *Aphis craccivora* Koch, by Lees (1967) on *Megoura viciae* (Buckton), by Sutherland (1969a) on the pea aphid and by Shaw (1970) on *Aphis fabae*.

Johnson and Birks (1960) pointed out that aphids on plants in poor condition tend to withdraw their stylets, and are more restless than on favorable plants. Sutherland (1967, 1969a) demonstrated that seedling tissue had an appetizing influence on isolated pea aphids. When Johnson (1965) reared cowpea aphids on leaf discs until mature and transferred them to old leaf discs during parturition, they produced more alate progeny than those transferred to fresh leaf discs. The immediacy of the response in alate production was explained as a response to taste rather than nutrition. Sutherland (1967, 1969b) observed pea aphids producing predominantly alate while on mature leaves and reverting to apterae production immediately upon being returned to seedlings. When Sutherland (1969b) reared pea aphids isolated on bean seedlings he obtained fewer alate offspring than when he reared them on mature bean leaves. He concluded that seedling tissue was amino acid rich and provided a more complete diet than did mature leaves of older plants. He further concluded that optimal nutrition resulted in greater production of apterous pea aphids. His findings agree with the widely held assumption that existing or impending depletion of nutrients or deterioration of the host plant lead to the production of alatae.

However, in contrast to that, Johnson (1965) found that by increasing starvation time of adult cowpea aphid there was a gradual shift from 100% alate production during a 4-hour and 1-day starvation period to 100% apterae production after a total starvation period of 3 days. Lees (1967) noted a decline in alate production after parent *Megoura viciae* were crowded off the plant for 3 days. Toba, et al. (1967) observed that the longer the spotted alfalfa aphids were starved the less alatae they produced.

It appears that increased alate production cannot be explained as a means to escape from failing food supply in all aphids. From continuing research with artificial diets, leaf tissue analysis and phloem-sap analysis a more reliable supplement of data is expected to correlate with those obtained from directly observing fecundity, reproduction rates, growth rates, wing formation, or size of aphids. Wing polymorphisms and alate production may thus come to aid as an additional sensitive measure in evaluating nonpreference and antibiosis.

Artificial Diets.--In recent years artificial diets have been developed for many phytophagous insects; among the media are those that support continuous cultures of polyphagous aphids like *Myzus persicae* (Sulzer), *Aphis fabae* (Scopoli), and *Neomyzus circumflexus* (Buckton) (Dadd and Mittler, 1966; Dadd and Krieger, 1967; Ehrhardt, 1968a) as well as of the oligophagous pea aphid *Acyrthosiphon pisum* (Harris) which has a more restricted range of hosts among legumes (Akey and Beck, 1971). Using such diets has elucidated some nutritional requirements (Dadd and Krieger, 1967, 1968; Ehrhardt, 1968b) and has made it possible to study other problems related to nutrition, e.g., feeding behavior (Mittler, 1967a), ingestion (Mittler, 1967b), and wing determination (Dadd, 1968). The artificial diet has permitted mass rearing of several hundred thousand European cornborer egg masses in the same season for artificial infestation techniques in large nurseries and in screening tests in the field (Guthrie, et al., 1965).

Factors Affecting Expression of Resistance

Insect Biotypes.--As insect populations are heterogeneous and dynamic, any stress or selection pressure may result in new, physiologically distinct strains. Polyphagous species are generally less subject to biotype formation than monophagous species because starvation as selection pressure is rarely achieved.

The alfalfa weevil *Hypera postica* (Gyllenhal) occurs in eastern and western strains that met in Kansas in 1971. It will be of importance to follow new developments closely since cross matings between the two strains demonstrated that they are partially intersterile. Eastern females crossed with western males produced infertile eggs. The reciprocal cross produced fertile eggs, but progeny were preponderantly female. Hybrid progeny backcrossed to parent populations produced viable eggs, as did hybrid by hybrid crosses (Blickenstaff, 1965). Incompatibility between the two strains was confirmed by White, et al. (1972), but they observed hybrid vigor in backcrosses. Interaction of the strains in the field and in the greenhouse should be observed to detect any change in ecological or behavioral traits in relation to different alfalfa varieties.

Six biotypes of the spotted alfalfa aphid (SAA), *Therioaphis maculata* (Buckton), have now been recognized in western United States. Among them, four are recent discoveries identified from a combination of studies including tests on the parent clones of "Moapa" and "Washoe" alfalfa, biological activity, and response to some organophosphate insecticides. The biotypes have been designated ENT-A, ENT-B, ENT-C, ENT-D, ENT-E, and ENT-F. Biotype ENT-A was collected near El Centro, California in 1958. The population reproduced and survived on 3 to 9 clones of the resistant cultivar Moapa and had high biological activity on 1 clone of the SAA and PA resistant 8-clone cultivar Washoe, and was susceptible to some organophosphate insecticides. Biotype ENT-B was the original population found in southwestern United States in 1954. It did not reproduce or survive on the parent clones of Moapa or Washoe; had high biological activity, and was susceptible to some organophosphate insecticides. Biotype ENT-C was discovered near Perryville, Arizona, in 1968. It reproduced and survived on 1 clone of Moapa and 1 clone of Washoe; had high biological activity and was resistant to some organophosphate insecticides. Biotype ENT-D was found in Kings County, California, in 1968. It reproduced and survived on 1 clone of Moapa and 1 clone of Washoe, had low biological activity, and was resistant to some organophosphate insecticides. Biotype ENT-E was found in San Bernardino, California, in 1968. It reproduced and survived on 5 clones of Moapa and 1 clone of Washoe, had very high biological activity, and was susceptible to some organophosphate insecticides. Biotype ENT-F was discovered near El Centro, California, in 1969. It reproduced and survived on 4 clones of Moapa but not on any clone of Washoe, had moderately high biological activity, and was susceptible to some organophosphate insecticides (Nielson, et al., 1970b).

Nielson, et al. (1971) evaluated fifty-two experimental alfalfa and cultivars for resistance to 4 biotypes (ENT-A, ENT-C, ENT-E, ENT-F) in greenhouse tests at Tucson, Arizona. Three alfalfas (Caliverde 65, developed by the University of California at Davis; T-3-12, a new experimental one developed jointly by University of California and Entomology Research Division, ARS, USDA), were rated highly resistant, 14 moderately resistant, 13 intermediate, 7 low resistant, 9 susceptible, and 6 highly susceptible. Differences in seedling survival were highly significant. Differences among biotypes were also highly significant. ENT-F was the most virulent, followed by ENT-E, ENT-A and ENT-C.

Vigilance and continuous testing against biotypes as they were discovered ensured development of varieties more resistant to the new biotype than the one damaged.

Biotypes of the pea aphid (PA) were observed in and studied in Canada (Cartier, 1957, 1959, 1960, 1964), Finland (Markkula, 1963), Germany (Muller,

1962), Poland (Wegorek, 1968), Switzerland (Meier, 1964). Cartier, et al. (1965) in studying two aphid populations from Quebec and Kansas, respectively, on ten alfalfa clones at three temperatures, concluded that the two populations differed. Cartier and Auclair (1965) investigated the behavior of different biotypes when exposed to different colors on an artificial diet.

Cartier (1964) studied 151 different parthenogenetical lines collected in fields in Quebec and Ontario. He distinguished three biotypes: A, B, and C. Biotype A was superior when cultivated on peas and on alfalfa. He assumed that biotypes A and B were dominant where alfalfa was grown. These biotypes differed also in other respects. Copulations took place between individuals of all three biotypes with normally appearing eggs produced. Auclair (1966) collected 7 clones of the pea aphid on alfalfa in southern New Mexico and reared them in the greenhouse on peas and on alfalfa. Growth and reproduction were generally higher on alfalfa than on peas. Alfalfa seedlings were more resistant than pea seedlings. As the plants aged several weeks, more aphids grew and reproduced on alfalfa than on peas. Such variations in plant susceptibility to the pea aphid indicate that the age and growth stage of the host plant must be carefully standardized for evaluation of resistance. Markkula (1963) found a green, more abundant, and a red pea aphid in Finland. Their reproductive periods and other biological properties differed on pea and on red clover. The pea plant reduced progeny of red aphids more than of green aphids. The red form has not yet been found in the United States, suggesting that only the green form was introduced from Europe.

Environmental Factors Modifying Resistance.--Changes in plant resistance to insects are associated with various growing conditions. The degree to which a plant is suitable as an insect's host can be modified by such environmental factors as light, temperature, moisture, and nutrients. The responses of the insects, including those to olfactory and gustatory stimuli, also, may be modified by environmental changes. Thus, a cultivar that exhibits resistance in one locality or environment may be susceptible in another.

That low temperatures may reduce resistance in alfalfa to the spotted alfalfa aphid was first noted by Howe and Smith (1957). Then Hackerott and Harvey (1959) demonstrated higher SAA survival at low (15.6°C) than at high (27°C), temperatures on resistant plants, but without difference in plant damage. Further observations on temperature-dependent modifications of resistance to the spotted alfalfa aphid were reported by McMurtry (1962), Isaak, et al. (1965), Schalk, et al. (1969) and Kindler and Staples (1970b). Laboratory studies demonstrated that plant nutrition affects resistance to the SAA. Resistant alfalfa deficient in phosphorus became more resistant; deficient in potassium, it became less resistant, while nitrogen deficiency had no effect (McMurtry, 1962). Kindler and Staples (1970a) reported identical results plus decreased resistance with deficiencies in calcium or potassium or excess magnesium or nitrogen. Sulphur levels had no effect. Plant age influenced resistance with older plants more resistant than seedlings (Howe and Pesho, 1960a). Photoperiod (McMurtry, 1962), relative humidity (Isaak, et al., 1963), and soil moisture (Kindler and Staples, 1970b) had little or no influence.

Development of Resistance to Alfalfa Insects

Alfalfa Seed Chalcid.--Methods are described for field screening and laboratory management of this insect (Nielson, 1967; Howe and Manglitz, 1961; Strong, 1960, 1962a; Watts, et al., 1967; Booth, 1969; Nielson and Schonhorst, 1965a). Certain cultivars, particularly Lahontan, have been reported by several investigators as having light infestations (Howe and Manglitz, 1961; Strong, 1962b; Nielson and Schonhorst, 1967). Basic studies on causes of resistance may be found in Kamm and Fronk (1964), who observed response of adult chalcids to 95 chemicals in alfalfa.

Alfalfa Weevil.--Methods are available for field and greenhouse screening. Laboratory management was summarized by Schroder, et al. (preparation). Work of Huggans and Blickenstaff (1964) made it possible to rear the alfalfa weevil continuously in the laboratory. Two cultivars, Team and Weevl-chek, have been released for resistance to the alfalfa weevil. Although neither has a high degree of resistance, both perform better than other varieties when infested with the weevil. In the case of Team, antibiosis and nonpreference are operating along with tolerance. In Weevl-chek tolerance is the only mechanism known to be operating.

Spotted Alfalfa Aphid (SAA).--Field screening and greenhouse management techniques are available (Harvey and Hackerott, 1956; Nielson, 1957; Howe and Smith, 1957; Graham, 1959; Daniels, 1960; Harvey, et al., 1960; Howe and Pesho, 1960a; Howe and Pesho, 1960b; Howe, et al., 1963; Messenger, 1964; Manglitz, et al., 1966; Schalk and Manglitz, 1969; Manglitz and Schalk, 1970). A combination of greenhouse and field screening works well with SAA because plants evaluated in the greenhouse will be resistant in the field and vice versa (Harvey, et al., 1960). Often excised plant parts have been used in such tests with apparently good results. However, Thomas, et al. (1966) demonstrated that the degree of resistance may be less in excised leaves than that in intact leaves. Large numbers of plants must be examined as only one resistant plant may be selected for each 3,000 susceptible plants tested (Padilla and Young, 1958). Many cultivars resistant to SAA have been developed. The USDA lists 14 that were released before December, 1969 (Anonym., 1969). Possibly all three modalities of resistance are involved (Jones, et al., 1968; Nielson and Currie, 1959; Kishaba and Manglitz, 1965; McMurtry and Stanford, 1960; Kindler and Staples, 1969; Kishaba and Manglitz, 1968; Marble, et al., 1959; Ortman, 1965; Kirchner, et al., 1970).

Pea Aphid (PA).--Methods of managing field populations of the pea aphid, evaluating resistance in the field, and rearing the insect in the greenhouse are closely similar to those for the spotted alfalfa aphid. Various methods to evaluate resistance to the pea aphid in alfalfa were investigated by Ortman, et al. (1960). Greenhouse and field evaluations for pea aphid resistance in alfalfa are positively correlated (Hackerott, et al., 1963). Several resistant varieties have been released: Washoe was the first (Peaden, et al., 1966), followed by Dawson (Kehr, et al., 1968), Mesilla (Melton, 1968), and Kanza (Sorensen, et al., 1969b). Each also is resistant to the SAA and certain diseases. Another variety, Team, which was developed primarily for alfalfa weevil resistance, also resists pea aphids (Barnes, et al., 1970). Heritable and stable resistance to the pea aphid is available, but as with SAA causes of resistance are not understood. Dahms and Painter (1940) first reported reduced fecundity of pea aphids on resistant alfalfa. Reduced fecundity and survival (Manglitz, et al., 1962; Carnahan, et al., 1963; Hackerott, et al., 1963) and perhaps reduced feeding (Ortman, 1965) seem characteristic of pea aphids on resistant plants. The specific amino acids found in pea aphids and their honey dew may differ from those in SAA; otherwise the behavior of the two aphids on resistant plants is similar (Ortman, 1965).

Sandmeyer, et al. (1971) compared longevity, reproduction and rate of nymphal development of 3 successive generations of the PA and the SAA reared on the same 6 alfalfa clones. Clones 1388 and 1-418 were highly susceptible, whereas N-556 and N-529 were highly resistant to both aphids. Clones C902 and N-466 were moderately resistant to PA but highly resistant to SAA. Reproductive rates, cumulative number of nymphs and length of reproductive periods varied from maximal on susceptible to low or 0 on highly resistant clones. Developmental rates of SAA appeared more directly affected by the various plants than PA. Molting difficulties appeared on clone N-522 for the third and fourth SAA instars, which may have caused the lag phase observed. These results indicate that probably different metabolisms of both plant and insect may govern responses of the 2 aphids.

Meadow Spittlebug.--Field screening and laboratory techniques are described by Wilson and Davis (1953, 1958, 1965) and Hill and Newton (unpublished data). Antibiosis and tolerance were determined and a simple preference test has been used effectively in Indiana (Wilson and Davis, 1965). Culver is the only cultivar so far developed specifically for resistance to the meadow spittlebug (Wilson and Davis, 1960). It performed well as a variety in Indiana, demonstrating both antibiosis and tolerance.

Potato Leafhopper.--Field screening and laboratory or greenhouse management are reported by Sorensen, et al. (1972). Jarvis and Kehr (1964, 1966); Schillinger, et al. (1964); and Newton and Barnes (1965). Webster, et al. (1968a) successfully selected for resistance on the basis of seedling survival in the growth chamber. Kindler and Kehr (1970) had similar results with seedlings screening in the greenhouse. However, when the selected plants were evaluated against leafhoppers in the field, they were no more resistant than unselected plants. Differences among cultivars in susceptibility to leafhopper attack were reported by Jewett (1929); Farrar and Woodworth (1939); Graber (1941); Davis and Wilson (1953); Hanson, et al. (1964); Webster, et al. (1968b); Moore (1968). Cultivars, such as Rhizoma, Rambler, Teton, Vernal, and Culver, that contain some *M. falcata* germplasm exhibit relatively high tolerance compared with unselected *M. sativa*. Resistant cultivars properly managed perform well with little damage from leafhoppers.

Other Alfalfa Insects.--Aamodt and Carlson (1938) found some cultivars, especially Grimm, able to flower despite lygus bug (*Lygus* spp.) injury, and Malcolm (1953) reported differences in lygus bug preference among 16 cultivars. Turkistan and Ladak harbored low populations compared with those on Ranger and Buffalo. Nielson and Schonhorst (1965b) sampled *Lygus* populations on many alfalfa cultivars in Arizona and found significant differences among entries. Differences in resistance among cultivars as measured by survival or damage to infested seedlings was demonstrated by Lindquist, et al. (1967). Several progenies from intravariety crosses of selected plants display higher seedling survival after infestation with *Lygus lineolaris* (Palisot de Beauvois).

Radcliffe and Barnes (1970) concluded that it should be possible to develop alfalfa cultivars with resistance to the alfalfa plant bug (*Adelphoris lineolatus* (Goeze)).

Horber (1965, 1972) demonstrated antibiosis, nonpreference, and tolerance to whitegrub, *Melolontha vulgaris* F., in roots of several alfalfa strains collected from severely infested fields.

Resistance to Various Forage Legume Insects

a. Red Clover (*Trifolium pratense* L.)

Jewett (1941) reported differences in resistance to the pea aphid among red clover varieties. A strain of red clover from Minnesota had significantly more aphids than other strains from Kentucky, Tennessee, and Virginia. He also found differences among plants of a given strain.

Wilcoxson and Peterson (1960) found Dollard variety of red clover more resistant than Wegener variety. They classified this resistance as nonpreference and antibiosis. Resistance to the pea aphid seemed to explain a much lower incidence of virus diseases in Dollard than in Wegener under field conditions at St. Paul, Minnesota. When mosaic and pea-stunt-viruses were inoculated mechanically, Dollard was just as susceptible as Wegener.

Markkula and Roukka (1970) exposed 10 red clover varieties to 3 pea aphid biotypes, a red biotype and two green biotypes originating on red clover. They reared aphids on broad bean before the test and transferred newly matured, wingless aphids to rearing cages on leaves of the same age. Ten to 20 aphids were confined singly and their progeny examined at weekly intervals.

All varieties were highly resistant to green biotype la, with fewer than 10 progenies. Some varieties were susceptible to red biotype lb. All varieties tested were susceptible to green biotype 16, their progeny varying from 79 to 88. Some varieties were heterogeneous; individual plants produced low, others high progeny numbers.

b. Alsike (*Trifolium hybridum* L.) and White Clover (*Trifolium repens* L.)

Markkula and Roukka (1971) compared alsike and white clover varieties for resistance to 3 pea aphid biotypes (la, lb, and 16) in a cage test to 10 diploid and tetraploid alsike clovers and to 12 white clovers.

Number of progeny produced indicated that all varieties of alsike and white clover were highly susceptible to biotype 16. All were considerably more resistant to biotype lb, and all were most resistant to biotype la. Some differences in fecundity of the same biotype also were noted on different clover varieties.

c. Sweetclover (*Melilotus* spp.) and *Trigonella* spp.

Wilson, et al. (1956) noted differences in sweetclover weevil adult feeding among various sweetclover varieties, but none were free of injury. A greenhouse technique to evaluate feeding preference on seedlings by the sweetclover weevil was developed by Connin, et al. (1958). In their greenhouse studies Manglitz and Gorz (1964) observed feeding on 18 species of *Melilotus*, on 12 of *Trigonella*, and on alfalfa but not on *Melilotus infesta* Guss. This species did not appear to have any deleterious effect upon the weevil, but did strongly deter its feeding. There seemed to be no relationship between coumarin content and weevil feeding since some hosts were accepted regardless of coumarin content.

Manglitz and Gorz (1961 and 1963) demonstrated that the sweetclover aphid *Therioaphis riehmi* Boerner may be controlled by using resistant plants, and that resistant plants are found among sweetclover introductions from many areas of the world. The resistance mechanism, which appears to be a type of nonpreference, is sufficiently effective to starve aphids confined to resistant plants (Kishaba and Manglitz, 1965).

Table 1.--Alfalfa varieties resistant to insects.

Insect	Alfalfa variety
Alfalfa weevil	Team, Weevl-chek, WL 215
Alfalfa seed chalcid	Lahontan
Pea aphid	Dawson, Kanza, Mesilla, Team, Washoe
Potato leafhopper	Cherokee, Culver, Team, Weevl-chek
Spittle bug	Culver
Spotted alfalfa aphid	Caliverde 65, Cody, Dawson, El Unico, Hayden, 183, Kanza, Lahontan, Mesa-Sirsa, Mesilla, Moapa, Sonora-70, Washoe, WL 504, WL 508

The host range of the sweetclover aphid was extended in their greenhouse studies (Manglitz and Gorz, 1964) to include 9 species of *Melilotus* and 8 of *Trigonella*. They reported that among most entries tested, older plants demonstrated greatest resistance.

Although the sweetclover weevil and the sweetclover aphid have a common group of host species, factors determining host suitability appeared to operate independently for the 2 insects.

The biology of the sweetclover root borer, *Walshia miscecolorella* (Chambers) and its injury to sweetclover have been studied recently by Manglitz, et al. (1971). They reported that the variety Denta (*Melilotus alba* Desv.) was less susceptible to root borer attack than the variety Goldtop (*M. officinalis* L. (Lam.)). The most resistant entries were an annual variety Israel (*M. alba*) and two introductions identified as Pl 314094 (*M. taurica* (Bieb.) Seringe) and Pl 318386 (*M. polonica* (L.) Desv.). The latter had significantly fewer root borer larvae per plant in replicated tests under natural field infestations in 1969 and 1970 than did *M. officinalis* (var. Goldtop). However, the top growth of *M. polonica* was damaged more than was the variety Denta by equal larvae per plant. Progeny of the interspecific of *M. alba* x *M. polonica* reacted to the root borer infestations in a manner similar to the *M. polonica* parent. The low coumarin variety Denta (*M. alba*) generally had fewer larvae than Goldtop but Denta was heavily infested in 1969 when root borer populations were exceptionally high. Observations on four homozygous genotypes in closely related and highly inbred lines differing in both coumarin contents and 8-glucosidase activities revealed that plants high in coumarin had significantly more larvae per plant than low-coumarin plants did. Thus, the apparent resistance of Denta may be related to its lower content of coumarin.

d. Lupine (*Lupinus* spp.)

Wegorek and Jasienska-Obrebska (1964) reported resistance to pea aphid in lupine varieties. On a yellow lupine variety, Gorzki, the population died or was barely maintained. Among blue lupine lines the most resistant was Wielkopolski Gorzki; aphids died after several days exposure on it. On Obornicki and Szybkopedny aphid populations developed only slowly. On white lupine varieties the aphid population remained extremely low or disappeared.

Leaf morphology and stem anatomy of seven lupine varieties belonging to 3 species were examined by Wegorek and Dunajska (1964) to explain differences in resistance to pea aphid. Very susceptible varieties had the least but longest hair per surface unit. Very resistant and resistant varieties had denser hair; however, not enough to protect the leaf surface from aphid feeding. Differences were also observed in stem anatomy, thickness, and localization of sclerenchymatic tissue and formation of several layers of cortex.

e. Vetches (*Vicia* spp.)

Albrecht (1940) observed striking varietal differences in resistance of vetches to pea aphid injury. Injury to lines rated resistant was confined entirely to stem tips that later recovered, whereas susceptible lines were severely attacked; their tips were killed and their stems were badly damaged or destroyed.

The most susceptible, a purple vetch (*Vicia atropurpurea* Desv.) was completely destroyed.

Three cultivated varieties (*Vicia sativa* L.), 1 hairy vetch (*V. villosa*), 1 smooth vetch (*V. villosa* var.), 1 woollypod vetch (*V. dasycarpa*) and 4 wild vetches (*V. melanops*, *v. angustifolia*, *V. hybrida*, *V. grandiflora*) were among those rated resistant.

f. Various Legumes

Hewitt (1969) screened four species of legumes and 13 species of grasses for resistance to feeding by the migratory grasshopper, *Melanoplus sanguinipes* (F.). Grasshoppers that fed on legumes gained the most weight and had lowest mortality. Relative toughness of 7 varieties among 4 legume species: birdsfoot trefoil

(Lotus corniculatus L.), cicer milkvetch (Astralagus cicer L.), crownvetch (Coronilla varia L.), and sainfoin (Onobrychis viciaefolia Scop.) did not correlate with amount of each eaten by the migratory grasshopper during 6 minutes feeding time. Significant differences in survival and weight gain as a result of living on one of the 7 varieties of legumes for 30 days were observed. The cicer milkvetch variety A-13107 and the birdsfoot trefoil variety Empire gave lowest grasshopper survival and weight gain.

The interrelationships among more than 70 small-seeded legumes with spotted alfalfa aphid, yellow clover aphid (Pterocallidium trifolii (Monell)) and the sweetclover aphid (Therioaphis riehmi Boerner) were studied by Peters and Painter (1957, 1958), largely in the greenhouse. Twelve species in 4 genera, Medicago, Melilotus, Trifolium and Trigonella, were immune or resistant to all 3 aphids.

Inheritance of Resistance to Insects by Forage Legumes

Although remarkable progress has been made in developing insect-resistant germplasm and cultivars, little is known about the genetic basis of resistance. In a limited study to determine inheritance patterns of spotted alfalfa aphid resistance in individual plants of Zia alfalfa, Glover and Melton (1966) concluded that inheritance was quantitative. Similarity of parent and progeny performance indicated relatively high heritability.

Jones, et al. (1950) indicated that 2 genes, a dominant and a recessive, conditioned resistance to the pea aphid. Based on disomic analysis, they suggested linkage with a crossover value of about 28%. Glover and Stanford (1966) explained inheritance of pea aphid resistance on the basis of a tetrasomically-inherited, dominant gene; modifying factors may also contribute to resistance.

Taylor (1956) found that potato leafhopper resistance in alfalfa correlated with pubescence. He also obtained evidence of heritable resistance in glabrous plants. Kehr, et al. (1970) reported that resistance to leafhopper yellowing in alfalfa appeared dominant to susceptibility, but they could not interpret how many genes were involved. Sufficient additive genetic variance was present for progress in breeding for resistance to leafhopper yellowing.

The inheritance of resistance in sweetclover Melilotus officinalis (L.) Lam., to the sweetclover aphid, Therioaphis riehmi (Boerner) was studied by Manglitz and Gorz (1961) who used mass infestation and confinement of aphids to measure resistance. F_1, F_2, and backcross progenies revealed that resistance was a dominant characteristic and that a single pair of alleles was segregating in most of the progenies observed. An additional pair of genes that acted complementarily appeared to be involved in several progenies derived from 1 resistant parent.

Breeding Insect Resistant Forage Legumes

a. Collecting germplasm.--Initiation of a breeding program depends on genes for resistance. They are generally found where adequate search is conducted among attacked plants. Success is proportional to the number and diversity of plants that can be studied, to the efficiency of the screening technique and to the skill in managing the infesting population of insects. Techniques are available or may be developed to assess all three modalities of resistance, and to detect and to exploit both low and high resistance. All possible sources of genetically different materials should be explored, but adapted lines from the problem area should receive first consideration. Maximum diversity of a crop occurs near centers of origin. Such areas often contain plants with wide differences in physiological characters and likely genes for insect resistance. Legumes from the native habitat of the insect also furnish sources of resistance because of natural selection. Alfalfa and other forage legumes contain vast variability, which should be used before attempts to induce mutations (Sorensen, et al. 1972).

Sources of high resistance to the seed chalcid have yet to be found. Cultivars mentioned as being resistant or being good sources for selecting resistance are: Hairy Peruvian, Zia, Ranger, Sirsa #9, and the species *Medicago agropyrotorum* Vass. emend. (Nielson and Schonhorst, 1967). Strong (1962b) and Rowley and Haws (1964) state that the best sources of resistance to the seed chalcid are among the nondormant types, particularly from the Afghanistan area; dormant types, particularly from western Europe, are the most susceptible. Alfalfa cultivars with high resistance to the alfalfa seed chalcid have not been developed, but judging from results to date, i.e., low but consistent heritable resistance, such a development seems quite possible (Sorensen, et al., 1972).

Dogger and Hanson (1963), and Pitre, et al. (1970) evaluated experimental and commercial cultivars for resistance to the alfalfa weevil and found differences in susceptibility. Low resistance was noted among 500 alfalfa introductions (Busbice, et al., 1967). *Medicago sativa* var. *gaetula* Urb. and *M. falcata* L. are probably the best known sources for resistance to oviposition (Campbell and Dudley, 1965).

Cultivars of Turkistan origin generally appeared to be good sources of resistance to the SAA (Harpaz, 1955; Hackerott, et al., 1958; Klement and Randolph, 1960; Howe and Pesho, 1960a). Lahontan, the first cultivar known to resist SAA, traces directly to Nemastan, an introduction from Turkistan (Smith, 1955). Lahontan was not developed for aphid resistance but its release in 1954 (Smith, 1958) coincided with the discovery of the SAA in the U.S. Some resistance has been found in early all sources examined. For example, the resistant cultivar Cody was developed from susceptible Buffalo (Sorensen, et al., 1961). Sources of resistance to the pea aphid were detected in Ladak alfalfa by Painter and Grandfield (1935) (see Figure 1). Flemish types seem to be the best source of resistance, followed by Turkistan alfalfas; however, some resistant plants were recovered from most alfalfa sources evaluated (Hackerott, et al., 1963; Nielson and Schonhorst, 1965b).

Sources of resistance to the meadow spittlebug occur in most of the *M. sativa* types. The Flemish varieties appear to have more tolerance than most of the commonly grown American varieties have (Wilson and Davis, 1958). However, *M. falcata* is even more promising for high resistance, including antibiosis. Sources of resistance to leafhoppers are obtainable from *M. falcata*. Cultivars such as Rhizoma, Rambler, Teton, Vernal, and Culver, which contain some *falcata* germplasm exhibit relatively high tolerance compared with unselected *M. sativa* types. Plants with varying degrees of resistance can be isolated from most varieties as demonstrated by the development of Cherokee (Dudley, et al., 1963).

b. Breeding procedures.--Breeding for insect resistance is complicated by variability of two organisms involved. Valuable information includes that about the genetic variance for resistance in the legume, genetic variance in the pest, consequent interactions between the genotypes, and about environmental conditions that affect plant resistance and insect behavior. Recurrent phenotypic selection, a form of mass selection, has been especially effective in developing resistance to the spotted alfalfa aphid (Harvey, et al., 1960), pea aphid (Carnahan, et al., 1963), potato leafhopper (Dudley, et al., 1963), and alfalfa weevil (Barnes, et al., 1970). It is useful in developing multiple-pest resistance (Painter, et al., 1965; Hanson, et al., 1972a). Mass selection conserves genetic diversity, increases the frequency of desirable genotypes, develops new genotypes, and enhances the success of selecting individuals that combine attributes needed in future cultivars (Sorensen, et al., 1972). Since it is necessary to recombine numerous parents to initiate the next generation, large plant populations must be examined (Hill, et al., 1969). Cumulative resistance may be achieved by combining components of resistance and by combining genes for a particular component. This accumulation of resistance factors makes the natural selection of biotypes able to infest or injure the resistant variety much more difficult (Painter, 1966). It is essential to devise and use screening and testing techniques that clearly distinguish between resistant and susceptible genotypes and to prevent escapes. A small percentage of escapes may invalidate a breeding effort. Initial screening based on seedling mortality is especially valuable because it measures the sum of the resistant mechanisms--e.g., nonpreference, antibiosis, and tolerance. However, seedling resistance as obtained in the greenhouse must be evaluated under field conditions to determine accuracy of selection and potential economic values. When Hanson, et al. (1972) selected for resistance in the laboratory at the seedling stage, or in the field without regard to vigor, yield decreased. Yield increased with recurring cycles when selection was made in the field and vigorous plants were selected as parents.

Strategic Breeding for Insect Resistant Forage Legumes

Given only limited means and time, what is wiser, to produce one highly resistant or several moderately resistant varieties? For years breeders and entomologists probably aimed too high, e.g., searching for highly resistant or nearly immune varieties while rejecting moderately resistant germplasm of tolerance type. Moderate mortality, combined with fewer progeny repeated several generations achieves the same results as the sterile-male technique, with fewer side effects than from insecticides.

There are many advantages to producing several moderately resistant varieties instead of one highly resistant one:

1. Use can be made of moderately resistant germplasm, occurring more frequently than near immunity among otherwise widely diverse germplasm.

2. Greater diversity in the gene pool allows breeders to adapt new varieties to local or parochial agronomic or ecologic requirements.

3. Varieties may be changed more frequently by releasing moderately resistant varieties at shorter intervals.

4. Monoculture or monopoly of one single variety may be avoided because one variety likely will take over only a small fraction of total acreage. New aggressive biotypes are less likely to be selected under these circumstances.

Hanson, et al. (1972) discussed directed, mass selection for developing multiple pest resistance and conserving germplasm in alfalfa, and suggested ways to conserve genetic diversity still available in alfalfa, to develop combined resistance to diseases and insect pests, and to reduce dependence on pesticides. They proposed an integrated program of worldwide collection, recombination, and mild selection to conserve and improve alfalfa germplasm resources. Key feature of such a program would be 7 gene pools developed by mass selection, representing 7 geographic regions and a wide range of environments, to provide improved source material for breeding regionally adapted, multiple pest-resistant varieties. It is reasonable to assume that varieties developed for multiple resistance by recurrent mass selection would be less vulnerable to attack by mutant forms of pathogens or insects than varieties developed by breeding procedures that give more attention to genetic uniformity. Such a program for conserving germplasm by collecting a wide range of ecotypes while they are still available in most forage legumes, e.g., red clover, alsike and white clover and other forage legumes and for screening the recombinants for multiple pest resistance would require international cooperative efforts.

Value of Host Plant Resistance for Insect Control

a. Contribution of insect resistant cultivars to crop improvements.--It is difficult to establish values for insect-resistant cultivars because insects damage susceptible legumes many ways and the damage also varies. Luginbill (1969) stated that $35 million was a conservative estimate of annual savings to growers from use of spotted alfalfa aphid-resistant cultivars. Resistance aids in stand establishment and maintenance, assures higher quality and yield, and lower insect and disease control costs. Under spotted alfalfa aphid attack, resistant cultivars showed excellent stand establishment while susceptible entries were eliminated (Smith, et al., 1958; Howe and Pesho, 1960a; Sorensen, et al., 1961). Infestations that killed mature stands of susceptible entries failed to kill resistant entries, and regrowth after cutting varied directly with plant resistance (Howe and Pesho, 1960b). In areas where the spotted alfalfa aphid occurred after the last cutting in the fall, winter survival paralleled resistance scored noted the previous fall (Carnahan, et al., 1963).

In alfalfa forage yield trials, where pea aphids damaged susceptible cultivars, resistant entries produced 2 to 3 times more forage than did susceptible cultivars (Peaden, et al., 1966; Sorensen, et al., 1969a) (see Figure 2). Also, pea aphid damage to the first cutting weakened plants so succeeding yields were reduced. Average increase in forage yields for resistant over susceptible plants was 211, 188, 107 and 114% for the first, second, third, and fourth cuttings, respectively. Resistant selections maintained a 78% increase over susceptible plants the first cutting the following year (Harvey, et al., 1971). Under epidemic infestations of spotted alfalfa aphids, forage yields of resistant cultivars exceeded those of susceptible ones 3 to 4 times (Howe and Smith, 1957). In field cages, heavy infestations of either spotted alfalfa aphids or pea aphids, reduced quality and quantity of forage of resistant alfalfa cultivars too, but losses were lower than those of susceptible entries (Harvey, et al., 1971; Kindler, et al., 1971). In 15- to 80-acre fields foliage damage by the spotted alfalfa aphid was 15 to 22 times greater on susceptible than on resistant cultivars (Barnes, 1963). Protein yields of resistant Kanza attacked by the pea aphid were almost double, and carotene yields triple those of the susceptible cultivars, Buffalo, Ranger, and Vernal (Sorensen, et al., 1969a). Loper (1968) reported higher coumestrol content in aphid-susceptible Vernal than in resistant Moapa and Washoe cultivars when all were subjected to aphid feeding.

b. Does insect resistance in legumes affect nutritive value of forage?--Because of increased apprehension to potentially harmful side effects of pest control on higher animals and man varietal resistance also should be scrutinized. Tolerance is less suspect than antibiosis or nonpreference. However, insects are much further removed phylogenetically from warm-blooded animals and man who may use insect resistant crops and their products than are rodents, dogs, and monkeys from which toxicological and teratological information usually are available and extrapolated for man. Since insects, like higher animals, react with wide range of physical and behavioral responses to the presence or absence of chemical stimuli, it is extremely difficult to infer reactions of man or domesticated animals directly from insect studies on resistant varieties. Resistance is localized in some tissues or plant parts. Sometimes they are biologically active only during a short time at certain developmental stages of the plant, eg., DIMBOA[1/], in the whorl of the young corn plant or alkaloids in the most active growth period, well in advance of harvest time.

Chemical constituents important in animal nutrition such as protein, carotene, and digestible dry matter were similar in susceptible alfalfa cultivars to

---

[1/] DIMBOA = 2,4=dihydroxy-7-methoxy-1,4-benzoxazin-3-one.

those resistant to the pea aphid and spotted alfalfa aphid (Kehr, et al., 1968; Kindler, et al., 1971). Also, chemical components of Team, resistant to weevil and pea aphid, were similar to those found in susceptible cultivars. Neither digestibility coefficients nor performance of yearling Holsteins differed significantly when fed resistant or susceptible varieties (Barnes, et al., 1970).

The quality and nutritive value of feed is improved through host plant resistance by avoiding loss of protein, carotene, and vitamin A. When under attack by the pea aphid, resistant Kanza yielded almost twice as much protein and three times as much carotene as the susceptible cultivars, Buffalo, Ranger, and Vernal (Sorensen, et al., 1969a). Loper (1968) reported higher coumestrol content in aphid susceptible Vernal than in resistant Moapa and Washoe cultivars, when all were subjected to aphid feeding.

Price of Host Plant Resistance

Because success in developing host plant resistance hinges on uninterrupted long-range programs, research is no on-and-off business to be financed by whims of politicians.

The public must be willing to invest money today to reap future benefits.

Success in developing resistant varieties was achieved with a relatively modest input in research efforts: Hanson (1961) reported the cost of developing Moapa alfalfa resistant to the spotted alfalfa aphid to be less than $30,000, the amount paid in salaries to part-time plant breeders and entomologists.

Luginbill (1969) compared the total estimated value of the research on resistant plants with its cost for 4 insect pests. About 115 professional man-years for the Hessian fly resistant wheat varieties, 92 for the sawfly, 119 for the spotted alfalfa aphid, and 136 for the European corn borer have cost about $20,000 per man-year for salary and other expenses, totaling $9.3 million. Federal, state and private agencies invested that sum over many years. Savings are estimated to be $308 million per year. After a variety or inbred line is developed, it may be grown successfully for about 10 years before it is discarded for biological, agronomical, or other reasons. Thus the cost of research $9.3), the annual savings ($308 million), and the 10 years use of a variety give a total net monetary value of about $3 billion or a 300:1 return on the investment. That value is based only on preventing damage and yield loss. It does not include bonus effects on subsequent crops by population control through eradication or suppression of insect pests and elimination of chemicals and their residues. Since so much has been achieved with so little money, a very large pay-off can be expected from modest but continued investments.

With presently increased concern about pollution and environment, simultaneous screening is needed to find resistance to diseases, insects, and nematodes among such diverse forage crops as legumes, small grains, and grasses so expenditures will have to be increased accordingly. Major items of investments for a thorough screening program include modern greenhouses and growth chambers, scanning electron microscopes, gas chromatographs, spectrophotometers, instruments for radioactive tracer technique, computers, etc.

For a long time resistance was attempted against only one insect at a time. To step up screening procedures, techniques are needed to screen for multiple resistance simultaneously in several crops against several insect pests, diseases, and nematodes.

One of the odds against resistance breeding is the pressure put on researchers to "publish or perish" because that favors short-term projects and "salami-sliced" publications. A handicap to overcome in the present organization of experiment stations is the lack of properly trained technicians and dependence on graduate students. Technicians could better do the tedious routine work of screening literally thousands of entries in the field, in the greenhouse and growth chamber. A well-trained technical staff also would have fewer interruptions and take better care of such expensive items as growth chambers and insect cultures.

Place of Resistant Forage Legumes in Pest Management Programs

It is important to stipulate for each pest management program economic injury levels for local areas and conditions (Ray F. Smith, 1967).

In some instances, rather high levels can be tolerated at least during certain periods in the cycle of legume development. Breeding highly resistant varieties may result in selecting new biotypes more aggressive to the resistant variety, as demonstrated by insecticide-resistant biotypes developed during continued use of highly potent insecticides. In both cases, selection pressure is high enough to screen the population for the most aggressive biotypes. With tolerant varieties, insect populations may reach densities high enough to allow outbreeding of the population and avoid new biotypes and thus make tolerant varieties more economical.

Growing a tolerant variety may not be sufficient by itself alone to solve a problem but combined with other means, e.g., biological control or a reduced spray program with specific insecticides, it may help avoid crop losses.

Significance of host plant resistance or tolerance in forage crops for pest management is obvious. It allows the host plant to maintain subeconomic levels of the pest species which in turn supports entomophagous forms. Low density populations of insects may be pivotal as a food supply for beneficial organisms needed later in the growth period in the same crop or in neighboring fields. The higher the economic injury level of a variety is, the higher the number of tolerated insects and the longer one can wait for natural enemies to take over without causing economic loss. Host plant tolerance is comparable to buying time for natural enemies. Neighboring crops often benefit from varieties tolerant to pea aphid or other insects that develop to high populations early in the season and, thus, attract and maintain high populations of beneficial ladybugs and syrphid flies. After each cutting of alfalfa fields, predators and parasites are forced to forage in neighboring crops, e.g., sorghum infested with greenbugs.

The defensive nature of tolerance should be emphasized because no mechanism so far has been recognized in insects to overcome or to fight back against this type of resistance, in contrast to antibiosis or nonpreference that often is overcome by new, aggressive biotypes. A grower may accept insects on a tolerant variety and see some damage during an outbreak. Even low to moderate resistance may permit a grower to omit treatments, start spraying later in the season, to protect natural enemies, or stop spraying well ahead of harvest and thus reduce residues on crops and in the environment. An example of the beneficial effect of moderate resistance or tolerance was provided by Campbell (1970) when he compared alfalfa varieties Team and Weevl-chek, both moderately tolerant to the alfalfa weevil, with the susceptible Cherokee. The tolerant varieties suffered less damage from a higher population than the susceptible variety (see Table 2). A control program integrating a moderately tolerant variety with an insecticide made it possible to delay spraying and omit one treatment (see Table 3), thus demonstrating that moderately resistant varieties are easier to protect with insecticides than are susceptible varieties.

Summary

Techniques and procedures were surveyed to evaluate and screen forage legumes, to develop insect resistant cultivars, to study modalities of resistance and factors determining its expression and persistence. Examples of successful development of cultivars resistant to several insect pests as well as diseases and nematodes were cited from alfalfa: seed chalcids (*Bruchophagus* spp.), alfalfa weevil (*Hypera postica* Gyllenhal), spotted alfalfa aphid (*Therioaphis maculata* Buckton), pea aphid (*Acyrthosiphon pisum* Harris), meadow spittlebug (*Phylaenus leucophthalmus* L.), potato leafhopper (*Empoasca fabae* Harris).

Resistance observed in various forage legumes to insect pests was discussed in red clover, alsike, white clover, vetches, and lupines to pea aphid; sweetclover to sweetclover weevil (*Sitona cylindricollis* Fahr.), sweetclover aphid (*Therioaphis riehmi* Boerner), and sweetclover rootborer (*Walshia miscecolorella* Chambers), and in several forage legumes to a migratory grasshopper (*Melanoplus sanguinipes* F.).

Examples of contributions to crop and environment improvement with insect resistant cultivars were given. Cost of developing insect resistance, and its place and value in a pest management program, were exemplified for specific cases.

Recurrent mass selection for developing multiple pest resistance and conserving diversity in germplasm, which has been effectively used in alfalfa, is also suggested for other legumes, along with a worldwide collection and recombination program.

There is still a pressing need for forage crops resistant to a host of insects and other pests. A rich diversity of genes remains in forage legumes as a largely untapped resource. It appears that the few concerted efforts to tap it in a cooperative approach to produce insect and pest resistant varieties has only scratched the surface. Since so much has been achieved with so little money, a very large pay-off can be expected from modest but continued investments in breeding multiple pest resistant forage legumes.

Table 2.--Alfalfa weevil damage on resistant (R) and on susceptible (S) alfalfa (Campbell, 1970) compared.

Alfalfa variety	No. larvae April 21	Alfalfa ht. (in.) April 29	% weevil damage May 5
Team (R)[a]	675.0	16.3	41.7
Weevl-chek (R)[b]	822.0	13.0	70.0
Cherokee (S)	511.7	12.0	97.7

[a]Developed by USDA, N.C., MD., and Va.

[b]Developed by Farmers Forage Research.

Table 3.--Integrated control with weevil-resistant alfalfa + insecticide (Campbell, 1970).

Variety and chemical	Lb. Al/a	% Weevil damage
Cherokee (Susc.)		
Methoxychlor (2x)	1.5	8.3
M + M[1] (2x)	1.5	8.3
Control	--	90.0
Team (Res.)		
Methoxychlor (1x)	1.5	11.7
M + M[1] (1x)	1.5	10.0
Control	--	48.3

[1] Malathion + methoxychlor.

References Cited

Aamodt, O. S. and J. Carlson. 1938. Grimm alfalfa flowers in spite of lygus injury. Wisconsin Agr. Exp. Sta. Bul. 440, Pt. 11:67.

Akeson, W. R., G. R. Manglitz, H. J. Gorz and F. A. Haskins. 1967. A bioassay for detecting compounds which stimulate or deter feeding by the sweet-clover weevil. J. Econ. Entomol. 60: 1082-1084.

Akeson, W. R., F. A. Haskins, H. J. Gorz and G. R. Manglitz. 1968. Water-soluble factors in Melilotus leaves which influence feeding by the sweetclover weevil. Crop Sci. 8: 574-576.

Akeson, W. R., G. L. Beland, F. A. Haskins and H. J. Gorz. 1969a. Influence of developmental stage on the resistance of Melilotus infesta to feeding by the sweetclover weevil. Crop. Sci. 9: 667-669.

Akeson, W. R., H. J. Gorz and F. A. Haskins. 1969b. Sweetclover weevil feeding stimulants: isolation and identification of glucose, fructose, and sucrose. Crop. Sci. 9: 810-812.

Akeson, W. R., G. L. Beland and G. R. Manglitz. 1969c. Nitrate as a deterrent to feeding by the sweetclover weevil. J. Econ. Entomol. 62: 1169-1172.

Akeson, W. R., F. A. Haskins and H. J. Gorz. 1969d. Sweetclover weevil feeding deterrent B: isolation and identification. Science 163: 293-294.

Akeson, W. R., F. A. Haskins and H. J. Gorz. 1970. Feeding response of the sweetclover weevil to various sugars and related compounds. J. Econ. Entomol. 63: 1079-1080.

Akey, D. H. and S. D. Beck. 1971. Continuous rearing of the pea aphid, Acyrthosiphon pisum on a holidic diet. Ann. Econ. Entomol. 64: 353-356.

Albrecht, H. R. 1940. Species and variety differences in resistance to aphid injury in vetch. J. Econ. Entomol. 33: 833-834.

Albrecht, H. R. and T. R. Chamberlain. 1941. Instability of resistance to aphids in some strains of alfalfa. J. Econ. Entomol. 34: 551-554.

Anonymous. 1969. The spotted alfalfa aphid--how to control it. USDA Leaflet 422 (revised): 8 pp.

Applebaum, S. W., J. Marco and Y. Birk. 1969. Saponins as possible factors of resistance of legume seeds to the attack of insects. J. Agr. Food Chem. 17: 618-620.

Auclair, J. L. 1966. Dissimilarities in the biology of the pea aphid Acyrthosiphon pisum (Homoptera: Aphidae), on alfalfa and peas in New Mexico. Ann. Entomol. Soc. Amer. 59: 780-786.

Barnes, O. L. 1963. Resistance o f Moapa alfalfa to the spotted alfalfa aphid in commercial-size fields in south-central Arizona. J. Econ. Entomol. 56: 84-85.

Barnes, D. K., C. H. Hanson, R. H. Ratcliffe, T. H. Busbice, J. A. Schillinger, G. R. Buss, W. V. Campbell, R. W. Hemken, and C. C. Blickenstaff. 1970. The development and performance of Team alfalfa, a multiple pest resistant alfalfa with moderate resistance to the alfalfa weevil. USDA ARS 34-115. 38 pp.

Beland, G. L., W. R. Akeson and G. R. Manglitz. 1970. Influence of plant maturity and plant part on nitrate content of the sweetclover weevil resistant species Melilotus infesta. J. Econ. Entomol. 63: 1037-1039.

Blanchard, R. A. and J. E. Dudley, Jr. 1934. Alfalfa plants resistant to the pea aphid. J. Econ. Entomol. 27: 262-264.

Blanchard, R. A. 1943. Insect resistance in forage plants. J. Amer. Soc. Agron. 35: 716-724.

Blickenstaff, C. C. 1965. Partial intersterility of eastern and western United States strain of the alfalfa weevil. Ann. Entomol. Soc. Amer. 58: 523-526.

Booth, G. M. 1969. Use of uric acid analysis to evaluate seed chalcid infestation in alfalfa seed. Ann. Entomol. Soc. Amer. 62: 1379-1382.

Busbice, T. H., D. K. Barnes, C. H. Hanson, R. R. Hill, Jr., W. V. Campbell, C. C. Blickenstaff and R. C. Newton. 1967. Field evaluation of alfalfa introductions for resistance to the alfalfa weevil, Hypera postica (Gyllenhal). USDA ARS 34-94. 13 pp.

Campbell, W. V. and J. W. Dudley. 1965. Differences among Medicago species in resistance to oviposition by the alfalfa weevil. J. Econ. Entomol. 58: 245-248.

Campbell, W. V. 1970. Alfalfa weevil investigations. Ann. Rept. North Carolina State University. 12 pp.

Carnahan, H. L., R. N. Peaden, F. V. Lieberman and R. K. Peterson. 1963. Differential reactions of alfalfa varieties and selections to the pea aphid. Crop Sci. 3: 219-222.

Cartier, J. J. 1957. Variations du poids des adultes virginipares apteres de deux races du puceron du pois, Acyrthosiphon pisum (Harris) (i.e., Macrosiphum pisi) au cours de 44 generations. Entomol. Soc. Quebec Ann. 2: 37-41.

Cartier, J. J. 1959. Recognition of three biotypes of the pea aphid from southern Quebec. J. Econ. Entomol. 52: 293-294.

Cartier, J.J. 1960. Growth, reproduction, and longevity in one biotype of the pea aphid, Acyrthosiphon pisum (Harr.) (Homoptera: Aphididae). Canadian Entomol. 92: 762-764.

Cartier, J. J. 1964. Les races du puceron du pois. Canadian Entomol. 96: 107-108.

Cartier, J. J. and J. L. Auclair. 1965. Effets des couleurs sur le comportement de diverses races du puceron du pois, Acyrthosiphon pisum (Harris), en elevage sur un regime nutritif de composition chimique connue. Proc. XII Int. Cong. Entomol. London 1964: 414.

Cartier, J. J., A. Isaak, R. H. Painter and E. L. Sorensen. 1965. Biotypes of pea aphid Acyrthosiphon pisum (Harris) in relation to alfalfa clones. Canadian Entomol. 97: 754-760.

Connin, R. V., H. J. Gorz and C. O. Gardner. 1958. Greenhouse techniques for evaluating sweetclover weevils' preference for seedling sweetclover plants. J. Econ. Entomol. 51: 190-193.

Dadd, R. H. 1968. Dietary amino acids and wing determination in the aphid, Myzus persicae. Ann. Entomol. Soc. Amer. 61: 1201-1210.

Dadd, R. H. and T. E. Mittler. 1966. Permanent culture of an aphid on a totally synthetic diet. Experientia 22: 832.

Dadd, R. H. and D. L. Krieger. 1967. Continuous rearing of aphids of the Aphis fabae complex on sterile synthetic diet. J. Econ. Entomol. 60: 1512-1514.

Dadd, R. H. and D. L. Krieger. 1968. Dietary amino acid requirements of the aphid Myzus persicae. J. Insect Physiol. 14: 741-764.

Dahms, R. G. and R. H. Painter. 1940. Rate of reproduction of the pea aphid on different alfalfa plants. J. Econ. Entomol. 33: 482-485.

Daniels, N. E. 1960. Field studies of spotted alfalfa aphid resistance. Texas Agr. Exp. Sta. Prog. Rept. 2153. 4 pp.

Davis, R. L. and M. C. Wilson. 1953. Varietal tolerance of alfalfa to the potato leafhopper. J. Econ. Entomol. 46: 242-245.

Dexter, S. T. 1941. A comparison of Hardigan and Ladak alfalfa in their reactions to leafhopper infestations. J. Amer. Soc. Agron. 33: 947-951.

Dogger, J. R. and C. H. Hanson. 1963. Reaction of alfalfa varieties and strains to alfalfa weevil. J. Econ. Entomol. 56: 192-197.

Dudley, J. W., R. R. Hill and C. H. Hanson. 1963. Effects of seven cycles of recurrent phenotypic selection on means and genetic variances of several characters in two pools of alfalfa germplasm. Crop Sci. 3: 543-546.

Ehrhardt, P. 1968a. Die Wirkung verschiedener Spurenelemente auf Wachstum, Reproduktion and Symbionten von Neomyzus circumflexus Buckt. (Aphididae, Homoptera, Insecta) bei kunstlicher Ernahrung. Z. Vergl. Physiol. 58: 47-75.

Ehrhardt, P. 1968b. Nachweis einer durch symbiontische Mikroorganismen bewirkten Sterin-synthese in kunstlich ernahrten Aphiden (Homoptera, Rhynchota, Insecta) Experientia 24: 82-83.

Farrar, N. D. and C. M. Woodworth. 1939. New strains of alfalfa studied for leafhopper resistance. Illinois Agr. Exp. Sta. Ann. Rept. 50: 167-168.

Glover, C. and B. A. Melton. 1966. Inheritance patterns of spotted alfalfa aphid resistance in Zia plants. New Mexico Agr. Exp. Sta. Res. Rept. 127. 4 pp.

Glover, D. V. and E. H. Stanford. 1966. Tetrasomic inheritance of resistance in alfalfa to the pea aphid. Crop Sci. 6: 161-165.

Graber, L. F. 1941. Recovery after cutting and differentials in the injury of alfalfa by leafhoppers (*E. fabae*). J. Amer. Soc. Agron. 33: 181-183.

Graham, H. M. 1959. Effects of temperature and humidity on the biology of *Therioaphis maculata* (Buckton). Univ. California Publ. Entomol. 16: 47-80.

Gregory, L. H. 1917. The effect of starvation on wing development of *Macrosiphum destructor*. Biol. Bull. 33: 296-303.

Gross, A. T. H. and G. A. Stevenson. 1964. Resistance in *Melilotus* species to the sweetclover weevil. Canadian J. Plant Sci. 44: 487-488.

Guthrie, W. D., E. S. Raun, F. F. Dicke, G. R. Pesho and J. W. Carter. 1965. Laboratory production of European corn borer egg masses. Iowa State J. Sci. 40: 65-83.

Hackerott, H. L. and T. L. Harvey. 1959. Effect of temperature on spotted alfalfa aphid reaction to resistance in alfalfa. J. Econ. Entomol. 52: 949-953.

Hackerott, H. L., T. L. Harvey, E. L. Sorensen and R. H. Painter. 1958. Varietal differences in survival of alfalfa seedlings infested with spotted alfalfa aphids. Agron. J. 50: 139-141.

Hackerott, H. L., E. L. Sorensen, T. L. Harvey, E. E. Ortman and R. H. Painter. 1963. Reactions of alfalfa varieties to pea aphids in the field and greenhouse. Crop Sci. 3: 298-301.

Hans, H. and A. T. Thorsteinson. 1961. The influence of physical factors and host plant odor on the induction and termination of dispersal flight in *Sitona cylindricollis* Fahr. Entomol. Expl. & Appl. 4: 165-177.

Hanson, C. H. 1961. Moapa alfalfa pays off. Crops & Soils 13: 11-12.

Hanson, C. H., G. O. Kohler, J. W. Dudley, E. L. Sorensen, G. R. Van Atta, K. W. Taylor, M. W. Pedersen, H. L. Carnahan, C. P. Wilsie, W. R. Kehr, C. C. Lowe, E. H. Standford, and T. A. Yungen. 1963. Saponin content of alfalfa as related to location, cutting, variety and other variables. USDA-ARS Crops Res. 34-44: 37.

Hanson, E. W., C. H. Hanson, F. I. Frosheiser, E. L. Sorensen, R. T. Sherwood, J. H. Graham, L. J. Elling, D. Smith and R. L. Davis. 1964. Reactions of varieties, crosses, and mixtures of alfalfa to six pathogens and the potato leafhopper. Crop Sci. 4: 273-276.

Hanson, C. H., T. H. Busbice, R. R. Hill, Jr., O. J. Hunt and A. J. Oakes. 1972. Directed mass selection in alfalfa for obtaining improved performance and retaining genetic diversity. J. Environ. Qual. 1: 106-111.

Harpaz, I. 1955. Bionomics of *Therioaphis maculata* (Buckton) in Israel. J. Econ. Entomol. 48: 668-671.

Harvey, T. L. and H. L. Hackerott. 1956. Apparent resistance to the spotted alfalfa aphid selected from seedlings of susceptible alfalfa varieties. J. Econ. Entomol. 49: 289-291.

Harvey, T. L. and H. L. Hackerott. 1967. Use of parasitized pea aphids to evaluate alfalfa for resistance. J. Econ. Entomol. 60: 573-575.

Harvey, T. L., H. L. Hackerott and E. L. Sorensen. 1971. Pea aphid injury to resistant and susceptible alfalfa in the field. J. Econ. Entomol. 64: 513-517.

Harvey, T. L., H. L. Hackerott, E. L. Sorensen, R. H. Painter, E. E. Ortman and D. C. Peters. 1960. The development and performance of Cody alfalfa, a spotted alfalfa aphid resistant variety. Kansas Agr. Exp. Sta. Tech. Bul. 114. 27 pp.

Hewitt, G. B. 1969. Twenty-six varieties of forage crops evaluated for resistance to feeding by Melanoplus sanguinipes F. Ann. Entomol. Soc. Amer. 62: 737-741.

Hill, R. R., Jr., C.H. Hanson, and T. H. Busbice. 1969. Effect of four recurrent selection programs on two alfalfa populations. Crop Sci. 9: 363-365.

Hill, R. R. and R. C. Newton. 1972. A method for mass screening alfalfa for meadow spittlebug resistance in the greenhouse during the winter. J. Econ. Entomol. 62: 621-623.

Horber, E. 1965. Isolation of components from the roots of alfalfa (Medicago sativa L.) toxic to white grubs (Melolontha vulgaris F.). Int. Cong. Entomol. XII Proc. pp. 540-541. 1964.

Horber, E. 1972. Alfalfa saponins significant in resistance to insects. pp. 611-627. In Insect and Mite Nutrition (ed. J. G. Rodriguez). North-Holland-Amsterdam. p. 702.

Horber, E., K. T. Leath, B. Berrang, V. Marcarian and C. H. Hanson. 1974. Biological activities of saponin components from DuPuits and Lahontan alfalfa. Entomol. Expl. & Appl. 17 (in press).

Howe, W. L. and G. R. Manglitz. 1961. Observations on the clover seed chalcid as a pest of alfalfa in eastern Nebraska. Proc. North Central Branch, Entomol. Soc. Amer. 16: 49-51.

Howe, W. L. and G. R. Pesho. 1960a. Influence of plant age on the survival of alfalfa varieties differing in resistance to the spotted alfalfa aphid. J. Econ. Entomol. 53: 142-144.

Howe, W. L. and G. R. Pesho. 1960b. Spotted alfalfa aphid resistance in mature growth of alfalfa varieties. J. Econ. Entomol. 53: 234-239.

Howe, W. L. and O. F. Smith. 1957. Resistance to the spotted alfalfa aphid in Lahontan alfalfa. J. Econ. Entomol. 50: 320-324.

Howe, W. L., W. R. Kehr, M. E. McKnight and G. R. Manglitz. 1963. Studies of the mechanisms and sources of spotted alfalfa aphid resistance in Ranger alfalfa. Nebraska Agr. Exp. Res. Bull. 210. 22 pp.

Hsiao, T. H. 1969. Chemical basis of host selection and plant resistance in oligophagous insects. Proc. 2nd Int. Symp. "Insect and Host Plant," Wageningen (1969) 777-788.

Huggans, J. L. and C. C. Blickenstaff. 1964. Effects of photoperiod on sexual development in the alfalfa weevil. J. Econ. Entomol. 57: 167-168.

Hunt, O. J., R. N. Peaden, M. W. Nielson and C. H. Hanson. 1971. Development of two alfalfa populations with resistance to insect pests, nematodes and diseases. I. Aphid resistance. Crop Sci. 11: 73-75.

Isaak, A., E. L. Sorensen and E. E. Ortman. 1963. Influence of temperature and humidity on resistance in alfalfa to the spotted alfalfa aphid and the pea aphid. J. Econ. Entomol. 56: 53-57.

Isaak, A., E. L. Sorensen and R. H. Painter. 1965. Stability of resistance to pea aphid and spotted alfalfa aphid in several alfalfa clones under various temperature regimes. J. Econ. Entomol. 58: 140-143.

Jarvis, J. L. and W. R. Kehr. 1964. Evaluating alfalfa for resistance to the potato leafhopper. Proc. North Central Branch Entomol. Soc. Amer. 19: 64-5.

Jarvis, J. L. and W. R. Kehr. 1966. Population counts vs. nymphs per gram of plant material in determining degree of alfalfa resistance to the potato leafhopper. J. Econ. Entomol. 59: 427-430.

Jewett, H. H. 1929. Leafhopper injury to clover and alfalfa. Kentucky Agr. Exp. Sta. Bull. 293: 157-172.

Jewett, H. H. 1932. The resistance of certain red clovers and alfalfas in leafhopper injury. Kentucky Agr. Exp. Sta. Bull. 329: 157-172.

Jewett, H. H. 1941. Resistance of strains of red clover to pea aphid injury. Kentucky Agr. Exp. Sta. Bull. 412: 42-55.

Johnson, B. and P. R. Birks. 1960. Studies on wing polymorphism in aphids. I. The developmental process involved in the production of the different forms. Entomol. Exp. & Appl. 3: 327-339.

Johnson, B. 1965. Wing polymorphism in aphids. II. Interaction between aphids. Entomol. Exp. & Appl. 8: 49-64.

Johnson, B. 1966. Wing polymorphism in aphids. III. The influence of the host plant. Entomol. Exp. & Appl. 9: 213-222.

Jones, B. F., E. L. Sorensen and R. H. Painter. 1968. Tolerance of alfalfa clones to the spotted alfalfa aphid. J. Econ. Entomol. 61: 1046-1050.

Jones, L. G., F. N. Briggs and R. A. Blanchard. 1950. Inheritance of resistance to the pea aphid in alfalfa hybrids. Hilgardia 20: 9-17.

Kamm, J. A. and W. D. Fronk. 1964. Olfactory response of the alfalfa-seed chalcid Bruchophagus roddi Guss., to chemicals found in alfalfa. Wyoming Agr. Exp. Sta. Bull. 413. 36 pp.

Kehr, W. R., G. R. Manglitz and R. L. Ogden. 1968. Dawson alfalfa--a new variety resistant to aphids and bacterial wilt. Nebraska Agr. Exp. Sta. Bull. 497. 23 pp.

Kehr, W. R., R. L. Ogden and S. D. Kindler. 1970. Diallel analysis of potato leafhopper injury to alfalfa. Crop Sci. 10: 584-586.

Kindler, S. D. and R. Staples. 1969. Behavior of the spotted alfalfa aphid on resistant and susceptible alfalfas. J. Econ. Entomol. 62: 474-478.

Kindler, S. D. and R. Staples. 1970a. Nutrients and the reaction of two alfalfa clones to the spotted alfalfa aphid. J. Econ. Entomol. 63: 938-940.

Kindler, S. D. and R. Staples. 1970b. The influence of fluctuating and constant temperatures, photoperiod, and soil moisture on the resistance of alfalfa to the spotted alfalfa aphid. J. Econ. Entomol. 63: 1198-1201.

Kindler, S. D. and W. R. Kehr. 1970. Field tests of alfalfa selected for resistance to potato leafhopper in the greenhouse. J. Econ. Entomol: 63: 1464-1467.

Kindler, S. D., W. R. Kehr and R. L. Ogden. 1971. Influence of pea aphids and spotted alfalfa aphids on the stand, yield of dry matter, and chemical composition of resistant and susceptible varieties of alfalfa. J. Econ. Entomol. 64: 653-657.

Kirchner, H. W., R. L. Misiorowski and F. L. Lieberman. 1970. Resistance of alfalfa to the spotted alfalfa aphid. J. Econ. Entomol. 63: 964-969.

Kishaba, A. N. and G. R. Manglitz. 1965. Nonpreference as a mechanism of sweetclover and alfalfa resistance to the sweetclover aphid and the spotted alfalfa aphid. J. Econ. Entomol. 58: 566-569.

Kishaba, A. N. and G. R. Manglitz. 1968. Substances from alfalfa biologically active against the spotted alfalfa aphid. USDA ARS 33-126. 12 pp.

Klement, W. J. and N. M. Randolph. 1960. The evaluation of resistance of seedling alfalfa varieties and strains to the spotted alfalfa aphid, *Therioaphis maculata*. J. Econ. Entomol. 53: 667-669.

Lees, A. D. 1967. The production of the apterous and alate forms in the aphid *Megoura viciae* Buckton, with special reference to the role of crowding. J. Insect Physiol. 13: 289-318.

Lindahl, T. L., W. P. Shalkop, R. W. Dougherty and others. 1957. Alfalfa saponins. Studies on their chemical, pharmacological, and physiological properties in relation to ruminant bloat. USDA Tech. Bul. 1161. 83 pp.

Lindquist, R. K. and E. L. Sorensen. 1970. Interrelationships among aphids, tarnished plant bugs and alfalfas. J. Econ. Entomol. 63: 192-195.

Lindquist, R. K., R. H. Painter and E. L. Sorensen. 1967. Screening alfalfa seedlings for resistance to the tarnished plant bug. J. Econ. Entomol. 60: 1442-1445.

Lindquist, R. K., E. L. Sorensen and R. H. Painter. 1968. Insect age and sex effects on screening alfalfa seedlings for resistance to *Lygus lineolaris*. Proc. North Central Branch, Entomol. Soc. Amer. 23: 41.

Loper, G. M. 1968. Effect of aphid infestation on the coumestrol content of alfalfa varieties differing in aphid resistance. Crop Sci. 8: 104-106.

Luginbill, P. 1969. Developing resistant plants. The ideal method of controlling insects. USDA Prod. Res. Rept. 111. 14 pp.

Malcolm, D. R. 1953. Host relationship studies of *Lygus* in south-central Washington. J. Econ. Entomol. 46: 485-488.

Manglitz, G. R. and H. J. Gorz. 1961. Resistance of sweetclover to the aphid. J. Econ. Entomol. 54: 1156-1160.

Manglitz, G. R. and H. J. Gorz. 1963. Sources of resistance to the sweetclover aphid in introduced species of *Melilotus*. USDA ARS-33-86. 8 pp.

Manglitz, G. R. and H. J. Gorz. 1964. Host range studies with the sweetclover weevil and the sweetclover aphid. J. Econ. Entomol. 51: 683-687.

Manglitz, G. R., W. R. Kehr and C. O. Calkins. 1962. Pea aphid resistant alfalfa now in sight. Nebraska Exp. Sta. Quar. 24: 5-6.

Manglitz, G. R., C. O. Calkins, R. J. Walstrom, S. D. Hintz, S. D. Kindler and L. L. Peters. 1966. Holocyclic strain of the spotted alfalfa aphid in Nebraska and adjacent states. J. Econ. Entomol. 59: 636-639.

Manglitz, G. R. and J. M. Schalk. 1970. Occurrence and hosts of *Aphelinus semiflavus* Howard (Hymenoptera: Eulophidae) in Nebraska. J. Kansas Entomol. Soc. 43: 309-314.

Manglitz, G. R., H. J. Gorz and H. T. Stevens, Jr. 1971. Biology of the sweet-clover root borer. J. Econ. Entomol. 64: 1154-1158.

Manglitz, G. R., H. J. Gorz, and F. A. Haskins. 1971. Resistance to the sweet-clover root borer as influenced by species and coumarin content. Special Rept. X-324 ERD-ARS-USDA. 8 pp.

Marble, V. L., J. C. Meldoen, H. C. Murray and F. P. Scholle. 1959. Studies on free amino acids in the spotted alfalfa aphid, its honeydew and several alfalfa selections in relation to aphid resistance. Agron. J. 51: 740-743.

Markkula, M. 1963. Studies on the pea aphid, *Acyrthosiphon pisum* Harris (Hom., Aphididae) with special reference to the differences in the biology of the green and red forms. Ann. Agr. Fenniae 2: 3-30.

Markkula, M. and K. Roukka. 1970. Resistance of plants to the pea aphid *Acyrthosiphon pisum* Harris (Hom., Aphididae). II. Fecundity on different red clover varieties. Ann. Agr. Fenniae 9: 304-308.

Markkula, M. and K. Roukka. 1971. Resistance of plants to the pea aphid *Acyrthosiphon pisum* Harris (Hom., Aphididae). IV. Fecundity on different alsike and white clover varieties. Ann. Agr. Fenniae 10: 111-113.

McMurtry, J. A. 1962. Resistance of alfalfa to spotted alfalfa aphid in relation to environmental factors. Hilgardia 32: 501-539.

McMurtry, J. A. and E. H. Stanford. 1960. Observations of feeding habits of the spotted alfalfa aphid on resistant and susceptible alfalfa plants. J. Econ. Entomol. 53: 714-717.

Mayome, B. 1963. Saponine delle leguminose e loro effeto nell'-alimentazione animale. Annale Della Sperimentazione Agrarie N.S. 17, 1-2, 2-20.

Meier, W. 1964. Uber einen Caudalhaarindex zur Charakterisierung von Klonen der Erbsenblattlaus *Acyrthosiphon pisum* Harris. Mitt. Schweiz. Entomol. Ges. 37: 1-41.

Melton, B. A. 1968. Mesilla alfalfa. New Mexico Agr. Exp. Sta. Bull. 530. 8 pp.

Messenger, P. S. 1964. The influence of rhythmically fluctuating temperatures on the development and reproduction of the spotted alfalfa aphid, *Therioaphis maculata*. J. Econ. Entomol. 57: 76.

Mittler, T. E. 1967a. Gustation of dietary amino acids by the aphid *Myzus persicae*. Entomol. Exp. & Appl. 10: 87-96.

Mittler, T. E. 1967b. Effect of amino acid and sugar concentrations on the food uptake of the aphid Myzus persicae. Entomol. Exp. & Appl. 10: 39-51.

Moore, G. D. 1968. Evaluation of feeding injury to alfalfa by the potato leafhopper, Empoasca fabae (Harris). Ph.D. Thesis, University of Minnesota. 119 pp.

Muller, F. P. 1962. Biotypen and Unterarten der "Erbsenlaus" Acyrthosiphon pisum (Harris). Z. Pfl Krankh. Pfl Pathol. Pfl Schutz 69: 129-136 (English sum.)

Newton, R. C. and D. K. Barnes. 1965. Factors affecting resistance of selected alfalfa clones to the potato leafhopper. J. Econ. Entomol. 58: 435-439.

Newton, R. B., R. R. Hill, Jr. and T. H. Elgin, Jr. 1970. Differential injury to alfalfa by male and female potato leafhoppers. J. Econ. Entomol. 63: 1077-1079.

Nielson, M. W. 1957. Sampling technique studies on the spotted alfalfa aphid. J. Econ. Entomol. 50: 385-389.

Nielson, M. W. 1967. Procedures for screening and testing alfalfa for resistance to the alfalfa seed chalcid. USDA ARS 33-120. 10 pp.

Nielson, M. W. and W. E. Currie. 1959. Effect of alfalfa variety on the biology of the spotted alfalfa aphid in Arizona. J. Econ. Entomol. 52: 1023.

Nielson, M. W. and M. H. Schonhorst. 1965a. Research on alfalfa seed chalcid resistance in alfalfa. Prog. Agr. Arizona 17: 20-21.

Nielson, M. W. and M. H. Schonhorst. 1965b. Screening alfalfas for resistance to some common insect pests in Arizona. J. Econ. Entomol. 58: 147-150.

Nielson, M. W. and M. H. Schonhorst. 1967. Sources of alfalfa seed chalcid resistance in alfalfa. J. Econ. Entomol. 60: 1506-1511.

Nielson, M. W., M. H. Schonhorst and H. Don. 1969. Arizona alfalfas resist new strain of the spotted alfalfa aphid. Prog. Agr. Arizona 21: 18-19.

Nielson, M. W., W. F. Lehman and V. L. Marble. 1970a. A new severe strain of the spotted alfalfa aphid in California. J. Econ. Entomol. 63: 1489-1491.

Nielson, M. W., H. Don, M. H. Schonhorst, W. F. Lehman and V. L. Marble. 1970b. Biotypes of the spotted alfalfa aphid in western United States. J. Econ. Entomol. 63: 1822-1825.

Nielson, M. W., M. H. Schonhorst, H. Don, W. F. Lehman and V. L. Marble. 1971. Resistance in alfalfa to four biotypes of the spotted alfalfa aphid. J. Econ. Entomol. 64: 506-510.

Ortman, E. E. 1965. A study of the free amino acids in the spotted alfalfa aphid and pea aphid and their honey dew in relation to the fecundity and honey dew excretion of aphids feeding on a range of resistant and susceptible alfalfas. Diss. Abstr. 25: 4315-4316.

Ortman, E. E., E. L. Sorensen, R.H. Painter, T. L. Harvey and H. L. Hackerott. 1960. Selection and elevation of pea aphid-resistant alfalfa plants. J. Econ. Entomol. 53: 881-887.

Padilla, A. R. and W. R. Young. 1958. El pulgon manchado de la alfalfa en Mexico, Therioaphis (Pterocallidium) maculata (Buckton). Mexican Sec. de Agr. y Ganad. Fol. Tec. 25. 32 pp.

Painter, R. H. 1951. Insect resistance in crop plants. Macmillan Co., New York. 520 pp.

Painter, R. H. 1966. Plant resistance as a means of controlling insects and reducing their damage. USDA ARS 33-110: 138-148.

Painter, R. H. and C. O. Grandfield. 1935. Preliminary report on resistance of alfalfa varieties to the pea aphid. Agron. J. 27: 617-674.

Painter, R. H., E. L. Sorensen, T. L. Harvey and H. L. Hackerott. 1965. Selection for combined resistance in alfalfa, Medicago sativa L., to pea aphid, Acyrthosiphon pisum (Harris) and spotted alfalfa aphid, Therioaphis maculata (Buckton). Int. Cong. Entomol. XII Proc. p. 531. 1964.

Paschke, J. D. 1959. Production of the agamic form of the spotted alfalfa aphid, Therioaphis maculata (Buckton) (Homoptera: Aphidae). Univ. California Publ. Entomol. 16: 125-180.

Peaden, R. N., H. L. Carnahan, O. J. Hunt and F. V. Lieberman. 1966. Washoe alfalfa. Nevada Agr. Exp. Sta. Cir. 64. 6 pp.

Pesho, G. R., F. V. Lieberman and W. F. Lehman. 1960. A biotype of the spotted alfalfa aphid on alfalfa. J. Econ. Entomol. 53: 146-150.

Peters, D. C. and R. H. Painter. 1957. A general classification of available small seeded legumes as hosts for three aphids of the "Yellow Clover Aphid Complex." J. Econ. Entomol. 50: 231-235.

Peters, D. C. and R. H. Painter. 1958. Studies on the biologies of three related legume aphids in relation to their host plants. Agr. Exp. Sta. Tech. Bul. 93. 44 pp.

Pitre, H. N., V. H. Watson and C. Y. Ward. 1970. Field evaluation of alfalfa cultivars for resistance to the alfalfa weevil in Mississippi--a preliminary study. Agron. J. 62: 678-679.

Radcliffe, E. B. and F. G. Holdaway. 1964. Sweetclover resistance to weevil attack. Minnesota Farm Home Sci. 22: 5-7.

Radcliffe, E. B. and D. K. Barnes. 1970. Alfalfa plant bug injury and evidence of plant resistance in alfalfa. J. Econ. Entomol. 63: 1995-1996.

Roof, M., E. Horber and E. L. Sorensen. 1972. Bioassay technique for the potato leafhopper, Empoasca fabae (Harris). Proc. North Central Branch Entomol. Soc. Amer. 27: 140-143.

Rowley, W. A. and B. A. Haws. 1964. Studies on alfalfa resistance to the seed chalcid Bruchophagus roddi Gussakovsky (Abstract). Utah Acad. Sci. Arts & Letters Proc. 41: 150.

Sandmeyer, E. E., O. T. Hunt, W. H. Arnett and C. P. Heisler. 1971. Relative resistance of six selected alfalfa clones to the pea aphid and spotted alfalfa aphid. J. Econ. Entomol. 64: 155-162.

Schalk, J. M. and G. R. Manglitz. 1969. Migration of an holocyclic strain of the spotted alfalfa aphid into Nebraska. J. Econ. Entomol. 62: 946-947.

Schalk, J. M., S. D. Kindler and G. R. Manglitz. 1969. Temperature and the preference of the spotted alfalfa aphid for resistant and susceptible alfalfa plants. J. Econ. Entomol. 62: 1000-1003.

Schillinger, J. A., F. C. Elliott and R. F. Ruppel. 1964. A method for screening alfalfa plants for potato leafhopper resistance. Michigan Agr. Exp. Sta. Bull. 46: 512-517.

Schroder, R. F. W., R. H. Ratcliffe, H. D. Byrne and C. C. Blickenstaff. Rearing the alfalfa weevil in the laboratory (in preparation).

Shaw, M. J. P. 1970. Effect of population density on alienicolae of _Aphis fabae_ Scop. I. The effect of crowding on the production of alatae in the laboratory. Ann. Appl. Biol. 65: 191-196.

Smith, O. F. 1955. Breeding alfalfa for resistance to bacterial wilt and the stem nematode. Nevada Agr. Exp. Sta. Bull. 188. 15 pp.

Smith, O. F. 1958. Lahontan alfalfa. Nevada Agr. Exp. Sta. Bull. 14, 5 pp.

Smith, O. F. and R. N. Peaden. 1960. A method of testing alfalfa plants for resistance to the pea aphid. Agron. J. 52: 609-610.

Smith, O. F., R. N. Peaden and R. K. Petersen. 1958. Moapa alfalfa. Nevada Agr. Exp. Sta. Cir. 15. 4 pp.

Smith, R. F. 1967. Recent developments in integrated control. 4th British Insecti. & Fungi. Conf. 2: 464-471.

Sorensen, E. L., R. H. Painter, E. E. Ortman, H. L. Hackerott and T. L. Harvey. 1961. Cody alfalfa. Kansas Agr. Exp. Sta. Cir. 381. 10 pp.

Sorensen, E. L., T. L. Harvey and H. L. Hackerott. 1969a. New alfalfa unappetizing to pea aphid. Crops and Soils 21: 22.

Sorensen, E. L., R. H. Painter, H. L. Hackerott and T. L. Harvey. 1969b. Registration of Kanza alfalfa. Crop Sci. 9: 487.

Sorensen, E. L., M. C. Wilson and G. R. Manglitz. 1972. Breeding for insect resistance. Chapter 17, pp. 371-390. _In_ C. H. Hanson (ed.): Alfalfa Science & Technology. p. 812. #15 Ser. Agron. Amer. Soc. Agron., Madison.

Stanford, E. H. and J. A. McMurtry. 1959. Indications of biotypes of the spotted alfalfa aphid. Agron. J. 51: 430-431.

Strong, F. E. 1960. Sampling alfalfa seed for clover seed chalcid damage. J. Econ. Entomol. 53: 611-615.

Strong, F. E. 1962a. Laboratory studies on the biology of the alfalfa seed chalcid _Bruchophagus roddi_ Guss. (Hymenoptera: Eurytomidae). Hilgardia 32: 229-249.

Strong, F. E. 1962b. The reaction of some alfalfas to seed chalcid infestations. J. Econ. Entomol. 55: 1004-1005.

Sutherland, O. R. W. 1967. Role of host plant in production of winged forms by a green strain of pea aphid, _Acyrthosiphon pisum_ Harris. Nature. 216: 387-88.

Sutherland, O. R. W. 1969a. The role of crowding in the production of winged forms by two strains of the pea aphid, _Acyrthosiphon pisum_. J. Insect Physiol. 15: 1385-1410.

Sutherland, O. R. W. 1969b. The role of the host plant in the production of winged forms by two strains of the pea aphid, _Acyrthosiphon pisum._ J. Insect Physiol. 15: 2179-2201.

Taylor, N. L. 1956. Pubescence inheritance and leafhopper resistance relationships in alfalfa. Agron. J. 48: 78-81.

Thomas, J. G., E. L. Sorensen, and R. H. Painter. 1966. Attached vs. excised trifoliolates for evaluation of resistance in alfalfa to the spotted alfalfa aphid. J. Econ. Entomol. 59: 444-448.

Thorsteinson, A.J. 1958. The chemotactic influence of plant constituents on feeding by phytophagous insects. Entomol. Exp. & Appl. 1: 23-27.

Thorsteinson, A. J. 1960. Host selection in phytophagous insects. Ann. Rev. Entomol. 5: 193-218.

Toba, H. H., J. D. Paschke and S. Friedman. 1967. Crowding as the primary factor in the production of the agamic alate form of Therioaphis maculata (Homop: Aphididae). J. Insect Physiol. 13: 381-396.

VanDenburgh, R. S., B. L. Norwood, C. C. Blickenstaff and C.H. Hanson. 1966. Factors affecting resistance of alfalfa clones to adult feeding and oviposition of the alfalfa weevil in the laboratory. J. Econ. Entomol. 59: 1193-1198.

Watts, J. G., C. B. Coleman and C. R. Glover. 1967. Colorimetric detection of chalcid-infested alfalfa seed. J. Econ. Entomol. 60: 59-60.

Weaver, C. R. and D. R. King. 1954. Meadow spittlebug. Ohio Agr. Exp. Sta. Res. Bull. 741. 99 pp.

Webster, J. A., E. L. Sorensen, and R. H. Painter. 1968a. Temperature, plant-growth stage, and insect-population effects on seedling survival of resistant and susceptible alfalfa infested with potato leafhoppers. J. Econ. Entomol. 61: 142-145.

Webster, J. A., E. L. Sorensen and R. H. Painter. 1968b. Resistance of alfalfa varieties to the potato leafhopper: Seedling survival and field damage after infestation. Crop Sci. 8: 15-17.

Wegorek, W. 1968. Specjalizacja pokarmowa mszyey grochowej (Acyrthosiphon pisum Harris). Prace Naukowe Instytutu Ochrony Roslin 10: 53-60 (English Sum.).

Wegorek, W. and Lidia Dunajska. 1964. Morphology and anatomy of some lupine varieties and susceptibility to green pea aphid (Acyrthosiphon pisum Harris) infestation. Biuletyn Instyt. Ochrony Roslin 26, 15 pp.

Wegorek, W. and E. Jasienska-Obrebska. 1964. Biuletyn Instyt. Ochrony Roslin 27, 26 pp.

White, C. E., E. J. Armbrust and J. Ashley. 1972. Crossmating studies of eastern and western strains of alfalfa weevil. J. Econ. Entomol. 65: 85-89.

Wilcoxson, R. D. and A. G. Peterson. 1960. Resistance of Dollard red clover to the pea aphid Macrosiphum pisi. J. Econ. Entomol. 53: 863-865.

Wilson, M. C. and R. L. Davis. 1953. Varietal tolerance to the meadow spittlebug. J. Econ. Entomol. 46: 238-241.

Wilson, M. C., R. L. Davis, B. A. Haws and H. L. Thomas. 1956. Attractiveness of sweetclover varieties to the sweetclover weevil. J. Econ. Entomol. 49: 444-446.

Wilson, M. C. and R. L. Davis. 1965. Host plant resistance research on *Philaenus spumarius* (L.) in alfalfa. Proc. 9th Int. Grassland Cong. Sao Paulo, Brazil 2: 1283-1286.

Wilson, M. C. and R. L. Davis. 1958. Development of an alfalfa having resistance to the meadow spittlebug. J. Econ. Entomol. 51: 219-222.

Wilson, M. C. and R. L. Davis. 1960. Culver alfalfa, a new Indiana variety developed with insect resistance. Proc. North Central Branch, Entomol. Soc. Amer. 15: 30-31.

Figure 1.--Seedling test with pea aphid on alfalfa. Resistant line KS10 was selected in 2 cycles from the susceptible Ladak. Comparisons included NS27 = Dawson and Nev-Synt. = Washoe. (Courtesy of Dr. E. L. Sorensen, USDA-Agronomist, Kansas Experiment Station, Manhattan, Kansas).

Figure 2.--Differences in height of susceptible Cody (left) and resistant Kanza (right) following a severe pea aphid infestation in the field. (Courtesy of Dr. E. L. Sorensen, USDA-Agronomist, Kansas Experiment Station, Manhattan, Kansas).

TECHNIQUES, ACCOMPLISHMENTS, AND FUTURE POTENTIAL OF HOST PLANT RESISTANCE TO *Diabrotica*

E. E. Ortman
Head, Department of Entomology
Purdue University
T. F. Branson and E. D. Gerloff
Northern Grain Insects Research Laboratory, Agr. Res. Serv.
USDA, Brookings, South Dakota

In the Corn Belt, the western corn rootworm, *Diabrotica virgifera* LeConte, and northern corn rootworm, *D. longicornis* (Say), rank high among the insect species that are economic pests of corn. The area in which the western corn rootworm is an economic problem continues to increase. The economic impact of corn rootworms has increased for a number of reasons. The practice of planting corn following corn is favorable for increasing the rootworm population. Resistance to the chlorinated hydrocarbons first identified in Nebraska in 1959 has spread rapidly and currently the entire population is considered to be resistant.

The northern and western corn rootworm has a similar life and seasonal cycle with a single generation per year. They overwinter as eggs in the soil. Egg hatch begins in early June and continues for several weeks. Economic losses which result from larval feeding are manifest in 2 ways: reduction in actual yield, and reduction in the harvestable yield due to lodged plants. First adult emergence ranges from late June in Missouri to mid-July in South Dakota. The beetles feed on silks, leaves, and pollen of corn and numerous other species of plants. Occasionally an economic loss occurs when adults feed on silk prior to pollination which interferes with kernel set. Egg laying commences soon after emergence and adults are present in fields until frost.

Techniques

Soil insects such as the corn rootworm larvae present unique problems in host plant resistance studies due to their subterranean habitat. The methodology used in studies on host plant resistance to larvae of the corn rootworm is dependent on the manner of expression of resistance (Painter, 1951): (1) the type of resistance which has a negative effect on the insect population (antibiosis and nonpreference), and (2) the type of resistance where a plant can grow and produce in the presence of an insect population which is generally damaging to the average plant population (tolerance). In our program, antibiosis and nonpreference are studied primarily in the laboratory and greenhouse, while the studies of tolerance are conducted in the field.

We have developed a laboratory technique utilizing seedling corn to evaluate a relatively large number of plants as potential sources of antibiosis. A given number of 1st instar larvae are placed on roots of seedling corn grown in growth pouches (Figure 1). The larvae are examined 10 days after initial infestation to determine their growth and percent survival on test plants. The decision to save a particular plant as a potential source of antibiosis is based on an adjusted scale relative to the development and survival of larvae on susceptible checks versus the test plants. Potentially resistant plants can be transplanted from growth pouches and thus maintain the genetic source. Primary limitations include first--time and funds to test a large number of plants, and second, the test is done on seedling versus the 3-4-week-old plant which is the earliest stage at which attack generally occurs in the field. In searching through different genetic stocks of Corn Belt material, we have not found an indication of an identifiable level of antibiosis. However, in exotic material from Central and South America, and the West Indies, there have been indications of a low level of antibiosis. A number of plants from diverse populations have been retained, pollinated, and the resulting seed grown and tested. In a 2nd cycle of selection, there was a 4-10% increase in the number of plants retained as compared with the initial plant population. Since the level and frequency of antibiosis is relatively low, it may take considerable time to develop a usable level. We are continually searching for other potential sources of antibiosis.

In the greenhouse, we have successfully infested plants potted in soil. This procedure is limiting due to the time and space required to screen a large number of plants. Therefore, we have used it only in studies on host range and advanced evaluation studies. In view of the relatively low incidence and level of antibiosis in the corn germplasm sources, the relatives of corn were investigated as a potential source of antibiosis. There is a marked difference in larval survival and subsequent adult emergence among the members of the tribe *Maydeae* as shown in Table 1 (Branson, et al., 1969). *Tripsicum dactyloides* has a high level of antibiosis or nonpreference to feeding by larvae of the western corn rootworm. The adult emergence after larval infestation has been ascertained for 2 intergeneric hybrids. Adult emergence was not reduced on *Zea mays* X *T. dactyloides* hybrid but there was a marked reduction in adult emergence with a *T. dactyloides* X *Z. mays* hybrid containing 3 *Tripsicum* genomes and 1 corn genome. Other hybrids, addition lines, and substitution lines involving *Z. mays* and *T. dactyloides* are being tested to elucidate the inheritance of resistance. It is recognized that attempts to utilize resistance found in near relatives of corn is a long-term program and potentially frought with considerable difficulty. However, it is being studied in view of the low level of antibiosis found in corn.

Corn has been shown to be the most favorable host for development by larvae of the northern and western corn rootworm as determined by their growth and survival (Table 1). A limited number of larvae survive on plants of common species such as wheat, barley, green foxtail, yellow foxtail, foxtail millet, and intermediate wheat grass (Branson and Ortman, 1966, 1967, 1971). Therefore, plants of species other than corn may serve to maintain a residual insect population. Larvae cannot survive on the roots of broadleaf plants commonly found in the Corn Belt such as alfalfa, soybean, red clover, sweet clover, sunflower, vetch, field bean, and flax and therefore these can be used safely in rotations with corn (Branson and Ortman, 1970). Oats and sorghum are also nonhosts for larvae of both species though the roots are equally as attractive to larvae as those of corn. The nonhost status is apparently due to different causes in the two crops (Branson, et al. 1969; Branson and Ortman, 1969). In sorghum, levels of hydrocyanic acid toxic to larvae of the rootworm result when larvae, feeding on sorghum roots, liberate a beta glucosidase which hydrolyzes endogenous dyanogenetic glucosides such as dhurrin. A feeding deterrent is apparently responsible for the nonhost status of oats.

A limited number of antibiosis and host range studies have been conducted in the greenhouse and field with artificial infestation. However, current techniques do not make it possible to handle a large number of plants. In greenhouse studies, the most consistent results have been obtained when infesting with newly emerged larvae as compared with eggs. In our experience, the variability of egg and larval distribution found in natural field populations makes it virtually impossible to conduct antibiosis tests in those conditions especially where heterogeneous or segregating populations are being studied.

The tolerance component of host plant resistance has been studied through field experiments (Eiben and Peters, 1962, 1965; Fitzgerald and Ortman, 1965; Melhus, et al., 1954; Ortman and Fitzgerald, 1964; 1964; Ortman and Gerloff, 1970; Ortman, et al., 1968; Russell, et al. 1971; Shank, et al., 1965; Zuber, et al., 1971). Development of a suitable field testing site is a most important element in this program because we are not able to artificially infest large field plots. First, it is important to identify those areas where rootworms have been a chronic and consistent problem. Methods have been developed to enhance natural field populations in a particular area to be used for plots. A trap crop of corn is planted at 2-3 times the normal planting rate in a field which is to serve as a test site the following season. This trap crop is planted late with a mixture of plant maturities in order to provide fresh silk and pollen to attract beetles and hold them during the oviposition period. Adult counts should be made to determine the oivposition potential. Field conditions which tend to maximize the egg laying potential include moise soil (therefore, the availability of irrigation is important), ground cover, cracks, and soil debris. A further check on the infestation potential in the field involves taking soil samples to determine egg abundance and uniformity of distribution throughout the field. A number of techniques and procedures have been described (Chandler, et al., 1966; Chiang, 1968; Lawson, 1964; Lawson and Weekman, 1963, 1966; Matteson, 1966). A desirable test field should have uniform distribution of eggs throughout the area. The experimental design and evaluation methods used are dependent upon the objective of the studies.

In the initial field screening test, we have used single-row plots replicated and randomized within similar genetic and maturity groupings, Evaluations are made 2-3 times during the season. The first evaluation is made at the time of maximum damage by larvae to the root system and before there has been time for extensive root growth after damage. The time for this evaluation generally corresponds to the time when the first adult is found in the field. Also at this time, an estimation of the level of larval infestation is made by digging a 7-in. cube of soil around the base of randomly selected plants and extracting the immature forms. Ten to 40 larvae per plant represent a range in which differential plant reactions can be studied.

A relative row-rating for line evaluation is based on an examination of each plant within the row for firmness of anchoring, uniformity of plant growth, and general appearance. To compensate for variable natural infestations, the row-rating is biased toward the poorest performing part of the row. Potential escapes are identified through replication and testing in several locations and different years. Plants and lines which are firm and show no obvious damage are dug and checked for antibiosis potential. The 1-9 rating scale which has been developed is viewed as representing 3 general performance categories: 1-3, acceptable; 4-6, marginal; and 7-9 unacceptable. This line rating system is a relative scale based on the performance of standards and the overall differences within a plot. The objective of the first rating is to assess the amount of damage incurred by the root system of a plant. A second evaluation is made after allowing time for root growth after damage. This second evaluation is generally made after tasseling or 2-4 weeks following the first evaluation. Again, the 1-9 relative rating scale is based on the differences within the plot and the performance of standards. The measure of tolerance is based on 2 considerations, namely the difference in rating between the first and second time of evaluation indicating improved anchorage and the level of the second rating. Most lines exhibit some root growth potential after damage but large differences in this capacity have been observed. A third and supplementary evaluation may be made at the normal harvest. If conditions exist which cause lodging, notes on this character are also taken. An example of the differences observed between lines is shown in Figure 2. Differences in root growth are illustrated in Figure 3. A liberal planting of susceptible checks within the plot area is essential in order to obtain a general index of the severity and uniformity of infestation within a field and in order to relate performance from 1 plot to another and from season to season. In addition to the row-rating when making progeny selections, we have used a rank rating within a particular germplasm group to determine which line should be carried forward in a breeding program. Another useful technique in evaluation of tolerance involves rating a line against itself under rootworm infested and controlled conditions. This is accomplished by maximizing insecticidal control on a paired row basis and then rating these lines. Where the insect population has been ascertained to be too great for the purpose of the experiment, we have been able to reduce the population to an acceptable testing level through use of varying levels of insecticide.

Tolerance is the primary form of resistance which has been identified in these field studies. Through extensive field studies using the methods just described, we have examined a large amount of very divergent plant material and achieved a major elimination of susceptible, nontolerant lines. Corn Belt inbreds identified as having best performance under rootworm infestation include SD10, B69, B57, Oh05, B14, and N38A. We have observed vast differences in the root growing capacity of corn lines grown under rootworm infestation. Those differences were observed in adapted types and therefore we felt significant as well as immediate progress could be made on the problem by the general improvement of the root growth potential. Lines having tolerance will serve a useful role in reducing root lodging through improved anchorage, and, therefore improve the harvestability. The tolerance component of resistance is practical when used under 2 conditions: (1) where the insecticides kill a major portion of the larval population; however, the surviving population may still cause damage to weak rooted lines, and (2) in areas of relatively light infestation, tolerance components may serve as an adequate measure to cope with the problem.

A study of the mechanism or characteristics of the tolerance component of resistance is complex because it involves the interaction of 3 dynamic entities-plant x insect x environment. The synchrony of the growth and development of roots with the growth and development of the rootworm larvae is a most important variable. The node of roots that is attacked by the larvae is dependent upon the time of hatch in relation to the time of planting and growth of the corn plant. Generally the larvae feed on the youngest and most succulent roots which are available. Feeding periods generally average about 4 weeks. Environmental hazards such as wind, rain, and drought compound the damage in rootworm fields. In the field studies just described, we find that there is a greater possibility of finding a higher level of tolerance in late maturing than in early maturing lines. This relationship between maturity and tolerance may be due to the prolonged presence of meristematic or juvenile tissue in late maturing lines. The apparent association between late maturity and tolerance needs to be broken by careful selection in order to maximize progress. For example, SD10 is an early maturing line which performs very well under rootworm stress. In a breeding program, it may be necessary to utilize some lines with marginal ratings in order to maintain diversity in germplasm and maturity.

A series of root studies under controlled conditions were initiated after we had demonstrated differences in performance of lines grown under rootworm infestation. The objectives of these studies were: (1) to develop quantitative criteria which can be used to describe the inherent differences among root systems; (2) to develop a precise characterization of the growth of the root system, and (3) to establish performance base lines for inbreds.

Root growth rate and total development are important characteristics which condition the expression of tolerance. The primary objective of root development studies is to quantitatively characterize root growth and identify the differentiating characteristics for a rapidly growing and extensive root system. The criteria developed to characterize the root system include number of nodes per root system; number of roots per node; rating of the root system for size, symmetry and angle of root growth; secondary or fibrous roots; root volume; dry weight; pounds of force required to remove the system from the soil; and an assessment of damage by the larvae. Correlations between several root system measurements are given in Table 2 with a regression equation to elucidate the contribution of root characteristics to firmness of anchorage as measured by the pounds of vertical force required to remove the root system from the soil. The inherent capacity of a plant to grow and produce roots has been studied in several ways. Root systems have been dug on a scheduled systematic basis throughout the growing season to characterize the root growth and development of lines which have shown differences in response to rootworm infestation. These studies have served to identify certain of the characteristics and mechanisms which contribute to the tolerance component of resistance. These studies have also shown how the tolerance component of resistance is influenced by environment. The growth of the root system follows the characteristic log growth curve (Figure 4). At our location, corn rootworm larvae cause maximum root damage during the month of July. At this time, less than 1/2 of the total dry weight of the root system has accumulated. In order for the plant to withstand considerable root damage, it must develop additional roots from either the nodes located in the stem base or from the remaining intact portion of the damaged root system. From a synchronization standpoint, the root growth which occurs after maximum damage is important in attaining high levels of tolerance. This growth would occur after the majority of the rootworm larvae have pupated and root damage is minimal. Under moderate rootworm infestation levels, root growth during as well as after maximum damage may account for a major portion of the expression of tolerance. Under field conditions, rootworm damage may be only one of a series of related stresses mediated through the root zone. Some plants have the capacity to grow roots at a very rapid rate provided there is a favorable environment. Figure 5 shows the root dry weight accumulation of 3 inbreds during 2 years. In 1968, soil moisture was more favorable than during the previous season. Inbred A34 did not grow a large root system either year while M14 was quite responsive to the more favorable conditions of 1968. By August 23, M14 nearly tripled the root dry weight it had accumulated the previous year. B69 demonstrated its superior root formation during either season. Presumably the more extensive root growth shown by B69 would endow the plant with the capacity to more effectively cope with other types of stresses which occur through the root zone. Generally our results indicate that plants having a more rapid growing and extensive root system in the absence of rootworm will also produce the best root growth under the rootworm infestation. Results show that the root characteristics which condition the expression of tolerance are quantitatively inherited. A root growth index was developed to compare the relationship of the root to the shoot: root volume X root dry weight divided by the top dry weight. The root system or foundation needs to be considered in relation to shoot or superstructure it is supporting. With this index, we search for a line having a rapid initial rate of development and reaching a high plateau. It would appear to be advantageous to combine inbreds that have a rapid initial development with those which attain a high level of prolonged development. Our data from artificial root damage studies indicate that production of fine fibrous roots as a result of damage may be important but not necessarily the most important factor in tolerance. Artificial root damage has also been used to apply a known uniform stress to a plant population. Several forms of artificial injury have been used. Results obtained in such tests show a similar plant response to that for plants grown under rootworm infestation. This known level of stress has an advantage in studying a given measure of injury and plant response. These studies also show that production of secondary or fibrous roots as a result of damage is second in importance to the overall capacity to grow nodal roots. Table 3 gives the ranking of some of the most tolerant inbreds based on several different evaluation criteria.

Accomplishments

Laboratory and greenhouse research has yielded the following: (1) development of methods for studying antibiosis with soil inhabiting insects, (2) identification of a low level of antibiosis in corn germplasm; (3) identification of a relatively high level of antibiosis in the near relatives of corn and initiation of a crossing program; (4) ascertainment of the host range of the larvae to determine use of crops in a rotation; and (5) knowledge of the feeding behavior of larvae and the influence of the secondary plant chemicals in limiting host range. In the area of field tests involving the plant tolerance component of resistance, we have achieved the following: (1) developed methods for field testing; (2) identified inbreds which have superior performance under rootworm infestation based on the tolerance components of resistance; (3) developed synthetics from which superior lines can be extracted; (4) developed techniques for quantitatively characterizing root systems; (5) characterized the tolerance component of resistance; and (6) shown the environment x insect x plant interaction effect on the expression of tolerance.

What are the propsects for resistance in the future? With the identification of tolerant corn lines and elucidation of the mechanisms that contribute to the tolerance component of resistance, the root system of the corn plant can be markedly improved. Tolerance, however, does not provide a means for controlling rootworm populations. This form of resistance will minimize losses in relatively low rootworm infestation areas and serve as an adjunct to other control measures. Though significant progress is being made, in order to maximize the potential for developing host plant resistance, we need to develop an improved capacity to manipulate the insect population in the laboratory and field. The work in antibiosis and nonpreference is continuing and represents a relatively long-term program to develop lines which will be available to the farmer. Low levels of antibiosis have been found in corn and there is reason to believe that it can be increased. We believe we have made an excellent start in the resistance program; however, we also feel the program is in its infancy and to reach the final goal, time, imagination, and perseverance will be needed. There is evidence of marked progress on a complex soil insect problem and this will continue to develop as the effort is increased in the area.

Table 1.--Known host range of larvae of the western corn rootworm (WCR) and northern corn rootworm (NCR). A negative host-plant relationship is indicated by "-," a blank indicates that the host-plant relationship has not been established, and a positive host-plant relationship is indicated by "+." A subscript "1" indicates a poor host, "2" indicates an intermediate host, and "3" indicates a superior host. (Classification of grasses is from Hitchcock, 1950).

	WCR	NCR
Tribe Hordeae		
Fairway wheatgrass, *Agropyron cristatum* (L.) Gaertn.	-	$+_1$
Tall wheatgrass, *A. elongatum* (Host) Beauv.	$+_1$	$+_1$
Intermediate wheatgrass, *A. intermedium* (Host) Beauv.	$+_1$	$+_1$
Slender wheatgrass, *A. trachycaulum* (Link) Malte	$+_1$	$+_1$
Pubescent wheatgrass, *A. trichophorum* (Link) Richt.	$+_1$	$+_1$
Canada wildrye, *Elymus canadensis* L.	-	$+_1$
Barley, *Hordeum vulgare* L.	$+_1$	$+_1$
Wheat, *Triticum aestivum* L.	$+_1$	$+_1$
Spelt, *T. spelta* L.	$+_1$	$+_1$
Tribe Festuceae		
Sand lovegrass, *Eragrostis trichodes* (Nutt.) Wood	$+_1$	-
Weeping lovegrass, *E. curvula* (Schrad.) Nees	$+_2$	$+_1$
Tribe Oryzeae		
Rice, *Oryza sativa* L.	$+_1$	$+_1$

Tribe Paniceae		
Proso millet, *Panicum miliaceum* L.	-	$+_1$
Foxtail millet, *Setaria italica* (L.) Beauv.	$+_1$	-
Yellow foxtail, *S. lutescens* (Weigel) Hubb.	$+_1$	$+_1$
Green foxtail, *S. viridis* (L.) Beauv.	$+_1$	$+_1$
Tribe Maydeae (Tripsaceae)		
Teosinte, *Euchlaena mexicana* Schrad.	$+_3$	
Mexican (perennial) teosinte, *E. perennis* Hitchc.	$+_3$	
Eastern gamagrass, *Tripsacum dactyloides* (L.) L.	$+_1$	
Florida gamagrass, *T. floridanum* Porter ex Vasey	$+_3$	
T. australe Cutler and Ander.	$+_3$	
T. laxum Nash	$+_2$	
T. latifolium Hitch.	$+_2$	
Job's tears, *Coix lacryma-jobi* L.	$+_2$	
Corn, *Zea mays* L.	$+_3$	$+_3$

Table 2.--Correlations between criteria used to evaluate root systems.

	Correlations				
	Pulling resistance	Root angle	Secondary roots	Visual rating	Root nodes
Root angle	0.000ns				
Secondary roots	.185*	0.709**			
Visual rating	.643**	- .229**	0.085ns		
Root nodes	.132*	.018ns	.089ns	0.374**	
Roots on N-1 node	.331*	- .077ns	.005ns	.485**	0.394**

Estimated pulling resistance = 5.89370 + 11.25767 (root angle) + 4.09966 (secondary roots) + 85.95524 (visual rating) - 10.19526 (root nodes) + 1.53055 (roots on N-1 node).

Table 3.--Inbred rank for various evaluation criteria.

Inbred	Field[1/] performance rating	Root rating[2/] Control	Root rating[2/] Damaged	Secondary roots[2/] Control	Secondary roots[2/] Damaged	Root Pulling[3/] wt.	Root Growth[4/] index	Root Vol.	Root Dry wt.
Zap15	1	4	1	12	9	1	3	1	2
B69	2	8	11	1	5	5	5	6	6
B57	3	7	9	19	19	6	7	8	10
SD10	4	2	2	9	1	3	1	3	3
CI21E	5	17	20	17	11	20	15	16	13
N38A	6	6	5	13	14	10	6	9	9
B14	7	5	6	7	6	11	9	10	7
Oh05	8	3	3	8	3	4	2	2	1
A251	9	15	16	18	18	12	16	18	21
A556	10	11	7	3	8	15	8	15	16
Mo22	11	19	18	22	22	16	11	7	12
A265	12	12	4	4	4	19	14	13	11
A73	13	1	13	10	15	2	10	5	5

R168	14	21	19	21	21	8	13	12	14
Ms107	15	9	8	11	2	13	4	4	4
Oh45	16	16	10	16	12	14	21	20	20
M14	17	20	21	14	10	17	17	14	17
B64	18	18	14	20	17	21	12	11	8
B37	19	14	12	6	13	9	19	17	18
N6	29	10	15	15	16	18	20	19	19
WF9	21	13	17	5	7	7	18	21	15
A34	22	22	22	2	20	22	22	22	22

1/A rating scale was applied to lines grown under rootworm infestation.
2/A rating scale was applied to normal and artificially damaged root systems.
3/Pounds of vertical force required to remove plant from the soil.
4/Root volume plus root dry wt. divided by shoot dry weight.

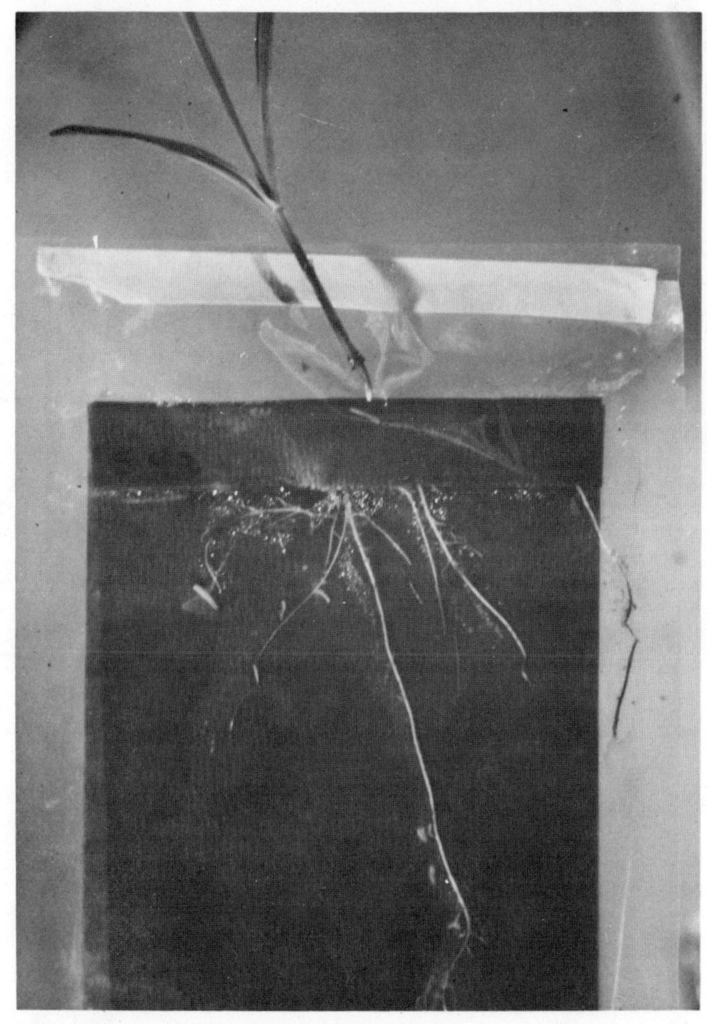

Figure 1.--Growth pouch with seedling corn plant for infestation with larvae of rootworm.

Figure 2.--Difference in line performance under rootworm infestation.

Figure 3.--Differential root growth by inbreds grown under rootworm infestation. Extensive brace root growth has occurred above damaged nodes on lower roots versus upper.

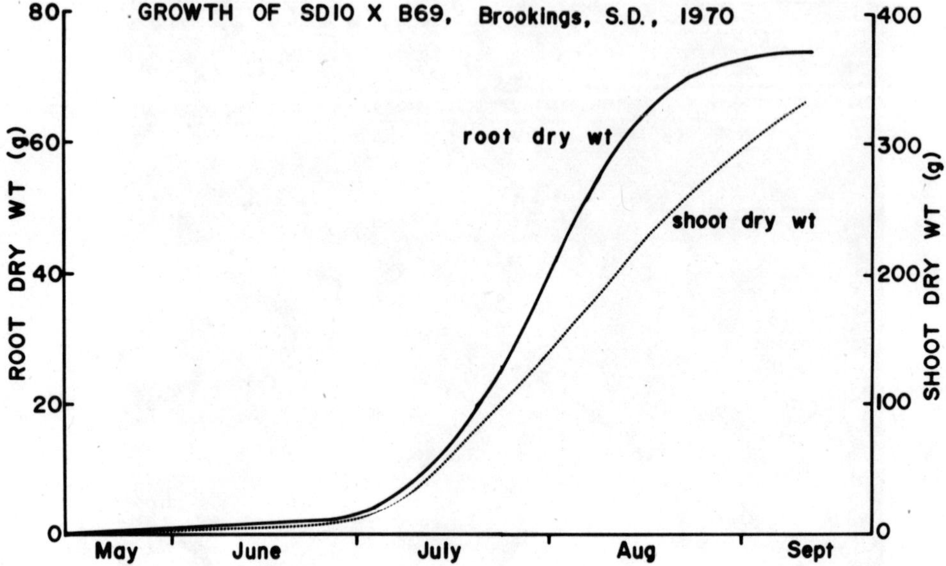

Figure 4.--Root and shoot dry weight accumulation of a single cross grown at Brookings, South Dakota, 1970.

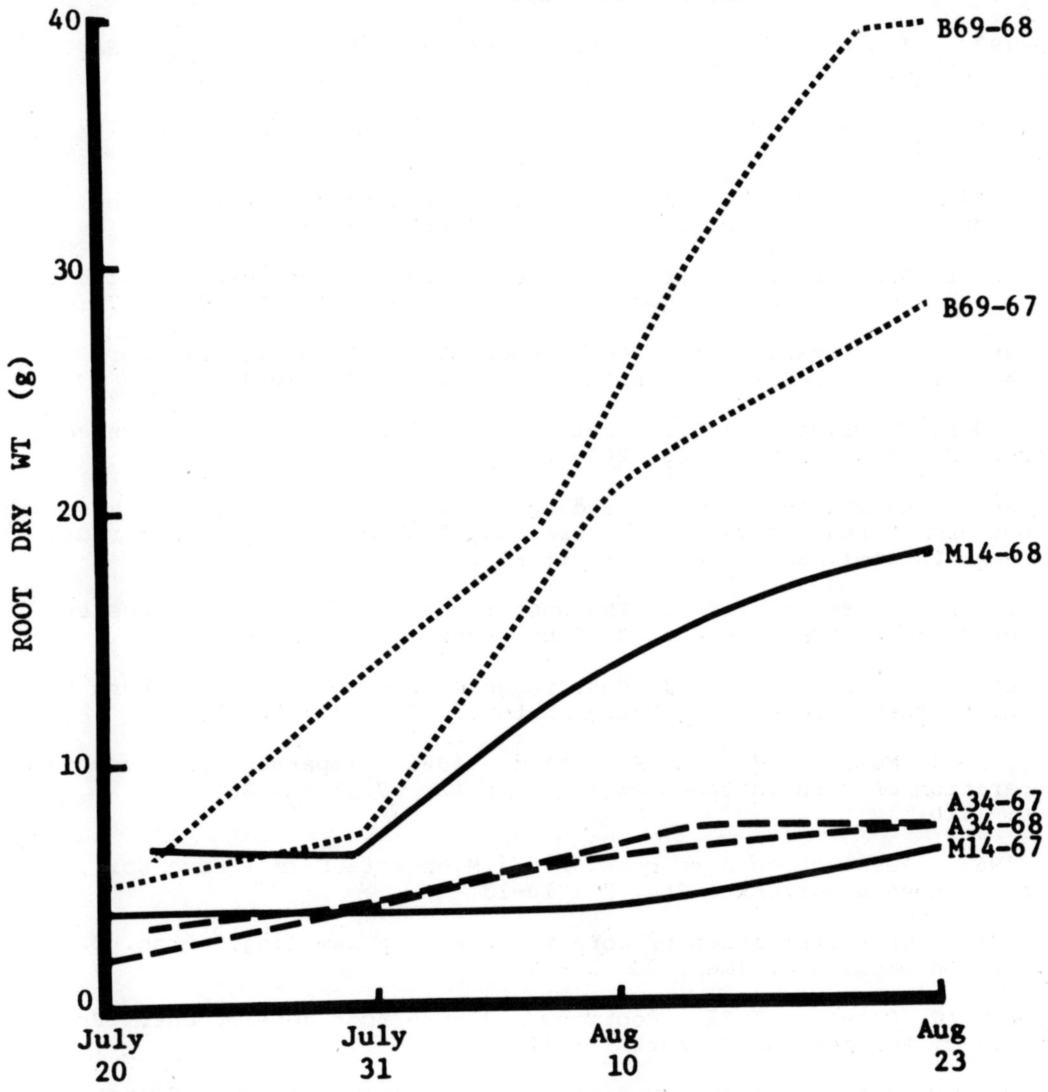

Figure 5.--Root dry weight accumulations of 3 inbreds for 2 years.

References Cited

Bigger, J. H. 1932. Short rotation fails to prevent attack of Diabrotica Say. J. Econ. Entomol. 25: 196-199.

Bigger, J. H. 1941. Breeding corn for resistance to insect attack. J. Econ. Entomol. 34: 341-347.

Branson, T. F. 1971. Resistance in the grass tribe Maydeae to larvae of the western corn rootworm. Ann. Entomol. Soc. Amer. 64: 861-863.

Branson, T. F., P. L. Guss and E. E. Ortman. 1969. Toxicity of sorghum roots to larvae of the western corn rootworm. J. Econ. Entomol. 62: 1375-1378.

Branson, T. F. and E. E. Ortman. 1966. Host range of larvae of western corn rootworm. Proc. N. Centr. Br. Entomol. Soc. Amer. 21: 40-41.

Branson, T. F. and E. E. Ortman. 1967. Host range of larvae of the western corn rootworm. J. Econ. Entomol. 60: 201-203.

Branson, T. F. and E. E. Ortman. 1969. Feeding behavior of larvae of the western corn rootworm: Normal larvae and larvae maxillectomized with laser radiation. Ann. Entomol. Soc. Amer. 62: 808-812.

Branson, T. F. and E. E. Ortman. 1970. The host range of larvae of the western corn rootworm: Further studies. J. Econ. Entomol. 63: 800-803.

Branson, T. F. and E. E. Ortman. 1971. Host range of larvae of the northern corn rootworm: Further studies. J. Kansas Entomol. Soc. 44: 50-52.

Chandler, J. H., G. J. Musick and M. L. Fairchild. 1966. Apparatus and procedure for separation of corn rootworm eggs from soil. J. Econ. Entomol. 59: 1409-1410.

Chiang, H. C. 1965. Research on corn rootworms. Minnesota Farm & Home Sci., Univ. of Minnesota Agr. Exp. Sta. 23: 10-13.

Chiang, H. C. 1968. Characteristics of corn rootworm egg sampling. Proc. N. Centr. Br. Entomol. Soc. Amer. 23: 19-20.

Eiben, G. J. and D. C. Peters. 1962. Rootworm and corn root development. Proc. N. Centr. Br. Entomol. Soc. Amer. 17: 124-126.

Eiben, G. J. and D. C. Peters. 1965. Varietal response to rootworm infestation in 1964. Proc. N. Centr. Br. Entomol. Soc. Amer. 20: 44-46.

Fitzgerald, P. J. and E. E. Ortman. 1965. Two-year performance of inbreds and their single crosses grown under corn rootworm infestation. Proc. N. Centr. Br. Entomol. Soc. Amer. 20: 46-47.

George, B. W. and A. M. Hintz. 1966. Immature stages of the western corn rootworm. J. Econ. Entomol. 59: 1139-1142.

Hagen, A. F. and F. N. Anderson. 1967. Nutrient imbalance and leaf pubescence in corn as factors influencing leaf injury by the adult western corn rootworm. J. Econ. Entomol. 50: 1071-1073.

Howe, W. L. and B. W. Goerge. 1966. Corn rootworms, pp. 367-383. In Insect Colonization and Mass Production. Academic Press, Inc., New York.

Kirk, V. M., C. O. Calkins and F. J. Post. 1968. Oviposition preferences of western corn rootworms for various soil surface conditions. J. Econ. Entomol. 61: 1322-1324.

Lawson, D. E. 1964. Rootworm egg distribution in corn fields. Proc. N. Centr. Br. Entomol. Soc. Amer. 19: 92.

Lawson, D. E. and G. T. Weekman. 1963. Sampling techniques for eggs *Diabrotica virgifera* LeConte. Proc. N. Centr. Br. Entomol. Soc. Amer. 18: 99.

Lawson, D. E. and G. T. Weekman. 1966. A method of recovering eggs of the western corn rootworm from the soil. J. Econ. Entomol. 59: 657-659.

Lonniquist, J. H. and T. A. Kiesselback. 1948. Corn rootworm studies. Nebraska Agr. Exp. Sta. Ann. Rept. 61: 18.

Matteson, J. W. 1966. Flotation techniques for extracting eggs of *Diabrotica* spp. and other organisms from soil. J. Econ. Entomol. 59: 223-224.

Matteson, J. W., C. O. Calkins and W. L. Howe. 1965. The effect of tillage practices on corn rootworm populations. Proc. N. Centr. Br. Entomol. Soc. Amer. 20: 46.

Melhus, I. E., R. H. Painter and F. O. Smith. 1954. A search for resistance to the injury caused by species of *Diabrotica* in the corns of Guatemala. Iowa State Coll. J. Sci. 29: 75-94.

Mendoza, C. E. and D. C. Peters. 1964. Species differentiation among mature larvae of *Diabrotica undecimpunctata howardi*, *D. virgifera*, and *D. longicornis*. J. Kansas Entomol. Soc. 37: 123-125.

Olson, L. A., G. J. Eiben and D. C. Peters. 1965. Crop sequence influences on rootworm damage. Proc. N. Centr. Br. Entomol. Soc. Amer. 20: 42-44.

Ortman, E. E. and P. J. Fitzgerald. 1964. Developments in corn rootworm research. Proc. 19th Ann. Hybrid Corn Ind. Res. Conf. 1964: 1-8.

Ortman, E. E. and P. J. Fitzgerald. 1964. Evaluation of corn inbreds for resistance to corn rootworms. Proc. N. Centr. Br. Entomol. Soc. Amer. 19: 92.

Ortman, E. E. and E. D. Gerloff. 1970. Rootworm resistance problems in measuring and its relationship to performance. Proc. 25th Ann. Corn & Sorghum Res. Conf. 1970: 161-174.

Ortman, E. E., D. C. Peters and P. J. Fitzgerald. 1968. Vertical-pull technique for evaluating tolerance of corn root systems to northern and western corn rootworm. J. Econ. Entomol. 61: 373-375.

Painter, R. H. 1951. Insect resistance in crop plants. McMillan, New York. 520 p.

Pruess, K. P., G. T. Weekman and B. R. Somerholder. 1968. Western corn rootworm egg distribution and adult emergence under two corn tillage systems. J. Econ. Entomol. 61: 1424-1427.

Reissig, W. H. and G. E. Wilde. 1971. Feeding responses of western corn rootworm on silks of fifteen genetic sources of corn. J. Kansas Entomol. Soc. 44: 479-483.

Russell, W. A., L. H. Penny, W. D. Guthrie and F. F. Dicke. 1971. Registration of maize germplasm inbreds. Crop Sci. 11: 140.

Shank, D. B., D. W. Beatty, P. J. Fitzgerald and E. E. Ortman. 1965. SD10 inbred corn for hybrids with resistance to corn rootworms. South Dakota Farm & Home Res. 16: 4-5.

Sifuentes, J. A. and R. H. Painter. 1964. Inheritance of resistance to western corn rootworm adults in field corn. J. Econ. Entomol. 57: 475-477.

Walter, E. V. 1965. Northern corn rootworm resistance in sweet corn. J. Econ. Entomol. 58: 1076-1078.

Zuber, M. S., G. J. Musick and M. L. Fairchild. 1971. A method of evaluating corn strains for tolerance to the western corn rootworm. J. Econ. Entomol. 64: 1514-1518.

TECHNIQUES, ACCOMPLISHMENTS AND FUTURE POTENTIAL OF BREEDING FOR RESISTANCE TO EUROPEAN CORN BORERS IN CORN

W. D. Guthrie
Entomologist - Professor
ARS-USDA
Iowa State University
Ankeny, Iowa

Summary

Investigations of corn borer resistance have become an integral part of the corn breeding programs in several states and some private seed companies. A considerable number of inbred lines used in hybrid combination are resistant or intermediate in resistance to leaf feeding by 1st-brood larvae. During the 1940's and 1950's, many hybrids were extremely susceptible to leaf feeding by a 1st-brood infestation. Today, most hybrids have a lower level of susceptibility. Many hybrids have at least an intermediate degree of resistance; some hybrids are resistant.

Inbred lines of corn that have been evaluated for resistance to sheath feeding by a 2nd-brood infestation differ in degree of susceptibility, but not in degree of resistance. B52 is the only inbred available to date with a high level of 2nd-brood resistance (antibiosis). Inbred lines and hybrids, however, differ in tolerance to a 2nd-brood infestation.

Rearing large numbers of corn borers on a meridic diet should accelerate research on host plant resistance by private seed companies. Our research, including larval rearing techniques and methods for measuring resistance, is made available to private seed corn companies; at least 2 companies plan to produce large numbers of egg masses for studies of 1st- and 2nd-brood resistance. We hope that this cooperative effort between USDA-ARS-state and private seed corn companies will eventually result in hybrids resistant to both a 1st- and 2nd-brood infestation.

The search for inbred lines of corn resistant to a 1st-brood infestation by the European corn borer, *Ostrinia nubilalis* (Hubner), has been in progress for many years as a cooperative effort by entomologists, corn breeders, and corn geneticists. As a result of this team approach, many inbred lines have been screened for resistant germ plasm (Stringfield, 1959; Guthrie, et al., 1960; Guthrie and Dicke, 1972), and the genetic basis for leaf feeding (1st-brood) resistance and methods of corn breeding to transfer resistance to susceptible inbred lines have been determined (Patch, et al., 1942; Schlosberg and Baker, 1948; Singh 1953; Ibrahim, 1954; Penny and Dicke, 1956, 1957; Guthrie and Stringfield, 1961a, b; Scott, et al., 1964; Scott and Dicke, 1965; Scott, et al., 1966; Penny, et al., 1967). Similar efforts to find sources of resistance to 2nd-brood borers have not proceeded as fast because the egg masses necessary to achieve the artificial infestations used in such studies have not been so readily available.

In the Corn Belt States, cultural practices have changed during the past several years. During the 1940's and 1950's, a considerable number of acres were planted to oats. Oats usually were planted on fields in corn the previous season. Most stalks were left on top of the ground. Larvae overwintering in the stalks were a source of moths for egg laying in June. Today, only a few acres are planted to oats; most Corn Belt acres are planted to corn and soybeans. Almost all corn stalks are plowed under; clean plowing of corn fields destroys large numbers of overwintering corn borers.

During the 1940's and 1950's, ear corn stored on farms was a source of moths for egg laying in June (many corn borer larvae overwinter in cobs). Today, almost all corn is shelled in the field, and the cobs are plowed under.

During the 1940's and 1950's, many hybrids were extremely susceptible to leaf feeding by a 1st-brood infestation. Today, most hybrids have a lower level of susceptibility. Many hybrids have at least an intermediate degree of resistance; some hybrids are resistant.

The combination of these changes in cultural practices and higher levels of resistance in hybrids have reduced 1st-brood populations during the past several years. First-brood corn borers, however, may still be a problem in local areas.

Heavy infestations by 1st-brood corn borers can decrease the yields of susceptible single crosses by some 30 bu/acre (Penny and Dicke, 1959); thus, one purpose in developing inbreds resistant to leaf feeding (1st brood) is to reduce 1st-brood populations to reduce losses. The biotic potential of the corn borer, however, is great when climatic conditions are favorable, and a large 2nd-brood infestation can develop from a small 1st-brood population. As a result, even a 50% reduction in the 1st-brood population achieved by the use of resistant hybrids would reduce damage by 2nd-brood borers. For example, an experimental inbred (W24 X B2)-2-38-1, highly resistant to leaf feeding, reduced borer populations 89% 30 days after egg hatch compared with a susceptible inbred line (WF9); another resistant inbred, Oh43, reduced borer populations 73% (Guthrie, et al., 1960). Even an inbred line with an intermediate degree of resistance would be effective in reducing 1st-brood populations.

In recent years, most farmers have planted single crosses or simulated single crosses instead of double cross hybrids. Some single crosses between a resistant and a susceptible line are intermediate in performance between the 2 parents (Dicke, 1954). The high resistance of Oh45 showed a high degree of dominance in crosses with 3 susceptible inbreds (Klun, et al., 1970). Also, the high resistance of CI.31A had a high degree of dominance in crosses with 4 susceptible inbreds (Penny and Dicke, 1959; Klun, et al., 1970). In general, single crosses with the following combinations of inbred lines are effective in reducing 1st-brood populations: Resistant X Resistant, Intermediate X Intermediate, Resistant X Intermediate, or Resistant X Susceptible. But, either dominance of resistance or incomplete dominance is necessary if the Resistant X Susceptible combination is to be effective, and most single crosses involving Intermediate X Susceptible combinations are susceptible. For example, inbreds B52, Hy, and W22 (intermediate in resistance to leaf feeding) in combination with 4 susceptible inbreds (B14A, B37, R101, WF9) were susceptible to leaf feeding (Klun, et al., 1970).

Moreover, inbred lines resistant to leaf feeding (1st brood) are not necessarily resistant to sheath feeding (2nd brood) (Guthrie, et al., 1970). Oh43 is resistant to a 1st-brood infestation (Guthrie, et al., 1960), but susceptible to a 2nd-brood infestation (Pesho, et al., 1965; Guthrie, et al., 1970). In contrast, inbred B52 is highly resistant to a 2nd-brood infestation (Pesho, et al., 1965; Guthrie, et al., 1970), but intermediate in resistance to a 1st-brood infestation (Klun and Brindley, 1966). Thus, resistance to whorl leaf feeding (1st brood) must be considered separately from resistance to sheath feeding (2nd brood) until we can develop methods of selecting for resistance to both broods in the same plant population (Guthrie, et al., 1970).

First- and 2nd-Brood Egg Production Methods

In investigations of European corn borer resistance, all plots are artificially infested with egg masses produced in the laboratory. Progress would be nil without artificial infestation techniques.

For many years, we have had a good source of moths for 1st-brood egg production; moths were collected from large emergence cages filled with infested corn stalks the previous fall (Guthrie, et al., 1965a).

Until recently, we have not had a good source of moths for 2nd-brood egg production; moths were obtained primarily by net-collecting in patches of grass or weeds near cornfields that had a 1st-brood infestation, or from infested, caged, green, sweet corn stalks (Guthrie, et al., 1965a).

During the past several years, we have had good success in rearing corn borer larvae individually in 3-dram vials on a plug of meridic diet; over 90% survival is often obtained (Guthrie, et al., 1965b, Lewis and Lynch, 1969). Two larvae per vial averages about 1 1/2 pupae per vial. This procedure is too slow for producing the number of moths needed for egg production in resistance research. During 1965-1972, we reared larvae on a meridic diet in plastic dishes (10-inch diam., 3 1/2-inch deep). The primary problem in the dishes was *Perezia pyraustae*. This microsporidian increased very rapidly in the dishes and drastically affected egg production and the quality of the eggs. Lewis and Lynch (1970) and Lynch and Lewis (1971) solved the *Perezia* problem in 1969 by adding Fumidil B to the diet.

We also had a mold problem on diet in dishes. Four species of mold have been troublesome: (1) *Aspergillus niger*, (2) *A. flavus*, (3) *Cladosporium avellaneum* and (4) *Penicillum* spp. The mold problem has largely been solved by using 4 different mold inhibitors (Table 1). The approximate cost of producing 1 million egg masses is given in Table 2.

Directions for Cooking Meridic Diet (15,243 g)

1. Add 8,125 ml water to agar and heat (194°F) until agar is cooked.

2. When agar is cooked, add 1,625 ml water cooled in a refrigerator to 38°F.

3. Add wheat germ, add 1,625 ml water cooled in a refrigerator to 38°F.

4. Stir until temperature of medium decreases to 136°F.

5. Pour dextrose, casein, cholesterol, salt mixture, vitamin supplement, ascorbic acid, aureomycin, and Fumidil B into a blender, add 1,625 ml water (at 74°F) and blend for 3 min. Pour this mixture into the cooker. Pour the mold inhibitors (methyl p hydroxybenzoate, propionic acid, formaldehyde, and sorbic acid) into the cooker. Stir until well mixed. This brings the temperature down to 122-126°F.

6. Pour approximately 3,800 g diet into a blender and blend for 1 min.

7. Pour 930 g diet into each dish.

The procedure used in cooking the diet is important. It is important to cool the diet to about 122°F (by adding water before adding any ingredients except agar and wheat germ). High heat destroys some of the value of the ingredients.

The top of the cooled diet is scratched with a fork so that the larvae will start feeding instead of crawling over the diet and on the side of the dish. Each dish is infested with 40 egg masses (1,000 larvae). The dishes are placed in an incubator operating at a temperature of 80-83°F and 75-80% RH.

Strips of corrugated paper (1-inch wide, treated in hot wax) are placed in the dishes for pupation. A mold inhibitor (sorbic acid, 5% by weight) is added to the hot wax. The sorbic acid in the wax helps considerably in controlling mold. More larvae crawl out of the dish into the corrugated strips if the lids and half the dish are painted black (Reed, et al., 1972, in press).

Twenty-one days after egg hatch, the corrugated strips containing pupae are hung in a large room (70°F and 80-85% RH) for moth emergence.

Moths are collected each day and placed in oviposition cages for egg production (Guthrie, et al., 1965b). A moth aspirator (made from an electric hair dryer) is used to facilitate collection. The top of each oviposition cage is made of 4-mesh hardware cloth. Two sheets of bleached waxed paper (6 x 24 inches) are placed on top of the hardware cloth. The moths deposit their eggs on the waxed paper between the openings in the hardware cloth. The sheets of waxed paper are removed and replaced with new paper each morning.

The oviposition room is operated at a temperature of 80°F during 18-hr each day and 65-68°F during 6-hr each day through a series of time clocks operating a heater and air conditioner. Relative humidity is maintained at about 85%. The cycling temperature is required to ensure adequate mating (Sparks, 1963).

Disks of waxed paper (1/2 inch in diam), each containing 1 corn borer egg mass, are cut from the waxed paper with a specially designed machine. The waxed paper disks, containing egg masses, are pinned onto 8 x 10-inch cellotex boards, 200 disks per board, and incubated to near hatching before being used for field infestations.

During 1965-69, moths originating from larvae reared on a meridic diet in plastic dishes were used for 1st-brood egg production to supplement egg production from wild moths collected in large emergence cages.

During the 1970, 1971, and 1972 seasons, we did not place corn stalks in the large emergence cages. We depended entirely on rearing larvae on the meridic diet. In 1970, 1,252,400 egg masses were produced for 1st-brood infestations and 308,000 egg masses, for 2nd-brood infestations. In 1971, 1,188,000 egg masses were produced for 1st-brood infestations and 447,000 egg masses, for 2nd-brood infestations.

In 1965-68, infection by the microsporidian Perezia pyraustae in the dishes increased to a high level, number of moths recovered per dish was low, egg production and quality of the eggs were poor, and a low level of larval establishment on corn plants resulted from the artificial infestations. In 1969, 1970, and 1971, number of moths recovered per dish was considerably higher than during the previous 4 years, egg production and the quality of the eggs were excellent, and a high level of larval establishment on corn plants resulted from the artificial infestations; we believe that the high performance was due primarily to control of P. pyraustae with Fumidil B in the diet. In the future, we may have disease problems, but, we hope that this procedure can be used year after year.

Larval Survival on Field Corn from Corn Borer Cultures Continuously Reared on a Meridic Diet

Data collected during the past 6 years show that the European corn borer reared continuously on a meridic diet cannot be used for screening inbred lines for resistance or in any type of resistance studies because the leaf-feeding ratings are too low for measuring resistance.

One culture has been reared 87 generations on a meridic diet. This culture was compared for level of larval establishment on inbred line WF9 (susceptible to leaf feeding, 1st brood) with a wild population during a 6-year period. In 1966, the culture had been reared for 34 generations on a meridic diet; in 1967, 45 generations; in 1968, 54 generations; in 1969, 65 generations; in 1970, 76 generations; and in 1971, 87 generations. In a 9-class rating scale, plants infested with this culture rated 1 to 3 (a very low level of larval establishment); plants infested with the wild population rated 7 to 9 (Huggans and Guthrie, 1970; Rathore and Guthrie, in press). Crosses between this culture and a wild population and backcrosses to each parent show that this loss in virulence is genetically controlled (Guthrie and Carter, 1972; Rathore, et al., in press). Cultures reared for at least 7 generations on a meridic diet perform as well as the wild type (Rathore and Guthrie, in press).

We start a new colony each spring. Several thousand wild borers are dissected from corn plants each fall (Guthrie, et al., 1972). These larvae are placed in a 40°F room; at weekly intervals for 5 weeks (starting on January 25), larvae are isolated individually in vials and placed in an 80°F room. These larvae pupate in February. The progeny from the moths are reared through 2 generations on a meridic diet containing 1,500 ppm Fumidil B (1 larva per vial for the 1st generation in March; 2 larvae per vial for the 2nd generation in April). This procedure eliminates P. pyraustae. Egg masses from moths originating from larvae reared 2 generations on a meridic diet are used to infest the dishes (500 ppm Fumidil B). Moths from the dish-reared material (3 generations on a meridic diet) are used for egg production in 1st-brood resistance research. Dishes are started on May 10 and continued 4 to 5 weeks for 1st-brood infestations. Corn planted during the 1st week of May at Ankeny and Ames, Iowa, is infested June 15-25 (45-52 days after planting, midwhorl stage of plant development). All egg masses are incubated to near the hatching point before being placed on corn plants. If plant growth and egg production are not synchronized, the egg masses can be held for 10 to 12 days at 60-62°F under high humidity. Egg masses usually are wrapped in moist paper and placed in plastic-lined boxes for incubation (4 days at 80°F).

Moths from the dish-reared material (4 generations on a meridic diet) are used for egg production in 2nd-brood research (Guthrie, et al., 1972). Dishes are started on June 21 and continued for 3 or 4 weeks. Corn planted during the last week of May at Ankeny is infested during July 25-August 15 depending on the maturity of the material.

The techniques of rearing corn borer larvae on a meridic diet has greatly accelerated research on 2nd-brood resistance.

Biology of 1st-Brood Larvae on the Corn Plant

A knowledge of the biology of an insect on the host plant is imperative in investigations of host-plant resistance. In the Corn Belt States, the corn borer has 2 broods each season. Biological relationships between the corn borer and the corn plant are not the same for both broods. During the period of egg deposition by the 1st brood, most dent corn in the Corn Belt States is in the whorl stage of plant development. The 1st- and 2nd-instar larvae feed primarily on the spirally rolled leaves in the whorl. Factors that inhibit 1st-brood borer establishment and survival on resistant lines are operative against the early larval instars. Most 1st-brood larval mortality on resistant inbred lines occurs during the 1st few days after egg hatch. This high rate of larval mortality reflects a high degree of antibiosis against the 1st- and 2nd-instar larvae of a 1st-brood infestation (Guthrie, et al., 1960). Therefore, 1st-brood resistance is actually leaf-feeding resistance. As the plant grows out of the whorl stage, the larvae develop to the 3rd and 4th instars; these larvae feed primarily on sheath and collar tissue. Resistance on some inbred lines also has been evaluated for sheath and collar feeding by 3rd- and 4th-instar larvae of the 1st brood (Guthrie, et al., 1960). Inbred lines susceptible to a 1st-brood infestation suffer extensive leaf, sheath, and collar damage. Cavities in the stalk are caused primarily by 5th-instar larvae.

Methods for Measuring 1st-Brood Resistance in Corn

Egg masses incubated to near the hatching point are dropped into the whorl of each plant for 1st-brood tests. In screening material for resistant germ plasm (late generation inbreds), 10 plants are infested with 4 egg masses or approximately 100 eggs per plant in 2 applications of 2 masses each during the midwhorl stage of development. Most inbred lines are about 25 inches in extended leaf height and hybrids, 35 inches, at this time. In genetic studies a large plant population is infested. Over 1-million egg masses have been used in a single season for field plot infestations (Guthrie, et al., 1972).

Rating Scale.--A 9-class rating scale, as described by Guthrie, et al. (1960) is used for evaluating damage from borer leaf feeding during the whorl stage of plant development. In the relative resistance scale, lines that rate 1 and 2 are considered highly resistant, lines that rate 3 and 4 are considered resistant, lines that rate 5 and 6 are considered intermediate in resistance, and lines that rate 7 to 9 are considered highly susceptible. Classification into a highly resistant, resistant, intermediate, or susceptible class is dependent upon the size and shape of leaf injuries, and rating within each class is determined by the number of holes or amount of feeding. The entries are rated on a plot or individual-plant basis, depending on the type of material under test, before pollination (about 3 weeks after egg hatch). This system preserves the resistant culture for pollination and progeny testing, and is particularly valuable in individual plant selection in segregating populations to study inheritance of resistant factors. This is an excellent method for evaluating a large amount of material. For example, in 1965, 40,000 individual plants and 6,000 plots were rated in 8 days by 1 man.

Cavity counts are not a good criterion for measuring resistance to a 1st-brood infestation in areas where 100% of the 1st-brood pupate to form a 2nd brood. Corn borer larvae are primarily external feeders for at least 20-25 days after egg hatch, and, if pupation occurs 30-35 days after egg hatch, the larvae have not been inside the stalk long enough to cause extensive damage to stalk tissue.

Plant Breeding Methods (1st Brood)

For several years, investigations on resistance to the 1st brood of the corn borer consisted of testing thousands of open-pollinated varieties of inbred lines for resistant germ plasm. At present, the practice of direct extraction of lines from special single and 3-way crosses and from synthetic varieties is being used extensively. In some instances, the backcross method in combination with various methods of intensification has been used successfully. A straight backcrossing procedure has not been successful.

Modified Backcrossing Procedure.--In Ohio, a modified backcrossing method was used in an attempt to improve agronomic characteristics of Oh45 (Guthrie and Stringfield, 1961a). Oh45 (designated as recurrent parent) is characterized by being sparsely husked, tightly husked, short shanked, and inadequate in pollen production--these characteristics being undesirable. Oh45 is resistant to a 1st-brood corn borer infestation. This line was outcrossed to single cross M14 X CI.187-2 (designated as donor parent), which is copiously husked, loose husked, well shanked, and adequate in pollen production, but highly susceptible to a 1st-brood infestation. A single cross, rather than an inbred line, was chosen as donor parent in the belief that the greater genetic diversity of the 3-parent population would provide improved selection over a 2-parent one. The objective of this program was to retain the corn borer resistance of Oh45 and recover the desirable morphological characters from the donor parent.

The outcross (M14 X CI.187-2) X Oh45 was then both selfed, [(M14 X CI.187-2) X Oh45]S_1, and backcrossed to Oh45, [(M14 X CI.187-2) X Oh45] X Oh45. The backcross progenies seemed to offer too little variation for effective selection, having too little of the donor morphology. We decided to continue selfing with selection in this population [M14 X CI.187-2) X Oh45]S_1 S_2 S_3 S_4 to establish the donor shank, husk, and pollen type along with Oh45 plant and grain type insofar as possible. After the S_4 generation, [M14 X CI.187-2) X Oh45]S_4, test crosses were made using both Oh45 and (M14 X CI.187.2) as test cross parents; i.e., [(M14 X CI.187-2) X Oh45 S_4] X Oh45 and [(M14 X CI.187.2) X Oh45 S_4] X (M14 X CI.187.2). This was to establish the genetic drift in each S_4 culture. Those that had drifted toward Oh45 should exhibit relatively low heterosis crossed with Oh45 and relatively high heterosis crossed with (M14 X CI.187-2). S_1, S_2, S_5, and S_6 cultures from the backcross to Oh45 were artificially infested during the next seasons, and appropriate discards were made. With this procedure, both the corn borer resistance of the line Oh45 and the morphological characters of the single cross M14 X CI.187-2 were recovered. Three of the best isolates were given permanent designations of Oh45A, Oh45B, and Oh45C. Oh45B has been released, and in Ohio tests in combination with Oh43 as the pollen parent, has performed well in double crosses.

Recurrent Selection.--The value of a recurrent-selection technique in selecting for resistance to leaf feeding by the European corn borer (1st brood) has been determined. This study was started in 1954 and concluded in 1966 (Penny, et al., 1967).

Recurrent selection is essentially a breeding method of concentrating genes for certain desirable characters for which selection is being practiced while maintaining a broad gene base for other characteristics in the population. This allows for the accumulation of desirable genes at numerous loci. The following 5 synthetic varieties were used in these studies:

1. Pa. early synthetic
2. Pa. intermediate synthetic
3. Pa. late synthetic
4. Synthetic A
5. Synthetic B

The 3 Pennsylvania varieties are closely related, all having been derived from the same base population developed in Pennsylvania for resistance to *Helminthosporium turcicum* (northern corn leaf blight). The breeding procedure followed for each of these populations was as follows:

First cycle (C_1) of breeding
- Year 1 -- Self 100 plants.
- Year 2 -- Plant the 100 plants, infest, save the best 10 for corn borer resistance.
- Year 3 -- Plant the 10 S_1 selections in paired rows to make all possible S_1 combinations = 45 S_1 X S_1 singles.

Second cycle (C_2) of breeding
- Year 4 -- Plant the 45 crosses, infest, and save 2+ ears and plant with the best corn borer rating from each of the 45 S_1 crosses for a total of 100.
- Year 5 -- Grow these 100 S_1's, infest, save 10 with the best resistance.
- Year 6 -- Intercross the best 10 to make 45 S_1 X S_1 combinations.

Third cycle (C_3) of breeding
- Year 7 -- Plant the 45 crosses, infest, and save 2+ ears from plants with the best corn borer rating from each of the 45 S_1 crosses for a total of 100.
- Year 8 -- Grow the 100 S_1's, infest, save 10 with the best resistance.
- Year 9 -- Intercross the best 10 to make 45 S_1 X S_1 combinations.

The above procedure can be repeated for as many times as wanted.

A group of S_1 lines from the original population (C_0) and the populations derived from three cycles (C_1, C_2, C_3) of the selection in each variety were artificially infested in 1965 and 1966. As shown in Table 3, 2 cycles of selection were sufficient to shift the frequencies of resistance genes to a high level in all varieties. Three cycles of selection produced essentially borer-resistant varieties.

The use of synthetic varieties as pools of germ plasm for the development of new inbred lines of corn by both public and private plant-breeding agencies has increased with the availability of more sources of good parental lines. Improvement of these varieties for resistance to various insects and diseases would be desirable. Recurrent selection seems a useful breeding procedure for this purpose.

Modified Recurrent Selection.--A modified recurrent selection technique is being used in Ohio. A 24-line synthetic and a Cash synthetic are used. In 1 generation, the families are planted ear-to-row and selfed. In the next generation, each family is planted ear-to-row and allowed to wind-pollinate (female parent). The pollinator parent for the wind block is a composite of all families ear-to-row in the wind block. Thus, a broad gene base should be maintained, and also, segregation for various characteristics should be high. These plots are infested with corn borer egg masses and inoculated with leaf blight and stalk rot organisms. In this study, we are attempting to maintain a broad gene base and, at the same time, concentrate genes for resistance to corn borers, stalk rot, and leaf blight. We hope to select lines with high yielding ability and resistance to environmental hazards.

Genetic Methods (1st Brood)

The breeding methods used to develop resistant crop varieties are determined by 2 factors: (1) mode of reproduction in the crop species and (2) the kind of gene action that conditions resistance in the host plant to the insect (Russell, (1972).

The search for sources of resistance must precede the study of inheritance of resistance. Techniques for determining the various degrees of resistance in plant material are needed before studies of the inheritance of resistance can be started. When sources of resistance have been found, the genetics of resistance must be determined. A knowledge of the genetic basis of resistance does not necessarily preclude starting a breeding program. Usually, the breeder and entomologist can speculate enough of the genetics of resistance so that some preliminary breeding can be done. Detailed breeding plans should not be completed, however, until some information is obtained on the genetics of the character involved (Russell, (1972).

The genetic basis of leaf-feeding (1st-brood) resistance has been studied during the past few years. Conventional methods have been used in these studies. The information obtained thus far indicates that the various resistant inbreds may carry different factors conditioning resistance. In these studies, susceptible X resistant crosses are made, and segregation in the F_2, F_3 and backcross populations is determined. A high level of infestation is imperative in this type of material; "escapes" are a problem. These studies indicate that resistance is dominant in some inbred lines and that susceptibility is dominant in others.

Segregation of F_2 and Backcross Populations from Susceptible X Resistant Crosses.--The segregation of F_2 and backcross populations of a susceptible X resistant cross, M14 (susceptible) X (MS1) (resistant), indicates at least 3 gene pairs are involved in resistance, with at least partial phenotypic dominance of susceptibility (Penny and Dicke, 1956). In a B14 (susceptible) X N32 (resistant) cross, 1 or 2 gene pairs for leaf-feeding resistance were indicated on the basis of individual plant segregations in F_2 and backcrosses (Penny and Dicke, 1956).

In another susceptible X resistant cross, WF9 (susceptible) X gl_7V_{17} (resistant and homozygous for 2 very closely linked genes), segregation of F_2 and backcross populations showed that resistance of gl_7V_{17} was conditioned by a single dominant gene. The resistant gene was linked with gl_7V_{17} genes of the resistant parent with the cross-over frequencies estimated at from 31 to 37% (Penny and Dicke, 1957).

Use of Test Crosses in Breeding for Corn Borer Resistance.--In a study of test crosses in breeding corn for resistance to the European corn borer, we found that segregation in a 24-line synthetic variety, as measured by the net variance, diminished after each selfing, but that a significant residue of segregation remained in the 5th selfed generation. The 24-line synthetic contained the following inbred lines: susceptible lines -- Oh02, Oh26, Oh26A, Oh28, Oh65, Oh67A, A, B8, L289, M14, WF9, 38-11, Ia. 159L1; intermediate lines -- Oh7B, Oh33, Oh51A, Hy, I205, K155, P8; resistant lines -- Oh40B, Oh41, W22, W23.

Two single-cross testers, Oh51A X Oh26D (susceptible) and Oh43 X Oh45A (resistant), were used to evaluate each S_2 culture. A susceptible double-cross tester (Oh51A X Oh26F) (Oh26A X Oh26D) was used to evaluate each S_3 and S_5 culture. We had concluded from the ease of transferring resistance by backcrossing with selection in the improvement of Oh45 (Guthrie and Stringfield, 1961a) that corn borer resistance was simply inherited. But, if there was an average of one effectual heterozygous locus in the S5, theoretically there should have been 2^5 or at least 32 effectual heterozygous loci 5 generations back in the S_0 (Guthrie and Stringfield, 1961b).

Type of Gene Action Involved in Leaf-Feeding Resistance.--Scott, et al., (1964) used F_2, F_3, and selfed backcross populations of CI.31A (resistant) X B37 (susceptible) plus individual F_2 plants of (CI.31A X B37) X CI.31A and individual F_2 plants of (CI.31A X B37) X B37. The data indicate that most genetic variance was of the additive type, although a portion of the genetic variance was of the dominant type. The inheritance of resistance to leaf feeding is probably not as complex as that of yield. Since additive genetic variance was the major type of gene action, an efficient breeding program would be one that allows for the accumulation of desirable genes. Thus, a recurrent-selection program or mass selection should be effective. The value of a recurrent-selection technique has been determined and was discussed in a previous section.

Translocations.--Reciprocal translocations have been used in identifying chromosome arms involved in resistance to a 1st-brood infestation. Corn has 10 pairs of chromosomes; thus to have all chromosome arms involved, at least 20 translocations are needed (Scott, et al., 1966).

Anderson, et al. (1949) and Longley (1950, 1958, 1961) have produced large numbers of chromosomal translocations in corn, over 500 are available where the chromosomes concerned are known. The approximate position of the breaks and the chromosome arms involved are known in many instances. Translocation is the transfer of a chromosome to a nonhomologous chromosome. If only one segment is transferred, the alteration is designated as simple. Such a change is very rare, however, and may not occur at all. Most translocations involve a mutual exchange of terminal segments of nonhomologous chromosomes and is therefore designated as reciprocal. For purposes of identification and reference, the chromosomes may be designated according to their centromeres. A translocation is indicated by the letter T followed by the identification number of the chromosome involved. An interchange between chromosomes 1 and 4 would be designated T1-4. Since chromosomes 1 and 4 could exchange many different segments, each exchange is given a letter to specify it. A specific translocation involving chromosomes 1 and 4 would be T1-4a, T1-4d, etc. Translocations are common results of irradiation, and they do occur naturally in species.

Reciprocal translocations cause semisterility. In corn, about 50% of the pollen grains abort and about 50% of the ovules abort. The reason for semisterility is that, from a translocation ring, disjunction may be of 3 types: (1) alternate, (2) adjacent I, and (3) adjacent II. Alternate disjunction results in viable gametes; adjacent I and adjacent II disjunctions result in sterile gametes.

Resistant inbreds CI.31A and B49 were outcrossed to heterozygous translocation stocks (homozygous translocation stocks could have been used; then, all the F_1's would have been semisterile). A total of 29 translocations were used.

The semisterile F_1 plants were crossed to susceptible inbred lines WF9 and M14. The seed from these crosses were planted, and the plants were artificially infested with corn borer egg masses and rated in classes 1 to 9 on an individual-plant basis. At harvest, each plant was classified as normal or semisterile.

Chromosomal translocations that show association with a gene(s) for corn borer susceptibility should give those plants classified as semisterile a higher (more susceptible) corn borer rating than that of the normal plant; i.e., if there is a gene for resistance in the translocated segment, the normal plants should be more resistant than are the semisterile plants. This indicates that a gene for resistance from the resistant parent (CI.31A) is operative in the normal plants, but not in the semisterile plants. Thus, the gene for resistance must be located in the region of the translocation. If there is no association of semisterility and susceptibility, the translocated segment is not involved in corn borer resistance.

The translocation data (Scott, et al., 1966) indicate that inbred CI.31A has genes for resistance to corn borer leaf feeding on the short arm of chromosomes 1, 2, and 4 and on the long arm of chromosomes 4 and 6. Inbred B49 has genes for resistance on these same chromosome arms (possibly allelic to those of CI.31A) plus an additional gene for resistance on the long arm of chromosome 8.

Chromosomal translocations as used by Scott, et al. (1966) for determining the number of genes conditioning a character has the following limitations: (1) linked genes would probably be identified as a single gene, (2) recessive genes for resistance would not be detected, and (3) unless a gene has enough potency in the heterozygous condition to be measured as a significant difference, it would not be detected.

The use of translocations for determining which chromosome arm(s) have certain genetic factors is a good procedure for qualitative characteristics with a high degree of dominance for the expression of this character. As a character is conditioned by more genes for its complete expression (quantitative character), or the dominance of expression of the character becomes less pronounced, or both, the detection of these genes becomes progressively more difficult. Add to these conditions plants that can "escape" injury, and it becomes apparent that possibly not all genes that actually contribute to resistance will be detected. Thus, the number of genes Scott, et al. (1966) located should be considered the minimum. Genes with the greatest potency, however, were probably located.

Since resistance in CI.31A and B49 is conditioned by several genes, a breeding method by which genes for resistance could be accumulated in a population (i.e., mass selection, recurrent selection, etc.) would be effective. The value of a recurrent-selection technique in breeding corn resistant to the European corn borer was discussed in a previous section.

<u>Mutable Loci in Corn (Ac-Ds)</u>.--The mutable system as discussed here was discovered by one of the world's most renowned maize geneticists, Barbara McClintock (1950, 1951, 1953, 1956a, 1956b, 1957, 1958,1959, 1961a, 1961b). Dollinger (1956), a former student of McClintock's and presently maize geneticist at the Ohio Agricultural Research and Development Center, Wooster, has discussed a

method by which a mutable system may be used in plant breeding. Dollinger started this study to incorporate stalk rot resistance into inbred line Oh28. In 1958, we added a study to attempt to induce mutation for corn borer resistance into otherwise superior susceptible lines (Oh28 and WF9). E. J. Dollinger is conducting the breeding aspect of this research.

A mutable system is one that contains a gene-like component or controlling element that modifies and controls the action of genes. The controlling elements discovered to date act in many respects as dominant genes. The controlling elements differ from dominant genes in that their effects are upon other gene loci, inducing them to mutate, rather than directly upon the expression of a character. They also differ in that they may move from one location to another within the chromosome complement. Mutations may be induced that resemble those occurring spontaneously and those induced by mutagenic agents such as radiation and certain chemicals. Of particular importance to the plant breeder is that, for the first time, we have a means of producing dominant mutation. The induced mutations may be stabilized by removing the controlling element.

The mutable system may cause chromosome breakage. The controlling elements, however, may exist in states that induce mutations, either dominant or recessive, that do not involve chromosome breakage. For a plant breeder concerned with a diploid, this is a distinct advantage over mutagenic agents such as radiation of various kinds, which produce, for the most part, mutations resulting from chromosome breakage. In diploids such as maize, mutations involving chromosome breakage usually result in sterility; in polyploid plants, this is not necessarily so. Diploids will tolerate very little, if any, loss of chromatin material.

Controlling elements may modify the expression of many different genes in the chromosome complement. They may modify the genic expression affecting widely different characters such as anthocyanin, starch formation, morphology, etc. They may act to control the expression of other genes at any particular location.

The mutable system used at Wooster is designated the Ac-Ds system. This is a 2-unit system. Each unit is a type of controlling element. Ac controls the action of Ds. Ds may exist in states that produce chromosome breakage and in states that induce mutations without breakage. Mutation may be either dominant or recessive. Ds controlled mutation occurs only when Ac is present. A Ds-induced mutation will be stable upon the removal of Ac. Ds may move from one location to another in the chromosome complement without losing its identity, thus it is able to modify the expression of many different genes.

A general outline of the method used at Wooster is as follows: inbreds Oh28 and WF9, highly susceptible to the European corn borer and stalk rot, were outcrossed to a genetic stock that carries the mutable Ac-Ds system. A backcrossing program for transferring Ds and Ac to each elite line was continued for 6 generations until each elite line was recovered, except that they now carried the mutable system. Each elite line was the recurrent parent, and the genetic stock, the donor or nonrecurrent parent. The presence of both Ac and Ds must be identified in each backcrossing generation. For the identification of Ac and Ds, a genetic stock with Ds closely linked to the I locus on the short arm of chromosome 9 was used. The I locus, an allele of C, inhibits anthocyanin formation in the endosperm. But, in the appropriate background, when Ac is present, Ds will induce mutation at I. The resulting phenotype is mosaic. The endosperms are nonanthocyanin with anthocyanin mutant areas. Such phenotypes can be readily recognized.

In each backcross generation, some progeny may carry Ds that has moved to a new location. Up to the final backcross, however, only those plants in which Ds has not moved from its position near the I locus are selected. When the final backcross is made, an appropriate cross to a T tester will show that a majority of the gametes contained Ds near the I locus.

Large numbers of the recovered Oh28 or WF9 carrying the mutable system are planted in an isolated crossing plot. This material is the detasseled female parent, and the original Oh28 or WF9, the pollinator. If a plant resistant to stalk rot or corn borers occurs in the detasseled material, then Ds has moved near a locus concerned with resistance and has induced a dominant mutation for resistance. The progeny from the resistant plant then are tested for the presence of Ac. In resistant plants that do not possess Ac, the resistance will be stable.

Corn borer susceptible Oh28 (rr,CC) was crossed to a

$$\frac{\text{I Ds Sh Bz Wx}}{\text{C ds sh bz wx}} \text{ or } \frac{\text{I Ds Sh Bz Wx}}{\text{C ds Sh Bz Wx}} \text{ Ac/ac, } A_1\text{, } A_2\text{, RR stock.}$$

Kernels were selected that were I Ds and Ac from this cross. These kernels were distinguished by an endosperm with a colorless background and with Ds-type, anthocyanin mosaic areas. The independent, complementary dominant genes C, A_1, A_2, and R are necessary for anthocyanin development in the endosperm. Most inbred lines of corn are homozygous recessive for r, and some for one or more of the other anthocyanin loci as well; thus, no anthocyanin is produced in these endosperms. The I locus, an allele of C, inhibits anthocyanin formulation in the presence of the other dominant genes. One dose of R, if the Mt gene is present, will produce an anthocyanin mosaic type of endosperm. But, the Ds type of mosaic, resulting from the mutation of I, often can be distinguished when superimposed on the R-type mottling. If Oh28 was crossed to a genetic inbred containing the R locus, the F_1 endosperm would be heterozygous for one dose of R. In this particular instance, however, the material is recessive for mt; thus, identification of the I Ds type of anthocyanin mosaic is quite easy.

Individual F_1 plants from kernels whose endosperms were I Ds/C ds, Ac/ac were backcrossed to Oh28 and to a C ds sh bz wx ac stock. If an individual plant crossed to the C ds sh bz wx tester shows the appropriate mosaics from chromosome 9, then this F_1 plant was heterozygous for I Ds and Ac. The backcross of this plant to Oh28 is selected for the next backcross. This process is then repeated for a number of generations until Oh28 germ plasm is recovered (heterozygous for Ac and Ds).

It is necessary to cross to a C ds sh bz wx stock during the backcrossing program because the elite line is recessive for rr and perhaps another anthocyanin loci. Thus, the presence of I Ds, Ac cannot be detected in a background of this type. The C ds sh bz wx stock is used only to detect the presence of the mutable system in an individual plant and was subsequently discarded. If the original elite line is rr, A_1A_1, A_2A_2, CC, the R gene from the donor parent can be carried during the early backcrossing generations, and the presence of I Ds and Ac can be detected directly in each backcross, thus eliminating the need to cross to the C tester. In later backcross generations, however, the R locus must be eliminated from the material.

Since a character such as stalk rot resistance is expressed after pollination, a method must be used that permits controlled pollination of any resistant plant in the Oh28, Ac/ac, Ds/ds material (corn borer ratings can be made before pollination, but we have used the same method in our corn borer research as in the stalk rot experiments). This may be accomplished by a crossing plot. The Oh28, I ds Ac/ac or Ds new location, C ds, Ac/ac plants are used as the female to be detasseled, and the original Oh28 line is used as the pollinator.

If a plant resistant to corn borers appears, Ds has moved to a location in the chromosome complement that affects resistance. It is not necessary to know the location of this locus. The resistant plant should have the genetic constitution

$$\frac{Ds}{ds}\frac{Ac}{ac}.$$

Several thousand of these plants have been artificially infested with corn borer egg masses and checked for mutations. It is imperative to have a high level of establishment in this type of study. "Escapes" are a distinct problem. Several hundred plants with a low level of leaf feeding, have been saved as possible mutants. The resulting progeny from the Oh28 backcross of the resistant plant; (i.e.,

Oh28 $\frac{Ds}{ds}\frac{Ac}{ac}$ X Oh28) will be as follows:

1. $\frac{Ac}{ac}\frac{Ds}{ds}$ corn borer resistant, unstable
2. $\frac{ac}{ac}\frac{Ds}{ds}$ corn borer resistant, stable
3. $\frac{Ac}{ac}\frac{ds}{ds}$ corn borer susceptible
4. $\frac{ac}{ac}\frac{ds}{ds}$ corn borer susceptible

Several hundred progeny, which indicated resistance in the isolated plot, have been checked to determine if any are mutants or all are "escapes." All progeny checked thus far are "escapes." The progeny are planted ear-to-row, and if a mutation has occurred, half the plants in a row should be resistant

$$\frac{Ac}{ac}\frac{Ds}{ds} \text{ and } \frac{ac}{ac}\frac{Ds}{ds}$$

and half the resistant plants should be stable

$$\frac{ac}{ac}\frac{Ds}{ds}.$$

Each individual resistant plant in a row may be selfed and crossed to a I Sh Bz Wx Ds ac tester. The susceptible plants in a row are discarded. A number of progeny from the resistant plant X I sh Bz Wx Ds ac tester cross for each individual plant are grown and crossed to a C sh bz wx, ds, ac line to test for the presence of Ac. Absence of chromosome 9 mosaics (anthocyanin, waxy mosaics) will indicate the absence of Ac. Those resistant and stable plants

$$\frac{ac}{ac}\frac{Ds}{ds}$$

then can be traced because remnant seed was saved from each ear. These plants may be used to make a resistant line for use in the breeding program.

Corn borer resistance of most inbred lines seems inherited quantitatively. In general, for most characters studied, the more intensive the studies, the more loci that have been found that affect the expression of the character. In some instances, action at certain loci may affect expression of the character drastically while action at other loci affects expression very little. It seems not unreasonable to assume that a change in gene action could be induced at a locus that might affect corn borer resistance to a large degree.

The corn breeding plan as outlined in this section was arranged to select induced resistance of a dominant type, since this type of resistance would be most useful. Recessive mutants, however, may be selected by selfing

Oh28 $\frac{I\ Ds\ Ac}{C\ ds\ ac}$ plants and then placing them in a crossing block.

Biology of 2nd-Brood Larvae on the Corn Plant

During the period of egg deposition by the 2nd brood, early planted corn in the Corn Belt States has tasseled and has completed the pollen-shedding stage of plant development; late planted corn is in the pollen-shedding stage during part of the 2nd-brood oviposition period. The 1st- and 2nd-instar larvae feed primarily on pollen accumulation at the axils of the leaves and on sheath, collar, ear shoots, husk, and silk tissue (Dicke, 1950; Guthrie, et al., 1969, 1970). First-, 2nd-, 3rd-, and 4th-instar larvae can develop satisfactorily on a pollen diet (Guthrie, et al., 1969); these 4 larval instars also feed extensively on sheath and collar tissue (Guthrie, et al., 1970). Therefore, 2nd-brood resistance is actually collar- and sheath-feeding resistance; but, the husks and silks are also favorite larval feeding sites through 18 days of age (Guthrie, et al., 1970).

More than 95% of 2nd-brood larval mortality occurs within 3 days after egg hatch on inbred lines resistant to a 2nd-brood infestation, indicating a high degree of antibiosis to 1st- and 2nd-instar larvae of a 2nd-brood infestation (Guthrie, et al., 1970).

In research on host-plant resistance, the word "brood" is meaningless. The growth stage of the plant being attacked is important.

Methods for Measuring 2nd-Brood Resistance

In research on 2nd-brood resistance, egg masses incubated to near the hatching point are pinned through the leaf midrib under the ear leaf and under the leaf above and below the ear during the active pollen-shedding stage as described by Pesho, et al. (1965). The infestations are made in 2 or 3 applications of 2 masses each spread 2 or 3 days apart. In genetic studies or when selections are made in segregating material, several applications of egg masses may be used to avoid "escapes." Usually 10 plants in each plot are infested. Variability is introduced by applying egg masses over time under varying bioclimatic conditions. Since increased survival of 2nd-brood larvae is associated with anthesis (Dicke, 1950; Guthrie, et al., 1969), however, egg masses are applied at a comparable stage of plant development rather than in a comparable environment.

Number of cavities in the stalk and ear shank is used in evaluating resistance or susceptibility. A cavity 1/2 to 1-inch long is counted as a cavity, a cavity 6 inches long is counted as 6 cavities (Pesho, et al., 1965). Cavity counts are made 50 to 60 days after egg hatch because corn borer larvae are primarily external feeders through 20-25 days of age. Less than 40% of the larvae are located in the stalk on most inbred lines 35 days after egg hatch (Guthrie, et al., 1970); 50-60 days is ample time for the larvae to cause extensive damage to stalk tissue.

Lesion counts also can be used as an index for 2nd-brood resistance or susceptibility. Lesions in the sheath are calculated on the basis of the number of and size of the lesions; i.e., a lesion 1-inch long is counted as 1 lesion, but a lesion 6 inches long is counted as 6 lesions. A lesion that girdles 1/3 of the collar is counted as 1 lesion, a lesion that girdles 2/3 of the collar is counted as 2 lesions, and a lesion completely girdling the collar is counted as 3 lesions. A lesion completely girdling the sheath at the point of attachment to the node is counted as 3 lesions (Guthrie, et al., 1970). Lesion counts are more time-consuming than are splitting the stalk and counting cavities.

Plant damage as an index of relative resistance in research on both 1st- and 2nd-brood resistance is used in preference to insect counts because many factors, including disease, predation, and parasitism, can result in the absence of viable insect forms at the time of examination even though extensive plant damage is present (Pesho, et al., 1965). Inbred lines highly susceptible to a 2nd-brood infestation may be so badly damaged that the plant is no longer suitable as a source of food; therefore, many larvae may leave the plant before it is examined.

Inbred lines highly susceptible to a 2nd-brood infestation suffer extensive sheath, collar, stalk, and shank damage; resistant inbred lines suffer little damage.

Second brood resistance research is much slower than 1st-brood resistance research because techniques are more cumbersome.

Corn Breeding Methods for Selecting for Both 1st- and 2nd-Brood Resistance in the Same Plant Populations

Resistance to leaf feeding (1st brood) has been easy to find (Stringfield, 1959; Guthrie, et al., 1960; Guthrie and Dicke, 1972). Based on the evaluation of 114 inbred lines (Pesho, et al., 1965) and 159 inbred selections from W. A. Russell's breeding nursery, sheath feeding resistance (2nd brood) occurs less frequently.

Inbred lines of corn evaluated for resistance to sheath feeding (2nd brood) differ in degree of susceptibility, but not in degree of resistance. B52 is the only inbred available to date with a high level of 2nd-brood resistance (antibiosis). Inbred lines or hybrids with more than 10 cavities per plant (10 inches of damage inside the stalk) are considered susceptible. Inbreds such as WF9 and Oh43 (20-23 cavities per plant) are highly susceptible to a 2nd-brood infestation. Inbreds W182-E and NN14 (28-30 cavities per plant) are even more susceptible than are WF9 and Oh43. All stalks of W182-E and NN14 are usually completed tunneled by 2nd-brood larvae. Inbred lines and hybrids also differ in tolerance to a 2nd-brood infestation.

From a total of 103 experimental inbred lines that have a high level of 1st-brood resistance, none had a satisfactory level of 2nd-brood resistance (Guthrie, et al., 1972). More sources of 2nd-brood resistant germ plasm need to be located.

Research is underway for evaluating breeding methods for selecting for both 1st- and 2nd-brood resistance in the same plant population.

In one experiment involving B52 X Oh43, selections are being made in the F_3, F_4, etc. generations. In a 2nd experiment, 5 F_4 lines derived from F_3 progenies that had excellent 2nd-brood resistance, but only intermediate 1st-brood resistance were backcrossed to Oh43 in 1971. In these 2 experiments, plants are infested during the midwhorl stage of growth and are selected for 1st-brood resistance in our Ames nursery. Selections with good 1st-brood resistance are infested during the pollinating stage of plant development in our 2nd-brood nursery at Ankeny. Entries with good 2nd-brood resistance are selected for further evaluation. In a 3rd experiment, 20 F_4 progenies from B52 X Oh43 were recombined into a synthetic; this population may be used in recurrent selection breeding for selecting for 1st- and 2nd-brood resistance in the same plant populations. The objective of these 3 experiments is to retain Oh43 resistance to the 1st brood and add B52 resistance to the 2nd brood.

In a 4th experiment, a synthetic variety was made by using inbreds B49, B50, B52, B54, B55, B57, B68, CI.31A, Mo17, and SD10. Inbreds B49 and CI.31A contribute high resistance to 1st-brood larvae. B50, B54, B55, B57, B68, and Mo17 contribute intermediate resistance. Inbred B52 contributes high resistance to 2nd-brood larvae. A small amount of 2nd-brood resistance is contributed by B49, B50, B55, B57, and B68. Inbred SD10 was included for early maturity and tolerance to the western corn rootworm, *Diabrotica virgifera* LeConte. A recurrent selection technique will be used for selecting for both 1st- and 2nd-brood resistance in this 10-line synthetic (Guthrie, et al., 1972). Recurrent selection is effective in selecting for 1st-brood resistance (Penny, et al., 1967).

Genetic Methods (2nd Brood)

Considerable information has been obtained on the genetic basis of leaf-feeding resistance (1st brood) by the European corn borer (Patch, et al., 1942; Schlosberg and Baker, 1948; Singh, 1953; Ibrahim, 1954; Penny and Dicke, 1956, 1957; Scott, et al., 1964, 1966).

During the past 4 years, we have determined to some extent the type of gene action involved in sheath feeding resistance (2nd brood). Generation mean studies, involving P_1, P_2, F_1, F_2, F_3, BC_1, BC_2, $BC_1(x)$ and $BC_2(x)$ populations show that the resistance of B52 is dominant or at least partly dominant in crosses with Oh43, B39, and L289. The dominance of B52 with WF9 was not as clear-cut. Oh 43, B39, L289, and WF9 are highly susceptible to 2nd-brood larvae (Guthrie, et al., 1972). Data from 45 diallel crosses among 10 inbred lines indicate that the high resistance of B52 is transmitted in hybrid combination (Scott, et al., 1967).

We will use reciprocal translocations to determine chromosome arms involved in sheat feeding resistance (2nd brood).

Table 1.--Ingredients for European corn borer diet. Ankeny, Iowa, 1972. (1 batch = 15,243 g).

Ingredient	Price	Quantity (1 batch)	Cost/1000 dishes (930 g diet/dish)
Water		13,000 g	
Agar[a]	$4.86/lb	280 g	$200.62
Wheat germ[a]	0.2775/lb	520 g	21.28
Dextrose[a]	0.25/lb	400 g	14.75
Casein[a]	1.97/lb	440 g	127.26
Cholesterol[a]	7.25/lb	32 g	34.15
Salt mixture #2[a]	1.04/lb	144 g	22.13
Vitamin supplement[a]	21.50/k	92 g	131.15
Ascorbic acid[b]	6.35/k	120 g	50.67
Aureomycin[c]	1.25/6.4 oz	9 teaspoon	13.00
Fumidil B[d]	29.20/9.5 g	6.9 g	139.16
Methyl p hydroxybenzoate[e]	2.35/100 g	75 ml	38.00
Propionic acid[f]	0.36/lb	86 ml	
Formaldehyde[g]	1.32/gal	7 ml	0.18
Sorbic acid[h]	4.50/500 g	40 ml	4.82
Total			$797.17

[a] Nutritional Biochemicals Corp., Cleveland, Ohio

[b] Merck & Co., Inc., 4545 Oleatha Avenue, St. Louis, Missouri

[c] Iowa Veterinary Supply Co., Box 616, Iowa Falls, Iowa

[d] Dadant & Sons, Hamilton, Illinois.

[e] Dissolve 56 g methyl p hydroxybenzoate in 200 ml 95% ethyl alcohol

[f] Mix 418 ml propionic acid with 82 ml distilled H_20; mix 42 ml phosphoric acid ($0.36/lb) with 458 ml distilled H_20; mix the propionic-distilled H_2O with the phosphoric-distilled H_2O.

[g] 40% formalin

[h] Dissolve 100 g sorbic acid in 500 ml 95% ethyl alcohol.

Table 2.--Approximate cost of producing 1,000,000 egg masses (excluding equipment).

Item	Hours	Wages @ $2.00/hr
Dissect 6000 larvae from corn stalks to start new culture	96	192.00
Clean 13,6000 vials	24	48.00
Plug 13,000 vials with cotton and sterilize	48	96.00
Cut plugs of diet for 13,600 vials	30	60.00
Place larvae in 13,600 vials for a source of moths for egg production for infesting dishes	30	60.00
Place pupae from 13,600 vials in cages (egg production for dishes)	6	12.00
Wax strips for 1000 dishes	32	64.00
Cook diet for 1000 dishes	100	200.00

Table 2. continued.

Item	Hours	Wages @ $2.00/hr
Punch 40,000 egg masses to infest 1000 dishes (40 masses/ dish)	60	120.00
Prepare 1000 dishes for incubation	22	44.00
Take corrugated strips out of dishes and wash 1000 dishes	30	60.00
Collect moths (100,000 females and 100,000 males) and place in oviposition cages	125	250.00
Punch out 1,000,000 egg masses for field infestations	1336	2,672.00
Place 1,000,000 egg masses on plants (1st brood)	500	1,000.00
Clean 500 oviposition cages	40	80.00
Diet ingredients		797.17
Pins (1,000,000 @$3.15/5000) (Union Pin Co., Winsted, Conn.)		630.00
Wax paper for oviposition (12,000 sheets 6 x 24") (72 sheets/ lb = 167 lb @ $32.40/cwt) (Carpenter Paper Co., Des Moines, Iowa)		54.00
Parawax for waxing corrugated strips (85 lb @ $0.20)		17.00
Corrugated paper for 1000 dishes (25 rolls 2 3/4" x 250' @ $1.50/roll) (Butler Paper Co., Des Moines, Iowa)		37.50
Total		6,493.67

Table 3.--Frequency distribution of mean corn borer leaf feeding ratings of S_1 lines in four populations of five synthetic varieties.

Cycle of selection	Class intervals of corn borer ratings[a/]								Mean rating
	1.0-2.0	2.1-3.0	3.1-4.0	4.1-5.0	5.1-6.0	6.1-7.0	7.1-8.0	8.1-9.0	
Synthetic A - rated in 1965									
C_0	8	8	15	17	17	10	9	0	4.8
C_1	24	28	19	7	4	2	0	0	3.0
C_2	49	20	9	5	1	0	0	0	2.3
C_3	67	12	3	1	1	0	0	0	1.9
Pennsylvania late synthetic - rated in 1965									
C_0	5	24	27	20	13	9	2	0	4.1
C_1	36	29	17	15	2	1	0	0	2.8
C_2	66	24	7	3	0	0	0	0	2.0
C_3	73	23	4	0	0	0	0	0	1.8
Pennsylvania early synthetic - rated in 1966									
C_0	0	2	7	22	18	27	16	8	6.1
C_1	6	10	25	26	20	8	5	0	4.6
C_2	16	44	24	8	7	1	0	0	3.2
C_3	17	52	21	8	1	0	0	0	2.9
Pennsylvania intermediate synthetic - rated in 1966									
C_0	0	2	4	19	33	27	10	5	6.0
C_1	6	19	27	26	13	5	4	0	4.2
C_2	10	36	31	15	7	1	0	0	3.5
C_3	20	44	22	11	2	1	0	0	3.0

Table 3.--continued.

Cycle of selection	Class intervals of corn borer ratings[a/] 1.0-2.0	2.1-3.0	3.1-4.0	4.1-5.0	5.1-6.0	6.1-7.0	7.1-8.0	8.1-9.0	Mean rating
				Synthetic B - rated in 1966					
C_0	0	0	6	18	28	30	7	11	6.2
C_1	5	19	25	26	14	8	2	1	4.3
C_2	11	32	31	13	9	4	0	0	3.5
C_3	18	42	24	13	3	0	0	0	3.1

[a/]Rated in classes 1 = least to 9 = highest infestation level.

Literature Cited

Anderson, E. G., A. G. Longley, C. H. Li and K. L. Retherford. 1949. Hereditary effects produced in maize by radiations from the Bikini atomic bomb. 1. Studies on seedlings and pollen of exposed generation. Genetics 34: 639-46.

Dicke, F. F. 1950. Response of corn strains to European corn borer infestations. Proc. N. Centr. Br. Entomol. Soc. Amer. 5: 47-49.

Dicke, F. F. 1954. Breeding for resistance to European corn borer. Proc. Ann. Hybrid Corn Industry Res. Conf. 9: 44-53

Dollinger, E. J. 1956. The possible uses of genes that induce mutation. Proc. Ann. Hybrid Corn Industry Res. Conf. 11: 92-102.

Guthrie, W. D. and G. H. Stringfield. 1961a. The recovery of genes controlling corn borer resistance in a backcrossing program. J. Econ. Entomol. 54: 267-270.

Guthrie, W. D. and G. H. Stringfield. 1961b. Use of test crosses in breeding corn for resistance to European corn borer. J. Econ. Entomol. 54: 784-787.

Guthrie, W. D. and S. W. Carter. 1972. Backcrossing to increase survival of larvae of a laboratory culture of the European corn borer on field corn. Ann. Entomol. Soc. Amer. 65: 108-109.

Guthrie, W. D. and F. F. Dicke. 1972. Resistance of inbred lines of dent corn to leaf feeding by 1st-brood European corn borers. Iowa State J. Sci. 46: 339-357.

Guthrie, W. D., F. F. Dicke and C. R. Neiswander. 1960. Leaf and sheath feeding resistance to the European corn borer in eight inbred lines of dent corn. Ohio Agr. Exp. Sta. Res. Bull. 860. 38 pp.

Guthrie, W. D., F. F. Dicke and G. R. Pesho. 1965a. Utilization of European corn borer egg masses for research programs. Proc. N. Centr. Br. Entomol. Soc. Amer. 20: 48-50.

Guthrie, W. D., E. S. Raun, F. F. Dicke, G. R. Pesho and S. W. Carter. 1965b. Laboratory production of European corn borer egg masses. Iowa State J. Sci. 40: 65-83.

Guthrie, W. D., J. L. Huggans and S. M. Chatterji. 1969. Influence of corn pollen on the survival and development of second-brood larvae of the European corn borer. Iowa State J. Sci. 44: 185-192.

Guthrie, W. D., J. L. Huggans and S. M. Chatterji. 1970. Sheath and collar feeding resistance to the second-brood European corn borer in six inbred lines of dent corn. Iowa State J. Sci. 44: 297-311.

Guthrie, W. D., W. A. Russell and C. W. Jennings. 1972. Resistance of maize to second-brood European corn borers. Proc. Ann. Corn and Sorghum Res. Conf. 27: 165-179.

Huggans, J. L. and W. D. Guthrie. 1970. Influence of egg source on the efficacy of European corn borer larvae. Iowa State J. Sci. 44: 313-353.

Ibrahim, M. A. 1954. Association tests between chromosomal interchanges in maize and resistance to the European corn borer. Agron. J. 46: 293-298.

Klun, J. A. and T. A. Brindley. 1966. Role of 6-methoxybenzoxazolinone in inbred resistance of host plant (maize) to first-brood larvae of European corn borer. J. Econ. Entomol. 59: 711-718.

Klun, J. A., W. D. Guthrie, A. R. Hallauer and W. A. Russell. 1970. Genetic nature of the concentration of 2,4-dihydroxy-7-methoxy 2H-1,4-benzoxazin-3(4H)-one and resistance to the European corn borer in a diallel set of eleven maize inbreds. Crop. Sci. 10: 87-90.

Lewis, L. C. and R. E. Lynch. 1969. Rearing the European corn borer, *Ostrinia nubilalis* (Hubner), on diets containing corn leaf and wheat germ. *Iowa State J.* Sci. 44: 9-14.

Lewis, L. C. and R. E. Lynch. 1970. Treatment of *Ostrinia nubilalis* larvae with Fumidil B to control infections caused by *Perezia pyraustae*. J. Invert. Pathol. 15: 43-48.

Longley, A. E. 1950. Cytological analysis of translocations in corn chromosomes resulting from ionizing radiation of the test Able atomic bomb and X-rays and of translocations from other sources. Rept. of Naval Med. Res. Sec. Joint Task Force One, on Biol. Aspects of Atomic Bomb Tests. Appendix #10: 1-60.

Longley, A. E. 1958. Breakage points for two corn translocation series maintained at the California Institute of Technology. USDA, ARS-34-4.

Longley, A. C. 1961. Breakage points for four corn translocation series and other corn chromosome aberrations. USDA, ARS-34-16: 1-40.

Lynch, R. E. and L. C. Lewis. 1971. Re occurrence of the microsporidian, *Perezia pyraustae* in the European corn borer, *Ostrinia nubilalis*, reared on *diet containing* Fumidil B. J. Invertebr. *Pathol. 17: 243-246*.

McClintock, B. 1950. The origin and behavior of mutable loci in maize. Proc. Nat. Acad. Sci. 36: 344-355.

McClintock, B. 1951. Chromosome organization and genic expression. Cold Springs Harbor Sym. Quan. Biol. 16: 13-47.

McClintock, B. 1953. Induction of instability at selected loci in maize. Genetics 38: 579-599.

McClintock, B. 1956a. Controlling elements and the gene. Cold Springs Harbor Sym. Quan. Biol. 21: 197-216.

McClintock, B. 1956b. Mutation in maize. Ann. Rept. Dept. Genetics, Carnegie Inst. Washington Yearbook: 323-332.

McClintock, B. 1957. Genetic and cytological studies in maize. Ann. Rept. Dept. Genetics, Carnegie Inst. Washington Yearbook: 393-401.

McClintock, B. 1958. The suppressor-mutator system of control of gene action in maize. Ann. Rept. Dept. Genetics, Carnegie Inst. Washington Yearbook: 415-429.

McClintock, B. 1959. Genetic and cytological studies of maize. Ann. Rept. Dept. Genetics, Carnegie Inst. Washington Yearbook: 452-456.

McClintock, B. 1961a. Some parallels between gene control systems in maize and in bacteria. American Naturalist 95: 265-277.

McClintock, B. 1961b. Further studies of the suppressor-mutator system of control of gene action in maize. Ann. Rept. Dept. Genetics, Carnegie Inst. Washington Yearbook: 469-476.

Patch, L. H., J. R. Holbert and R. T. Everly. 1942. Strains of field corn resistant to the survival of the European corn borer. USDA Tech. Bull. 823. 22 pp.

Penny, L. H. and F. F. Dicke. 1956. Inheritance of resistance to leaf feeding of the European corn borer. Agron. J. 48: 200-203.

Penny, L. H. and F. F. Dicke. 1957. A single gene-pair controlling segregation for European corn borer resistance. Agron. J. 49: 193-196.

Penny, L. H. and F. F. Dicke. 1959. European corn borer damage in resistant and susceptible dent corn hybrids. Agron. J. 51: 323-326.

Penny, L. H., G. E. Scott and W. D. Guthrie. 1967. Recurrent selection for European corn borer resistance in maize. Crop Sci. 7: 407-409.

Pesho, G. R., F. F. Dicke and W. A. Russell. 1965. Resistance of inbred lines of corn (*Zea mays* L.) to the second brood of the European corn borer (*Ostrinia nubilalis* (Hubner). Iowa State J. Sci. 40: 85-98.

Rathore, Y. S. and W. D. Guthrie. 1972. Survival on field corn of European corn borer larvae reared for different generations on a meridic diet. J. Econ. Entomol. (in press).

Rathore, Y. S., W. D. Guthrie and S. W. Carter. 1972. Inheritance of decreased survival on field corn of European corn borer larvae from cultures continuously reared on a meridic diet. Ann. Entomol. Soc. Amer. (in press).

Reed, G. L., W. B. Showers, J. H. Huggans and S. W. Carter. 1972. Improved procedures for mass rearing the European corn borer. J. Econ. Entomol. (in press).

Russell, W. A. 1972. A breeder looks at host-plant resistance for insects. Proc. N. Centr. Br. Entomol. Soc. Amer. 27: 77-87.

Schlosberg, M. and W. A. Baker. 1948. Tests of sweet corn lines for resistance to European corn borer larvae. J. Agr. Res. 77: 137-156.

Scott, G. E. and F. F. Dicke. 1965. Types of gene action of resistance in corn to leaf feeding of the European corn borer. Crop Sci. 5: 487-489.

Scott, G. E., F. F. Dicke and G. R. Pesho. 1966. Location of genes conditioning resistance in corn to leaf feeding of the European corn borer. Crop Sci. 6: 444-446.

Scott, G. E., A. R. Hallauer and F. F. Dicke. 1964. Types of gene action conditioning resistance to European corn borer leaf feeding. Crop Sci. 4: 603-606.

Scott, G. E., W. D. Guthrie and G. R. Pesho. 1967. Effect of second-brood European corn borer infestations on 45 single-cross corn hybrids. Crop Sci. 7: 229-230.

Singh, R. 1953. Inheritance of maize of reaction to the European corn borer. Indian J. Genetics and Plant Breeding 13: 18-47.

Sparks, A. N. 1963. Preliminary studies of factors influencing mating of the European corn borer. Proc. N. Centr. Br. Entomol. Soc. Amer. 18: 95.

Stringfield, G. H. 1959. Maize inbred lines of Ohio. Ohio Agr. Exp. Sta. Res. Bull. 831. 67 pp.

TECHNIQUES, ACCOMPLISHMENTS, AND FUTURE POTENTIAL OF BREEDING FOR RESISTANCE IN CORN TO THE CORN EARWORM, FALL ARMYWORM, AND MAIZE WEEVIL; AND IN SORGHUM TO THE SORGHUM MIDGE

B. R. Wiseman, W. W. McMillian and N. W. Widstrom
Southern Grain Insects Research Laboratory
ARS, USDA, Tifton, Georgia

The host plant resistance program at the Southern Grain Insects Research Laboratory involves a multidisciplinary approach. The team consists primarily of a geneticist and two entomologists. The broad goals established are: (1) screen the world collection of corn for resistance to corn earworm, *Heliothis zea* (Boddie) and fall armyworm, *Spodoptera frugiperda* (J. E. Smith), and the world collection of sorghum for sorghum midge, *Contarinia sorghicola* (Coquillett), resistance, (2) determine the mechanisms of resistance involved (nonpreference, antibiosis, and tolerance) and elucidate the basis or cause of the resistance, and (3) develop usable resistant germ plasm for release to the public.

Some of the basics of our host plant resistance research have been (1) to develop techniques for mass production of the insects researched, (2) to develop infestation procedures for field evaluations, and (3) to develop measurement systems for separating plants differing in insect damage, i.e., the resistant from the susceptible.

Infestation Techniques

Mass production of lepidopterous insects by the Southern Grain Insects Research Laboratory is more than adequate for our infestation needs. This requirement has been met by our capable personnel in the rearing section of the laboratory.

Corn.--*Corn earworm*.--Artificial infestations have been made using all ages of instars, adults, and eggs. Most of our plot work infestations have been made using first-instar larvae or eggs. Newly hatched larvae are taken to the field in an ice cooler and transferred to silk masses with a camel's hair brush at three larvae per silk mass. A small polyethylene test tube filled with water and strapped to the hand was stoppered with a cork with a small hole bored in its center. This was used as a source of moisture for the brush so that the larvae may be easily picked up and transferred to the silk masses.

Infestations using corn earworm eggs were initiated by Widstrom and Burton in 1970. They used eggs suspended in 0.25% agar solution and injected into the silk mass with a hypodermic syringe. Double and single applications of 10, 20, and 30 eggs were tested against 3 larvae per silk and an uninfested check. More recently we have refined these techniques and are now using a 0.2% agar solution and a single application into the tip of the silk mass of 30 to 35 eggs suspended in 0.4 milliliters of agar solution. We have adopted a pressure applicator that is normally used for dispensing hand lotion. Egg infestations may be effectively made any time from 100% silk to 7 days following. Other applicators tested were a chromatographic sprayer, a syringe, and a squeeze bottle. All of them were more time consuming and/or far less accurate than the pressure applicator. Damage from egg infestation levels and techniques adopted have been at least equal to that obtained by three larvae per silk mass.

Laboratory infestations for evaluating insect feeding responses (feeding stimulant) have been accomplished as illustrated in Figure 1 and reported by McMillian, et al. (1970). Extracts of corn kernels, leaves, or silks, reconstituted in a ratio of one-gram lyophilized residue per six milliliters distilled water, are placed on Oxalis violacea (L.) leaves (Wiseman, et al., 1969) or filter paper at the rate of 0.1 milliliter extract per substrate. Treatments are rated

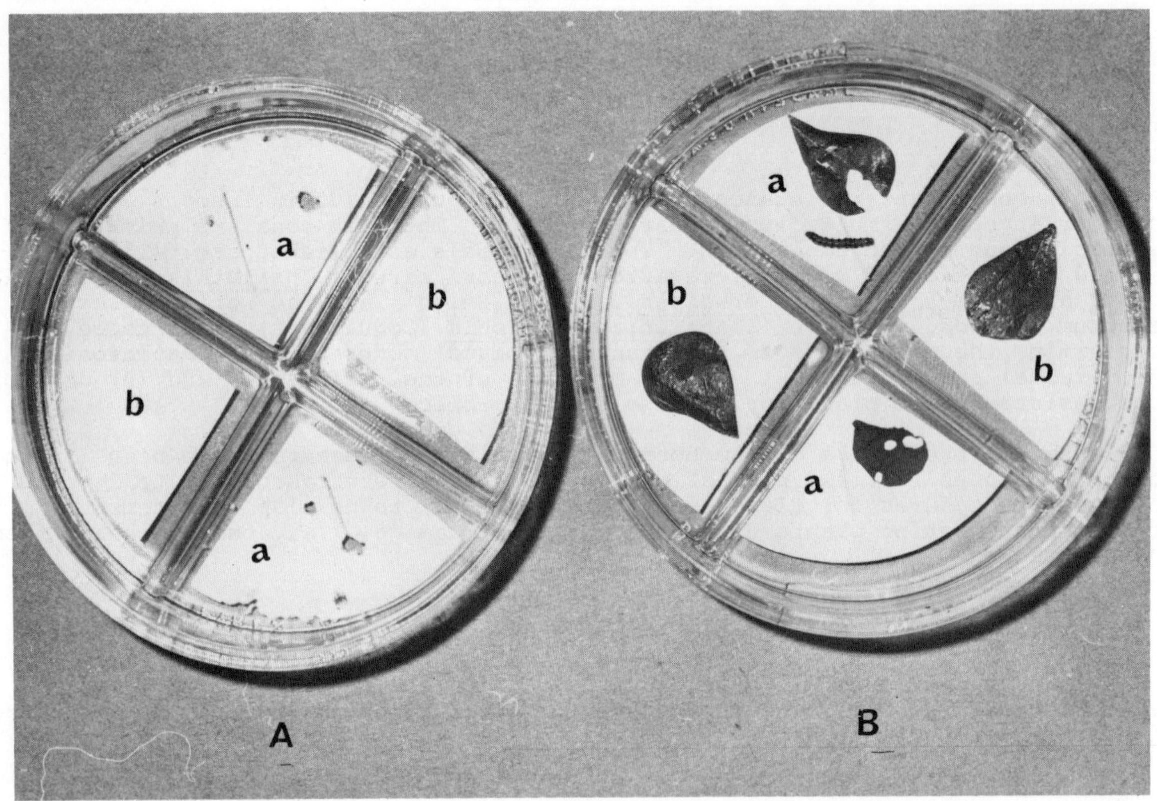

Figure 1.--Typical dishes set up for bioassaying showing responses to phagostimulative substances. A, used for fourth, fifth, and sixth instars; B, used for first, second, and third instars. (Carriers extract-treated, a, and water-treated, b.).

18 hours after exposure to corn earworm larvae (McMillian, et al., 1967).

The use of corn earworm adults as a means of determining differences between corn lines or plant chemicals has not been particularly rewarding. However, at present we are studying oviposition preference of adults for corn in the whorl stage. Nonpreference for oviposition to whorl-stage corn could have drastic and lasting effects on the buildup of corn earworm populations.

Fall armyworm.--Artificial infestations have been made using first-instar larvae applied to whorl-stage or seedling corn (Wiseman, et al., 1966). This technique is very laborious, and large-scale rapid screening is virtually impossible. Just recently McMillian and Wiseman (1972) developed a technique whereby they separated fall armyworm egg masses into single eggs. Adequate hatch was obtained when these eggs were applied to laboratory diets and on seedling corn in the greenhouse. This technique could enhance our resistance work with the fall armyworm immensely if details for field applications of eggs can be worked out. The technique for separating fall armyworm egg masses is as follows:

1. Use Scott® paper towel for adults to oviposit on.
2. Cut material into two-inch squares.
3. Place paper sections into flask in a quantity that 200 milliliters of 0.05 molar KOH will cover.
4. Shake vigorously for 4 minutes.
5. Add 200 milliliters of 0.05 molar HCl.
6. Shake vigorously for 2 minutes.
7. Add 40 milliliters of pH 7 buffer solution.
8. Shake vigorously for 1 minute.
9. Filter eggs - rinse for 20 minutes.
10. Drain off most of water and add eggs to 0.2% agar solution.
11. Suspend eggs in concentration desired.
12. Apply eggs to diet cups or plants.

Maize weevils.--Artificial infestation of maize weevils, *Sitophilus zeamais* Motschulsky, has been accomplished since 1967, using the birdhouse method developed by McMillian, et al. (1968a). This method utilizes shelled corn and laboratory-reared weevils at the rate of 1000 weevils per birdhouse placed in test plots at one for each 1000 square feet; adequate populations build up so that 100% infestation is assured.

Sorghum.--Sorghum midge.--Infestations have been accomplished by very early plantings of an ultra-susceptible sorghum line, usually PI-29166. The sorghum midge are allowed to build up initial populations so that screening of the world sorghum collection can be subjected to midge attack throughout the flowering period. This situation is a type of "artificial" infestation and has proven very satisfactory for our needs.

Resistance Measurements

Corn Earworm.--The method of measuring damage to corn by the corn earworm has usually been a visual rating scale or an estimate of kernel loss. Widstrom (1967) evaluated several corn earworm damage measurements. A system was devised called the revised centimeter scale where 0 = no damage, 1 = silk damage, 2 = ear tip damage to a depth of one centimeter, and 3 to n = damage increased by one unit for each additional centimeter depth of penetration. This revised centimeter scale for measuring corn earworm damage has been used to date in rating plots of corn for resistance.

Other measurements of resistance in corn to the corn earworm have been egg counts and percent infestation.

Laboratory measurements of responses of corn earworm larvae to extracts of corn have been made as the amount of previously treated *Oxalis violacea* (L.) (Wiseman, et al., 1969) or filter paper (Figure 2) consumed by using a grid divided into square millimeters as reported by McMillian et al. (1966, 1967, 1970).

A photoelectric counter to monitor olfactory response of corn earworm moths was developed by Starks, et al. (1966) to detect behavioral responses to corn plant extracts. Flight activities of the moths at the test areas are recorded when

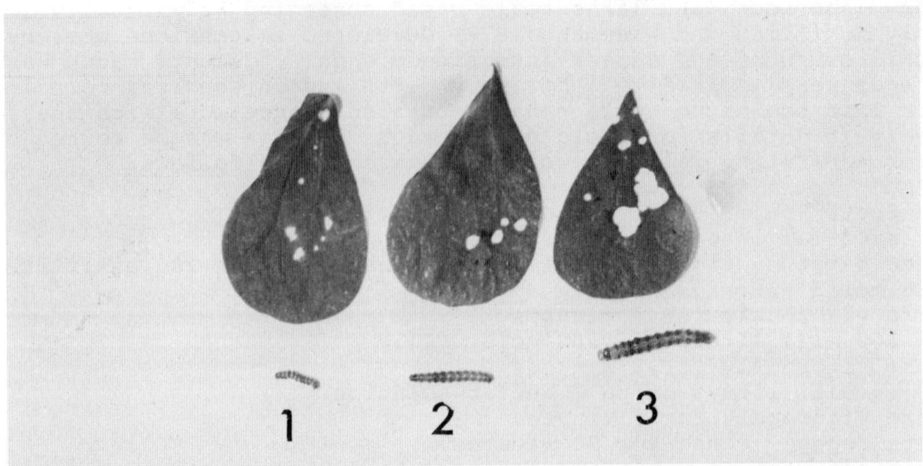

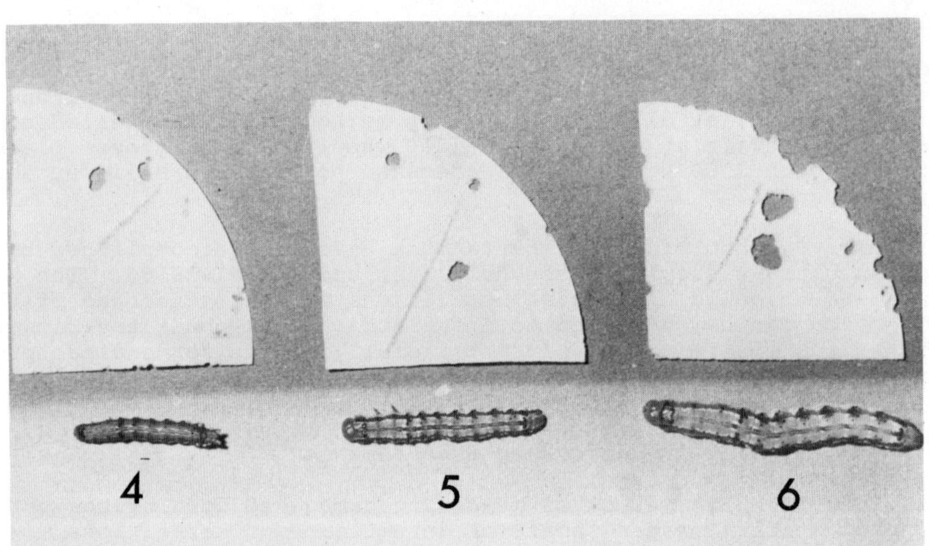

Figure 2.--Illustrations to show relative sizes of six instars of corn earworm and relative area consumed by each instar.

the moths pass through the provided light beam. The number of times the beam to the photo-cell is broken at a particular extract site is considered a function of its attractiveness.

Plant damage caused by insect feeding in relation to the ability of insects to use their food source was researched by McMillian, et al. (1966) and Wiseman, et al. (1970a). McMillian, et al. (1966) adapted a method whereby chromic oxide was introduced into diet material and fed to larval insects. Then the excreta was analyzed for chromic oxide by the colorimetric method and the percent use was calculated as follows:

$$\% = \frac{1 - A/B}{C} \times 100$$

where: A = percent of Cr_2O_3 in diet (by analysis, dry basis)
B = percent of Cr_2O_3 in excreta (dry basis)
C = fractional part of sample available for use.

Fall Armyworm.--Fall armyworm injury has been rated using a visual scale of 0-10 as devised by Wiseman, et al. (1966) where 0 = no damage; 1 = small amount of pin-hole type injury; 2 = several pin holes; 3 = small amount shot-hole type injury with 1 or 2 lesions; 4 = several shot-hole type injuries and a few lesions; 5 = several lesions; 6 = several lesions, shot-hole injury and portions eaten away; 7 = several lesions and portions eaten away with some areas dying; 8 = several portions eaten away and areas dying; 9 = the whorl almost or completely eaten away and several lesions with more areas dying; and 10 = plant dead, dying, or almost completely destroyed. We have used this system to date on both seedling corn in the greenhouse and whorl stage corn in the field. McMillian and Starks (1967) used a slightly different visual rating scale of one to nine when they evaluated sorghum for fall armyworm resistance.

Maize Weevil.--Ratings of maize weevil damage have been made using a visual scale for infestation or damage of one to ten as outlined by Wiseman, et al. (1970 b) when 1 = 0-10% of the kernels damaged and 2-10 = 11-100% of the kernels damaged in 10% increments.

Sorghum Midge.--Visual ratings of sorghum midge damage (Wiseman and McMillian, 1970) (Figure 3) are usually made on ten individual sorghum heads per plot using a scale 0-10, with 0 = no damage, 1 = $> 0 \leq 10\%$ of head damaged, and 2-10 = $> 10 \leq 100\%$ of head damaged.

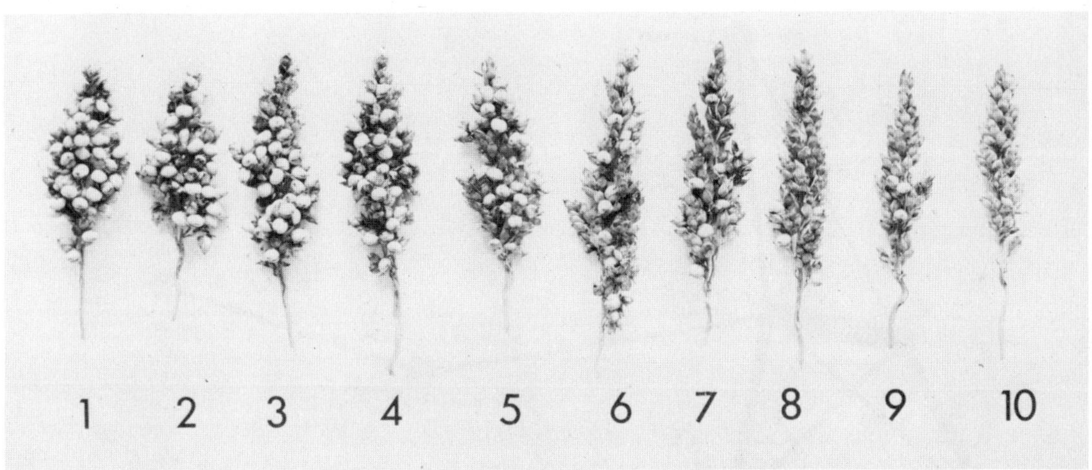

Figure 3.--Rating scale for sorghum midge damage illustrated by damaged spikelets where: 0 = no damage, 1 = $> 0 \leq 10\%$ of head damaged, and 2-10 = $> 10 \leq 100\%$ of head damaged.

Accomplishments

Corn.--Resistance in maize to the corn earworm has been known for many years. Listed below are some of the characteristics that have been associated with resistance or may be useful in resistance studies (Starks and McMillian, 1967; McMillian, et al., 1966; 1967; McMillian and Wiseman, 1972; McMillian and Starks, 1966; Wiseman, et al., 1967b).

Resistance Character*	Corn Line
Poor larval growth (CEW)	380, Oh26F, Tx727, 245, SC335
Low feeding stimulant (CEW)	L501, C17
Low larval survival (CEW)	380, PI 217413, 166, 81-1, 221, Asgrow 101W
Husk length (CEW)	81-1
Husk tightness (CEW)	Zapalote Chico
Low utilization (CEW)	166
Silk penetration (CEW)	SC97
Low kernel feeding (CEW)	F6, M119
Low egg count on silk (CEW)	Zapalote Chico
Oviposition nonpreference (FAW)	Antigua 2D
Leaf feeding nonpreference (FAW)	Antigua 2D

*Insects resisted are corn earworm (CEW) and fall armyworm (FAW)

Widstrom, et al. (1970) found that of 14 plant characters tested among 36 inbreds, only husk characters and feeding stimulant (Figure 4) provided information closely enough related to damage to be used in identifying genotypes resistant to injury by the corn earworm. Wiseman, et al. (1970b) reported that husks of commercial hybrids provided significant protection against several insect species (Table 1) and contributed to increased yields. Among some of the hybrids evaluated, both husk and kernel resistance were found for the earworm, maize weevil, and pink scavenger caterpillar, Sathrobrota rileyi (Walsingham).

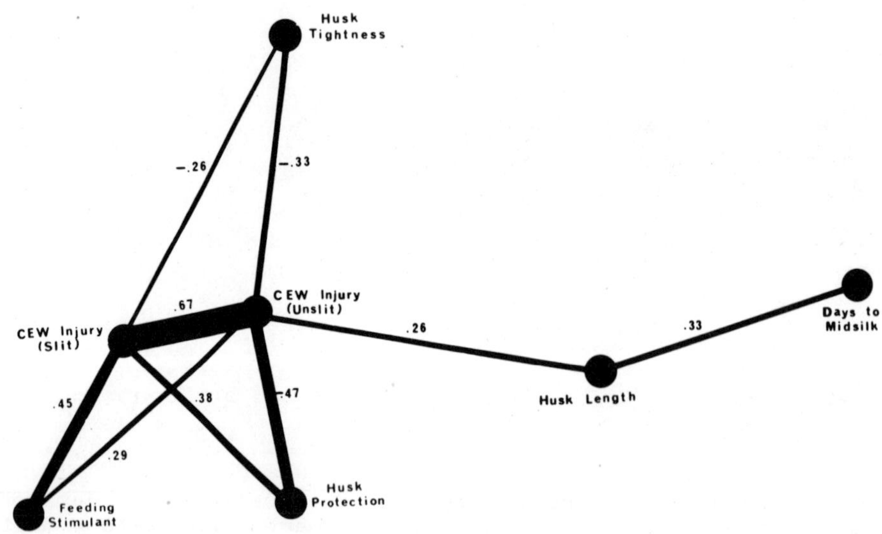

Figure 4.--Relationship between corn earworm injury and several selected plant characters measured among 36 inbred lines and illustrated by average "r" values listed on lines of length inversely proportional to "r" and of width directly proportional to r^2.

Table 1.--Husk and/or kernel resistance to a corn insect complex.[1/]

Hybrid	Corn earworm Husk-kernel	Maize weevil Husk-kernel	Pink scavenger caterpillar Husk-kernel
Taylor 196A	R - S	R - R	R - S
Funk G-732	S - S	R - S	R - S
Dixie 18	R - S	R - S	R - S
Coker 811A	R - R	R - R	R - R

[1/] Husk resistance was measured as insect damage leaving the husks intact; kernel resistance was measured as insect damage when the husks were slit. R = resistance, S = susceptible.

Later Wiseman, et al. (1972) found that resistant hybrids Dixie 18 and 471-U6 x 81-1 were damaged less than the susceptible hybrids, Asgrow 200B and Ioana, and yet supported fully equal numbers of earworm larvae. Wiseman and McMillian (1973) studied the behavior of earworm larvae on two susceptible sweet corns and concluded that the quantity and/or quality of silk available in one resulted in decreased earworm damage during the first 8 days. However, after 8 days the hybrid with early earworm protection was the most susceptible. Therefore, the silks are most likely the influencing factor associated with the resistance of Dixie 18 and 471-U6 x 81-1, whereas the lack of silk quantity and/or quality are the contributing factors in Asgrow 200B and Ioana susceptibility.

In laboratory investigations to complement the field studies, McMillian, et al. (1972a) developed an economical freeze-dryer with high capacity enabling the host plant resistance team to process adequate plant material for evaluating insect responses to extracts of the corn plant. McMillian, et al. (1966, 1967) and Jones, et al. (1972) conducted extensive laboratory studies on the extracts of corn silks and/or kernels to the earworm. The following summarizes findings on the feeding stimulant extracted from 10-day-old sweet corn kernels:

1. Feeding stimulant is contained in a water extract.
2. Earworm feeding stimulant response differs among plant species, corn lines, plant parts, and ages of the corn plant.
3. Feeding stimulation is correlated with field damage (r=0.71*), *significant at 5% level probability.
4. Feeding stimulant contains a complex of 4 major sugars and 7 amino acids as well as small amounts of several unidentified components.
5. Early instars (first through third) prefer extract of fresh silk and immature kernels (5-day) to more mature kernels (10 through 15-day).
6. Late instars (fourth through sixth) prefer extract of more mature kernels (10 through 15-day) to pollinated silks.
7. First through third instars respond equally to sugars and amino acid fractions.
8. Fourth instars prefer the amino acid fraction.
9. Fifth and sixth instars prefer the sugar fraction.
10. Synthetic mixtures of identified sugars and amino acids achieved 70% of the activity of the natural extract.
11. A synchronization of larval feeding preferences for certain plant chemicals with larval movement on the corn ear and plant and larval maturity has been indicated.

Widstrom, et al. (1972a) in studies of genetic parameters for earworm injury in corn populations with Latin American germ plasm found that heterosis was weak in the hybrid population. Estimates of heritability and genetic response to selection were highest in those populations which also had the largest estimate for dominance and the highest level of earworm resistance. It was concluded that prospects for successful selection within these populations were good.

The host plant resistance team released, through Plant Science Research Division of ARS, a dent corn synthetic GT-CEW-RS-8 in 1970. The base population prior to selection was composed of 423 single crosses among 34 adapted lines with some earworm resistance. The synthetic was derived from 8 cycles of recurrent selection in the base population for resistance to ear damage by the corn earworm. The synthetic has fair to good ear height and yielding ability.

The use of resistant plants or plant materials as an adjunct to other control measures has greatly increased efficacy in reducing corn earworm losses in sweet corn. The feeding stimulant has been incorporated with an insecticide and damage was reduced significantly over the insecticide alone. In fact, more than 8 times more insecticide would be required to equal the control achieved with the feeding stimulant plus insecticide (McMillian, et al., 1968b).

A resistant sweet corn hybrid has been used in combination with 7 applications of insecticide under artificial infestation (Figure 5) (McMillian, et al., 1972b), resulting in 52% more damage-free ears than the susceptible hybrid plus insecticide. Under natural infestation the resistant hybrid plus insecticide had 7% more damage-free ears than the susceptible hybrid plus insecticide. Thus, sweet corn hybrids with equal and similar type of resistance should require less insecticides and even possibly fewer applications at lower rates to achieve a high level of damage-free ears.

Resistance studies with the fall armyworm have been less rewarding than studies involving the corn earworm. Wiseman, et al. (1966) first found indications of resistance among the Antigua corns. Also, Wiseman, et al. (1967a, b) showed that fall armyworm larvae highly preferred corn to *Tripsacum dactyloides* (L.), which is considered a near relative of corn. Antigua 2D was the least preferred of the corns studied by Wiseman, et al. (1967a, b). Wiseman, et al. (1973a) found that in an intermediate resistant Antigua corn, both a higher level of resistance and susceptibility could be induced by the use of a complete fertilizer or by individual fertilizer components. In another paper, Wiseman, et al. (1973b) showed that foliar applications of the recommended rate of zinc could produce detrimental effects to the fall armyworm having fed on the treated foliage.

Widstrom, et al. (1972b) in studies with 8 maize inbreds and their F_1 progeny found that heterosis contributed substantially to the mean level of resistance among F_1 progenies to fall armyworm leaf feeding injury. This resistance is most likely a case of tolerance rather than nonpreference or antibiosis as exhibited in the studies above. Selection among the lines and their progeny would depend on the accumulation of additive gene effects for resistance to fall armyworm damage.

Sorghum.--Resistance in sorghum to sorghum insects has not been studied as extensively as other crop-insect relationships. Resistance in sorghum to the sorghum midge in the United States was first reported by Wiseman and McMillian in 1968. Based on a visual rating of one through five, nine lines were consistently less damaged. After more extensive evaluation, Wiseman and McMillian in 1970 found that the resistance mechanism in sorghum was of a nonpreference nature and that ODC 19 (select) was the most resistant of the lines studied, whereas SPI 29166 and CI 938 were the most preferred sorghum lines.

The first sorghum line, SGIRL-MR-1, resistant to the sorghum midge [Wiseman, et al. 1973c)] was released in 1971. SGIRL-MR-1 was developed as follows: Since 1964, the host plant resistance team at the Southern Grain Insects

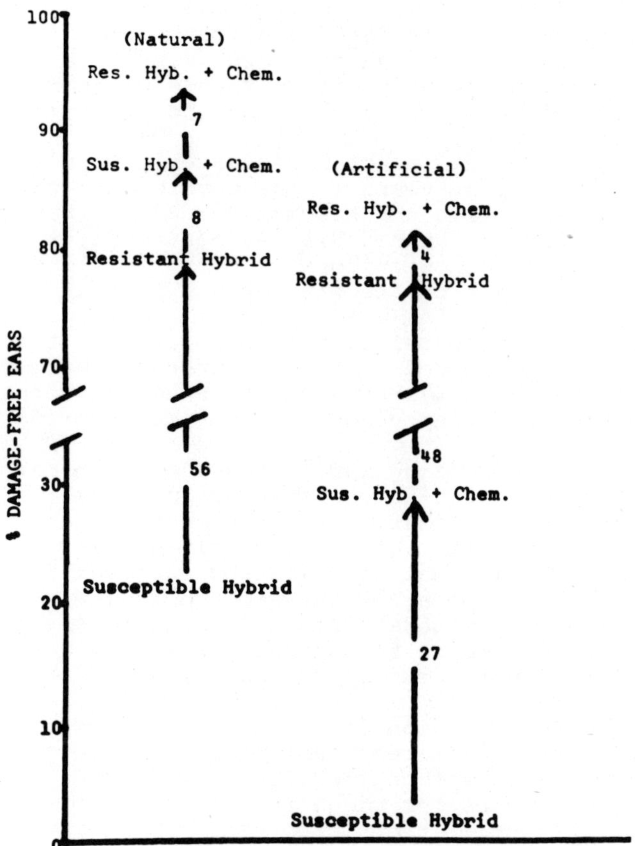

Figure 5.--Percentage damage-free ears resulting from combinations of resistant or susceptible sweet corn hybrids with insecticide and natural or artificial infestation of insects. 1970.

Research Laboratory continuously evaluated and selected within ODC-19 (select) for sorghum midge resistance. The least damaged heads were selected and exposed to a heavy midge population in successive years. SGIRL-MR-1 is the product of 7 years of this type selection and exhibits nonpreference-type resistance. In field tests it rates highly resistant, receiving significantly less damage than ODC-19 and averages 50% of the damage sustained by Ga. 615. SGIRL-MR-1 is a restorer (R) line.

Widstrom, et al. (1972c) reported gene effects conditioning resistance to the sorghum midge and sorghum webworm, Celama sorghiella (Riley). Parental, F_2, F_3, and selfed backcross populations were evaluated for resistance. Analysis of generation means indicated highly significant additive gene effects for both crosses and insects. Dominance effects were significant only for the cross SGIRL-MR-1 x 130. Dominance conditioned susceptibility to insect injury.

In a program closely related to host plant resistance, Wiseman and McMillian (1969) showed that heavy sorghum midge damage could be avoided by observing proper planting dates. These dates for South Georgia are considered effective up to about May 20 or a flowering date before July 10 through 15. In the absence of commercial grain sorghum hybrids with adequate midge resistance, serious considerations should be given to early plantings to avoid losses from midge damage.

Potential for Breeding Insect Resistance in Corn and Sorghum.--Considerable progress has been made in past years in the development of resistant corn hybrids to the corn earworm. McMillian and Wiseman (1972) have estimated that for every one dollar spent by the United States Department of Agriculture on research pertaining to earworm resistance in corn, a corn yield increase valued at 20 dollars was obtained. This estimate was for the 20-year period, 1950 to 1970.

In their review of the relationship of corn and the corn earworm McMillian and Wiseman (1972) have noted numerous sources of earworm resistant germ plasm, such as the sweet corn hybrid 471-U6 x 81-1, and field corns, Zapalote Chico and Antigua 2D. These have not been used in commercial hybrids to any extent.

The potential for breeding insect resistance in corn to the corn earworm lies in the fact that we must take resistant germ plasm such as Zapalote Chico and put it into a usable form that, as an end result, farmers will benefit from the higher level of resistance. Other, possibly greater potentials, exist in areas of research and development of resistance to leaf feeding insects early in the crop season. The development of earworm and fall armyworm nonpreference for oviposition combined with a moderate amount of antibiosis for early season resistance has great potential and in this area, resistance could have lasting effects on the development of future earworm and fall armyworm populations.

The potential for breeding for resistance in sorghum to the sorghum midge is almost unlimited. The first resistant line released in the United States did not occur until 1971. The vast collection of the world sorghum germ plasm has been only partially evaluated for midge resistance. Some of the greatest potentials depend on advancement of techniques, such as rearing, infestations, and refinement of measurements for resistance.

Lastly, potentials exist in using multiple pest control approaches of corn and sorghum insects such as resistant corn hybrids and less insecticide, resistant sorghum and early plantings, and resistant corns and management of fertilizer or cropping practices. The use of intermediate levels of resistance in crops with multiple approaches of insect control has advantages such that higher levels of control are sometimes achieved than when similar measures are used with high resistance in crops. Tolerance to insects attacking corn and sorghum may be used in conjunction with other control programs as an advantage rather than discounted as not being particularly valuable in resistance programs. Combinations of the mechanisms of resistance (nonpreference, antibiosis, and tolerance) in lesser amounts could exist or be developed so that very high levels of resistance would occur in corns or sorghums. Thus, the insects discussed herein would not be as damaging in the future as they have in the past.

References Cited

Jones, R. L., W. W. McMillian and B. R. Wiseman. 1972. Chemicals in kernels of corn that elicit a feeding response from larvae of the corn earworm. Ann. Entomol. Soc. Amer. 65: 821-824.

McMillian, W. W. and K. J. Starks. 1966. Feeding responses of some noctuid larvae (Lepidoptera) to plant extracts. Ann. Entomol. Soc. Amer. 59: 516-519.

McMillian, W. W., K. J. Starks and M. C. Bowman. 1966. Use of plant parts as food by larvae of the corn earworm and fall armyworm. Ann. Entomol. Soc. Amer. 59: 863-864.

McMillian, W. W. and K. J. Starks. 1967. Greenhouse and laboratory screening of sorghum lines for resistance to fall armyworm larvae. J. Econ. Entomol. 60: 1462-1463.

McMillian, W. W., K. J. Starks and M. C. Bowman. 1967. Resistance in corn to the corn earworm, *Heliothis zea*, and the fall armyworm, *Spodoptera frugiperda* (Lepidoptera: Noctuidae). I. Larvae feeding responses to corn plant extracts. Ann. Entomol. Soc. Amer. 60: 871-873.

McMillian, W. W., N. W. Widstrom and K. J. Starks. 1968a. Rice weevil damage as affected by husk treatment within methods of artificially infesting field corn plots. J. Econ. Entomol. 61: 918-921.

McMillian, W. W. and B. R. Wiseman. 1972. Host plant resistance: A twentieth century look at the relationship between *Zea mays* L. and *Heliothis zea* (Boddie). Univ. of Florida Monograph Ser. No. 2, 131 pp.

McMillian, W. W., A. N. Sparks, B. R. Wiseman and E. A. Harrell. 1972a. An economical high capacity freeze-dryer. J. Georgia Entomol. Soc. 7: 64-87.

McMillian, W. W. and B. R. Wiseman. 1972. Separating egg masses of the fall armyworm. J. Econ. Entomol. 65: 900-902.

McMillian, W. W., B. R. Wiseman and A. A. Sekul. 1970. Further studies on the responses of corn earworm larvae to extracts of corn silks and kernels. Ann. Entomol. Soc. Amer. 63: 371-378.

McMillian, W. W., B. R. Wiseman, N. W. Widstrom and E. A. Harrell. 1972b. Resistant sweet corn hybrid plus insecticide to reduce losses from corn earworm. J. Econ. Entomol. 65: 229-231.

McMillian, W. W., J. R. Young, B. R. Wiseman and A. N. Sparks. 1968b. Arrestant-feeding stimulant as an additive to Shell SD-8447 for control of corn earworm damage in sweet corn. J. Econ. Entomol. 61: 642-644.

Starks, K. J. and W. W. McMillian. 1967. Resistance in corn to the corn earworm and fall armyworm. II. Types of field resistance to the corn earworm. J. Econ. Entomol. 60: 920-923.

Starks, K. J., P. S. Callahan, W. W. McMillian and H C Cox. 1966. A photoelectric counter to monitor olfactory response of moths. J. Econ. Entomol. 59: 1015-1017.

Widstrom, N. W. 1967. An evaluation of methods for measuring corn earworm injury. J. Econ. Entomol. 60: 791-794.

Widstrom, N. W. and R. L. Burton. 1970. Artificial infestation of corn with suspension of corn earworm eggs. J. Econ. Entomol. 63: 443-446.

Widstrom, N. W., W. W. McMillian and B. R. Wiseman. 1970. Resistance in corn to the corn earworm and fall armyworm. IV. Earworm injury to corn inbreds related to climatic conditions and plant characteristics. J. Econ. Entomol. 63: 803-808.

Widstrom, N. W., B. R. Wiseman and W. W. McMillian. 1972a. Genetic parameters for earworm injury in corn populations with Latin American germ plasm. Crop Sci. 12: 358-359.

Widstrom, N. W., B. R. Wiseman and W. W. McMillian. 1972b. Resistance among some maize inbreds and single crosses to fall armyworm injury. Crop Sci. 12: 290-292.

Widstrom, N. W., B. R. Wiseman and W. W. McMillian. 1972c. Some gene effects conditioning resistance to midge and webworm injury in sorghum. Georgia Agron. Abstr. 15: 1-2.

Wiseman, B. R., D. B. Leuck and W. W. McMillian. 1973a. Effects of fertilizers on resistance of Antigua corn to fall armyworm and corn earworm. Florida Entomol. 56: 1-7.

Wiseman, B. R., D. B. Leuck and W. W. McMillian. 1973b. Effects of crop fertilizer on feeding of larvae of fall armyworm on excised leaf sections of corn foliage. J. Georgia Entomol. Soc. 8: 136-141.

Wiseman, B. R. and W. W. McMillian. 1968. Resistance in sorghum to the sorghum midge, Contarinia sorghicola (Coquillett) (Diptera: Cecidomyiidae). J. Georgia Entomol. Soc. 4: 15-22.

Wiseman, B. R. and W. W. McMillian. 1969. Relationship between planting date and damage to grain sorghum by the sorghum midge, Contarinia sorghicola (Diptera: Cecidomyiidae), in 1968. J. Georgia Entomol. Soc. 4: 55-58.

Wiseman, B. R. and W. W. McMillian. 1970. Preference of sorghum midge among selected sorghum lines, with notes on overwintering midges and parasite emergence. USDA ARS Prod. Res. Rept. 122. 8 pp.

Wiseman, B. R. and W. W. McMillian. 1973. Response of instars of the corn earworm, Heliothis zea (Lepidoptera: Noctuidae), to two susceptible sweet corn hybrids. J. Georgia Entomol. Soc. 8: 79-82.

Wiseman, B. R., W. W. McMillian and M. C. Bowman. 1970a. Retention of laboratory diets containing corn kernels or leaves of different ages by larvae of the corn earworm and the fall armyworm. J. Econ. Entomol. 63: 731-732.

Wiseman, B. R., W. W. McMillian and R. L. Burton. 1969. Feeding response of larvae of the corn earworm (Lepidoptera: Noctuidae) to water extracts of 16 host plants. J. Georgia Entomol. Soc. 4: 15-22.

Wiseman, B. R., W. W. McMillian and N. W. Widstrom. 1970b. Husk and kernel resistance among maize hybrids to an insect complex. J. Econ. Entomol. 63: 1260-1262.

Wiseman, B. R., W. W. McMillian and N. W. Widstrom. 1972. Tolerance as a mechanism of resistance in corn to the corn earworm. J. Econ. Entomol. 65: 835-837.

Wiseman, B. R., W. W. McMillian and N. W. Widstrom. 1973c. Registration of a sorghum line resistant to the sorghum midge. Crop Sci. 13: 398.

Wiseman, B. R., R. H. Painter and C. E. Wassom. 1966. Detecting corn seedling differences in the greenhouse by visual classification of damage by the fall armyworm. J. Econ. Entomol. 59: 1211-1214.

Wiseman, B. R., R. H. Painter and C. E. Wassom. 1967a. Preference of first-instar fall armyworm larvae for corn compared with _Tripsacum dactyloides_. J. Econ. Entomol. 60: 1739-1742.

Wiseman, B. R., C. E. Wassom and R. H. Painter. 1967b. An unusual feeding habit to measure differences in damage to 81 Latin American lines of corn by the fall armyworm, _Spodoptera frugiperda_ (J. E. Smith). Agron. J. 59: 279-281.

USE OF HOST PLANT RESISTANCE IN PEST MANAGEMENT OR ERADICATION SCHEMES

Johnie N. Jenkins
Research Geneticist
Host Plant Resistance, USDA, ARS
Mississippi State, Mississippi

Abstract

This report described the use of low levels of host plant resistance in pest management and eradication programs. The major ecosystems of cotton were described and the concept of the key pest for each ecosystem was discussed. The ecosystem in the southeastern Rain grown cotton area where the boll weevil is the key pest was used as the example.

Frego bract is a morphological mutant in cotton that confers a low level of resistance to the boll weevil. The use of Frego bract as one component of an integrated pest management program was shown to reduce boll weevils 69% and 79%, respectively, when used with and without a diapause program. The results were based on data from 22 fields in Yalobusha County, Mississippi in 1971. In addition to the reduction in boll weevil population, the beginning of weekly applications of insecticides for boll weevil was delayed 4 weeks longer in Frego than in non-Frego fields. In the Frego, but not in the non-Frego fields, we were able to go through the peak period of activity of the bollworm complex, *Heliothis* spp. before we needed boll weevil insecticides. The details of this experiment are described and reported in the following reference:

Jenkins, J. N., W. L. Parrott and J. C. McCarty, Jr. 1973. The role of a boll weevil resistant cotton in pest management research. J. Environ. Qual. 2: 337-340.

GENETIC INTERRELATIONSHIPS BETWEEN HOST AND ORGANISM AND INFLUENCE ON RESISTANCE

R. L. Gallun
Entomologist and Professor
Department of Entomology
USDA, ERD, Purdue University

After hearing the previous speakers, you may realize by now that a considerable amount of research has been conducted to develop crop plants that are resistant to insect pests. Today, many insect resistant cultivars are being grown in the United States and almost every major crop has resistant cultivars that are preventing damage from insect pests. I would hazard to guess that there are more than 100 different cultivars resistant to many different insect species, and these cultivars are probably grown on more than 50 million acres in the United States today.

When we speak of resistance, we are speaking of something that is real. It is heritable, meaning that it can be transferred from parent plant to its offspring. Because of this genetic heritability, crops can be protected by the addition of resistance to specific insect pests. The late Dr. Painter defined 3 kinds of resistance in crop plants called mechanisms of resistance (Painter, 1951). These are tolerance, nonpreference and antibiosis, and they influence the permanence of resistance that is inherent in the plant.

With tolerance, the plant responds to insect attack by repairing damaged tissue, growing new parts, or having enough foliage or plant vigor so that there is no significant injury to the plant. With this mechanism, there is no apparent change in the insect's life cycle or well-being. The insect continues to feed and survive while the plant responds by compensating for the injury. A good example of tolerance is the interrelationship that exists between the northern corn rootworm and corn plant as Dr. Ortman has already mentioned to you. Here the corn plant responds to root feeding by the larvae by growing new rootlets or by having the capacity for a greater root system than the less resistant plants. There is little reduction in the size of insect population; it continues to go on its merry way with both the plant and insect apparently satisfied with the arrangement.

The remaining two mechanisms work differently. Here the insect responds to the plant instead of the plant responding to the insect. In nonpreference, the insect may be attracted to the plant from a distance for feeding or oviposition but when arriving at the plant, it may not prefer it for food or oviposition and will move away to another host plant that it prefers better.

What is responsible for this unsatisfactory situation to the insect? This varies considerably with the insect and the plant. There may be certain chemical compounds that the plant emits or are found in the plant tissue that the insect does not care for. This is demonstrated by the boll weevil research being conducted at the USDA lab in Starkville. Boll weevils in the lab will not feed on their natural host cotton squares or bolls because of a feeding repellent applied to these plant parts, but they will feed on bean and okra seedlings not their native food plant, because of a feeding stimulant sprayed on these plants (Maxwell, et al., 1968; Jenkins, et al., 1963).

The morphology of the plant also plays an important role in protecting the crop from insect damage. The thickness of plant epidermis may prevent feeding or oviposition, the thickness or solidness of stem may also influence oviposition as is the case with the wheat stem sawfly on wheat (Wallace, 1966). The size of a plant part such as Frego bract inhibits egg laying on cotton as is the case with the boll weevil (Jenkins and Parrott, 1971). Wheat leaves with hairy surfaces prevent the cereal leaf beetle from ovipositing and hence saves the plant from defoliation (Gallun, et al., 1966; Schillinger and Gallun, 1968). These are morphological. There are also cases where color of plant influences feeding or oviposition by certain insects. Red color in cotton foliage has been reported to be less preferred by the boll weevil than green cotton leaves (Istey, 1928). The pea aphid prefers blue green varieties to yellow green varieties for feeding (Cody, 1941).

All these nonpreference mechanisms do have an important part in resistance programs, and I believe they should be utilized to their utmost capabilities. Insect populations may be reduced in size somewhat because of there not being as many preferred alternate hosts or susceptible varieties, but the insects are not killed on the plant and there is no immediate selection for races and biotypes.

The mechanism antibiosis is a mechanism that does select for insect biotypes because of the adverse effect the plant has on the insect. The feeding insects may die, produce fewer eggs or living young, or the growth stages of the insect may be affected and smaller and less healthy insects will develop. This mechanism can work well with parasites and predators because it tends to reduce insect populations to a level that is more appropriate for biological or integrated control. For controlling insects, it is the ideal way to reduce populations to nonsignificant levels as far as economical insect damage is concerned and of the 3 mechanisms, it is utilized the most. Antibiosis does have one drawback, however, and that is that it does apply strict selection pressure for the buildup of biotypes, races, or strains that can survive on the heretofore resistant plants. This mechanism has been utilized more than nonpreference and tolerance mainly because of the adverse effect it has upon the insect and because of its availability. Many genes have been identified that condition antibiosis in crop plants, most resistance being monogenic and dominant in character. Plants having single genes for resistance are generally more vulnerable to biotype buildup than plants having more than one gene for resistance. How these biotypes develop is dependent upon the genetics of the host plant, the genetics of the insect, and the kind of selection pressure applied to the insect population by the plant.

Biotypes are individual variants in an insect population that differ genetically from the majority of the insects in the population in that they have the capability of surviving on and damaging crop plants that are resistant to most of the insects in the population. When antibiosis occurs, the avirulent type insects are killed or reduced in numbers and only the virulent strains that are left in the population survive and interbreed. If crop plants of the same genetic resistance are grown over large areas for a great many years, the only insects left that can survive are the new race or biotypes and they increase in numbers from generation to generation until they buildup into epidemic populations and become major crop pests.

This type of interaction between plant and pest is probably better known with the pathogens and resistance to diseases. The most known cases have been the breakdown of resistance in wheat to stem rust and most recently the breakdown of T-cytoplasm resistance in corn to the southern corn leaf blight. In insects, the same thing occurs but generally not over such large areas and not as fast.

Let's look at a few examples of resistance in crop plants and insect biotypes:

1. The greenbug, *Schizaphis graminum* (Rond.), attacks small grains and sorghums with damage being quite extensive under heavy infestation. Plants are killed in different parts of the field which results in reduced stands. This insect occurs most generally in Texas and Oklahoma, but is found in almost every state where wheat is grown. Resistance to the greenbug is mainly due to the mechanism antibiosis although some varieties show tolerance. In barley, a single dominant gene conditions resistance where in wheat, a single recessive gene conditions resistance (Gardenhire, 1965; Curtis, et al., 1960). No wheat varieties have been released that are resistant to the greenbug, but the resistant barleys named Will and Kerr have been released and have been growing in Oklahoma for a number of years. There are now 3 known biotypes of the greenbug (Wood, 1971). Originally two wheats, Dickinson sel 28A and CI 9058, were resistant to the field strain of greenbug, biotype A. Then in the greenhouse, a new strain developed that could live on these heretofore resistant wheats. This strain was labeled biotype B. Just recently, another biotype, biotype C, was identified that is able to survive and cause damage to sorghum. Before this biotype, the greenbug was not known to cause damage to sorghum although it was occasionally found on sorghum. So here are 3 biotypes of greenbug, and their capabilities of surviving on wheat and sorghum are dependent upon the types of resistance utilized. I understand there are already lines of sorghum that have been found to be resistant to biotype C making it possible for the doubling of the number of races already known.

2. Another pest of sorghum, the corn leaf aphid, *Rhopalosiphum maidis* (Fitch), has 4 or more biotypes depending upon the genetics of the sorghum and the genetics of the insect (Singh and Painter, 1964).

3. The spotted alfalfa aphid, *Therioaphis maculata* (Buckton), a major pest of alfalfa, has 6 biotypes based on survival on different alfalfa varieties and clones (Nielson, et al., 1970).

4. Another aphid pest, the raspberry aphid, *Amphorophora rubi* (Kalt), transmits a virus disease of raspberry in the United States, Canada and England. Researchers in England have isolated 4 biotypes that are virulent to one or more different resistant raspberry varieties. These raspberries differ from one another by single dominant genes for resistance and the biotypes differ from each other by single genes for virulence (Briggs, 1965; Keep, et al., 1969).

You have noticed that the biotypes I spoke about are forms of aphids. Aphids, because of their ability to produce young without fertilization of the female, can develop offspring of the same genetic constitution as the parent and as prolific as aphids are, large populations of the same kind of aphid can develop over a short period of time.

There is, however, another insect pest, the Hessian fly, *Mayetiola destructor* (Say), a dipterous insect pest of wheat and barley that has this same kind of genetic capability. However, it is unique from aphids in that it is of a more advanced order of insects and population increases of this race are due to the production of offspring by fertilized females thereby maintaining the genetic variability in the population.

Let's examine the Hessian fly and its interrelationship between biotypes and host plant. The adult males resemble a mosquito in size and shape. The female adult before ovipositing has a swollen red abdomen full of eggs. The eggs are laid on the wheat leaves and the newly hatched red colored larvae migrate down the leaf to the growing point. In seedling wheat, the larvae feed between the leaf sheaths at the base of the plant. In more mature wheat, the feeding area is behind the leaf sheath at the node. Larvae in the red stage on resistant plants die within 4 days after they begin to feed. Larvae on susceptible plants continue to feed for at least 15 days and then pupate to form adults which emerge to start the cycle over again (Gallun and Langston, 1963).

Damage to wheat occurs in the fall and in the summer. In the greenhouse, a susceptible plant is a dark green stunted plant and a resistant plant is a normal elongated plant, lighter green in color.

Wheat plants respond to Hessian fly biotypes by either reacting susceptible or resistant, depending upon the specific gene for resistance in the wheat plant and the specific gene or genes for virulence in the insect (Tables 1 and 2) (Gallun and Hatchett, 1968). Three differentials determine the occurrence of 8 races. Turkey is universally susceptible and Ribeiro is universally resistant, so only Seneca, Monon, and Knox 62 act as differentials.

When reciprocal crosses are made between the Great Plains race and other races, the resulting progenies react as the Great Plains race in that they are unable to attack Seneca, Monon, and Knox 62. For example, when using the Great Plains race (SSMMKK) which cannot attack Seneca, Monon, and Knox 62, is crossed with race A (ssMMKK) which can attack Seneca but cannot attack Monon and Knox 62, the progeny will resemble the Great Plains race (SsMMKK) and will not be able to attack Seneca, Monon, and Knox 62 because of the avirulent dominant genes at the 3 loci. The same with crosses between the Great Plains race and race D. Here a race D (ssmmkk) has recessive genes for virulence to all 3 wheats, but when crossed with the Great Plains race (SSMMKK), will produce progeny that resembles the Great Plains race (SsMmKk) in that it will not attack the 3 wheats. This is because the alleles for avirulence that come from the female parent were dominant to the recessive alleles for virulence that come from the male parent. This and other crosses demonstrated to us that avirulence or inability to attack is dominant to the ability to attack (Hatchett and Gallun, 1970). This means that virulence in the insect (or ability to attack the plant) is controlled by recessive gene pairs for virulence to Turkey and dominant gene pairs for avirulence to Arthur 71.

When races B and C are crosses (ssmmKK x ssMMkk), something else occurs (Table 3). We get a new race, race A (ssmMkK) and you can see why. Race B has recessive virulent genes for attacking Seneca and Monon, but not Knox 62. Race C has virulent recessive genes for attacking Seneca and Knox 62 but not Monon. The F_1 progeny receive virulent recessive genes from both parents making them capable of attacking Seneca, but the dominant gene for avirulence to Monon comes from the male side and the dominant gene for avirulence to Knox 62 comes from the female and hence the progeny phenotypically reacts like A in that it only can attack Seneca but not Monon and Knox 62.

Results from this and other crosses between races give us genetic evidence that races of Hessian fly function because of independent single recessive genes for virulence and dominant genes for avirulence. A biotype can attack a wheat having a specific gene for resistance only if the Hessian fly biotype has both recessive alleles for virulence. If one of the alleles is a dominant allele, the insect cannot attack the wheat having the matching dominant gene for resistance.

These are a few things we have learned from our genetic studies that have aided us in breeding resistant wheats. By knowing the genetics of the insect and the genetics of the wheat, we can do a better job of understanding the interrelationships that occur.

Today, there are 29 wheat varieties that are resistant to this insect, and resistance is controlled by one or more of 6 dominant genes (Gallun and Reitz, 1971). In 1969, these varieties were estimated to have been grown on over 8 1/2 million acres in 34 states. In certain states such as Indiana where wheats having the same source of resistance are being grown on almost the total wheat acreage in the state, biotypes became noticeable and farmers became aware that their resistant wheats were becoming susceptible.

In Indiana, the first resistant variety named Dual was released to growers in 1955. In 1959, Monon was released, and by 1962 Redcoat and Reed had been released. All 4 wheats had the same genetic source of resistance to Hessian fly, the H_3 gene, and by 1962 over 70% of the wheat acreage in Indiana was planted to these wheats. This meant that there was extreme selection pressure for the development of a specific race of Hessian fly that could develop on these wheats, and in 1962 race B became prevalent in the field. This race could live on wheats that had no genes for resistance plus those wheats that had the H_3 gene for resistance such as Dual, Monon, Redcoat, and Reed. What was needed in 1962 was a variety that had resistance to race B. It so happened that the wheat variety Knox 62 was released that same year and this variety did have resistance to race B. It was not a case of luck but of planning since our previous research had shown us the genetic capabilities of new races so that they could be isolated and utilized in the laboratory to evaluate and develop wheats having a kind of resistance to this race. Prior to when race B became prevalent in the field, crosses involving wheats that were resistant to laboratory race B were made leading to the development of Knox 62. In 1966, another wheat, Benhur, also resistant to race B was released. This may sound that we had the race situation licked, but nature is hard to put down and when augmented by the farmer's non-acceptance of certain wheats, the story was not finished. Knox 62, although extremely immune to race B, was not accepted by wheat growers for some reason and it never did increase in acreage. Benhur, the other variety although a top yielder, also was not accepted. One contributing factor to these wheats not being accepted was the popularity of Monon wheat. Monon wheat which was released in 1960 had such a good reputation for high yielding capacity that even though susceptible to race B, it was grown extensively by the wheat growers. In 1969, this variety was grown on over 50% of Indiana wheat acreage. Other wheats having the H_3 gene for resistance made up approximately 40% more of the acreage, so race B could maintain itself very well. This year we released a new variety called Arthur 71 (Caldwell, et al., 1972). It has a different gene for resistance and is resistant to all known races of Hessian fly. Since resistance in this wheat is controlled by a single dominant gene, it should only be a matter of time before some new race would develop on it, although as yet we don't have one in the lab. If this does happen, the number of possible races will double from 8 now to 16. This new wheat although resistant to all 8 races has a temperature sensitive gene and at temperatures of 75°F and higher, the plant loses its resistance to Hessian fly. This complicates matters in the field when it comes to determining if biotypes are developing on the plant. We do have another wheat in the making which has the same kind of resistance but is not as temperature sensitive. It will be released in the near future. As you can see, breeding a crop for resistance to insect pests isn't all that simple.

If we could regulate the kind of wheat that is grown in a state, I'm sure we could eliminate the Hessian fly or at least reduce populations to such a low level as to make them ineffective. For instance, all our wheats have single genes for resistance and they control biotypes having single genes for virulence. Field surveys have shown that the H_3 wheats that make up over 90% of the wheat acreage in Indiana suppressed races A and C that were prevalent before the release of these wheats, and only race B exists in great numbers along with a few race D types. If we could switch to wheats having the H_6 genes for resistance like Knox 62 and Benhur and completely blanket the state with these varieties, then we could suppress race B and since A and C were already reduced to very few numbers, only race D would be able to survive. Then when race D would start to increase, we could switch to H_5 resistance wheats and suppress race D. If biotypes would build up on the H_5 wheats, we could go back to the H_3 or H_6 wheats again. I am positive it would work, but regulations of this sort are not forthcoming in Indiana. The farmers are an independent lot and our resistant wheats have helped keep fly populations at a low level, so as yet they are not really hurt from Hessian fly.

It certainly would be a challenge to see if we could eliminate Hessian fly by regulating the use of varieties.

From what little I have shown you about Hessian fly races, you can see that variability exists within an insect species and you can appreciate the complexities that occur when developing insect resistant cultivars.

If you are faced with biotypes, all is not lost. Insect biotypes can be used to distinguish between different genes for resistance in world collections of crop plants. They can also be utilized in the breeding programs to evaluate lines for resistance and pick out segregating material. They can also be used to determine if two or more different genes exist in one plant, and they are very instrumental for genetic studies.

It is still preferable to work with the mechanisms nonpreference and tolerance. These mechanisms do not select for biotypes and resistant varieties should hold up in the field. Regardless of what mechanism is used, it is only the well-funded research programs that produce resistant cultivars, and then only when there is excellent cooperation between scientists of the different disciplines, entomology, plant pathology, agronomy, and genetics. Without the team approach, I'm afraid all we will be doing is screening plants for resistance and not producing varieties.

Table 1.--Wheat reactions to races of Hessian fly.

Hessian fly race	Turkey	Seneca H_1H_2?	Monon H_3	Knox 62 H_6	Ribeiro H_5
GP	S	R	R	R	R
A	S	S	R	R	R
B	S	S	S	R	R
C	S	S	R	S	R
D	S	S	S	S	R
E	S	R	S	R	R
F	S	R	R	S	R
G	S	R	S	S	R

Table 2.--Race genotypes based on virulence to wheat differentials.

Hessian fly race	Turkey	Seneca H_1H_2?	Monon H_3	Knox 62 H_6	Ribeiro H_5
GP	tt	SS	MM	KK	RR
A	tt	ss	MM	KK	RR
B	tt	ss	mm	KK	RR
C	tt	ss	MM	kk	RR
D	tt	ss	mm	kk	RR
E	tt	SS	mm	KK	RR
F	tt	SS	MM	kk	RR
G	tt	SS	mm	kk	RR

Table 3.--Interracial crosses.

Race phenotypes	Race genotypes	Progeny genotype	Progeny phenotype
GP X A	SSMMKK x ssMMKK	SsMMKK	GP
GP X D	x ssmmkk	SsMmKk	GP
A X B	ssMMKK x ssmmKK	ssMmKK	A
A X D	x ssmmkk	ssMmKk	A
B X D	ssmmKK x ssmmkk	ssmmKk	B
C X D	ssMMkk x ssmmkk	ssMmkk	C
B X C	ssmmKK x ssMMkk	ssmMKk	A
E X A	SSmmKK x ssMMKK	SsmMKK	GP

References Cited

Briggs, J. B. 1965. The distribution, abundance and genetic relationships of four strains of the Rubus aphid (Amphorophora rubi (Kalt)) in relation to raspberry breeding. J. Hort. Sci. 49(2): 109-117.

Caldwell, R. M., J. J. Roberts, R. E. Finney, G. E. Shaner, F. L. Patterson and R. L. Gallun. 1972. Arthur 71 soft red winter wheat improved for resistance to leaf rust and Hessian fly. Purdue Univ. Agr. Exp. Sta. Res. Prog. Rept. 406: 1-3.

Cody, C. E. 1941. Color preferences of the pea aphid in western Oregon. J. Econ. Entomol. 34: 584.

Curtis, B. C., A. M. Schlehuber, and E. A. Wood. 1960. Genetics of greenbug resistance in two strains of common wheat. Agron. J. 52: 599-602.

Gallun, R. L. and R. Langston. 1963. Feeding habits of Hessian fly larvae on P32-labeled resistant and susceptible wheat seedlings. J. Econ. Entomol. 56(5): 702-706.

Gallun, R. L., R. Ruppel and E. H. Everson. 1966. Resistance of small grains to the cereal leaf beetle. J. Econ. Entomol. 59(4): 827-829.

Gallun, R. L. and J. H. Hatchett. 1968. Interrelationship between races of Hessian fly, Mayetiola destructor (Say), and resistance in wheat. Proc. 3rd Int. Wheat Genet. Symp. Canberra, Australian Acad. Sci.

Gallun, R. L. and L. P. Reitz. 1971. Wheat cultivars resistant to races of Hessian fly. Agr. Res. Serv., USDA, Prod. Res. Rept. 134: 1-16.

Gardenhire, J. H. 1965. Inheritance and linkage studies on greenbug resistance in barley, Hordeum vulgare L. Crop Sci. 5: 28-29.

Hatchett, J. H. and R. L. Gallun. 1970. Genetics of the ability of the Hessian fly, Mayetiola destructor Say, to survive on wheats having different genes for resistance. Ann. Entomol. Soc. Amer. 63: 1400-1407.

Istey, D. 1928. The relation of leaf color and leaf size to boll weevil infestation. J. Econ. Entomol. 21(4): 553-559.

Jenkins, J. N., F. G. Maxwell, J. C. Keller and W. L. Parrott. 1963. Investigations of the water extracts of Gossypium, Abelmoschus, Cucumis, and Phaseolus for an arrestant and feeding stimulant for Anthonomus grandis Boh. Crop Sci. 3: 215-219.

Jenkins, J. N. and W. L. Parrott. 1971. Effectiveness of frego bract as a boll weevil resistance character in cotton. Crop Sci. 11: 739-743.

Keep, E., R. L. Knight and J. H. Parker. 1969. Fruit breeding. Further data on resistance to the Rubus aphid, Amphorophora rubi (Kltb.) Rept. E. Malling Res. Sta. X21.

Maxwell, F. G., J. N. Jenkins and J. C. Keller. 1963. A boll weevil repellent from the volatile substance of cotton. J. Econ. Entomol. 56(6): 894-895.

Nielson, M. W., H. Don, M. H. Schonhorst, W. F. Lehanan and V. L. Marble. 1970. Biotypes of the spotted alfalfa aphid in varieties in western United States. J. Econ. Entomol. 63: 1822-1825.

Painter, R. H. 1951. Insect Resistance in Crop Plants. Macmillan, New York.

Schillinger, J. A. and R. L. Gallun. 1968. Leaf pubescence of wheat as a deterrent to the cereal leaf beetle, Oulema melanopus. Ann. Entomol. Soc. Amer. 61(4): 900-903.

Singh, S. R. and R. H. Painter. 1964. Reaction of four biotypes of corn leaf aphid, Rhopalosiphum maidis (Fitch), to differences in host plant nutrition. Proc. XII Int. Cong. Entomol. London.

Wallace, L. E. and F. H. McNeal. 1966. Stem sawflies of economic importance in grain crops in the United States. ARS-USDA Tech. Bul. 1350: 1-50.

Wood, E. A., Jr. 1971. Designation and reaction of three biotypes of the greenbug cultured on resistant and susceptible species of sorghum. J. Econ. Entomol. 64: 183-185.

BIOCHEMICAL BASES OF RESISTANCE OF PLANTS TO PATHOGENS

Alois A. Bell
Director and Research Leader
USDA, ARS, National Cotton Pathology Research Laboratory
College Station, Texas

Development of infectious diseases of plants requires a potential host, a potential pathogen, and a conducive environment. When these are present, the ability of a plant to prevent, restrict, or retard disease development is called _disease resistance_. The ability of the plant to remain healthy by avoiding the pathogen, environment conducive to disease, or both is called _disease escape_ (or _klendusity_). Orton (1908) originally recognized 3 degrees of resistance: immunity, resistance, and endurance (now called tolerance). _Immunity_ refers to the ability of a plant to remain completely free of the disease. _Tolerance_ refers to the ability of a plant to endure invasion of a pathogen without much symptom expression (particularly in viral infections) or damage (particularly in fungal infections). _Susceptibility_ is the opposite of resistance. It includes qualities that permit disease development and limit the plant's ability to overcome or withstand injurious effects of disease.

Disease escape must be clearly distinguished from disease resistance and should be avoided in biochemical studies. Plants that escape disease frequently have no more biochemical resistance than plants that fail to escape the disease, and they are susceptible when infection occurs. Yet, plant characters that allow disease escape can be useful in disease-control programs. Okra leaf and nectariless characters in cotton reduce boll rots by modifying the microenvironment of the plant canopy and by reducing insect movement and feeding on the plant; insect wounds provide the initial entry for many boll rot organisms. Earliness in cotton reduces losses from _Phymatotrichum_ root rot and _Verticillium_ wilt, which occur late in the growing season; early varieties also have less boll rot in southwestern United States because they are harvested before fall rains occur.

Resistance may be either specific or general (also called vertical and horizontal, respectively (Van der Plank, 1968). Resistance expressed to some, but not to other, strains of a pathogen is called _specific_, while resistance expressed similarly to all strains of a pathogen is called _general_. Strains of pathogens that vary in their potential to cause disease are said to vary in _virulence_ (or aggressiveness); virulence also may be specific or general. Strains of a pathogen clearly distinguished by differential susceptible or resistant reactions of a group of closely related plants (usually species or varieties) are called _physiological races_. A given plant may have both specific and general resistance to a disease. Specific resistance generally is expressed against obligate (or near obligate) parasites, but also has been demonstrated against the bacterium, _Xanthomonas malvacearum_ in cotton (Brinkerhoff, 1970) and Hessian fly in wheat (Gallun and Reitz, 1971). Resistance to facultative parasites usually is general. Biochemical mechanisms that distinguish susceptible and resistant varieties might be different for general and specific resistance.

Responses that distinguish susceptible and resistant reactions may be directed primarily against growth and development of the pathogen, action of toxins and enzymes of the pathogen, or both. Resistance to diseases caused by obligate parasites (for example, rusts and viruses) appears to be directed against growth or multiplication of the pathogen. Resistance to certain diseases caused by facultative fungal parasites (for example, *Helminthosporium*, *Periconia*, and *Alternaria* sp.) (Scheffer and Pringle, 1967; Scheffer and Samaddar, 1970 and Scheffer and Yoder, 1972) is directed primarily against the fungal toxins. In the latter instance, the plant must also resist pathogen development, but this ability usually resides equally in susceptible and resistant varieties and is not critical in distinguishing them.

Resistance, tolerance, and susceptibility are arbitrary, relative terms. A plant in a given environment may be susceptible to one but resistant to another strain of the same pathogen; under a slightly different environment it may be resistant to both strains. A plant susceptible to a given disease during the first week of its life may become and remain resistant during the second week and thereafter. One tissue or organ may be susceptible but another resistant in the same plant. Biochemical expressions of resistance to most diseases probably occur in both susceptible and resistant plants and vary only in their speed, magnitude, and localization. Different tissues in a single plant may vary more in their resistance responses than the same tissue in susceptible and resistant varieties. Thus, biochemical studies of resistance should use genetically pure and stable plants and pathogens, only appropriate tissues, and carefully defined and reproducible environments that clearly distinguish susceptible and resistant responses. The time-course of biochemical events should be related with that of disease development.

Mechanisms and nature of plant disease resistance have been discussed extensively in several books (Goodman, et al., 1967; Metlitskii and Ozeretskovskaya, 1968; Van der Plank, 1968; Wood, 1967), and numerous review articles (Akai, 1959; Akai, et al., 1967; Akai, et al., 1971; Albersheim, et al., 1969; Allen, 1959; Beckman, 1964; Bell, 1972; Bingefors, 1971; Brown, 1934, 1965; Buddenhagen and Kelman, 1964; Byrde, 1963; Chester, 1933; Cruickshank, 1963, 1966, Cruickshank and Perrin, 1963; Daly, 1972; DeVay, et al., 1967; Diener, 1963; Dimond, 1967, 1970; Dropkin, 1969; Ellingboe, 1968; Farkas and Kiraly, 1962; Farkas and Stahmann, 1966; Fawcett and Spencer, 1969; Fuchs, 1971; Hadwiger and Schwochau, 1969; Klement and Goodman, 1967; Kosuge, 1969; Krusberg, 1963; Kuc, 1964; 1968; Martin, 1964; Matta, 1971; Muller, 1958, 1959, 1963; Mundry, 1963; Oku, 1967; Price, 1962; Rhode, 1965; Rohringer and Samborski, 1968; Ross, 1966; Scheffer and Pringle, 1967; Scheffer and Samadder, 1970; Scheffer and Cowling, 1967; Schwochau and Hadwiger, 1970; Shaw, 1963, 1967; Stahmann, 1965, 1967; Stoessl, 1970; Thatcher, 1942; Tomiyama, 1963, 1971; Tomiyama, et al., 1967; Uritani, 1963, 1971; Uritani, et al., 1967; Walker, 1924; 1963; Walker and Stahmann, 1955; Ward, 1905; Wingard, 1941; Yarwood, 1967). This review briefly summarizes preinfectional and postinfectional characteristics that contribute to plant disease resistance. Only a few selected literature citations are used to illustrate each characteristic.

Preinfectional Resistance

Certain defense mechanisms of the plant occur without previous infection and thus are *preinfectional* (also called preformed, preexisting, mechanical, or passive). These include many general resistance mechanisms of plants to saprophytic and facultatively parasitic microorganisms. They frequently are better developed in wild than in domestic plants. Preinfectional mechanisms are limited in value in breeding programs, since energy converted to defense mechanisms usually occurs at the expense of growth and reproduction; thus, yield and quality may be reduced. Preinfectional resistance of plants to disease may be due to anatomical barriers, nutritional limitations, resistance to microbial enzymes and toxins, or antimicrobial substances.

Anatomical Barriers.--The epidermis, comprised of the cuticle and outer walls of the epidermal cells, serves as both a chemical (Martin, 1964; Martin, et al., 1957; Roberts, et al., 1961) and physical (Akai, et al., 1967; Dickinson, 1960; Flentje, 1958; Martin, 1964; and Wood, 1960) barrier to exclude microorganisms from plant tissues. The cutin and waxes in the cuticle constitute a hydrophobic surface, which repels water and prevents its accumulation as a film on the plant surface (Davies, 1961); this surface also restricts diffusion of nutrients from tissues to the surface of the plant. Thus, the cuticle limits foliar retention of microorganisms, germination of fungal spores, and multiplication and penetration of bacteria. A wound through the epidermal cell wall is necessary to establish viral, many bacterial, and some fungal infections (Dickinson, 1960; Esau, 1966; Flentje, 1958; and Wood, 1960).

For fungi that directly penetrate the epidermis, cuticle thickness and epidermal resistance to mechanical penetration have been related to disease resistance (Dickinson, 1960; Martin, 1964). For example, varietal resistance of barberry to *Puccinia graminis* (Melander, 1927) coffee to *Colletotrichum coffeanum* (Nutman and Roberts, 1960) and strawberry to *Sphaerotheca macularis* (Perrin, 1964) has been related to cuticle characteristics. The greater thickness of cuticle in mature versus young leaves and stems also relates to the greater disease resistance of the mature tissue (Robinson, 1969; Schieferstein and Loomis, 1959). High temperatures, high light intensity and low humidity enhance cuticle thickness (Martin, 1964) and generally enhance resistance to fungi that directly penetrate the epidermis.

Some fungi and bacteria invade through natural openings in the epidermis such as stomata, lenticels, and hydathodes. The structure and function of these have been related to resistance (Hart, 1924; 1931; Walker, 1924). McLean (1921) suggested that varietal resistance in mandarin oranges to *Pseudomonas citri* was due to the size of the stomatal opening. Likewise, small size as well as early cork formation in lenticels has been related to varietal resistance to bacterial diseases. Romig and Caldwell (1968) found that differential development of leaf rust on peduncles, sheaths, and blades of wheat was caused by differences in stomatal exclusion, determined by thickness and function of guard cells.

The relative occurrence of cells with thickened secondary walls (for example, sclerenchyma, collenchyma, and xylem and phloem parenchyma), has also been related to resistance. Hart (1931) suggested that bundles of such cells restricted the development of stem-rust pustules in wheat. Such cells in leaf veins effectively restrict "angular leaf spot" diseases to the area between veins. Numerous other relationships between anatomical structures of the host and disease resistance are cited in early reviews of disease resistance (Appel, 1915; Butler, 1918; Coons, 1937; Freeman, 1911; Walker, 1924; and Ward 1905).

While anatomical barriers are undoubtedly involved in the general resistance of plants, their importance in varietal resistance generally has been based on casual observations or simple correlations involving a few varieties. Critical proof of involvement (concurrent transfer of the anatomical feature and resistance in a breeding program, induction of susceptibility by removing or bypassing the barrier, and absence of toxic chemicals) is usually lacking. Thus, the exact importance of these features remains to be determined.

Nutritional Limitations.--Massee in 1905 defined an immune plant as one in which positive chemotactic substances, necessary for facilitating the entrance of the germ tubes of a given parasitic fungus into its tissues, were absent. He suggested that studies of the nature of resistance should be concerned with the exact constituents of the cell sap. Subsequent studies, however, showed that most obligate fungal parasites penetrated susceptible and resistant hosts with similar ease, and Massee's theory was discounted.

Studies by Brown (1922, 1934) caused renewed interest in the role of nutrition in virulence and resistance. He demonstrated exosmosis of electrolytes into water droplets placed on foliage and flower petals of certain plants; drops on petals had increased capacity to bring about germination of spores of Botrytis sp. Exosmosis of nutrients into drops on petals was unaffected by presence of the fungus until after penetration (6-8 hr) when it increased rapidly. The amount of exosmosis from petals was directly related to the speed and percentages of infection. Water drops on leaves also increased in electrical conductivity, but either they did not affect germination, or they inhibited germination in proportion to the increase of conductivity. Kovacs (1955) found that germination of spores of Cercospora beticola were inhibited on resistant beet leaves; spores of the non-pathogen, Alternaria tenuis, also were inhibited. Leaf washings and exudates from numerous other plants also are primarily fungitoxic.

Exudates from roots, seeds, or hypocotyls have been studied extensively for effects on seedling and soilborne disease (Schroth and Hildebrand, 1964; Spender, 1962). Kerr and Flentje (1957) reported that exudates from radish roots, when applied to washed strips of radish cuticles, stimulated organization of hyphal masses and penetration by virulent, but not by avirulent, strains of Pellicularia filamentosa. Glucose or peptone did not mimic this effect. The stimulatory compound was not identified.

Many investigators have shown large increases of fungal and bacterial flora in the rhizosphere and next to germinating seeds (Schroth and Hildebrand, 1964; Spencer, 1962), but attempts to demonstrate a nutritional basis for varietal resistance have been frustrated by concurrent exudation of toxic compounds. Timonin (1941), for example, found a greater population of microorganisms in the rhizosphere of the Fusarium-resistant flax variety Bison than in that of the susceptible variety Novelty under field conditions. Sterile root exudates from the susceptible variety, however, were more stimulatory to microflora growth, and sterile exudates from the resistant variety were toxic to Fusarium and Helminthosporium sp. Exudates from the resistant variety contained 80 ppm KCN, which inhibited Fusarium oxysporum f. lini, but stimulated Trichoderma viride, a soil saprophyte. Buxton (1957) examined root exudates of three varieties of peas that showed differing degrees of susceptibility to three races of F. oxysporum f. pisi and concluded that exudates inhibited germination of races to which the varieties were resistant. Similar results have been obtained by others trying to relate nutrition to resistance.

To resolve this dilemma, Garber (1956) proposed a nutrition-inhibition hypothesis of pathogenicity (and resistance). According to this theory, the nutritional environment of the microorganism may be adequate or inadequate, and the inhibitory environment effective or ineffective. Only an adequate-nutrient environment, coupled with an ineffective inhibitory environment, results in virulence or susceptibility. Evidence that supports the concept that nutrient supply is critical to susceptibility is as follows:

1) Mutations that impose major nutrient requirements, such as nucleotides or amino acids (except proline and sometimes methionine), on Venturia inaequalis (Boone, et al., 1958; Kline, et al., 1958), Ustilago maydis (Holliday, 1961), Verticillium albo-atrum (Hastie, 1970), and V. dahliae (Puhalla, unpublished), result in a loss of virulence. These nutrients are present in the host but usually in complex forms within the cytoplasm. Auxotrophic mutations for minor nutrients such as the vitamins, nicotinic acid or inositol, cause slight or no loss in virulence.

2) Facultative pathogens invariably produce pectinase and usually cellulase enzymes, which allow them to degrade and use cell-wall components (Albersheim, et al., 1971; Bateman and Millar, 1966; Brown, 1965 and Roberts and Martin, 1963). Inability to produce pectinase naturally or due to mutation results in loss of virulence.

3) Host-specific toxins and other phytotoxins that appear to be involved in disease development invariably cause cellular leakage of nutrients (Brown, 1965; Romanko, 1959; Scheffer and Pringle, 1967; Scheffer and Yoder, 1972). These toxins are essential for overcoming host resistance (Scheffer and Yoder, 1972).

4) Tissues like bark and heartwood have extremely high C/N ratios. Only basidiomycetes grow in the presence of such ratios and invade trees through these tissues (Scheffer and Cowling, 1967).

5) Cellular pH and osmotic concentrations restrict the number of pathogens that attack the plant. Only acid-tolerant fungi such as *Monilinia*, *Botrytis*, or *Penicillium* sp. attack the highly acid fruits of plants, and only osmotic-tolerant fungi such as *Aspergillus* or *Penicillium* sp. cause diseases of stored seed.

6) Numerous environmental conditions such as moisture, temperature, and supply of N, K, and Ca markedly affect disease resistance (Bell, 1972; Wingard, 1941). Those conditions (low temperature, high water, high N, low K and Ca), which favor enhanced levels of soluble nutrients, particularly nitrogenous compounds, in plant tissues, usually increase susceptibility to soilborne and seedling diseases.

In spite of much evidence that adequate nutritional supply is necessary for virulence (or susceptibility), few if any instances of varietal resistance caused by inadequate nutrient supply have been shown. The genetic potential to use the host as a substrate must be a primary requirement for pathogenicity; thus, resistance of plants to pathogens probably depends on development of effective inhibitory environments.

Resistance to Enzymes.--Hydrolytic enzymes appear important for pathogenesis by facultative fungi and bacteria (Albersheim, et al., 1969; Bateman and Miller, 1966; Brown, 1965; Wood, 1960). These compounds dissolve and disorganize plant cells and tissues, making nutrients available to the pathogen. They also may kill plant cells and thus disrupt development of postinfectional resistance mechanisms. Brown (1934) proposed that plant resistance to facultative fungi was largely a result of insensitivity or of regulation by the host of the pathogen's enzyme production. He proposed four categories of resistance: a) plant composition is unsuitable for growth or production of active substances by the fungus, because of nutritional limitations or toxic barriers, b) plant composition allows ready fungal growth, but not secretion of an appreciable quantity of active substances by the fungus, c) plant composition favors fungal growth and allows secretion of enzymes, but secretion or activity of enzymes is limited, and d) the active principle of the fungus is unable to affect the tissue of the plant.

As an example of (b), *Botrytis allii* (an onion pathogen) penetrated and grew slightly in apple but did not cause disease. Likewise, the fungus grew readily in apple extract but did not produce detectable pectinase. Addition of nitrogenous compounds allowed the fungus to attack apple and to secrete pentinase when grown in apple extracts. Resistance of apple appeared due to the adverse effects of a high C/N ratio on production of pectinase.

Brown (1934) proposed that resistance was usually due to (c). For example, potato is resistant to *Botrytis cinerea* but susceptible to *Pythium debaryanum*. On potato decoctions, *Botrytis* produced considerable pectinase and *Pythium* very little but on living or dead potato tubers *Pythium* produced considerable enzyme, probably because the pectic substances in the latter substrate induced its production. The activity of the enzyme from *Botrytis* was favored by acid conditions and was sensitive to salts, while that from *Pythium* was favored by neutral or alkaline conditions and was insensitive to salts. Conditions in the live potato tuber favored the production and activity of pectinase from *Pythium*. Pectinase from *Botrytis* was also quickly absorbed by subturgid potato cells, and little rot developed; injection of tissues with water prevented such absorption and allowed rot to occur. Pectinase from *Pythium* attacked subturgid cells, and Brown (1934) suggested that some substance secreted by Pythium allowed adequate water exchange between the fungus and host.

Several other factors regulating enzymes have also been related to disease resistance:

1) Rate of enzyme formation by pathogen. Lapwood (1957) compared host-parasite behavior of four virulent and three avirulent strains of Erwinia aroideae, which made similar quantities of macerating enzymes in culture during 3 days of incubation. In potato decoction, the growth curves of the virulent strains began to separate upward from those of the avirulent strains after about 4 hr and continued to diverge until about the 12th hr; by 24 hr, the curves had converged and populations were similar. Production of macerating enzyme was demonstrated for virulent but not avirulent strains after 4 hr, and great differences occurred after 8 hr; by 24 hr all strains produced high enzyme activity. Initial attack on potato tissues occurred at 2 to 3 hr by virulent and at 4 to 8 hr by avirulent strains, but only virulent strains continued to progress. The speed of enzyme production, therefore, is considered of critical importance in the host-parasite relationship.

2) Calcium pectate content of tissues. Bateman and Lumsden (1964, 1965, 1966) reported that calcium pectate is not degraded by endopolygalacturonase (endoPG) of Rhizoctonia solani. Calcium ions accumulated in and around developing Rhizoctonia lesions, and these tissues were more difficult to macerate enzymatically than similar tissue from healthy plants. Shear and Drake (1971) also reported localized accumulation of calcium in cork cells, which limit development of apple-scab lesions. Soaking bean hypocotyls in calcium ions caused them to be more resistant to Rhizoctonia (Bateman, 1964). Old (3 or more weeks) bean tissues were more resistant than young (less than 2 weeks) tissues to endoPG and the disease (Bateman and Lumsden, 1965). Increased resistance with aging was related to large increases of calcium and conversion of pectins to pectates.

3) Differential induction and repression of enzyme synthesis. The syntheses of pectinase, cellulase, and other hydrolytic enzymes by pathogenic microorganisms frequently are regulated by concentrations of substrates and products. For example, pectinase synthesis is stimulated by pectin and inhibited by galacturonic acid, cellulase is stimulated by cellulose or cellobiose and inhibited by glucose (Albersheim, et al., 1969; Bateman and Millar, 1966; Goodman, et al., 1967; Wood, 1967). Thus, differences in host composition can markedly affect the production of enzymes. Albersheim, et al., 1969, have proposed that carbohydrate structures of plant cell walls, through their different sensitivities to enzymes and effects on enzyme production, may determine varietal resistance. Deese and Stahmann (1962) reported that Verticillium albo-atrum produced pectinesterase only on stems of susceptible varieties. Polygalacturonase production on stem sections also was directly related to varietal susceptibility. Mussell and Green (1970) reported similar results with V. albo-atrum and Fusarium oxysporum in susceptible and resistant stem sections of cotton and tomato. English and Albersheim (1969) reported that α-galactosidase production by various physiological races of Colletotrichum lindemuthianum was always greatest when strains were grown on susceptible host wall materials. Possibilities of differential growth rates of the fungi or inactivation of the enzymes, however, were not eliminated in any of the above studies.

In few, if any, instances has resistance to an enzyme clearly been shown to be the cause of varietal resistance to natural populations of pathogens. Nevertheless, resistance to enzymes undoubtedly is involved in total resistance and might be regulated by many indiscrete genes.

Resistance to Toxins.--Phytotoxins produced by microorganisms cause leakage, disruption, and death of host cells and are critical determinants of pathogenesis by certain facultative fungal parasites (Braun and Pringle, 1948; Brown, 1965; Scheffer and Pringle, 1967; Scheffer and Yoder, 1972; Wood, et al.,

1972). Varietal resistance to pathogens producing host-specific phytotoxins is shown to the toxin alone (Braun and Pringle, 1958; Scheffer and Pringle, 1967; Scheffer and Yoder, 1972). Since toxic action on cells requires only minutes, predisposition to phytotoxins appears to be critically important in determining susceptibility (Scheffer and Yoder, 1972). Romanko (1959) found that toxin of *Helminthosporium victoriae* was inactivated totally in resistant but not in susceptible oat varieties. He suggested that resistant plants contain some mechanism for inactivating the toxin. However, Scheffer, et al. (1967) and Scheffer and Yoder, 1972), working with several host-specific toxins, found that active toxin could not be recovered after absorption by either susceptible or resistant varieties. They proposed that phytotoxins might have to be absorbed on specific receptors or sensitive sites to be active and that a lack of receptors might be the cause of varietal resistance.

Differences in varietal resistance to disease have also been related to reactions of plants to fungal filtrates, e.g., *Verticillium* wilt of lucerne (Carr, 1971). Filtrates in this instance contain several toxins, and no single toxin showing the host-specificity has been isolated. Many facultative fungi, such as *Fusarium* sp., *Aspergillus* sp., *Helminthosporium* sp., *Sclerotinia* sp., *Alternaria* sp., and bacteria, each produce numerous toxins (Herout, 1971; Turner, 1971; Wood et al. (1972), which may vary qualitatively and quantitatively between species and strains. Host specificity toward these pathogens may be determined to a great extent by the plant's resistance to the composite of microbial toxins and enzymes, but this is difficult to evaluate.

Antimicrobial substances.--Ward (1905), in his classical paper in 1905, concluded: "the matter (disease resistance) has nothing to do with anatomy, but depends entirely on physiological reactions of the protoplasm of the Fungus and the cells of the host. In other words, infection, and resistance to infection, depend on the power of the Fungus-protoplasm to overcome the resistance of the cells of the host by means of enzymes or toxins; and, reciprocally, on that of the protoplasm of the cells of the host to form antibodies which destroy such enzymes or toxins, or to excrete chemotactic substances which repel or attract the Fungus-protoplasm." Ward further emphasized that external factors (temperature, moisture, and nutrition) greatly affect the susceptibility of individual plants to their pathogens. Following his suggestion, numerous workers (Butler, 1918; Chester, 1933; Price, 1932) attempted to find chemotactic substances, antibodies, and acquired immunity (increased resistance to a disease resulting from previous infection by the same pathogenic agent, or one of its strains) in plants. The production of specific antibodies was never found in plants. However, agglutinins, precipitins, lysins, enzyme denaturants, and antimibrobial substances in plants were demonstrated (Bawden, 1953; Bell, et al., 1962; Byrde, 1963; Cook and Taubenhaus, 1911; Cruickshank and Perrin, 1964; DeBaun and Nord, 1951; Farkas and Kiraly, 1962; Fawcett and Spencer, 1969; Kosuge, 1969; Kuc, 1964; Mandels and Reece, 1965; Rennerfelt and Nacht, 1955; Scheffer and Cowling, 1967; Sehgal, 1961; Spencer, 1962; Van Fleet, 1972; Virtanen, et al., 1957; Walker and Stahmann, 1953; and Whittaker, 1970). These materials were generally nonspecific and frequently occurred in healthy as well as infected plants.

Preinfectional antimicrobial compounds have usually been found in dead tissues of living plants (such as heartwood, bark, or bulb scales) or in specialized living cells (such as endodermis and hypodermis, resin ducts, or gland cells). These compounds function in the biochemical ecology of higher plants competing with other plants, pests, and pathogens (Whittaker, 1970). Most toxic chemicals found in dead tissue, glands, or specialized cells also are formed by other living plant cells under stress of infection; therefore, the biochemistry of their synthesis will be discussed under postinfectional chemical barriers.

Antimicrobial Compounds in Dead Tissues.--Cook and Taubenhaus (1911) demonstrated that tannins were toxic to fungi at the concentrations found in bark, cork, and heartwood. They extracted tannins from cork and found that fungal attack occurred unless the tannin was replaced. The work of Walker, Link, et al. (1929, 1933, 1924, 1963 and 1955) with onion further established the importance of polyphenols as preinfectional chemical barriers. Varieties with white bulbs were uniformly susceptible, while those with red bulbs were resistant to *Colletotrichum circinans*. Cold-water extracts from the dry outer scales of red, but not white, bulbs were highly toxic to spores and mycelia of the fungus. When the dry outer scales were removed, previously resistant, colored bulbs were readily infected. The fungitoxic chemicals in red scales were identified as catechol (Link, and Walker, 1933) and protocatechuic acid (Link, et al., 1929) (Figure 4).

Numerous chemicals have been found in fungitoxic concentrations in bark and heartwood (DeBaun and Nord, 1951; Erdtman, 1952; Fawcett and Spencer, 1969; Rennerfelt and Nacht, 1955 and Scheffer and Cowling, 1967). These include condensed tannins, hydrolyzable tannins, lignins, tropolones (thujaplicin and nootkatin), stilbenes (pinosylvin and 2,4,3',5'-tetrahydroxy stilbene), terpenoids (chamic acid and carvacrol), flavanoids (taxifolin and quercetin), and quinones (juglone and plumbagin); see Figure 1 for examples. The toxicity and concentrations of the materials, together with high C/N ratios, account for much of the relative resistance of heartwoods to decay.

Antimicrobial Compounds in Live Cells and Glands.--Water extracts or juice expressed from young plant leaves or shoots usually contain antimicrobial materials (Bawden, 1954; Sehgal, 1961; Spencer, 1962; Van Fleet, 1972; Virtanen, et al., 1957). In many of these, the toxic substances are formed from nontoxic precursors by rapid enzymatic reactions; these are discussed under postinfectional mechanisms. In others, the toxic substances seem to be isolated in vacuoles of certain living cells, in resin ducts, or in glands. Such substances include alkaloids, terpenoids, polyacetylenes, and possibly proteins.

In 1938 Greathouse and Watkins demonstrated that roots of *Mahonia trifoliolata* and *M. swaseyi* contained from 1.33 to 2.48% of the alkaloid, berberine (Figure 2). This was more than 65 times the concentration required to inhibit growth of *Phymatotrichum omnivorum* completely in culture. They suggested the alkaloid was responsible for the high level of resistance that these plants had to *Phymatotrichum* root rot. Berberine was distributed in walls of tracheids and vessels in the xylem and in smaller amounts in lumina of cells in the wood rays. In extracambial tissues, the alkaloid occurred in a nearly continuous zone of parenchyma cells surrounding the phloem and below the periderm. The bast fibers were impregnated with berberine, and small amounts were frequently observed in the periderm. The fact that berberine was most concentrated in a zone of cells just below the periderm was given special significance, since this is the zone of cells first attacked by the fungus in susceptible hosts. Greathouse (1939) later showed a similar relationship between the alkaloids sanguinarine (Figure 2), chelerythrine, and protopine and resistance of *Sanguinaria canadensis* to *Phymatotrichum* root rot.

Greathouse and Rigler (1940) later studied 62 alkaloids from 15 families and more than 50 species. Toxicity of each alkaloid to *P. omnivorum* was generally associated with the relative resistance of the plant from which it was isolated. They concluded that alkaloids constitute an important factor in root resistance to *P. omnivorum*.

The steroidal glyco-alkaloids (Figure 2), tomatine (D-xylo-(1-3-)-[D-gluco-(1-2)]-D-gluco-(1-4)-D-galacto-tomatidine) in tomato, and solanine (L-rhamno-(1-4)-D-galacto-(1-4)-D-gluco-solanidine) and chaconine (L-rhamno-(1-4)-[L-rhamno-(1-2)-]-D-gluco-solanidine) in potato, are fungitoxic and have been implicated in resistance to diseases caused by *Fusarium* sp. (Arneson and Durbin, 1968; Fawcett, and Spencer, 1969; Stoessl, 1970). Detoxification of tomatine by pathogenic *Septoria lycopersici*, but not by nonpathogenic *S. linicola* or *S. lactucae*, has been demonstrated (Arneson and Durbin, 1967).

Other nitrogen-containing compounds occur as preinfectional inhibitors in young grasses. Avenacin (Figure 2), a glyco-triterpenoidal amino-acid ester, occurs at concentrations of 155 mg/kg fresh weight in oat roots (Burkhardt, et al. 1964) and is toxic to many fungi at 3-50 μg/ml (Maizel, et al., 1964). The oat pathogen, _Ophiobolus graminis avenae_, but not avirulent strains of _O. graminis_, detoxified avenacin by hydrolyzing one of the sugar moieties (Turner, 1961).

Extracts or juice from young barley plants are fungitoxic (Spencer, 1962; Stoessl, 1969; Virtanen, et al., 1957). Ludwig, et al. (1960) observed that young barley seedlings were not infected by the pathogen, _Helminthosporium sativum_, in the first 5 days after emergence from seed. A group of closely related compounds, hordatines, apparently were responsible for the antifungal activity in young plants (Stoessl, 1970). Hordatines are formed by a peroxidase-catalyzed dimerization of cinnamyl amides formed with the organic base agmatine. Hordatine A (Figure 2), its glucoside, and its methyl ether are equally fungitoxic at about 10^{-5} M. Loss of resistance to _H. sativum_ in young barley seedlings has been correlated with accumulation of Ca^{++} in the plants. Fungitoxic activity of plant sap or hordatine A was inactivated by addition of Ca^{++} or Mg^{++}.

Many terpenes have antimicrobial activity (Fawcett and Spencer, 1969; Robinson, 1969; Stoessl, 1970) and are frequently found in gland secretions, resins, and essential oils of plants. Gossypol (Figure 3) occurs in subepidermal glands of leaves and shoots and in root periderm of cotton (Bell, 1967). The cyclic ketones, humulone and lupulone (Figure 3) occur in soft resin of hop cones (Lewis, et al., 1959; Salle, et al., 1949), and limonene (Figure 3) occurs in essential oils of citrus (Stoessl, 1970). These compounds have broad-spectrum fungicidal activity, but their roles in specific cases of disease resistance have not been elucidated.

The acetylenic compounds are among the most potent antifungal compounds found in plants. Capillin (Figure 3), which occurs in essential oils of _Artemisia capillaris_ (Imai, 1956) and _Chrysanthemum frutescens_ (Boglmann and Kleine, 1961), inhibits _Trichophyton purpureum_ at 0.25 μg/ml (Imai, et al., 1956) and _Penicillium italicum_ at 4 μg/ml (Tanaka, 1961). Another acetylenic compound, wyerone (Figure 3), is responsible for toxicity of extracts from _Vicia faba_ (Deverall and Vessey, 1969). Fresh broad-bean seedlings yielded 10 mg wyerone/kg fresh tissue. The compound completely inhibited _Alternaria brassicicola_ at 3 μg/ml. Nematicidal compounds in roots of _Tagetes_ are terthienyl compounds (Uhlenbroek and Bijloo, 1958, 1959), and the first acetylenic compound isolated from plants, carlina oxide (Figure 3), is bactericidal (Robinson, 1969). More than 300 acetylenic compounds have been isolated from fungi and higher plants (Robinson, 1969). Those that contain hydroxyl, carbonyl, or furan groups usually are antimicrobial. Acetylenic compounds are particularly abundant in the Compositae, Umbelliferae, and Araliaceae and are produced mainly by the endodermis or epithelial cells forming ducts (Van Etten and Bateman, 1971; Van Fleet, 1971).

Several observations indicate that preinfectional proteins are important in disease resistance.

1) Many plants contain proteins that inhibit mechanical transmission of viral diseases when wound saps from infected plants are used to inoculate healthy plants of a different species (Bawden, 1954). These inhibitors usually are not effective when wound saps are used to inoculate healthy plants of the same species. The inhibitor from pokeweed has been identified as a glycoprotein.

2) Albersheim and Anderson (1971) found proteins in cell walls of bean hypocotyls that inhibit enzymes secreted by fungal pathogens. The ability of these proteins from differential bean varieties to inhibit endopolygalacturonases from α, γ, and δ races of _Colletotrichum lindemuthianum_ was nonspecific (1972).

3) Abeles, et al. (1971) have shown that plants contain gluconase and chitinase enzymes and have suggested that these may have an antibiotic role in attacking cell walls of invading pathogens.

4) Doubly, et al. (1960) found that a specific antigen in each of 4 races of _Melampsora lini_ also was found only in lines of flax that were susceptible to each particular race. Rust races were avirulent when they lacked a common antigen with their host. Since each race was grown on its corresponding susceptible host, rust spores used for antigen preparation might have been contaminated with host tissues to account for the results. However, the same results subsequently were repeated with spores of all races produced on the variety Bison (Flor, 1971). Common antigen relationships between susceptible hosts and their pathogens have since been shown for several diseases (DeVay and Charudattan, 1972; DeVay, et al., 1967). The common antigen of _Fusarium oxysporum_ or _Verticillium albo-atrum_ and cotton is a soluble polysaccharide-protein complex, whereas that of _Ustilago maydis_ and corn is present in ribosomes (Wimalajeewa and DeVay, 1971). Other common antigens have not been characterized. The role of common antigens in disease resistance is not known.

Attempts to explain resistance on the basis of preinfectional toxic compounds have generally been frustrated by inadequate concentrations of the compounds or their occurrence only in dead or a few localized cells. Recent studies have shown that synthesis of preinfectional toxins is greatly increased and occurs generally in all or most cells when they are infected or chemically stressed. Pinosylvin (Hillis and Inoue, 1968; Shain, 1967), safynol (Allen and Thomas, 1971; Thomas and Allen, 1971), gossypol-related terpenoids (Bell, 1967, 1968) and solanine and chalconine (Allen and Kuc, 1968; Locci and Kuc, 1967) are preinfectional toxins, but their behavior is similar to that of the phytoalexins, which are induced only by infection.

Postinfectional Resistance

While preinfectional mechanisms account for much of the general resistance to nonpathogens, they usually do not prevent pathogens from penetrating and establishing themselves in their hosts (Akai, 1959; Akai, et al., 1967; Dickinson, 1960; Ellingboe, 1968; Flentje, 1958; Matta, 1971; Suzuki, 1965). Establishment of pathogens induces numerous biochemical changes in the host, some of which cause containment or death of the pathogen. These, therefore, are _postinfectional_ (also called induced, physiogenic, physiological, or biochemical) _resistance mechanisms_ and include both anatomical and chemical barriers to disease development.

Anatomical Barriers.--Infected plants form a number of characteristic structures that appear to contain the growth or spread of the pathogen (Beckman, 1966; Muller, 1959; Suzuki, 1965; Van Fleet, 1972). Many of these same structures also are formed in mechanically injured tissues, but they are not present in healthy plants. Some anatomical defense structures result from differentiation of tissues or deposition of substances (for example, callose, suberin, lignin, gums) in advance of and around the pathogen. Other structures or responses are cellular and include swelling of cell walls, hyphal sheaths, tyloses, permeability changes, and hypersensitivity.

Formation of Periderm and Cork Layers.--Infections of stems, roots, and young fruits frequently result in the formation of cork layers, or periderm, beyond the developing pathogens. These cells are differentiated from parenchyma cells, which frequently are formed from new, stimulated meristematic activity in the region of the infection. Cork or periderm cells are tightly arranged and have thickened walls strengthened by deposits of lignin and suberin. They appear to prevent spread of the pathogen, contain pathogenic enzymes and toxins within the infection site, and restrict flow of nutrients and water from plant cells to the infection site. This type of defense response is commonly found with scab

(Jones, 1931), anthracnose (Stanghellini, and Aragaki, 1966), or leaf-spot (Cunningham, 1928) diseases. Effectiveness of periderm, or cork-layer, formation depends largely on the speed of its development and the degree of compactness and impregnation of the cells with lignin and suberin.

Formation of Abscission Layers.--Abscission layers are formed from adjoining layers of parenchyma cells, which swell and become thin-walled, while the pectic materials of the middle lamella between them are dissolved. These layers most commonly form between leaf petioles and stems, but also may form in localized areas of young leaves, particularly in stone fruits. For example, the symptoms of "shot-hole" diseases (Samuel, 1927) result from abscission layers as resistance responses. A layer of cork cells usually is formed just behind the abscission layer, which confines the fungus to the infected tissues. Severe infection of leaves, for example, in bacterial blight of cotton (Brinkerhoff, 1970) frequently results in defoliation, which prevents spread of the pathogen from leaves to the remainder of the plant.

Formation of Gums.--Intense deposits of gums frequently develop in cells and intercellular spaces around infected tissues (Akai, 1959; Suzuki, 1965). Gums also are secreted through pits of xylem parenchyma cells into vessels and tracheids and function in resistance to wilt diseases (Beckman, 1964; Dimond, 1970). The gums exuded by stone-fruit trees are composed mainly of pentosans. Wound gums from various plants also give positive tests for lignin, and many contain other yellow or brown granulated substances. Gums of cotton are rich in antimicrobial terpenoids and polyphenols (unpublished), and present a chemical as well as anatomical barrier to infection. The speed and extent of gum deposits partly account for resistance of rice varieties to *Helminthosporium* leaf spot (Akai, 1959).

Formation of Swollen Walls and Hyphal Sheaths.--Fungal hyphae, even before they initiate penetration, frequently induce thickening of epidermal cell walls and slight protuberances into the lumen of the host cell (Akai, 1959; Akai, et al. 1967, 1971; Pierson and Walker, 1954). Later, the protuberances develop into a sheath of materials between the plant cell wall and cytoplasmic membrane and envelope the penetrating hyphal tip (Akai, 1959; Bell, 1972). Many of the attempted fungal infections abort, with concurrent lysis of the hyphal tip, in swollen walls or hyphal sheaths, which appear to represent structural as well as possibly chemical barriers to microbial penetration. The swollen epidermal walls usually become lignified. Some contain high levels of suberin. Protuberances and hyphal sheaths may be composed primarily of callose (callosities) lignin-related compounds (lignitubers), or cellulose. The last type is found mostly around haustoria of rust, downy mildew, or smut fungi. Many investigators have proposed that hyphal sheaths are produced mainly from cell walls of the host. However, the ultrastructure studies of Griffiths (1971) show that lignitubers, formed in response to *V. dahliae*, result from aggregation and lysis of small vesicles pinched off from the protoplasmic membrane. The contents of these vesicles might interact with enzymes of the wall to form the gummy materials that accumulate in hyphal sheaths.

Formation of Tyloses.--Beckman (1966) proposed that resistance to vascular infections involves a three-step mechanism: a) screening out of mobile microbial cells from the transpiration stream by perforation plates or their remnants in the xylem vessels, b) formation of gels in the vessel to immobilize cells of the pathogen, and c) overgrowth of vasicentric parenchyma cells to form tyloses, which eventually seal off the infected portion of the vessel. Tyloses, which protrude through half-bordered pits of the xylem vessel, have thin walls made of cellulose and other carbohydrates. Beckman (1971) proposed that pasticizing of plant walls and formation of tyloses may be conditioned by diurnal fluctuations in pH and the presence of metabolic acids that remove calcium from cell walls and permit their extension. The speed of formation of tyloses has been related to varietal resistance of numerous plants to vascular diseases (Beckman, 1964, 1966; Bell, 1972; Dimond, 1970). Gums frequently are deposited with tyloses to seal the vessel completely and possibly form a chemical as well as anatomical barrier.

Permeability Changes.--Thatcher (1943) found that increased permeability of plant cells was consistently associated with susceptible host-parasite relationships. Marked decreases in permeability to urea occurred in host cells surrounding infection sites and were associated with resistance of Mindum wheat to stem rust. Treating Mindum with chloroform vapors increased both permeability of these cells and disease susceptibility. Decreases in permeability to water were found in cork cells that restricted necrotic lesions caused by *Phoma lingam* in swede root. Infection of cotton with *V. dahliae* caused marked increases in nutrient content of xylem fluids collected from susceptible host-parasite combinations and decreases in those from resistant combinations (Bell, 1972). This might be caused by the differential effects on permeability. Pathogens producing host-specific toxins cause much greater increases of permeability in susceptible than in resistant cells (Scheffer and Samaddar, 1970). The nature of changes in permeability have not been determined, but they probably involve structural changes in the plasma membrane.

Hypersensitivity.--Hypersensitivity is characteristic of the specific varietal resistance of plants to pathogens (Muller, 1959). However, the same type of resistance is also exhibited against many nonpathogens (Cruickshank, 1963; Matta, 1971). Ward, while studying rust of brome grass in 1902, was the first to recognize clearly that hypersensitivity was a defense mechanism. Susceptible and resistant brome varieties were penetrated equally well by the rust pathogen, and differences in behavior of the fungus occurred only after direct contact between the host and pathogen. In the susceptible host "the parasite slowly taxes its host and even stimulates the cells for some greater activity," while in the resistant host "the tissues turned brown and died, the destructive action of the infecting tubes having killed the cells too rapidly." Stakman (1915) borrowed the medical term "hypersensitivity" (literally meaning sensitivity of a host to a pathogen beyond the norm) to describe this resistance mechanism. Muller (1959) defines hypersensitivity as "all morphological and histological changes that, when produced by an infectious agent, elicit the premature dying off (necrosis) of the infected tissue as well as inactivation and localization of the infectious agent."

Early studies of hypersensitivity involved only the obligate fungal parasites. However, it soon became apparent that resistance to facultative fungi such as *Phytophthora infestans* and *Venturia inaequalis* also involved hypersensitivity (Muller, 1959). Hypersensitive resistance now has been found for viruses (Diener, 1963; Mundry, 1968; Ross, 1966), bacteria (Klement and Goodman, 1967), and nematodes (Bingefors, 1971; Dropkin, 1969; Endo, 1971; Krusberg, 1963 and Rohde, 1960).

The ultrastructure of the hypersensitive response has been studied recently (Goodman and Plurad, 1971; Heath, et al., 1971; Paulson and Webster, 1972). During the hypersensitive reaction of tobacco to bacteria, there was widespread damage to membranes of subcellular organelles (Goodman and Plurad, 1971). Plasmalemma, tonoplast, bounding and internal membranes of chloroplasts and mitochondria, and the external membrane of microbodies were all profoundly deranged. Damage to membranes coincided with and was probably the cause of tissue collapse and rapid loss of electrolyte-laden water. A much greater loss of semipermeability and electrolytes in hypersensitive than in susceptible cells, particularly from 12-96 hours after infection, has been noted frequently (Muller, 1959).

In cowpea, the first sign of hypersensitivity to rust was deposition of callose-containing material on the host cell wall around the point of entry of the haustorium; this reaction did not occur in compatible reactions (Heath, 1971). Next, the host plasmalemma around the haustorial body became convoluted and was associated with aggregations of phospholipid-like material. Subsequently, simultaneous necrosis of both host cell and haustorium occurred, or the haustorium was enclosed in a callose sheath possibly derived from the activity of dilated rough endoplasmic reticulum.

The hypersensitive reaction of tomato roots to nematodes first was characterized by increased numbers of ribosomes, proliferation of endoplasmic reticulum, and stainability of cytoplasmic ground substances (Paulson and Webster, 1972). Concurrently, dense osmiophilic inclusions disappeared from vacuoles. Distinctness of cell membranes then was lost, and mitochondria and Golgi bodies disappeared. The fibrillar structure also disappeared, and electron-dense inclusions appeared in the nucleoplasm. Organized arrays of ribosomes appeared on the outer membrane of the nuclear envelope. Plastid stroma lost their granular structure, but starch grains persisted. Changes were restricted to cells close to the nematode.

Early concepts of the nature of hypersensitivity, which prevailed even into the 1960's, are reflected in synonyms used for this defense mechanism--suprasensitivity, hypersusceptibility, hyperenergy, necrogenous defense, and necrotic defense. Hypersensitivity was thought to be a rapid host response (death) caused by the extra-aggressive mechanisms (enzymes, toxins, etc.) of the pathogen. The death and desiccation of the host was thought to retain the pathogen by depriving it of nutrition and water. Some workers even stressed that crosses between the hypersensitive and truly resistant hosts should be avoided, since the progeny were apt to be extremely susceptible because of the influence of the hypersensitive parent.

Recently, Tomiyama (1963) concluded that hypersensitive resistance of potato to P. infestans involves a 2-phase response: (a) a sequence of physiological phenomena that result in rapid cell death within a few minutes or at most a few hours after cells are penetrated; and (b) a defense mechanism, which appears to develop an inhibitory environment and repair in cells adjoining the dead invaded cell. Cell death apparently did not inhibit fungal growth, since both compatible and incompatible races of the fungus grew at equal rates for several hours after cell death.

Hypersensitive death of cells appears to depend on metabolic activity. The uncoupling toxins, sodium azide and 2,4-dinitrophenol, greatly delayed hypersensitive death of potato cells when applied after penetration by incompatible strains (1971). Hypersensitivity also occurred more slowly when freshly-cut surfaces were inoculated than when similar cells were inoculated five or more hours after cutting. The latter cells had a more greatly accelerated metabolic activity than the former. Cycloheximide (an inhibitor of protein synthesis on 80S ribosomes) inhibits the hypersensitive reaction of tobacco to bacteria, even though it has no effect on bacterial growth (Pinkas and Novacky, 1971).

A number of metabolic alterations that appear to be characteristic of diseased tissues occur more rapidly and intensely in hypersensitive reactions than in compatible reactions (Uehara, 1964). The following metabolic changes are clearly established during hypersensitive reactions.

1) Oxidative metabolism increases greatly 12-96 hr after inoculation (Bell, 1964; Farkas, et al., 1960; Kuck, 1967; Millerd and Scott, 1962; Williams and Kuc, 1969). This increase apparently is caused by enhanced activity of the pentose phosphate pathway of respiration (Bell, 1964; Farkas, et al., 1960; Uritani, 1963). Synthesis of the key enzymes, glucose-6-phosphate dehydrogenase and 6-phosphogluconate hydrogenase, is greatly increased in hypersensitive cells (Uritani, 1963; Uritani, et al., 1967). Synthesis of NADP, the cofactor of these enzymes, is enhanced (Uritani, et al., 1967). As soon as necrosis is complete, as seen from visible browning, oxygen uptake rapidly declines and becomes less than that of healthy cells.

2) The content of polyphenols derived from the hydroxycinnamicacid pathway or the shikimate pathway are increased during the hypersensitive reaction (Farkas and Kiraly, 1962; Kosuge, 1969; Kuc, 1964). The key enzymes in polyphenol synthesis, phenylaline ammonia-lyase (PAL) and tyrosine ammonia-lyase (Hanson, et al. 1967), and enzymes of the shikimic-acid pathway are newly synthesized, or their

rate of synthesis is greatly enhanced in either hypersensitive or wounded tissue (Cruickshank and Perrin, 1964; Hadwiger, 1968; Hadwiger and Schwochau, 1969, 1970; Stahmann, 1965, 1967; Uritani, 1963). Shikimate:NADP oxidoreductase was formed after a 2-hr lag phase, and PAL and dehydroshikimate hydrolyase were detected after about a 6-hr lag phase in wounded tissue (Uritani, 1963). Various protein and RNA synthesis inhibitors prevented PAL synthesis, implying that the enzyme was synthesized de novo during the hypersensitive reaction (Hadwiger, 1968; Hanson, et al. 1967; Schwochau and Hadwiger, 1970; Uritani, 1963).

3) Peroxidase and phenolase enzyme contents and activities are increased during the hypersensitive reaction (Cruickshank and Perrin, 1964; Farkas and Kiraly, 1962; Farkas, et al., 1960, 1965; Kosuge, 1969; Kuc, 1964, 1967; Stahmann, 1967; Tamiyama, et al., 1967; Tamiyama and Stahmann, 1964; Uritani, 1963, 1971; Uritani, et al., 1967). Increases in activity of both enzymes frequently are associated with the synthesis of new isozymes that are not found in healthy tissues (Farkas and Stahmann, 1966; Lovrekovich, et al., 1968; Stahmann, 1967; Tamiyama, et al., 1967; Tamiyama and Stahmann, 1964; Uritani, 1971; Weber and Stahmann, 1964). The increase of peroxidase and phenolase enzymes also appears to be due to de novo synthesis rather than to activation of preformed enzymes. In cut tissues, new peroxidase and phenolase were formed after lag periods of less than 24 hr and 24 hr or more, respectively (Uritani, 1971). Synthesis of protoporphrin IX, the prosthetic group of peroxidase, was stimulated concurrently with peroxidase synthesis (Uritani, 1963).

4) The number and activity of mitochondria (Asahi, et al., 1966), microsomes (Uritani and Stahmann, 1961), and ribosomes (Paulson and Webster, 1972) were increased in cells adjoining the hypersensitive reaction. These changes were accompanied by increased breakdown and consumption of stored carbohydrate (Tamiyama, 1971) and increased activity of enzymes such as amylase, phosphorylase, and hexokinase (Uritani, 1963). Activity of cytochrome oxidase and succinic dehydrogenase was greater per mitochondrion (Asahi, et al., 1966).

5) Secondary metabolites with antimicrobial activity are deposited in killed cells and in live cells surrounding the hypersensitive reaction (Cruickshank, 1963; Gaumann, et al., 1950; Muller, 1956, 1958, 1959, 1963; Muller and Borger, 1940; Uritani, et al., 1963, 1967). These compounds are specific for the host plant producing them. Unlike antibodies, they are nonproteins and nonspecific for the organism inciting them. Muller (1959) suggested calling these compounds phytoalexins.

The various responses associated with hypersensitivity are consistent with the hypothesis that hypersensitivity rapidly forms chemical barriers to contain the pathogen and its metabolites. Chemical barriers are manifested in two stages: a) the production of wound toxins (responses 1 through 4) within minutes or hours of the penetration and death of the cells, and b) the mobilization of stored carbohydrates and energy to deposit phytoalexins in the killed and surrounding cells from 8-48 hr after penetration. Phytoalexin synthesis appears to be a unique response in the hypersensitive reaction, since it frequently is not induced by wounding or ethylene treatments, which consistently stimulate the other responses similar to those of infection (Uritani, 1963, 1971). Many of the anatomical structures induced by infection might be more important for maintaining high, localized concentrations of toxic chemicals than for mechanical confinement of the organism (Muller, 1959).

Wound Toxins.--Wound saps or juices expressed from many plant tissues quickly form antimicrobial (Beijersbergen and Lemmers, 1972; Fawcett and Spencer, 1969; Sehgal, 1961; Spencer, 1962; Stoessl, 1970; Virtanen, et al., 1957; Walker and Stahmann, 1955) and enzyme-denaturing (Bell, et al., 1962; Byrde, 1963; Mandels and Reese, 1965; Williams, 1963; Wood, 1967) substances. The toxins apparently are formed passively by the mixing of compartmentalized enzymes and substrates. Boiling water, or alcohol, extracts of tissues usually do not contain toxins. Also, various protein and RNA inhibitors do not inhibit the formation of these compounds, and living cells are not required for their formation. Wound toxins are of three general types--oxidation products of dihydroxy polyphenols, hydrolysis and oxidation or rearrangement products of nontoxic glycosides and esters, and thiosulphinates formed from sulphoxides.

Oxidation Products of Polyphenols.--Phenolic compounds are among the most abundant and varied compounds in plants (Geissman, 1962; Harborne, 1964). Some of the more common polyphenolic compounds and their derivatives from the shikimic-acid pathway are shown in Figure 4. Various derivatives from hydroxycinnamic acids are shown in Figure 7. Any given plant part may contain many different types of phenolic compounds. Contents vary qualitatively and quantitatively between species plants of the same genotype, tissues of the same plant, or even individual cells. Phenolic contents also vary markedly with age and under different environments. Thus, phenolics have nearly unlimited potential in accounting for the many differences that occur in disease resistance.

Cook, et al. in 1911 were the first to show experimentally that tannins were fungitoxic and to suggest they were involved in disease resistance of green apple and pear fruit. They further observed "that tannin as such does not exist in any part of the normal, uninjured fruit previous to maturity, except possibly a small amount in the peel, but exists as a poly-atomic phenol, which upon injury is acted upon by the oxidase and forms a tannin or tannin-like body having the property of precipitating proteid matter, and at the same time forming germicidal fluid." Injury greatly increased the oxidase content in cells next to the injury. In uninjured fruits, oxidase activity was concentrated just below the peel and around the core. Activity was greatest in young fruits, decreased with aging, and disappeared in fully ripe fruit.

Numerous workers (Cruickshank and Perrin, 1964; Farkas and Kiraly, 1964; Farkas, et al., 1965; Kosuge, 1969; Kuc, 1963, 1964, 1967; Muller, 1958; Rohringer and Samborski, 1968; Rubin and Artsikhovskaya, 1964; Tomiyama, et al., 1967; Uritani, et al., 1967; Williams and Kuc, 1969) have since shown that toxic activities of polyphenols depend largely on their oxidation by enzymes such as peroxidase and *o*- or *p*-diphenyl oxidoreductase (polyphenol oxidase). Major products of these reactions are *o*- and *p*-quinones, which act as powerful enzyme and metabolic inhibitors (Webb, 1966) because of their reactions with sulfhydryl and amino groups. The *o*-quinones also readily undergo self-condensation, frequently with a half-life of less than one minute. As a consequence, they seldom can be demonstrated in tissues by direct evidence. Some of the less polar polyphenols (catechol) are themselves antimicrobial, but this toxicity probably is caused by formation of the *o*-quinone within the microorganism.

The effectiveness of polyphenols in disease resistance depends on: a) quantity and type of polyphenols in healthy tissue, b) speed and quantity of polyphenol synthesis induced by infection, c) quantity and type of oxidase in healthy tissue, d) speed and quantity of oxidase synthesis induced by infection, e) location of polyphenols and oxidases, f) sensitivity of pathogens and their enzymes to polyphenols and oxidation products, and g) cellular environment (pH, temperature, presence of sulfhydryl, amino, or reducing compounds that greatly affect oxidase activity and quinone formation; and preinfectional phenols that affect the induction of enzymes that synthesize more phenolic compounds). A complete study of the role of polyphenols in resistance should consider all these points.

Varietal resistance has been related to polyphenol content on numerous occasions. Condensed tannins (complex polyphenols formed from flavolans like +-catechin, Figure 4) and hydrolyzable tannins (complex polyphenols formed from glucose and gallic acid, Figure 4) have been implicated in resistance of cotton, strawberry, and apricot to *Verticillium* wilt (Bell, 1972) and the resistance of chestnut to *Endothia parasitica* (Nienstaedt, 1953). The polyphenols,

3,4-dihydroxyphenylalanine (DOPA) and 3-hydroxytyramine, have been implicated in the resistance of banana to Fusarium wilt (Mace, 1964; Mace and Solit, 1966) and of sugarbeet to Cercospora leaf spot (Harrison, et al., 1961; Rautela and Payne, 1970). Protocatechuic acid and catechol contents of onion scales were strongly correlated with anthracnose resistance in onion (Link, et al., 1929; Link and Walker, 1933, Walker and Link, 1935; Walker and Stahmann, 1955). Concentrations of hydroxycinnamic acids and their esters with quinic acid, shikimic acid, and glucose have been related to resistance of: potato to scab (Johnson and Schaal, 1952, 1957); Verticillium wilt (Lee and LeTourneau, 1958; McLean, 1921; McLean, et al., 1961; Patil, et al., 1964), and late blight (Sokolava, et al., 1961; Tomiyama, et al., 1967; Virtanen, et al., 1957); wheat to rust (Walker and Stahmann, 1955); apples and pears to storage rots (Byrde, et al., 1959; Hulme and Edney, 1960; Williams, 1963) and scab (Kirkham, 1958; Williams and Kuc, 1969); carrots to storage rots (Hampton, 1962); coffee to canker (Echandi and Fernandez, 1962); and Nicotiana glutinosa to virus (Farkas, et al., 1960).

The importance of oxidase enzymes is indicated in several studies. Infiltration of N. glutinosa leaves with ascorbic acid (which reduces quinones) resulted in fewer and larger local lesions from virus infections (Farkas, et al., 1960; Parish, et al., 1962), and infected plants consumed 50% more ascorbate than healthy plants. Polyphenols with phenolase or peroxidase, but not polyphenols alone, rapidly inactivate pectinase and to a lesser extent cellulase enzymes (Byrde, 1963; Mandels and Reese, 1965; Williams, 1963; Wood, 1967). Deverall (1964) found that varietal resistance of French beans to halo-blight was associated with a marked increase in oxygen consumption by leaf homogenates. A substrate that appeared to stimulate oxygen uptake was isolated and had properties of a galactolipid. Lipoxygenase (lipoxidase), which specifically oxidizes linoleic or linolenic acid (a major component of galactolipids), was found in the homogenate. Mace (1964a, b) concluded that vascular browning associated with resistance of banana to Fusarium oxysporium f. cubense probably was caused by phenol oxidases and peroxidases of both the plant and fungus. Kedar (1959) concluded that the correlation between general resistance of potato to Phytophthora infestans and peroxidase activity of crude sap of healthy plants was good enough to use in field selection programs. Peroxidase activity of different organs of the potato plant also was correlated with their resistance to P. infestans (Fehrmann and Dimond, 1967). The wheat variety, Transfer, bears leaf-rust resistance that resulted from incorporation of a segment of an Aegilops umbellulata chromosome (Macdonald and Smith, 1971). This segment carries gene(s) for production of a unique peroxidase, which appeared to be suppressed in the parent variety but was expressed in Transfer, possibly by deletion of a suppressor element.

Only potato varieties immune to late blight showed marked increases of phenoloxidase activity directly after fungal penetration (Rubin and Artsikhovskaya, 1964). Concurrent inhibition of dehydrogenase occurred, presumably by products of phenol oxidation. However, tyrosinase can inactivate alcohol dehydrogenase by direct oxidation (Lobarzewski, 1962), and DOPA and catechol accelerate this reaction. Resistance of sugar beets affected with Cercospora leaf spot was correlated with both the speed and magnitude of increased peroxidase and o-diphenol oxidase activities (Rautela and Payne, 1970). Similar relationships between increased peroxidase activity and varietal resistance are reported for bean infected with Pseudomonas phaseolicola (Imaseki, et al., 1968) and tomato infected with Fusarium oxysporum f. lycopersici (Van Fleet, 1971). The ability of nonpathogenic organisms and ethylene to induce disease resistance in sweet potatoes to C. fimbriata was related to their ability to induce new peroxidase and polyphenol oxidase synthesis (Imaseki, et al., 1968; Stahmann, 1967). Heat-killed cells of Pseudomonas tabaci injected into tobacco leaves increased peroxidase activity and induced resistance to subsequent inoculation with the live bacterium (Lovrekovich, et al., 1968). Certain physiological diseases, for example, scald of apple and blackheart of potato, have symptoms similar to those of the hypersensitive reaction. Varieties susceptible to these physiological diseases have higher phenolic contents, higher phenolase activity, and lower dehydrogenase activity than resistant varieties.

In preparations from healthy plants, phenolase is bound to chloroplasts, mitochondria, and other particulate fractions, from which it is extracted with Triton X-100. Latent phenolase in broad-bean extracts is activated by acid or alkali (Kenten, 1957), anionic wetting agents (Kenten, 1958), or pectic substances released by enzymes of pathogens (Deverall and Wood, 1961). The cambium that gives rise to the cork of oak bark contains both polyphenolixidase and β-glucosidase, which are probably responsible for tannin synthesis. β-Glucosidase and esterase enzymes also are found in association with oxidases in other cells.

Peroxidase occurs abundantly in secondary walls, in which lignification or polymerization of polyphenols occurs (Harkin, 1967). Both peroxidase and polyphenolase are prominent in endodermal cells, particularly in the root (Van Fleet, 1971). During development of the epidermis, peroxidase first is distributed in the cytoplasm, later it becomes localized in a mosaic in the tangential walls next to the stele and in plasmodesmatal tubules connecting epidermal and adjoining cells The enzyme appears to be bound in a lipid-like polymer, and binding can be blocked by ethylene (Van Fleet, 1971). Peroxidase is also found abundantly in cells of the duct canals, bundle sheaths, hypodermis, and meristems, phloem companion cells, wound-regeneration cells, and scattered parenchyma cells throughout the plant. In cotton parenchyma cells, the enzyme appears to be bound to the outer part of the plasma membrane and occurs in particulate bodies in the cytoplasm (unpublished).

The localization of phenoloxidases and peroxidases in diseased tissues has not been studied extensively. Mace (1964a, b) found both enzymes concentrated in particulate bodies in cytoplasm of banana and *F. oxysporium* f. *cubense*. In sugar beet leaves affected with *Cercospora* leaf spot, however, peroxidase was mainly in the soluble fraction, while *o*-diphenol oxidase was in the chloroplasts and mitochondria (Rautela and Payne, 1970).

Polyphenols usually are located in tonoplasts (vacuoles) of the same cells that contain the oxidase enzymes in other structures (Rautela and Payne, 1970; Reeve, 1951; Van Fleet, 1971). Cell injury apparently can liberate the polyphenols and allow them to be quickly oxidized. Beckman, et al. (1972) found that the bulbous part of stalked glands on stems and leaves of many plants contained intense amounts of polyphenols; Veech (unpublished) has shown intense polyphenolase activity, particularly in the stalks, in the same glands in tomato. Thus, rupture of the fragile gland should readily generate quinones on the surface of the stem or leaf. The endodermis and hypodermis also have intense polyphenol as well as oxidase contents. Greater phenolic contents in resistant than in susceptible varieties may be because of greater percentages of phenol-containing cells (McLean, et al., 1961; Van Fleet, 1972), or phenolic content per cell. Individual cells in cherry and peach fruits undergo a 90-fold increase in polyphenol content from the beginning of cell enlargement until fruits reach about half-mature size (Reeve, 1959). During ripening, the phenolic content per cell drops to about 15% of that in green fruit. Similar changes occur in polyphenol content in other fruits. Resistance of green fruits to various rots usually parallels the phenol content.

Microorganisms vary considerably in their sensitivity to different phenolic compounds (Bell, 1967; Cook and Taubenhaus, 1911; Farkas and Kiraly, 1962; Fawcett, et al., 1969; Fawcett and Spencer, 1969; Walker and Link, 1935). The order of sensitivity of fungal pathogens to polyphenols is: obligate parasites>facultative saprophytes>facultative parasites>saprophytes. Thus, spore germination of mildew and rust fungi frequently is inhibited by polyphenols, while the saprophytes *Aspergillus* and *Penicillium* readily make use of moderate concentrations of polyphenols as carbon sources.

The toxicity of polyphenols is inversely related to their polarity (for example, *o*-dihydroxybenzene>3,4 dihydroxybenzaldehyde>3,4 dihydroxycinnamic acid> 3,4 dihydroxybenzoic acid) and directly related to their oxidation potential (for example, *o*-dihydroxybenzene>*p*-dihydroxybenzene>*m*-dihydroxybenzene). The toxicity of phenolics, particularly acids, increases markedly as pH is lowered from 6 to 3. The effect of various cell components on toxicity of phenolics have not been elucidated. Amino acids, peptides, and reducing compounds probably adversely affect the toxic efficiency of phenolics, since quinones readily react with sulfhydryl and amino groups. The possible relationships between toxic efficiency of phenolic compounds and varietal resistance have not been studied.

Antimicrobial Compounds from Glycosides or Esters.--Many plants contain nontoxic glycosides (usually glucosides) that form highly toxic, labile compounds in wound sap. Formation of the toxic entity depends on the action of glycosidase (usually β-glucosidase), which occurs in cell walls and cytoplasmic particles of plants (Chkanikov, et al., 1969; Mace, 1973), or membranes and cytoplasmic particles of fungi (Kaplan, Tacreiter, 1966; Mace, 1973). The glycosidases and their substrates are spatially separated in the same cell. Their distribution in tissues is similar to that of the polyphenols and oxidases.

The glycosidic-bound toxins appear to be a higher form of defense evolution. They are unique to certain taxonomic groups of plants; and the toxins are more potent than phenols and have greater half-lives than the o-quinones. These compounds, in general, can be placed into five major groups (Figure 5): phenolic glycosides (particularly 1,4-dihydroquinone glycosides), benzoxazinone glucosides, cyanohydrin (or mandelonitrile) glucosides, isothiocyanata glucosides, and lactone glucosides or glucose esters.

Glucosides of 1,4-hydroquinones, 1,4-hydronaphthoquinones, and anthranols occur in the plant families Gramineae and Rosaceae; Juglandaceae, Ericaceae, and Boraginaceae; and Rubiaceae, Rhamnaceae, and Polygonaceae, respectively (Robinson, 1969; Van Fleet, 1971). β-Glucosidase and an oxidase are found spatially separated in the same cells (Fan Fleet, 1972). Thus, quinones are quickly formed from the glucosides by cellular injury and concomitant decompartmentalization. In healthy plants the glucosides and activating enzymes have been found in the endodermis (Robinson, 1969), but they probably occur in other cells as well.

Mace and Hebert (1963) reported that anaerobic treatments that control loose smut in wheat and barley grains also induce loss of 2-methoxy-hydroquinone glucoside, with an apparent accumulation of the toxic quinone in tissue. The hydroquinone glucoside, arbutin (Figure 5, No. 1) has been implicated in the resistance of pears to fireblight (Hildebrand and Schroth, 1965; Powell and Hildebrand, 1970; Smale and Keil, 1966). Other hydroquinone glucosides also yield antimicrobial quinones--juglone in walnut; trichocarpin in poplar; plumbagin in European leadwort; and 2-methoxy-1,4-naphthoquinone in balsam (Fawcett and Spencer, 1969; Robinson, 1969). Most of these have ED_{50} concentrations of 5-25 μg/ml against pathogenic fungi. Their roles in disease resistance have not been elucidated.

Nearly all flavanoids, except flavan-3-ols and flavan-3,4-diols, exist as glycosides, and many of the aglycones exhibit antimicrobial activity (Geissman, 1962; Harborne, 1964). Many of these glycosides have free o-dihydroxy groups and, like the glucoside arbutin (Powell and Hildebrand, 1970), they might be oxidized directly to toxic quinones. The content of flavanoids in plants changes markedly with different conditions of stress. However, their role in resistance is largely unexplored. One exception is the dihydrochalcone glucoside, phloridzin, which appears to be involved in resistance of apple to scab (Kuc, 1967; Williams and Kuc, 1969). Phloridzin is hydrolyzed by β-glucosidase, and the aglucone is oxidized by a tyrosinase to yield fungitoxic compound(s) (Noveroske, 1964). A heat-stable metabolite isolated from the apple-scab fungus caused lysis of appleleaf cells; this compound might cause mixing of enzymes and the glucoside in the natural host-parasite relationship.

Virtanen, et al. (1955, 1957, 1965) found glucosides (Figure 5, No. 2) in rye, corn, and wheat, which are hydrolyzed by β-glucosidase to yield the unstable benzoxazinones. The latter readily lose formic acid and rearrange to benzoxazolinones, which are isolated readily from expressed juice. The glucoside of 2,4-dihydroxy-1,4-benzoxazine-3-one (DIBOA: Figure 5, No. 2, R = H) occurs in rye. The glucoside of a similar compound with a 7-methoxy group (DIMBOA: Figure 5, No. 2, R = OCH_3) occurs in wheat and corn. Recently a 6,7-dimethoxy and a 4-deoxy, 7-methoxy derivative have been isolated from Corn (Klun, et al., 1970; Tipton, et al., 1967) and may be related to insect and disease resistance. The aromatic ring of these compounds is derived from quinic acid, and the heterocyclic ring is derived from ribose (Reimann and Byerrum, 1964). The origin of the hydroxyamate moity is unknown. All free aglucones and benzoxazolinones, except for 4-deoxy derivatives, are fungitoxic (Stoessl, 1970; Virtanen, 1965; Virtanen and Hietala, 1955). ED_{50} values range from 50-500 μg/ml for benzoxazolinones and 1-20 μg/ml for benzoxazinones.

Decrease of DIMBOA glucoside in cornstalks with age is correlated with loss of resistance to stalk rots (BeMiller and Pappelis, 1965). Likewise DIMBOA glucoside content in wheat has been related to varietal resistance to rust (ElNaghy and Linko, 1962; ElNaghy and Shaw, 1966). In corn the dominant gene, Bx, mediates production of the hydroxamates, so that homozygous recessive lines are deficient in DIMBOA glucoside (Couture, et al., 1971). When bxbx genes were incorporated into either resistant or susceptible genotypes, the percentage of leaf area infected by *Helminthosporium turcicum* was significantly increased, suggesting that cyclic hydroxamate glucosides play an important role in normal resistance to this fungus. Resistance of 12 inbred strains of corn to the same fungus also was closely related to their contents of hydroxamate glucosides (Molot and Anglade, 1968).

Cyanogenic glucosides (Figure 5, No. 3) are formed from decarboxylated, aliphatic or aromatic amino acids and glucose (Robinson, 1969). These compounds occur in various species of the Leguminosae, Rosaceae, and Gramineae. They are best known for their toxicity to animals (Montgomery, 1969). Wound saps of plants usually contain both β-glucosidase, which hydrolyzes the glucosides to glucose and a cyanohydrin, and oxynitrilase, which converts the latter to an aldehyde and HCN (Figure 5, No. 3). Cyanohydrins also decompose spontaneously above pH 7, but are much more stable below pH 6. All of the hydrolytic products are fungitoxic, but the cyanohydrins are much more toxic to fungi than the aldehydes (Bell, 1970).

Valenta and Sisler (1962) correlated resistance of lima bean varieties to downy mildew with the toxicity of expressed juice from stems toward zoospores of the pathogen. The toxicity later was shown to develop only after the juice was expressed (Bell, 1967). Part of the toxicity was attributed to hydrolysis of the cyanohydrin glucosides, that give rise to 4-hydroxy- and 3-methoxy- 4-hydroxybenzaldehyde (Bell, 1970). The glucosides were formed only in aerial parts of lima bean after seedlings emerged from the soil and were much more prevalent in leaves than in stems.

The resistance of flax varieties to *Fusarium* wilt has been related to the amount of KCN exuded by roots (Timonin, 1941). Presumably, the cyanide originated from the cyanogenic glucosides known to occur in flax (Montgomery, 1969).

The isothiocyanates are formed from sulfur-linked glucosides (Figure 5, No. 4), which occur abundantly in the Cruciferae and are found in a few species of other plant families (Sehgal, 1961). The enzyme thioglucosidase (myrosinase) cleaves glucose from the aglucone, which then rearranges to form the isothiocyanate (Virtanen, 1965: Figure 5, No. 4). Synthesis of isothiocyanate glucosides appears to proceed by homologation of amino acids with acetic acid (Chisholm and Wetter, 1967). The mustard oils, which contain allyisothiocyanate formed from the glucoside, sinigrin, have been known for more than 100 years to have antimicrobial activity (Virtanen, 1965). Various isothiocyanates have since been shown to be fungitoxic at concentrations of 5-80 μg/ml (Drobnica, et al., 1967;

Virtanen, 1965). Contents of isothiocyanate glucosides have been correlated with clubroot resistance in cabbage, but later work refuted the involvement of these compounds in resistance (Walker and Stahmann, 1955). The causal fungus is an intracellular parasite, which caused no tissue maceration. Thus, the thioglucosidase and glucoside probably would not be combined to form the isothiocyanates. Davis (1964) was unable to establish any clear relationship between resistance of cabbage and toxicity of phenylethylisothiocyanate to isolates of *Fusarium oxysporum*. Toxicity of cabbage-root homogenates had the same specificity as the isothiocyanate but also was unrelated to fungal virulence.

Tulipalin is a highly fungitoxic lactone (Figure 5, No. 5) found in wound sap of tulip and other species of the Liliaceae (Bergman, 1966; Bergman, et al., 1967). The lactone probably is synthesized from γ-methyleneglutamic acid, which also occurs in tulip (Bergman, et al., 1967). Tulipalin and β-hydroxytulipalin are hydrolyzed from the nontoxic glucose esters tuliposide A and B (Figure 5, No. 5; R = H and OH, respectively), which occur in intact tissues (Kovacs, 1964; Tscheshe et al., 1969, 1968). Tuliposides readily break down nonenzymatically at pH 7.5 or above. Tulipalin is lethal to many fungi at 100 μg/ml. Since the lactones are inactivated by cysteine and other thiols, their activity probably is caused by interference with sulfhydryl groups (Sehgal, 1961).

Bergman (1966) found that tulipalin content of tulip bulbs was correlated with their resistance to rots. Water-soluble substances from pistils (presumably containing tulipalins) were inhibitory to many fungi but not to tulip pathogens (Schonbeck, 1967). The nonpathogen *Botrytis cinerea* caused a more rapid release of tuliposides from infected tissue than the pathogen *Botrytis tulipae* (Schonbeck and Schoeder, 1972). Also, lactones were formed in the presence of *B. cinerea*, while *B. tulipae* caused cleavage of the tuliposides to the corresponding acids, which have very low activity. *B. cinerea* was more sensitive to the inhibitory substances than *B. tulipae*.

The lactone protoanemonin (5-methylene-2-oxodihydrofuran) is released from a glucoside, ranunculin, in foliage of the Ranunculaceae. On standing, the lactone polymerizes to the less toxic anemonin and other products (Holden, et al., 1947; Sehgal, 1961). Protoanemonin is toxic to many bacteria and fungi at 3-20 μg/ml, but its role in resistance has not been determined.

Other antimicrobial unsaturated lactones that have been found in plants include: parasorbic acid (delta-hexano-lactone) in ash; plumericin in roots of *Plumeria multiflora*; and the coumarins, scopoletin and umbelliferone, in many plants (Fawcett, Spencer, 1969; Sehgal, 1961; Uritani, 1963). These lactones are fungitoxic at concentrations of 100-1,000 μg/ml. Whether they are all derived from glucosides has not been ascertained. The concentrations of coumarins and their glycosides increase dramatically in infected plants such as sweet potato, tobacco, potato, and cotton (Bell, 1972; Fawcett, Spencer, 1969; Van Fleet, 1971). However, these concentrations still are usually less than those required for appreciable fungitoxic activity.

The saponins, glycosides of steroids (C_{27}) or triterpenoids (C_{30}), are distributed in more than 400 species and 80 families of the plant kingdom (Birk, 1969). They are extremely variable in structure, with five or more different types occurring in single species (alfalfa and soybeans). Many saponins possess antifungal properties. The hydrolysis of at least one saponin, sarsaparilloside, forms an aglycone that is more toxic than the glycoside (Stoessl, 1970). The role of saponins in plant disease resistance is unexplored.

Thiosulfinates from Sulfoxides.--The possible therapeutic effects of onion and garlic were recognized by the early Romans and Egyptians (Virtanen, 1965). Garlic and onion juice have since been known to contain potent bacteriocidal and fungicidal activities (Kovacs, 1964; Virtanen, 1965). The major toxicant in garlic juice is allicin (allyl-2-propene-1-thiosulphinate), which is formed from alliin (S-allyl-L-cysteine sulfoxide) by the enzyme, alliinase, by a lyase-type reaction. Allicin is present to the extent of 0.4% in garlic and is toxic to some fungi at 1 μg/ml (Fawcett and Spender, 1969).

Volatile toxins in onion juice were identified as methyl- and propyl-thiosulphinates (Virtanen, 1965). These are less toxic than allicin, but the total toxicity of onion juice also depends on a number of related organic sulfides, which have been described (Virtanen, 1965). In total, these may constitute 4-5% of the dry weight of onion. The antibiotic effect of thiosulphinates appears to be caused by their binding of essential sulfhydryl groups (Sehgal, 1961). Walker and Stahmann (1955) found that pungent onion varieties were more resistant to smudge or neck rot than corresponding mild varieties. This seems to be the only attempt to relate thiosulfinates to disease resistance.

Phytoalexins,--These anitmicrobial compounds occur only in trace amounts in living cells of healthy plants but are synthesized extensively after infection or inoculation, particularly with avirulent microorganisms. Phytoalexins are not formed in wound sap or dead tissue because their synthesis is energy-dependent. Since each phytoalexin is unique to a closely related group of plants, their synthesis appears to represent an evolutionary advance in resistance mechanisms.

The original phytoalexin theory proposed by Muller (1956, 1958, 1963; Muller and Borger, 1940) and restated by Cruickshank (1963, 1966; Cruickshank and Perrin, 1964) included the following postulates: phytoalexins are antifungal substances, formed or activated in hypersensitive tissue only when host cells come into contact with the fungal parasite; the reaction occurs only in living cells; phytoalexins are chemicals and may be regarded as products of necrobiosis of the host cell; they are nonspecific in toxicity, but fungal species are differentially sensitive; the basic response in susceptible and resistant hosts is similar, but the speed of phytoalexin formation is greater in resistant hosts; the reaction is confined to colonized and immediately adjacent tissue; and the sensitivity of the host cell to the fungus is specific, it is genotypically determined, and it determines the speed of host reaction. The original phytoalexin theory was based primarily on the response of potato tubers, pea pods, bean pods, and orchid rhizomes to various fungi. Further elucidation of the theory was based mostly on behavior of pisatin in pea and phaseollin in bean as model phytoalexin systems.

Our current understanding of phytoalexins reveals that several modifications of the original theory are needed.

1) Phytoalexin synthesis may be induced by various agents other than fungi. These include bacteria (Bell and Stipanovic, 1972; Cruickshank and Perrin, 1971; Stholasuta, et al., 1971), viruses (Bailey, Ingham, 1971; Klarman and Hammerschlag, 1972), toxic chemicals (Bell, 1967; Cruickshank and Perrin, 1963; Hadwiger, 1972; Hadwiger and Martin, 1971; Hadwiger and Schwochau, 1969, 1971; Hess and Hadwiger, 1971; Schwochau and Hadwiger, 1970; Uehara, 1963; Uritani, et al., 1960), microbial metabolites (Cruickshank and Perrin, 1968; Schwochau and Hadwiger, 1970; Varns, et al., 1971), and physical treatments such as chilling or ultraviolet irradiation (Bell and Christiansen, 1968; Hadwiger and Schwochau, 1971). Most effective inducers cause cell damage, as evidenced by cell leakage or necrosis. Cell injury inflicted by cutting induces small detectable concentrations of phytoalexins in the absence of microorganisms.

2) Phytoalexins and closely related compounds also occur in cells or tissues of healthy plants. These include heartwood (Hillis and Inoue, 1968; Scheffer and Cowling, 1967; Shain, 1967; Stoessl, 1970), bark or peel (Allen and Kuc, 1968; Bell, 1967, 1972; Stoessl, 1970), epithelial cells and resin ducts (Langcake, et al., 1972; Van Fleet, 1972), subepidermal and hair glands (Beckman, et al., 1972; Bell, 1967), and endodermis (Van Fleet, 1972). Early investigators failed to recognize phytoalexins in healthy plants because they studied only fleshy parenchyma cells, mostly of edible vegetable crops.

3) Phytoalexins in the susceptible and resistant reactions of a given plant may vary qualitatively (Bell and Stiponovic, 1972) and in the speed of their destruction or detoxification (Higgins and Millar, 1969; Wit-Elschove, et al., 1971), as well as in the speed of their formation.

4) Phytoalexins may be general antibiotics (Bell, 1967; Cruickshank and Perrin, 1968) and not just antifungal antibiotics.

The original concept of phytoalexins was too rigid and needs to be expanded. A more realistic definition of phytoalexins is antibiotics produced by anabolic biosynthesis in plant cells irritated by microorganisms or other agents.

Compared to wound toxins, phytoalexins a) form more slowly (usually from 12-96 hr) after infection, b) accumulate in much higher concentrations (frequently 1-2% of dry weight in lesions involving storage parenchyma, endocarp, or cambial tissues), c) are generally less toxic to microorganisms (ED_{50} = 25-1,000 μg/ml), d) are more persistent (half-life of days or weeks in many host tissues), and) are less phytotoxic. The mode of action of phytoalexins against microorganisms may be different from that of wound toxins, which affect mostly sulfhydryl or amino groups. The phytoalexin, phaseollin, appears to affect permeability of fungal membranes (VanEtten and Bateman, 1971).

Phytoalexin synthesis usually occurs in connection with the hypersensitive response (Muller, 1959; Uritani, 1963). It appears to result primarily from de novo synthesis of enzymes as a result of gene activation (Hadwiger and Schwachau, 1969; Schwachau and Hadwiger, 1970). Phytoalexins are derived biochemically from acetate or hydroxycinnamates by various condensation, rearrangement, and oxidative reactions.

Acetate-Derived Phytoalexins.--Terpenoid phytoalexins (Figure 6) include ipomeamarone, ipomeanine, and related compounds in sweet potato (Uritani, 1971; Uritani, et al, 1967); gossypol (Bell, 1967, 1969), hemigossypol (Bell, 1972; Bell and Stiponovic, 1972; Zaki, et al., 1972), vergosin (Bell and Stiponovic, 1972; Zaki, et al., 1972), gossylic acid lactone (Bell and Stipanovic, unpublished) and related compounds (Bell, 1967, 1969) in cotton, rishitin in potato (Sato, et al., 1971; Sato and Tomiyama, 1971; Varns, et al., 1971), tomato (Sato, et al, 1969); and phytuberin in potato. The antimicrobial, preinfectional terpenes that occur abundantly in heartwoods, resins, and essential oils of many plants also might act as phytoalexins, but they have not been studied in diseased tissues.

Terpenoid phytoalexins have been related to both general and specific disease resistance. Ipomeamarone was less toxic to pathogenic than to related nonpathogenic fungi of sweet potato (Nonaku, 1966). Rishitin was studied in the specific resistance of potato to *Phytophthora infestans* and in general resistance to nonpathogens (Ishiyawa, et al., 1969; Sato, et al., 1971; Sato and Tomiyama, 1969; Tomiyama, 1963; Varns, et al., 1971; Varns and Kuc, 1971; Varns, et al., 1971). It was equally toxic to all races of *P. infestans* (Ishiyawa, et al., 1969; Tomiyama, 1971), and the C-3 hydroxyl was essential for activity. Most of the rishitin was present in the brown infected and immediately adjacent cells. The compound was first found at 7-8 hr after inoculation, when mycelial growth also first became inhibited (Tomiyama, 1971); concentrations of rishitin increased linearly after 12 hr, reaching a maximum in 2-3 days, and then decreased rapidly (Sato, et al., 1971; Sato and Tomiyama, 1969). Rishitin reached a completely inhibitory concentration of 100 μg/g at the same time that lesion development was ceasing (Sato, et al., 1971).

Substantial amounts of rishitin and phytuberin were synthesized in tuber slices from various potato varieties inoculated with avirulent strains of *P. infestans* or with nonpathogens of potato (Tomiyama, 1971; Varns and Kuc, 1971). However, slices from various susceptible host-fungus combinations produced little or no detectable phytoalexins. Mycelial contents of all strains of *P. infestans* obtained by sonification induced necrosis and similar accumulation of the phytoalexins in tuber slices of either susceptible or resistant potatoes. Susceptibility was related to a delay or absence of phytoalexin synthesis. Varns, et al. (1971), however, were unable to show a similar relationship in sprouts and suggested that caution be exercised in extrapolating results from tubers to foliage and stems.

Phytoalexins in cotton have been related to disease resistance in several ways:

1) The resistance of *Gossypium barbadense*, compared with susceptible *Gossypium hirsutum*, to *Verticillium dahliae* was associated with a faster rate of accumulation of the gossypol-related phytoalexins (Bell, 1969, 1972; Zaki, et al., 1972) and with a much greater proportion of vergosin in the total phytoalexin content (Bell and Stipanovic, 1972; Zaki, et al., 1972).

2) Tolerant *G. hirsutum* varieties inoculated with mildly virulent strains accumulated phytoalexins more rapidly than those inoculated with severe defoliating isolates (Bell, 1969; Zaki, et al., 1972). Virulent and avirulent strains were equally sensitive to the phytoalexins.

3) Infection of susceptible cotton varieties with *V. dahliae* is confined primarily to the stele, vascular bundles of petioles, and leaf tissues. Resistance of specific tissues to fungal infection was related to speed and extent of phytoalexin synthesis. Cambial tissues and root cortex, boll endocarp, and xylem parenchyma formed concentrations of phytoalexins completely toxic to *V. dahliae* in 24, 48 and 72-96 hr, respectively; leaf tissues formed only trace concentrations of phytoalexins (Bell and Stipanovic, 1972).

4) Marked increases in *Verticillium* wilt resistance with increasing temperatures from 23-30°C were correlated with an increase in the ratio of the rate of phytoalexin synthesis to the rate of conidial production (Bell and Presley, 1969).

5) The degree of induced resistance in cotton to *Verticillium* wilt by injections of avirulent, killed, or temperature-inhibited conidia also was strongly related to the amount of phytoalexin induced by these treatments (Bell and Presley, 1969).

6) In cotton varieties inoculated with races of *Xanthomonas malvacearum*, phytoalexins accumulated in similar patterns for 12-36 hr in susceptible and resistant host-bacteria combinations (Bell, unpublished). Beyond 36 hr, little or no additional biosynthesis occurred in susceptible combinations. The capsular slime from the bacterium also induced appreciable phytoalexin synthesis in resistant but little or no synthesis in susceptible varieties. Virulence of these bacteria may be related to their ability to prevent phytoalexin synthesis.

The steroid alkaloids in potato and tomato have similarities to phytoalexins (Allen, 1970; Allen and Kuck, 1968; Langcake, et al., 1972; Tomiyama, et al., 1968; Locci, and Kuc, 1967; Mohanakumarane, et al., 1969). α-Solanine and α-chaconine, major fungitoxins in healthy potato peels (Allen, 1970), also are synthesized at cut surfaces of the tuber (Allen, 1970; Allen and Kuc, 1968; Locci and Kuc, 1967). Infection with nonpathogens or avirulent strains of *Phytophthora infestans* stimulated production of solanidine, α-solanine, and α-chaconine to about the same degree as wounding (Locci and Kuc, 1967; Tomiyama, et al., 1968). Little or no increase of the steroidal alkaloids occurred in susceptible potatoes after infection.

Concentrations of α-tomatine in roots of tomato varieties resistant to *Pseudomonas solanacearum* were much higher than in those of susceptible varieties (Mohamakumaran, et al., 1969). Infection with the bacterium caused the tomatine content to double only in resistant varieties. *Fusarium oxysporum* f. *lycopersici* caused marked increases of α-tomatine content in roots and stems 2-3 days after inoculation, but responses were similar in resistant and susceptible hosts (Langcake, et al., 1972). Various fungal pathogens of tomato were considerably less sensitive to α-tomatine than were nonpathogens (Arneson and Durbin, 1968).

The acetylenic compounds, safynol and dehydrosafynol in safflower (Allen and Thomas, 1971a, b, c; Thomas and Allen, 1970, 1971) and wyerone in broad bean (Fawcett, et al., 1971), behave like phytoalexins, even though they occur in small concentrations in healthy plants. Wyerone increased from 0.1 μg/g to 45 μg/g 4 days after inoculation with Botrytis fabae. Safynol increased from 0.83 and 0.55 μg/g to 33.7 and 22 μg/g at 2 and 4 days, respectively, and dehydrosafynol increased from 0.01 μg/g to 2.96 μg/g at 2 days, after inoculation of safflower hypocotyls with Phytophthora drechsleri (Allen and Thomas, 1971c; Thomas and Allen, 1970). The ED_{50} of safynol and dehydrosafynol to P. drechsleri were 12 and 1.7 μg/ml, respectively. Van Fleet (1971) reported that polyacetylenes in healthy tissues occur in the endodermis and epithelial cells of ducts. Induced safynol occurred mostly in tissues outside the vascular ring.

In the resistant safflower variety, Biggs, safynol concentrations increased rapidly in the first 48 hr and remained relatively constant at 30 μg/g fresh weight during the next 48 hr after inoculation with P. drechsleri. In a susceptible variety, concentrations increased rapidly in the first 24 hr to 17 μg/g fresh weight, but then decreased during the next 72 hr (Allen and Thomas, 1971a). Differences in safynol content of susceptible and resistant varieties were even greater on a dry-weight basis, since loss of water was greater in the susceptible variety.

Plants of the Biggs variety held in photoperiods of 0, 8, 16, and 24 hr/24 hr after inoculation with P. drechsleri were susceptible, moderately susceptible, and resistant, respectively. At 48 hr after inoculation, safynol concentrations were 11.2, 22.3, 27.0, and 30.5 μg/g fresh tissue, and dehydrosafynol concentrations were 0.13, 0.94, 2.51, and 3.73 μg/g fresh tissue, respectively (Thomas and Allen, 1971). In another experiment, Biggs plants inoculated with the nonpathogen P. megasperma var. sojae contained 15 μg safynol and 7.5 μg dehydrosafynol/g fresh tissue at 4 days after inoculation and were resistant to subsequent infection by P. drechsleri, even when plants were grown in darkness for 4 to 6 days.

Hydroxycinnamate-Derived Phytoalexins.--Numerous antibiotic compounds derived from the hydroxycinnamic acids (Figure 7) are stimulated by infection. The free acids and their coumarins, lignins, and esters (for example, chlorogenic acid) occur in many different plant families. These compounds are readily induced by wounding or infection, but they have weak antibiotic activity. Other compounds, like the stilbenes, dihydrophenanthrenes, pterocarpins (formed from chalcones), and amides (Figure 7), are more toxic, and each occurs in a single or few plant families.

The first phytoalexin to be identified in this group was the dihydrophenanthrene, orchinol, in orchids (Gaumann, et al., 1950, Figure 6). A related compound, hircinol (Uhlenbroeck and Bijloo, 1959, Figure 6), was later found. The stilbenes have been known for many years to occur in heartwood (Schieferstein and Loomis, 1959). Pinosylvin (Figure 1) also is formed rapidly by infected living tissues of pine (Hillis and Inoue, 1968; Shain, 1967). The most studied group of phytoalexins is the pterocarpins (or chromanocoumarans), Figure 6, which include: pisatin in pea (Perrin and Bottomley, 1962); phaseollin (Perrin, 1964), and a related compound (Smith, et al., 1971) in bean; hydroxyphaseollin in soybean (Keen, et al., 1971); medicarpin in alfalfa and clover (Higgins and Smith, 1972; Smith, et al., 1971); and trifolirhizin, maachiain, and maachiain-β-o-glucoside in red clover (Higgins and Smith, 1972). At least 10 other fungitoxic pterocarpins are known to occur in heartwood of various trees and probably will be shown to act as phytoalexins (Geissman, 1962; Harborne, 1964; Stoessl, 1970). Medicarpin occurs both in alfalfa and in heartwood of several tropical leguminous trees (Smith, et al., 1971). Recently, the fungitoxic cinnamylamides, N-(p-coumaroyl)-2-hydroxy-putrescine and N-(feruloyl)-2-hydroxyputrescine, have been shown to accumulate in infected wheat leaves (Samborski and Rohringer, 1970) and might act as phytoalexins (Stoessl, et al., 1969).

Involvement of pterocarpin phytoalexins in general disease resistance to facultative fungal parasites is suggested by numerous experiments with pea (Cruickshank, 1962, 1965; Nonaka, 1967; Perrin and Cruickshank, 1969; Uehara, 1964) bean (Cruickshank and Perrin, 1971; Perrin and Bottomley, 1962), alfalfa (Higgins, 1972; Higgins and Millar, 1968), and *Vicia faba* (Deverall and Vessey, 1969). Phytoalexin synthesis was induced by both virulent and avirulent pathogenic fungi and by nonpathogens. After infection of resistant hosts, phytoalexin accumulation began at 6-9 hr and progressed almost linearly for 48-72 hr. An ED_{50}, or greater, concentration of phytoalexin was formed within 48-72 hr after inoculation, and usually within 96 hr a completely fungistatic concentration was formed. In susceptible hosts, accumulation also began at 6-9 hr after inoculation, but completely fungistatic concentrations failed to develop within a week or more after inoculation. Either pathogens were more insensitive than nonpathogens to the phytoalexins, or phytoalexin accumulation was much slower or did not persist when plants were inoculated with the pathogens. Slow accumulation of phytoalexin in response to infection by pathogens frequently was shown to result from degradation of phytoalexins by the microorganism (Deverall and Vessey, 1969; Higgins, 1972; Higgins and Millar, 1968; 1970; Nonaka, 1967; Turner, 1961; Wit-Elshove, 1968; Wit-Elshove and Fuchs, 1971).

Heath and Wood (1971) reported that in leaf lesions caused by *Ascochyta pisi* and *Mycosphaerella pinoides*, pisatin continued to accumulate in infected tissue until it finally was present in inhibitory concentrations in brown tissue beyond the region colonized by the pathogen (Heath and Wood, 1971); this coincided with retarded lesion development. Pierre and Bateman (1967) also proposed that phytoalexins of bean contain lesions caused by *Rhizoctonia solani*. Phytoalexins were most concentrated in lesions, but they also accumulated in tissues near the lesion during the course of pathogenesis. Concentrations present in young, water-soaked lesions were sufficient to inhibit linear growth of *R. solani*. Since lesions become delimited in this stage, the phytoalexins appeared to contribute to the induced resistance of the host to progressive invasion by the pathogen. Pierre (1971) also examined the nature of general varietal resistance of beans to *Fusarium* and *Thielaviopsis* root rot. In each case greater quantities of phytoalexins accumulated in resistant than in susceptible varieties. The greater restriction of lesion sizes in the resistant variety was attributed to this difference in concentration.

With physiologically specialized pathogens, induction of phytoalexin synthesis is delayed or fails to occur in susceptible, but not resistant, host-parasite combinations. Large amounts of phaseollin were found in bean tissue that showed hypersensitive reactions to an incompatible race of the bacterium *Pseudomonas phaseolicola* (Stholasuta, et al., 1971). Very little phaseollin was found in leaves developing halo-blight symptoms typical of a susceptible reaction. Likewise, bean varieties hypersensitive to rust formed significant amounts of phaseollin, but none could be detected in susceptible varieties showing chlorotic flecks (Bailey and Ingham, 1971). Bean and soybean developed appreciable concentrations of phaseollin and hydroxyphaseollin in necrotic tissue incited by infection with tobacco necrosis virus (Bailey and Ingham, 1971; Klarman and Hammerschlag, 1972). However, no phytoalexin has been found with systemic viruses that cause no necrosis.

In bean varieties differentially resistant to races of *Colletotrichum lindemuthianum*, phaseollin accumulated much earlier during hypersensitive resistant responses than during susceptible responses in which a lesion eventually formed (Bailey and Deverall, 1971). Phaseollin accumulation was limited to infected tissue that was visibly brown. Phaseollin synthesis was not incited by the initial, extensive intracellular growth of compatible races of C. lindemuthianum; after several days, however, host cells collapsed and became brown, and some phaseollin synthesis occurred. Rahe, et al. (1969) also found phaseollin and other phenolic accumulation closely correlated with browning in the same host-parasite combinations; browning occurred earlier in resistant combinations.

Hydroxyphaseollin apparently is involved in the specific resistance of soybeans to Phytophthora sojae. Klarman and Gerdemann (1963) demonstrated that only the resistant variety, Harosoy 63, produced a phytoalexin, presumably hydroxyphaseollin (Keen, 1971), in response to this fungus. However, the susceptible variety, Harosoy, produced similar amounts of the phytoalexin in response to the nonpathogens, Phytophthora megasperma and Phytophthora cactorum. The phytoalexin was equally toxic to the three fungi. Susceptibility was induced in Harosoy 63 by inserting a wick through the inoculation site and continuously flushing with distilled water to remove the phytoalexin (Klarman and Gerdemann, 1963). Heat treatment (44°C for 1 hr) of inoculated resistant soybeans abolished resistance and decreased hydroxyphaseollin production to 10% of that in inoculated control plants (Chamberlain, 1970; Keen, 1971). Greater plant age increased both resistance and potential for hydroxyphaseollin accumulation (Paxton and Chamberlain, 1969).

The nature of accumulation of hydroxyphaseollin in Phytophthora-infected soybeans was further elucidated in recent studies (Gray, et al., 1967; Keen, 1971; Keen, et al., 1971). Some phytoalexin synthesis was found in susceptible host-pathogen combinations, especially in those showing intermediate degrees of susceptibility. The degree of resistance always was directly related to the speed and magnitude of phytoalexin accumulation, particularly during the 72 hr after inoculation. Thus, the resistance conferred to soybean by the dominant Rps and Rps2 allelomorphs appears associated with ability to accumulate phytoalexin rapidly in response to the pathogen, P. megasperma var. sojae (Keen, 1971).

Antimicrobial compounds found in various other infected plants have been called phytoalexins (Fawcett and Spencer, 1969; Stoessl, 1970). These compounds have not been identified, however, and data are inadequate to show whether the compounds are formed as wound toxins, phytoalexins, or products of the infecting microorganism. Two cases illustrate the need for caution in recognizing a compound as a phytoalexin. Numerous studies indicated that 6-methoxy-3-methyl-8-hydroxy-dihydroisocoumarin (IC) was the natural phytoalexin of carrot; and the behavior of this compound in host-pathogen relationships between Ceratocystis fimbriata and carrot was frequently studied (Fawcett and Spencer, 1969; Kosuge, 1969; Stoessl, 1970). Recent studies (Curtis, 1968; Stoessl, 1969), however, show that 6-methoxy-3-methyl-8-hydroxyisocoumarin is a natural metabolite of C. fimbriata, and identification of IC as a phytoalexin is doubtful. Fawcett and Spencer (1969) isolated the toxic phenolic acids, 4-hydroxy-benzoic acid and vanillic acid, from apple infected with Sclerotinia fructigena. However, they pointed out that these were false phytoalexins because they are produced by metabolism of host-produced chlorogenic and quinic acid by fungal rather than by host enzymes.

Conclusions

Much of the general resistance of plants to microorganisms appears to depend on their preinfectional characteristics such as anatomical barriers, predisposition to microbial enzymes and toxins, and chemical barriers. These decrease the availability of nutrients and water at plant-microbe interfaces or exert physical and chemical resistance to penetration and invasion of the microorganism. Organisms that cause infectious disease generally have developed means of overcoming or circumventing the preinfectional resistance mechanisms. Specific resistance of plants to these pathogens appears to depend largely on postinfectional development of antimicrobial compounds. Many of the induced anatomical responses probably enhance the effectiveness of the antimicrobial compounds by preventing their diffusion and thereby increasing their concentrations at localized sites.

Attempts to correlate resistance with the quantity or quality of wound toxins or phytoalexins in whole tissues frequently have been unsuccessful. In cotton infected with V. dahliae, the concentrations of antimicrobial compounds in stems is consistently greater in susceptible than in resistant varieties two or more weeks after infection (Bell, 1969, 1972). The concentrations in this instance reflect the percentage of cells infected and producing delayed defense responses. Plant varieties with marked differences in specific resistance usually produce the same antimicrobial compounds in response to various nonpathogens and races of pathogens. The speed of accumulation of antimicrobial compounds after infection appears to be more important in disease resistance than their quality or the final quantity produced.

A model consistent with various resistance responses is illustrated in Figure 8. According to the model, resistance depends on the speed of accumulation of antimicrobial substances relative to the speed of colonization by the pathogen. The pathogen is contained only after the development of a completely inhibitory environment. If an ED_{100} concentration of antimicrobial substances is developed before secondary colonization, infection is confined to the initially penetrated cell, and the plant is perfectly resistant (immune). If an ED_{100} concentration fails to develop, the disease becomes progressive, and the host is susceptible. If only slight secondary colonization occurs in advance of ED_{100} concentrations (as illustrated in Figure 8), the disease is slowly progressive, but the host is tolerant. In some instances, ED_{100} concentrations develop in advance of colonization only after several sequences of successful penetration of cells; cankers, leaf spots, or other confined diseases result. In a few instances, nutrient and water deprivation may become so acute that colonization is halted without an ED_{100} concentration of antimicrobial substances.

The speed of attaining an ED_{100} concentration of antimicrobial substances is controlled by several genetically-controlled characteristics of the host and pathogen. Sequentially in the host, the preinfectional concentrations, the concentrations formed quickly from cell injury (that is, formed from preformed substrates and enzymes), and concentrations formed by de novo synthesis after infection, may contribute to the total accumulation of antimicrobial substances. In some instances, for example, plants bearing alkaloids in root bark, the preinfectional concentrations are so high that induced synthesis of antimicrobial substances is not necessary for resistance. Plants that form wound toxins from preformed glucosides have not been shown to contain appreciable concentrations of preinfectional antimicrobial substances or phytoalexins; thus, wound toxins alone may contain many infections. In potatoes, legumes, and cotton, particularly in storage and endocarp tissues, the only effective means of accumulating antimicrobial substances appears to be from phytoalexin synthesis. Qualitative and quantitative variations of antimicrobial substances occur among varieties and determine the speed of reaching ED_{100} concentrations.

The pathogen also affects accumulation of antimicrobial substances in several ways. Various facultative fungal parasites that cause leaf spots and stem cankers can degrade or detoxify phytoalexins. These same fungi use phenolic metabolism and oxidation to form melanin pigments in hyphae and spores. Fungi that do not produce melanin or pigments have not been shown to degrade phytoalexins. These, particularly the obligate parasites, cause no appreciable injury to host tissues in early stages of infection, and phytoalexin synthesis is greatly delayed or does not occur. Whether the delay is due to "immunological tolerance" of the host to the pathogen, or to an active suppression of induced resistance mechanisms, is not known. Other pathogens secrete powerful enzymes or phytotoxins, which might prevent infection-induced synthesis of antimicrobial substances by total disruption of cell metabolism. The quantitative distribution of phytoalexins and therefore their efficiency, also varies somewhat according to the specific pathogen (Pierre, 1971), and might depend on redox changes created by the pathogen.

The speed of pathogen colonization also depends on several genetically-controlled characteristics of both the host and pathogen. The genetic potentials of the pathogen to utilize rapidly host components as nutrients, and to produce enzymes and toxins that cause cell leakage, contribute to the speed of colonization. The plant's ability to convert reserves, such as starch, lipids, and globulins, to callose, suberin, cutin, lignin, tannin, and ammonia probably is an important means of depriving the pathogen of nutrients, in addition to its value in creating mechanical barriers. The synthesis of phenolic compounds and their oxidation to quinones and tannins occur universally in plants and are probably most important for inactivation of hydrolytic enzymes and toxins. Recent reports indicate that proteinaceous, preinfectional inhibitors of enzymes occur in plants and may affect colonization. The susceptibility of the host to specific microbial toxins is extremely important and probably determines host resistance to some extent by affecting pathogen colonization. Plants also secrete lytic enzymes, which might affect rates of colonization (Abeles, et al., 1971).

Pathogenic organisms appear to have evolved in two directions to overcome disease resistance. Facultative parasites have developed simple nutritional requirements and potent enzymes and toxins that allow them to colonize plant tissues rapidly. Pathogens in this group usually induce the synthesis of wound toxins and phytoalexins with the same efficiency as nonpathogens or saprophytes. Pathogens frequently degrade the antimicrobial substances of their host and prevent their accumulation. In other instances, pathogens are less sensitive than nonpathogens or saprophytes to the antimicrobial compounds. Many facultative pathogens are contained in lesions and cankers, probably because their metabolites or massive injury finally induces high concentrations of antimicrobial substances in advance of pathogen development. Varietal resistance to facultative parasites usually depends on slightly faster rates or greater extent of accumulation of antimicrobial substances, occurrence of antimicrobial substances with greater activity, and more rapid or efficient inactivation of pathogenic enzymes and toxins.

Physiologically specialized pathogens (obligate parasites and facultative saprophytes) have developed increasingly efficient mechanisms for preventing the infection-induced resistance responses of plants. These organisms usually are much more sensitive to infection-induced antimicrobial compounds than are saprophytes or facultative parasites. As a consequence, whenever they do induce the normal resistance response, they are contained in a small volume of necrotic tissue (hypersensitive reaction). Thus, the major difference between resistant and susceptible plant varieties is the speed of the infection-induced biochemical responses. With some obligate pathogens, such as viruses causing systemic invasion infection-induced host responses appear to be absent during the infection. With facultative fungal saprophytes, infection-induced responses are delayed for only a few days after infection. When hypersensitive resistance occurs in response to specialized pathogens, the speed of infection-induced resistance approaches that shown to facultative parasites and saprophytes. Understanding specific resistance to physiologically specialized pathogens appears to depend on understanding the mechanism of infection-induced resistance.

Hadwiger and Schwochau (1969); Schwochau and Hadwiger (1970) have proposed "an induction hypothesis" to explain infection-induced resistance of plants. When genes for resistance in the host and genes for avirulence in the pathogen are dominant, the genes of the pathogen direct synthesis of fungal metabolites that activate (de-repress) the resistance genes in the host. These activated host genes alter cellular organization of the host, resulting in a hypersensitive response and destroying the symbiosis necessary for disease development. Hopper, et al. (1972) recently proposed a similar model to explain plant resistance to bacterial diseases.

Evidence that supports the induction hypothesis includes: a) synthesis of new proteins and enzymes (phenylaline ammonia lyase, glucose-6-phosphate dehydrogenase, and peroxidase) associated with induced disease resistance can be detected a few hours after infection (Hadwiger, 1968; Hadwiger, et al., 1970; Hess, et al., 1971; Stahmann, 1967; Uritani, 1963; von Broembsen, et al., 1972); b) radioactive precursors are readily incorporated into polyphenols and phytoalexins only after infection, indicating that these compounds are newly synthesized (Hadwiger, 1966; Hess, et al., 1971; Keen, et al., 1972); c) various antibiotics produced by saprophytic fungi affect genetic transcription and translation (Franklin and Snow, 1971), and these induce and inhibit phytoalexin synthesis in plants at low and high concentrations, respectively (Hadwiger, 1972; Schwochau and Hadwiger, 1970); d) various chemicals that intercalate with DNA (Franklin and Snow, 1971) induce phytoalexin synthesis (Hadwiger and Martin, 1971; Hadwiger and Schwochau, 1971; Hess and Hadwiger, 1971); e) various chemicals known to react with DNA or affect histones (repressors are thought to be composed of histones and RNA) also induce phytoalexin synthesis (Hadwiger and Schwochau, 1970); and, f) phytoalexin synthesis is activated by ultraviolet light, which is known to affect genetic transcription by causing thymine-thymine bonding in DNA (Hadwiger and Schwochau, 1971). Microbial metabolites exhibiting some specificity for induction of phytoalexin synthesis have been reported (Cruickshank and Perrin, 1968). However, no microbial metabolite showing the predicated specificities required by the induction theory has been found.

Frank and Paxton (1971) proposed a modified induction theory for soybean resistance to Phytophthora megaspermae var. sojae. They suggested that the host resistance gene induces the fungus to produce a compound, which in turn induces phytoalexin synthesis. The fungus does not produce this compound in susceptible varieties.

Kuc (1968) proposed that induced resistance responses are a part of a general response of cells subjected to stress by infection agents or wounding. Membrane damage may be particularly important in inciting such responses. Evidence that supports this concept includes: a) death of the initially penetrated cell precedes phytoalexin synthesis in adjoining cells by more than 5 hr in the hypersensitive response of potato (Tomiyama, 1963); b) extensive membrane injury, as shown by leakage of electrolytes and organic substances usually precedes phytoalexin synthesis in hypersensitive reactions (Goodman and Plurad, 1971; Muller, 1959); c) ultrastructure studies reveal extensive membrane injury associated with hypersensitive responses (Goodman and Plurad, 1971; Paulson and Webster, 1972); d) toxic chemicals that disrupt cell structure and metabolism also induce phytoalexin synthesis at low concentrations (Bell, 1967; Uehara, 1963; Uritani, et al., 1960); e) chilling and physical injury of cells induce resistance responses similar to those in diseased plants (Bell and Christiansen, 1968; Hirai, et al., 1968); f) sloughing-off of the primary root cortex and deposition of heartwood in healthy perennials involve disruption of cell structure and metabolism and initiate the same responses as infection (Bell, 1967; Stoessl, 1970); g) obligate parasites produce few (if any) injurious metabolites, such as phytotoxins (Wood, et al., 1972) and enzymes (Albersheim, et al., 1969; Bateman and Millar, 1966), and they induce phytoalexins slowly or not at all in susceptible hosts (Bailey and Ingham, 1971); and h) progressively more saprophytic fungi secrete progressively greater numbers of hydrolytic enzymes (Albersheim, et al., 1969; Bateman and Millar, 1966) and toxins (Herout, 1971; Wood, et al., 1972) and more readily induce phytoalexin synthesis (Cruickshank and Perrin, 1963;1971). Current evidence favors the concept of genetic activation of postinfectional resistance by inducers or metabolic disturbances in host tissues.

The selective delay or absence of infection-induced resistance in plants infected with virulent physiological races of pathogens presents an intriguing problem to plant pathologists. The occurrence of common antigens between such pathogens and their hosts (DeVay and Charudattan, 1972; DeVay, et al., 1967) suggests plants might have an "immunological tolerance" to these organisms. The glycopeptides and glycolipids found in membranes and cell walls of various pathogens could be important for such tolerance. These compounds might act as insulators between the pathogen wall and cytoplasmic membranes of the host and thereby prevent electrical or chemical disturbances of the membrane that result in infection-induced resistance.

The highly specialized pathogens also might suppress directly the metabolism of the host required for postinfectional resistance. The fungal product, cycloheximide (Pinkas and Novacky, 1971), which inhibits protein synthesis on 80S ribosomes, and cytokinins (Novacky, 1972), which are produced by microorganisms and plants, inhibited hypersensitivity of tobacco to bacteria. Susceptible reactions were mostly unchanged by either treatment. Bacterial populations in treated tissues were equal to or greater than those in untreated tissues. Pathogens might produce similar compounds to suppress infection-induced resistance. These compounds might include reactants for host inducers, substitute repressors, hormones to stimulate synthesis of the host repressors, inhibitors of translation, inhibitors of RNA or protein synthesis, or specific inhibitors of enzymes involved in wound toxin and phytoalexin synthesis. Understanding the nature of induced disease resistance and its prevention by specialized pathogens is of utmost importance in determining our future approaches to plant disease control.

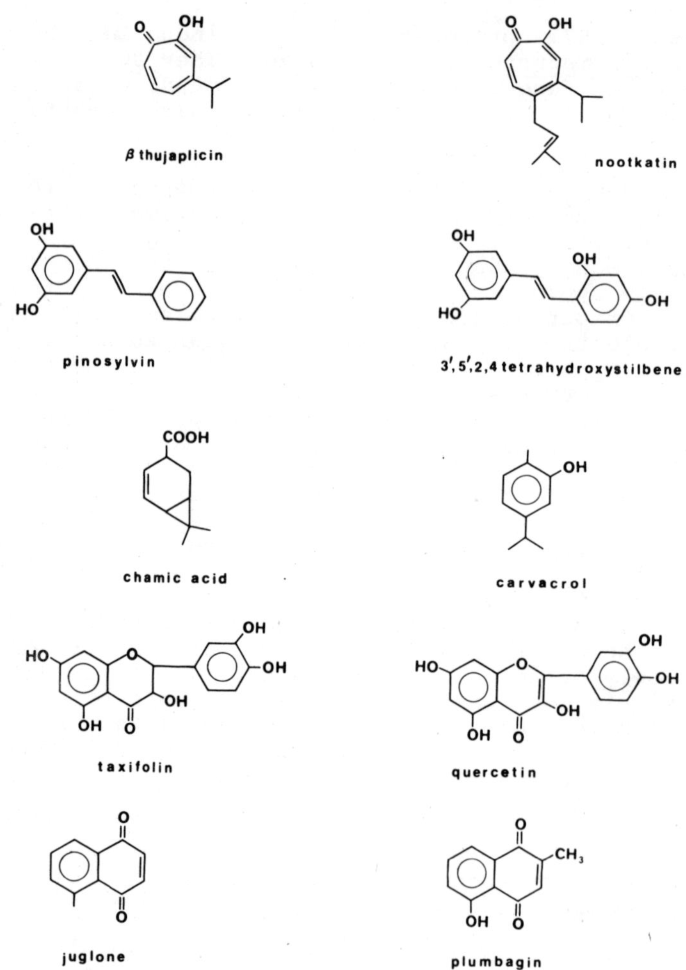

Figure 1.--Examples of fungitoxic compounds found in heartwood.

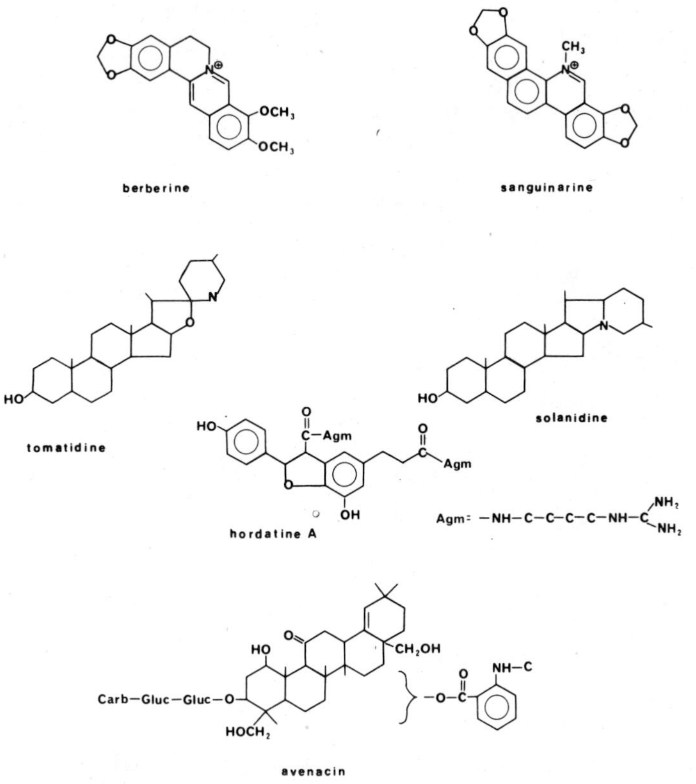
OCH3
OCH3
berberine
CH3
sanguinarine
HO
tomatidine
HO
solanidine
HO
C—Agm
C
Agm
OH
hordatine A
Agm= —NH—C—C—C—C—NH—C
NH2
NH2
HO
CH2OH
Carb—Gluc—Gluc—O
HOCH2
—O—C
NH—C
avenacin

Figure 2.--Examples of alkaloids, steroidal alkaloids and other nitrogenous compounds found in fungitoxic concentrations in healthy plants.

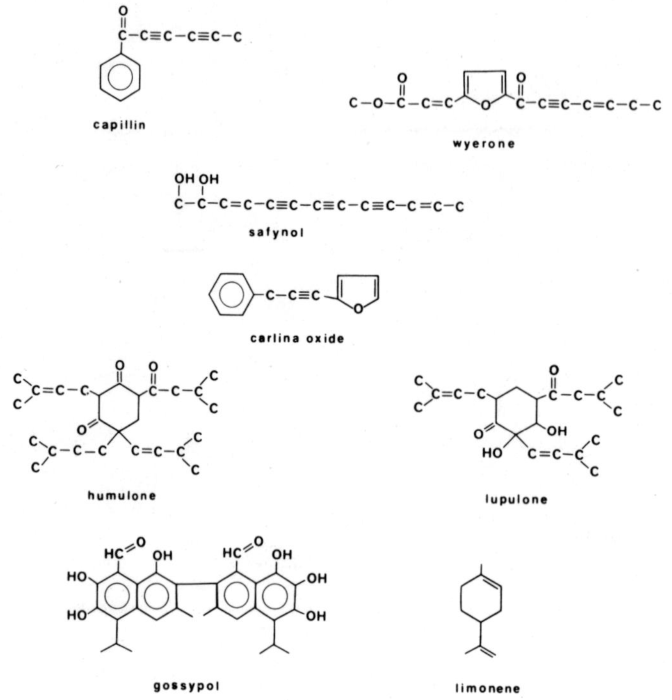

capillin

wyerone

safynol

carlina oxide

humulone

lupulone

gossypol

limonene

Figure 3.--Examples of fungitoxic acetylenes, terpenoids and related compounds found in endodermis, ducts and glands of healthy plants.

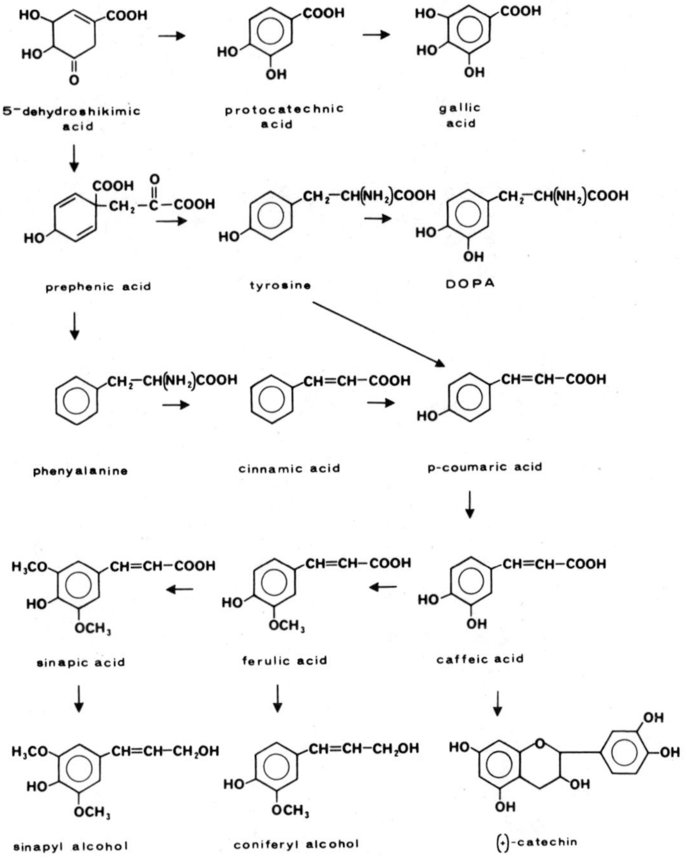

Figure 4.--Relationships and biosynthetic derivation of various polyphenols from the shikimic acid pathway.

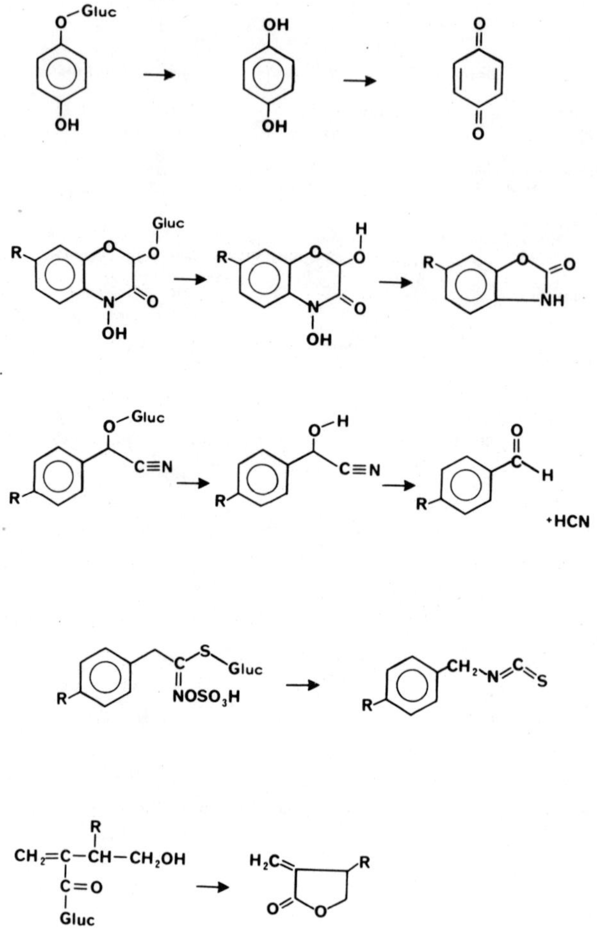

Figure 5.--Formation of wound toxins from glucosides or glucose esters. (1) hydroquinones and paraquinones, (2) benzoxazinones and benzoxazolinones, (3) cyanohydrins (or mandelonitriles), aldehydes and cyanide, (4) isothicyanates, and (5) unsaturated lactones.

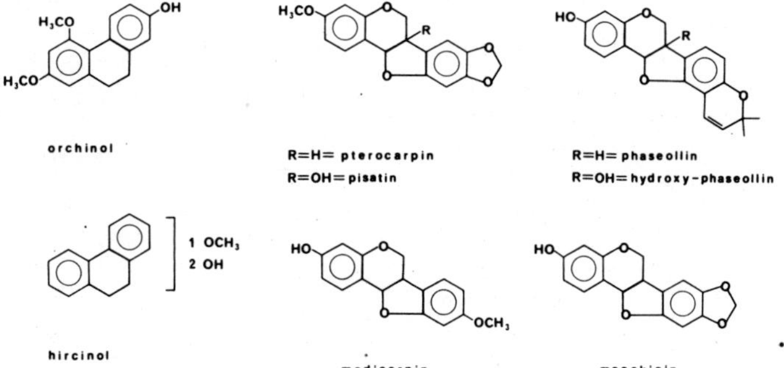

Figure 6.--Examples of phytoalexins derived from acetate (terpenoids) or hydroxycinnamates (phenanthrenes and pterocarpins).

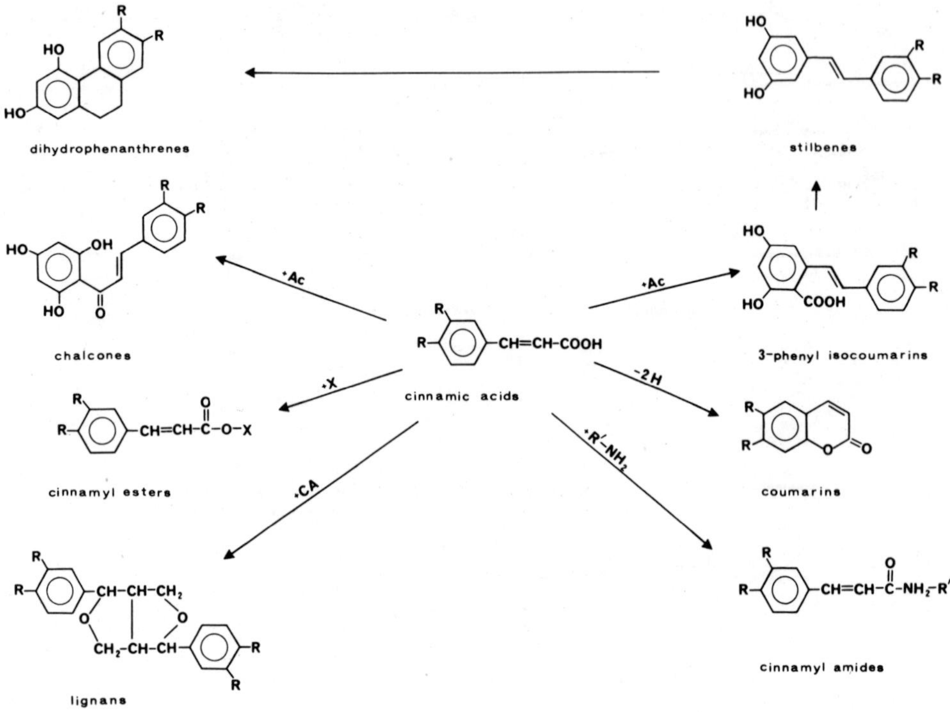

Figure 7.--Various types of antimicrobial compounds derived from hydroxycinnamic acids. R = H, OH or OCH_3 groups, R'-NH_2 = organic bases, AC = acetate, X = glucose, quinic acid or shikimic acid, CA = hydroxycinnamic acids.

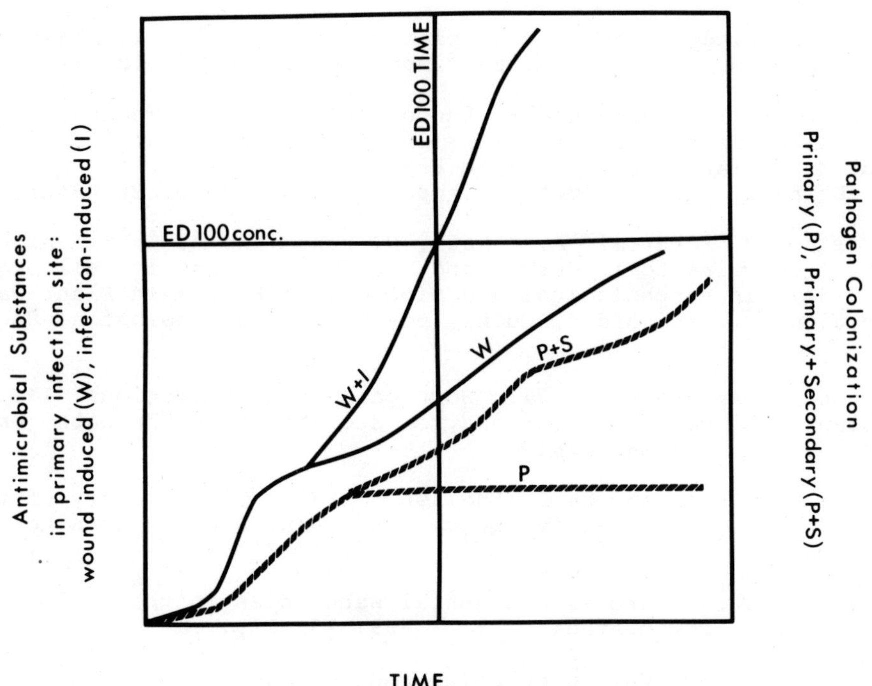

Figure 8.--A model illustrating the factors which contribute to the effectiveness of induced disease resistance.

Literature Cited

Abeles, F. B., R. P. Bosshart, L. E. Forrence, and W. H. Habig. 1971. Preparation and purification of glucanase and chitanase from bean leaves. Plant Physiol. 47: 129-134.

Akai, S. 1959. Histology of defense in plants, pp. 391-434. In Plant Pathology, Vol. 1 (J. G. Horsfall and A. E. Dimond, eds.). Academic Press, New York.

Akai, S., M. Fukutomi, N. Ishida and H. Kunoh. 1967. An anatomical approach to the mechanism of fungal infections in plants, pp. 1-20. In The Dynamic Role of Molecular Constituents in Plant-Parasite Interaction (C. J. Mirocha and I. Uritani, eds.). Amer. Phytopathological Soc., St. Paul, Minnesota.

Akai, S., O. Horino, M. Fukutomi, A. Nakata, H. Kunoh and M. Shiraishi. 1971. Cell wall reaction to infection and resulting change in cell organelles, pp. 329-347. In Morphological and Biochemical Events in Plant-Parasite Interaction (S. Akai and S. Ouchi, eds.). Phytopathological Soc. of Japan, Tokyo.

Albersheim, P. and A. Anderson. 1971. Host pathogen interactions. III. Proteins from plant cell walls inhibit enzymes secreted by plant pathogens. Proc. Nat. Acad. Sci. 68: 1815-1819.

Albersheim, P., T. M. Jones and P. D. English. 1969. Biochemistry of the cell wall in relation to infective processes. Ann. Rev. Phytopathology 7: 171-194.

Allen, E. H. 1970. The nature of antigunfal substances in the peel of Irish potato tubers. Phytopathol. Z. 69: 151-159.

Allen, E. H. and J. Kuc. 1968. α-Solanine and α-Chaconine as fungitoxic compounds in extracts of Irish potato tubers. Phytopathology 58: 776-781.

Allen, E. H. and A. Thomas. 1971. Time course of safynol accumulation in resistant and susceptible safflower infected with *Phytophthora drechsleri*. Physiol. Plant Pathol. 1: 235-240.

Allen, E. H. and C. A. Thomas. 1971. *Trans-trans*-3,11-tridecadiene-5,7,9-triyne-1,2-diol, an antifungal polyacetylene from diseased safflower (*Carthamus tinctorius*). Phytochem. 10: 1579-1582.

Allen, E. H. and C. A. Thomas. 1971. A second antifungal polyacetylene compound from *Phytophthora*-infected safflower. Phytopathology 61: 1107-1109.

Allen, P. J. 1959. Physiology and biochemistry of defense, pp. 435-467. In Plant Pathology, Vol. 1 (J. G. Horsfall and A. E. Dimond, eds.). Academic Press, New York.

Anderson, A. and P. Albersheim. 1972. Aspects of the specificity of the plant proteins which inhibit pathogen secreted endopolygalacturonases. Phytopathology 62: 744 (Abstr.).

Appel, O. 1915. Disease resistance in plants. Science 41: 773-782.

Arneson, P. A. and R. D. Durbin. 1967. Hydrolysis of tomatine by *Septoria lycopersici*: a detoxification mechanism. Phytopathology 57: 1358-1360.

Arneson, P. A. and R. D. Durbin. 1968. The sensitivity of fungi to α-tomatine. Phytopathology 58: 536-537.

Arneson, P. A. and R. D. Durbin. 1968. Studies on the mode of action of tomatine as a fungitoxic agent. Plant Physiol. 43: 683-686.

Asahi, T., Y. Honda and I. Uritani. 1966. Increase of mitochondrial fraction in sweet potato root tissue after wounding or infection with Ceratocystis fimbriata. Plant Physiol. 41: 1179-1184.

Bailey, J. A. and B. J. Deverall. 1971. Formation and activity of phaseollin in the interaction between bean hypocotyls (Phaseolus vulgaris) and physiological races of Colletotrichum lindemuthianum. Physiol. Plant Pathol. 1: 435-449.

Bailey, J. A. and J. L. Ingham. 1971. Phaseollin accumulation in bean (Phaseolus vulgaris) in response to infection by tobacco necrosis virus and rust Uromyces appendiculatus. Physiol. Plant Pathol. 1: 451-456.

Bateman, D. F. 1964. An induced mechanism of tissue resistance to polygalacturonase in Rhizoctonia-infected hypocotyls of bean. Phytopathology 54: 438-445.

Bateman, D. F. and R. D. Lumsden. 1965. Relation of calcium content and nature of the pectic substances in bean hypocotyls of different ages to susceptibility to an isolate of Rhizoctonia solani. Phytopathology 55: 734-738.

Bateman, D. F. and R. L. Millar. 1966. Pectic enzymes in tissue degradation. Ann. Rev. Phytopathology 4: 119-146.

Bawden, F. C. 1954. Inhibitors and plant viruses. Adv. Virus Res. 2: 31-57.

Beckman, C. H. 1964. Host responses to vascular infection. Ann. Rev. Phytopathology 2: 231-252.

Beckman, C. H. 1966. Cell irritability and localization of vascular infections in plants. Phytopathology 56: 821-824.

Beckman, C. H. 1971. The pasticizing of plant cell walls and tylose formation--a model. Physiol. Plant Pathol. 1: 1-10.

Beckman, C. H., W. C. Mueller and W. E. McHardy. 1972. The localization of stored phenols in plant hairs. Physiol. Plant Pathol. 2: 69-74.

Beijersbergen, J. C. M. and C. B. G. Lemmers. 1972. Enzymic and non-enzymic liberation of tulipalin A. (α-methylene butyro-lactone) in extracts of tulip. Physiol. Plant Pathol. 2: 265-270.

Bell, A. A. 1964. Respiratory metabolism of Phaseolus vulgaris infected with alfalfa mosaic and southern bean mosaic virus. Phytopathology 54: 914-922.

Bell, A. A. 1967. Formation of gossypol in infected or chemically irritated tissues of Gossypium (cotton) species. Phytopathology 57: 759-764.

Bell, A. A. 1967. Formation of fungitoxins in wound sap of Phaseolus lunatus. Phytopathology 57: 1111-1115.

Bell, A. A. 1969. Possible relationship between activation of phytoalexin synthesis and peroxidase enzymes in cotton. Phytopathology 59: 1018 (Abstr.)

Bell, A. A. 1969. Phytoalexin production and Verticillium wilt resistance in cotton. Phytopathology 59: 1119-1127.

Bell, A. A. 1970. 4-Hydroxybenzaldehyde and vanillin as toxins formed in leaf wound sap of Phaseolus lunatus. Phytopathology 60: 161-165.

Bell, A. A. 1972. Nature of disease resistance. In Verticillium wilt of Cotton. USDA ARS 34-137 (in press).

Bell, A. A. and M. N. Christiansen. 1968. Gossypol synthesis in chilled cotton tissues. Phytopathology 58: 883 (Abstr.)

Bell, A. A. and C. R. Howell. 1971. Comparative physiology of Verticillium albo-atrum isolates. II. Induction of phytoalexins in cotton, pp. 84-85. Proc. Beltwide Cotton Prod. Res. Conf., Atlanta, Georgia.

Bell, A. A. and J. T. Presley. 1969. Temperature effects upon resistance and phytoalexin synthesis in cotton inoculated with Verticillium albo-atrum Phytopathology 59: 1141-1146.

Bell, A. A. and J. T. Presley. 1969. Heat-inhibited or heat-killed conidia of Verticillium albo-atrum induce disease resistance and phytoalexin synthesis in cotton. Phytopathology 59: 1147-1151.

Bell, A. A. and R. D. Stipanovic. 1972. Chemistry and nature of fungitoxic compounds in diseased cotton. Proc. Beltwide Cotton Prod. Res. Conf., Memphis, Tenn. pp. 87-88.

Bell, T. A., J. L. Etchells, C. F. Williams and W. L. Porter. 1962. Inhibition of pectinase and cellulase by certain plants. Botan. Gaz. 123: 220-223.

BeMiller, J. N. and A. J. Pappelis. 1965. 2,4-dihydroxy-7-methoxy-1,4-benzoxazin-3-one glucoside in corn. Relation of water-soluble, 1-butanol-soluble glucoside fraction content of pith cores and stalk rot resistance. Phytopathology 55: 1237-1240.

Bergman, B. H. H. 1966. Presence of substance in the white skin of young tulip bulbs which inhibits the growth of Fusarium oxysporum. Netherlands J. Plant Pathol. 72: 222-230.

Bergman, B. H. H.& J. C. M. Biejersbergen. 1968. A fungitoxic substance extracted from tulips and its possible role as a protectant against disease. Netherlands J. Plant Pathol. 74 (Suppl. 1): 157-162.

Bergman, B. H. H., J. C. M. Biejersbergen, J. C. Overeem and A. Kaars-Sijpestein. 1967. Isolation and identification of γ-methylene-butyrolactone, a fungitoxic substance from tulips. Rec. Trav. Chim. 86: 709-714.

Bingefors, S. 1971. Resistance to nematodes and the possible value of induced mutations, pp. 209-235. In Mutation Breeding for Disease Resistance. International Atomic Energy Agency, Vienna, Austria.

Birk, Y. 1969. Saponins, pp. 169-210. In Toxic Constituents of Plant Foodstuffs (I. Liener, ed.). Academic Press, New York.

Bohlmann, F. and K. M. Kleine. 1961. Polyacetylene compounds. XXXV. The polyynes from Chrysanthemum frutescens and Artemisia dracunculus. Chem. Ber. 95: 39-46.

Boone, D. M., D. M. Kline and G. W. Keitt. 1958. Venturia inaequalis (Cke.) Wint. XIII. Pathogenicity of induced biochemical mutants. Amer. J. Bot. 44: 791-796.

Braun, A. C. and R. B. Pringle. 1958. Pathogen factors in the physiology of disease--toxins and other metabolites, pp. 88-99. In Plant Pathology Problems and Progress 1908-1958 (C. S. Holton, G. W. Fischer, R. W. Fulton, H. Hart and S. E. A. McCallan, eds.) The Univ. Wisconsin Press, Madison.

Brinkerhoff, L. A. 1970. Variation in Xanthomonas malvacearum and its relation to control. Ann. Rev. Phytopathology 8: 85-110.

Brown, W. 1922. Studies in the physiology of parasitism. VIII. On the exosmosis of nutrient substances from the host tissue into the infection drop. Ann. Bot. 36: 101-119.

Brown, W. 1934. Mechanisms of disease resistance in plants. Trans. British Mycol. Soc. 19: 11-33.

Brown, W. 1965. Toxins and cell-wall dissolving enzymes in relation to plant disease. Ann. Rev. Phytopathology 3: 1-18.

Buddenhagen, I. and A. Kelman. 1964. Biological and physiological aspects of bacterial wilt caused by Pseudomonas solanacearum. Ann. Rev. Phytopathology 2: 203-230.

Burkhardt, H. J., J. V. Maizel and H. K. Mitchell. 1964. Avenacin, an antimicrobial substance isolated from Avena sativa. Structure. Biochem. 3: 426-431.

Butler, E. J. 1918. Immunity and disease in plants. Agr. J. India 1918: 10-28.

Buxton, E. W. 1957. Some effects of pea root exudates on physiologic races of Fusarium oxysporum Fr. f. pisi (Linf.) Snyder & Hansen. Trans. British Mycol. Soc. 40: 145-154.

Byrde, R. J. W. 1963. Natural inhibitors of fungal enzymes and toxins in disease resistance, pp. 31-47. In Perspectives of Biochemical Plant Pathology (S. Rich, ed.). Connecticut Agr. Exp. Sta. Bul. 663.

Byrde, R. J. W., A. H. Fielding and A. H. Williams. 1959. The role of oxidized phenols in the varietal resistance of apples to brown rot, pp. 95-99. In Phenolics in Plants in Health and Disease (J. B. Pridham, ed.). Pergamon Press, New York.

Cadman, C. H. 1960. Inhibition of plant virus infection by tannins, pp. 101-105. In Phenolics in Plants in Health and Disease (J. B. Pridham, ed.). Pergamon Press, New York.

Carr, A. J. H. 1971. The role of wilt toxins produced by the lucerne strain of Verticillium albo-atrum, p. 10. Proc. Intern. Verticillium Symp., Wye College, near Ashford, Kent, England.

Chamberlain, D. W. 1970. Temperature ranges inducing susceptibility to Phytophthora megasperma var. sojae in resistance soybeans. Phytopathology 60: 293-294.

Chester, K. S. 1933. The problem of acquired physiological immunity in plants. Quart. Rev. Biol. 8: 129-154, 275-324.

Chisholm, M. D. and L. R. Wetter. 1967. The biosynthesis of some isothiocyanates and oxazolidinethiones in rape (Brassica campestris L.). Plant Physiol. 42: 1726-1730.

Chkanikov, D. I., G. A. Tarabrin, A. M. Shabanova and P. F. Konstantinov. 1969. Localization of β-glucosidase in the cells of higher plants. Soviet Plant Physiol. 16: 261-264.

Cook, M. T., H. F. Bassett, F. Thompson and J. J. Taubenhaus. 1911. Protective enzymes. Science 33: 624-629.

Cook, M. T. and J. J. Taubenhaus. 1911. The relation of parasitic fungi to the contents of the cells of the host plants. I. The toxicity of tannin. Delaware Col. Agr. Exp. Sta. Bul. No. 91. 77 p.

Coons, G. H. 1937. Progress in plant pathology: Control of disease by resistant varieties. Phytopathology 27: 622-632.

Couture, R. M., D. G. Routley and G. M. Dunn. 1971. Role of cyclic hydroxamic acids in monogenic resistance of maize to Helminthosporium turcicum. Physiol. Plant Pathol. 1: 515-521.

Cruickshank, I. A. M. 1962. Studies on phytoalexins. IV. The antimicrobial spectrum of pisatin. Australian J. Biol. Sci. 15: 147-159.

Cruickshank, I. A. M. 1963. Phytoalexins. Ann. Rev. Phytopathology 1: 351-374.

Cruickshank, I. A. M. 1965. Phytoalexins in the luguminosae with special reference to their selective toxicity. Deut. Akad. Landwirtschaftswiss. Tangunsber. 74: 313-332.

Cruickshank, I. A. M. 1966. Defense mechanisms in plants. World Rev. Pest Control 5: 161-175.

Cruickshank, I. A. M. and D. R. Perrin. 1963. Studies on phytoalexins. VI. Pisatin. The effect of some factors on its formation in Pisum sativum L. and the significance of pisatin in disease resistance. Australian J. Biol. Sci. 16: 111-128.

Cruickshank, I. A. M. and D. R. Perrin. 1964. Pathological function and phenolic compounds in plants, pp. 511-544. In Biochemistry of Phenolic Compounds (J. B. Harborne, ed.). Academic Press, New York.

Cruickshank, I. A. M. and D. R. Perrin. 1968. The isolation and partial characterization of monicolin A, a polypeptide with phaseollin-inducing activity from Monilinia fructicola. Life Sci. 7: 449-458.

Cruickshank, I. A. M. and D. R. Perrin. 1971. Studies on phytoalexins. XI. The induction, antimicrobial spectrum and chemical assay of phaseollin. Phytopathol. Z. 70: 209-229.

Cunningham, H. S. 1928. A study of the histologic changes induced in leaves by certain leaf spotting fungi. Phytopathology 18: 717-751.

Curtis, R. F. 1968. 6-Methoxymellein as a phytoalexin. Experientia 24: 1187-1188.

Daly, J. M. 1972. The use of near-isogenic lines in biochemical studies of the resistance of wheat to stem rust. Phytopathology 62: 392-400.

Davies, R. R. 1961. Wettability and the capture, carriage and deposition of particles by raindrops. Nature 191: 616-617.

Davis, D. 1964. Host fungitoxicants in selective pathogenicity of *Fusarium oxysporum*. Phytopathology 54: 290-293.

DeBaun, R. M. and F. F. Nord. 1951. The resistance of cork to decay by wood-destroying molds. Arch. Biochem. Biophys. 33: 314-319.

Deese, D. C. and M. A. Stahmann. 1962. Formation of pectic enzymes by *Verticillium albo-atrum* on susceptible and resistant tomato stem tissues and on wheat bran. Phytopathol. Z. 46: 53-70.

DeVay, J. E. and R. Charudattan. 1972. Common antigens between host and pathogen in relation to disease susceptibility, pp. 82-83. Proc. Beltwide Cotton Prod. Res. Conf., Atlanta, Ga.

DeVay, J. E., W. C. Schnathorst and M. S. Foda. 1967. Common antigens and host-parasite interactions, pp. 313-328. *In* The Dynamic Role of Molecular Constituents in Plant Parasite Interaction. Amer. Phytopathological Soc. St. Paul, Minn.

Deverall, B. J. 1964. Studies on the physiology of resistance to the halo-blight disease of French beans, pp. 127-130. *In* Host Parasite Relations in Plant Pathology (Z. Kiraly and G. Urizsy, eds.). Publ. Res. Inst. for Plant Prot. Budapest.

Deverall, B. J. and J. C. Vessey. 1969. Role of a phytoalexin in controlling lesion development in leaves of *Vicia faba* after infection by *Botrytis* spp. Ann. Appl. Biol. 63: 449-458.

Deverall, B. J. and R. K. S. Wood. 1961. Chocolate spot of beans (*Vicia faba* L.)-interactions between phenolase of host and pectic enzymes of the pathogen. Ann. Appl. Biol. 49: 473-487.

Dickinson, S. 1960. The mechanical ability to breach the host barriers, pp. 203-232. *In* Plant Pathology, Vol. 2 (J. G. Horsfall and A. E. Dimond, eds.). Academic Press, New York.

Diener, T. O. 1963. Physiology of virus-infected plants. Ann. Rev. Phytopathology 1: 197-218.

Dimond, A. E. 1967. Physiology of wilt disease, pp. 100-120. *In* The Dynamic Role of Molecular Constituents in Plant Parasite Interaction. (C. J. Mirocha and I. Uritani, eds.). Amer. Phytopathological Soc., St. Paul, Minn.

Dimond, A. E. 1970. Biophysics and biochemistry of the vascular wilt syndrome. Ann. Rev. Phytopathology 8: 301-322.

Doubly, J. A., H. H. Flor and C. O. Clagett. 1960. Relation of antigens of *Melamposora lini* and *Linum usitatissimum* to resistance and susceptibility. Science 131: 229.

Drobnica, L., M. Zemanova, P. Nemec, K. Antos, P. Kirstian, A. Stullerova, V. Knoppova and P. Nemec, Jr. 1967. Antifungal activity of isothiocyanates and related compounds. Appl. Microbiol. 15: 701-709.

Dropkin, V. H. 1969. Cellular responses of plants to nematode infections. Ann. Rev. Phytopathology 7: 101-122.

Echandi, E. and C. E. Fernandez. 1962. Relation between chlorogenic acid content and resistance to coffee canker incited by *Ceratocystis fimbriata*. Phytopathology 52: 544-546.

Ellingboe, A. H. 1968. Inoculum production and infection by foliage pathogens. Ann. Rev. Phytopathology 6: 317-330.

ElNaghy, M. A. and P. Linko. 1962. The role of 4-O-glucosyl-2,4-dihydroxy-7-methoxy-1,4-benzoxazin-3-one in resistance of wheat to stem rust. Physiologia Pl. 15: 764-771.

ElNaghy, M. A. and M. Shaw. 1966. Correlation between resistance to stem rust and the concentration of a glucoside in wheat. Nature 210: 417-418.

Endo, B. Y. 1971. Nematode-induced syncytia (giant cells). Host-Parasite relationships of Heteroderidae, pp. 91-118. In Plant Parasitic Nematodes, Vol. 2 (B. M. Zuckerman, W. F. Mai, and R. A. Rohde, eds.). Academic Press, New York.

English, P. D. and P. Albersheim. 1969. Host pathogen interactions. I. A correlation between α-galactosidase production and virulence. Plant Physiol. 44: 217-224.

Erdtman, H. 1952. Chemistry of some heartwood constituents of conifers and their physiological and taxonomic significance. Prog. in Organic Chemistry, London: 22-63.

Esau, K. 1966. Anatomy of plant virus infections. Ann. Rev. Phytopathology 5: 45-76.

Farkas, G. L. and Z. Kiraly. 1962. Role of phenolic compounds in the physiology of plant diseases and disease resistance. Phytopathol. Z. 44: 106-150.

Farkas, G. L., Z. Kiraly and F. Solymosy. 1960. Role of oxidative metabolism in the localization of plant viruses. Virology 12: 408-421.

Farkas, G. L., F. Solymosy and L. Lovrekovich. 1965. The role of altered enzyme levels in the regulation of metabolic pattern in diseased plant tissues. Tagungsberichte Nr. 74: 71-81.

Farkas, G. L. and M. A. Stahmann. 1966. On the nature of changes in peroxidase isoenzymes in bean leaves infected by southern bean mosaic virus. Phytopathology 56: 669-677.

Fawcett, C. H., R. D. Firn and D. M. Spencer. 1971. Wyerone increase in leaves of broad bean (Vicia faba L.) after infection by Botrytis fabae. Physiol. Plant Pathol. 1: 163-166.

Fawcett, C. H. and D. M. Spencer. 1969. Natural antifungal compounds, pp. 637-669. In Fungicides, Vol. 2 (D. C. Torgenson, ed.) Academic Press, New York.

Fawcett, C. H. and D. M. Spencer. 1970. Plant chemotherapy with natural products. Ann. Rev. Phytopathology 8: 403-418.

Fehrmann, H. and A. E. Dimond. 1967. Peroxidase activity and Phytophthora resistance in different organs of the potato plant. Phytopathology 57: 69-72.

Flentje. N. T. 1958. The physiology of penetration and infection, pp. 76-87. In Plant Pathology Problems and Progress 1908-1958 (C. S. Holton, G. W. Fischer, R. W. Fulton, H. Hart, and S. E. A. McCallan, eds.). The Univ. Wisconsin Press, Madison.

Flor, H. H. 1971. Current status of the gene for gene concept. Ann. Rev. Phytopathology 9: 275-296.

Frank, J. A. and J. D. Paxton. 1971. An inducer of soybean phytoalexin and its role in the resistance of soybeans to Phytophthora rot. Phytopathology 61: 954-958.

Franklin, T. J. and G. A. Snow. 1971. Biochemistry of Antimicrobial Action. Academic Press, New York. 163 pp.

✓ Freeman, E. M. 1911. Resistance and immunity in plant diseases. Phytopathology 1: 109-115.

Fuchs, W. H. 1971. Physiological and biochemical aspects of resistance to disease, pp. 5-16. In Mutation Breeding for Disease Resistance. International Atomic Energy Agency, Vienna, Austria.

Gallun, R. L. and L. P. Reitz. 1971. Wheat cultivars resistant to races of Hessian fly. USDA ARS Prod. Res. Rept. No. 134.

Garber, E. D. 1956. A nutrition-inhibition hypothesis of pathogenicity. Amer. Naturalist 90: 183-194.

Gaumann, E., R. Braun and G. Bazzigher. 1950. Uber induzierte Abwehrreaktionen bei Orchideen. Phytopathol. Z. 17: 36-62.

Geissman, T. A., ed. 1962. The Chemistry of Flavonoid Compounds. The Macmillan Co., New York. 666 pp.

Goodman, R. N., Z. Kiraly and M. Zaitlin. 1967. The Biochemistry and Physiology of Infectious Plant Disease. Van Nostrand-Reinhold, Princeton, New Jersey. 354 pp.

Goodman, R. N. and S. B. Plurad. 1971. Ultrastructural changes in tobacco undergoing the hypersensitive reaction caused by plant pathogenic bacteria. Physiol. Plant Pathology 1: 11-15.

Gray, G., W. L. Klarman and M. Bridge. 1967. Relative quantities of antifungal metabolites produced in resistant and susceptible soybean plants inoculated with Phytophthora megasperma var. sojae and closely related nonpathogenic fungi. Canadian J. Bot. 46: 285-288.

Greathouse, G. A. 1939. Alkaloids from Sanguinaria canadensis and their influence on growth of Phymatotrichum omnivorum. Plant Physiol. 14: 377-380.

Greathouse, G. A. and N. E. Rigler. 1940. The chemistry of resistance of plants to Phymatotrichum root rot. V. Influence of alkaloids on growth of fungi. Phytopathology 30: 475-485.

Greathouse, G. A. and G. M. Watkins. 1938. Berberine as a factor in the resistance of Mahonia trifoliata and M. swaseyi to Phymatotrichum root rot. Amer. J. Bot. 25: 743-748.

Griffiths, D. A. 1971. The development of lignitubers in roots after infection by Verticillium dahliae Kleb. Canadian J. Microbiol. 17: 441-444.

Hadwiger, L. A. 1966. The biosynthesis of pisatin. Phytochem. 5: 523-525.

Hadwiger, L. A. 1968. Changes in plant metabolism associated with phytoalexin production. Netherlands J. Plant Pathol. 74 (Suppl. 1) 163-169.

Hadwiger, L. A. 1972. Increased levels of pisatin and phenylalanine ammonia lyase activity in Pisum sativum treated with antihistaminic, antiviral, antimalarial, tranquilizing, or other drugs. Biochem. Res. Comm. 46: 71-79.

Hadwiger, L. A., S. L. Hess and S. von Broembsen. 1970. Stimulation of phenylalanine ammonia-lyase activity and phytoalexin production. Phytopathology 60: 332-336.

Hadwiger, L. A. and A. R. Martin. 1971. Induced formation of phenylalanine ammonia lyase and pisatin by chlorpromazine and other phenothiazine derivatives. Biochem. Pharmacol. 20: 3255-3261.

Hadwiger, L. A. and M. E. Schwochau. 1969. Host resistance--an induction hypothesis. Phytopathology 59: 223-227.

Hadwiger, L. A. and M. E. Schwochau. 1970. Induction of phenylaline ammonia lyase and pisatin in pea pods by polylysine, spermidine or histone fractions. Biochem. Biophys. Res. Comm. 38: 683-691.

Hadwiger, L. A. and M. E. Schwochau. 1971. Ultraviolet light-induced formation of pisatin and phenylalanine ammonia lyase. Plant Physiol. 27: 588-590.

Hadwiger, L. A. and M. E. Schwochau. 1971. Specificity of deoxyribonucleic acid intercalating compounds in the control of phenylalanine ammonia lyase and pisatin levels. Plant Physiol. 47: 346-351.

Hampton, R. E. 1962. Changes in phenolic compounds in carrot root tissue infected with Thielaviopsis basicola. Phytopathology 52: 413-415.

Hanson, K. R., M. Zucker and E. Sondheimer. 1967. The regulation of phenolic biosynthesis and the metabolic roles of phenolic compounds in plants, pp. 68-93. In Phenolic Compounds and Metabolic Regulation (B. J. Finkle and V. C. Runeckles, eds.). Appleton-Century-Crofts, New York.

Harborne, J. B., ed. 1964. Biochemistry of Phenolic Compounds. Academic Press, New York. 618 pp.

Harkin, J. M. 1967. Lignin--a natural polymeric product of phenol oxidation, pp. 243-321. In Oxidative Coupling of Phenols (A. R. Battersby and A. I. Taylor, eds.). Dekker, New York.

Harrison, M., M. G. Payne and J. O. Gaskill. 1961. Some chemical aspects of resistance to Cercospora leaf spot in sugar beets. J. Amer. Soc. Sugar Beet Tech. 10: 457-468.

Hart, H. 1929. Relation of stomatal behavior to stem-rust resistance in wheat. J. Agr. Res. 39: 929-948.

Hart, H. 1931. Morphologic and physiologic studies on stem rust resistance in cereals. USDA & Minnesota Agr. Exp. Sta. Tech. Bul. 266. 76 pp.

Hastie, A. C. 1970. The genetics of asexual phytopathogenic fungi with special reference to Verticillium, pp. 55-62. In Root Diseases and Soil-Borne Pathogens (T. A. Toussoun, R. V. Bega and P. E. Nelson, eds.). University of California Press, Berkeley.

Heath, M. C. and I. B. Heath. 1971. Ultrastructure of an immune and a susceptible reaction of cowpea leaves to rust infection. Physiol. Plant Pathol. 1: 277-287.

Heath, M. C. and R. K. S. Wood. 1971. Role of inhibitors of fungal growth in the limitation of leaf spots caused by Ascochyta pisi and Mycosphaerella pinodes. Ann. Bot. 35: 475-491.

Herout, V. 1971. Biochemistry of sesquiterpenoids, pp. 53-94. In Aspects of terpenoid chemistry and biochemistry (T. W. Goodwin, ed.). Academic Press, New York.

Hess, S. L. and L. A. Hadwiger. 1971. The induction of phenylalanine ammonia lyase and phaseollin by 9-aminoacridine and other deoxyribonucleic acid intercalating compounds. Plant Physiol. 48: 197-202.

Hess, S. L., L. A. Hadwiger and M. E. Schwochau. 1971. Studies on biosynthesis of phaseollin in excised pods of Phaseolus vulgaris. Phytopathology 61: 79-82.

Higgins, V. J. 1972. Role of the phytoalexin medicarpin in three leaf spot diseases of alfalfa. Physiol. Plant Pathol. 2: 289-300.

Higgins, V. J. and R. L. Millar. 1968. Phytoalexin production by alfalfa in response to infection by Colletotrichum phomoides, Helminthosporium turcicum, Stemphylium loti, and S. botryosum. Phytopathology 58: 1377-1383.

Higgins, V. J. and R. L. Millar. 1969. Comparative abilities of Stemphylium botryosum and Helminthosporium turcicum to degrade a phytoalexin from alfalfa. Phytopathology 59: 1493-1499.

Higgins, V. J. and R. L. Millar. 1969. Degradation of alfalfa phytoalexin by Stemphylium botryosum. Phytopathology 59: 1500-1506.

Higgins, V. J. and R. L. Millar. 1970. Degradation of alfalfa phytoalexin by Stemphylium loti and Colletotrichum phomoides. Phytopathology 60: 269-271.

Higgins, V. J. and D. G. Smith. 1972. Separation and identification of two pterocarpanoid phytoalexins produced by red clover leaves. Phytopathology 62: 235-238.

Hildebrand, D. C. and M. N. Schroth. 1964. Arbutin-hydroquinone complex in pear as a factor in fire blight development. Phytopathology 54: 640-645.

Hillis, W. E. and T. Inoue. 1968. The polyphenols formed in Pinus radiata after Sirex attack. Phytochem. 7: 13-22.

Hirai, T., Z. Hidaka and I. Uritani, eds. 1968. Biochemical regulations in diseased plants and injury. Nat. Inst. Agr. Sci., Tokyo. 351 pp.

Holden, M., B. C. Seegal and H. Baer. 1947. The range of antibiotic activity of protanemonin. Proc. Soc. Exptl. Biol. Med. 66: 54-60.

Holliday, R. 1961. The genetics of Ustilago maydis. Genet. Res. Camb. 2: 204-230.

Hopper, D. G., K. R. Gholson and L. A. Brinkerhoff. 1972. A biochemical lock-and-key model for bacterial plant diseases. Proc. Amer. Chem. Soc. (in press).

Hulme, A. C. and K. L. Edney. 1960. Phenolic substances in the peel of Cox's orange pippin apples with reference to infection by G. perennans, pp. 87-94. In Phenolics in Plants in Health and Disease (J. B. Pridham, ed.). Pergamon Press, New York.

Imai, K. 1956. Studies on the essential oil of Artemisia capillaris Thunb. III. Antifungal activity of the essential oil. Structure of antifungal principle, capillin, J. Pharm. Soc. Japan 76: 405-408.

Imai, K., N. Ikeda, K. Tanaka and S. Sugaivara. 1956. Studies on the essential oil of Artemisia capillaris. Thunb. III. Antifungal activity of the essential oil. 2. Isolation of the antifungal principle. J. Pharm. Soc. Japan 76: 400-404.

Imaseki, H., T. Asahi and I. Uritani. 1968. Investigations on the possible inducers of metabolic changes in injured plant tissues. pp. 189-201. In Biochemical Regulation in Diseased Plants or Injury (T. Hirai, ed.). Kyoritsu Printing Co., Ltd., Tokyo.

Ishiyawa, N., K. Tomiyama, N, Katsui, A. Murai and T. Masamune. 1969. Biological activities of rishitin, an antifungal compound isolated from diseased potato tubers, and its derivatives. Plant Cell Physiol. 10: 183-192.

Johnson, G. and L. A. Schaal. 1952. Relation of chlorogenic acid to scab resistance in potatoes. Science 115: 627-629.

Johnson, G. and L. A. Schaal. 1957. Chlorogenic acid and other orthodihydricphenols in scab-resistant Russet Burbank and scab-susceptible Triumph potato tubers of different maturities. Phytopathology 47: 253-255.

Jones, A. P. 1931. The histogeny of potato scab. Ann. Appl. Biol. 18: 313-333.

Kaplan, J. G. and W. Tacreiter. 1966. The β-glucosidase of the yeast cell surface. J. Gen. Physiol. 50: 9-23.

Kedar, N. 1959. The peroxidase test as a tool in the selection of varieties resistant to late blight. Amer. Potato J. 36: 315-324.

Keen, N. T. 1971. Hydroxyphaseollin production by soybeans resistant and susceptible to Phytophthora megasperma var. sojae. Physiol. Plant Pathol. 1: 265-275.

Keen, N. T., J. J. Sims, D. C. Erwin, E. Rice and J. E. Partridge. 1971. 6α-Hydroxyphaseollin: an antifungal chemical induced in soybean hypocotyls by Phytophthora megasperma var. sojae. Phytopathology 61: 1084-1089.

Keen, N. T., A. I. Zaki and J. J. Sims. 1972. Biosynthesis of hydroxyphaseollin and related isoflavanoids in disease-resistant soybean hypocotyls. Phytochem. 11: 1031-1039.

Kenten, R. H. 1957. Latent phenolase in extracts of broad bean (Vicia faba L.) leaves. I. Activation by acid and alkali. Biochem. J. 67: 300-307.

Kenten, R. H. 1958. Latent phenolase in extracts of broad bean (Vicia faba L.) leaves. II. Activation by anionic wetting agents. Biochem. J. 68: 244-251.

Kerr, A. and N. T. Flentje. 1957. Host infection of Pellicularia filamentosa controlled by chemical stimuli. Nature 179: 204-205.

Kirkham, D. S. 1958. Host factors in the physiology of disease, pp. 110-118. In Plant Pathology Problems and Progress 1908-1958 (C. H. Holton, G. W. Fischer, R. W. Fulton, H. Hart and S. E. A. McGallan, eds.). The Univ. Wisconsin Press, Madison.

Klarman, W. L. and J. W. Gerdemann. 1963. Induced susceptibility in soybean plants genetically resistant to Phytophthora sojae. Phytopathology 53: 863-864.

Klarman, W. L. and J. W. Gerdemann. 1963. Resistance of soybeans to three Phytophthora species due to the production of a phytoalexin. Phytopathology 53: 1317-1320.

Klarman, W. L. and F. Hammerschlag. 1972. Production of the phytoalexin, hydroxyphaseollin, in soybean leaves inoculated with tobacco necrosis virus. Phytopathology 62: 719-721.

Klement, Z. and R. N. Goodman. 1967. The hypersensitive reaction to infection by bacterial plant pathogens. Ann. Rev. Phytopathology 5: 17-44.

Kline, D. M., D. M. Boone and G. W. Keitt. 1958. Venturia inaequalis (Cke.) Wint. XIV. Nutritional control of pathogenicity of certain induced biochemical mutants. Amer. J. Bot. 44: 797-803.

Klun, J. A., C. L. Tipton, J. F. Robinson, D. L. Ostrem, and M. Beroza. 1970. Isolation and identification of 6,7-dimethoxy-2-benzoxazolinone from dried tissues of Zea mays (L.) and evidence of its cyclic hydroxamic acid precursor. Agr. Food. Chem. 18: 663-665.

Kosuge, T. 1969. The role of phenolics in host response to infection. Ann. Rev. Phytopathology 7: 195-222.

Kovacs, A. 1955. Uber die ursachen der unterschiedlichen resistenz der zuckerrubensorten gegen Cercospora beticola Sacc. Phytopathol. Z. 24: 283-298.

Kovacs, G. 1964. Studies on antibiotic substances from higher plants, with special reference to their plant pathological importance. Denmark Vet. Og Landbohojsk. Arsskr.: 47-92.

Krusberg, L. R. 1963. Host response to nematode infection. Ann. Rev. Phytopathology 1: 219-240.

Kuc, J. 1963. The role of phenolic compounds in disease resistance, pp. 20-30. In Perspectives of Biochemical Plant Pathology (S. Rich, ed.). Connecticut Agr. Exp. Sta. Bul. 663.

Kuc, J. 1964. Phenolic compounds and disease resistance in plants, pp. 63-81. In Phenolics in Normal and Diseased Fruits and Vegetables (V. C. Runeckles, ed.). Proc. Plant Phenolics Group of North America Symp. Imperial Tobacco Co., Montreal, Quebec.

Kuc, J. 1967. Shifts in oxidative metabolism during pathogenesis, pp. 183-202. In The Dynamic Role of Molecular Constituents in Plant-Parasite Interaction. (C. J. Mirocha and I. Uritani, eds.). Amer. Phytopathological Soc., St. Paul, Minnesota.

Kuc, J. 1968. Biochemical Control of disease resistance in plants. World Rev. Pest Control 7: 42-55.

Langcake, P., R. B. Drysdale and H. Smith. 1972. Post-infectional production of an inhibitor of Fusarium osysporum f. lycopersici by tomato plants. Physiol. Plant Pathol. 2: 17-25.

Lapwood, D. H. 1957. On the parasitic action of certain bacteria in relation to the capacity to secrete pectolytic enzymes. Ann. Bot. 21: 167-184.

Lee, S. F. and D. LeTourneau. 1958. Chlorogenic acid content and Verticillium wilt resistance of potatoes. Phytopathology 48: 268-274.

Lewis, J. C., G. Alderton, G. F. Barley, J. F. Carson, D. M. Reynolds and F. Stitt. 1959. Antibacterial agents from hops. USDA Bur. Agr. Ind. Chem., Mimeo. Circ. Ser. AIC 231: 1-15.

Link, K. P., H. R. Angell and J. C. Walker. 1929. The isolation of protocatechuic acid from pigmented onion scales and its significance in relation to disease resistance in onions. J. Biol. Chem. 81: 369-375.

Link, K. P. and J. C. Walker. 1933. The isolation of catechol from pigmented onion scales and its significance in relation to disease resistance in onions. J. Biol. Chem. 100: 379-383.

Lobarzewski, J. 1962. The inactivation of alcohol dehydrogenase by tyrosinase. Annales Universitatis Mariae Curie--Sklodowska, Lublin-Polonia: Sect C 16: 155.

Locci, R. and J. Kuc. 1967. Steroid alkaloids as compounds produced by potato tubers under stress. Phytopathology 57: 1272-1273.

Lovrekovich, L., H. Lovrekovich and M. A. Stahmann. 1968. The importance of peroxidase in the wildfire disease. Phytopathology 58: 193-198.

Ludwig, R. A., E. Y. Spencer and C. H. Unwin. 1960. An antifungal factor from barley of possible significance in disease resistance. Canadian J. Bot. 38: 21-29.

McLean, F. T. 1921. A study of the structure of the stomata of two species of citrus in relation to citrus canker. Bull. Torrey Botan. Club 48: 101-106.

McLean, J. G., D. J. LeTourneau and J. W. Guthrie. 1961. Relation of histochemical tests for phenols to *Verticillium* wilt resistance of potatoes. Phytopathology 51: 84-89.

Macdonald, T. and H. H. Smith. 1971. Biochemical variation associated with an *Aegilops umbellulata* chromosome segment incorporated in wheat, pp. 17-24. *In* Mutation Breeding for Disease Resistance. International Atomic Energy Agency, Vienna, Austria.

Mace, M. E. 1963. Histochemical localization of phenols in healthy and diseased banana roots. Phytiologia Plant. 16: 915-925.

Mace, M. E. 1964. Phenol oxidases and their relation to vascular browning in *Fusarium*-invaded banana roots. Phytopathology 54: 840-842.

Mace, M. E. 1964. Peroxidases and their relation to vascular browning in banana roots. Phytopathology 54: 1033-1034.

Mace, M. E. 1964. Phenols and their involvement in *Fusarium* wilt pathogenesis, pp. 13-19. *In* Phenolics in Normal and Diseased Fruits and Vegetables (V. C. Runeckles, ed.). Imperial Tobacco Co., Montreal.

Mace, M. E. 1973. Histochemistry of β-glucosidase in isolines of *Zea mays* susceptible or resistant to northern corn leaf blight. Phytopathology 63: 243-245.

Mace, M. E. and T. T. Hebert. 1963. Naturally occurring quinones in wheat and barley and their toxicity to loose smut fungi. Phytopathology 53: 692-700.

Mace, M. E. and E. Solit. 1966. Interactions of 3-indoleacetic acid and 3-hydroxytyramine in *Fusarium* wilt of banana. Phytopathology 56: 245-247.

Maizel, J. V., H. J. Burkhardt, and H. K. Mitchell. 1964. Avenacin, an antimicrobial substance isolated from Avena sativa. Isolation and antimimicrobial activity. Biochem. 3: 424-426.

Mandels, M. and E. T. Reese. 1965. Inhibition of cellulases. Ann. Rev. Phytopathology 3: 85-102.

Martin, J. T. 1964. Role of cuticle in the defense against plant disease. Ann. Rev. Phytopathology 2: 81-100.

Martin, J. T., R. F. Batt and R. T. Burchill. 1957. Defense mechanism of plants against fungi. Fungistatic properties of apple leaf wax. Nature 180: 796-799.

Massee, G. 1905. On the origin of parasitism in fungi. Trans. Roy. Phil. Soc. London, Ser. B 197: 7-24.

Matta, A. 1971. Microbial penetration and immunization of uncongenial host plants. Ann. Rev. Phytopathology 9: 387-410.

Melander, L. W. and J. H. Craigie. 1927. Nature of resistance of Berberis spp. to Puccinia graminis. Phytopathology 17: 95-114.

Metlitskii, L. V. and O. L. Ozeretskovskaya. 1968. Plant immunity. Plenum Press, New York. 114 p.

Millerd, A. and K. Scott. 1962. Respiration of the diseased plant. Ann. Rev. Plant Physiol. 13: 559-574.

Mohanakumaran, N., J. C. Gilbert and I. W. Buddenhagen. 1969. Relationship between tomatin and bacterial wilt resistance in tomato. Phytopathology 51: 14 (Abstr.).

Molot, P. M. and P. Anglade. 1968. Resistance commune des lignees de mais a l'helminthosporiose (Helminthosporium turcicum Pass.) et a la pyrale (Ostrinia nubilalis Hbn.) en relation avec la presence d'une substance identifiable as la 6-methoxy2(3)-benzoxazolinone. Annls. Epiphyt. 19: 75-95.

Montgomery, R. D. 1969. Cyanogens, pp. 143-157. In Toxic Constituents of Plant Foodstuffs (I. Liener, ed.). Academic Press, New York.

Muller, K. O. 1956. Einige einfache versuche zum nachweis von phytoalexin. Phytopathol. Z. 27: 237-254.

Muller, K. O. 1958. Relationship between phytoalexin output and the number of infections involved. Nature 182: 167-168.

Muller, K. O. 1958. Studies on phytoalexins I. The formation and the immunological significance of phytoalexin produced by Phaseolus vulgaris in response to infections with Sclerotinia fructicola and Phytophthora infestans. Australian J. Biol. Sci. 11: 275-300.

Muller, K. O. 1959. Hypersensitivity, pp. 469-519. In Plant Pathology, Vol. 1 (J. G. Horsfall and A. E. Dimond, eds.). Academic Press, New York.

Muller, K. O. 1963. The phytoalexin concept and its methodological significance. Recent Adv. Bot. 1: 396-400.

Muller, K. O. and H. Borger. 1940. Experimentelle untersuchungen uber die phytophthora-resistenz der kartoffel; zugleich ein beitrag zum problem der "erworbenen resistenz" in pflanzenreich. Arb. Biol. Reichsanstalt. Land-u. Fortwirtsch. Berlin-Dahlem 23: 189-231.

Mundry, K. W. 1963. Plant virus-host cell relations. Ann. Rev. Phytopathology 1: 173-196.

Mussell, H. W. and R. J. Green, Jr. 1970. Host colonization and polygalacturonase production by two tracheomycotic fungi. Phytopathology 60: 192-195.

Nienstaedt, H. 1953. Tannin as a factor in the resistance of chestnut, Castanea spp., to the chestnut blight fungus, Endothia parasitica (Murr) A and A. Phytopathology 43: 32-38.

Nonaka, F. 1966. On the selective toxicity of ipomeamarone towards the phytopathogens. Agr. Bull. Saga Univ. 22: 39-49.

Nonaka, F. 1967. Inactivation of pisatin by pathogenic fungi. Agr. Bull. Saga Univ. 24: 109-121.

Novacky, A. 1972. Suppression of the bacterially induced hypersensitive reaction by cytokinins. Physiol. Plant Pathol. 2: 101-104.

Noveroske, R. L., J. Kuc and E. B. Williams. 1964. β-Glucosidase and phenoloxidase in apple leaves and their possible relation to resistance to Venturia ineaqualis. Phytopathology 54: 98-103.

Nutman, F. J. and F. M. Roberts. 1960. Investigations on a disease of Coffea arabica caused by a form of Colletrotrichum coffeanum. I. Some factors affecting infection by the pathogen. Trans. British Mycol. Soc. 43: 489-505.

Oku, H. 1967. Role of parasite enzymes and toxins in development of characteristic symptoms in plant disease, pp. 237-255. In The Dynamic Role of Molecular Constituents in Plant-Parasite Interaction (C. J. Mirocha and I. Uritani, eds.). Amer. Phytopathological Soc., St. Paul, Minnesota.

Orton, W. A. 1908. The development of farm crops resistant to disease, pp. 453-464. In Yearbook of the Department of Agriculture, 1908.

Parish, C. L., M. Zaitlin and A. Seigel. 1963. A study of necrotic lesion formation by tobacco mosaic virus. Virology 26: 413-418.

Patil, S. S., R. L. Powelson and R. A. Young. 1964. Relation of chlorogenic acid and free phenols in potato roots to infection by Verticillium albo-atrum. Phytopathology 54: 531-535.

Paulson, R. E. and J. M. Webster. 1972. Ultrastructure of the hypersensitive reaction in roots of tomato, Lycopersicon esculentum L., to infection by the root-knot nematode, Meloidogyne incognita. Physiol. Plant Pathol. 2: 227-234.

Paxton, J. D. and D. W. Chamberlain. 1969. Phytoalexin production and disease resistance in soybeans as affected by age. Phytopathology 59: 775-777.

Pellizzari, E. D., J. Kuc and E. B. Williams. 1970. The hypersensitive reaction in Malus species: Changes in the leakage of electrolytes from apple leaves after inoculation with Venturia inaequalis. Phytopathology 60: 373-376.

Peries, O. S. 1962. Studies on strawberry mildew, caused by Sphaerotheca macularis (Wallr. ex Fries) Jaczewski. II. Host parasite relationships on foliage of strawberry varieties. Ann. Appl. Biol. 50: 225-233.

Perkins, D. D. 1949. Biochemical mutants in the smut fungi Ustilago maydis. Genetics 34: 607-626.

Perrin, D. R. 1964. The structure of phaseolin. Tetrahedron Ltrs. 29-35.

Perrin, D. R. and W. Bottomley. 1962. Studies on phytoalexins. V. The structure of pisatin from Pisum sativum L. J. Amer. Chem. Soc. 84: 1919-1922.

Perrin, D. R. and I. A. M. Cruickshank. 1969. The antifungal activity of pterocarpans toward Monilinia fructicola. Phytochem. 8: 971-978.

Pierre, R. E. 1971. Phytoalexin induction in beans resistant or susceptible to Fusarium and Thielaviopsis. Phytopathology 61: 322-327.

Pierre, R. E. and D. F. Bateman. 1967. Induction and distribution of phytoalexins in Rhizoctonia-infected bean hypocotyls. Phytopathology 57: 1154-1160.

Pierson, C. F. and J. C. Walker. 1954. Relation of Cladosporium cucumerinum to susceptible and resistant cucumber tissue. Phytopathology 44: 459-465.

Pinkas, Y. and A. Novacky. 1971. The differentiation between bacterial hypersensitive reaction and pathogenesis by the use of cycloheximide. Phytopathology 61: 906 (Abstr.).

Powell, C. C., Jr. and D. C. Hildebrand. 1970. Fire blight resistance in Pyrus: involvement of arbutin oxidation. Phytopathology 60: 337-340.

Price, W. C. 1932. Acquired immunity to ring-spot in Nicotiana. Contr. Boyce Thompson Inst. 4: 359-403.

Price, W. C. 1964. Strains, mutation, acquired immunity and interference, pp. 93-117. In Plant Virology (M. K. Corbett and H. D. Sisler, eds.). Univ. of Florida Press, Gainesville. 527 pp.

Rahe, J. E., J. Kuc, C. Chuang and E. B. Williams. 1969. Correlation of phenolic metabolism with histological changes in Phaseolus vulgaris inoculated with fungi. Netherlands J. Plant Pathol. 75: 58-71.

Rautela, G. S. and M. G. Payne. 1970. The relationship of peroxidase and orthodiphenol oxidase to resistance of sugarbeets to Cercospora leaf spot. Phytopathology 60: 238-245.

Reeve, R. M. 1951. Histochemical tests for polyphenols in plant tissues. Stain Technology 26: 91-96.

Reeve, R. M. 1959. Histological and histochemical changes in developing and ripening peaches. III. Catechol tannin content per cell. Amer. J. Bot. 46: 645-650.

Reimann, J. E. and R. U. Byerrum. 1964. Studies on the biosynthesis of 2,4-dihydroxy-7-methoxy-2H-1,4-benzoxazin-3-one. Biochem. 3: 847-851.

Rennerfelt, E. and G. Nacht. 1955. The fungicidal activity of some constituents from heartwood of conifers. Svensk. Botan. Tidskr. 49: 419-432.

Roberts, M. F. and J. T. Martin. 1963. Withertip disease of limes (Citrus aurantifolia) in Zanzibar. III. The leaf cuticle in relation to infection by Gloeosporium limetticola Clausen. Ann. Appl. Biol. 51: 411-413.

Roberts, M. F., J. T. Martin and O. S. Peries. 1961. Studies on plant cuticle. IV. The leaf cuticle in relation to invasion by fungi, pp. 102-110. Ann. Rept. Long Ashton Res. Sta. 1960.

Robinson, T. 1969. The organic constituents of higher plants. Burgess Publishing Co., Minneapolis, Minnesota.

Rohde, R. A. 1960. Mechanisms of resistance to plant-parasitic nematodes, pp. 447-453. In Nematology: Fundamentals and Recent Advances with Emphasis on Plant Parasitic and Soil Forms (J. N. Sasser and W. R. Jenkins, eds.). Univ. North Carolina Press, Chapel Hill, North Carolina.

Rohde, R. A. 1965. The nature of resistance in plants to nematodes. Phytopathology 55: 1159-1162.

Rohringer, R. and D. J. Samborski. 1968. Aromatic compounds in host-parasite interaction. Ann. Rev. Phytopathology 5: 77-86.

Romanko, R. R. 1959. A physiological basis for resistance of oats to Victoria blight. Phytopathology 49: 32-36.

Romig, R. W. and R. M. Caldwell. 1964. Stomatal exclusion of Puccinia recondita by wheat peduncles and sheaths. Phytopathology 54: 214-218.

Ross, A. F. 1966. Systemic effects of local lesion formation, pp. 127-150. In Viruses of Plants (A. B. R. Beemster and J. Dijkstra, eds.). North Holland.

Rubin, B. A. and E. V. Artsikhovskaya. 1963. Biochemistry and Physiology of Plant Immunity. The Macmillan Co., New York. 358 pp.

Rubin, B. A. and E. V. Artsikhovskaya. 1964. Biochemistry of pathological darkening of plant tissues. Ann. Rev. Phytopathology 2: 157-178.

Rudolph, K. and M. A. Stahmann. 1964. Interactions of peroxidases and catalases between Phaseolus vulgaris and Pseudomonas phaseolicola (halo blight of bean. Nature 204: 474-475.

Salle, A. J., J. G. Jann and M. Ordanik. 1949. Lupulon, an antibiotic extracted from the strobiles of Humulus lupulus. Proc. Soc. Exptl. Biol. Med. 70: 409-411.

Samaddar, K. R. and R. P. Scheffer. 1971. Early effects of Helminthosporium victoriae toxin on plasma membranes and counter-action by chemical treatments. Physiol. Plant Pathol. 1: 319-328.

Samborski, D. J. and R. Rohringer. 1970. Abnormal metabolites of wheat: occurrence, isolation and biogenesis of 2-hydroxy-putrescine amides. Phytochem. 9: 1939-1945.

Samuel, G. 1927. On the shot-hole disease caused by Cladosporium carpophilum and on the "shothole" effect. Ann. Bot. 41: 375-404.

Sato, N., K. Kitazawa and K. Tomiyama. 1971. The role of rishitin in localizing the invading hyphae of Phytophthora infestans in infection sites at the cut surfaces of potato tubers. Physiol. Plant Pathol. 1: 289-295.

Sato, N. and K. Tomiyama. 1969. Localized accumulation of rishitin in the potato tuber tissue by an incompatible race of Phytophthora infestans. Ann. Phytopathol. Soc. Japan 35: 202-217.

Sato, N., K. Tomiyama, N. Katsui and T. Masamune. 1968. Isolation of rishitin from tomato plants. Ann. Phytopathological Soc. Japan 34: 344-345.

Scheffer, R. P. and R. B. Pringle. 1967. Pathogen-produced determinants of disease and their effects on host plants, pp. 217-236. In The Dynamic Role of Molecular Constituents in Plant-Parasite Interaction (C. J. Mirocha and I. Uritani, eds.). Amer. Phytopathological Soc., St. Paul, Minnesota.

Scheffer, R. P. and K. R. Samaddar. 1970. Host-specific toxins as determinants of pathogenicity. Rec. Adv. Phytochem. 3: 123-142.

Scheffer, R. P. and O. C. Yoder. 1972. Host-specific toxins and selective toxicity, pp. 251-272. In Phytotoxins in Plant Disease (R. K. S. Wood, A. Ballio and A. Graniti, eds.). Academic Press, New York.

Scheffer, T. C. and E. B. Cowling. 1967. Natural resistance of wood to microbial deterioration. Ann. Rev. Phytopathology 4: 147-170.

Schieferstein, R. H. and W. E. Loomis. 1959. Development of the cuticular layer in angiosperm leaves. Amer. J. Bot. 46: 625-635.

Schonbeck, F. 1967. Untersuchungen uber bluteninfektionen. V. Untersuchungen an tulpen. Phytopathol. Z. 59: 205-224.

Schonbeck, F. and C. Schroeder. 1972. Role of antimicrobial substances (tuliposides) in tulips attacked by Botrytis spp. Physiol. Plant Pathol. 2: 91-99.

Schroth, M. N. and D. C. Hildebrand. 1964. Influence of plant exudates on root-infecting fungi. Ann. Rev. Phytopathology 2: 101-132.

Schwochau, M. E. and L. A. Hadwiger. 1970. Induced host resistance--a hypothesis derived from studies of phytoalexin production. Rec. Adv. Phytochem. 3: 181-189.

Sehgal, J. M. 1961. Antimicrobial substances from flowering plants. I. Antifungal substances. Hindustan Antibiot. Bul. 4: 3-29.

Shain, L. 1967. Resistance of sapwood in stems of loblolly pine to infection by Fomes annosus. Phytopathology 57: 1034-1045.

Shaw, M. 1963. The physiology and host-parasite relations of the rusts. Ann. Rev. Phytopathology 1: 259-294.

Shaw, M. 1967. Cell biological aspects of host parasite relations of obligate fungal parasites. Canadian J. Bot. 45: 1205-1220.

Shear, G. M. and C. R. Drake. 1971. Calcium accumulation in apple fruit infected with Venturia inaequalis (Cooke) Wint. Physiol. Plant Pathol. 1: 313-318.

Smale, B. C. and H. L. Keil. 1966. A biochemical study of the intervarietal resistance of Pyrus communis to fireblight. Phytochemistry 5: 1113-1120.

Smith, D. A., H. D. Van Etten and D. F. Bateman. 1971. Isolation of substance II, an antifungal compound from Rhizoctonia solani-infected bean tissue. Phytopathology 61: 912 (Abstr.).

Smith, D. G., A. G. McInnes, V. J. Higgins, and R. L. Millar. 1971. Nature of the phytoalexin produced by alfalfa in response to fungal infection. Physiol. Plant Pathol. 1: 41-44.

Sokolava, V. E., O. N. Savelieva and G. A. Solovieva. 1961. The toxic effects of caffeic and quinic acids on the fungus Phytophthora infestans. Dokl. Akad. Nauk SSSR 136: 723-726.

Spencer, D. M. 1962. Antibiotics in seeds and seedling plants, pp. 125-146. In Antibiotics in Agriculture (M. Woodbine, ed.). Butterworth, London and Washington, D.C.

Stahmann, M. A. 1965. The biochemistry of proteins of the host and parasite in some plant diseases. Tagungsberichte Nr. 74: 9-40.

Stahmann, M. A. 1967. Influence of host-parasite interactions on proteins, enzymes and resistance, pp. 357-372. In The Dynamic Role of Molecular Constituents in Plant-Parasite Interaction. (C. J. Mirocha and I. Uritani, eds.). Amer. Phytopathological Soc., St. Paul, Minnesota.

Stakman, E. C. 1915. Relation between Puccinia graminis and plants highly resistant to its attack. J. Agr. Res. 4: 193-200.

Stanghellini, M. E. and M. Aragaki. 1966. Relation of periderm formation and callose deposition to anthracnose resistance in papaya fruit. Phytopathology 56: 444-450.

Stholasuta, P., J. A. Bailey, V. Severin and B. J. Deverall. 1971. Effect of bacterial inoculation of bean and pea leaves on the accumulation of phaseollin and pisatin. Physiol. Plant Pathol. 1: 177-183.

Stoessl, A. 1969. 8-Hydroxy-6-methoxy-3-methylisocoumarin and other metabolites of Ceratocystis fimbriata. Biochem. Biophys. Res. Comm. 35: 186-192.

Stoessl, A. 1970. Antifungal compounds produced by higher plants. Rec. Adv. Phytochem. 3: 143-180.

Stoessl, A., R. Rohringer and D. J. Samborski. 1969. 2-Hydroxyputrescine amides as abnormal metabolites of wheat. Tetrahedron Ltrs. 33: 2807-2810.

Suzuki, N. 1965. Histochemistry of foliage diseases. Ann. Rev. Phytopathology 3: 265-286.

Tanaka, K. 1961. Effect of capillin, an antifungal substance, on mycelial growth of Penicillium italicum. Takamine Kenkyusho Nempo 13: 112-116.

Thatcher, F. S. 1942. Further studies of osmotic and permeability relations in parasitism. Canadian J. Res. 20: 283-311.

Thatcher, F. S. 1943. Cellular changes in relation to rust resistance. Canadian J. Res. 21: 151-172.

Thomas, C. A. and E. H. Allen. 1970. An antifungal polyacetylene compound from Phytophthora-infected safflower. Phytopathology 60: 261-263.

Thomas, C. A. and E. H. Allen. 1971. Light and antifungal polyacetylene compounds in relation to resistance of safflower to Phytophthora drechsleri. Phytopathology 61: 1459-1461.

Timonin, M. I. 1941. The interaction of higher plants and soil microorganisms. III. Effect of by-products of plant growth on activity of fungi and actinomycetes. Soil. Sci. 52: 395-413.

Tipton, C. L., J. A. Klun, R. R. Husted and M. O. Pierson. 1967. Cyclic hydroxamic acids and related compounds from maize. Isolation and characterization. Biochem. 6: 2866-2870.

Tomiyama, K. 1963. Physiology and biochemistry of disease resistance. Ann. Rev. Phytopathology 1: 295-324.

Tomiyama, K. 1971. Cytological and biochemical studies of the hypersensitive reaction of potato cells to *Phytophthora infestans*, pp. 387-401. *In* Morphological and Biochemical Events in Plant-Parasite Interaction (S. Akai and S. Ouchi, eds.). The Phytopathological Soc. of Japan, Tokyo.

Tomiyama, K., N. Ishizaka, N. Sato, T. Masamune and N. Katsui. 1968. "Rishitin" a phytoalexin-like substance. Its role in the defense reaction of potato tubers to infection, pp. 287-292. *In* Biochemical Regulation in Diseased Plants or Injury (T. Hirai, ed.). Kyoritsu Printing Co., Tokyo, Japan.

Tomiyama, K., R. Sakai, T. Sakuma and N. Ishizaka. 1967. The role of polyphenols in the defense reaction in plants induced by infection, pp. 165-182. *In* The Dynamic Role of Molecular Constituents in Plant-Parasite Interaction (C. J. Mirocha and I. Uritani, eds.). Amer. Phytopathological Soc., St. Paul, Minnesota.

Tomiyama, K. and M. A. Stahmann. 1964. Alteration of oxidative enzymes in potato tuber tissue by infection with *Phytophthora infestans*. Plant Physiol. 39: 483-490.

Tschesche, R., F. J. Kammerer and G. Wulff. 1969. Uber die struktur der antibiotisch aktiven substanz der tulpe. Chem. Ber. 102: 2057-2071.

Tschesche, R., F. J. Kammerer, G. Wulff, and F. Schonbeck. 1968. Uber die antibiotisch wirksamen substanzen der tulpe (*Tulipa gesneriana*). Tetrahedron Ltrs.: 701-706.

Turner, E. M. C. 1961. An enzymic basis for pathogenic specificity in *Ophiobolus graminis*. J. Exptl. Bot. 12: 169-175.

Turner, W. B. 1971. Fungal metabolites. Academic Press, New York. 446 pp.

Uehara, K. 1963. On the production of phytoalexin by metallic salts. Bull. Hiroshima Agr. Coll. 2: 41-44.

Uehara, K. 1964. Relationship between the host specificity of pathogen and phytoalexin. Ann. Phytopathological Soc. Japan 29: 103-110.

Uhlenbroek, J. H. and J. D. Bijloo. 1958. Investigations on nematicides. I. Isolation and structure of a nematicidal principle occurring in *Tagetes* roots. Rec. Trav. Chim. Pays-Bas. 77: 1004-1009.

Uhlenbroek, J. H. and J. D. Bijloo. 1959. Investigations on nematicides II. Structure of a second nematicidal principle isolated from *Tagetes* roots. Rec. Trav. Chim. Pays-Bas. 78: 382-390.

Urech, J., B. Fechtig, J. Nuesch and E. Vischer. 1963. Hircinol, eine antifungische wirksame substanz aus knollen von *Laroglossum hircinum* (L.) Rich. Helv. Chim. Acta 46: 2758.

Uritani, I. 1963. The biochemical basis of disease resistance induced by infection, pp. 4-19. In Perspectives of Biochemical Plant Pathology. Connecticut Agr. Exp. Sta. Bul. 663 pp.

Uritani, I. 1971. Protein changes in diseased plants. Ann. Rev. Phytopathology 9: 211-234.

Uritani, I., T. Asahi, T. Minamikawa, H. Hyodo, K. Oshima and M. Kojima. 1967. The relation of metabolic changes in infected plants to changes in enzymatic activity, pp. 342-356. In The Dynamic Role of Molecular Constituents in Plant-Parasite Interaction. (C. J. Mirocha and I. Uritani, eds.). The Amer. Phytopathological Soc., St. Paul, Minnesota.

Uritani, I., H. Nomura and T. Teramura. 1967. Comparative analysis of terpenoids in roots of *Ipomoea* species induced by inoculation with *Ceratocystis fimbriata*. Agr. Biol. Chem. (Tokyo) 31: 385.

Uritani, I. and M. A. Stahmann. 1961. Changes in nitrogen metabolism in sweet potato with black rot. Plant Physiol. 36: 770-782.

Uritani, I., M. Uritani and H. Yamada. 1960. Similar metabolic alterations induced in sweet potato by poisonous chemicals and by *Ceratostomella fimbriata*. Phytopathology 50: 30-34.

Valenta, J. R. and H. D. Sisler. 1962. Evidence for a chemical basis of resistance of lima bean plants to downy mildew. Phytopathology 52: 1030-1037.

Van der Plank, J. E. 1968. Disease Resistance in Plants. Academic Press, New York. 206 pp.

Van Etten, H. D. and D. F. Bateman. 1971. Studies on the mode of action of the phytoalexin phaseollin. Phytopathology 61: 1363-1372.

Van Fleet, D. S. 1971. Enzyme localization and the genetics of polyenes and polyacetylenes in the endodermis. Adv. Front. Plant Sci. 26: 109-143.

Van Fleet, D. S. 1972. Histochemistry of plants in health and disease. Rec. Adv. Phytochem. 5: 165-195.

Varns, J. L., W. W. Currier and J. Kuc. 1971. Specificity of rishitin and phytuberin accumulation by potato. Phytopathology 61: 968-971.

Varns, J. L. and J. Kuc. 1971. Suppression of rishitin and phytuberin accumulation and hypersensitive response in potato by compatible races of *Phytophthroa infestans*. Phytopathology 61: 178-181.

Varns, J. L., J. Kuc and E. B. Williams. 1971. Terpenoid accumulation as a biochemical response of the potato tuber to *Phytophthora infestans*. Phytopathology 61: 174-177.

Virtanen, A. I. 1965. Studies on organic sulphur compounds and other labile substances in plants. Phytochem. 4: 207-228.

Virtanen, A. I. and P. K. Hietala. 1955. 2(3)-Benzoxazolinone, an anti-*Fusarium* factor in rye seedlings. Acta Chem. Scand. 9: 1543-1544.

Virtanen, A.I., P. K. Hietala and O. Wahlroos. 1957. Antimicrobial substances in cereals and fodder plants. Arch. Biochem. Biophys. 69: 486-500.

Von Broembsen, S. L. and L. A. Hadwiger. 1972. Characterization of disease resistance responses in certain gene-for-gene interactions between flax and Melampsora lini. Physiol. Plant Pathol. 2: 207-215.

Walker, J. C. 1924. On the nature of disease resistance in plants. Trans. Wisconsin Acad. Sci. 21: 225-247.

Walker, J. C. 1963. The physiology of disease resistance, pp. 1-25. In West Virginia Agr. Exp. Sta. Bul. 488T.

Walker, J. C. and K. P. Link. 1935. Toxicity of phenolic compounds to certain onion bulb parasites. Botan. Gaz. 96: 468-484.

Walker, J. C. and M. A. Stahmann. 1955. Chemical nature of disease resistance in plants. Ann. Rev. Plant Physiol. 6: 351-366.

Ward, H. M. 1902. On the relations between host and parasite in bromes and their brown rust, Puccinia dispersa (Eriks.). Ann. Bot. 16: 233-315.

Ward, H. M. 1905. Recent researches on the parasites of fungi. Ann. Bot. 19: 1-54.

Webb, J. L. 1966. Enzyme and metabolic inhibitors. Academic Press, New York. 1028 pp.

Weber, D. J. and M. A. Stahmann. 1964. Ceratocystis infection in sweet potato: its effect on proteins, isozymes and acquired immunity. Science 146: 929-931.

Whittaker, R. H. 1970. The biochemical ecology of higher plants, pp. 43-70. In Chemical Ecology (E. Sondheimer and J. B. Simeone, eds.). Academic Press, New York, 336 pp.

Williams, A. H. 1963. Enzyme inhibition by phenolic compounds, pp. 87-95. In Enzyme Chemistry of Phenolic Compounds (J. B. Pridham, ed.). Pergamon Press, London.

Williams, E. B. and J. Kuc. 1969. Resistance in Malus to Venturia inaequalis. Ann. Rev. Phytopathology 7: 223-246.

Wimalajeewa, D. L. S. and J. E. DeVay. 1971. The occurrence and characterization of a common antigen relationship between Ustilago maydis and Zea mays. Physiol. Plant Pathol. 1: 523-535.

Wingard, S. A. 1941. The nature of disease resistance in plants. I. Bot. Rev. 7: 59-109.

Wit-Elshove, A. de. 1968. Breakdown of pisatin by some fungi pathogenic to Pisum sativum. Netherlands J. Plant Pathol. 74: 44-47.

Wit-Elshove, A. de and A. Fuchs. 1971. The influence of the carbohydrate source on pisatin breakdown by fungi pathogenic to pea (Pisum sativum). Physiol. Plant Pathol. 1: 17-24.

Wood, R. K. S. 1960. Chemical ability to breach the host barriers. pp. 233-272. In Plant Pathology, Vol. 2 (J. G. Horsfall and A. E. Dimond, eds.). Academic Press, New York.

Wood, R. K. S. 1960. Pectic and cellulolytic enzymes in plant disease. Ann. Rev. Plant Physiol. 11: 299-322.

Wood, R. K. S. 1967. Physiological Plant Pathology. Blackwell Scientific Publications, Oxford. 570 pp.

Wood, R. K. S., A. Ballis and A. Graniti, eds. 1972. Phytotoxins in Plant Disease. Academic Press, New York. 530 pp.

Yarwood, C. E. 1967. Response to parasites. Ann. Rev. Plant Physiol. 18: 419-438.

Zaki, A. I., N. T. Keen and D. C. Erwin. 1972. Implications of vergosin and hemigossypol in the resistance of cotton to _Verticillium albo-atrum._ Phytopathology 62: 1402-1406.

Zaki, A. I., N. T. Keen, J. J. Sims and D. C. Erwin. 1972. Vergosin and hemigossypol--antifungal compounds produced in cotton plants inoculated with _Verticillium albo-atrum_. Phytopathology 62: 1398-1401.

BIOCHEMICAL BASIS OF RESISTANCE OF PLANTS TO PATHOGENS AND INSECTS: INSECT HORMONE MIMICS AND SELECTED EXAMPLES OF OTHER BIOLOGICALLY ACTIVE CHEMICALS DERIVED FROM PLANTS

Jerome A. Klun
Entomologist
USDA, ARS, ERD
Iowa State University
Ankeny, Iowa

I. Evolutionary Perspective

The interactions of plants with the phytophagous insects are as diverse and complex as life itself. In the evolutionary sense, the plants and the insects generally are best considered as dynamic and independently evolving systems that interact at their periphery. The interaction of the two systems can lead to co-adaptation; the insects evolve with respect to the plants, and the plants, with respect to the insects. It must be realized, however, that the major selective pressures that result in evolutionary change in the respective systems are not necessarily a consequence of the plant-insect interaction. Other selective pressures such as environmental factors, microorganisms, and competition are probably the greatest impetus for evolutionary change. This evolutionary concept can explain the diversity of associations between insects and plants (Dethier, 1970).

The plant evolves in an atmosphere of a multitude of selection pressures, among which predation by insects is only one. In response to these pressures, several courses of escape are available to the plant: emigration, morphological or biochemical changes. Often the plant will respond in defense of itself by quantitatively or qualitatively altering its chemical arrays to counteract the threat. Whatever the impetus for the change in chemical composition of the plant, the insect using that plant as a host must somehow circumvent the change if it poses a threat to the insect's survival. In this conceptual view of the plant-insect interaction, the insect is relegated to a somewhat passive role. That is, for the most part, the insect has little influence on the evolutionary trends of the plants and must be resourceful and respond evolutionarily to chemical innovations of the plants.

II. The Plant-Insect Chemical Interface

The successful interaction of an insect with its host plant is dependent on synchrony of a complex set of environmental, visual, tactile, and chemical factors that mediate the behavior and physiology of the insect. Modification of any one or more of these factors profoundly influences the suitability of any plant as a host for any insect predator. These factors and heterogeneity in the botanical world explain why all plants are not acceptable hosts for all phytophagous insects. It is not within the scope of this presentation to consider the multiplicity of interacting factors that influence plant-insect relationships. These factors have been expertly reviewed by Painter (1951), Beck (1965), Thorsteinson (1960), and

Dethier (1970). It is pertinent, however, that many mechanistic aspects of the resistance or susceptibility of plants to insects can be explained, in part, by behavioral and/or physiological responses of insects to chemicals produced by plants. These chemicals are of considerable consequence since they play a role in the linkages between plants and insects (Whittaker and Feeny, 1971). Some plant chemicals are insect repellents, feeding deterrents, or toxins, and others are attractants, feeding stimulants, or arrestants. The chemical identity of many of these substances is unknown, and often complex blends (Dethier, 1970; Hedin, et al. 1968), of chemicals are responsible for the observed behavioral or physiological effects on the insects. The complexity of the chemical interface and the wealth of plant chemical stimuli that await elucidation pose an enormous scientific challenge to students of the plant-insect interaction.

Among the many chemicals produced by plants are substances called secondary chemicals. These chemicals are distinguished from the primary chemicals (lipids, carbohydrates, proteins, nucleic acids) in that they have no obvious metabolic function but are elaborated by plants as by-products of metabolic systems producing other substances (Whittaker, 1970). The secondary chemicals can be classified into five major classes on the basis of their biosynthetic origin: terpenoids, steroids, acetogens, alkaloids, and phenyl propanes (Figure 1).

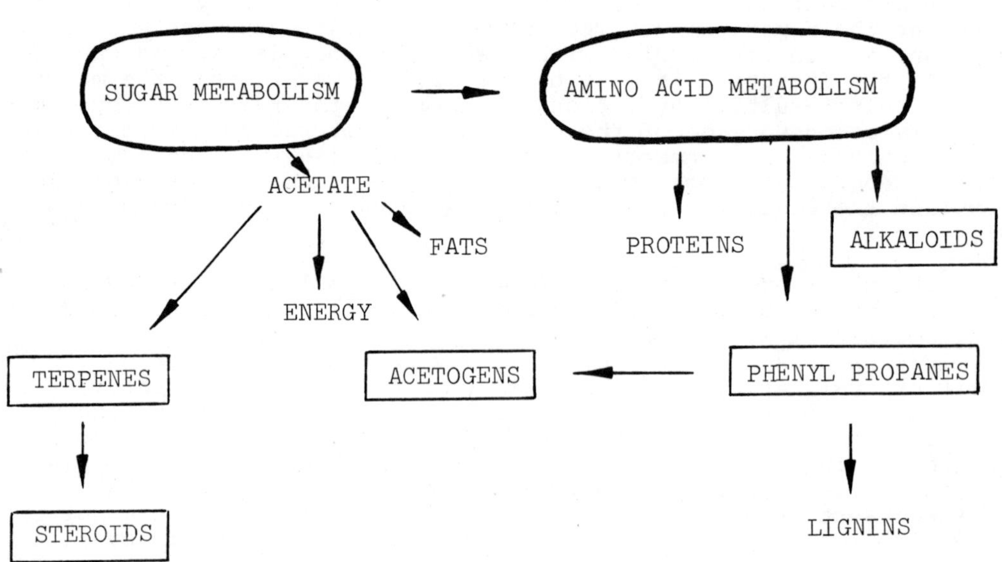

Figure 1.--Biosynthetic relationship of the five major classes of secondary chemicals (enclosed in rectangles) to other metabolic pathways of the plants.

Whittaker (1970) and Whittaker and Feeny (1971) have reviewed the chemistry of the five classes of secondary chemicals and considered their significance in the interaction of a diversity of life forms.

The secondary chemicals of plants are characteristically toxic, unpalatable or offensive to many would-be herbivores. Thus, such chemicals are of considerable significance to the phytophagous insects (Dethier, 1970, 1954; Fraenkel, 1959; Ehrlich and Raven, 1965).

III. The Juvenile Hormones and Ecdysones--Hormonal Interaction of the Plants and Insects

In recent years, an intriguing story involving the terpenoids and steroids of certain plants and insect endocrinology has unfolded. To tell this story, I must briefly review the hormonal control of insect growth, development, and reproduction.

Insect development is a stepwise process regulated by hormones (Schneiderman and Gilbert, 1964). For example, in the holometabolous insects, four distinct developmental stages are recognized: egg, larva, pupa, and adult (Figure 2). The process of insect growth and differentiation is controlled mainly by three endocrine tissues of the insects: the brain, corpus allatum, and prothoracic gland.

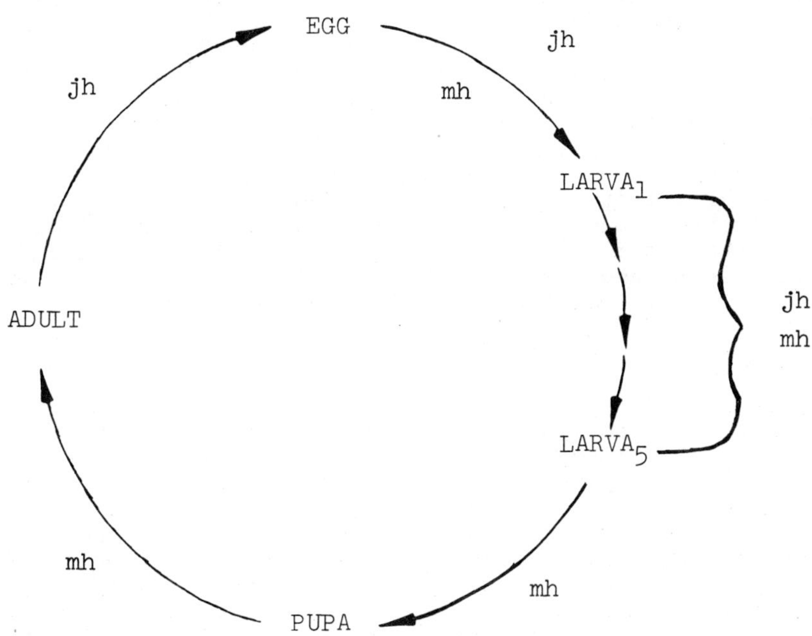

Figure 2.--Hormonal control of insect growth, development, and reproduction. jh = juvenile hormone, mh = molting hormone.

Endocrinological events occurring within the egg are poorly understood; there, however, is evidence that the egg stage is highly susceptible to hormonal applications (Bowers, 1971a; Riddiford, 1970). The larval stage of insect development characteristically involves sequential increases in size and weight. This growth occurs without morphological differentiation even though each larval instar has the genetic capacity for differentiation. When the time comes for a larva to undergo a molt, the brain activates the prothoracic gland, which in some way effects an increase of molting hormone (ecdysone) titer within the insect body. It was believed for many years that the prothoracic gland was the principal secretory source of molting hormone. Nakanishi, et al. (1972) have shown, however, that molting hormone is produced by tissues other than the prothoracic gland. Aside from this endocrinological puzzle, the presence of molting hormone in the insect body triggers gene activity in certain cells (Williams, 1970). If juvenile hormone is secreted by the corpus allatum before the molting hormone is released, only larval characters are generated in the molt (Siddall, 1970). If juvenile hormone is not released in the insect body before the molting hormone release, the molt results in differentiation and leads to development of the adult form (Williams, 1970). Thus, the molting hormone and juvenile hormone act in concert during the larval stage, and the juvenile hormone, as its name implies, functions to maintain larval or immature characters, permitting growth, but not maturation. In the adult stage, the juvenile hormones are called into action once again and play a significant role in insect ovarian development, sex pheromone production, and termination of diapause (Bowers, 1971a, c; Pan and Wyatt, 1971).

From this brief resume of insect endocrinology, it is clear that the molting and juvenile hormones, by their absence or presence, control important aspects of the physiology and biochemistry of each stage in the insect life cycle. Morphological abnormalities result if either hormone is supplied to the insect at a time inconsistent with its normal physiological program. For example, if an insect is treated with juvenile hormone at the pupal stage, the expected process of maturation to the adult is disrupted, and an intermediate form results with both adult and pupal characters. Similarly, if an insect is treated with excess molting hormone at certain stages of its development, abnormal morphs result. An insect experiencing such anomalous metamorphosis is incapable of further development and subsequently dies, providing a basis for hopes that hormonal-based insecticides may be developed. In recent years, there has been a flurry of research in this area (Pfiffner, 1971; Bowers, 1971b; Siddall, 1970, 1971). Correspondingly, significant advancements have been made in our knowledge of the chemistry of the insect hormones. The events leading to the elucidation of identities of the insect juvenile and molting hormones have been reviewed extensively by Pfiffner (1971); Rees (1971), and Williams (1970).

A. Juvenile Hormones

It was through the research of Roller, et al. (1967) that the juvenile hormone of the giant silkworm moth, *Hylaphora cecropia*, was identified as methyl-10-11-epoxy-7-ethyl-3,11-dimethyl,10,11-*cis*,2-*trans*,6-*trans* tridecadienoate (Figure 3 [1]). Subsequently, a second juvenile hormone [2] was isolated and identified from the giant silkworm moth by Meyer, et al. (1968). The molecule was identical in biological activity and structure to Roller's compound [1] except for a methyl substitution on C-7 instead of an ethyl group. Although these compounds are widely accepted as natural juvenile hormones, Bowers (1971a) appropriately pointed out that these substances were isolated and identified from the adult male *Cecropia* moth and not from the immature stages. Therefore, reasonable doubt existed if these juvenile hormones were truly the hormonal elements involved in regulation of larval growth. Most recently, however, this doubt was laid to rest by the isolation of a third juvenile hormone, 10,11-epoxyfarnesenic acid methyl ester [3], from larvae of the tobacco hornworm, *Manduca sexta* (Johannson) (Siddall, 1972). The compound identified from these larvae differs only slightly from those isolated from *Cecropia* in that it possesses a C-12 chain length. Bowers, et al. (1965) predicted, by inference, that the general structure of the juvenile hormones isolated from adult *Cecropia* males by Roller, et al. (1967) and Meyer, et al. (1968) would be chemically similar to 10,11-epoxyfarnesenic acid methyl ester [3]. Nearly 7 years later, the juvenile hormone of tobacco hornworm larvae turned out to be identical to the structure predicted by Bowers' group.

The structure of the juvenile hormones suggests their terpenoid origin via a farnesol derivative. If a farnesol [4] is a biosynthetic precursor of juvenile hormone [3], conversion to [3] would require epoxydation at C-10, 11, oxidation to

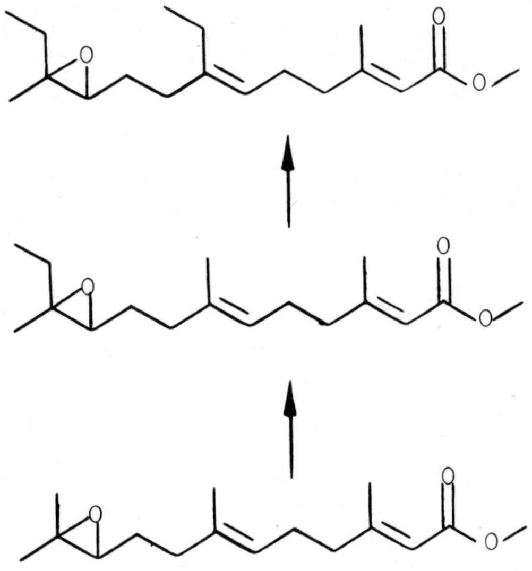

[1]

[2]

[3]

Figure 3.--Juvenile hormones isolated and identified from the giant silkworm moth.

to the acid, and formation of the methyl ester. Pfiffner (1971) suggested that juvenile hormone [2] isolated from the Cecropia moth by Meyers' group may be a precursor of [1] and that biomethylation (uncommon) of [2] may take place leading to [1]. Analogously, I wonder if [3] may be a biosynthetic precursor of [1] and [2] (Figure 3). This viewpoint is speculative, of course, and at present, little is known regarding the biosynthetic origin of juvenile hormone in the insect (Metzler, et al., 1972).

[4] [3]

Before isolation and elucidation of the first Cecropia juvenile hormone, a discovery was made that certain sesquiterpenes native to conifers possessed morphogenic activities analogous to the insect juvenile hormones (Slama and Williams, 1966). Bowers, et al. (1966) subsequently isolated one of these active factors from the balsam fir, identified it as the methyl ester of todomatic acid, and assigned it a trivial name, juvabione [5]. The morphogenic activity of juvabione was uniquely specific toward the Pyrrhocoris bugs and certain other hemipterans. A second morphogenic agent of the bugs, dehydrojuvabione [6], identical to juvabione [5] except for occurrence of a double bond at C-4 in the side chain, was identified later from the balsam fir by Cerny, et al. (1967). Juvabione and dehydrojovabione are analogous to the natural juvenile hormones [1,2,3], in that both are α-β-unsaturated methyl esters. The unique specificity of the juvabiones toward the Pyrrhocoridae is probably due to the cyclohexene ring (Siddall, 1970). Aside from the juvabiones (5,6], no other insect juvenile hormone mimics have been identified from plants, and it is difficult to speculate on the occurrence and distribution of such morphogenic agents in the botanical world or the significance of such compounds as they might restrict the host range of the phytophagous insects. However, research on the molecular requirements for juvenile hormone activity has shown that assorted terpenoid synthetics of various structure and functionality are active as morphogenic agents (Figure 4) (Bowers, 1971a, b; Siddall, 1971). The morphogenic potency of these synthetics varies according to the insect treated, and

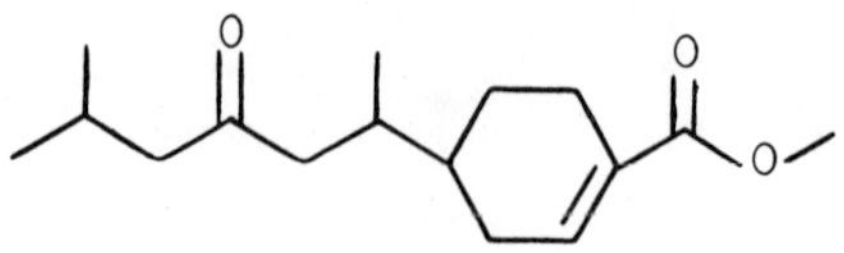

JUVABIONE [5]

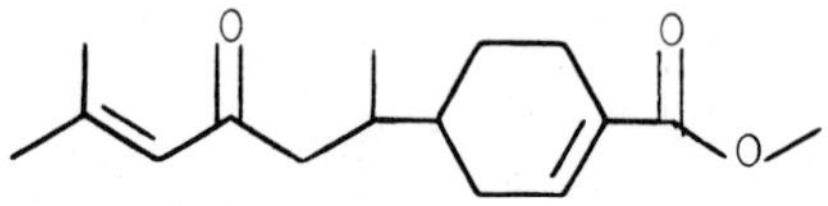

DEHYDROJUVABIONE [6]

some are more potent than the Cecropia juvenile hormone. Moreover, the insect endocrine system is susceptible to disruption by assorted compounds with terpenoid characters. Considering the versatility of the terpenoid biosynthetic pathways of the plants, it is plausible that juvenile-hormone-type compounds may lie undiscovered in the botanical world.

B. The Steroidal Molting Hormones

Butenandt and Karlson (1954) isolated 25 mg of molting hormone (α-ecdysone) from 500 kg of silkworm pupae (*Bombyx mori*). The structure of the hormone is shown in Figure 5 [7]. Events leading to structural elucidation of α-ecdysone have been reviewed by Rees (1971) and Horn (1971). After the isolation and identification of the ecdysone from *Bombyx mori*, the compound was identified from the Moroccan locust (*Dociostaurus morocanus*, Thunberg) (Stam, 1959), tobacco hornworm (*Manduca sexta*, Johannson) (Kaplanis, et al., 1966) and the oak silkworm moth (*Antherea pernyi*) (Horn, et al., 1966).

In addition to α-ecdysone, two other molting hormones have been isolated from the insects: 20-hydroxyecdysone [8] and 20,26-dihydroxyecdysone [9]. Note that the stereochemistry of the steroid nucleus is the same for all three compounds and that major differences occur in position and number of hydroxyls in the side chain (C_{20}-C_{26}). The absolute configurations of the side chain are known only for ecdysone (Siddall, 1970). All three ecdysones have been isolated and identified from tobacco hornworm pupae (Kaplanis, et al., 1966; Thompson, et al., 1967). Thompson, et al. (1967) speculated that α-ecdysone may be actively metabolized to 20-hydroxyecdysone and 20,26-dihydroxyecdysone in the tobacco hornworm by sequential hydroxylation (Figure 5) and that this metabolism may be part of a metabolic pathway that leads to deactivation of the ecdysones. Subsequent research (Robbins, et al., 1971) on steroid metabolism in the insects has provided evidence of this hydroxylation pathway in support of Thompson and coworkers' hypothesis.

In recent years, it has become obvious that the plant kingdom is a replete source of steroids (Siddall, 1970; Horn, 1971; Rees, 1971) that possess molting hormone activity toward the insects. Indeed, molting hormones evidently are distributed throughout the plant kingdom, having been demonstrated to occur in the

STRUCTURE	RELATIVE POTENCY (MOSQUITOES)
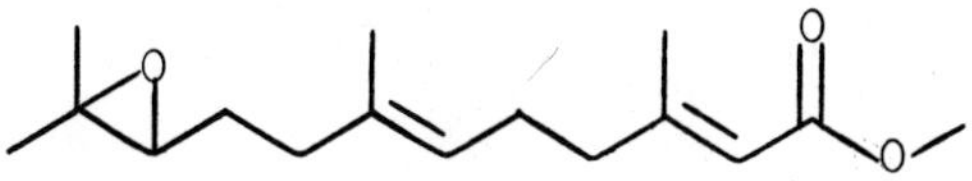	1.0
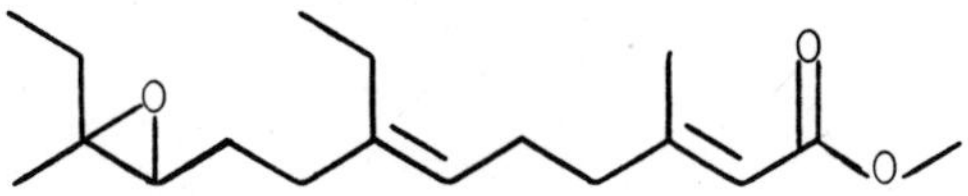	1.4
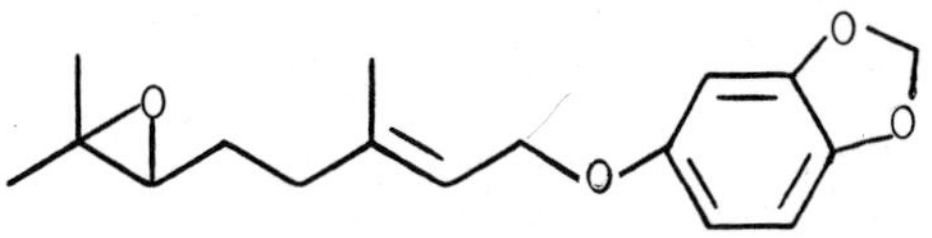	70
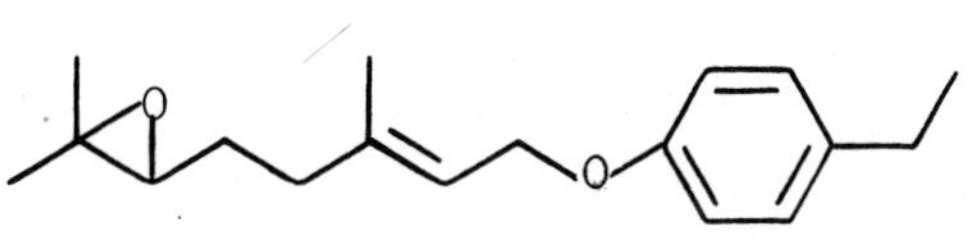	450
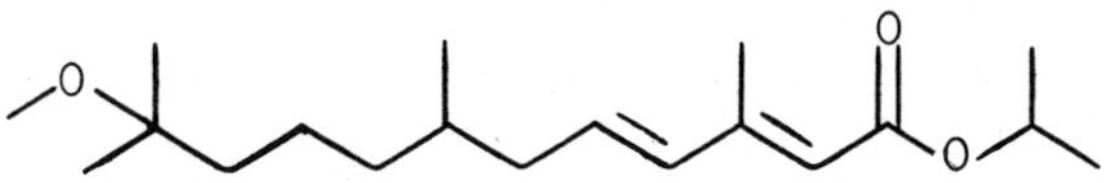	1900

Figure 4.--Structures of compounds having juvenile hormone activity (from Siddall, 1971).

Pteridophytes, Gymnosperms, and Angiosperms (both subclasses). At least 28 different active steroids have been identified from the plants, and the concentration of the insect molting hormones in some plants is often greater than that in insect sources (Horn, 1971). Among the 28 molting hormones identified from plants are the established insect hormones α-ecdysone [7] and 20-hydroxyecdysone [8]. 20-Hydroxyecdysone is the most ubiquitous of the phytoecdysones. Kaplanis, et al. (1967) found both α-ecdysones and 20-hydroxyecdysone in the same plant, a bracken fern (*Pteridium aquilinum*).

In general, the phytoecdysones bear a striking structural similarity to the insect ecdysones. Nearly all the identified phytoecdysones possess 2, 3β, and 14-α-hydroxyls, an unsaturated (Δ^7)-6-ketone system, and *cis* fusion between the A

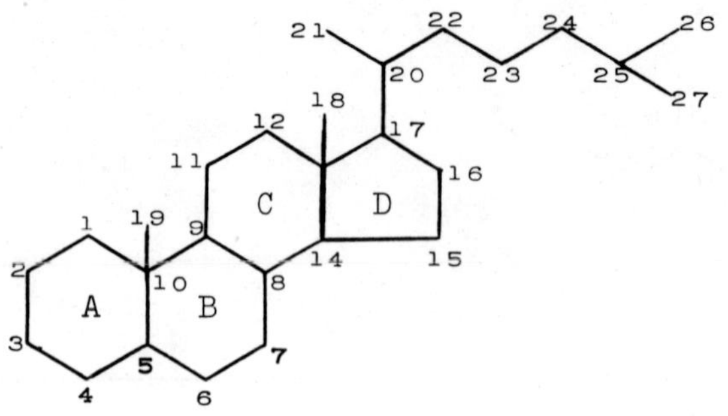

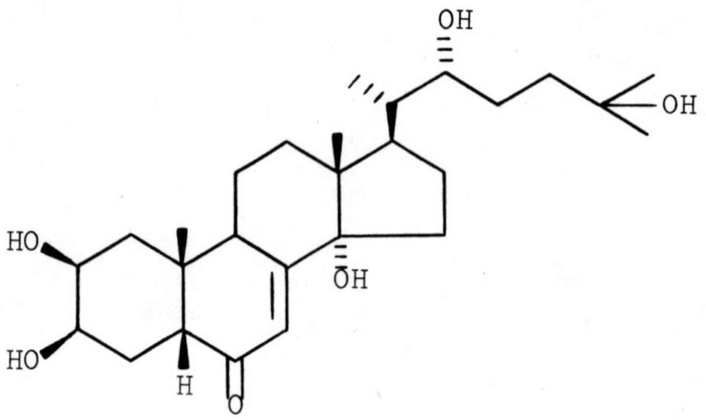

20-HYDROXYECDYSONE [8]

20, 26-DIHYDROXYECDYSONE [9]

Figure 5.--Structure of molting hormones identified from the insects.

and B rings. Major structural differences among the phytoecdysones therefore are attributed to the various substituents of the cholestane side chain. For example, two of the most potent phytoecdysones are polypodine B [10] and ponasterone A [11]. In certain standard assays for insect molting hormone activity, these compounds are more potent than α-ecdysone (Williams, 1970; Siddall, 1970; Rees, 1971).

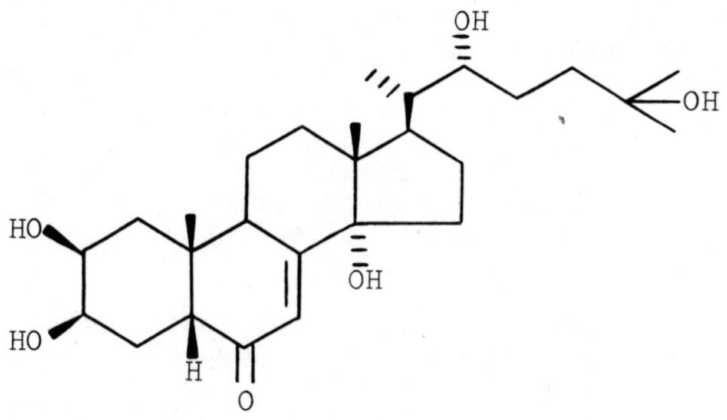

α-ECDYSONE

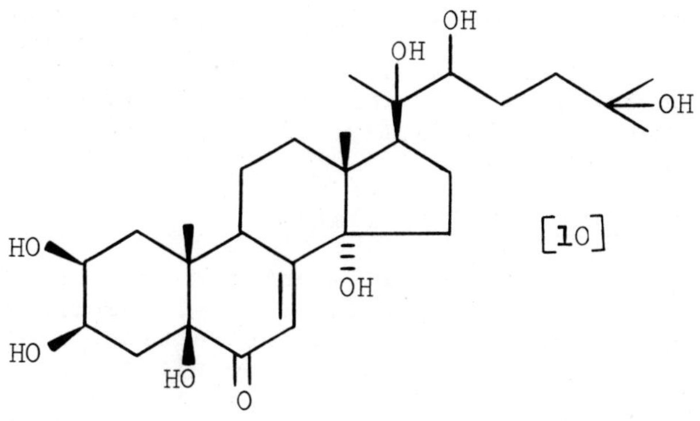

[10] POLYPODINE B

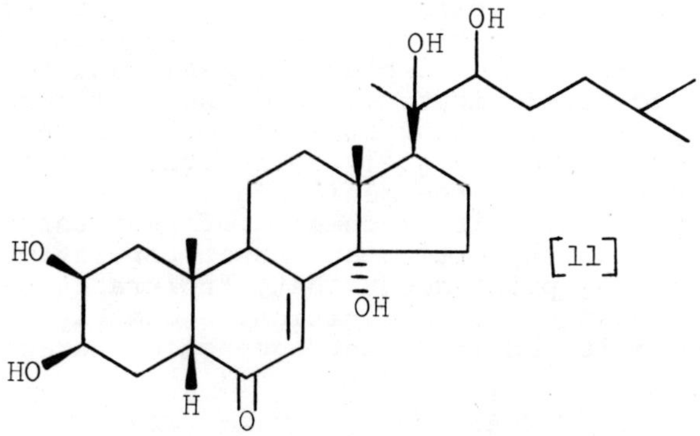

[11] PONASTERONE A

The physiological function of phytoecdysones in plants is unknown; therefore, these steroids must be considered secondary chemicals. Given the occurrence and distribution of the phytoecdysones in the plant kingdom, the question arises: are these compounds of any significance in the plant-insect interaction? In this connection, most recent studies (Rees, 1971) show that nearly all the phytoecdysones (Ponasterone A [11] excluded; Williams, 1970) are without effect on insects when administered orally. It also is clear that the ecdysones are not readily absorbed through the insect cuticle. Therefore, in bioassays for molting hormone activity, the ecdysones are either applied topically in an appropriate solvent or injected intersegmentally. Under such conditions, the phytoecdysones have profound morphogenic activities. But, attempts to speculate on the significance of phytoecdysones in the plant-insect interaction on the basis of such laboratory assays are as tenuous as attempting to speculate on the significance of certain glycosides, used as arrow-tip poisons, on the evolutionary trends of the elephant. In short, at present, the real function(s) of ecdysones in plants is totally uncertain.

Nonetheless, it is abidingly curious that plants and insects have enzyme systems that produce identical complex steroids, and it is possible that interactions of the insect endocrine system and plant steroids, other than the phytoecdysones, may occur. For example, in contrast to the inactivity of orally administered phytoecdysones, Robbins, et al. (1968) have shown that three synthetic steroids (Figure 6 [12, 13, 14]), analogous to α-ecdysones, are absorbed from the intestinal tract of insects and inhibit larval growth and development. These compounds have low molting hormone activity _per se_ and are suspected metabolic intermediates in α-ecdysone biosynthesis. The mechanisms by which these compounds interfere with normal growth and development are not clear. But, this finding opens the additional possibility that certain steroids that could inhibit or alter normal steroid (ecdysone) metabolism in the insect or interfere with mechanisms that regulate molting hormone titers in the insect body may exist in plants.

IV. Biochemical Basis of Insect Resistance in Crop Plants.

The impetus for study of most plant-insect interactions is based largely on implications for agriculture. From an agronomic viewpoint, it is desirable to understand those factors that influence the plant-insect interaction. In particular, knowledge of the interaction can be invaluable in the development of superior plant varieties with greater insect resistance. Historically, agronomic varieties with good insect resistance have been developed through the application of efficient plant breeding and entomological methods and in the absence of a detailed knowledge of the mechanisms of insect resistance. However, an increased understanding of the bases of resistance could contribute measurably to additional agronomic advancement.

Beck (1965) defines plant resistance as, "The collective heritable characteristics by which a plant species, race, clone, or individual may reduce the probability of successful utilization of that plant as a host by an insect species, race, biotype, or individual." The mechanisms of resistance have been classified in three categories: First, antibiosis--in which a plant is resistant by exerting an adverse effect on insect growth and development. Second, nonpreference--in which a plant displays resistance by exerting an adverse effect on insect behavior. These two categories are quite arbitrarily and vaguely delineated. Experimentally, it is often very difficult to attribute the manifestation of resistance solely to antibiosis or nonpreference. The third mechanism of resistance is tolerance--in which a plant is capable of supporting an insect population without loss of vigor or crop yield. Beck (1965) pointed out that, "tolerance cannot be considered resistance in the strict sense of its definition." Clearly, however, plant tolerance to insects is a desirable and important agronomic characteristic.

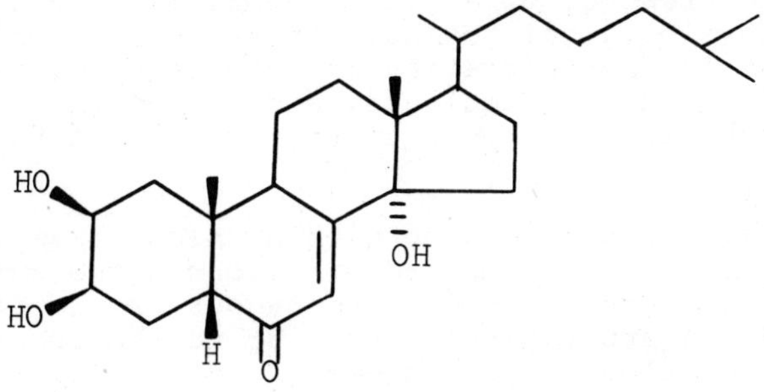

[12]

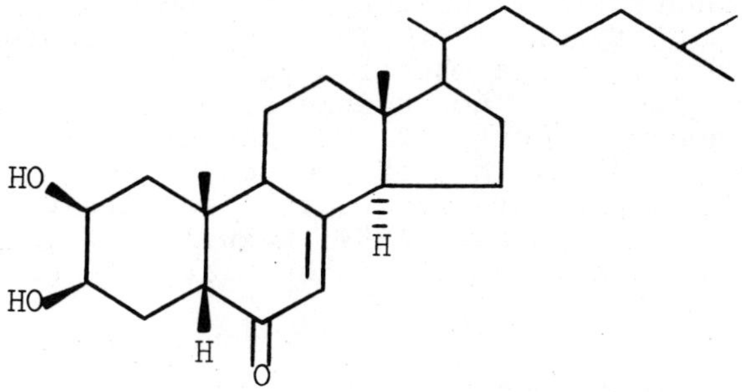

[13]

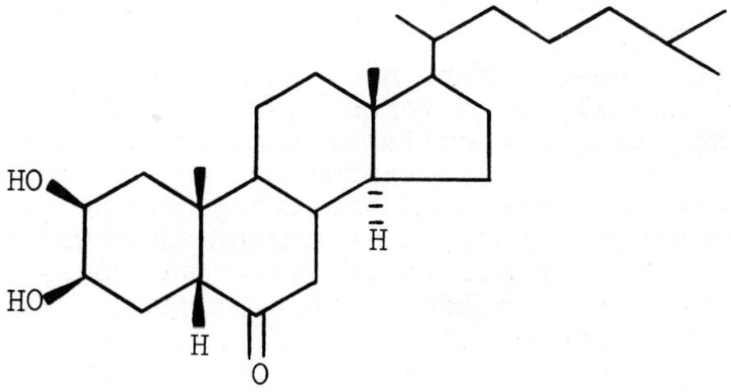

[14]

Figure 6.--Synthetic steroids analogous to α-ecdysones that inhibit larval growth and development in feeding tests.

The chemical basis of plant resistance can be viewed from two perspectives. A plant may be resistant owing to the presence of a plant chemical(s) such as feeding deterrents, repellents, or physiological toxins. On the other hand, a plant may be resistant because of the low concentration or absence of certain chemicals such as required nutrients, feeding stimulants, and attractants. The objective, then, is to define these chemical elements as they relate to the host-plant resistance phenomenon.

Recently Da Costa and Jones (1971) provided us with a unique example of how the presence or absence of a single set of compounds influences the suitability of cucumbers (*Cucumis sativus* (L.)) as hosts for two phytophagous arthropods. They found that tetracyclic triterpenoids (cucurbitacins), peculiar to the cucurbitaceae are feeding attractants for cucumber beetles (Crysolimadeae) and that cucurbitacin-free cucumber genotypes were beetle-resistant through nonpreference. Field tests showed, however, that cucurbitacin-free genotypes were susceptible to the two-spotted mite, *Tetranychus urticae* (Koch) while genotypes containing cucurbitacins were mite-resistant. Thus, the cucurbitacins function ambivalently as plant protectants on one hand, and feeding attractants, on the other.

Another unique example of the significance of a plant chemical in the plant-insect interaction has been demonstrated in cotton. All species of cotton produce a dimeric sesquiterpene called gossypol [15]. The compound occurs in the seed and green plant parts of cotton and is toxic when fed to poultry and swine. To increase the feed value of the cottonseed for nonruminants, plant breeders bred strains of glandless cotton with low gossypol content. The concentration of gossypol in these glandless strains was 1/3-1/4 that of the normal and morphologically distinct glanded cotton. Plant breeders were satisfied with their accomplishment until entomologists (Bottger, et al., 1964; Maxwell, et al., 1965) discovered that breeding gossypol out of cotton deprived the plant of much of its

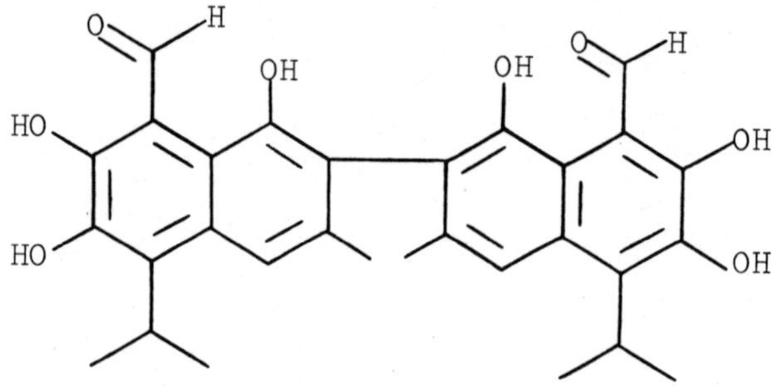

[15]

resistance (antibiosis) to known cotton insect pests and, at the same time, made the plant susceptible to insects not normally considered pests of cotton. This result adequately demonstrates the importance of collaboration of entomologists and plant breeders in any plant-breeding program. Ironically, after discovering the significance of gossypol as a plant protectant and contrary to the plant breeders' initial desire to decrease gossypol content in cotton, Bottger and Patana (1966) suggested that it would be desirable to increase the gossypol content of cotton, because such an increase could render cotton virtually immune to four of its most important lepidoptera insect pests.

Plant chemicals that provide the basis of resistance to insects need not be complex molecules such as the cucurbitacins or gossypol. Most recently, Burton, et al. (1972) found that the relatively simple compound, benzyl alcohol [16] was the chemical factor responsible for the resistance of Omugi barley to greenbug, *Schizaphus graminum* (Rondani). Significantly, benzyl alcohol is readily absorbed and transported systemically by greenbug-susceptible varieties of barley

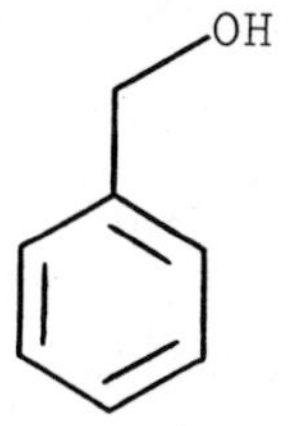

OH

[16]

and sorghum making them resistant to this aphid. Thus, it may be possible to directly apply this natural resistance factor against the greenbug without having to breed benzyl alcohol content into barley and sorghum. The potential realized here is an excellent example of the value of host-plant resistance studies.

A key to partial explanation of dent corn resistance to the 1st-brood European corn borer lies in the reaction scheme shown in Figure 7. The first

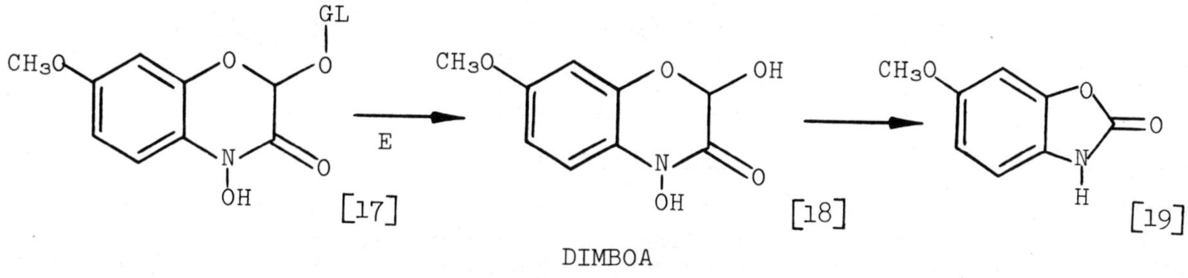

CH_3O GL O O N OH O [17] E CH_3O O OH N OH O [18] DIMBOA CH_3O O O N H [19]

Figure 7.--Formation of DIMBOA in corn from a glucoside precursor.

compound in the sequence is a glucoside [17] that exists in the uninjured corn plant (Wahlroos and Virtanen, 1959). When the plant tissues are damaged by feeding activity of the insect, the glucoside is hydrolyzed by a plant enzyme to an aglucone, 2,4-dihydroxy-7-methoxy-1,4-(2H)-benzoxazin-3-one (DIMBOA) [18], at the site of larval feeding. It has been determined through bioassays and quantitative studies of this compound in corn that DIMBOA, a cyclic hydroxamic acid, is a significant factor in the resistance of corn to the 1st-brood European corn borer (Klun, et al., 1967).

DIMBOA is chemically labile and slowly decomposes to MBOA]19] (Wahlroos and Virtanen, 1959), which is chemically stable, biologically inactive, and plays no role in plant resistance to the borer (Klun, et al., 1967). In addition to DIMBOA, two other minor cyclic hydroxamic acid constituents (Figure 8) have been identified from corn (Tipton, et al., 1967; Hofman and Hofmanova, 1969; Klun, et al., 1970). These benzoxazinones exist in the intact plant as glucosides and are hydrolyzed to labile aglucones when plant tissues are crushed.

The reactions in Figures 7 and 8 demonstrate an important element that must be considered in investigations of the chemical basis of plant resistance to insects. This element is the response of plants to tissue damage. Many plants contain precursors from which physiologically active labile substances are formed enzymatically when plant tissues are damaged (Virtanen, 1962). The chemical entities responsible for the resistance of plants to insects can easily escape detection if consideration and control of such enzymic and chemical degradation reactions are neglected.

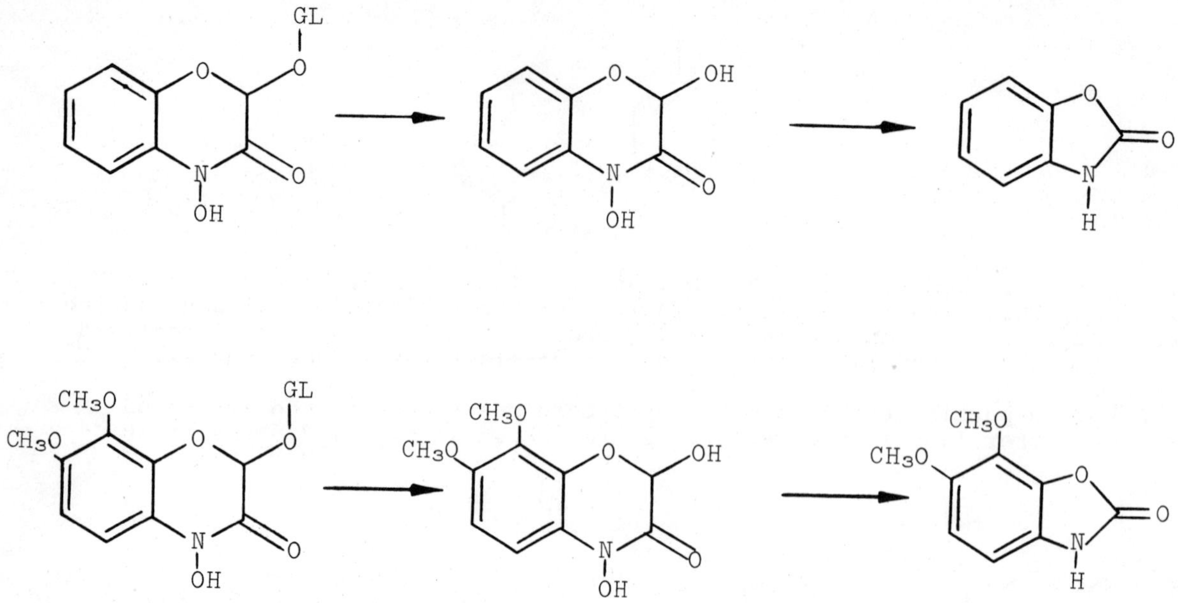

Figure 8.--Benzoxazinones other than DIMBOA identified from corn.

Before the identification of DIMBOA as a chemical factor in the resistance of corn to the European corn borer, extensive studies were conducted to determine the biology of the interaction of the insect with the plant. Such studies are a prerequisite and are indispensible to any chemical investigation of plant resistance. For example, resistance to the borer is expressed at three stages of plant development. At the seedling stage, most strains of corn are resistant to the borer. At the whorl stage, some varieties lose their resistance, and others retain it; this resistance is termed 1st-brood borer resistance (Guthrie, et al., 1960). At the whorl stage of plant development, larvae feed predominantly on leaves at the center of the whorl. Leaf lesions incurred by resistant varieties are characteristically few and small as compared with the extensive feeding damage sustained by the susceptible varieties (Guthrie, et al., 1960; Chiang and Hodson, 1953; Dicke and Penny, 1954). Thus, the target tissue for chemical study of 1st-brood borer resistance was established to be the whorl tissue of the plant. Resistance occurs again at the pollen-shedding stage of plant development. This resistance is termed 2nd-brood resistance. The chemical basis of this resistance is unknown, but Klun and Robinson (1969) established that this resistance is due to factors other than DIMBOA. Paradoxically, a variety borer resistant at the whorl stage of plant development is not necessarily resistant at the pollen-shedding stage.

These observations attest to the dynamic nature of the plant-insect interaction and accentuate the importance of the coordination of plant tissue sampling for chemical study with the biology of the insect. As obvious and logical as it may seem, many investigators frequently neglect to analyze the proper plant tissue at the right stage of plant growth as it relates to the synchrony of the plant-insect interaction and expression of plant resistance.

To establish a plant-borne chemical as a factor in plant resistance requires that biological assays of the plant-isolated chemicals be conducted and the physiological or behavioral effects of the plant isolate characterized. Meaningful interpretation of laboratory assay results, in terms of the plant-insect interaction, however, is difficult because the artificial environment of a laboratory assay can produce artifactual behavioral or physiological responses that in no way represent the natural responses of the insect to the plant chemical. Correspondingly, interpretation of bioassay results must be cautiously related to the insect-plant interaction.

Laboratory bioassays of DIMBOA (Klun, et al., 1967) show that the compound causes inhibition of development and mortality of borers feeding on artificial diets treated with the compound. Larvae fed tissues of the resistant varieties in the laboratory respond similarly (Reed, et al., 1972). Laboratory and field assays indicate that DIMBOA functions as a feeding deterrent and repellent (Robinson and Klun, 1972). Therefore, 1st-brood borer resistance is explained by the nonpreference mechanism. In addition, quantitative analyses of the tissues of corn for DIMBOA via analysis for the degradation product of DIMBOA, MBOA [19] show that the resistance of seedling corn is accounted for by the high concentration of the compound found in all varieties of seedlings (Klun and Robinson, 1969) (Figure 9). Varieties that maintain high concentrations of DIMBOA in the whorl tissue at later stages of plant development are borer resistant. Thus, the degree of borer resistance exhibited by a variety is directly related to the concentration of DIMBOA in its tissue (Figure 10).

It is possible to identify 1st-brood borer resistance genotypes on an individual-plant basis by chemical analysis of an aliquot sample of whorl tissue for DIMBOA. This development may have practical significance because such analysis could facilitate identification of borer-resistant genotypes in a corn breeding program. Currently, the visual rating system (Guthrie, et al., 1960; Guthrie, 1972) is being used by plant breeders to select for borer resistance. The efficiency of this system is highly variable, and application of the visual rating system requires extensive time, personnel and facilities. As a result, only a few corn breeders are able to carry out programs that include screening for resistance to the European corn borer.

On the other hand, the objective chemical analysis for DIMBOA could be of great value to the plant breeders because it would eliminate some of the variables in a selection program and increase the effectiveness and efficiency for identifying genotypes resistant to the 1st-brood borer. Therefore, we are currently evaluating the application of the DIMBOA chemical analysis to inbred-line breeding. Starting with the 200 entries of the F_2 (CI.31A X WF9) X (CI.31A X WF9), we have gone through 3 cycles of inbreeding and have applied the visual leaf-feeding rating method and chemical analyses in selection of resistant genotypes from this base population. After 3 cycles, 22 borer-resistant lines were selected by chemical analysis for DIMBOA, and 122 lines were selected by the visual leaf-feeding rating system. Significantly, all lines selected by the chemical method are among those lines selected by the visual rating system. Although it is somewhat premature to judge conclusively, these results may indicate the superior efficiency with which resistance can be concentrated via application of chemical analysis for DIMBOA as a selection tool. Alternately, some of the 122 borer-resistant lines selected by the visual rating system, which may not contain high concentrations of DIMBOA, could represent new sources of 1st-brood borer resistance.

The genetic nature of DIMBOA concentration and resistance to the 1st-brood borer in a diallel set of 11 inbreds has been studied (Klun, et al., 1970). Additive genetic effects are of primary importance for the expression of leaf-feeding resistance and high DIMBOA concentration.

The physiological function of the cyclic hydroxamic acids in the grasses is unknown. In addition to being involved in the resistance of corn to the European corn borer however, they also are of significance in the resistance of cereal grasses to fungi (BeMiller and Pappelis, 1965; ElNaghy and Linko, 1962; Molot and Anglade, 1968; Couture, et al., 1971) and resistance to 2-chloro-s-triazine herbicides (Tipton, et al., 1971; Roth and Knusli, 1961; Hamilton, 1964).

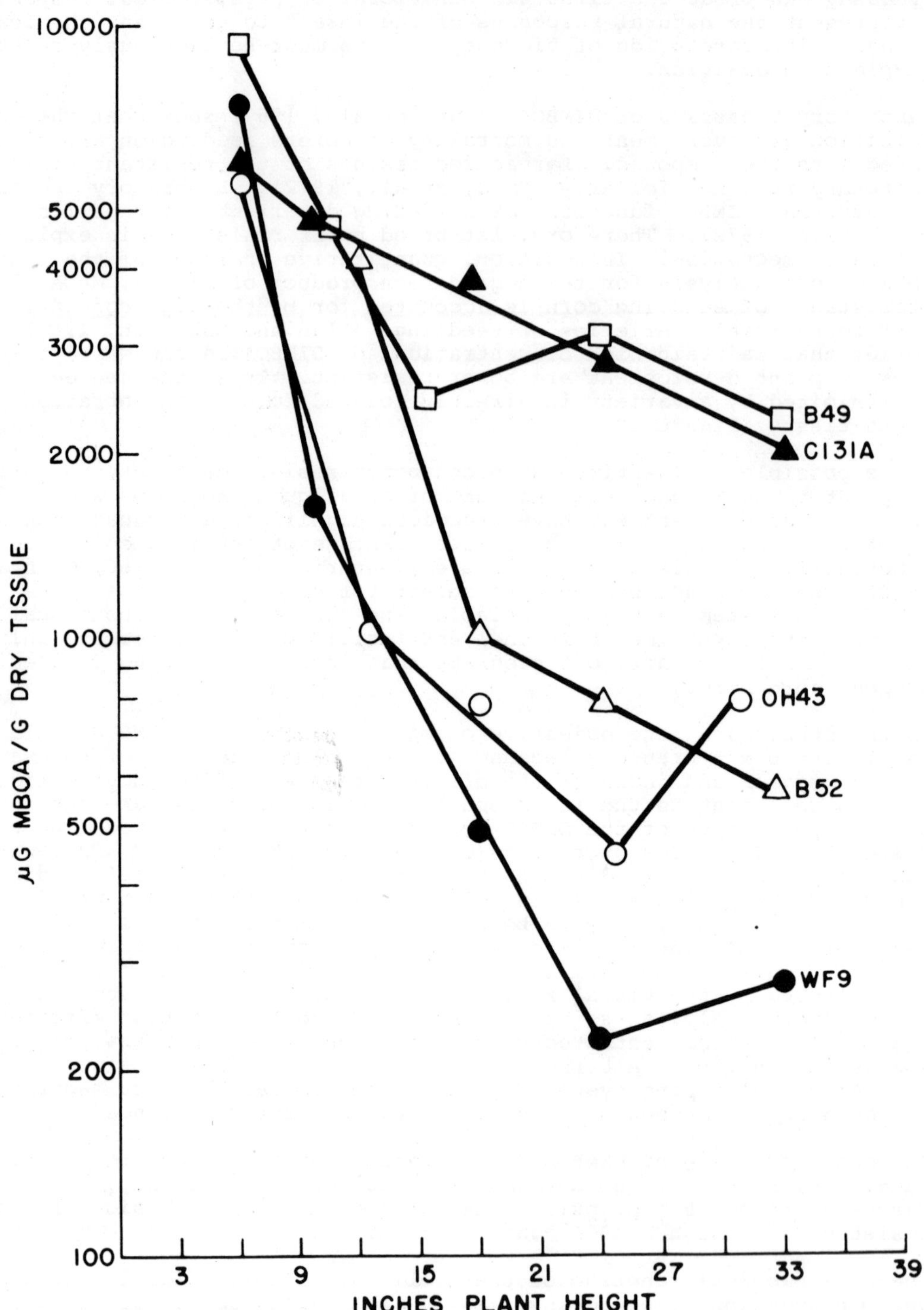

Figure 9.--Concentration changes of DIMBOA in corn leaves in five inbred lines from seedling stage to whorl stage of plant development.

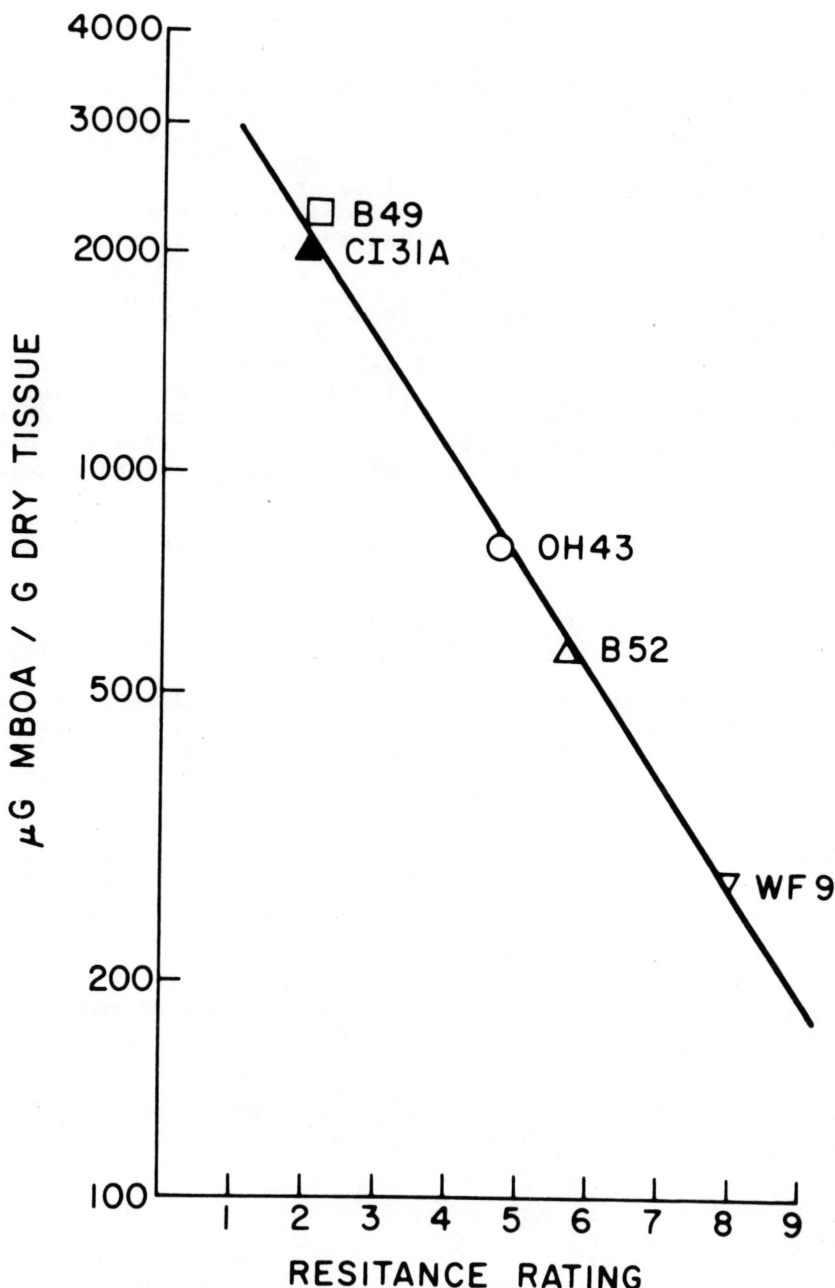

Figure 10.--Relation between DIMBOA concentration and plant resistance in five inbred lines of dent corn. Abscissa is resistance rating (1 indicates a corn borer resistant inbred, 5 intermediate, and 9 susceptible).

Thus, the cyclic hydroxamic acids ostensibly serve as plant protectants against a diversity of deleterious agents.

V. Conclusion

Study of plant resistance to insects at the chemical level is still in its infancy. From even a limited review of literature it is clear that most plant-resistance phenomena are chemically undefined. Where the identity of biologically active plant chemicals has been established (e.g., the phytoecdysones), their role in the plant-insect interaction is often undefined or casual. In short, there is need for research on the chemistry of plant chemicals and the role of identified biologically active plant chemicals in the plant-insect interaction.

References Cited

BeMiller, J. N. and A. J. Pappelis. 1965. 2,4-Dihydroxy-7-methoxy-1,4-benzoxazin-3-one glucoside in corn. I. Relation of water-soluble, 1-butanol-soluble glucoside fraction content of pith cores and stalk rot resistance. II. Isolation of 6-methoxy-2(3)-benzoxazolinone fraction as a measure of glucoside content and tissue differences of glucoside content. Phytopathology 55: 1237-1243.

Beck, S. D. 1965. Resistance of plants to insects. Ann. Rev. Entomol. 10: 207-232.

Bottger, G. T. and R. Patana. 1966. Growth, development, and survival of certain lepidoptera fed gossypol in the diet. J. Econ. Entomol. 59: 1166-1168.

Bottger, G. T., E. T. Sheehan and M. J. Lukefahr. 1964. Relation of gossypol content of cotton plants to insect resistance. J. Econ. Entomol. 57: 283-285.

Bowers, W. S. 1971a. Chemistry and biological activity of morphogenic agents. Mitt. Schweiz. Entomol. Gissel. 44: 115-130.

Bowers, W. S. 1971b. Insect hormones and their derivatives as insecticides. Bull. World Health Organ. 44: 381-389.

Bowers, W. S. 1971c. Juvenile hormones, pp. 307-332. In Naturally Occurring Insecticides. Marcel Dekker, Inc., New York.

Bowers, W. S., M. J. Thompson and E. C. Uebel. 1965. Juvenile and gonadotropic hormone activity of 10,11-epoxyfarnesenicacid methyl ester. Life Sci. 4: 2323-2331.

Bowers, W. S., H. M. Fales, M. J. Thompson and E. C. Uebel. 1966. Juvenile hormone: identification of an active compound from balsam fir. Science 154: 1020-1022.

Burton, R. L., K. J. Starks, P. S. Juneja and R. K. Gholson. 1972. Agr. Res. (Washington, D.C.) 20: 5.

Butenandt, A. and P. Karlson. 1954. Uber die isolierung eines metamorphose-hormons der insectin in kristallisierter form. Z. Naturforsch. Teil B 96: 389-391.

Cerny, V., L. Dolejs, L. Labler and F. Sorm. 1967. Dehydrojuvabione--a new compound with juvenile hormone activity from balsam fir. Tetrahedron Ltrs. 1053-1057.

Chiang, H. C. and A. C. Hodson. 1953. Leaf injury caused by first-generation corn borer in field corn. J. Econ. Entomol. 46: 68-73.

Couture, R. M., D. G. Routley and G. M. Dunn. 1971. Role of cyclic hydroxamic acids in monogenic resistance of maize to Helminthosporium turcicum. Physiol. Plant Pathol. 1: 515-521.

Da Costa, C.P. and C. M. Jones. 1971. Cucumber beetle resistance and mite susceptibility controlled by the bitter gene in Cucumis sativus L. Science 172: 1145-1146.

Dethier, V. G. 1954. Evolution of feeding preferences in phytophagous insects. Evolution 8: 33-54.

Dethier, V. G. 1970. Chemical interactions between plants and insects, pp. 83-102. In Chemical Ecology, Academic Press, New York.

Dicke, F. F. and L. H. Penny. 1954. Breeding for resistance to European corn borer. Proc. Ann. Hybrid Corn Ind.-Res. Conf. 9: 44-53.

Ehrlich, P. R. and P. H. Raven. 1965. Butterflies and plants: a study in coevolution. Evolution 18: 586-608.

ElNaghy, M. A. and P. Linko. 1962. The role of 4-0-glucosyl-2,4-dihydroxy-7-methoxy-1,4-benzoxazin-3-one in resistance of wheat to stem rust. Physiol. Plant. 15: 764-771.

Fraenkel, G. S. 1959. The raison d'etre of secondary plant substances. Science 129: 1466-1470.

Guthrie, W. D. 1972. Techniques, accomplishments and future potential of breeding for resistance to European corn borers in corn. Proc. Inst. Biol. Contr. Plant Insect and Diseases. (in press).

Guthrie, W. D., F. F. Dicke and C. R. Neiswander. 1960. Leaf and sheath feeding resistance to the European corn borer in eight inbred lines of dent corn. Ohio Agr. Exp. Sta. Res. Bull. 860.

Hamilton, R. H. 1964. A corn mutant deficient in 2,4-dihydroxy-7-methoxy-1,4-benzoxazine-3-one with an altered tolerance of atrazine. Weeds 12: 27-30.

Hedin, P. A., A. C. Thompson and J. P. Minyard. 1968. Constituents of the cotton bud. Formulation of a boll weevil feeding stimulant mixture. J. Agr. Food Chem. 16: 505-513.

Hofman, J. and O. Hofmanova. 1969. 1,4-Benzoxazine derivatives in plants. Sephadex fractionation and identification of a new glucoside. European J. Biochem. 8: 109-112.

Horn, D. H. S. 1971. The ecdysones, pp. 333-459. In Naturally Occurring Insecticides, Marcel Dekker, Inc., New York.

Horn, D. H. S., E. J. Middleton and J. A. Wunderlich. 1966. Identity of the moulting hormones of insects and crustaceans. Chem. Commun. 339-341.

Kaplanis, J. N., M. J. Thompson, W. E. Robbins and B. M. Bryce. 1967. Insect hormones: Alpha ecdysone and 20-hydroxyecdysone in bracken fern. Science 157: 1436-1438.

Kaplanis, J. N., M. J. Thompson, R. T. Yamamoto, W. E. Robbins and S. J. Louloudes. 1966. Ecdysones from the pupa of the tobacco hornworm, Manduca sexta (Johannson). Steroids 8: 605-623.

Klun, J. A. and J. F. Robinson. 1969. Concentration of two 1,4-benzoxazinones in dent corn at various stages of development of the plant and its relation to resistance of the host plant to the European corn borer. J. Econ. Entomol. 62: 214-220.

Klun, J. A., C. L. Tipton, and T. A. Brindley. 1967. 2,4-Dihydroxy-7-methoxy-1,4-benzoxazin-3-one (DIMBOA), an active agent in the resistance of maize to the European corn borer. J. Econ. Entomol. 60: 1529-1533.

Klun, J. A., W. D. Guthrie, A. R. Hallauer and W. A. Russell. 1970. Genetic nature of 2,4-dihydroxy-7-methoxy-1,4(2H)-benzoxazin-3-one concentration in a diallel set of eleven maize inbreds. Crop Sci. 10: 87-90.

Klun, J. A., C. L. Tipton, J. F. Robinson, D. L. Ostrem and M. Beroza. 1970. Isolation and identification of 6,7-dimethoxy-2-benzoxazolinone from dried tissue of _Zea mays_ (L.) and evidence of its cyclic hydroxamic acid precursor. J. Agr. Food Chem. 18: 663-665.

Maxwell, F. G., H. N. LaFever and J. N. Jenkins. 1965. Blister beetles on glandless cotton. J. Econ. Entomol. 58: 792-793.

Metzler, M., D. Meyer, K. H. Dahm, H. Roller and J. B. Siddall. 1972. Biosynthesis of juvenile hormone from 10-epoxy-7-ethyl-3,11-dimethyl-2,6-tridecadienoic acid in the adult _Cecropia_ moth. Z. Naturforsch. 27b: 321-322.

Meyer, A. S., H. A. Schneiderman, E. Hanzmann and J. H. Ko. 1968. The two juvenile hormones from the _Cecropia_ silk moth. Proc. Natl. Acad. Sci. U.S. 60: 853-860.

Malot, P. M. and P. Anglade. 1968. Resistance commune des lignees de mais a d'helminthospoiose (_Helminothosporium turcicum_ Pass.) et a la pyrale (_Ostrinia nubilalis_ Hbn.) en relation avec la presence d'une substance identifiable a la 6-methoxy-2(3)-benzoxazolinone. Ann. Epiphyt. (Paris) 19: 75-95.

Nakanishi, K., H. Moriyama, T. Okauchi, S. Fujioka and M. Koreeda. 1972. Biosynthesis of α and β-ecdysones from cholesterol outside the prothoracic gland in _Bombyx mori_. Science 176: 51-52.

Painter, R. H. 1951. Insect Resistance in Crop Plants. Macmillan, New York. 520 pp.

Pan, M. L. and G. R. Wyatt. 1971. Juvenile hormone induces vitellogenin synthesis in the monarch butterfly. Science 174: 503-505.

Pfiffner, A. 1971. Juvenile hormones, pp. 95-135. _In_ Aspects of Terpenoid Chemistry and Biochemistry, Academic Press, New York.

Reed, G. L., T. A. Brindley and W. B. Showers. 1972. Influence of resistant corn leaf tissue at two stages of plant development on the biology of the European corn borer. Ann. Entomol. Soc. Amer. 65: 658-662.

Rees, H. H. 1971. Ecdysones, pp. 181-222. _In_ Aspects of Terpenoid Chemistry and Biochemistry, Academic Press, New York.

Riddiford, L. M. 1970. Prevention of metamorphosis by exposure of insect eggs to juvenile hormone analogs. Science 167: 287-288.

Robbins, W. E., J. N. Kaplanis, J. A. Svoboda and M. J. Thompson. 1971. Steroid metabolism in insects. Ann. Rev. Entomol. 16: 53-72.

Robbins, W. E., J. N. Kaplanis, M. J. Thompson, T. J. Shortino, C. F. Cohen, and S. C. Joyner. 1968. Ecdysones and analogs. Effects on development and reproduction of insects. Science 161: 1158-1159.

Robinson, J. F. and J. A. Klun. 1972. Unpublished data.

Roller, H., K. H. Dahm, C. C. Sweely and B. M. Trost. 1967. The structure of the juvenile hormone. Angewandte Chemie, Int. Ed. Engl. 6: 179-180.

Roth, W. and E. Knusli. 1961. Beitrag zur kennis der resistenzphanomene einzlner pflanzen gegenuber dem phytotoxischen wirkstoff simazin. Experentia 17: 313-313.

Schneiderman, H. A. and L. I. Gilbert. 1964. Control of growth and development in insects. Science 143: 325-333.

Siddall, J. B. 1970. Chemical aspects of hormonal interactions, pp. 281-306. In Chemical Ecology, Academic Press, New York.

Siddall, J. B. 1971. Insect hormone mimics near market. Chem. Eng. News. Nov. 29. 33-34.

Siddall, J. B. 1972. Personal communication. Zoecon Corp., Palo Alto, California 94304.

Slama, K. and C. M. Williams. 1966. The juvenile hormone. V. The sensitivity of the bug, Pyrrhocoris apterus, to a hormonally active factor in American paper pulp. Biol. Bull. 130: 235-246.

Stam, M. D. 1959. An. Real Soc. Esp. Fis. Quim. (Madrid) B55: 171 (From Pfiffner 1971).

Thompson, M. J., J. N. Kaplanis, W. E. Robbins and R. T. Yamomoto. 1967. 20,26-Dihydroxy-ecdysone, a new steroid with moulting hormone activity from the tobacco hornworm, Manduca sexta (Johannson). Chem. Commun. 650.

Thorsteinson, A. J. 1960. Host selection in phytophagous insects. Ann. Rev. Entomol. 5: 193-218.

Tipton, C. L., R. R. Husted and F. H. C. Tsao. 1971. Catalysis of simazine hydrolysis by 2,4-dihydroxy-7-methoxy-1,4-benzoxazin-3-one. J. Agr. Food Chem. 19: 484-486.

Tipton, C. L., J. A. Klun, R. D. Husted and M. D. Pierson. 1967. Cyclic hydroxamic acids and related compounds from maize. Isolation and characterization. Biochemistry 6: 2866-2870.

Virtanen, A. I. 1962. Some organic sulfur compounds in vegetables and fodder plants and their significance in human nutrition. Agew. Chem. Intern. Ed. English 1: 299-305.

Wahlroos, O. and A. I. Virtanen. 1959. The precursors of 6MBOA in maize and wheat plants: Their isolation and some of their properties. Acta Chem. Scand. 13: 1906-1908.

Whittaker, R. H. 1970. The biochemical ecology of higher plants, pp. 43-70. In Chemical Ecology, Academic Press, New York.

Whittaker, R. H. and P. P. Feeny. 1971. Allelochemics: Chemical interactions between species. Science 171: 757-770.

Williams, C. M. 1970. Hormonal interactions between plants and insects, pp. 130-132. In Chemical Ecology, Academic Press, New York.

BIOCHEMICAL BASIS OF RESISTANCE OF PLANTS TO INSECTS ANTIBIOTIC FACTORS OTHER THAN PHYTOHORMONES

Ted N. Shaver
Biochemist, ERD, ARS, USDA
Brownsville, Texas

The insecticidal activity of substances contained in plant tissue has been recognized for many years, some of the better known examples of which are rotenone from *Derris* roots and pyrethroids from *Chrysanthemum* flowers. In recent years, there have been increasing instances of investigations designed to study the role that toxic plant components might contribute in providing resistance of a given plant to insect infestations.

The term antibiosis as one of the 3 mechanisms for insect resistance in crop plants was proposed by Painter (1941) to designate those cases in which a host plant or species demonstrated an adverse effect on the life history of the insect. This adverse effect on the life of the insect can take the form of death of larvae, nymphs, or eggs; small size; reduced fecundity, fertility, and longevity and abnormal length of life. These physiological effects can be caused either by the presence of a toxin or metabolic inhibitor in the plant or by insufficient quantities of a specific nutrient required by the insect for growth, reproduction or metamorphosis.

Although the study of the biochemical basis for antibiosis is mainly a laboratory problem, a considerable amount of field work must be done prior to attempting laboratory studies. The response of a particular insect to its host plant is dependent upon the biology of both. The plant and the insect are dynamic systems subject to changes from external stresses; the life history of each should be studied, and the biological characteristics of the resistance under investigation should be thoroughly understood. Before the problem is brought into the laboratory, techniques must be developed for maintaining the insect in the absence of the host. This necessitates the development of a culturing medium and rearing techniques for the insect.

A bioassay technique must be developed to follow the fractionation of the resistance factor in the case of antibiosis due to a toxin or metabolic inhibitor, and the bioassay procedure should be based on one of the facets of insect growth and development affected by the resistance factor. Most commonly measured parameters are effects on mortality, growth, and reproduction. However, there have been cases where fractionation of a resistance factor was monitored using an organism other than the target insect. For example, Beck and Stauffer (1957) used the mold *Penicillium chrysogenum* as an assay organism to determine activity during the fractionation of the resistance factor contained in corn and active against European corn borer.

Many bioassay techniques have been developed to detect resistance in laboratory or small-cage tests in which insects are maintained directly on the plant. These tests are extremely valuable and essential in screening programs where large numbers of plants must be tested to locate sources of resistance or in a breeding program to detect resistant progeny, and much important information concerning the relationship of the plant and insect can be obtained by careful planning and observation. However, the insect must be maintained separately from the plant on a nutritionally adequate substrate before it can be determined whether the resistance is nutritional in nature or if it involves a toxic compound.

A bioassay to detect antibiosis was developed by Glover and Stanford (1966) for testing alfalfa plants for resistance to the pea aphid, _Acyrthosiphon pisum_ (Harris). They used a quantitative score of the surviving aphids per adult day as an index of plant resistance when considered in reference to a susceptible check variety or clone.

Larval growth was used by Schillinger (1966) as an index for screening _Triticum_ spp. for resistance to the cereal leaf beetle, _Oulema melanopus_ (L.), by caging late 1st- or early 2nd-instar larvae on seedling plants. Notes were taken daily on the number of larvae alive and dead and their location within the test chamber; at the end of 75 hr, larvae were weighed to determine percentage weight gain. Todd and Canerday (1968) used larval development as an index for antibiosis to the cowpea curculio, _Chalcodermus aeneus_ Boheman, in southern peas.

An antibiotic type of resistance to corn earworm, _Heliothis zea_ (Boddie), was reported by Walter (1957) and Wann and Hills (1966). Antibiosis was later demonstrated in sweet corn hybrid lines and hybrids by Chambliss and Wann (1971) by periodic growth measurements of corn earworm larvae feeding on resistant and susceptible types. One 1st-instar larva was placed on silks of each ear 1-3 days after silks emerged from the husks. Resistant lines were distinguished from susceptible ones by a significant increase in larval mortality. Resistant lines also retarded growth, decreased depth to which larvae penetrated the ear, and delayed pupation of the insect. They discounted the importance of the observation that lines showing the greatest inhibition also had relatively tight husks and long silk channels beyond the ear tip since their technique of opening the husks to follow larval growth minimized these physical barriers.

Todd, et al. (1971) reared biotype B greenbugs, _Schizaphis graminum_ (Rondani) on chemically defined diets containing commercially available phenolic and flavonoid compounds and found that compounds having ortho-hydroxyl groups including catechol, tannic acid, quercetin, chlorogenic acid, and protocatechuic acid were detrimental to greenbug growth; the number and survival of progeny were drastically reduced when these compounds were included in the diet. Cis-caffeic acid caused drastic reduction growth with no reproduction from greenbugs feeding on it, whereas trans-caffeic acid caused some reduction in weight gain with no effect on reproduction. Also, another group of compounds were benzoic or cinnamic acid derivatives having a para-hydroxyl group. Although these compounds were less toxic, they effectively reduced survival of the progeny. Less than 20% of the greenbugs survived feeding on diets containing vanillic, sinopic, syringic, gentisic, or ferulic acids. They concluded that since many of these compounds are constituents of barley leaves, it is likely that at least part of the resistance to greenbugs of some barley varieties is attributable to the presence of these phenolic and flavonoid substances in quantities sufficient to retard the insect's growth and reproduction.

The inadequacies of young corn plants as larval hosts for the European corn borer was reported by Beck (1956) in that all genetic lines are resistant to larvae as young plants. Results of various workers indicated that newly hatched borer larvae placed in whorls of small corn plants undergo a high mortality, with only about 20% surviving the 1st 72 hr. Although some genetic lines of corn become more susceptible as they approach tasseling, other lines maintain juvenile resistance to a certain degree as the plant matures. Beck and Stauffer (1957) demonstrated that aqueous extracts of whorl tissue was toxic and caused death or poor growth of larvae on synthetic diets. Fractionation of the extracts yielded ether-soluble substances, termed Resistance Factor A (RFA), and ether-insoluble substances, termed Resistance Factor B(RFB). Later, Beck (1957) demonstrated the presence of 2 resistance factors in the ether-soluble fraction, necessitating the use of Resistance Factor C (RFC) to designate the 2nd ether-soluble fraction. Resistance Factor A was identified as 6-methoxybenzoxazolinone (Loomis, et al., 1957; Smissman, et al., 1957) and RFC as 2,4-dihydroxy-7-methoxy-1,4-benzoxozine 3-one.

Since most bioassay techniques are designed to obtain qualitative or quantitative measurement of the effect of a particular host plant on an insect, proper plant sampling techniques are an essential part of the overall program. This involves critical evaluation of field and laboratory observations on the feeding habits of the insect and the site of attack of the insect at various stages of development. Thus, the plant parts normally used by the insect for food can be evaluated under bioassay conditions. The importance of a knowledge of the feeding habits of the insect under investigation was pointed out by Beck (1957) who reported that although the leaves of a European corn borer-resistant corn variety contained large quantities of resistance factor, the corn was at the growth stage in which borers do not feed on the leaves. Also very young tassel buds of some corn inbreds contained high resistance factor concentrations; however, at a growth stage when borer larvae feed on the tassel, they contained little resistance factor.

Thurston and Webster (1962) reported that resistance in *Nicotiana* species to the green peach aphid, *Myzus persicae* (Sulzer) appears to result from the production of a toxic material produced by the aerial parts of the plants, since a material exuded from the leaf hairs was demonstrated as toxic to several species of aphids. Later, Thurston, et al. (1966) found that these secretions contained nicotine as the major alkaloidal constituent. Alkaloids were found in the trichome secretions of 7 *Nicotiana* species tested. Although nicotine content of some susceptible varieties is high in a total plant analysis, the aphid feeds in the phloem while nicotine is transported in the xylem (Guthrie, et al., 1962).

If possible, plant samples should be taken at about the same time each day, since the concentration of many types of compounds are not comparable under dark and light conditions. For example, Loomis, et al. (1957) found about 4 times as much resistance factor for the European corn borer, *Ostrinia nubilalis* Hubner, in corn seedlings exposed to light for 12 hours preceding analysis than in those held in the dark.

A research program was initiated in the early 1960's to find sources of resistance in cotton to the bollworm, *H. zea* (Boddie), and tobacco budworm, *H. virescens* (F.), and among the first characters studied were glabrous, nectariless, and high gossypol contents of the cotton squares. In a sense, the nectariless character is an example of antibiosis due to inadequate food intake by the adult. Commercial upland cotton contains both floral and extrafloral nectaries. One extrafloral nectary is located on the midrib of each true leaf and six are located on the flower or boll. The floral nectary is located at the base on the inner side of the calyx, but the importance of these as feeding sites for lepidopterous insects have been discounted since the flowers are closed during their feeding period. Lukefahr and Martin (1964) demonstrated that both longevity and fecundity were reduced by 50% in moths receiving only water compared with moths fed a sucrose solution. In cage tests by Lukefahr and Rhyne (1960), populations of cotton leafworm, *Alabama argillacea* (Hubner), and cabbage looper, *Trichoplusia ni* (Hubner) were 7 to 10 times as high on a commercial cotton containing nectaries than on an experimental strain containing no extrafloral nectaries.

The cotton plant contains numerous subepidermal glands in the above-ground parts. These glands, which are commonly referred to as pigment glands, contain gossypol, a polyphenolic compound toxic to larvae of bollworm and tobacco budworm (Bottger, et al., 1964; Lukefahr and Martin, 1966; Shaver and Lukefahr, 1969). On the basis of laboratory data, it was estimated that the amount of gossypol in cotton squares would have to exceed 1.2% to effectively inhibit larval development in *Heliothis* spp. Lukefahr and Houghtaling (1969) tested an experimental cotton containing 1.7% gossypol in laboratory and cage tests and demonstrated the feasibility of using gossypol as a source of insect resistance in cotton.

In a laboratory test to determine whether the amount of gossypol in the experimental high gossypol line, XG-15, was sufficient to be the major factor in growth suppression of bollworm larvae, Shaver, et al. (1970) compared the growth of larvae reared on XG-15 square powder diets with that on diets prepared from a glandless strain of cotton to which gossypol had been added to provide concentrations of gossypol of 0.06, 0.40, 0.74, and 1.76%. In addition, the M-8 glandless and XG-15 square powders were mixed to give gossypol concentrations of 0.14, 0.23, 0.40, 0.73, and 1.70%. Square powders (10g) were blended in 1.5% agar solution (80 ml) containing aureomycin and methyl parasept, and test diets were infested with 3-day-old bollworm larvae. Larvae were weighed on the 7th and 14th days, and mortality was recorded on the 14th day. Bollworm larvae demonstrated no difference in larval weight for survival whether the gossypol was added as gossypol acetic acid or as XG-15 square powder (Table 1). Therefore, it was concluded that the effect of XG-15 on growth and survival in bollworm larvae can be attributed to its gossypol content.

Shaver and Garcia (1973) reported the gossypol content of the component parts of cotton flower buds, including petals, sepals, stamen, stigma + style, and ovary; the petals usually contained at least twice the concentration of gossypol as any of the other parts (Table 2). This large variation in gossypol in individual parts of cotton buds could lead to problems in relating the biological effects of a cotton strain on insects to gossypol content of whole buds. This suggests that a thorough study of the feeding habits of *Heliothis* larvae of various ages be made to determine if they show a preference for a particular part of the cotton flower bud.

Shaver and Parrott (1970) showed that larval age had an effect on the toxicity of gossypol to the bollworm, tobacco budworm, and pink bollworm. The amount of gossypol required to affect larval weight and survival, length of time in larval stage, and adult emergence increased with increasing larval age when exposed to gossypol. Food utilization and consumption were studied by Oliver, et al. (1970) who found that all ages of larvae tested gained less weight on diets made from glanded cotton compared with those fed a diet prepared from glandless cotton due to less efficient food conversion and a reduction in feeding.

Numerous flavonoid compounds are contained in cotton in vegetative parts and in the seed (Parks, 1965). Lukefahr and Martin (1966) and Shaver and Lukefahr (1969) assayed several of these compounds by incorporating graded levels of each into the insect diet and observing the growth and survival of *Heliothis* larvae placed on the diets. Each of the cotton flavonoids tested exhibited antibiotic activity against larvae, but no work has been done on concentration of individual flavonoids in cotton buds. Therefore, it is difficult to assess the importance that this type of compound might play in a host plant resistance project.

Although the 3 resistant factors in cotton strains mentioned previously have proven effective in tests thus far, other sources of resistance with different mechanisms of action would be desirable to supplement their action. A project was initiated in 1967 to search for new sources of antibiosis for tobacco budworm and bollworm. Wild types of *Gossypium hirsutum* L. collected largely from dooryard plantings in southern Mexico, Guatemala, and El Salvador were selected for screening since they contain a diversity of germ plasm previously untested in host plant resistance projects.

A bioassay technique capable of detecting plants containing growth inhibitors or toxic substances was required. Initially, cotton lines were bioassayed by using growth rates of Heliothis larvae when they were fed a reconstituted square powder diet. However, the amount of plant material available for testing was limited since most of the wild types of cotton have a short-day photoperiodic response and are relatively unproductive, so the number of larvae used in each test was low. This, coupled with poor growth rate and large variations in individual weights of larvae on each of the entries, made interpretation of the results difficult. Also, most of the nutrition in squares is furnished the larvae by the anthers, especially in the early instars, and there are large differences in the number of anthers in these lines. Thus, we needed a bioassay that would eliminate nutrients in the different cotton lines and would measure only the antibiotic chemicals that might be present.

The bioassay used presently at our laboratory involves measuring larval response on diets containing extracts of square powder and response of larvae on detached fresh squares. In the first phase of the bioassay, lyophilized square powders are extracted with ether and then acetone, these extracts are combined and coated onto alphacel by evaporation under reduced pressure in a rotary evaporator, and the alphacel preparations are incorporated into a casein-wheat germ diet (extract of 1 g of squares in 10 ml medium). Test diets are infested with 3-day-old tobacco budworm larvae, and the larvae are weighed after feeding for 6 days. Larval weights are compared with weights of larvae feeding on diets containing extracts of a susceptible check.

In the 2nd phase of the bioassay, 3-day-old tobacco budworm larvae are caged in 2-oz cups with a freshly picked square, and the squares are replaced daily with fresh ones. Larvae are weighed after feeding for 5 and 10 days, and weights are compared with those larvae feeding on a susceptible check. Normally, lines chosen for the fresh square assay are those with less than 1% gossypol that caused a 50% or more reduction in larval weight compared with the check in the ether-acetone extract bioassay. The fresh square bioassay must be performed near growing plants so that a supply of fresh squares will be available; wild races of cotton must be planted in the greenhouse since the growing season in the Cotton Belt does not coincide with the photoperiod required for flowering. This, coupled with the time required for fresh square assays, preclude the use of fresh squares as the primary screening tool in our program. However, we feel that the fresh-square assay is a valuable means of further testing lines selected for resistance on the basis of the assay with extracts of square powder since it is more closely related to the natural feeding habits of the tobacco budworm, and the larvae can be selective in their feeding on component parts of the square as they are in nature.

Of the first 256 selections of the race stock collections tested, 14 have met the requirements of producing larvae at least 50% smaller than those on the standard and those containing 1% or less gossypol in the squares. Currently, we are conducting studies toward isolation and identification of the resistance factors and transferring these characters into an agronomically acceptable plant. In attempts to isolate the active component of one of these lines, designated as Texas 144, we have been able to obtain several active fractions by column chromatography and liquid-liquid extractions, but we have been unable to obtain an active fraction containing no gossypol.

The gossypol content of the active fractions was not sufficient to account for the effect on larval growth based on the effect of gossypol added to the diet as gossypol:acetic acid at equal concentrations. Also, the biological effect of an amount of gossypol added to the diet as gossypol-acetic acid could be quite different from the biological effect of the same amount of gossypol added as an extract of square tissue. The colorimetric methods used for estimation of gossypol concentration are not specific for gossypol. The reduced larval weights of insects feeding on diets containing extracts of Texas 144 or those feeding on fresh squares of this line could be due to additive effects of gossypol and an unknown or "X" factor. Therefore, the difference in response of larvae to Texas 144 and M-8 glanded could be due to a greater amount of "X" factor in Texas 144, a greater amount of gossypol in Texas 144, or a combination of the two.

To determine the effect on larval weight of gossypol added as gossypol: acetic acid or as extracts from square tissue, buds of M-8 glandless (0.1% gossypol) and M-8 glanded (0.5%) were lyophilized and milled through a 40-mesh screen. Fourteen 5-g samples of M-8 glandless buds and 10 samples of M-8 glanded buds were supplemented with gossypol:acetic acid so that the gossypol contents of the 5-g samples would increase by increments of 0.1% from the original amount to 1.4% gossypol. These samples were extracted with ether in a continuous extractor, and each extract was coated onto 1.75 g of alphacel in a rotary evaporator. A series of gossypol samples were coated directly onto alphacel to correspond to the 0.1-1.4% gossypol levels in extracts. Also, a 5-g sample of Texas 144 (0.9% gossypol) and 247-1 (1.7%) were each extracted with ether and tested in the same manner. Test diets were infested with 3-day-old tobacco budworm larvae, and the larvae were weighed after feeding on the diet for 8 days. The results indicate that the response of tobacco budworm larvae to gossypol is about the same whether added to the diet as gossypol:acetic acid or as bud extracts containing gossypol. Also, the 8-day larval weight from extracts of 247-1 falls in line with extrapolated values obtained from gossypol and extracts of M-8 glanded and glandless plus gossypol. However, the extracts of Texas 144 caused a greater reduction in larval weight than could be expected from its gossypol content, although the 0.9% gossypol was sufficient to cause some reduction in larval weight. The curve of larval weight (mg) vs. percentage gossypol shows a sharp drop in larval weight with increasing gossypol from about 0.8% to 1.2%.

One of the most important considerations in the study of the biochemical basis for resistance is that the study be based on the biology of both the plant and the insect. The presence of a chemical in a resistant plant and its absence in a susceptible one is not evidence that the chemical is the cause of resistance unless evidence is presented to demonstrate that the chemical acts as a toxin to the insect and is present in the plant at the site of insect feeding. Although it is possible to develop an insect resistant plant without knowing the exact mechanism of resistance, a knowledge of the chemical basis of resistance would be useful in many cases. This is especially true when the insect is difficult to maintain under laboratory or cage conditions and a chemical analysis is necessary for selection of breeding material.

Table 1.--Development of bollworm larvae on square powder diet consisting of various combinations of glandless and heavily glanded cottons and gossypol.

Gossypol (mg/ml) in diet by adding		Larval wt. (mg) on 14th day		% survival after 14 days	
XG-15	Gossypol	XG-15	Gossypol	XG-15	Gossypol
0.14	0.06	386	376	100	100
.23	--	186	--	86	--
.40	.40	162	186	63	76
.73	.74	88	78	33	16
1.70	1.76	--	--	0	0

Table 2.--Gossypol content of component parts of cotton buds.

Cotton line	% gossypol in				
	Petals	Sepals	Stamen	Stigma + style	Ovaries
M-8 glanded	1.03	0.42	0.47	0.54	0.40
Pima glanded	1.55	.71	.82	.54	.12
XG-15	2.64	.91	.94	1.01	1.35

Beck, S. D. 1956. The European corn borer, Pyrausta nubilalis (Hubn.), and its principal host plant. II. The influence of nutritional factors on larval establishment and development on corn plant. Ann. Entomol. Soc. Amer. 49: 582-588.

Beck, S. D. 1957. The European corn borer, Pyrausta nubilalis (Hubn.), and its principal host plant. VI. Host plant resistance to larval establishment. J. Insect Physiol. 1: 158-177.

Beck, S. D. and J. E. Stauffer. 1957. The European corn borer, Pyrausta nubilalis (Hubn.), and its principal host plant. III. Toxic factors influencing larval establishment. Ann. Entomol. Soc. Amer. 50: 166-170.

Bottger, G. T., E. T. Sheehan and M. J. Lukefahr. 1964. Relation of gossypol content of cotton plants to insect resistance. J. Econ. Entomol. 57: 283-285.

Chambliss, O. L. and E. V. Wann. 1971. Antibiosis in earworm resistant sweet corn. J. Amer. Soc. Hort. Sci. 96: 273-277.

Glover, D. V. and E. H. Stanford. 1966. Tetrasomic inheritance of resistance in alfalfa to the pea aphid. Crop Sci. 6: 161-165.

Guthrie, F. E., W. V. Campbell and R. L. Baron. 1962. Feeding sites of the green peach aphid with respect to its adaptation to tobacco. Ann. Entomol. Soc. Amer. 55: 42-46.

Jenkins, J. N., F. G. Maxwell and W. L. Parrott. 1964. A technique for measuring certain aspects of antibiosis in cotton to the boll weevil. J. Econ. Entomol. 57: 679-681.

Loomis, R. S., S. D. Beck and J. F. Stauffer. 1957. The European corn borer, Pyrausta nubilalis (Hbn.), and its principal host plant. V. A chemical study of host plant resistance. Plant Physiol. 32: 379-385.

Lukefahr, M. J. and J. E. Houghtaling. 1969. Resistance of cotton strains with high gossypol content to Heliothis spp. J. Econ. Entomol. 62: 588-591.

Lukefahr, M. J. and D. F. Martin. 1964. The effects of various larval and adult diets on the fecundity and longevity of the bollworm and cotton leafworm. J. Econ. Entomol. 57: 233-235.

Lukefahr, M. J. and D. F. Martin. 1966. Cotton plant pigments as a source of resistance to the bollworm and tobacco budworm. J. Econ. Entomol. 59: 176-179.

Lukefahr, M. J. and C. Rhyne. 1960. Effects of nectariless cottons on populations of three lepidopterous insects. J. Econ. Entomol. 53: 242-244.

Oliver, B. F., F. G. Maxwell and J. N. Jenkins. 1970. Utilization of glanded and glandless cotton diets by the bollworm. J. Econ. Entomol. 63: 1965-1966.

Painter, R. H. 1941. The economic value and biologic significance of plant resistance to insect attack. J. Econ. Entomol. 34: 358-367.

Parks, C. R. 1965. Floral pigmentation studies in the genus Gossypium. I. Species specific pigmentation patterns. Amer. J. Bot. 52: 309-316.

Schillinger, J. A. 1966. Larval growth as a method of screening Triticum spp. for resistance to the cereal leaf beetle. J. Econ. Entomol. 59: 1163-1166.

Shaver, T. N. and J. A. Garcia. 1973. Gossypol content of cotton flower buds. J. Econ. Entomol. 66: 327-329.

Shaver, T. N. and M. J. Lukefahr. 1969. Effect of flavonoid pigments and gossypol on growth and development of the bollworm, tobacco budworm, and pink bollworm. J. Econ. Entomol. 62: 643-646.

Shaver, T. N., M. J. Lukefahr and J. A. Garcia. 1970. Food utilization, ingestion and growth of larvae of the bollworm and tobacco budworm on diets containing gossypol. J. Econ. Entomol. 63; 1544-1546.

Shaver, T. N. and W. L. Parrott. 1970. Relationship of larval age to toxicity of gossypol to bollworms, tobacco budworms, and pink bollworms. J. Econ. Entomol. 63: 1802-1804.

Smissman, E. E., J. B. LaPidus and S. D. Beck. 1957. Isolation and synthesis of an insect resistance factor from corn plants. J. Amer. Chem. Soc. 79: 4697-4698.

Thurston, R., W. T. Smith and B. P. Cooper. 1966. Alkaloid secretion by trichomes of *Nicotiana* species and resistance of aphids (*Myzus persicae*). Entomol. Expl. & Appl. 9: 428-432.

Thurston, R. and J. A. Webster. 1962. Toxicity of *Nicotiana gossei* Domin. to *Myzus persicae* (Sulzer). Entomol. Exptl. & Appl. 5: 233-238.

Todd, J. W. and T. D. Canerday. 1968. Resistance of southern peas to the cowpea curculio. J. Econ. Entomol. 61: 1327-1329.

Todd, G. W., A. Getchum and D. C. Cress. 1971. Resistance in barley to greenbug, *Schizaphis graminum*. 1. Toxicity of the phenolic and flavonoid compounds and related substances. Ann. Entomol. Soc. Amer. 64: 718-722.

Walter, E. V. 1957. Corn earworm lethal factor in silks of sweet corn. J. Econ. Entomol. 50: 105-106.

Wann, E. V. and W. A. Hills. 1966. Earworm resistance in sweet corn at two stages of ear development. Proc. Amer. Soc Hort. Sci. 89: 491-496.

INSECT PLANT ATTRACTANTS, FEEDING STIMULANTS, REPELLENTS, DETERRENTS, AND OTHER RELATED FACTORS AFFECTING INSECT BEHAVIOR

Paul A. Hedin, Fowden G. Maxwell and Johnie N. Jenkins
Chemist, USDA Boll Weevil Research Laboratory,
Head, Department of Entomology
Geneticist, USDA Boll Weevil Research Laboratory
Mississippi State, Mississippi

For the purpose of this discussion, an "attractant" is defined as a stimulus to which the insect responds by orienting movements toward the apparent source. "Repellents" elicit an oriented response away from the apparent source. An "arrestant" is a stimulus that causes the insect to cease locomotion in close contact with the apparent source. Although arrestants and attractants may have the mechanical effect of causing aggregation, they should not be called aggregants. The term "aggregant," if used at all, should be reserved for insect-to-insect factors, such as behavioral pheromones. The terms incitant and suppressant have been introduced by Beck (1965) as feeding has been demonstrated to be separable into two phenomena, that of initiation and maintenance. An "incitant," therefore, describes a stimulus that evokes a biting or piercing reaction; and conversely, a stimulus tending to prevent this response is designated as a "feeding suppressant." Stimuli tending to promote continuous feeding are "feeding stimulants" and those preventing continuous feeding or hastening termination are termed "feeding deterrents."

A series of tables has been compiled which include the type of response, the insect, the host, the compounds or classes of compounds if known, and the investigators. Materials from hosts which acted as growth inhibitors on the insect were not included, and only those materials which influence attraction, feeding, and oviposition on the host were considered. Since there are over 350,000 described plant feeding species and this review contains only about 200 references, many with only preliminary evidence, it is easy to see how little is known about the basic underlying factors controlling or contributing to insect-plant interrelationships.

Feeding Stimulants

Glycosides were reported as feeding stimulants for several insects. Sinagrin (and other glycosides) was reported as a feeding stimulant for 2 species of imported cabbage worms (Verschaffelt, 1910), the diamondback moth (Thorsteinson, 1953), and the cabbage butterfly (David and Gardiner, 1966). The cyanogenic glycosides, phaseolunatin and lotaustrin were reported for the Mexican bean beetle (Klingenberg and Bucher, 1960). The flavanoid glycoside, 7-α-L-rhamnosyl-6-methoxyluteolin, was reported for the Chrysomelid beetle (Zielske, et al., 1972). Several flavonoid glycosides, quercetin, quercetin-7-glucoside, quercetin-3'-glucoside and cyanidin-3-glucoside, were reported as feeding stimulants for the boll weevil (Hedin, et al., 1968).

A number of phenols were also reported. They include methyl chavicol for the black swallow tail (Dethier, 1941), chlorogenic acid for the Colorado potato beetle (Hsaio and Fraenkel, 1968), salicin for the willowbeetle (Kearns, 1931), p-hydroxyacetophenone, o-hydroxybenzyl alcohol, and p-hydroxybenzaldehyde for the scolytid beetle (Baker and Norris, 1968), colotropin for a grasshopper (Euw, et al. 1967), anethole and anisic aldehyde for the black swallowtail (Erlich and Raven, 1967), and hypericin for *Calliphora brunsvicensis* (Rees,1966). A few phospholipids, fatty acids, and glycerides were reported. They include a partially characterized fraction for the Colorado potato beetle (Yamamoto and Fraenkel, 1959), lecithin and phosphatidyl inositol for the twostriped grasshopper (Thorsteinson and Nyar, 1963), lecithins for the cabbage looper (Gothilf and Beck, 1967), linoleic and linolenic acids for the imported fire ant (Vinson, et al., 1967), triolein for the prairie grain wireworm (Davis, 1965), and 3 other species of wireworms (Thorpe, et al., 1946), and palmitic acid for the confused flour beetle (Loschiavo, 1965). Sugars and terpenes were not as prevalent as may have been expected. Sucrose, glucose and raffinose were reported for the red cotton bug (Saxena, 1964), and mono- and diccharides for the cabbage looper (Gothilf and Beck, 1967). Pheophytins a and b were reported for the boll weevil (Temple, et al., 1968), curcubitacins for the spotted cucumber beetle (Chambliss and Jones, 1966; DaCosta and Jones, 1971), and a partially characterized terpene containing essential oil (Stride, 1965), for *Epilachna fulvosignata*. Examination of Table 1 points out very strikingly that in most cases feeding stimulants are comprised of chemicals which fall into what has been grouped as secondary plant substances (Fraenkel, 1959). These substances are largely thought to contain no primary function in the plant or in the insect.

When the known feeding stimulants were classified by chemical structure, 20 were glycosides, 15 were acids, 8 were flavonoid aglycones, 6 were carbonyls, 5 each were phospholipids and terpenoids, with 17 miscellaneous responses. When these feeding stimulant classes were further subdivided by insect order, somewhat more specific preferences were suggested. Among the Lepidoptera, 8 glycosidic feeding stimulants were reported, along with 3 aglycones which could become glycosidated; there were 9 miscellaneous responses. Among the Coleoptera, 10 acids, 6 glycosides, 4 flavonoid aglycones, and 4 terpenes were reported. There were also 8 miscellaneous classes. In the order Homoptera, both compounds reported were glycosides. In the order Orthoptera, acids, and their esters and salts, were most prevalent. These compounds accounted for 8 of the 14 listed responses.

In reports on insect feeding stimulants, none studied has provided unambiguous evidence that a single compound acts individually. Furthermore, when a "glycoside," "sugar," or "purine" has been implicated, it has often been questionable whether the activity could have been maintained after rigorous purification. In addition, when pure compounds have been demonstrated to give activity (Hamamura, et al., 1962; Robbins, et al., 1965), several compounds might have been isolated from the same source, each of which could have elicited some activity or synergized the activity of the compound. The concept of host-plant specificity implies that insects can discriminate flavors. Therefore, though insects almost certainly do sometimes respond to a single, dominant compound when it is present, there probably is a much larger number of situations where no dominant compound exists. Consequently, an adequate response more likely requires a complicated profile of compounds.

An insect which feeds as a result of a complicated, and probably sequential presentation of stimuli is the silkworm. Volatile substances including citral, linalyl acetate, linalool and terpinyl acetate attract the larvae to the mulberry leaves. Biting factors such as β-sitosterol-β-glucoside, lupeol, isoquercetrin, morin, & 2',2,4'5,7-pentahydroxyflavone initiate plant consumption. Swallowing factors such as sitosterol, sugar, silica, cellulose, and potassium phosphate were demonstrated. From these results, the substances controlling the feeding behavior of silkworm larvae were described, but these substances are not found exclusively in mulberry leaves, and are in fact rather common in green leaves. The preference for mulberry leaves may depend on the amounts and proportions of these compounds, and on the absence of repellents. A repellent effect could be demonstrated by adding raw soybean cake or powdered milk to a mulberry leaf preparation. Extraction of these additives with methanol removed the repellent components (Hamamura, 1970).

In work conducted at the Boll Weevil Research Laboratory which illustrates some of the complexities of feeding stimulant research, feeding-stimulant components extractable with petroleum ether, chloroform, acetone, and chloroform-methanol from cotton buds appear to belong to several major groups that cause feeding activity in the boll weevil (Hedin, et al., 1966; Struck, et al., 1968a, b; Hedin, et al., 1968; Temple, et al., 1968). Also, the physical and chemical properties of these active fractions clearly differentiate between nonpolar and polar active constituents in the cotton bud. However, attempts to purify the major groups frequently result in a dissipation of feeding activity that cannot be fully regenerated by recombination, indication of a breakdown in the chemical structure of some components during purification. This inability to regenerate full feeding activity by recombination is especially true for many of the TLC systems studied. Also, the labile nature of some components is further indicated by the gradual loss of the feeding activity of some compounds stored in certain solvent systems. For example, an extract will show only token feeding activity after storage in diethyl ether for 12 hr, but activity can be retained at the same level for 2-3 weeks in either chloroform or methanol.

It is significant that the activity of lypolytic enzymes in diethyl ether is high. Even though no significant feeding-stimulant activity can be attributed to the lipid components derived from cotton buds, the increase in the concentration of hydrogen ions caused by their chemical degradation may have an adverse effect on the feeding-stimulant components. These considerations suggest that lability of some feeding-stimulant components may occasionally contribute to the observed loss of activity during procedures of fractionation. More importantly, however, these studies indicate that a multicomponent system is necessary for optimum boll weevil feeding (Struck, et al., 1968a, b; Temple, et al., 1968).

Since recombination or fortification of fractions by sugars and buffers often rejuvenated part of the activity, efforts were directed to formulating an active feeding mixture from known cotton constituents, common metabolites, and compounds inducing primary mammalian sensations of taste and odor (Hedin, et al., 1968). Of 286 compounds bioassayed individually, 52 elicited substantial activity, and 14 of these previously had been reported in cotton. They include gossypol, α-ketoglutaric acid, malonic acid, vanillin, formic acid, lactic acid, ℓ-malic acid, quercetin, β-sitosterol, succinic acid, valine, quercimeritrin, quercetin-3'-glucoside, and cyanidin-3-glucoside. The insect was found to express preference for sweet, sour, and cooling taste properties, but odor preferences were difficult to establish. Sixteen carboxylic acids, 8 alcohols, 8 carbonyls, 8 phenols, and 10 amides or amines were among the most stimulatory. When the active components were factored on the bases of taste and molecular size, it became apparent that sweet substances having molecular weights above 200 were consistently well accepted. Since most of these compounds were di- or triterpenoids or steroids, hydroxylated, and much less sweet than the sugars, their activity may be associated with their predicted low rate of desorption. The most favored molecular size for sour and salty compounds was below 150. Bitter deterrent compounds were concentrated in the 100-200 range. Pungent compounds of 150-200 were most deterrent.

Maxwell and Jenkins (F. G. Maxwell and J. N. Jenkins, USDA, Mississippi State, Miss., personal communication) made an exhaustive review of biologically active substances in host plants affecting insect behavior. From these data a frequency table was prepared relating feeding stimulation and/or attractants and deterrents and/or repellents to molecular structure. The leading attractive compounds were: 8 acids attractive to 1 unattractive; monoterpene hydrocarbons, 9 to 2; di- or triterpenoids and steroids, 23 to 1; esters and alcohols, each 7 to 0; and nucleotides and tannins, each 5 to 0. The leading repellent compounds were: 14 alkaloid glycosides unattractive to 1 attractive; and lactones, 7 to 1. Flavonoid, cyanogenic, and other unclassified glycosides were normally stimulatory, 17 to 6. The tastes of the stimulatory substances appeared to be sour, cooling, semisweet, and salty, and the stimulatory odors appeared to be floral, musky, pepperminty, and comphoraceous. Repellent substances appeared to have selected bitter tastes and strongly flavored (pungent?) characteristics accorded by the lactones. The similarity of these results to those that we observed with the boll weevil is striking.

Formulation of mixtures was initiated on an empirical basis (Hedin, et al., 1968). Within a short time, some 330 mixtures had been tested with limited success. Consequently, the mixtures were inspected for components which were most often present when good feeding occurred. From this list, 17 compounds were selected including 6 which were known to be present in cotton. They were formulated into 4 to 8 component mixtures with some consideration given to taste class and intensity. Subsequently, several 8-component mixtures were selected which gave feeding responses equal to cotton bud water extracts. The best was a mixture of β-sitosterol, 15-pentadecanolide, cineole, N,N-dimethyl-aniline, vanillin, mannitol, rhamnose and phosphate buffer pH 7.0. This mixture also elicited slightly more insect punctures than did the water extracr when rolled plugs of each were placed side by side.

There has been no extensive effort to evaluate these mixtures in field trials for several reasons. Cottonseed oil "foots" and cottonseed meal contain most of the components of the water or organic solvent extracts, and are a cheaper and more convenient source for bait preparation. Secondly, the baits are not competitive during the period of greatest plant growth. Finally, the insects are overwhelmingly attracted to males or the synthetic pheromone in the presence of growing cotton or baits.

McKibben, et al. (1971) investigated eight acids including adipic, fumaric, malic, phosphoric, succinic, citric, lactic and tartaric acids, as food acidulants. When added to cottonseed oil baits used to attract the weevil, they caused a 41-67% increase in the amount ingested but there were no significant differences among the acids tested.

Plant Attractants

Fraenkel (1959) has discussed the role of secondary plant substances in the development of host plant-insect relationships. Compounds produced by plants as waste, metabolic intermediates, defense mechanisms, insect attractants or stimulants influence the establishment of highly specific plant-insect ecologies. He suggested that these secondary plant substances, known to exist in many plants without having any apparent role in plant function, may have developed from past mutations and natural selections. The development of these substances would afford the plant protection against increased insect damage or other enemies. Certain insects, less susceptible to the undesirable effects of the compounds, may then have adapted to the secondary substances as a stimulus for feeding or propagation.

With regard to the mechanisms of odor perception, Dethier (1970) reported that whereas taste receptors tend to be individually specific to restricted classes of compounds, olfactory receptors may be either specific or nonspecific. The antennal olfactory receptors of *Manduca sexta* (Johan.) respond to a wide variety of odors and resemble "generalists." In the absence of stimulation they are spontaneously active. Stimulation may increase or decrease the rate of firing depending on the identity of the stimulus and the particular receptor. Thus, a given odor may increase the rate of firing of one receptor and decrease the rate of firing of another while a different odor may have exactly the opposite effect. The receptors have overlapping but not congruent action spectra. As a consequence, divers odorous compounds, by their differential effects on the several receptors, cause different compound patterns of activity to be generated when all receptor input is viewed as a whole.

When the known attractants were classified by chemical structure (Table 2), 21 were terpenes, 7 were alcohols, 4 each were esters, acids, and sulfur containing, and 2 were phenolics. There were 7 miscellaneous responses including 3 which possessed little or no volatility and probably were misclassified as attractants. When these attractants were further subdivided by insect odor, somewhat more specific preferences were suggested. Among the Coleoptera, 17 terpenes, 3 ketones, 2 alcohols, 2 acids, and 4 miscellaneous responses were reported. A parallel with the compounds reported as sex attractants for insects of this order is therefore suggested. These data are also predicted because a number of tree-infesting beetles were in the group. Among the Lepidoptera, 5 esters, 2 glycosides (probably improperly identified as attractants), 2 acids, and 4 alcohols were reported. A parallel is also obvious with the sex attractants of this insect order; mostly esters and their constituent alcohols and acids were reported. Terpenes were also reported as attractants in the orders Hymenoptera, Diptera, and Isoptera.

Plant attractants elicit responses by several mechanisms and are subject to various conditions. Some appear to act individually, both under laboratory and field conditions. Perhaps the attraction of the oriental fruitfly by methyl eugenol (Steiner, 1952) is the best example, where eradication of this insect from several islands has been achieved. In some other instances, such as the Japanese beetle, two components, methyl cyclohexanepropionate and eugenol, are found to be attractive under field conditions (McGovern, et al., 1970). In still other instances, a fairly complicated mixture of components is required, but some partial response can be obtained from individual components in the mixture. Some examples of these are the silkworm (Hamamura, 1970), the onion maggot (Matsumoto and Thorsteinson, 1968), the lacewing (Sakan, et al., 1970), the Pales weevil (Thomas and Hertell, 1969), the western pine beetle (Vite and Pitman, 1969), the checkered beetle (Harwood and Rudinsky, 1966), the codling moth (Dethier, 1947; Masden and Falcon, 1969), and the boll weevil (Minyard, et al., 1969).

In this laboratory, Minyard, et al., (1969) found β-bisabolol to be present in the highest concentration of any polar compound in the cotton bud essential oil. It was not previously reported in nature, but has subsequently been found by our group to be present in several other malvaceous oils (Thompson, et al., 1971). However, β-bisabolol alone is only about 50% as attractive in the laboratory olfactometer bioassay as a hot water extract of cotton. By fractionation and bioassay of the cotton bud essential oil, several other components were identified which were attractive in their own right, and improved the activity of β-bisabolol; they included β-carophyllene, ℓ-limonene, α-D-pinene, and β-caryophyllene oxide (Minyard, et al., 1969). Subsequently, α-bisabolol (Hedin, et al., 1971) and bisabolene oxide (Hedin, et al., 1972) have been identified as other contributing attractants.

However, boll weevils of both sexes are much more attracted to males than to cotton. Hardee, et al. (1969) showed a definite preference of overwintered and late season insects for male weevils compared with that of fruiting cotton and speculated that insects find cotton by random flight or by response to "short range" plant attractants.

Nevertheless, this mixture of components isolated from cotton bud essential oil and of a synthetic mixture of terpenes increased the response of insects in the field to a dye marker bait two-fold during late August and early September, but did not elicit a significant increase during the late season phase of the experiment (McKibben, et al., 1971). Failure to obtain increased feeding with the added attractant in the second phase was attributed to the increased feeding with all treatments in October; this suggests decreased discrimination by the insect as the food supply decreased.

It is not too difficult to identify by screening, compounds which show attractancy for insects. Beroza and Green (1963) published on several thousand compounds, many of which showed some attractancy for various insects. Also, it is not too difficult to isolate from plants by steam distillation, essential oils which are attractive. Thompson, et al. (1971) obtained the essential oils from a number of Malvaceous plants, and also from some Angiosperms and Gymnosperms. Several of the oils were as attractive as cotton oil, and the insects actually favored essential oils of 2 alternate host plants, *Hibiscus militaris* Cav. and *H. lasiocarpus* Cav. in comparative preference tests with cotton oil. Thus, in the laboratory, the essential oils in plants did not control insect preference exclusively. Morphological and environmental conditions may influence the concentrations of volatile components released by plants to the atmosphere. Thompson et al. (1971) attempted to assess the extent of these differences by a GLC study of the condensate from the atmosphere around cotton plants in the field. The total essential oil from the condensate was found to contain 50-60% β-bisabolol and β-bisabolene, and 30-40% of geraniol myrtenal, nerolidol and β-caryophyllene oxide. The identification of some of these components as attractants, and now the recognition that they are major components in the cotton aroma of the field, give added credence to the expectation that they may play some role in the movement of boll weevils to cotton in the field.

Feeding Deterrents

Beck (1965) defines a feeding deterrent in terms of factors which do not support maintenance of feeding, or which cause cessation of feeding. Munakata (1970) classifies feeding deterrents as synonymous with antifeedants and gustatory repellents, and further defines them as chemicals that inhibit feeding, but do not kill the insect direct. Thorsteinson (1960) attributes host selection by phytophagous insects to the presence or absence of attractants or repellents.

Alkaloid glycosides have been among the most frequent compounds reported as feeding deterrents. Work by Kuhn and coworkers (Kuhn and Low, 1957, and earlier), and by Shreiber (1958) showed the presence of several of these compounds in *Solonaceae* and implicated them as deterrents against the Colorado potato beetle. More recently, Harley and Thorsteinson (1967) have shown the deterrency of this class against twostriped grasshopper larvae. Lichtenstein, et al. (1962) isolated 2-phenylethyl-isothiocyanate from the turnip as an antifeedant of vinegar flies. Two resistance factors in corn plants, 6-methoxy-benzoxazolinone (Smissman, et al., 1957b) and 2,4-dihydroxy-7-methoxy-1,4-benzoxazine-3-one (Klun, et al., 1967) were shown to be effective against the European corn borer. Rudman and Gay (1961) showed that 2 anthraquinones present in teak heartwood inhibited termite activity. Several other interesting compounds include juglone (Gilbert, et al., 1967), azadirachten (Gill and Lewis, 1971), nepetalactone (Eisner, 1964), cocculolidine and isoboldine (Wada and Munakata, 1968), and 2 shiromodiol acetates (Wada, et al. 1968). When the known feeding deterrents were classified by chemical structure (Table 3), 18 related alkaloids or alkaloid glycosides affecting 5 insects were reported. The other classes and their frequencies were 3 lactones, 4 quinones affecting 2 insects, 4 heterocyclic ring compounds, 1 isothiocyanate, and 1 acid. While 7 orders of insects were studied, all of the reports except 7 involved Coleoptera.

Although Beck (1965) and Munakata (1970) stress the concept that feeding deterrents do not directly kill the insect, most of those reported are in fact biological poisons. Two representative compounds for which LD_{50} values were given in Merck Index were the alkaloid glycoside, tomatin (25 mg/kg i.p. and 500 mg/kg oral in mice) and nornicotine (23.5 mg/kg i.p. in rats). The alkaloid aglycones would be expected to be more toxic. Most of these compounds are bitter to humans, and some similar perception apparently occurs with insects.

Repellents

Beck (1965) defines a repellent as eliciting an oriented response away from the apparent source. The property of volatility is thus inferred for the candidate compound. Another group of repellents are the defense secretions. However, since they are biosynthesized by insects rather than by plants, they have not been included in this discussion. Some of the compounds reported (Table 4) were capsaicin (Shreiber, 1958), myrcene and limonene (Smith, 1966), valeric and isovaleric acids (Goodhue, 1963), and butyric acid and coumarin (Kamm and Fronk, 1964). Several others which were reported as repellents have no appreciable vapor pressure and therefore should have been reported as feeding deterrents. Any repellency probably should have been attributed to impurities in the isolated fraction. They include tomatin (Shreiber, 1958), anacardic acid (Wolcott, 1946), tannic acid (Bennett, 1965), and shikimic acid, pyridoxine, succinic acid, malic acid, betaine, and xanthophyl (Kamm and Fronk, 1964).

Eight appreciably volatile compounds from 5 orders were reported; they include 3 acids, 2 phenols, 2 terpenes and 1 alcohol (Table 4). The limited nature of work on naturally occurring insect repellents is apparent. However, since many plants which are resistant to insects have characteristic odors, it is possible that a number of others exist. The vastly improved capability for identification of volatile compounds, i.e., integrated gas chromatography-mass spectrometry and high sensitivity nuclear magnetic resonance spectrometry, could facilitate expanded research on repellents in the next few years.

Oviposition Stimulants

Many of the compounds reported as oviposition stimulants (Table 5) were also identified as feeding stimulants or attractants (Tables 1 and 2), but not necessarily for the same insect. More oviposition stimulants were reported for Diptera than any other order. This may reflect the relative simplicity of the bioassay as compared with that for other orders. The cabbage root fly was stimulated by sinagrin and β-phenylethylamine (Traynier, 1965). Matsumoto and Thorsteinson (1968) reported that the onion maggot was stimulated, not surprisingly, by n-propyl disulfide, methyl disulfide, n-propyl mercaptan, and n-propyl alcohol. Sulfides and mercaptans are found in high concentrations in onions. The carrot rust fly was stimulated by methyl isoeugenol and trans-1,2-dimethoxy-4-propenyl-benzene (Berueter and Staedler, 1971). The mosquito was reported by Perry and Fly (1967) to be stimulated by 3 esters; ethyl acetate, methyl propionate, and methyl butyrate.

In Lepidoptera, a partially characterized glycoside for the tobacco hornworm (Yamamoto and Fraenkel, 1960), allyl-isothiocyanate for the diamond back moth (Gupta and Thorsteinson, 1960) and two partially characterized growth regulators for the European corn borer (Schurr, 1970) were reported.

In Coleoptera, lecithin was reported for the Colorado potato beetle (Grison, 1958). In Orthoptera, 4 terpenes were reported for the desert locust (Carlisle, et al., 1965). In Hymenoptera, saponins were reported for the multivoltive bruchid (Applebaum, et al., 1965). All data on the oviposition stimulants are summarized in Table 5.

There were also reports on feeding incitants, flight termination stimulants, swallowing factors, mating factors, growth factors, and an oviposition deterrent; these data are included in Table 6. β-Sitosterol, β-sitosterol glucoside, lupeol, isoquercitrin and 2'3,4',5,7-pentahydroxyflavone were reported as feeding incitants for the silkworm larvae (Hamamura, et al., 1961; Nyar and Fraenkel, 1962; Goto, 1965; Hamamura, et al., 1962; Hayashiya, 1966).

Phaseolunatin and lotaustrin, cyanogenic glycosides, were reported as feeding incitants for the Mexican bean beetle (Klengenburg and Bucher, 1960). Coumarin was reported as a flight termination agent for the sweet clover weevil (Heidewig and Thorsteinson, 1961). Cellulose, silica and potassium phosphate were reported as swallowing factors for the silkworm larvae (Hamamura, et al., 1962), and protein, leucine, and guanosine-5'-monophosphate as swallowing factors for the mosquito (Yamamoto and Jensen, 1967).

Trans-2-hexenal was reported as a mating factor for the Polyphemus moth (Riddiford and Williams, 1967; Riddiford, 1967). Chlorogenic acid was described as a growth factor for the silkworm larvae (Kato and Yamada, 1966), and some partially characterized sapogenic aglycones were reported as oviposition deterrents for the multivoltive bruchid. The compounds reported for these miscellaneous categories are the same as or similar to those reported earlier as attractants, feeding stimulants, and deterrents.

Table 1.--Feeding stimulants.

Insect					
Scientific name	Common name	State	Plant	Compound	Reference
Pieris rapae L.	Imported cabbage-worm	Larvae	Cruciferae	Sinigrin	Verschaffelt (1910)
Pieris brassica L.	Imported cabbage-worm	Larvae	Cruciferae	Sinigrin & related glucosides, glucocapparin, glucoiberin	Verschaffelt (1910) David & Gardner (1966)
Papilio polyxenes asterius Stoll	Black swallowtail	Larvae	Umbelliferae	Carvone, methyl chavicol, coriandrol	Dethier (1941)
Leptinotarsa decemlineata (Say)	Colorado potato beetle	Larvae & adult	Potato leaves	Phospholipid, chlorogenic acid, glycoside, $C_{17}H_{29}O_{10}N$	Hsiao (1966) Hsiao & Fraenkel (1968a) Hsiao & Fraenkel (1968b) Chauvin (1952) Yamamoto & Fraenkel (1959) Fraenkel et al. (1960)
Bombyx mori L.	Silkworm	Larvae	Mulberry	glycoside: $C_{30}H_{62}O$(?)	Fraenkel (1959)
Hypera postica (Gyllenhal)	Alfalfa weevil		Alfalfa	Unknown	Yamamoto (1963)
Diabrotica virgifera L. LeConte	Western corn rootworm		Corn, *Zea mays* L.	Unknown	Derr et al. (1964)
Diabrotica longicornus (Say)	Northern corn rootworm		Corn, *Zea mays*	Unknown	Derr et al. (1964)
Musca domestica L.	Housefly		Yeast	Guanine, monophosphate	Robbins et al. (1965)
Anthonomus grandis Boh.	Boll weevil	Adult	Cotton *Gossypium* sp.	glycoside, flavonoids	Keller et al. (1962) Maxwell et al. (1963) Jenkins et al. (1963)
Anthonomus grandis Boh.	Boll weevil	Adult	Cotton	Gossypol, α-ketoglutaric acid, malonic acid, vanillin, formic acid, lactic acid, ℓ-malic acid, quercetin, β-sitosterol, succinic acid, valine, quercetin-7-glucoside, quercetin-3'-glucoside, cyanidin-3-glucoside	Hedin et al. (1968)

Table 1.--Continued

Insect					
Scientific name	Common name	State	Plant	Compound	Reference
Anthonomus grandis Boh.	Boll weevil	Adult	Cotton Gossypium sp.	Pheophytin a, Pheophytin b	Temple et al. (1968)
Galerucella xanthomelaena (Shiffermuller)	Elm leaf beetle	Larvae	Ulmus americanan L., American Elm	Unknown	Keller et al. (1965)
Gastrophysa cyanea (Melch)	Dock beetle		Dock, Rumex obtusifolius L.	Unknown	Keller et al. (1965)
Danaus plexiggus (L.)	Monarch butterfly	Larvae	Milkweed, Ascelpius syriaca L.	Unknown	Dethier (1937)
Anasa tristis (DeGeer)	Squash bug	Adult	Pumpkin, Curbita pepo L.	Glycoside	Keller et al. (1965)
Heliothis zea	Corn earworm	Larvae	Corn, Zea mays L.	Unknown	Starks et al. (1965)
Choristoneura fumiferana (Clem.)	Spruce budworm	Larvae	White spruce, Picea glauca (Moench)	Unknown	Heron (1965)
Laspeyresia caryana (Fitch)	Hickory shuckworm	Larvae	Pecan, Carya illinoensis (Wang) K. Koch	Unknown	Howell & Maxwell (1965)
Scolytus multistriatus (Marsham)	Smaller elm bark		Elm	Unknown	Loschiavo et al. (1963)
Protoparce sexta (Johan.)	Tobacco hornworm	Larvae	Tomato	Glycoside	Yamamoto & Fraenkel (1960)
Plutella maculipennis (Curt)	Diamond-back moth	Larvae	Cruciferae	Sinigrin, Sinalbin Glucocheirolin	Thorsteinson (1953)
Epilachna varivestis Mulsant	Mexican bean beetle	Adult	Phaseolus sp.	Phaseolunatin Lotaustrin	Klingenburg & Bucher (1960)
Epilachna fulvosignata Rche. var. suahelorum (Weise)		Larvae	Solanum campylacanthum L.	essential oil	Stride (1965)
Ceratomia catalpae (Boisduval)	Catalpa worm	Larvae	Catalpa Catalpa bignonoides	Unknown Catalposide	Fraenkel (1960) Nyar & Fraenkel (1962)

Table 1.--Continued.

Insect			Plant	Compound	Reference
Scientific name	Common name	State			
Melanoplus bivetattus (Say)	Twostriped grasshopper		Corn, soybeans, other plants	Lecithin, phosphatidyl inositol	Thorsteinson & Nyar (1963)
Camnula pellucida (Scudder)	Clear-winged grasshopper		Corn, soybeans, other plants	Lecithin, phosphatidyl inositol, amylacetate	Thorsteinson & Nyar (1963) Parker (1924)
Cheimotobia brumata	Clear-winged grasshopper		*Rosaceae*, *Ericaceae* *Salicinae*	Tannins	Lagerheim (1900)
Nygmia phaearhoea Donavan	Browntailed moth	Larvae	Oak	Tannins	Grevillins (1905)
Malacosoma neustria Fab.	Tent caterpillar	Larvae	Oak	Tannins	Grevillins (1905)
Euproctis chrysorrhoea			Oak	Tannins	Grevillins (1905)
Periophorus padi			Rosaceae	Glycoside	Verschaffelt (1910)
Malacosoma americana (Fab.)	Eastern tent caterpillar	Larvae	Rosaceae	Amygdalin	Verschaffelt (1910)
Gastroidea vividula			*Polygonaeae rumix* sp.	Oxalic acid	Verschaffelt (1910)
Hylemia antiqua (Meig.)	Onion maggot		Onion	Allyl sulfide	Peterson (1924)
Phyllodecta vitallinae	Willow beetle		Willow var.	Salicin	Kearns (1931)
Diabrotica undecimpunctata howardi (Barb.)	Spotted cucumber	Adults	*Cucurbitaceae* sp.	Curcurbitacins	Chambliss & Jones (1966) DaCosta & Jones (1971)
Trichoplusia ni	Cabbage looper	Larvae	Laboratory diets	proteins, sugars, wheat germ,oil, salts, phospholipids, lecithins	Gothilf & Beck (1967)
Poekilocerus bufonius (Klug)	Grasshopper		Milkweed, *Asclepius syrica* (L.)	Calotropin	Euw, et al. (1967)
Brevicoryne brassicae	Cabbage aphid			Mustard oil glycosides	Wensler (1963)

Table 1.--Continued.

Insect					
Scientific name	Common name	State	Plant	Compound	Reference
Agasicles sp. (Nov.)	Chrysomelid beetle	Adult	Alligatorweed *Alternanthera phylloxeroides*, Amaranthaceae	7-alpha-L-rhamnosyl-6-methoxyluteolin	Zielske, et al. (1972)
Scolytus multistriatus (Marsham)	Scolytid beetle	Adult	Elm, *Ulmus americana*	*p*-hydroxyacetophenone *o*-hydroxybenzyl alcohol, *p*-hydroxybenzaldehyde	Baker & Norris (1968)
Ansitio plagiata	Geometrid	Larvae	*Hypericum*	Cuticular waxes	Harris (1967)
Calophasia lanula	Noctuid		Linaria	Unknown	Harris (1963)
Heliothis virescens	Tobacco budworm	Larvae	Corn kernels & silks, cotton buds & flowers	Unknown	Guerra & Shaver (1968)
Hypera postica (G.)	Alfalfa weevil		Alfalfa	Adenine salts, nucleotides	Hsiao (1969)
Scolytus multistriatus (Marsham)	Smaller European Elm Bark Beetle		*Ulmus americana* (L.)	(+)-catechin-5-alpha-D-xylopyranoxide, Lupeyl cerotate	Doskotch, et al. (1970)
Papilio polyxenes asterius (Stoll.)	Black swallowtail	Larvae	Anise, Coriander celery, angelica, citrus	Anethole, anisic aldehyde	Ehrlich & Raven (1967)
Agriotes sputator, *A. obscuris*, *A. lineatus*	Wireworms			Triolein	Thorpe, et al. (1946)
Spodoptera frugiperda (Smith)	Fall armyworm	Larvae	Corn, cotton, tomato, sorghum	Unknown	McMillan & Starks (1966)
Heliothis zea (Boddie)	Corn earworm	Larvae	Corn, cotton, tomato, sorghum	Unknown	McMillan & Starks (1966)
Heliothis virescens (F.)	Tobacco budworm	Larvae	Corn, cotton, tomato, sorghum	Unknown	McMillan & Starks (1966)
Tribolium confusum (J. DuVac)	Confused flour beetle		Wheat germ, yeast	Unknown Palmitic acid	Loschiavo (1965)
Solenopsis saevissima (U.V. Richteri)	Imported fire ant		Boll weevil, cabbage looper, ground beef	Linoleic acid, linolenic acid	Vinson, et al. (1967)

Table 1.--Continued.

Insect					
Scientific name	Common name	State	Plant	Compound	Reference
Chorthippus curtipennis	Grasshopper		General	Ascorbic acid, thiamine, betaine, monosodium glutamate	Thorsteinson (1955b)
Melanoplus bivattatus (Say)	Twostriped grasshopper		General	Amides, anisic acid, benzoic acid, ammonium salts, pentylacetate Plant phospholipids wheat germ oil	Thorsteinson (1960)
Shistocerca gregaria	Desert locust		Bran, olives, peanuts, miaouli	Unknown	Chauvin & Mentzer (1951)
	Grasshoppers, several species		Host plants	Unknown	Mulkern (1967)
Pieris brassicae (L.)	Cabbage butterfly		*Brassica* sp., *Capparidaceae* *Nasturtium trogacolum majus* (L.)	glucotropacolin glucoapparin, glucoiberin, glucocheirolin, progortrin, glucosinalbin, sinigrin, glucoerucin	David & Gardiner (1966)
Clenicera aeripennis destructor (Brown)	Prairie grain wireworm	Larvae	Germinating rye seed	Water extract triolein	Davis (1961, 1965)
Calliphora brunsvicensis			*Hypericum* sp. (St. Johns wort)	Hypericin	Rees (1966)
Phrydiuchus toparius (Germ.)			Salvia sp.	Unknown	Andres (1964)
Gastrophysa cyanea (Melsheimer)		Beetle	*Rumex crispis*	Water extract	Force (1966)
Calophasia lunula (Hufn.)			Toadflax *Linaria vulgaris*	Pyridine extract	Harris (1963)
Pectinophora gossypiella	Pink bollworm	Larvae	*Gossypium* sp.	volatiles	Keller & Sheets (1968) Parrott (1968)
Dsydercus koenigii (F.)	Red cotton bug		Cottonseed	sucrose, glucose raffinose	Saxena (1964)
Heliothis zea (Boddie)	Corn earworm	Larvae	16 host plants	water extracts	Wiseman, et al. (1969) Allen & Pate (1966)

Table 2.--Attractants.

Insect					
Scientific name	Common name	State	Plant	Compound	Reference
Pieris rapae L.	Imported cabbage-worm	Larvae	Cruciferae	Allyl isothiocyanate	Hovanitz & Chang (1963)
Papilio polyxenes asterius	Black swallowtail	Adult	Umbelliferae Stoll	Carvone, methyl chavieol, Coriandrol	Dethier (1941)
Bombyx mori L.	Silkworm	Adults Larvae	Mulberry, Morus alba L.	2-hexenol 3-hexenol β-gamma-hexenol α-β-hexenal	Guenther (1949) Watanabe (1958)
Bombyx mori L.	Silkworm	Larvae	Mulberry, Morus alba L.	Citral, terpinyl-acetate, linalyl acetate, linalool	Hamamura (1959) Hamamura et al. (1962) Horie (1962) Ito (1961)
Anthonomus grandis (Boh.)	Boll weevil	Adult	Cotton, Gossypium sp.	Unknown	Keller, et al. (1963)
Anthonomus grandis (Boh.)	Boll weevil	Adult	Cotton, Gossypium sp.	β-bisabolol, β-carophyllene oxide, ℓ-pinene, limonene, β-carophyllene trimethyl amine, ammonia	Minyard, et al. (1969) Folsom (1931)
Chilo suppressalis (Walker)	Rice stem borer		Rice, Oryza sativa L.	Oryzanone	Munakata, et al. (1959)
Listroderes costirustris oblignus Klug	Vegetable weevil		Crucifera and Umbelliferae	leaf alcohol 3-hexenol	Matsumoto & Sugiyama (1960)
				mustard oil, isothio-cyanates; methyl, ethyl, allyl, isobutyl, n-butyl, phenyl, benzyl, β-phenylethyl, α-naphthyl	
			Sweet clover, Melilotus officianalis L.	Coumarin	Matsumoto (1962)
Laspeyresia caryanae Fitch	Hickory shuckworm	Larvae	Pecan, Carya illinoensis (Wang) K. Koch	Unknown	Howell & Maxwell (1965)
Protoparce sexta (Johan.)	Tobacco hornworm	Adult	Jimson weed, Datura stromonium	Amyl solicylate	Morgan & Lyon (1928)

Table 2.--Continued.

Insect					
Scientific name	Common name	State	Plant	Compound	Reference
Plutella maculipennis (Curt.)	Diamond-back moth	Adult	Cruciferae	Progoitrin, glucocouringlin, glucoerucin, glucotropacolin, gluconastertium, gluconapin	Nyar & Thorsteinson (1963)
Oncopeltus faciatus (Dallas)	Milkweed bug	Adult	Milkweed seed	Unknown	Feir & Beck (1963)
Cosmopolites sordidus (Cheur.)	Banana weevil		Banana	Unknown	Cuille (1950)
Gastroidae vividula			Polygonaeae rumix sp.	Oxalic acid	Vershaeffelt (1910)
Malacosoma americana (Fab.)	Eastern tent caterpillar		Rosaceae	Amygdalin	Vershaeffelt (1910)
Hylemia antiqua (Meig.)	Onion maggot		Onion	Allyl sulfide, n-propyl disulfide, n-propyl, mercaptan, methyl disulfide, isopropyl mercaptan	Peterson (1924) Matsumoto & Thorsteinson (1968)
Drosophila melanogaster (Meig.)	Fruit fly		Banana	Volatiles	Wright (1965)
Dacus diversus (Coq.)	Fruit fly	Male adults	Fruit	Oil of citronella	Howlett (1912)
Dacus zonatus (Saund.)	Fruit fly	Male adults	Fruit	Oil of citronella	Howlett (1912)
Dacus dorsalis (Hendel.)	Oriental fruit fly	Adults	Fruits	Methyl eugenol	Howlett (1915)
Cryphalus fulvus (Niyima)	Pine bark beetle		Pine bark, Pinus densiflora L.	Benzoic acid	Yasunaga, et al. (1962)
Reticulitermes flavipes	Termite		Decayed wood by Lenzites trabea	Essential oil	Watanabe & Cassida (1963)
Dendroctonus pseudodotsugae (Hopkins)	Douglas fir beetle	Adult	Douglas fir, Pseudotsuga menziezei Mirb. Franco	Alpha pinene	Heikkenenard & Hrutfiord (1965)
Musca domestica	House fly	Adult	Mushroom, Amarita muscaria L.	1,3-Diolein	Muto & Sugawara (1965)

Table 2.--Continued.

Insect					
Scientific name	Common name	State	Plant	Compound	Reference
Adris tyrannua amurensis (Standinger)			Grapes	Neutral volatile	Saito & Munakata (1970)
Chrysopa septempunctata (Wesmael)	Lacewings		Matatabi, *Actinidia polygama* (Miq.)	Iridodiol; metatabiol, 5-hydroxymatatabiether, 7-hydroxydehydromatatabiether, allomatatabiol	Sakan, et al. (1970)
Blastophagus piniperda (L.)	Pine beetle	Adult	*Pinus densiflora* *Pinus silvestris*	Benzoic acid α-terpineol	Kangas, et al. (1965)
Gnathotrichus sulcatus (LeConte)	Ambrosia beetle	Adult	Western hemlock *Tsuga heterophylla* (R.) Sargent	Ethanol	Cade, et al. (1970)
Erioischia brassicae (Bouche)	Cabbage root fly	Adult	Cabbage	Volatiles	Traynier (1967)
Shistocerca gregaria (Forskil)	Desert locust	Adult	Grasses	Ammonium dihydrogen phosphate, D-ammonium hydrogen phosphate, ammonium sulfide	Kennedy & Moorhouse (1969) Haskell, et al. (1961)
Hypera postica (Gyllenhal)	Alfalfa weevil	Adult	Alfalfa	Unknown	Byrne & Steinauer (1966)
Hylobius pales (Herbst.)	Pales weevil	Adult	Loblolly pine stem	Monoterpenoid(s), eugenol, anethole, α-pinene	Thomas & Hertell (1969)
Ips pini	Pine engraver beetle	Adult	*Pinus resinosa*	Terpenes	Seybert & Gara (1970)
Dendroctonus pseudodotsugae (Hopkins)	Douglas fir	Adult female	Douglas fir, *Pseudotsuga menziezei* Mirb. Franco	Terpenes	Johnson & Belluschi (1969)
Dendroctonus brevicomis (LeConte)	Western pine beetle	Adult	*Pinus ponderosa* (Laws)	"Oleoresin," α-pinene	Vite & Pitman (1969)
Scolytus multistriatus (Marsham)	Smaller elm bark beetle	Adult	Decaying hardwood	Syringaldehyde, vanillin	Meyer & Morris (1967)
Popilia japonica	Japanese beetle	Adult		Geraniol Citronellol	Wilde (1957)

Table 2.--Continued.

Insect					
Scientific name	Common name	State	Plant	Compound	Reference
Enoclerus sphegeus (Fab.)	Checkered beetle	Adult	Douglas fir, Ponderosa pine, grand fir	α-pinene β-pinene limonene	Harwood & Rudinsky (1966)
Enoclerus undatulus (Say)	Checkered beetle		same	same	same
Laspeyresia pomonella (L.)	Codling moth	Adult		esters, oil of cloves, oil of citronella, alcohols, acids, propionate, pine tar oil, Anethole, terpene alcohols Terpinyl acetate	Eyer & Rhodes (1931) Dethier (1947 Masden & Falcon (1960)
Grapholitha (=Laspeyresia) molesta Busck	Oriental fruit			Terpinyl acetate	Brunson (1955)
Diabrotica undecimpunctata howardi (Barber)	Spotted cucumber beetle	Adult	Cucumis melo (L.) Cucurbitae foetidissima (HBK.)	Cucurbitacins	Benepal & Hall (1967)
Kalotermes flavicollis (Fab.); Zootermopsis nevadensis (Hagen); Heterotermes indicola (Wasmann) Reticulitermes	Termites		Wood infected Basidomycetes	Vanillic acid p-hydroxybenzoic acid p-coumaric acid protocatechuic acid ferulic acid	Becker (1964)
Leptinotarsa decemlineata (Say)	Colorado potato beetle		Potato leaves	Alcoholic extract	Hsaio & Fraenkel (1968b)
Pectinophora gossypiella	Pink bollworm		Gossypium sp.	Volatiles	Keller & Sheets (1968)
Bruchophagus roddi (Gus.)	Alfalfa seed chalcid	Adult	Alfalfa	β-Carotene, niacin, Vit. D_2, cholesterol, diethylstilbesterol, DL-aspartic acid, L-proline, histidine, Pangamic acid	Kamov & Frank (1964)
Dysdercus koengii (F.)	Red cotton bug	Adult	Cotton	Ether extract	Saxena (1964)

Table 2.--Continued.

Insect			Plant	Compound	Reference
Scientific name	Common name	State			
Hylotrupes bajulus (Gyll.) & *H. ater* Payk. (Col., Scolytidae)	Bark beetles	Female adults	*Pinus* sp. and others	α-pinene	Pettunen (1957)
Sitophilus oryzae (L.)	Rice weevil	Adult	Rice grains	Ether extract, methyl ketones	Hondo, et al. (1969)
Costelytra zealandica (White)	Grass grub beetle	Adult	Elder, *Sambucus niger* (L.)	Essential oil	Osborne & Hoyt (1968)
Cerambycid sp.	Wood boring beetles	Adult	Wood	Turpentine, smoke	Gardiner (1957)
Ostrinia nubilalis (Hubner)	European corn	Adult	Corn	Essential oil	Moore (1928)
Apis mellifera (L.)	Honey bees	Adult	General	Geraniol	Free (1962)

Table 3.--Feeding deterrents.

Insects					
Scientific name	Common name	State	Plant	Compound	Reference
Leptinotarsa decemlineata (Say)	Colorado potato	Adult & larvae	Potato, *S. demisum* L.	Demissine	Schreiber (1958) Kuhn & Gauhe (1947)
Leptinotarsa decemlineata (Say)	Colorado potato	Adult & larvae	Potato *S. tuberosum*	dihydro-alpha-solanin	Kuhn & Low (1955)
			S. chacoense	Leptines	Kuhn & Low (1957) Sturckow & Low (1961)
			S. caulescens	Solacaulin	Kuhn & Low (1957)
			S. dulcamare	Soladulcin	Schreiber (1958)
			S. polyademium	Tomatin	Kuhn & Low (1955) Schreiber (1958)
			S. schreiteri	Solanin	Kuhn & Low (1955) Schreiber (1958)
			S. punae	Solanin	Kuhn & Low (1955) Schreiber (1958)
			S. acaulia	Solacaulin	Kuhn & Low (1955) Schreiber (1958)
Anthonomus grandis Boh.	Boll weevil	Adult	Rose of Sharon (calyx)	Unknown	Maxwell, et al. (1965)
Listroderes costirustris obliguus Klug	Vegetable weevil		Sweet clover, *Melilotus officianalis* L.	Coumarin	Matsumoto (1963)
Ostrinia nubilalis (Hubner)	European corn		Corn	6-Methoxybenzo-xazolinone, 2,4-dihydroxy-2-methoxy-1,4-benzoxazine-3-one	Beck & Smissman (1960) Smissman, et al. (1957a, b) Klun, et al. (1967)
Oraesia excavata	Fruit-piercing moth	Adult	*Cocculus trilobus* (DC)	Cocculolidine, isoboldine	Wada & Munakata (1968)
Oraesia emarginata (Fab.)	Fruit-piercing moth		*Cocculus trilobus* (DC)	same	same
Abraxas miranda (Butler)			*Cocculus trilobus* (DC)	same	same
Prodenia litura F.			*Cocculus trilobus* (DC)	same	same
Prodenia litura F.			*Clerodendron tricotomum* (Thunb.)	Clerodendrin A	Munakata (1970)
			Parabenzoin trilobum Nakoi	Shiromodiol di-acetate, Shiromodiol monoacetate	Wada, et al. (1968)

Table 3.--Continued.

Insect					
Scientific name	Common name	State	Plant	Compound	Reference
Trichoplusia ni	Cabbage looper	Larvae	Laboratory diets	Ascorbic acid	Gothilf & Beck (1967)
Two ant species	Phytophagous insects		Catnip, Nepeta cataria	Nepetalactone	Eisner (1964)
Melanoplus bivittatus	Twostriped grasshopper	Larvae	Catnip, Nepeta cataria	Nornicotine dipicrate solanine, tomatine, digitonin, saponin, santonin, indicane diosegin, hecogenin, lukenine	Harley & Thorsteinson (1967)
Scolytus multistriatus (Marsham)	Bark beetle	Adult	Carya ovata bark	Juglone	Gilbert, et al. (1967)
Schistocerca gregaria	Desert locust	Adult	Nelm tree, Azadirachta indica melia, Azerdarach scilla maritima	Azadirachtin, glycoside	Gill & Lewis (1971) Butterworth & Morgan (1971) Bhatia (1940) Chauvin & Mentzer (1951)
Various			Solanum luteum Solanum lycopersicum	Unknown	Bongers (1970
Heliothis zea (Boddie)	Corn earworm	Larvae	Corn silks	Unknown	Straub & Fairchild (1970)
Homoptera Empoasca fabae (Harris)	Leafhopper	Nymphs	Solanum sp.	Tomatine, other alkaloids & alkaloid glycosides	Dahlman (1965)
Various termites			Teak heartwood	2-Methyl-, 2 hydroxy-methyl-, and 2-formyl-anthraquinones	
Drosophila melonagaster (Meig)	Vinegar fly	Adults	Turnip Brassica rapa L.	2-phenylethyl-isothiocyanate	Lichtenstein, et al. (1962)

Table 4.--Repellents.

Insect			Plant	Compound	Reference
Scientific name	Common name	State			
Leptinotarsa decemlineata (Say)	Colorado potato beetle	Adult & larvae	Tomato, S. esculentum Mill.	Tomatin	Schreiber (1958)
Leptinotarsa decemlineata (Say)	Colorado potato	Adult & larvae	Pepper, S. capsicum	Capsaicin	Schreiber (1958)
Leptinotarsa decemlineata (Say)	Colorado potato	Adult & larvae	Tobacco, Nicotiana tobacum L.	Nicotine	Buhr, et al. (1954) Trouvelot (1958)
Heliothis zea (Boddie)	Corn earworm	Adult	Corn, Zea mays L.	Essential oil	Starks, et al. (1965)
Protoparce sexta (Johan.)	Tobacco hornworm	Adult	Nicandria sp.	Alcohol $C_{22}H_{27}O$	Fraenkel, et al. (1960)
Reticulitermes sp.	Termite	Adult	Cashew, Anacardium occidentale L.	Anacardic acid	Wolcott (1946)
Dendroctomus brevicomis LeConte	Western pine beetle	Adult	Ponderosa pine	Myrcene, limonene	Smith (1966)
	Locust	Adult	Neem seedcake	sulfur containing	Sinha and Gulati (1966)
Shistocerca gregaria	Desert locust	Adult	General, corn	N-valeric acid iso-valeric acid Unknown	Haskell, et al. (1961) Goodhue (1963) Mulkern (1967)
Hypera postica (Gyllenhal)	Alfalfa weevil		Alfalfa	Tannic acid	Bennett (1965)
Brucho phagus roddi (Gus.)	Alfalfa seed	Adult	Alfalfa	Butyric acid, coumarin shikimic acid, pyridoxine, succinic acid, xanthophyll, malic acid, betaine	Kamon & Frank (1964)

Table 5.--Oviposition stimulant.

Insect					
Scientific name	Common name	State	Plant	Compound	Reference
Leptinotarsa decemlineata (Say)	Colorado potato beetle	Adult	*S. tuberosum*	Lecithin	Grison (1958)
Protoparce sexta (Johan.)	Tobacco hornworm	Adult	Tomato	glycoside, $C_{17}H_{29}O_{10}N$	Yamamoto & Fraenkel (1960) Fraenkel, et al. (1960)
Plutella maculipennis (Curt)	Diamond-back moth	Adult	*Brassica nigra* L. & other Cruciferae	Allyl isothiocyanate	Gupta & Thorsteinson (1960)
Erioischia brassicae (Boughe)	Cabbage root		Swede, *Brassicae napus* L.	Sinigrin, beta-phenylethylamine, allyl ___	Traynier (1965)
Hylemia antiqua (Meig.)	Onion maggot	Adult	Onion	N-propyl disulfide, methyl disulfide, N-propyl mercaptan, N-propyl alcohol	Matsumoto & Thorsteinson (1968)
Shistocerca gregaria	Desert locust	Adult	*Commiphora myrrha*	Alpha-pinene, beta-pinene, limonene, eugenol.	Carlisle, et al. (1965)
Psila rosae	Carrot rust fly	Adult	Carrot leaves	Methyl isoeugenol	Berentes & Staedler (1971)
			Carrot	trans-1,2-dimethoxy-4-propenyl-benzene	Berentes & Staedler (1971)
Ostrinia nubilalis	European corn borer	Female adults	*Zea mays*	"Gibberellin-like" acytokinin, others	Schurr (1970)
Aedes aegypti (L.)	Mosquito	Adult	--	ethylacetate, methyl propionate methyl butyrate	Perry & Fay (1967)
Heliothis virescens (F.)	Robacco budworm	Adult	Tobacco	Unknown	Deutsch (1968)
Cryptorrhynchus lapathi L.	Poplar leaf perforator	Adult	Poplar	Unknown	Cadahia (1965)
Callosobruchus bruchus chinensis (L.)	Multivolive Bruchid		Soybeans	Saponins, SBSE, Urease	Applebaum, et al. (1965)

Table 6.--Feeding incitants.

Insect					
Scientific name	Common name	State	Plant	Compound	Reference
Bombyx mori L.	Silkworm	Larvae	Mulberry	Beta-sitosterol, beta sitosterol glucoside, lupeol	Hamamura, et al. (1961) Nyar & Fraenkel (1962) Goto (1965)
Epilachna varivestis (Mulsant)	Mexican bean	Adult	Phaseolus sp.	Phaseolunatin Lotaustrin	Klingenburg & Bucher (1960)
Bombyx mori L.	Silkworm	Larvae	Mulberry	isoquercitrin	Hamamura, et al. (1962)
Bombyx mori L.	Silkworm	Larvae	Mulberry	2',3,4',5,7-penta-hydroxy-flavone	Hayashira (1965)
Leptinotarsa decemlineata	Colorado potato		Potato leaves	Alcoholic extract	Hsaio & Fraenkel (1968b)
Dysidercus koenigii (F.)	Red cotton bug		Cottonseed	lipids	Saxena (1964)
Flight termination					
Sitona cylindricollis Fab.	Sweet clover weed	Adult	Melilotus officianalis L.	Coumarin	Heidewig & Thorsteinson (1961)
Bombyx mori (L.)	Silkworm	Larvae	Mulberry	Cellulose, water extract silica potassium phosphate	Hamamura, et al. (1962
Musca domestica (L.)	House fly		Laboratory diet	Protein, 1-leucine, guanosine-5'-mono-phosphate	Yamamoto & Jensen (1967)
Mating factor					
Polyphemus	Moth		Oak leaves	trans-2-hexenal	Reddiford & Williams (1967) Reddiford (1967)
Growth factor					
Bombyx mori (L.)	Silkworm	Larvae		chlorogenic acid	Kato & Yamada (1966)
	Alfalfa aphid	Nymphs	Alfalfa	Unknown	Kindler & Staples (1969)
Oviposition deterrent					
Callosobruchus chinensis (L.)	Multivoltive Bruchid		Soybeans	Sapongenin	Applebaum, et al. (1965)

References Cited

Akeson, W. R., H. J. Gorz, F. A. Haskins, G. R. Manglitz. 1968. A water soluble factor in Melilotus officinalis leaves which stimulates feeding by the adult sweet clover weevil. J. Econ. Entomol. 61: 1111-1112.

Allen, G. E. and T. L. Pate. 1966. The potential role of a feeding stimulant used in combination with the nuclear polyhedrosis virus of Heliothis. J. Invertebr. Pathol. 8: 129-131.

Andrea, L. 1964. Summary of observations on the host range and development of Phrydiuchus toparius Germ. and Phrydiuchus sp. (Curculionidae) on Salvia spp. (Labitae). USDA Spec. Rept.

Applebaum, S. W., B. Gestetner and Y. Bik. 1965. Physiological aspects of host specificity in the Bruchidae. IV. Developmental incompatability of soybeans for Callosobruchus. J. Insect Physiol. 11: 611-616.

Baker, J. and D. Norris. 1968. Behavioral responses of the smaller European elm bark beetle, Scolytus multistriatus, to extracts of non-host tree tissues. Entomol. Expl. & Appl. 11: 464-469.

Beck, S. D. 1965. Resistance of plants to insects. Ann. Rev. Entomol. 10: 207-232.

Beck, S. D. and E. E. Smissman. 1960. The European corn borer, Pyraustia nubilalis, and its principal host plant. VIII. Laboratory evaluation of host resistance to larval growth and survival. Ann. Entomol. Soc. Amer. 53: 755-762.

Becker, von Gunther. 1964. Termiten-anlockende Wirkung einiger bei Basidiomyceten-Angriff in Holz entstehender Verbindungen. Holzforschung 18(6): 168-172.

Benepal, P. S. and C. V. Hall. 1967. Biochemical composition of plants of Cucurbita foetidissima Hbdk. and Cucumis melo L. as related to cucumber beetle feeding. Amer. Soc. Hort. Sci. 91: 353-359.

Bennett, S. E. 1965. Tannic acid as a repellent and toxicant to alfalfa weevil larvae. J. Econ. Entomol. 58(2): 372.

Beroza, M. and N. Green. 1963. Materials tested as insect attractants. Agr. Handbook No. 239 (USDA). 148 pp.

Berentes, J. and E. Staedler. 1971. An oviposition stumulator for the carrot rust fly from carrot leaves. Z. Naturforsch. B. 26: 339-340.

Brunson, M. H. 1955. Effect on the Oriental fruit moth of parathion and EPN applied to control the plum curculio on peach. J. Econ. Entomol. 48: 390-392.

Butterworth, J. H. and E. D. Morgan. 1971. Investigation of the locust feeding inhibition of the seeds of the Nelm tree, Azadirachta indica. J. Insect Physiol. 17: 969-977.

Byrne, H. D. and A. L. Steinauer. 1966. The attraction of the alfalfa weevil, Hypera postica (Coleoptera, Curculionidae) to alfalfa. Entomol. Soc. Amer. 59: 303-309.

Buhr, H. 1954. Beobachtungen uber Parisitindefall an Pfropfungen und Chimaren von Pflnzen. Zuchter 24(7/8) 185-193.

Cadahia, A. 1965. Preferencias clonales del gorgjo perforador del chopo Cryptorrhynchus lapathi L. (Coleoptera, Curculionidae). Boln. Serv. Plagas 8(16): 115-125.

Cade, S., B. Hruitfiord, and R. Gara. 1970. Identification of primary attractant for Gnathotrichus sulcatus isolated from western hemlock logs. J. Econ. Entomol. 63: 1010-1015.

Carlisle, D. B., P. E. Ellis and E. Betts. 1965. The influence of aromatic shrubs on sexual maturation in the desert locust Schistocerca gregaria. J. Insect Physiol. 11: 1541-1558.

Chambliss, O. Y. and C. M. Jones. 1966. Cucurbitacins: Specific insect attractants in Cucurbitaceae. Science 153: 1392-1393.

Chauvin, R. 1952. Nouvelles recherches sur les substances qui attirent le doryphore (L. decemlineata Say) vers la pome de terre. Annales de L'I.N.R.A. 3: 303-308.

Chauvin, R. and C. Mentzer. 1951. Contribution a l étude des substances naturelles et de synthese repulsives pour les acridians. Bull. Offic. Natl. Anti-Acrid., 1: 5-14.

Cuille, J. 1950. Recherches sur le chavancon du bananier. Inst. Fruits et Agrumes Coloniaux, Ser. Tech. 4.

DaCosta, C. P. and C. M. Jones. 1971. Cucumber beetle resistance and mite susceptibility controlled by the bitter gene in Cucumis sativus L. Science 172: 1145-1146.

Dahlman, D. L. 1965. Responses of Empoasca fabae (Harris) to selected alkaloids and alkaloidal glycosides of Solanum species. Ph.D. thesis. Dissertation Abstr. 6245.

David, W. A. L. and B. O. C. Gardiner. 1966. Mustard oil glucosides as feeding stimulants for Pieris brassicae larvae in a semi-synthetic diet. Entomol. Expl. & Appl. 9: 247-255.

Davis, G. R. F. 1961. The biting response of larvae of the prairie grain wireworm, Ctenicera aeripennis destructor (Brown) (Coleoptera: Elateridae) to various extracts of germinating rye seed. Canadian J. Zool. 39: 299-303.

Davis, G. R. F. 1965. The effect of structural analogs of methionine and glutamic acid on larvae of the prairie grain wireworm, Ctenicera destructor. Arch. Intern. Physiol. Biochem. 73(2): 177-187.

Derr, R. F., D. D. Randall and R. W. Kieckhefer. 1964. Feeding stimulant for western and northern corn rootworm adults. J. Econ. Entomol. 57(6): 963-965.

Dethier, V. G. 1937. Gustation and olfaction in lepidopterous larvae. Biol. Bull. 72: 7-23.

Dethier, V. G. 1941. Chemical factors determining the choice of food plants by Papilio larvae. Amer. Naturalist 75: 61-73.

Dethier, V. G. 1947. Chemical insect attractants and repellents. The Blakiston Co., Philadelphia. 289 pp.

Dethier, V. G. 1970. Some general considerations of insect responses to the chemicals in food plants. In Control of Insect Behavior by Natural Products, ed. by D. L. Wood, R. M. Silverstein, M. Nakajima. Academic Press, New York., pp 21-28.

Deutsch, R. G. 1968. Ovipositional preferences of the tobacco budworm, Heliothis virescens (Fabricius). M.S. Thesis, Texas A&M Univ. 35 pp.

Dosthotch, R., S. Chatterji, J. Peacock. 1970. Elm bark derived feeding stimulants for the smaller European elm bark beetle. Science 167: 380-382.

Ehrlich, P. R. and P. H. Raven. 1967. Butterflies and plants. Scien. Amer. 216(6): 104-113.

Eisner, T. 1964. Catnip: Its raison d'etre. Science 146: 1318-1320.

Euw, J., L. Fishelson, J. Parsons, T. Rechstein and M. Rothschild. 1967. Cardenolides (heart poisons) in a grasshopper feeding on milkweeds. Nature 214: 35-38.

Eyer, J. R. and H. Rhodes. 1931. Preliminary notes on the chemistry of codling moth baits. J. Econ. Entomol. 24: 702-711.

Feir, D. and S. D. Beck. 1963. Feeding behavior of the large milkweed bug, Oncopeltus fasciatus. Ann. Entomol. Soc. Amer. 56: 224-229.

Folsom, J. W. 1931. A chemotropometer. J. Econ. Entomol. 24: 827-833.

Force, D. C. 1966. Reactions of the green dock beetle, Gastrophysa cyanea (Coleoptera: Chrysolelidae), to its host and certain nonhost plants. Ann. Entomol. Soc. Amer. 59(6): 1119-1125.

Fraenkel, G. 1959. The raison d'etre of secondary plant substances. Science 129: 1466-1470.

Fraenkel, G., J. K. Nayer, O. Nalbandov and R. T. Yamamoto. 1960. Further investigations into the chemical basis of the insect-host-plant relationship. Proc. Intern. Congr. Entomol. 11th Vienna 3: 122-126.

Free, J. B. 1962. The attractiveness of geraniol to foraging honey bees. J. Apic. Res. 1: 52-54.

Gardiner, L. M. 1957. Collecting wood-boring beetle adults by turpentine and smoke. Canada Dept. Agr. Forest Biol., Bimonthly Prog. Rept. 13: 1-2.

Gilbert, B., J. Baker and D. Norris. 1967. Juglone (5-hydroxy-1,4-naphthoquinone) from Carya ovata, a deterrent to feeding by Scolytus multistriatus. J. Insect Physiol. 13: 1453-1459.

Gill, J. and C. Lewis. 1971. Systemic action of an insect feeding deterrent. Nature 232: 402-403.

Goodhue, D. 1963. Feeding stimulants required by a polyphagous insect, Schistocerca gregaria. Nature 197: 405-406.

Gothilf, S. and S. D. Beck. 1967. Larval feeding behavior of the cabbage looper, Trichoplusia ni. J. Insect Physiol. 13(7): 1039-1053.

Grevillius, A. Y. 1905. Zur Kenntniss der Biologie des Goldafters (Euproctis chrysorrhoea Hbn.). Bot. Centralbl. Beihefte 18 Abt. II. 22.

Grison, P. A. 1958. L'influence de la plant hote sur la fecondite de l'insecte phytophage. Entomol. Exptl. & Appl. 1: 73-93.

Guenther, E. 1949. The essential oils. Vol. 2, D. van Nostrand Co., New York.

Guerra, A. and T. Shaver. 1968. A bioassay technique for screening feeding stimulants for larvae of tobacco budworm. J. Econ. Entomol. 61: 1393-1399.

Gupta, P. D. and A. J. Thorsteinson. 1960. Food plant relationships of the diamond-back moth (Plutella maculipennis (Curt.)), II. Sensory regulation of the adult female. Entomol. Exptl. & Appl. 3: 305-314.

Hamamura, Y. 1959. Food selection by silkworm larvae. Nature 183: 1746-1747.

Hamamura, Y., K. Hayashiya and K. Naito. 1961. Food selection by silkworm larvae, Bombyx mori. II. Betasitosterol as one of the biting factors. Nature 190(4779): 880-881.

Hamamura, Y., K. Hayashiya, K. Naito, K. Matsuura, and J. Nishida. 1962. Food selection by silkworm larvae. Nature 194: 754-755.

Hamamura, Y. 1970. The substances that control the feeding behavior and growth of the silkworm, Bombyx mori L., In Control of Insect Behavior by Natural Products, ed. by D. L. Wood, R. M. Silverstein, M. Nakajima. New York: Academic Press, pp. 55-80.

Hardee, D. D., W. H. Cross and E. B. Mitchell. 1969. Male boll weevils are more attractive than cotton plants to boll weevils. J. Econ. Entomol. 62: 165-169.

Harris, P. 1963. Host specificity of Calophasia bunula (Hbn.) (Lepidoptera: Noctuidae). Canadian Entomol. 95: 105.

Harris, P. 1967. Host specificity of Calophasia lunula (Hbn.) Canadian Entomol. 99: 1304-1310.

Harwood, W. G. and J. A. Rudinsky. 1966. The flight and olfactory behavior of checkered beetles (Coleoptera: Cleridae) predatory on the Douglas fir beetle. Tech. Bull. 95, Agr. Exp. Sta., Corvallis, Oregon. 36 pp.

Harley, K. and A. J. Thorsteinson. 1967. The influence of plant chemicals on the feeding behavior development and survival of the twostriped grasshopper, Melanoplus bivittatus. Canadian J. Zool. 45: 305-319.

Haskell, P. T., W. J. Paskin and J. E. Moorhouse. 1961. Laboratory observations on factors affecting the movements of hoppers of the desert locust. J. Insect Physiol. 8: 53-78.

Hedin, P. A., A. C. Thompson, and R. C. Gueldner. 1966. Constituents of the cotton bud. III. Factors that stimulate feeding by the boll weevil. J. Econ. Entomol. 59: 181-185.

Hedin, P. A., L. R. Miles, A. C. Thompson and J. P. Minyard. 1968. Constituents of the cotton bud. Formulation of a boll weevil feeding stimulant mixture. J. Agr. Food Chem. 16: 505-513.

Hedin, P. A., A. C. Thompson, R. C. Gueldner and J. P. Minyard. 1971. Isolation of alpha-bisabolol from the cotton bud. Phytochemistry 10: 1693-1694.

Hedin, P. A., A. C. Thompson, R. C. Gueldner and J. M. Ruth. 1972. Isolation of bisabolene oxide from the cotton bud. Phytochemistry (in press).

Heidweg, H. and A. J. Thorsteinson. 1961. The influence of physical factors and host plant odor on the induction and termination of dispersal flights in Sitona cylindricollis Fahr. Entomol. Exptl. & Appl. 4: 165-177.

Heikkenen, H. J. and B. F. Hriutfiord. 1965. Dendroctonus pseudotsygae: A hypothesis regarding its primary attractant. Science 150: 1457-1459.

Heron, R. J. 1965. The role of chemotactic stimuli in the feeding behavior of spruce budworm larvae on white spruce. Canadian J. Zool. 43: 247-269.

Honda, H., I. Yamamoto and R. Yamamoto. 1969. Attractant for rice weevil, Sitophilus zeamais. (Motschulsky) from rice grains. Appl. Entzool. 4: 23-31.

Horie, Y. 1962. Effects of various fractions of mulberry leaves on the feeding of the silkworm, Bombyx mori L. J. Sericult. Sci. Japan 31: 258-264.

Hovanitz, W. and V. C. S. Chang. 1963. Comparison of the selective effect of two mustard oils and their glucosides to Pieris larvae. J. Res. on the Lepidoptera 2(4): 281-288.

Howell, G. S., Jr., F. G. Maxwell and R. B. Nevins. 1965. Responses of the hickory shuckworm to certain extracts of pecan. Ann. Entomol. Soc. Amer.

Howlett, F. M. 1912. The effect of oil of citronella on two species of Dacus. Trans. Entomol. Soc. London 1912: 412-418.

Howlett, F. M. 1915. Chemical reactions of fruit flies. Bull. Entomol. Res. 6: 297-305.

Hsiao, T. H. 1966. The host plant specificity of the Colorado potato beetle, Leptinotarsa decemlineata (Say). Ph.D. Dissertation, Univ. of Illinois. 1966. 199 pp.

Hsiao, T. 1969. Adenine and related substances as potent feeding stimulants for the alfalfa weevil, Hypera postica. J. Insect Physiol. 15: 1785-1790.

Hsiao, T. H. and G. Fraenkel. 1968a. Isolation of phagostimulative substances from the host plant of the Colorado potato beetle. Ann. Entomol. Soc. Amer. 61(2): 476-484.

Hsiao, T. H. and G. Fraenkel. 1968b. The role of secondary plant substances in the food specificity of the Colorado potato beetle. Ann. Entomol. Soc. Amer. 61(2): 485-493.

Ito, T. 1961. Nutrition of the silkworm, Bombyx mori. IV. Effects of ascorbic acid. Bull. Sericult. Exptl. Sta. (Tokyo) 17: 119-136.

Jenkins, J. N., F. G. Maxwell, J. C. Keller and W. L. Parrott. 1963. Investigations of the water extracts of Gossypium, Abelmoschus, Cucumis and Phaseolus for an arrestant and feeding stimulant for Anthonomus grandis Boh. Crop Sci. 3: 215-219.

Johnson, N. and P. Belluschi. 1969. Host finding behavior of the Douglas fir beetle. J. Forestry 67: 290-295.

Kamon, J. A. and W. D. Frank. 1964. Olfactory response of the alfalfa seed chalcid Bruchophagus voddi Guss., to chemicals found in alfalfa. Univ. of Wyoming Agr. Exp. Sta. Bull. 413. 35 pp.

Kangas, E., V. Pertunnen, H. Oksanen and M. Rinne. 1965. Orientation of Blastophagus pineperda L., to its breeding material. Attractant effect of alpha terpineol isolated from pine rind. Ann. Entomol. Fenn. 31: 61-73.

Kato, M. and H. Yamada. 1966. Silkworm requires 3,4-dihydroxybenzene structure of chlorogenic acid as a growth factor. Life Sci. 5: 717-722.

Kearns, H. G. H. 1931. Ann. Rept. Agr. Hort. Sta., Long Ashton. 199.

Keller, J. C., F. G. Maxwell and J. N. Jenkins. 1962. Cotton extracts as arrestants and feeding stimulants for the boll weevil. J. Econ. Entomol. 55(5): 800-801.

Keller, J. C., F. G. Maxwell, J. N. Jenkins and T. B. Davich. 1963. A boll weevil attractant from cotton. J. Econ. Entomol. 56(1): 110-111.

Keller, J. C. and T. B. Davich. 1965. Response of 5 species of insects to water extracts of their host plants. J. Econ. Entomol. 58(1): 165.

Keller, J. C. and L. W. Sheets. 1968. Personal communication.

Kennedy, J. and J. Moorehouse. 1969. Laboratory observations on locust responses to windborne grass odor. Entomol. Exptl. & Appl. 12: 487-503.

Kindler, S. and R. Staples. 1969. Behavior of the spotted alfalfa aphid on resistant and susceptible alfalfas. J. Econ. Entomol. 62: 474-478.

Klingenburg, M. and J. Bucher. 1960. Biological oxidations. Ann. Rev. Biochem. 29: 669-708.

Klun, J., C. Tipton and T. Brindley. 1967. 2,4-Dihydroxy-7-methoxy-1,4-benzoxazin-3-one (DIMBOA), an active agent in the resistance of maize to the European corn borer. J. Econ. Entomol. 60: 1529-1533.

Kuhn, R. and A. Gauhe. 1947. Uber die Bedeutung des Demissins fur die resistance von Solanum demissum gegen die larven des Kartoffelkafers. Z. Naturforsch. 2: 407-409.

Kuhn, R. and I. Low. 1955. Resistance factors against Leptinotarsa decemlineata (Say), isolated from the leaves of wild Solanum species, in Origins of Resistance to Toxic Agents, 122-132. Academic Press, New York.

Kuhn, R. and I. Low. 1957. Nene alkaloidglykoside in den Blattern von Solanum chacoense. Angew. Chem. 69: 236.

Lagerheim, G. 1900. Zur frage der schutzmittel der pflanzen gegen raupenfrass. Entomol. Tidskv. 21: 209-232.

Lichtenstein, E. P., F. M. Strong and D. G. Morgan. 1962. Identification of 2 phenylethyl-isothiocyanate as an insecticide occurring naturally in the edible part of turnips. J. Agr. Food Chem. 10: 30-33.

Loschiavo, S. R. 1965. The chemosensory influence of some extracts of brewers' yeast and cereal products on the feeding behavior of the confused flour beetle, Tribolium confusum. Ann. Entomol. Soc. Amer. 58: 526-588.

Loschiavo, S. R., S. D. Beck and D. M. Norris. 1963. Behavioral responses of the smaller European elm bark beetle, Scolytus multistriatus, to extracts of elm bark. Ann. Entomol. Soc. Amer. 56: 764-768.

Loschiavo, S. R. 1965. The chemosensory influence of some extracts of brewers' yeast and cereal products on the feeding behavior of the confused flour beetle, Tribolium confusum (Coleoptera: Tenebrionidae). Ann. Entomol. Soc. Amer. 58(4): 576-588.

Madsen, H. F. and L. A. Falcon. 1960. DDT-resistant codling moth on pears in California. J. Econ. Entomol. 53: 1083-1085.

Matsumoto, Y. A. 1962. A dual effect of coumarin, olfactory attraction and feeding inhibition on the vegetable weevil adult, in relation to the uneatability of sweet clover leaves. Japan J. Appl. Entomol. Zool. 6: 141-149.

Matsumoto, Y. and S. Sugiyama. 1960. Studies on the host plant determination of the leaf-feeding insects. V. Attraction of leaf alcohol and some aliphatic alcohols to the adult and larvae of the vegetable weevil. Ber. Ohara Inst. Landwirtsch. Biol. 11: 359-364.

Matsumoto, Y. and A. J. Thorsteinson. 1968. Effect of organic sulfur compounds on oviposition in the onion maggot, Hylemya antiqua. Appl. Ent. Zool. 3: 5-12.

Maxwell, F. G., J. N. Jenkins, J. C. Keller and W. L. Parrott. 1963. An arrestant and feeding stimulant for the boll weevil in water extracts of cotton plant parts. J. Econ. Entomol. 56(4): 449-454.

Maxwell, F. G., W. L. Parrott, J. N. Jenkins, H. N. Lafever. 1965. A boll weevil feeding deterrent from the calyxes of an alternate host, Hibiscus syriacus. J. Econ. Entomol.

McGovern, T. P., M. Beroza, P. H. Schwartz, D. W. Hamilton, J. C. Ingangi and T. L. Ladd. 1970. Methyl cyclohexanepropionate and related chemicals as attractants for Japanese beetles. J. Econ. Entomol. 63: 276.

McKibben, G. H., P. A. Hedin, T. B. Davich, R. J. Daum and M. W. Laseter. 1971. Addition of food acidulants to increase attractiveness to boll weevils of bait containing cottonseed oil. J. Econ. Entomol. 64: 583-585.

McMillian, W. W. and K. J. Starks. 1966. Feeding responses of some Noctuid larvae (Lepidoptera) to plant extracts. Ann. Entomol. Soc. Amer. 59(3): 516-519.

Meyer, H. J. and P. M. Norris. 1967. Vanillin and syringaldehyde as attractants for Scolytus multistriatus. Ann. Entomol. Soc. Amer. 60(4): 858-859.

Minyard, J. P., D. D. Hardee, R. C. Gueldner, A. C. Thompson, G. Wiygul and P. A. Hedin. 1969. Constituents of the cotton bud. Compounds attractive to the boll weevil. J. Agr. Food Chem. 17: 1093-1097.

Moore, R. H. 1928. Odorous constituents of the corn plant in their relation to the European corn borer. Proc. Oklahoma Acad. Sci. 8: 16-18.

Morgan, A. C. and S. C. Lyon. 1928. Notes on amylsalicylate as an attractant to the tobacco hornworm moth. J. Econ. Entomol. 21: 189-191.

Mulkern, G. B. 1967. Food selection by grasshoppers. Ann. Rev. Entomol. 12: 59-78.

Munakata, K., T. Saito, S. Ogawa and S. Ishii. 1959. Oryzanone, an attractant of the rice stem borer. Bull. Agr. Chem. Soc. Japan, 23(1): 65-67.

Munakata, K. 1970. Insect antifeedants in plants. In Control of Insect Behavior by Natural Products, ed. by L. Wood, R. M. Silverstein and N. Nakajima. Academic Press, New York, pp. 179-187.

Nuto, T. and R. Sugawara. 1965. The house fly attractants in mushrooms. I. Extraction and activities of the attractive components in Amanita muscaria (L.) Fr. Agr. Biol. Chem. (Japan) 29: 949-955.

Nayar, J. K. and G. Fraenkel. 1962. The chemical basis of host selection in the Mexican bean beetle, Epilachna varvivestris. Ann. Entomol. Soc. Amer. 56: 174-178.

Nayar, J. K. and A. J. Thorsteinson. 1963. Further investigations into the chemical basis of insect-host relationships in an oligophagous insect, Plutella maculipennis (Curtis). Canadian J. Zool. 41: 923-929.

Osborne, G. O. and C. P. Hoyt. 1968. Preliminary notes on a chemical attractant for the grass grub beetle (Costelytra zealandica (White)) from the flowers of elder (Sambucus nigra L.). N. Zealand J. Sci. 11: 137-139.

Parker, J. R. 1924. Observations on the clear-winged grasshopper (Camnula pellucides). Minnesota Agr. Exp. Sta. Bull. 214. 44 pp.

Parrott, W. L., T. N. Shaver and J. C. Keller. 1968. A feeding stimulant for the pink bollworm in water extracts of cotton. J. Econ. Entomol.

Perry, A. S. and R. W. Fay. 1967. Correlation of chemical constituents and physical properties of fatty acid esters with oviposition response of Aedes aegypti. Mosquito News 27(2): 175-183.

Perttunen, V. 1957. Reactions of two bark beetle species, Hylurgops palliatus Gyll. and Hyastesater Payk. (Col., Scolytidae) to the terpene alpha-pinene. Ann. Entomol. Fennici 23: 101-110.

Peterson, A. 1924. Some chemical attractive to adults of the onion maggot (Hylemia antiqua Meig.) and the seed corn maggot (Hylemia cilicrura Rond.). J. Econ. Entomol. 17: 87-94.

Rees, C. J. C. 1966. A study of the mechanism and functions of chemoreception especially in some phytophagous insects in Calliphora. Ph.D. Dissertation, Univ. of Oxford, England.

Reddiford, L. 1967. Trans-2-hexenal: Mating stimulant for Polyphemus moths. Science 158: 139-140.

Reddiford, L. M. and C. M. Williams. 1967. Volatile principle from oak leaves: Role in sex life of the Polyphemus moth. Science 155: 589-590.

Robbins, W. E., M. J. Thompson, R. T. Yamamoto, and T. J. Shortino. 1965. Feeding stimulants for the female house fly, Musca domestica L. Science 147(3658): 628-630.

Rudman, P. and F. J. Gay. 1961. Causes of the natural durability in timber. V. The role of extractives in the resistance of Eucalyptus microcorys to attack of the subterranean termite, Nasutitermes exitiosis. Hotzforschung 15: 50-53.

Saeto, T. and K. Munakata. 1970. Insect attractants of vegetable origin. In Control of Insect Behavior by Natural Products, ed. by D. L. Wood, R. M. Silverstein and N. Nakajima. Academic Press, New York. pp. 225-235.

Sakan, T., S. Issac and S. Hyeon. 1970. The chemistry of attractants for Chrysopidae from Actinidia polygama (Mig.) in Control of Insect Behavior by Natural Products, ed. by D. L. Wood, R. M. Silverstein and N. Nakajima. Academic Press, New York, pp. 237-247.

Saxena, K. N. 1964. Control of the orientation and feeding behavior of red cotton bud, Dysdercus koenigii (F.) by chemical constituents of the plant. Proc. XII Int. Cong. Entomol., London. p. 294.

Schreiber, K. 1958. Uber einige Inhaltsstoffe der Solanaceen und ihre Bedeutung fur die Kartoffelkaferresistenz. Entomol. Exptl. & Appl. 1: 28-37.

Schurr, K. 1970. Host Plant Interactions. Adv. Front & Plant Sci. 25: 147.

Seybert, J. and R. Gara. 1970. Notes on flight and host selection behavior of the pine engraver, Ips pini. Ann. Entomol. Soc. Amer. 63: 947-950.

Smissman, E. E., J. P. Lapidus and S. D. Beck. 1957a. Corn plant resistant factor. J. Organ. Chem. 22: 220.

Smissman, E. E., J. P. Lapidus and S. D. Beck. 1957b. Isolation and synthesis on an insect resistance factor from corn plants. J. Amer. Chem. Soc. 79: 4697.

Smith, R. H. 1966. The monoterpene composition of Pinus ponderosa xylern resin and of Dendroctonus brevicomis pitch tubes. For. Sci. 12: 63-68.

Starks, K. J., W. W. McMillian, A. A. Sekul and H. C. Cox. 1965. Corn earworm larval feeding response to corn silk and kernel extracts. Ann. Entomol. Soc. Amer. 58(1): 74-76.

Steiner, L. F. 1952. Methyl eugenol as an attractant for oriental fruit fly. J. Econ. Entomol. 45: 241-248.

Straub, R. and M. Fairchild. 1970. Laboratory resistance studies in corn to the corn earworm. J. Econ. Entomol. 63: 1901-1903.

Stride, G. O. 1965. Studies on the chemical basis of host plant selection in the genus Epilachna (Coleoptera, Coccinellidae). I. A volatile phago-stimulant in Solanum campylacanthum for Epilachna fulvosignata. J. Insect Physiol. 11: 21-22.

Struck, R. F., J. Frye, Y. F. Shealy, P. A. Hedin, A. C. Thompson and J. P. Minyard. 1968. Constituents of the cotton bud. IX. Further studies on a polar boll weevil feeding stimulant complex. J. Econ. Entomol. 61: 270-274.

Struck, R. F., J. Frye, Y. F. Shealy, P. A. Hedin, A. C. Thompson and J. P. Minyard. 1968. Constituents of the cotton bud. XI. Studies of a feeding stimulant complex from flower petals for the boll weevil. J. Econ. Entomol. 61: 664-667.

Sturckow, B. and I. Low. 1961. Die wirkung einiger Solanum - alkaloidglykoside auf den Kartoffelkafer, Leptinotarsa decemlineata Say. Entomol. Exptl. & Appl. 4: 133-142. North-Holland Publ. Co., Amsterdam.

Temple, C., E. C. Roberts, J. Frye, R. F. Struck, Y. F. Shealy, A. C. Thompson, J. P. Minyard and P. A. Hedin. 1968. Constituents of the cotton bud. XIII. Further studies on a nonpolar feeding stimulant for the boll weevil. J. Econ. Entomol. 61: 1388-1393.

Thomas, H. and G. Hertell. 1969. Responses of the boll weevil to natural and synthetic host attractants. J. Econ. Entomol. 62: 383-386.

Thompson, A. C., D. A. Baker, R. C. Gueldner and P. A. Hedin. 1971. Identification and quantitative analysis of the volatile substances emitted by maturing cotton in the field. Plant Physiol. 48: 50-52.

Thorsteinson, A. J. 1953. The chemotactic responses that determine host specificity in an oligophagous insect (*Plutella maculipennis* (Curt.)). Canadian J. Zool. 31: 52-72.

Thorsteinson, A. J. 1955. The experimental study of the chemotactic basis of host specificity in phytophagous insects. Canadian Entomol. 87: 49-57.

Thorsteinson, A. J. 1960. Host selection in phytophagous insects. Ann. Rev. Entomol. 5: 193-219.

Thorsteinson, A. J. and J. K. Nyar. 1963. Plant phospholipids as feeding stimulants for grasshoppers. Canadian J. Zool. 41: 931-935.

Traynier, R. M. M. 1965. Chemostimulation of oviposition by the cabbage rootfly, *Erioischia brassicae* (Bouche). Nature 207(4993): 201-208.

Trouvelot, B. 1958. L'Immuniate Chez. les Solanacus a L'egard du Doryphore. Entomol. Exptl. & Appl. 1: 9-13.

Verschaeffelt, E. 1910. The cause determining the selection of food in some herbivorous insects. Proc. Acad. Sci. Amsterdam 13(1): 536-542.

Vite, J. and G. Pitman. 1969. Insect and host odors in the aggregation of the western pine beetle. Canadian Entomol. 101: 113-117.

Vinson, S. B., J. L. Thompson and H. B. Green. 1967. Phagostimulants for the imported fire ant, *Solenopsis saevissima* var. *richteri*. J. Insect Physiol. 13(11): 1729-1736.

Wada, K., Y. Enomoto, K. Matsui, K. Munakata. 1968. Insect antifeedants from parabenzointrilobum. I. Two new sesquiterpenes, shiromodiol diacetate and monoacetate. Tetrahedron Lettrs. No. 45: 4673-4676.

Wada, K. and K. Munakata. 1968. Naturally occurring insect control chemicals. Isoboldine, a feeding inhibitor and cocculalidine, an insecticide in the leaves of *Cocculus tribolus* DC. J. Agr. Food Chem. 16: 471-474.

Watanabe, T. 1958. Substances in mulberry leaves which attract silkworm larvae (*Bombyx mori*). Nature 182: 325-26.

Watanabe, T. and J. E. Cassida. 1963. Response of *Reticulitermes flavipes* to fractions from fungus-infected wood and synthetic chemicals. J. Econ. Entomol. 56: 300-307.

Wensler, R. 1962. Mode of host selection by an aphid. Nature 195: 830-831.

Wilde, J. de. 1957. Vergeten Hoofdstukken Uit de Phytopharmacie. Ghent Landbouw Hogeschoal Mededelingen. 22(3): 335-347.

Wiseman, B. R., W. W. McMillian and R. L. Burton. 1969. Feeding response of larvae of the corn earworm to water extracts of 16 host plants. J. Georgia Entomol. 4: 15-22.

Wright, R. H. 1965. Finding metarchons for pest control. Nature 207(4992): 103-104.

Yamamoto, R. T. and G. Fraenkel. 1959. Common attractant for the tobacco hornworm, Protoparce sexta (Johan.), and the Colorado potato beetle, Leptinotarsa decemlineata Say. Nature 184: 206-207.

Yamamoto, R. T. and G. Fraenkel. 1960a. Assay of the principle gustatory stimulant for the tobacco hornworm, Protoparce sexta, from solanaceous plants. Ann. Entomol. Soc. Amer. 53: 499-503.

Yamamoto, R. T. and E. Jenson. 1967. Ingestion of feeding stimulants and protein by the female house fly, Musca domestica (L.). J. Insect Physiol. 13: 91-98.

Zielske, A., J. Simons, and R. Silverstein. 1972. A flavone feeding stimulant in alligatorweed. Phytochemistry 11: 393-396.

THE DEVELOPMENT AND PROGRESS OF INSECT MICROBIAL CONTROL

A. M. Heimpel
Plant Protection Institute, ARS, USDA
Beltsville, Maryland

Disease symptoms in insects, particularly in beneficial insects such as the honey bee and the silkworm were noted as early as 2700 B.C. in China and later in ancient Greek times. The first truly scientific approach to the study of an insect disease was the investigation of the white muscardine disease of silkworms (known in Italy as 'calcino') caused by *Beauveria bassiana*.

The classical work by Pasteur, on the 'pebrine' disease of the silkworm caused by the microsporidial *Nosema bombycis* Nageli is famous and is cited in most microbiological textbooks.

The Japanese have contributed much to our knowledge about the diseases of the silkworm and the European Corn Borer Commission, coordinated experiments on control of *Ostrinia nubilalis* (Hubner) between several countries including Canada, France and Hungary. One of these scientists, A. Paillot (1933) wrote one of the first comprehensive books in insect pathology on his research in insect pathology entitled "L'infection Chez les Insectes." Many other scientists added, piece-meal, to our general knowledge on this subject. In this country several extremely competent and dedicated scientists spent part of or most of their lives on studies of insect diseases and the possibility of microbial control of insects. These included Snow, Forbes, Fawcett, Berger, Watson, White and Glaser.

However, the main thrust in this field was initiated by the late Edward A. Steinhaus, whose efforts literally created an organized field of endeavor. In 1947, Steinhaus collected all of the scattered references to insect diseases in one volume entitled Insect Microbiology. Two years later (1949) he wrote a text-book, Principles of Insect Pathology, to aid in teaching the first *bona fide* course in insect pathology given in the world. Today, many laboratories studying microbial control and insect pathology, throughout the world, are populated by Steinhaus' students.

It is some 87 years since Elie Metchnikoff and Isaak Krassilstschik built their first pilot plant to grow the fungus *Metarrhizium anisopliae* in an attempt to control the sugar beet curculio *Cleonus punctineutris*. Following their successful experiments using this fungus as a control agent, some 58 years passed without any significant contributions to the field of mass production of microbial agents.

In 1942, Dutky published a paper describing a method of producing a fastidious, anaerobic sporeformer, *Bacillus popilliae*, for use in effecting permanent control of larvae of the Japanese beetle, *Popillia japonica*. The method consisted of injecting beetle larvae with relatively small numbers of *B. popilliae* spores, allowing the bacterium to germinate, multiply and sporulate in the insects, and then harvesting them by crushing and drying the infected insects at an appropriate time. The dried insect material was then extended with talc and the product was ready for use. More than 200,000 lbs. of this preparation was distributed on 194,000 sites in 14 states and the District of Columbia some two decades ago, in a Federal-State cooperative control project. These figures do not include the thousands of pounds of commercial preparation bought and used by private individuals to protect their lawns and gardens.

It soon became obvious that this method of production was limited in two ways. (1) Production was very low compared to that obtainable by a fermentation method; (2) as control progressed on the eastern seaboard it became increasingly difficult to find large concentrations of beetles to dig up for bacterial production. Today manufacturers of the spore dust have to travel from 300 to 500 miles to find beetle larvae. In due time, the Northern Utilization Laboratory at Peoria, Illinois was contracted to attempt to develop a fermentation method to produce *B. popilliae*. These efforts have met with little success over the past few years, however, work still continues.

At present, the use of commercial preparations of *B. popilliae* on pastures has been forbidden by the Pesticide Regulation Division, Environmental Protection Agency (EPA) pending proof of its safety for vertebrates. The necessary experiments to apply for Exemption from Tolerance are currently in hand, sponsored by the Agricultural Research Service, USDA.

In 1956, E. A. Steinhaus, R. A. Fisher and J. M. Sudarsky met (of the Pacific Yeast Products--later to become Bioferm Corporation and finally to be absorbed by International Minerals and Chemical Corp.) and from this meeting developed the first large-scale fermentation of *Bacillus thuringiensis* tradename THURICIDE. Other companies including Nutrilite Products, Inc. (BIOTROL); Rohm and Haas (BAKTHANE); Merke, Sharpe, and Dohm (Agritol) and Grain Processing Co. (PARASPORIN) and Abbott Laboratories all developed and registered products. Other countries, France, Germany, Czechoslovakia and Russia, throughout the world developed *Bacillus thuringiensis* products as well.

In the United States, three companies are currently producing this material. These are International Minerals and Chemical Co., Nutrilite Products, Inc., and Abbott Laboratories.

Methods of fermentation, tests of efficacy against a variety of insects and safety tests which included human volunteer tests, research on methods of standardization, occupied the time from 1956 to 1958. Extensive studies were conducted by our European colleagues in France, Germany, Czeckoslovakia, Russia and in England. In 1958 the Food and Drug Administration (FDA) granted a temporary exemption from tolerance and the Pesticide Regulation Division (PRD) of the United States Department of Agriculture (USDA) issued an experimental permit.

However, the trials and tribulations continued for manufacturers of *Bacillus thuringiensis* all over the world. It became obvious that many factors were involved in growing toxic preparations of these bacteria. Conditions and components in the media could vastly influence the efficacy of the product. Standardization was incorrectly based on spore count. It was only two years ago that the PRD of EPA seriously considered using a bioassay based on the International unit as a substitute for spore count, a recommendation made by a learned panel at the Colloquium on Insect Pathology at Wageningen, Holland in 1966. Despite all of these annoyances, research has finally reached a point of knowledge where there is confidence that reliable, stable products can be produced and these products can be tailored if necessary, to challenge any specific target insect.

This then is to be considered a successful microbial control agent, commercially available and I predict that it will be used more and more extensively to protect food and fiber crops in the near future.

Let us turn to another group of microorganisms that show enormous promise as insect control agents; these are the nuclear polyhedrosis viruses and the granulosis viruses of insects. As you can see, they are unique viruses compared to the animal and plant viruses. They are generally quite specific for an insect species or a closely related group of species. These viral pathogens are fastidious and grow only in living insect tissues.

Previous attempts to use viruses to control noxious insects in the past have been so successful and promising that the thoughtful person wonders why they were not developed long ago as control agents. Perhaps one explanation for this lack of development is the fact that these pathogens could only be produced on living insects, reared in the laboratory or insectary on natural foliage, during a limited season of the year, at what could only be considered a prohibitive cost. Otherwise, naturally infected insects in the field were collected or populations of insects were artificially infected and collected in the field to provide a bulk of these microorganisms. The latter methods were unpredictable and provided virus preparations mixed with an astonishing number of other extraneous biotypes, competitive disease organisms, etc. They were dirty preparations and would never be acceptable for formulation under current regulations.

Obviously, a method of rearing insects on semi-artificial or artificial media was required to permit virus production 365 days of the year. In the late '50's and early '60's, Dr. C. M. Ignoffo had been working in Brownsville, Texas for the Agricultural Research Service, USDA, in an attempt to produce nuclear polyhedrosis viruses of the cabbage looper and the cotton bollworm (corn earworm) (*Heliothis zea*), in larvae of these insects reared on semi-artificial media.

Eventually he developed methods of virus production that basically are those now employed by International Minerals and Chemical Corporation and Nutrilite Products Incorporated.

However, mass rearing of Lepidoptera is an extremely difficult problem. One of the main troubles is competitive disease in the rearings. Recent developments in alternative methods of producing insect viruses seem very encouraging. An attempt to grow insect viruses in bacterial protoplasts, by the Midwest Research Institute was reported recently by Wells and Heimpel (1970). This preliminary contract work, done for the USDA, did little more than encourage the Midwest Research Institute to continue investigations under the auspices of the McLaughlin Gormley King Co., with other single-celled organisms, after the expiration of the USDA contract. Although their results are confidential at present, they are very optimistic concerning this possible production method.

Again, Goodwin, et al. (1970) have demonstrated that the fall armyworm and the cabbage looper lines of tissue culture can be infected and can produce appreciable amounts of virus. This may mean that the possibility of producing insect viruses in insect tissue culture is at hand and may be feasible in the near future. Of course virus produced in tissue culture is preferable to that produced in the insect because it should be cheaper to produce; there would not be the problem of competitive disease and the virus preparation would be bacteria free and would contain a minimum of contaminating insect debris.

Production of virus in living insects using the nuclear polyhedrosis virus of the cotton bollworm, *H. zea*, was begun by the two previously named companies in the early '60's. And, five years ago petitions were submitted to the regulatory agency requesting labeling of these products. These petitions contained a tremendous amount of data concerning the efficacy and safety of this virus. Details on safety testing have been published by Heimpel (1971). Although a final decision is yet forthcoming from the Environmental Protection Agency, we are hopeful that these materials will be released again under a temporary exemption from tolerance and experimental permits this year.

There are many viruses that are available and await development, furthermore, although most of my remarks are quite naturally restricted to this type of activity in the United States, there are a large number of viruses that undoubtedly will be developed in the near future, that show promise as potential control agents of noxious insects in many countries throughout the world.

Another notable group of insect pathogens is the fungi. Various attempts in the past have been made to produce *Metarrhizium anisopliae*, however, it is a curious point that these efforts have not persisted and no extended use of this effective organism has been continued.

Schaerffenberg, in Austria, developed a method of producing *Beauveria bassiana* on a bran medium. This method definitely has potential commercially since the scientists at Nutrilite Products, Inc., Lakeview, California, independently developed a similar method for producing the same fungus. However, where Schaerffenberg claimed exceptional results using the fungus spore suspension to control the potato beetle, tests using the Nutrilite product of *B. bassiana* against a variety of insects were not successful enough to label the fungus potentially useful.

Again, attempts have been made to produce quantities of nematodes and protozoa, particularly the microsporidia, for tests against insects. Most of these efforts hardly reached the pilot plant stage. The results of tests indicate that these microorganisms leave much to be desired with regard to efficacy in the field. The methods of production also will require extensive research to make them feasible on a commercial scale as well as cost-wise.

In summary, in the past two decades, enormous strides have been made in our ability to produce large quantities of bacterial, viral and fungal pathogens of insects. These products should be appearing on the market within the next few years in ever increasing numbers.

Development of other types of microbial pathogens has been slow but recent increase of interest in their testing and production is very encouraging.

References

Dutky, S. R. 1942. US Dept. Agr. Bur. Entomol. Plant Quar. ET-192. 10 pp.

Goodwin, R. H., J. L. Vaughn, J. R. Adams and S. J. Louloudes. 1970. J. Invertebr. Pathol. 16: 284-288.

Paillot, A. 1933. In: L'Infection Chez les insectes. Trevoux et Patissier, Paris, France.

Steinhaus, E. A. 1947. In: Insect Microbiology. Comstock Publishing Co., Inc., Ithaca, New York.

Steinhaus, E. A. 1949. In: Principles of Insect Pathology. McGraw-Hill Book Co., Inc. New York, N.Y. 759 pp.

Wells, F. E. and A. M. Heimpel. 1970. J. Invertebr. Pathol. 16: 301-304.

MICROBIOLOGICAL CONTROL OF INSECTS: BACTERIAL PATHOGENS

T. A. Angus
Insect Pathology Research Institute
Canadian Forestry Service
Sault Ste. Marie, Ontario, Canada

May I first of all express my thanks to the Committee for their kind invitation to speak today both for the honor you do me and the Insect Pathology Research Institute where I am employed. Most of what I shall present is derived from the work of others and has been reported in detail elsewhere or in reviews of various kinds. The general theme of this Summer Institute and particularly the subject assigned me, bacteriological control of insects, indicates the emphasis expected. Nevertheless I should like to digress and discuss some generalities before I embark on the specifics of my text.

There is a tendency to regard as bizarre or unusual those kinds of microorganisms which are found associated with insects. In my opinion this is a mistaken attitude and whenever I have an opportunity I do what I can to dispel it. My reason for alluding to this is that I feel that many potentially useful pathogens have been missed by entomologists simply because they do not think of the microorganisms found associated with insects as being potentially useful.

As you will all be aware one of the dominant if not the dominant group of animals on earth today is the Insecta. They are ubiquitous, cosmopolitan and at times prodigiously abundant. I have read that there are perhaps as many as a million species of insects which is many times more species than there are in the rest of the entire animal kingdom. The various species of insects exploit a bewildering array of natural substrates as saprophytes, as carnivores, as predators, and parasites. They are found in an astonishing variety of ecological and environmental extremes.

Again by way of general remarks, the Insecta are a very old group of animals. A Collembolid has been found fossilized in material which has been estimated to be over 300,000,000 years old. From this kind of estimate Grosser speculates that insects must have been flourishing a hundred million years before the earliest known mammals. Without straining at zeros, it is obvious that Man is a Johnny-come-lately. It is fairly certain that bacteria had made the transition from saprophyte to pathogen in the Insecta long before it occurred in other kinds of animals. The point I wish to make is that the very antiquity of this relationship makes it inappropriate to either designate or regard it as unusual. The fact that we know a great deal more about the bacterial pathogens of Man than of insects simply is a reflection of our bias.

In the earliest days of microbiology almost all study was devoted to eradication, prevention, and prophylaxis of the infectious diseases of humans and then after that, of the animals and plants useful to Man. Because the silkworm and the honey bee were useful, their diseases were the first noted in insects. Indeed there are references in Aristotle, Pliny and Virgil to diseases of bees. There was, of course, no knowledge and certainly no experimental proof that microorganisms were the cause of these diseases until in about 1834 Bassi demonstrated that a fungus caused a disease in *Bombyx mori*. In a sense the germ theory of disease had its origins in his work. Although Pasteur originally began with a protozoan disease of silkworm, he later worked with the disease we now call endemic flacherie and which is caused by a bacterial species. Thus Pasteur, one of the giants of microbiology, who started out as a chemist was in a sense introduced to microbiology as a result of some time spent as an insect pathologist.

As Man learned that many of his diseases and those of his domesticated animals and plants were caused by microorganisms he became interested in how pathogens were spread from one individual to another and found that many species of insects and arthropods were implicated. Horsfall in his most interesting text indicates that for thousands of years man was practically powerless to intervene effectively in recurring epidemics of arthropod-borne diseases. He further points out that in many cases this shaped many of his cultural and economic activities. Horsfall discusses a number of important bacterial pathogens which are transmitted by arthropods.

In nearly all of these associations the arthropod is really only a disseminating vector, a sort of peripatetic inoculating needle. If one considers the various kinds of sites used by insects for oviposition or the kinds of substrates they utilize for food it is easy to understand how insects can transfer microorganisms. The graphic posters prepared by the World Health Organization for use in certain underdeveloped parts of the world indicate that the ancient problem of arthropod-borne diseases is still a very real one for many members of the human race.

The demonstration that microorganisms and especially bacteria are the cause of disease and that insects can act as spreaders of these pathogens led to a great deal of work on the bacteria associated with insects. As a result a great many species of bacteria have been found associated in one way or another with various kinds of insects. For instance, certain kinds of aquatic insects utilize microorganisms directly as food. Various kinds of organic material such as feces or animal remains are broken down by bacteria to smaller compounds, many of which are attractive to or are utilized by insects. As the insects feed on such material they ingest bacteria. It is not surprising therefore that a great many of the species of bacteria found in insects are merely adventitious and transitory.

Understandably bacteriological studies of insects have been very largely concerned with pest species: i.e., those which have been implicated as vectors, carriers, or secondary hosts of pathogens of humans and domestic animals; those which attack men directly such as the mosquitoes and biting flies; and the much larger group of insect species which are in competition with man for the use of a wide variety of cereals, fibres, fruits and so on.

Although insect pathology had its beginning in the "medical entomology-public health bacteriology" aspects of the relationship between bacteria and insects, lately another motivation has been provided by the concept of "microbial control." In it we have simply turned the usual rationale of medical entomology and public health upside down, and our purpose in studying insect diseases is not how to prevent or to cure them but how to utilize them in our behalf. Microbial control as a concept was first enunciated about 125 years ago and since that time, it has enjoyed cycles of popularity and neglect.

The control of insects by use of chemicals began in North America just about 100 years ago. In that time we have gone from the simple use of Paris Green to a bewildering array of various kinds of natural and man-made chemicals. Through the use of such compounds we have accomplished a level of insect control that would have been unthinkable. Undoubtedly the development of the chlorinated hydrocarbons and the organic phosphates constituted a major breakthrough in chemical control of insects. Some agricultural writers claim that the reduction in insect damage and the improvement in crop yields and quality that we have achieved would not have been possible without the use of these modern chemical insecticides. Equally impressive has been the reduction in human misery as the result of the use of chemicals to control insect vectors of malaria, typhus and various kinds of dysentery.

It is, however, common knowledge, and I do not propose to dwell on it at length, that the use of modern chemical insecticides has proved to be a kind of two-edged sword. The involvement of nontarget species, biological concentration at the top of food chains, the problem of persistent residues, the development of resistance in some pest species, all of these have been discussed in detail elsewhere. My purpose in mentioning them again is to indicate that many responsible authorities argue that it is essential that we lessen our dependence on broad spectrum chemicals for insect control and that we must explore any avenue that holds promise in this direction. As a result, we are interested again in the possibilities of using insect diseases as a means of controlling the abundance of certain pest species. From observation we know that bacterial epizootics do occur naturally in insect populations and that under certain conditions it is possible to initiate such an epizootic. The problem is how to do this in circumstances and at a cost attractive to Man as farmer or forester.

The purpose then of this rather long preamble is to put my subject into its proper perspective as a very small part of a larger field of knowledge. The rather specialized approach that we are discussing today arose out of these larger considerations and should not be pursued in isolation for if we do so, we do so at our scientific peril.

Before beginning to discuss specific bacterial pathogens, two other subjects should be touched upon. The first is to agree on some sort of common definition of what is an insect pathogen. As indicated earlier, a wide variety of bacteria have been isolated from a great number of insect species. In many, if not most of these cases the bacteria are either adventitious or simply transient contaminants of the insect body or of the food it has been utilizing. Bucher has proposed a classification which has been very widely accepted. Briefly, he groups pathogens under three headings. The obligate pathogens are those which are found associated with a specific insect disease. Characteristically they have a very narrow host range, are readily transmitted and very often require quite specialized conditions for growth. In addition, in nature they are probably restricted to only a few species of particular insects. The facultative pathogens are those that have some mechanism for damaging or invading a susceptible tissue but are not truly obligate pathogens. They can be cultured in artificial media and sometimes actually multiply within the gut of the host insect before they invade the hemocoele. Potential pathogens are those kinds of bacteria which normally do not multiply in the gut but if they are injected or in some way gain access to the hemocoele, they can become established there. They will grow on artificial media and are not generally associated with a specific disease of a specific insect species.

When Bucher's criteria are applied to the bacteria that have been isolated from insects, only a few can be classed as insect pathogens, and more to the point only a few of them can be considered as having any potential for use in microbial insecticides.

The term microbial insecticide is, I think, pretty well self-explanatory. A chemical insecticide is one that is based on the use of a particular chemical or group of chemicals, thus a microbial insecticide is one which utilizes some kind of bacterial, viral, fungal or protozoan species.

The second point which requires some comment is what constitutes an ideal microbial insecticide, or as it has been described before, what are the attributes of an ideal microbial insecticide. Quite a long time ago Bucher listed a number of attributes that he thought were desirable. Some time after this, Heimpel and Angus slightly modified this, and in turn about five or six years ago, I attempted a further amendment.

"Previous attempts to utilize bacteria as microbial insecticides have revealed certain attributes that materially affect successful use. First of all the prospective pathogen should be virulent at least to the extent that it consistently causes a disease serious enough to inhibit the competitive activity of the pest insect. Variations in virulence when they do occur should not be so great as to affect materially a recommended dosage or require frequent reassay. The pathogen should not be markedly sensitive to the environmental hazards to which it will be exposed such as desiccation and sunlight, to the way in which it will be introduced (such as a spray or dust), or the suspending medium used (oil, water, stickers, emulsifying agents). It should also be persistent, in the sense that it will remain viable or infectious until it gains access to the target insect. In general terms, it is preferable that the pathogen be rapid in its action for it is the feeding activity of most agricultural or forest insects that makes them pest species; this is not a rigorous requirement, however. It is important that the pathogen be fairly specific for the insect pest it is used against and inactive against the host plant or useful insect species such as parasites and pollinators. It is of extreme importance that the pathogen be harmless under the conditions of use for vertebrates and especially for mammals. Finally, it must be possible to produce the pathogen in quantity at an economically acceptable cost in a form that is practicable and aesthetically satisfactory. Taken as a whole, this is a rigorous set of requirements that excludes most isolates from consideration."

Of the hundreds of bacterial species that have been isolated from or found associated with insects, only a very few have been actively considered or utilized as microbial insecticides. The best known of these are found in three families: the Enterobacteriaceae, the Micrococcaceae, and the Bacillaceae; some species occur in Pseudomonodales.

About the turn of the century d'Herelle reported that he had isolated from diseased grasshoppers a bacterial species that he called _Coccobacillus acridiorum_. He claimed that the release of this organism resulted in very large reductions in grasshopper populations. Many attempts were made to duplicate his findings, and the results were at best ambiguous. Subsequently, his isolate was found to be very unstable in respect to virulence and there was some confusion as to the exact identity of the strains used. All of this considerable body of work was reviewed a few years ago by Bucher. He concluded that d'Herelle's strain was a type of Enterobacteriaceae which are widely found in the gut of grasshoppers. In Bucher's opinion, _Coccobacillus acridiorum_ should not be classed as a true pathogen.

Another species of Enterobacteriaceae is represented by _Serratia marcescens_. Bacteriologically speaking, it is regarded as a ubiquitous saphrophytic species found in water, soil, milk and foods. It is also frequently isolated from diseased and dead insects. It is not nutritionally fastidious, it is a facultative anaerobe, is strongly proteolytic, and can grow in a fairly wide pH range. Thus, it is capable of multiplying in the gut of many insect species. In addition, if it is injected into the hemocoel it will cause disease in a wide range of insects but only causes disease in a few species if taken by mouth. In the literature, many insect hosts are reported but, returning to an earlier theme, probably in many of these the occurrence of the bacterium is accidental or adventitious. In many of the instances recorded, _S. marcescens_ is a pathogen of insects when they are being reared under laboratory conditions or are subjected to crowding or other stresses.

Some years ago Stevenson isolated a bacterium causing mortality in grasshoppers. This was later identified as a non-chromogenic strain of Serratia. There is no doubt that many unidentified strains of bacteria earlier isolated from insects belong in this group. Using Bucher's classification, these kinds of Serratia which are isolated from insects should be regarded as facultative pathogens.

Some years ago Bucher in Canada did a very exhaustive study of the occurrence of Pseudomonas aeruginosa in grasshoppers in Western Canada. This bacterium often causes serious mortality in laboratory rearings and for a number of years was thought to have some potential as a usable pathogen under field conditions. Bucher classes it as a potential pathogen, i.e., as a species which can multiply in the insect gut. Ruptures often occur in the gut of grasshoppers, and Bucher suggests that P. aeruginosa gains access to the insect hemocoel more by accident than as a result of some invasive mechanism. In investigating the potential of P. aeruginosa to initiate epizootics under field conditions it was found that it is very sensitive to sunlight and to desiccation. Even when protected by special coatings, releases of this isolate failed to establish themselves in grasshopper populations. As a result work on this particular bacterium has been discontinued in Canada.

The three species just discussed are non-sporulating bacterial pathogens and although a wide variety of this kind of bacterium has been isolated from insects, none of them really have shown a great deal of promise as potentially useful species for use in microbial insecticides. Generally speaking, the non-sporulating strains have a limited leaf life, that is to say that they are very sensitive to drying and sunlight. It has also been noted among this kind of bacteria that there is quite a variability of virulence.

A much more serious objection to the use of these non-sporulating types arises out of the fact that they are very often pathogenic for vertebrates. Occasionally they are minor pathogens of mammals (usually in unusual circumstances) but this is sufficient to raise serious doubts about the wisdom of attempting to utilize them as microbial insecticides.

Doane, working in New England with the gypsy moth, has isolated a strain of Streptococcus faecalis from naturally infected larvae. When broth cultures of this isolate were applied as sprays from a mist blower a considerable reduction of defoliation was noted. Doane reports that the most striking sign of infection with the Streptococcus is the shortened and shrunken appearance of the larvae and he suggests that the disease be named gypsy moth brachyosis. Brachyosis is the name proposed by Bucher to describe the shortened appearance seen in tent caterpillar larvae infected with bacteria.

The strain isolated by Doane is no doubt a facultative pathogen, but before this isolate can seriously be considered as a basis for a microbial insecticide, a great deal of work will have to be done on its potential pathogenicity for other life forms. Bergey's Manual indicates that the sources of this bacterium are human feces and the intestine of many warm-blooded animals. It is occasionally encountered in urinary infections, in the blood stream, and in heart lesions in subacute bacterial endocarditis. It is sometimes associated with European foulbrood of bees, and has been associated with mild outbreaks of food poisoning. In summary, although this gypsy moth pathogen is a very interesting one, it is too soon to give any reliable estimate of its potential usefulness as the basis for a microbial insecticide.

Turning now to the spore-forming bacterial pathogens, they can conveniently be divided into two groups, the aerobic and the anaerobic species. Dealing first with the anaerobic types, species of the genus Clostridium have occasionally been found in diseased insects. The spore-forming anaerobes are a difficult group of bacteria to work with and require very specialized techniques so that their relative scarcity in insects may simply reflect an absence of investigators. Two very interesting strains, Clostridium brevifaciens and Clostridium malacosoma were isolated by Bucher from diseased larvae of Malacosoma; these are the tent caterpillars. If spores of C. brevifaciens are fed to tent caterpillar larvae, they will germinate in the gut and multiply. After about 72 hours of incubation the infected larvae demonstrate a characteristic accordion-like shortening of body and eventually die. This condition was called brachyosis by Bucher. The other organism mentioned, Clostridium malacosoma, induces similar symptoms. Although these organisms are fastidious, they can be grown in highly specialized media. Similar kinds of Clostridia have been isolated from other Lepidoptera, notably the Essex skipper, Thymelicus lineola, and from the pine processionary moth. In the latter insect, the bacterium can sometimes cause disease in epizootic proportions.

Although attempts have been made to utilize Bucher's isolates under field conditions, a difficulty is that they sporulate only sparsely, even on a medium which supports abundant vegetative growth. This makes it difficult to accumulate sufficiently large amounts of material to give satisfactory field tests.

Turning now to the aerobic spore-formers, some very promising isolates exist and indeed some of these are available as commercial preparations. Pride of place historically belongs to Bacillus popilliae, and a closely related strain, Bacillus lentimorbus, which cause a lethal septicemia in a wide range of Scarabeid larvae. The commercial preparations based on Bacillus popilliae are used against larvae of the Japanese beetle, Popillia japonica, which is a pasture and lawn pest in eastern North America. Since the larval form is a soil insect, it has been very difficult to deal with it in the usual ways. Once introduced, Bacillus popilliae spores persist in the soil. In the course of feeding, larvae ingest these spores which germinate in the gut and give rise to vegetative cells which eventually penetrate the gut wall and enter the hemocoel where they multiply abundantly. Eventually the vegetative cells sporulate and the thick-walled refractile spores occur in such numbers that when viewed through the semi-translucent integument of the insect, display the appearance which gives rise to the name milky disease. The infected larvae die in the soil and disintegrate, releasing the spores which are then available for further cycle of ingestion, disease, death and disintegration. The milky-disease organism can be grown in its vegetative stage on laboratory media, but it does not sporulate readily. Since the infective spore is the ideal inoculum stage, this presents difficulties for commercial exploitation. The commercial preparations presently available are prepared by individual inoculation of larvae which at the point of death are dried, ground and extended with suitable mineral fillers. Although each infected larvae contains billions of spores at death, the milky diseases are not rapidly acting and from ingestion to death may take from two to four weeks, depending on soil temperatures.

Bacillus popilliae is available commercially under two names and these products are effective for the purposes for which they have been developed. Their more widespread use is limited by a number of factors among which are that the preparations are only cleared for certain kinds of use, but I understand this is under review. The method of production limits the annual output and thus it is not possible to achieve economies of scale. Also, the control imposed is one that develops only slowly over a whole or perhaps several seasons. It is in many ways an admirable pathogen because it is highly selective and poses no risk at all to other forms. It is self-perpetuating and thus once used, it is not necessary to repeat the treatment for many years, if at all. A number of milky disease organisms have been isolated from larvae of soil beetles in Europe and in Australia. In a modest way, the story of Bacillus popilliae is a success story.

I should like to turn now to the aerobic spore-forming bacteria which have been studied very extensively over many years. A group that has received much attention are those strains which are either closely related to or derived from

Bacillus cereus which is a ubiquitous soil form very widely distributed in soil, dust, milk and on plant surfaces. Bacillus cereus has been isolated under many names from diseased and dead larvae. A number of these isolates are pathogenic when ingested and insects from quite a wide range of species have been studied. Bacillus cereus grows readily on synthetic media and also sporulates readily on such media. This has made it possible to accumulate relatively large amounts of experimental material for testing and indeed quite a few isolates have been tested under field conditions. There is quite a range of pathogenicity depending on the isolate so that some of the isolates can be classed as potential pathogens and some others as facultative pathogens. Strains of Bacillus cereus have been isolated from a wide range of Lepidoptera including the coddling moth which is a pest of apples, the larch sawfly which is a pest of Canadian forests, and from spruce budworm which is probably the most serious forest pest in North America. Lately it has been reported as causing a disease condition in the tobacco worm.

Much work has been devoted to Bacillus cereus because it is readily cultivated and thus easily studied. It is also attractive because it forms a resistant endospore. A drawback of Bacillus cereus is that some strains produce a cereolysin which can cause tissue damage in some mammals. Occasionally isolates of Bacillus cereus have been associated with certain human disease conditions although these are rare and the attendant circumstances are obscure. Nevertheless, it is obvious that this might make it difficult to find easy acceptance of the Bacillus cereus as the basis for microbial insecticides.

The usefulness and many advantages of a closely related species Bacillus thuringiensis, had tended to push the work with Bacillus cereus off the stage. There has been a tendency to infer that whatever Bacillus cereus can do, Bacillus thuringiensis can do better. This is only partially true because there are a number of insect species, for example, some kinds of sawfly larvae, which are susceptible to disease with Bacillus cereus that do not respond to treatment with Bacillus thuringiensis. I do not believe that we have heard the last of Bacillus cereus. It is so similar and some aspects of the disease it causes are so similar to those caused by Bacillus thuringiensis that anyone working with B.t. should be constantly aware of these common features.

It should be mentioned that a number of bacterial taxonomists are of the opinion that Bacillus thuringiensis is merely a special kind of Bacillus cereus and they continue to insist that they should be grouped together. The taxonomic wrangle over Bacillus cereus and Bacillus thuringiensis continues to this day. Dr. Heimpel and I have always recognized the close relationship of these two strains, but have long been advocates of species status for Bacillus thuringiensis partially on the grounds of utility and usage.

Turning now to the thuringiensis story, it dates back to Pasteur's time, but I cannot take the time to provide you with a blow-by-blow account. In the last 20 years or so, about a thousand papers have been published dealing with the various aspects of research on Bacillus thuringiensis. I know that you will be relieved to hear that my review of this work will be cursory and of necessity incomplete. Frankly, this work has been reviewed so often that I am almost embarrassed to attempt yet another summary of it. If anything that I have to say here catches your fancy and you wish to pursue the matter, there are available quite a number of excellent review articles. Bacillus thuringiensis has caught the interest of bacterial taxonomists, bacterial physiologists, crystallographers, biochemists, people working in protein chemistry, insect physiologists, and economic entomologists. These various groups have studied Bacillus thuringiensis from their own point of view and thus, the published papers cover a very wide range of disciplines, some of which are remote from economic entomology.

Within the context of today's subject, the Bacillus thuringiensis strains are of interest because under normal conditions of growth, as a concomitant of sporulation, they produce proteinaceous crystalline parasporal inclusions which are the source of a compound which is toxic for the larvae of Lepidoptera and without appreciable effect on other life forms. Also, within the frame of reference of our present discussion, it is equally important that the *thuringiensis* varieties are readily adaptable to culture in modern large volume fermentation equipment.

If spores or vegetative cells of *Bacillus thuringiensis* are injected into the hemocoel of an insect, abundant growth soon takes place and this leads directly to a fatal septicemia. The fate of spores or cells taken by mouth depends on gut conditions which embrace pH, oxidation-reduction potential in the gut, degree of anaerobiosis, presence of antibiotics of plant origin, the digestive enzymes of the host insect and so on. The importance of the parasporal inclusion is that the protein that comprises it appears to act as a protoxin yielding a toxin that damages the midgut cells of susceptible Lepidoptera species in such a way as to inhibit feeding and subsequently causing other changes favoring the growth of the pathogen. The susceptible Lepidoptera include species affected by either the crystal alone or the spore alone. There is a gradation of response and in most species the greatest mortality is caused by preparations which are a mixture of these entities. The toxicity of the crystal protein seems to be limited almost exclusively to Lepidoptera.

The rapid inhibition of feeding makes *B. thuringiensis* very attractive as a biological control agent. In contrast, the diseases caused by viruses, fungi or protozoans are slow to develop and so there can be considerable defoliation during the incubation period of the disease. The toxinosis induced in susceptible Lepidoptera by the pre-formed toxin is analogous in a crude sort of way to botulism or to staphylococcus food-poisoning. It should not be confused with diseases like anthrax or cholera where there is an initial inoculation or ingestion of the pathogens, then bacterial growth, then toxin accumulation and finally the toxinosis. With *Bacillus thuringiensis* based insecticides, the process begins with the ingestion of spores and the toxic crystals. The latter cause an immediate enterotoxinosis which paves the way for a lethal septicemia arising from the spores.

There are a number of most interesting studies dealing with the structure of the crystalline parasporal inclusion body, its chemical nature and its effect on the mid-gut of susceptible Lepidoptera. Unfortunately time will not permit any extended reference and I can only draw this work to your attention.

The whole *B.t.* story is not complete, but enough is known to make it possible for substantial amounts of *B.t.* products to be produced, marketed and used as effective agricultural and forestry products.

In the last hour we have touched on many subjects and you have been most patient, for which I thank you. It would be unseemly for me (because of my personal involvement), to write a testimonial, but I hope you will permit me to advance the opinion that the development of the milky disease and *B.t.* formulations has been a laudable endeavor and worth the effort. I certainly hope it encourages others to persist in this kind of study for I believe that much useful work remains to be done.

References

Hannay, C. L. 1956. In: 6th Symp. Soc. Gen. Microbiol., London, April 1956 (E.T.C. Spooner and B. A. D. Stocker, eds.), University Press, Cambridge, 1956, pp. 318-340.

Heimpel, A. M. and T. A. Angus. 1956. Proc. 10th Intern. Congr. Entomol., Montreal, 1956, Vol. 4, p. 711 (1958).

Heimpel, A. M. and T. A. Angus. 1960. Bacteriol. Rev. 24: 266.

Krieg, A. 1961. Mitt. Biol. Bundesanstalt Land-Forstwirtsch. Berlin-Dahlem 103: 79.

Heimpel, A. M. and T. A. Angus. 1963. In: Insect Pathology, An Advanced Treatise (E. A. Steinhaus, ed.), Vol. 2, Academic Press, New York, 1963, pp. 21-73.

Heimpel, A. M. 1963. Adv. Chem. 41: 64.

Cameron, J. M. 1963. Ann. Rev. Entomol. 8: 265-286.

Heimpel, A. M. 1965. World Rev. Pest Control 4: 150. NA

Angus, T. A. 1965. Bacteriol. Rev. 29: 364. 448.28 B131

Rogoff, M. H. 1966. In: Advances in Applied Microbiology (W. H. Umbreit, ed.), Vol. 8, Academic Press, New York, pp. 291-313.

Heimpel, A. M. 1967. In: Ann. Rev. of Entomol. (R. F. Smith and T. E. Mittler, eds.), Vol. 12, Annual Reviews, 1967, pp. 287-322.

Krieg, A. 1967. Mitt. Biol. Bundesanstalt Land-Forstwirtsch. Berlin-Dahlem 125: 106.

Angus, T. A. 1968. World Rev. Pest Control 7: 11. NA

Rogoff, M. H. and A. A. Yousten. 1969. In: Ann. Rev. Microbiol. (C. E. Clifton, S. Raffel and M. P. Starr, Eds.), Vol. 23, Annual Reviews, 1969, pp. 357-386.

Angus, T. A. 1971. In: Naturally Occurring Insecticides (M. Jacobson and D. G. Crosby, eds.). Dekker, New York, 1971, pp. 463-497.

Burges, H. D. and N. W. Hussey, eds. 1971. Microbial Control of Insects and Mites. Academic Press, London. Chaps. 1,3,10, 11, 12, 13, 22, 24 and appendices.

MICROBIOLOGICAL CONTROL OF INSECTS: VIRAL PATHOGENS

C. M. Ignoffo
Biological Control of Insects Research Laboratory
USDA, ARS
Columbia, Missouri

Introduction

Most people think of viruses as plagues. Insect virologists, however, are working on the concept that viruses may actually be used to benefit man! Can a specific insect virus be used to biologically control an insect pest? It is evident that other means of controlling insects must be found. Our present-day potent, nonspecific chemical insecticides were, and still are, fulfilling an important role. They also, however, produce environmental effects beyond their original purpose. Largely because of these effects we are attempting to develop insect viruses into effective, safe, specific, biodegradable insecticides--man's effort to apply nature's developed product to man's best advantage.

About 450 different viruses have been isolated from insects and mites. This total probably represents about 30% of all diseases associated with insects and mites (Ignoffo, 1968a).

Use of viruses for insect control is the general theme I wish to develop today. This concept will be evaluated later as to the effectiveness of insect viruses; effects of insect viruses on humans, other animals, and plants; and how insect viruses may be produced.

Morphological Relationship to Other Viruses

Prior to this evaluation, I wish to introduce you to insect viruses by explaining their morphological relationship to other animal and plant viruses, classification, and the insects they attack.

Viral diseases of all animals and plants are induced by nucleic acid contained in a virus particle called a virion. The virion may be a filament, rod, sphere or complex as exemplified by bacteriophages, i.e., viruses which attack bacterial cells (Figure 1). Virions of some insect viruses are similar in general morphology to the known vertebrate, plant and invertebrate viruses (Table 1).

The citrus red mite virion, which is small, spherical and contains RNA, is morphologically similar to many plant viruses. Virions of the iridescent virus of the crane fly and rice stem borer, large DNA-containing spheres, morphologically resemble reptile and fish viruses. Cytoplasmic polyhedroses of arthropods are induced by small, spherical virions which contain RNA, and are embedded in polyhedral shaped inclusion bodies. Virions of the cytoplasmic polyhedrosis virus (CPV) are morphologically similar to *Reovirus*. The reoviruses include avian, mammalian and plant viruses. Virions of insect-pox (*Entomopoxvirus*) are in the same group as pox viruses of mammals and birds. Virions of granuloses and nuclear polyhedroses are rod-shaped and contain DNA. They are morphologically unique, i.e., not similar in shape or form to any of the described plant or animal viruses. The insect-pox, granuloses, nuclear polyhedroses and cytoplasmic polyhedroses are all characterized by the presence of an inclusion body (Figures 2, 4, 6, 8). No plant or vertebrate viruses described to date produce inclusion bodies similar to those produced by arthropod viruses. Approximately 90% of the presently described arthropod viruses are occluded in a proteinaceous, polyhedral shaped inclusion body. The virions (Figures 3, 5, 7, 9) are embedded in the protein matrix of the inclusion, generally at random, and without any apparent disruption of the lattice structure.

Morphological similarity in no way implies similarity in ability of virions of arthropod viruses to cause disease in other animals or plants. The present classification system is based on structural relationships, and consequently viruses of plants and other animals are placed in the same taxonomic groups.

Naming and Classification of Viruses

Arthropod viruses were once named and classified according to whether the virus particle was or was not occluded in an inclusion body; the shape of the inclusion body when present; the site of viral replication; and the shape of the virion (Holmes, 1948; Steinhaus, 1953). The present system is based on virion characteristics, i.e., virion shape, outline, and symmetry; type and percent of nucleic acid; strandedness of nucleic acid; and molecular weight of the nucleic acid (Wildy, 1971). Other criteria, i.e., symptomology, site of viral replication, chemical sensitivity, and serology, is used to supplement characteristics of the virion. There are six groups of arthropod viruses (Table 2). Latinized binomials are used to name all viruses. The suffix -virus is used for all generic names.

Kinds of Arthropods Susceptible to Viruses

All major orders of insects are susceptible to viruses (Table 3). About 500 arthropod species have at least one virus disease and more than 450 virus species are described. Most viruses (83%) have been isolated from caterpillars of moths and butterflies. This is not surprising since many of our economic pests are in this group. Likewise, most described viruses (ca. 90%) have inclusion bodies. Inclusion bodies are large and easily observed using light microscopy. In the future more of the smaller non-inclusion viruses will undoubtedly be described. Viruses of sawflies and flies account for 14% of the described arthropod viruses. The remaining 3% are equally divided among viruses of beetles, grasshoppers and mites.

Characteristics of Arthropod Viruses

Nuclear polyhedrosis

Nucleopolyhedrosis viruses (NPV), which account for 41% of described arthropod viruses, develop in cell nuclei (Table 4). Cells derived from ectoderm, mesoderm, endoderm, are all susceptible to infection. Infection generally results in death and rupture of the integument. Virions are occluded singly or in bundles in polyhedral-shaped inclusion bodies (PIB). The PIB range in diameter from 0.2 to 15 μ (Figure 2). Virions are rod-shaped (diameter 40 to 80 mμ, length 230-420 mμ) contain DNA (double stranded) with a molecular weight to 50 to 100 X 10^6 (Figure 3). NPV have been reported from Lepidoptera, Hymenoptera, Diptera and Orthoptera.

Cytoplasmic polyhedrosis

Cytoplasmic polyhedrosis viruses (CPV) develop in cytoplasm of mid-gut epithelial cells. Infection generally is debilative, resulting in slow larval growth and reduction in adult longevity and fecundity. Virions are occluded singly. The diameter of PIB range from 0.5 to 15.0 μ (Figure 4). Virions are spherical in outline, average 60 mμ in diameter, and contain double stranded RNA (Figure 5). CPV have been isolated from Lepidoptera, Neuroptera and Diptera.

Granuloses

Granulosis inclusion viruses (GIV) replicate in nuclei or cytoplasm immediately surrounding the nuclei of adipose, tracheal or epidermal cells. Infection results in death and symptoms are similar to the NPV. Inclusion bodies are oval-shaped, contain one or rarely two virions per inclusion body, and measure 200 X 400 mμ (Figure 6). Virions are similar to the NPV, i.e., rod-shaped and contain DNA (Figure 7). GIV have been reported only from species of Lepidoptera.

Pox viruses

Pox viruses are a recently described group. Pox viruses replicate in cytoplasm of susceptible cells, especially adipose tissue. Inclusion bodies are large (2 to 8 μ) and oval to polygonal in outline (Figure 8). Virions contain DNA, are ovoid-cuboid in shape (Figure 9) and measure 250 X 300 X 200 mμ. Pox viruses have been isolated from Lepidoptera, Diptera and Coleoptera.

Polymorphic viruses

Polymorphic viruses are presently represented by only one disease, i.e., disease of the cabbage worms (Pieris spp., Lepidoptera). Large (5 to 15 μ) irregular shaped inclusion bodies are found in cytoplasm of adipose and blood cells (Figure 10). Virions of this virus have not been described.

Non-inclusion viruses

About 10% of the described arthropod viruses are not occluded in inclusion bodies (Figures 11, 12, 13). Virions of non-inclusion viruses (NIV) are generally beyond limits of resolution of the light microscope (<200 mμ). NIV contain the bullet-shaped (180 X 70 mμ) RNA, virions of the Drosophila sigmavirus (Figure 11); isometric, small (diameter 20 to 25 mμ) DNA virions of the densonucleosis (Figure 12); icosahedral (diameter 130 mμ) DNA virions of iridescent viruses (Figure 13); and all non-occluded virions with cubic symmetry.

Pest Control Using Viruses

Now that a background on arthropod viruses has been developed, we will go back to the question formulated earlier. Can arthropod pests be controlled using viruses? Or more specifically: (1) can virus be produced; (2) are viruses safe; and (3) are viruses effective against field populations of pests.

Production of insect viruses

We have at our disposal many virus diseases which can be used against arthropod pests (Table 5) (Anderson and Ignoffo, 1967; Ignoffo, 1970). Some pest species have 3 to 4 different types of virus diseases. Species of the budworm-bollworm complex (Heliothis spp.), cosmopolitan pests which cause severe damage to many different crops, are susceptible to NPV, CPV, and a GIV. The same can be said for the sawfly complex (Diprion and Neodiprion spp.), armyworm complex (Pseudaletia, Spodoptera, and Prodenia spp.) and the cutworm complex (Loxagrotis, Agrotis, and Peridroma spp.).

Insect viruses, as well as all other viruses, can only be grown in living systems (Ignoffo, 1967). Various hosts from mites to caterpillars have been used to produce viruses in the laboratory and by industry (Table 6).

Previously, insects were collected in the field and fed foliage contaminated with virus. Dying insects were collected and processed into virus for use the following year. Recently developed techniques and semi-synthetic diets permit continuous virus production on an annual basis; caterpillars are laboratory-reared in large numbers and the viruses grown in the caterpillars (Ignoffo, 1966).

Essential nutrients and vitamins are mixed and prepared into a solid diet. Caterpillars are allowed to develop 5-7 days on the diet. Virus is then sprayed on the diet and the caterpillars develop for another 6-8 days; death generally occurs 7-9 days after spraying. One dead caterpillar produces 6-36 billion inclusion bodies. Dead caterpillars are collected, liquified, screened, and processed into a dry powder. This powder is standardized for activity and purity, formulated and then packaged for eventual field use. Generally less than 1/10 oz of the technical virus is needed to treat 1 acre of cropland.

Industry, in spite of many problems, has developed semi-automated systems for producing insect viruses (Ignoffo, 1967, 1972a; Greer, et al., 1971). Pilot plants which produce a million caterpillars per month are operational (Table 7). Larger plants have been designed to produce about 1 million caterpillars per week. Production of insect viruses at the plant-scale, however, is not presently operational.

Other techniques for producing viruses

Tissue culture or fermentation may be the virus production technology of the future (Ignoffo, 1967; Ignoffo and Hink, 1971). Insect cell lines are established and insect viruses will replicate in these lines. Vertebrate viruses are routinely produced by this method on a large scale and therefore it is possible to adapt current technology of producing vertebrate viruses to production of insect viruses. Production cost may not be as expensive as vertebrate viruses since levels of sanitation needed for vertebrate viruses may not be needed to produce insect viruses.

Fermentation also may be used to produce insect viruses. Living bacteria or yeasts have been used as the "host" for viral replication instead of caterpillars or tissue cells (Wells, 1970). Large numbers of bacterial or yeast cells can be grown in fermentors. Host cells can be specially treated to accept the insect virus inoculum. Cells, containing replicated insect virus, can then be concentrated, either physically or chemically, dried, standardized, and formulated into a finished product.

Safety of insect viruses

Once produced, insect viruses should be tested to see what effects they have on man, other animals, and plants. Indirect evidence, i.e., natural occurrence, presence on marketed vegetables or persons handling the virus, indicate that insect viruses are safe. Indirect evidence, however, should and must be corroborated by direct experimentation (Table 8). Experimental evidence relating to possible toxicity, pathogenicity, allergenicity, carcinogenicity, teratogenicity, specificity and mutability of the viruses, therefore, must be accumulated for both animals and plants which may be exposed to field applications of the viruses (Ignoffo, 1968b, 1972b; Heimpel, 1971). Safety in all living systems and for all times can never be absolutely guaranteed. However, each individual virus should be evaluated for safety prior to large-scale field use. Continual use, after initial safety is established, should be carefully monitored for several years to insure that detrimental accumulated effects will not develop. Based on present knowledge, however, insect viruses are generally safe, specific, and innocuous to other living organisms.

Within the last few years industry has developed the Heliothis NPV for use against bollworms-budworms on cotton (Ignoffo, 1972a, 1972b). This virus has been fed to many different insects, been tested on other invertebrates and on vertebrates including man, and repeatedly applied over wide areas to many different crops without any reported ill effects to users, wildlife, beneficial insects, or plants to which it was applied (Ignoffo, 1968b, 1972b); only species of Heliothis were found to be susceptible (Table 9). Test animals were fed or injected with the virus, or were made to inhale or receive, topical applications of the virus. No adverse effects were ever observed. This is a remarkable level of specificity when compared to the specificity of many other viruses.

A temporary exemption of a requirement for tolerance, based upon a series of 8 types of tests (Table 10), was granted in 1970 (Ignoffo, 1972b). This significant development has established general guidelines and protocols for evaluating future viruses. Costs for these tests were estimated at less than $125,000.

Effectiveness of insect viruses

All presently known insect viruses must be eaten by the insect pest if control is to be effective. Therefore, two factors are of utmost importance if viruses are to be successfully used against insect pests. First, application must be early enough in the life cycle of the pest or crop to minimize damage, and second, consistent, uniform plant coverage must be obtained.

In order to control an insect pest like loopers on a cole crop, for example, its NPV is mixed with water and sprayed on plants. Caterpillars feed on the plants and consequently swallow the viral inclusion bodies, which dissolve in their stomachs. Seconds later, virions or infective subunits are released. These pass through the gut wall of the caterpillar and infect the nuclei of susceptible cells in which replication occurs. The virus continues to grow in susceptible cells until the caterpillar eventually dies.

The occurrence of natural epidemics which completely wipe out populations of insects are striking demonstrations of the effectiveness of insect viruses. These natural epidemics can occur annually with some pests, e.g., the cabbage looper. In fact it is often impossible to grow cabbage loopers in the laboratory on raw cabbage or lettuce bought at the supermarket. Enough natural virus can still be on the leaves to kill about a million young cabbage looper larvae. The example of the cabbage looper is not an isolated case. Natural epidemics are observed in field populations of sawflies, the gypsy moth, tent caterpillars, the alfalfa caterpillar, as well as with many other arthropod pests. Unfortunately, most of this "natural" control comes after the damage is done! Plants must be sprayed with virus when the caterpillars are still young if extensive damage is to be prevented. Natural epidemics are not the only example of the effectiveness of insect viruses. Nearly 40 different insect viruses have been successfully used to control field populations of various pest insects (Ignoffo, 1967, 1968a, b; Stairs, 1971). All major viral groups and types of pests are included in examples as shown in Table 11.

Within the past 10 years, 5 different viruses have been produced by American or foreign commercial concerns and either sold or made available for experimental control of insect pests (Ignoffo, 1970). All viruses developed so far are of the nucleopolyhedrosis type. In a series of similar tests conducted in 6 cotton states in 1969, one NPV (Viron H)1/ significantly increased yields

1/Mention of a proprietary product does not constitute endorsement by the USDA.

over untreated fields and gave control of bollworm-budworms comparable to the standard insecticide (Table 12). These tests were all conducted at 240 X 10^9 PIB/acre, which represents the amount of virus obtained from 25 to 40 mature, virus-killed bollworms. Use of additives to inhibit inactivation by sunlight increased yields and reduced variability (Ignoffo, et al., 1972).

In 1970, the Heliothis NPV was federally registered and granted a temporary exemption for the requirement of tolerance (Ignoffo, 1972b). A permanent exemption was granted on May 23, 1973. This was the first time an insect virus was ever registered as a pesticide. The temporary permit authorized use of more than 2 million acre-treatments of virus in 19 cotton-growing states from coast to coast, from Illinois south to the Gulf.

Conclusion

Realistically, viral insecticides to be bought and used by growers must be effective, economical, and offer a distinct advantage over other materials. In most instances, they must compete directly with commonly-used chemical insecticides which have built up a following, as well as a technology for field application. In general, insect viruses show promise as living insecticides. They are effective, can be as available and competitive as insecticides, are not poisonous, and do not cause diseases in other animals and plants.

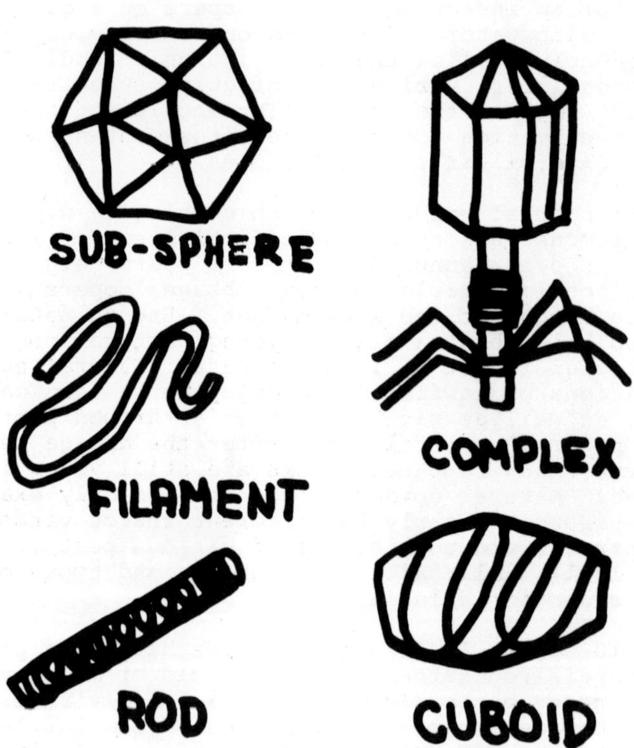

Figure 1.--Shapes of virions which cause diseases in bacteria, plants, insects, animals and man.

Figure 2.--Scanning electron microphotograph of purified inclusion bodies of an insect nucleopolyhedrosis virus (10,000 X).

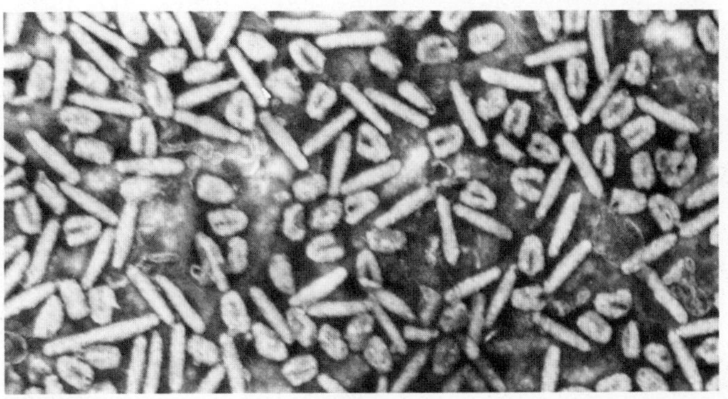

Figure 3.--Rod-shaped virions of an insect nucleopolyhedrosis virus (30,000 X).

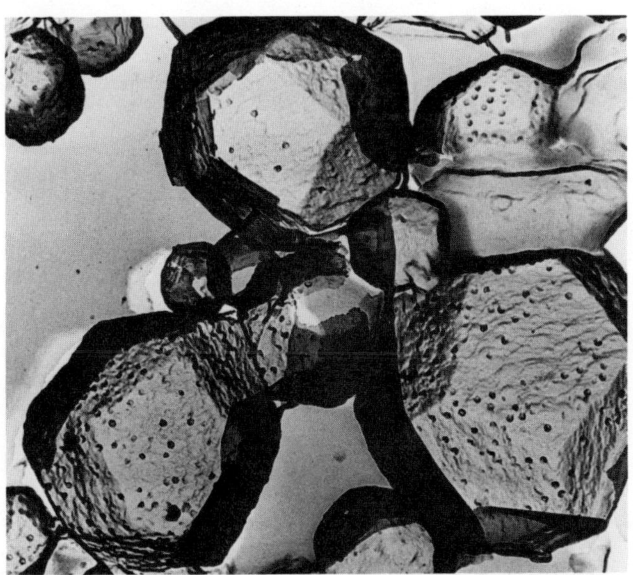

Figure 4.--Carbon replica of purified inclusion body of an insect cytoplasmic polyhedrosis virus (15,000 X).

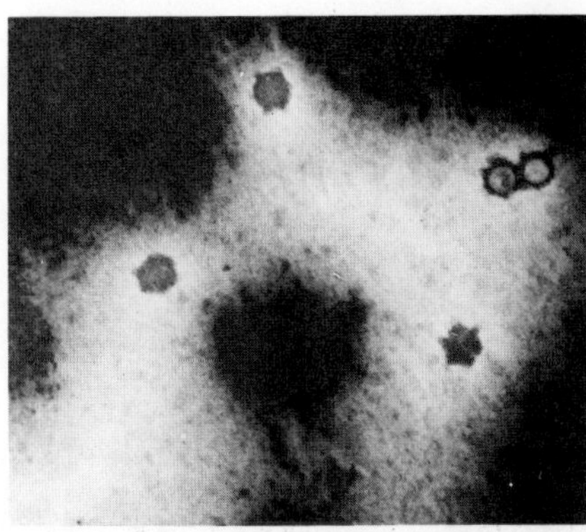

Figure 5.--Negative stain of sub-spherical virions of an insect cytoplasmic polyhedrosis virus (650,000 X).

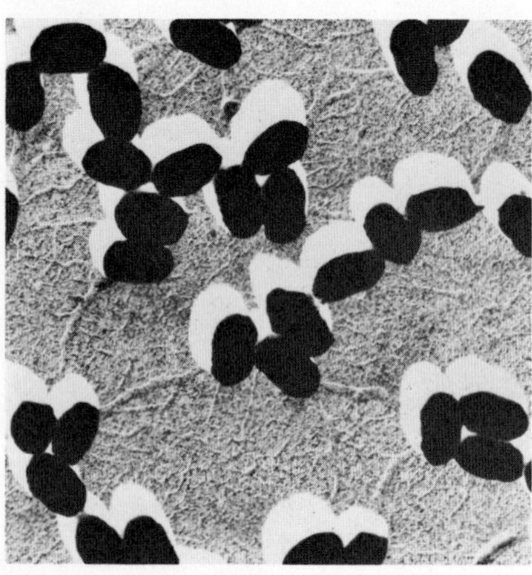

Figure 6.--Shadowed purified inclusion bodies of an insect granulosis virus (20,000 X).

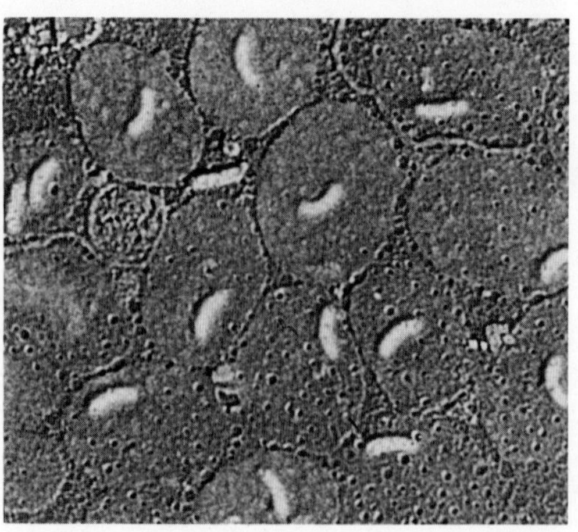

Figure 7.--Alkali-released, rod-shaped virions of an insect granulosis virus (30,000 X).

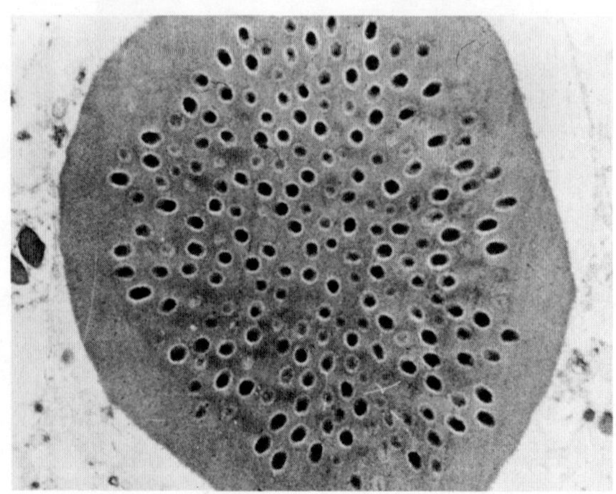

Figure 8.--Cross-section through an inclusion body of an insect pox virus (25,000 X). Courtesy M. Bergoin, Laboratoire de Cytopathologie, Gard, France.

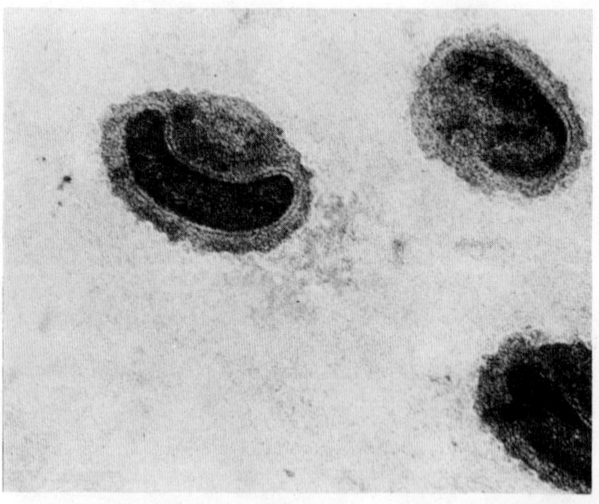

Figure 9.--Ovoid-cuboid shaped virions of an insect pox virus (100,000 X) Courtesy of M. Bergoin.

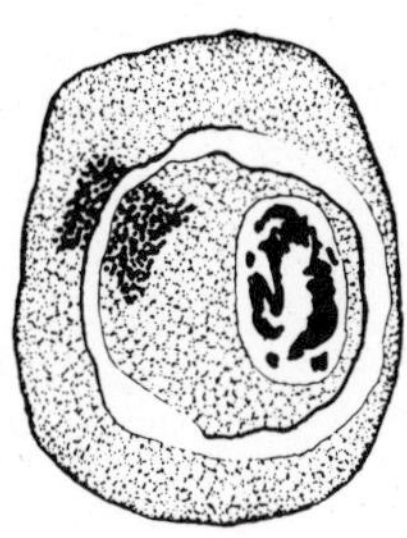

Figure 10.--Irregular-shaped body of an insect polymorphic inclusion virus (2,000 X). From E. A. Steinhaus,, University of California, Irvine, California, USA (deceased).

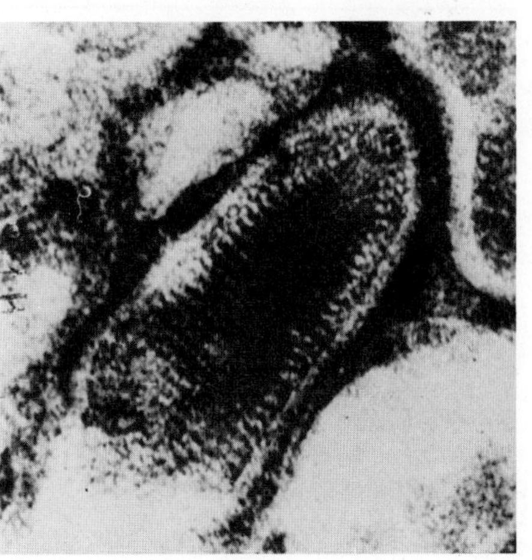

Figure 11.--Bullet-shaped virions of an insect non-inclusion virus (350,000 X). Courtesy of C. Vago, Laboratoire de Cytopathologie Gard, France.

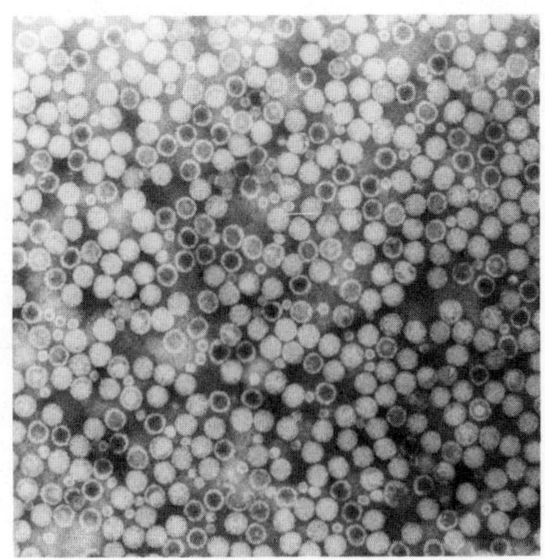

Figure 12.--Sub-spherical shaped virions of an insect non-inclusion virus (120,000 X). Courtesy of C. Vago.

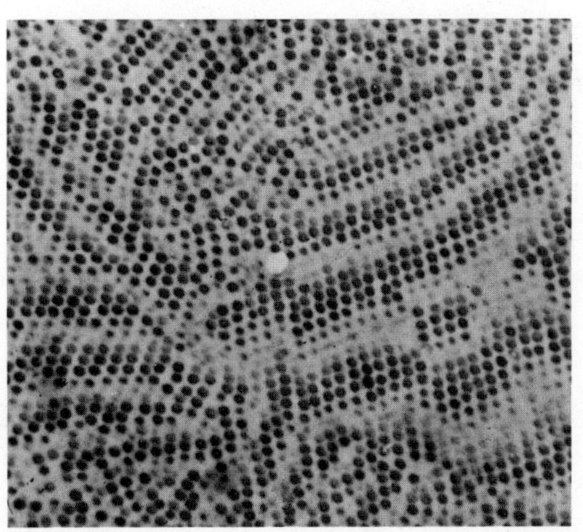

Figure 13.--Paracrystalline pattern of sub-spherical shaped virions of an insect non-inclusion virus (12,000 X).

Table 1.--Virion shape and size and nucleic acid type of viruses isolated from bacteria, plants, vertebrates and invertebrates.

Disease or virus	Nucleic acid	Virion size (mμ)
Complex shape		
Bacteriophage	DNA	Head, tail, 100
Spherical or sub-spherical shape		
Insect Iridiovirus	DNA	130
Mammalian Herpesvirus	DNA	125
Cauliflower mosaic	DNA	50
Bacterial coliphage	DNA	25
Insect cytoplasmic polyhedrosis	RNA	60
Tomato bushy stunt	RNA	30
Colorado tick fever	RNA	75
Cuboid-brick shape		
Insect pox	DNA	260 X 320 X 210
Vaccinia	DNA	100 X 200 X 300
Rod or bullet shape		
Insect nuclear polyhedrosis	DNA	60 X 320
Tobacco mosaic	RNA	20 X 300
Insect granulosis	DNA	60 X 320
Rabies	RNA	70 X 175
Filament shape		
Bacterial coliphage	DNA	800 X 5
Potato virus Y	RNA	750 X 4
Mammalian myxovirus	RNA	100 X 18

Table 2.--Current approved groups and names of arthropod viruses.

Virus type	Type host species	Virus genus
RNA Viruses		
Cytoplasmic polyhedrosis	*Bombyx mori* (L.)	undecided
Drosophila CO_2-virus	*Drosophila melanogaster* Meigen	*Sigmavirus*
DNA Viruses		
Nuclear polyhedrosis	*Bombyx mori* (L.)	*Baculovirus*
Granulosis	*Choristoneura fumiferana* (Clemens)	*Baculovirus*
Densonucleosis	*Galleria mellonella* (L.)	*Densovirus*
Pox-Spheroidosis	*Melolontha melolontha* (L.)	*Entomopoxvirus*
Iridescent	*Tipula paludosa* Meigen	*Iridovirus*

Table 3.--Percentage of the described viruses isolated from various types of arthropods.[1/]

Arthropod types	Percent of described viruses
Moths and butterflies	83
Sawflies	10
Flies and mosquitoes	4
Beetles	1
Spider mites	1
Grasshoppers	1

[1/] Ignoffo, 1968a.

Table 4.--General characteristics of the various types of viruses associated with arthropods.

Virus type	Estimated number	Nucleic acid	Shape	
			inclusion	virion
Inclusion viruses				
Nuclear polyhedroses	138	DNA	polyhedron	rod
Cytoplasmic polyhedroses	121	RNA	polyhedron	sphere
Granuloses	42	DNA	granule	rod
Pox viruses	12	DNA	polyhedron	oval
Polymorphic viruses	1	?	irregular	
Non-inclusion viruses				
Spherical viroses	12	RNA	none	sphere
Iridescent	6	DNA	none	sphere
Rod viroses	2	RNA	none	rod
Densonucleosis	1	DNA	none	sphere

Table 5.--Some selected examples of major economic arthropod pests in the United States having associated viruses.

Grain, Grasses, Forage and Fiber Crops	
Grasshoppers	Armyworms
Bud- and bollworms	Pink bollworm
Alfalfa caterpillar	Saltmarsh caterpillar
Fruit, Vegetables and Truck crops	
Cutworms	Redbanded leafroller
Imported cabbageworms	Citrus red mite
Codling moth	Cabbage looper
Forest, Ornamental and Shade Trees	
Sawflies	Spruce budworm
Gypsy moth	Webworms
Tent caterpillars	Tussock moths

Table 5.--continued.

Household, Stored Products, Man and Animals	
Clothes moths	Almond moth
Mosquitoes	Greater wax moth
Midges	Gnats

Table 6.--Representative yields of arthropod viruses produced on a laboratory scale.[1/]

Target pest	Food	Billions of viral units/gram of host[2/]
	Nuclear polyhedrosis	
Bollworm	diet	30
Cabbage looper	diet	43
	broccoli-cotton	26
Gypsy moth	diet	30
	Cytoplasmic polyhedrosis	
European pine processionary moth	pine	5
Japanese pine caterpillar	pine	30
	Granulosis	
Codling moth	diet	20
Redbanded leafroller	apple leaves	10
	Entomopox	
Spruce budworm	diet	1
	Non-Inclusion virus	
Citrus red mite	lemons	1

[1/] Ignoffo and Hink, 1971.

[2/] All viral units are inclusion bodies except median-infective dose of citrus red mite non-inclusion virus.

Table 7.--Levels of Heliothis NPV produced at various phases of its development.[1/]

Development phase	Thousand host larvae/month	Estimated cost/ larvae (¢)
Laboratory	54	7.0
Pilot-plant	1,000	4.8
Commercial plant	4,200	2.0

[1/] Ignoffo, 1972a.

Table 8.--Types of insect viruses tested for safety to various vertebrates.[1/]

Test animals	Viruses tested
Mammals: white mouse, white rat, guinea pig, rabbit, dog, human	nuclear polyhedrosis cytoplasmic polyhedrosis granulosis non-inclusion
Birds: chicken embryo, turkey embryo, chicken, sparrow, dove, mallard, quail	nuclear polyhedrosis cytoplasmic polyhedrosis granulosis
Amphibians: frog	nuclear polyhedrosis
Fishes: blue gill, black bullhead, killifish, rainbow trout, sheepshead minnow, spotfish, white sucker	nuclear polyhedrosis

[1/] Ignoffo, 1968b.

Table 9.--Plants, insects, other invertebrates, and vertebrates used to determine the specificity of the *Heliothis* NPV[1/]

Vertebrates

Man, monkey, dog, rabbit, guinea pig, white rat, white mouse, chicken, chicken egg, quail, sparrow, mallard, killifish, spotfish, rainbow trout, bluegill, black bullhead, white sucker, sheepshead minnow

Insects

Heliothis (5 species), tobacco hornworm, tomato hornworm, greater wax moth, cabbage looper, beet armyworm, fall armyworm, southern armyworm, Lucerne moth, tobacco cutworm, cabbageworms, beetworm, diamondback moth, smaller citrus dog, Oriental tussock moth, common cutworm, honey bee, house fly, rice leafhopper, Daikon leaf beetle, carmine spider mite

Invertebrates

Grass shrimp, brown shrimp, oyster

Plants

Cotton, corn, sorghum, bean, soybean, snapbean, kidney bean, tomato, tobacco, radish

[1/] Ignoffo 1968a, 1972a; Greer, et al., 1971.

Table 10.--Tests which established that the *Heliothis* NPV was safe and could be granted a temporary exemption from the requirement of tolerance.[1/]

Type of test	Animal system	Estimated cost ($)[2/]
Acute toxicity-pathogenicity		
Per os diet	rat or mouse, birds, fishes, oyster, shrimps	6,000
Inhalation	rat	750
Dermal	rats, rabbit, guinea pig	750
Intraperitoneal injection	rat or mouse	725
Sub-cutaneous injection	rat	500
Sensitivity-irritation		
Eye	rabbit	500
Skin	rabbit, man	6,500
Sub-acute toxicity-pathogenicity		
Diet	monkey, dog, rat or mouse	25,000
Inhalation	monkey, dog, rat or mouse	18,000
Sub-cutaneous injection	monkey, dog, rat or mouse	18,000
Teratogenicity	rat or mouse	3,000
Carcinogenicity	rat or mouse	30,000
Replication potential	man, primates, tissue culture	6,000
Phytotoxicity	agricultural crops	1,750
Invertebrates-specificity	beneficial and other arthropods	4,000

[1/] Ignoffo, 1972a, b.

[2/] Costs of tests conducted from 1965-1968, inclusive.

Table 11.--Selected examples of viruses used to experimentally control arthropod pests.

Target pest and crop attacked	Virus type	Reference
Vegetables		
Cabbage looper	nuclear polyhedrosis	Hofmaster & Ditman, 1961
Cabbageworms (*Pieris* spp.)	granulosis	Biliotti, et al., 1956
Orchard crops		
Codling moth	granulosis	Falcon, et al., 1968
Citrus red mite	non-inclusion	Munger & Gilmore, 1960
Forage crops		
Alfalfa caterpillar	nuclear polyhedrosis	Steinhaus & Thompson, 1949
Fiber and grain crops		
Boll- and budworms (*Heliothis* spp.)	nuclear polyhedrosis	Ignoffo, et al., 1965
Cotton leafworm	nuclear polyhedrosis	Abul-Nasr, 1959
Forest and shade trees		
Gypsy moth	nuclear polyhedrosis	Rollinson, et al., 1965
European pine processionary moth	cytoplasmic polyhedrosis	Grison, et al., 1959
Spruce budworm	entomopox	Bird, et al., 1971

Table 12.--Seed cotton yields from plots treated with Viron H® in tests conducted in Alabama, Arkansas, California, Mississippi, North Carolina, and Texas (1969)[1/].

Formulations[2/]	Pounds seed cotton/acre Average	Range
Viron H[3/]	1878	575 - 3278
Viron H + carbon	2038	970 - 3192
Viron H + IMC 90001[4/]	2140	980 - 3585
Standard insecticide[5/]	2179	1095 - 3648
Check	1647	392 - 2799

[1/] Ignoffo, et al., 1972.

[2/] 5 to 7 tests/formulation; [3/] Viron H used at 40 LE/acre.

[4/] Sunlight Protectant of Int. Minerals and Chemical Corporation.

[5/] Standard insecticide was insecticide commonly used on cotton in specific state.

References Cited

Abul-Nasr, S. 1959. Further tests on the use of a polyhedrosis virus in the control of the cotton leafworm Prodenia litura Fabricius. J. Insect Pathol. 1: 99-106.

Anderson, R. F. and C. M. Ignoffo. 1967. Microbial insecticides, pp. 172-182, In H. J. Peppler (ed.), Microbial Technology. Univ. Food Corp., 454 pp.

Bird, F. T., C. T. Sanders, J. M. Burke. 1971. A newly discovered virus disease of the spruce budworm Choristoneura biennis (Lepidoptera: Tortricidae). J. Invertebr. Pathol. 18: 159-161.

Biliotti, E., P. Grison and D. Martouret. 1956. L'utilisation d'une maladie a virus comme methode de lutte biologique contre Pieris brassicae L. Entomophaga 1: 35-44.

Falcon, L. A., W. R. Kane and R. S. Bethell. 1968. Preliminary evaluation of a granulosis virus for control of the codling moth. J. Econ. Entomol. 61: 1208-1213.

Greer, F., C. M. Ignoffo and R. F. Anderson. 1971. The first viral pesticide: A case history. Chemtech., June 1971, pp. 342-347.

Grison, P., C. Vago and R. Maury. 1959. La lutte contre la processionnaire du pin "Thaumetopoea pityocampa" Schiff dans le masif du Ventoux; essai d'utilisation practique d'un virus specifique. Rev. For. Fr. (Nancy) 5: 353-370.

Heimpel, A. M. 1971. Safety of insect pathogens for man and vertebrates, pp. 469-489, In H. D. Burges and N. W. Hussey (eds.), Microbial Control of Insects and Mites. Academic Press, New York, 861 pp.

Hofmaster, R. N. and L. P. Ditman. 1961. Utilization of a nuclear polyhedrosis virus to control the cabbage looper on cole crops in Virginia. J. Econ. Entomol. 54: 921-923.

Holmes, F. O. 1948. Order virales, the filterable viruses, pp. 1225-1228. In Bergey's Manual of Determinative Bacteriology, 6th ed. Baltimore: Williams and Wilkins Co., 1094 pp.

Ignoffo, C. M. 1966. Insect viruses, Chap. 36, pp. 501-530. In C. N. Smith (ed.), Insect Colonization and Mass Production. Academic Press, New York. 618 pp.

Ignoffo, C. M. 1967. Possibilities of mass-producing insect pathogens. Proc. Inter. Colloq. Insect Pathol. & Microbial Contr. Wageningen, The Netherlands, Sept., 1966, North-Holland Publ. Co., Amsterdam, pp. 91-117.

Ignoffo, C. M. 1968a. Viruses-living insecticides, pp. 129-167. In K. Maramorosch (ed.), Current Topics in Microbiology and Immunology. Springer-Verlag, New York, Inc., 192 pp.

Ignoffo, C. M. 1968b. Specificity of insect viruses. Bull. Entomol. Soc. Amer. 14: 265-276.

Ignoffo, C. M. 1970. Microbial insecticides: no-yes; now-when! Proc. Tall Timbers Conf. Ecol. Animal Control Habitat Manage. 2: 41-57.

Ignoffo, C. M. 1972a. Concept to commercialization: the development of a viral insecticide. Colloq. Proc. Exper. Parasitol. 33: 380-406.

Ignoffo, C. M. 1972b. Toward registration of a viral insecticide. Misc. Publ. Entomol. Soc. Amer. (in press).

Ignoffo, C. M. and W. F. Hink. 1971. Propagation of arthropod pathogens in living systems, pp. 541-580. In H. D. Burges and N. W. Hussey (eds.), Microbial Control of Insects and Mites. Academic Press, New York, 861 pp.

Ignoffo, C. M., A. J. Chapman and D. F. Martin. 1965. The nuclear polyhedrosis virus of *Heliothis zea* (Boddie) and *Heliothis virescens* (Fabricius). III. Effectiveness of the virus against field populations of *Heliothis* on cotton, corn, and grain sorghum. J. Invertebr. Pathol. 7: 227-235.

Ignoffo, C. M., J. R. Bradley, Jr., F. R. Gilliland, Jr., F. A. Harris, L. A. Falcon, L. V. Larson, R. L. McGarr, P. P. Sikorowski, T. F. Watson and W. C. Yearian. 1972. Field studies on stability of the *Heliothis* nucleopolyhedrosis virus at various sites throughout the cotton belt. Environ, Entomol. 1: 388-390.

Munger, F. and J. E. Gilmore. 1960. Mass production of the citrus red mite. J. Econ. Entomol. 53: 964-966.

Rollinson, W. D., F. B. Lewis and W. E. Waters. 1965. The successful use of a nuclear-polyhedrosis virus against the gypsy moth. J. Invertebr. Pathol. 7: 515-517.

Stairs, G. R. 1971. Use of viruses for microbial control of insects, pp. 97-124. In H. D. Burges and N. W. Hussey (eds.), Microbial Control of Insects and Mites. Academic Press, New York, 861 pp.

Steinhaus, E. A. 1953. Taxonomy of insect viruses. Ann. New York Acad. Sci. 56: 517-537.

Steinhaus, E. A. and C. G. Thompson. 1949. Preliminary field tests using a polyhedral virus to control the alfalfa caterpillar. J. Econ. Entomol. 42: 301-305.

Wells, F. E. 1970. Insect virus replication and reproduction. Patent No. 700831. Republic of South Africa.

Wildy, P. 1971. Classification and nomenclature of viruses. Monographs in virology. Basel: S. Karger AG. Vol. 5, 81 pp.

PROTOZOAN PATHOGENS

William R. Kellen
Market Quality Research Division, ARS, USDA
Fresno, California

Insects are known hosts for a wide variety of pathogenic protozoa. Lipa (1963) presented a general review of the insect pathogenic amoebae, flagellates, and ciliates; a similar review by Weiser (1963) covered the pathogenic Sporozoa, including members of the Eucoccida, Haplosporida, Microsporida, and Neogregarinida. A more recent general discussion of pathogenic protozoa and their possible application for insect control was prepared by McLaughlin (1971). The present review will be limited to the protozoan orders Microsporida and Neogregarinida because they have received the greatest attention of researchers as possible agents for the microbial control of insect pests.

Microsporida

Biology

The members of the order Microsporida are characterized by the formation of resting spores which contain a polar filament and a single infective sporoplasm. All are obligate intracellular pathogens. They are known to infect invertebrates and to a lesser degree lower vertebrates, but one representative has also been isolated from the nervous tissue of mammals. The Microsporida are relatively common pathogens of both beneficial and pestiferous insects.

The majority of the over 200 known species from insects belong to the family Nosematidae, which is comprised of 8 genera. The genera are distinguished by the number of spores (1, 2, 4, 8, etc.) formed in the sporoblast during the sporogonic (sexual) part of the life cycle. Because of differences in the interpretation of certain details in the formation of spores, however, the taxonomic validity of some genera recently has been challenged (Burges, et al., 1971).

As many species of Microsporida have relatively few distinct morphological differences, diagnostic characteristics of new species have been largely based on a combination of characters, including spore size and shape, tissue specificity, and host range. Even these characters, however, are not always conclusive. For example, Hazard and Lofgren (1971) presented data to show that tissue specificity, which was long considered to be a very stable character, can vary from host to host. They infected 4 species of mosquitoes with a *Nosema* which was originally isolated from the common malaria mosquito, *Anopheles quadrimaculatus* Say. In its original host the pathogen was highly virulent, invading 12 different tissues and organs, 4 of them heavily. In contrast, the southern house mosquito, *Culex pipiens quinquefasciatus* Say, was invaded very lightly by the Nosema; only 9 tissues and organs were involved and one of them only rarely. A third host, *Aedes salinarious* Coquillett, was invaded rather lightly, but the same tissues were involved as with *A. quadrimaculatus*. Finally, the pathogen only invaded the nervous tissue of the yellowfever mosquito, *Aedes aegypti* (Linnaeus). The authors concluded, of course, that the site of infection is not always a good criterion for species separation. In a similar study, Weiser and Coluzzi (1966) showed that *Pleistophora culisetae* Weiser and Coluzzi invaded the fat, gut, Malpighian tubes, muscles, and blood cells of the mosquito, *Culiseta longiarelolata* (Macquart), but it only invaded the fat body of the northern house mosquito, *Culex pipiens pipiens* Linnaeus. It is evident from these studies that degrees of tissue specificity must be determined separately for each pathogen, as broad generalizations cannot be made. The same care must be taken in the selection of all such diagnostic characters.

Until recently many workers assumed without conclusive data that microsporidans were host specific or restricted to closely related host species. Steinhaus and Hughes (1949), however, clearly showed that such is not always the case. They demonstrated that under laboratory conditions at least 10 different insect species in 3 different orders (Lepidoptera, Hymenoptera, and Neuroptera) were susceptible to peroral transmission of *Nosema destructor* Steinhaus and Hughes, which was originally described from the potato tuberworm, *Phthorimaea operculella* (Zeller). Furthermore, Kellen and Lindgren (1969) reported 5 species of Lepidoptera (4 genera) and 2 species of beetles as being susceptible to *Nosema heterosporum* Kellen and Lindgren which was first isolated from the Indian meal moth, *Plodia interpunctella* (Hubner). Similarly, Burges, et al. (1971) showed that *Nosema oryzaephili* Burges, Canning, and Hurst, a virulent pathogen of the saw-toothed grain beetle, *Oryzaephilus surinamensis* (Linnaeus), had a host range among storage insects that included 5 species of beetles and 3 moths, involving 5 insect families. It must be remembered, however, that such laboratory data do not necessarily indicate the extent of the host range as it occurs under natural field conditions. Weiser (1963) has suggested that selection for host range is dependent upon having potential hosts occupy a common biotype with a common food source, so that frequent transmission is facilitated; he concluded that close taxonomic relationship is of secondary importance.

The formation of resistant spores is the terminal stage of development of Microsporida. When the spore is ingested, chemical activity of the gut stimulates the extrusion of a hollow polar filament through which the sporoplasm passes. In this way the infective agent is released in the tissue to initiate further development and multiplication. In recent years electron microscope studies of spores have been conducted to elucidate the ultrastructure in an effort to clarify the mechanism of germination as well as to establish additional criteria for species determination (Sprague and Vernick, 1968).

Trans-ovum transmission of Microsporida is as common as peroral transmission in many species of insects. Ovaries and associated tissues of sublethally diseased females may become infected and eggs which harbor pathogens both internally and externally may be laid. Internal (transovarian) transmission is an especially important factor in maintaining pathogens in overwintering adult populations of insects.

Kramer (1959) reported that infected adult females of the European corn borer, *Ostrinia nubilalis* (Hubner), transmitted *Glugea pyraustae* (Paillot) to half of their offspring. The tissues of the accessory glands and ovarian tubes harbored every stage of the pathogen. External contamination of eggs consisted of clusters of spores which became lodged in the shallow fovea on the surface of the chorion. As larvae fed upon empty egg shells soon after hatching, young larvae acquired infections during this first feeding. Schizonts (asexual forms) and spores were also observed within about 50% of the eggs examined, so that many larvae became initially infected as embryos.

In a study of the transovarian transmission of *Nosema plodiae* Kellen and Lindegren in the Indian meal moth, Kellen and Lindegren (1973) found that 12.5% of 856 eggs laid by infected moths harbored pathogens. The highest level of transmission by an individual female was 14 infected eggs of 27 laid

(51.8%); the lowest level of transmission observed was one of 43 eggs (2.3%). All stages of N. plodiae were not transmitted with equal frequency, however, and eggs usually harbored predominantly only one stage of the pathogen. Approximately 82% of the infected eggs contained schizonts almost exclusively; about 6.5% contained mainly spores, and the remainder contained about equal number of spores and schizonts. Histological data indicated that infections probably started in ovarian nurse cells and subsequently were transferred to associated oocytes. Similarly, Brooks (1968) reported that eggs of the corn earworm, Heliothis zea (Boddie), were heavily infected with binucleate schizonts of Nosema heliothidis Lutz and Splendor, but spores were detected infrequently. Finally, Kellen and Wills (1962) reported that eggs of the mosquito, Culex tarsalis Coquillett, harbored only mononucleate and binucleate schizonts of Thelohania californica Kellen and Lipa. Typical spores of this pathogen were not observed in infected females. In contrast to the above relationships, however, Kellen and Lindegren (1969) noted that Nosema heterosporum rarely invaded the ovaries of the Indian meal moth, and transovarian transmission did not occur.

In addition to peroral and transovarian transmission, Microsporida can also be transmitted by inoculation through the cuticle. This may occur in nature through the agency of parasitic wasps which transfer spores on ovipositors which have become contaminated by previous attacks on infected larvae (Tanada, 1963).

Microsporida as Microbial Control Agents

Microbial insecticides may be applied as protectants much as chemical insecticides are used, or they may be introduced into insect populations to act as suppressing or stressing agents to assist other natural factors to regulate insect populations. The potential use of protozoan pathogens generally lies in the latter category.

One way in which Microsporida act to regulate populations is by reducing adult life span and fecundity. Thomson (1958) presented evidence to show that Glugea fumiferanae (Thomson) had a very significant depressing effect on the fecundity of the spruce budworm, Choristoneura fumiferana (Clemens), and the pathogen was credited with creating a general reduction in at least one population in Canada (Thomson, 1960). Kramer (1959) reported that European corn borers infected by Glugea pyraustae have significantly shorter adult life spans than uninfected adults. Moreover, in laboratory tests infected females produced fewer eggs and frequently were sterile. Veber and Jasic (1961) noted that the fecundity of moths of the silkworm, Bombyx mori (Linnaeus), with sublethal infections of Nosema bombycis Naegli was highly reduced. When reared on a diet containing 5 x 10^3 spores/g as 5th instar larvae, egg production of moths was reduced about 25%; when moths were fed the same concentration of spores as early 4th instar larvae, however, fecundity was reduced about 55%. In a similar study, Kellen and Lindgren (1971) reported that 25% of the eggs of Indian meal moths infected with Nosema plodiae did not hatch; moreover, 27% of the surviving progeny were infected transovarially.

Although pathogenic protozoa frequently give rise to chronic disease which may not cause mortality until late in the larval stage, there is evidence that protozoans can be effective regulatory agents when their total influence is evaluated. Franz and Huger (1970) investigated the role of Nosema tortricis Weiser in the collapse of an outbreak of the green tortrix, Tortrix viridana Linnaeus, in oak forests of central Germany. In a preliminary examination in 1968, they found that 29% of the 1st instar larvae collected from oak twigs were infected with the Nosema. A subsequent collection of over 1500 larvae showed that the level of infection had risen to 59%; moreover, 60% of the adults examined were also naturally infected. Because of transovarian transmission by infected females, the initial level of infection was about twice as high the following year. Two subsequent larval collections were made. The first consisted predominantly of 3rd instar larvae of which about 80% were infected; the final sample (2 weeks later) showed the typical post-epizootic phase of the disease; the incidence of infection, which was density dependent, had dropped to 24%. Franz and Huger (1970) recommended: (1) that the health status of insect pest populations should be determined regularly and (2) that when the population nears a critical threshold the levels of naturally occurring disease should be carefully considered before control measures are taken.

In addition to naturally occurring epizootics of Microsporida, it has been demonstrated that it is possible to introduce pathogens at optimal times to control populations before they are well established. For example, Henry (1971) conducted a field study to determine the feasibility of controlling grasshoppers in Montana with applications of _Nosema locustae_ Canning. Spores in a bran were applied at rates of 50, 100, 200, 400, and 800 spores/in^2 in plots of 10 acres each. The experimental plots were subsequently sampled to determine the incidence of infection and the reduction in population due to _Nosema_ disease. The average incidence of infection among the treated groups of insects of the several species increased from about 3.6% after 3 weeks to 34% after 6 weeks, but there was no significant difference in the 5 rates of spore application tested. Five of the species of grasshoppers were the most abundant and had the highest levels of infection. After 6 weeks the combined density of the 3 most predominant species was reduced by 34.5%. The density of one of them, _Melanoplus gladstoni_ Scudder, was reduced by 56.7%; the high level of reduction was attributed to the fact that this species developed later and was, therefore, at an early nymphal stage of development when the _Nosema_ spores were applied. Henry stated that in addition to the reduction in population density due to mortality, an additional reduction would probably occur because of reduced fecundity. Although the lowest effective dose for control was not determined, it was noted that a single infected adult grasshopper produced sufficient spores of _N. locustae_ to treat about 3 acres. Considering that there have been very few well documented field trials with pathogenic protozoa, the results reported by Henry are very encouraging.

Neogregarina

The Neogregarina include a variety of forms with diversified host-pathogen relationships. Like members of the Microsporida they are obligate pathogens. The species belonging to the genus _Mattesia_ (Schizocystidae) have received predominant attention as possible microbial control agents. They are pathogens of Lepidoptera and Coleoptera.

The life cycle of _Mattesia_ includes a micronuclear and macronuclear schizogony and a sporogonic cycle terminating in the formation of twin spores. Germination occurs in the gut of the susceptible host resulting in the release of 8 sporozoites from each spore. Sporozoites are motile and penetrate the fat tissue where the life cycle is repeated.

Considerable effort has been made to test the feasibility of using _Mattesia_ as an agent for controlling the boll weevil, _Anthonomus grandis_ Boheman. McLaughlin (1967) treated plants in 1/16-acre field cages with a bait containing spores of _Mattesia grandis_ McLaughlin and a feeding stimulant from cottonseed oil. At the end of the test he recorded that 55% of the weevils were diseased; moreover, the diseased population subsequently produced about one-half as many adults as the untreated check population. It was concluded that spores plus a feeding stimulant might be effective for use in large open fields where weevil movement is unrestricted. There was a 97% increase in cotton yield in the treated plots. In a subsequent field test which also included the application of spores of a microsporidan, McLaughlin, et al. (1969) observed that it was possible to reduce overwintering populations of the boll weevil by applying baits to populations entering diapause. If such a reduction were coupled with an induced epizootic of the spring population, control would be feasible. Two advantages of microbial control were apparent. It provided preservation of other natural enemies of the weevil which are not susceptible to the pathogens and an alternative method to chemical control to be used if resistance to chemicals should develop.

It is evident that insect pathogenic protozoa have potential for practical application in many control programs. Future applications of microbial control technology will utilize this potential.

References Cited

Brooks, W. M. 1968. Transovarian transmission of Nosema heliothidis in the corn earworm Heliothis zea. J. Invertebr. Pathol. 11: 500-512.

Burges, H. D., E. C. Canning and J. A. Hurst. 1971. Morphology, development and pathogenicity of Nosema oryzaephili n. sp. in Oryzaephilus surinamensis and its host range among granivorous insects. J. Invertebr. Pathol. 17: 419-432.

Franz, J. M. and A. M. Huger. 1970. Microsporidia causing the collapse of an outbreak of the green tortrix (Tortrix viridana L.) in Germany. Proc. IV Int. Colloq. Insect Pathol. 48-53.

Hazard, E. I. and Lofgren, C. S. 1971. Tissue specificity and systematics of a Nosema in some species of Aedes, Anopheles, and Culex. J. Invertebr. Pathol. 18: 16-24.

Henry, J. E. 1971. Experimental application of Nosema locustae for control of grasshoppers. J. Invertebr. Pathol. 18: 389-394.

Kellen, W. R. and W. Wills. 1962. The transovarian transmission of Thelohania californica Kellen and Lipa in Culex tarsalis Coquillett. J. Insect Pathol. 4: 321-326.

Kellen, W. R. and J. E. Lindegren. 1969. Host-pathogen relationships of two previously undescribed Microsporidia from the Indian meal moth, Plodia interpunctella (Hubner), (Lepidoptera, Phycitidae). J. Invertebr. Pathol. 14: 328-335.

Kellen, W. R. and J. E. Lindegren. 1971. Modes of transmission of Nosema plodiae Kellen and Lindegren, a pathogen of Plodia interpunctella (Hubner). J. Stored Prod. Res. 7: 31-34.

Kramer, J. P. 1959. Some relationships between Perezia pyraustae Paillot (Sporozoa, Nosematidae) and Pyrausta nubilalis (Hubner) (Lepidoptera, Pyralidae). J. Insect Pathol. 1: 25-33.

Lipa, J. J. 1963. Infections caused by protozoa other than sporozoa, In Insect Pathology, An Advanced Treatise (E. A. Steinhaus, ed.), 2: 335-361. Academic Press, New York.

McLaughlin, R. E. 1967. Development of the bait principle for boll weevil control II. Field-cage tests with a feeding stimulant and the protozoan Mattesia grandis. J. Invertebr. Pathol. 9: 70-77.

McLaughlin, R. E., T. C. Cleveland, R. J. Daum and M. R. Bell. 1969. Development of the bait principle for boll weevil control IV. Field tests with a bait containing a feeding stimulant and the sporozoans Glugea gasti and Mattesia grandis. J. Invertebr. Pathol. 13: 429-441.

McLaughlin, R. E. 1971. Use of protozoans for microbial control of insects. In Microbial Control of Insects and Mites (H. D. Burges and N. W. Hussey, eds.), pp. 151-172. Academic Press, New York.

Sprague, V. and S. H. Vernick. 1968. Light and electron microscope study of a new species of Glugea (Microsporida, Nosematidae) in the 4-spined stickleback Apeltes quadracus. J. Protozool. 15: 547-571.

Steinhaus, E. A. and K. Hughes. 1949. Two newly described species of Microsporidia from the potato tuberworm, Gnorimoschema operculella (Zeller) (Lepidoptera, Gelechiidae). J. Parasitol. 35: 67-75.

Tanada, Y. 1963. Epizootiology of infectious diseases. In Insect Pathology, An Advanced Treatise (E. A. Steinhaus, ed.), 2: 423-475. Academic Press, New York.

Thomson, H. M. 1958. The effect of a microsporidian parasite on the development, reproduction and mortality of the spruce budworm, Choristoneura fumiferana (Clem.). Canadian J. Zool. 36: 499-511.

Thomson, H. M. 1960. The possible control of a budworm infestation by a microsporidian disease. Canadian Dept. Agr. Bi-Mo. Progr. Rept. 16: 1.

Veber, J. and J. Jasic. 1961. Microsporidia as a factor in reducing fecundity in insects. J. Insect Pathol. 3: 103-111.

Weiser, J. 1963. Sporozoan infections. In Insect Pathology, An Advanced Treatise (E. A. Steinhaus, ed.), 2: 423-275. Academic Press, New York.

Weiser, J. and M. Coluzzi. 1966. Plistophora culisetae in the mosquito Culiseta longiareolata Macqu. - further remarks. Proc. Ist Int. Congr. Parasitol., Rome 1: 596-597.

Kellen, W. R. and J. E. Lindegren. 1973. Transovarian transmission of Nosema plodiae in the Indian meal moth, Plodia interpunctella. J. Invertebr. Pathol. 21: 248-254.

FUNGAL PATHOGENS AND THEIR USE IN THE MICROBIAL CONTROL OF INSECTS AND MITES

C. W. McCoy
Agricultural Research and Education Center
Institute of Food and Agricultural Sciences
University of Florida
Lake Alfred, Florida

Introduction

Excellent reviews of entomogenous fungi and their use as microbial agents have been published in the last decade (Muller-Kogler, 1965; Madelin, 1966, 1968; Roberts and Yendol, 1971). Students and researchers alike should familiarize themselves with these works before embarking on a problem relating to fungi and microbial control.

The late Dr. E. A. Steinhaus (1949) stated: "Entomogenous fungi, in nature and without any help from man, cause a regular and tremendous mortality of many insect pests in many parts of the world and do, in fact, constitute an efficient and extremely important natural control factor." The literature contains numerous reports that support this statement, for example, Baird (1958) lists 41 successful attempts to control 28 species or groups of insects with fungi. Nevertheless, the opinion of many insect pathologists is, fungi will not be used successfully in the field as a reliable control for insects. There are numerous reasons why this negative feeling prevails. Since the eighteenth century, when *Metarrhizium anisopliae* (Metch.) Sor. was used for sugarbeet curculio control, partial or complete failure in controlling insects with fungi has usually resulted; in fact, few fungi listed by Baird as successful, to my knowledge none in the U.S. are utilized in control programs today. Also, our meager knowledge of fungal epizootiology and host ecology does not provide answers to questions such as 1) are fungi truly regulatory agents and 2) would economic loss be greater if fungi were absent in a given ecosystem? Finally and most basic to the problem, fungi are unique from other microorganisms in that they cause mortality by entering the host, usually via the host integument. This mode of invasion has rigid tolerances in the micro-environment of the host; simply infection will not occur if favorable weather conditions do not prevail. Unfortunately, past researchers have either failed to recognize the importance of micro-climate or have been unsuccessful in their efforts to manipulate it to achieve insect control. It is interesting to note that most attempts to control insects with fungi have not occurred in subtropical and tropical regions, where micro-climate would appear to be most favorable.

I would like to discuss briefly the various fungal pathogens having received noteworthy attention as microbial agents in the past, outline the general mode of infection for fungi, discuss some factors such as micro-climate involved in fungal epizootiology, and conclude by presenting the ways fungi might be used in insect control in the future.

Entomopathogenic Fungi

Parasitic fungi of insects and other invertebrates have been reported in all four major classes of true fungi excluding the slime molds (Roberts and Yendol, 1971). Those fungal genera most frequently associated with insect disease are listed in bold type in Table 1.

Phycomycetes

This primitive fungal group is almost wholly aquatic and is known collectively as water molds or aquatic phycomycetes. They show the closest resemblances to protozoa, in that they produce motile spores furnished with flagella and in many of the simpler forms (e.g., chytrids), the vegetative structure is not mycelial.

Table 1.--List of the major taxa containing entomogenous species (the main genera are shown in bold type).

PHYCOMYCETES		BASIDIOMYCETES
Coelomomyces		Septobasidium
Myiophagus		Uredinella
Entomophthora[a]		
Massospora		

ASCOMYCETES	FUNGI IMPERFECTI[c]	
Ascosphaera (Pericystis)	Acrostalagmus	
Calonectria	Aegerita	
Cordyceps	Aschersonia	
Hypocrella	Aspergillus	
Myriangium	Beauveria	
Nectria	Cephalosporium	Paecilomyces
Ophiocordyceps	Gibellula	Penicillium
Podonectria	Hirsutella	Spicaria
Sphaerostilbe	Hymenstilbe	Sorosporella
Torrybiella	Isaria	Synnematium
Laboulbeniales[b]	Metarrhizium	Tetracrium
	Microcera	Verticillium

Table slightly modified from Roberts and Yendol, 1971.

[a]Divided into more than one genus by some authors.

[b]Approximately 125 genera.

[c]Many species in the genera listed are asexual states of ascomycetous insect-parasites.

Coelomomyces

Excellent reviews on the genus Coelomomyces have been reported by Couch and Umphlett (1963).

The genus, belonging to the order Blastocladiales, has been reported mainly as an obligate parasite of mosquitoes. However, it has been reported from sandflies, chironomids, and possibly blackflies too (Garnham and Lewis, 1959). These fungi prefer the immature stages of their host, usually first instar larvae, but occasionally infect adults. The genus appears to have worldwide distribution having been found in nearly all mosquito breeding habitats.

Like other representatives of the order Chytridiales, *Coelomomyces* spp. have a zoosporic stage and therefore depend on the presence of free water at some stage of their life cycle as a medium for zoospore locomotion. Parasitism culminates in the formation of resting sporangia within the host.

Coelomomyces spp. would appear to have a promising future as microbial agents in aquatic environments, if practical mass production procedures can be developed. Significant steps toward this goal have been achieved by Couch (1968) in North Carolina, where he and his associates have monitored the process of sporangial germination in the laboratory and in independent studies, brought about both laboratory and field infections of *Anopheles quadrimaculatus* by *C. punctatus*.

Laird (1967) in his well-known Tokelau Islands pilot project showed conclusively, that diseased mosquito populations introduced from one region into a non-diseased population in another region could become established. However, the value of this control endeavor from the standpoint of population suppression remains in question.

Entomophthora

Comprehensive accounts of the entomophthorales have been presented by MacLeod (1963) and Madelin (1966).

Fungi of the family Entomophthoraceae appear to be the most advanced evolutionarily of the Phycomycetes. These organisms have a worldwide distribution attacking many species of Homoptera, Hemiptera, Diptera, Lepidoptera as well as three families of Acari and one family of millipeds.

Members of the genus *Entomophthora* have short, thick-walled hyphae which develop within the host; reproduce asexual by means of modified sporangia called conidia which are shot-off singly from the apex of club-shaped conidiophores formed outside the host. Conidial germination must take place very soon after discharge, as the ability to germinate is rarely retained for more than 2 weeks. Secondary and tertiary conidia can be formed from germinated primary and secondary conidia, respectively, and then be discharged. Sexual reproduction is by the union of hyphal bodies to form thick-walled zygospores or parthenogenetically formed azygospores. Rhizoids are often formed to secure a diseased host to a given substrate and under unfavorable conditions modified hyphal bodies called chlamydospores may be formed within the host.

Because of their potential economic importance, numerous attempts have been made to culture many species of *Entomophthora* on an artificial medium (e.g., Gustafsson, 1969). In many cases, success has been achieved quite easily; however, the essentially obligate forms have presented considerable difficulty. Where isolation and cultivation of the fungus has been successful, conidia are often nonviable. Some species, however, produce resting spores quite easily (e.g., *E. virulenta* and *E. exitialis*), and these have been cultivated successfully on an artificial medium (e.g., Muller-Kogler, 1967). Gustafsson (1969) has successfully cultivated different *Entomophthora* species on soil media in air-tight containers and extended conidial survival by 4 to 5 months. In large-scale field studies, Hall and Dunn (1958) tried spraying resting spores and vegetative stages of the more virulent species for control of the alfalfa aphid. Also, they introduced small vegetative cultures in containers directly into the field where the sporulating cultures would subsequently contact the host. These methods of pathogen dissemination were either unsuccessful or too laborious for practical application.

Ascomycetes

The Ascomycetes are a large class in which insect parasitism is represented in diverse genera (Madelin, 1966). Apart from the Laboulbeniales which are highly specialized ectoparasites, the vegetative phase of parasitic Ascomycetes is confined to the body of the host. The largest single genus of such endoparasites is *Cordyceps* (Fries) Link.

Members of the genus *Cordyceps* are found on Diptera, Hymenoptera, Coleoptera, Lepidoptera, Hemiptera, Isoptera, and Orthoptera among insect orders and on spiders (McEwen, 1963). Immature stages appear to be the most common host. The genus is characterized by a stroma, an aerial outgrowth arising from the host. The stroma is produced by the sclerotium, a compact mycelial mass within the body of the host. The perithecia develop on the fertile portion of the stromata.

The stromata of the genus hosts infected with *Cordyceps* can be found in moist soil and decaying logs, but also on the bark and foliage of trees infested with host insects.

Whether or not infection by *Cordyceps* is a significant factor in the natural control of any insect population remains to be determined.

Rather little information has appeared in recent years on other endoparasitic ascomycetes, particularly of fungi in the genera found on scale insects and mealybugs. A review of the scale insect fungi has been reported by Steinhaus (1949). The more important genera are *Sphaerostilbe* (conidial stage believed to be *Fusarium*), *Nectria*, *Podonectria*, *Hypocrella* (conidial stage in *Aschersonia*), and *Myriangium*. *Sphaerostilbe auranticola* and *Nectria diploa* attack Florida red scale. *Nectria vilis* has been found on snow scale and purple scale and *Podonectria coccicola* has been reported from chaff and purple scale on citrus in Florida (e.g., Berger, 1942).

Basidiomycetes

The entomogenous habit is represented among the Basidiomycetes in only the genera *Septobasidium* and *Uredinella* (Madelin, 1966). Other than a few reports of parasitism and observation on behavior of *Septobasidium* spp., these fungi have apparently received little experimental study.

Fungi Imperfecti

The Fungi Imperfecti comprise many species in a variety of unrelated genera that are parasitic to insects. However, the majority of publications on insect mycoses discuss the imperfect genera *Beauveria* and *Metarrhizium*. Insect parasitism in this group was reviewed extensively by Madelin (1963, 1966). Only the major parasites will be considered separately here.

The Fungi Imperfecti are mainly hyphomycetous; conidia are generally produced on free or aggregated conidiophores arising from mycelium growing from the surface of their substratum. They can enter intact insects only by the penetration of hyphae arising from spores in the insects' immediate environment.

The entomogenous fungus most widely used for microbial control has been *Beauveria bassiana*, one of 14 species of the genus attacking insects (MacLeod, 1954). It has been used as a biological insecticide against a wide range of subterranean and subaerial pests (such as, European corn borer larvae, Colorado potato beetle, wireworms, and eye gnats) with varying degrees of success. In Europe and

Russia, Beauveria has been used in combination with sublethal doses of different insecticides in control work (Telenga, 1958). A number of techniques including submerged culture have been developed for the cultivation of B. bassiana on a more or less large scale for production of spores (Martignoni, 1964; Pristavko and Goral, 1967). The major shortcoming to the use of Beauveria as a biological insecticide concerns its effect on man. Allergenic reactions have been reported after harvesting B. bassiana spores (Hall, 1954; York, 1958).

The most common species of Metarrhizium is Metarrhizium anisopliae, the cause of the green muscardine disease. According to Veen (1968), the host range of M. anisopliae exceeds 200 species of insects, attacking mainly larvae of soil inhabiting forms. Its length of conidial viability at intermediate humidities around 45% is rather poor and is relevant to its use in microbial control (Clerk and Madelin, 1965). M. anisopliae can also be readily cultured on a suitable liquid and solid media and no doubt, will receive more consideration as a microbial insecticide in the future.

The genus Hirsutella is currently classified in the Fungi Imperfecti (Mains, 1951), although some reports have listed it incorrectly in the Basidiomycetes (Muma, et al., 1961; Lipa, 1971). Twenty-six species are known to be entomogenous (Mains, 1951) with H. thompsonii being the only species reportedly attacking Acari (Fisher, et al., 1949; Baker and Neunzig, 1968).

MacLeod (1959) conducted extensive nutritional studies on the spruce budworm pathogen, H. gigantea, while McCoy and Kanavel (1969) and McCoy, et al. (1971) isolated H. thompsonii from the citrus rust mite and cultivated it on both solid and liquid media. In adapting the mass production method for H. thompsonii to submerged culture, the authors successfully produced large quantities of mycelia and used it in a fragmented form with adjuvants rather than conidial powder as a field inoculum for controlling the citrus rust mite in Florida.

Many fungus species formerly placed in Isaria and Spicaria are now in the genus Paecilomyes (Brown and Smith, 1957); however, the rather common insect pathogen S. rilyei is still unassigned (Behnke and Paschke, 1966). Paecilomyes farinosus is the most common species causing disease in cerambycids, sawflies, and pine shoot moth larvae. Spicaria rilyei is the cause of disease in many field crop pests such as cabbage looper, Heliothis sp., velvetbean caterpillar, and soybean looper (e.g., Allen, et al., 1971).

In Florida, insects belonging to the families Coccidae and Aleyrodidae are subject to attacks by fungi. Of the Fungi Imperfecti, Aegerita, Aschersonia, Cephalosporium, and Verticillium have been reported on various species of scale insects and whitefly (e.g., Fawcett, 1908).

Mode of Infection

Although the mode of infection among fungi differ somewhat, disease is generally induced as outlined in Figure 1. The conidium or spore, zoospores in the case of many phycomycetes, is the infective unit. When environmental conditions are suitable, the conidium germinates on the host integument, producing a germ tube that either penetrates directly or forms an appressorium with an infection peg that penetrates the insect epicuticular layer (Figure 2). Studies by Gabriel (1968) and David (1968) suggest that enzyme activity as well as mechanical force are involved in cuticle penetration. The respiratory or alimentary tract has been reported as an invasion site (Muller-Kogler, 1965). Within the hemocoel of the host, hyphal bodies are generally produced which usually float free and apparently multiply in the hemocoel (Prasertphon and Tanada, 1968). Some fungal strains produce sufficient toxins at this time to cause death, though no vital organs have been invaded. Fat body is generally the preferred site for hyphal invasion. After death or in some cases before, the mycelia continue to develop within the body until the insect is virtually filled with mycelia. Conidiophores

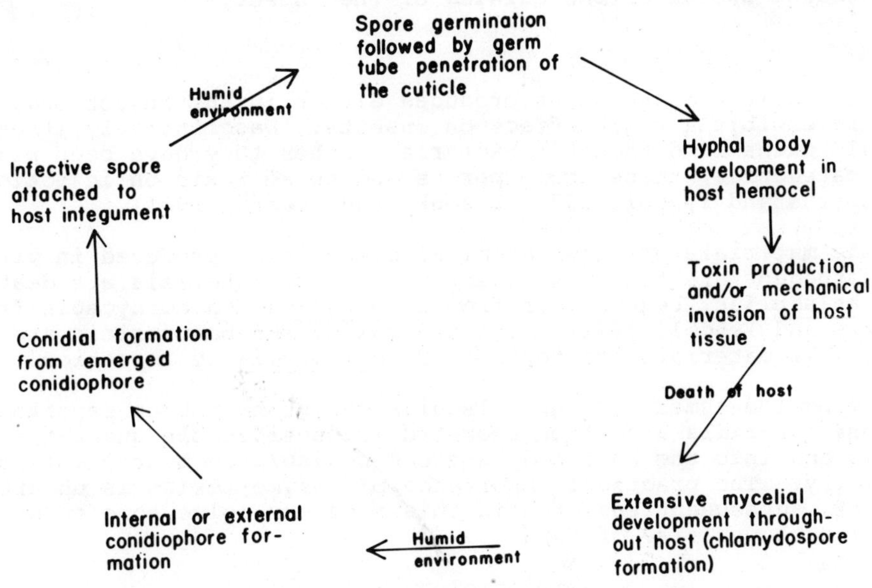

Generalized scheme of the mode of infection by entomogenus fungi

Figure 1 (above); and Figure 2 (below)

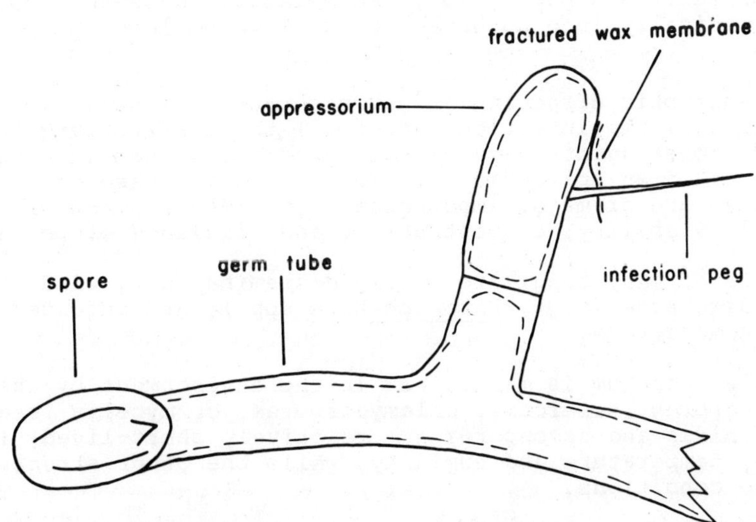

Appressorium of a *Paecilomyces farinosus* spore on a wax membrane (redrawn from Robinson, 1966).

are then produced which erupt through the cuticle during suitable environmental conditions and produce spores on the outside of the insect.

Toxins

In fungi, distinct metabolites produced either in the insect body or in artificial culture exhibit a toxic effect on insects. Descriptively, these toxins are not classical toxins like those in bacteria. Rather they have been placed in two overlapping categories; those toxic per os and those toxic on injection (Muller Kogler, 1965; Roberts and Yendol, 1971; Lysenko and Kucera, 1971).

Four toxic materials of known chemical composition produced in vitro by several insect pathogens are toxic on injection. These materials are destruxins A and B from *M. anisopliae*, beauvericin from *B. bassiana* and cordycepin from *C. militaris* (Roberts and Yendol, 1971). The destruxins and beauvericin are toxic to mosquito larvae. All materials are toxic to *G. mellonella* by injection.

Various fungal enzymes of high molecular weight have been reported as toxic to insects. These materials are often secreted in considerable quantities into the culture media and into the host body and can be isolated by reliable methods of protein chemistry. The practical importance of enzyme toxins is uncertain at this time, however, subsequent research in this area may solve some questions about the mechanisms of pathogenicity of fungi.

Epizootiology

General information on the important role of microbial pathogens of insects in the population dynamics and general ecology of insects can be found in the classical paper by Steinhaus (1954) and the reviews of Tanada (1963), Muller-Kogler (1965), and MacLeod, et al. (1966).

Entomogenous fungi may occur in insect populations as an enzootic or epizootic disease. An enzootic disease is one that has a low incidence but is constantly present in a population. An epizootic disease is simply one which may flare up sporadically into outbreaks which involve large proportions of the insect population.

Any epizootic affecting an invertebrate population is concerned with three primary agents: 1) the infectious agent, 2) the invertebrate host, and 3) the environment. These agents have certain attributes that interact to determine the characteristics of an epizootic (Figure 3). In the case of the infectious agent, four attributes are of major importance in the development of an epizootic. These are dispersal, viability, concentration, and virulence of the inoculum.

Spores of many fungi are widely disseminated by wind (Hirst, et al., 1967). Rain, spore discharge (e.g., *Entomophthora* spp.), and infected host movement account for some spread.

Viable inoculum is maintained in the environment by the production of conidia, ascospores, sclerotia, chlamydospores, or mycelia fragments in or on the host. Conidia and ascospores are relatively short-lived, longevity being influenced by temperature and humidity, while the other structures survive longer under adverse conditions.

Virtually nothing is known concerning inoculum concentration, specifically with regard to the minimum number of spores required to induce disease. Counts from environments where epizootics occurred have not been reported and LD_{50} values obtained in laboratory experimentation tend to be rather high and difficult to relate back to the field (Roberts and Yendol, 1971).

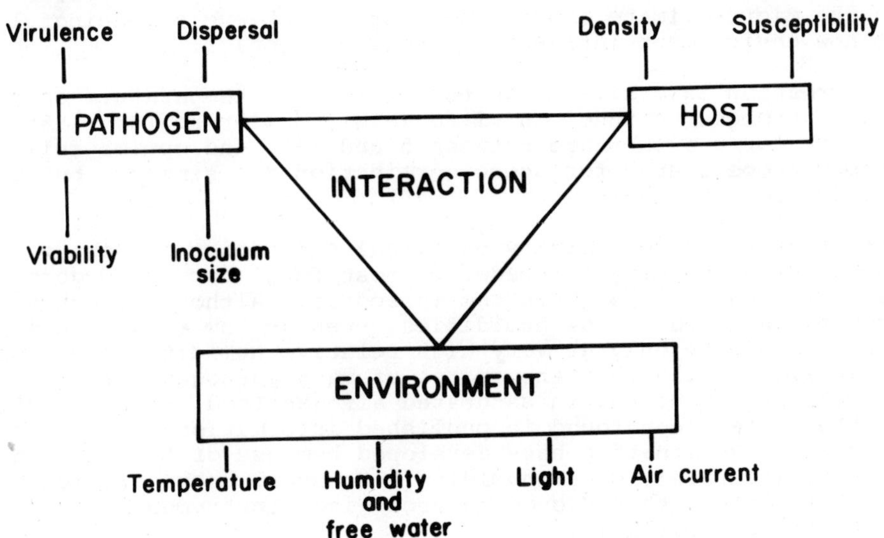

Various attributes of the primary agents that interact to initiate an epizootic in the field

Figure 3.

Finally, within single fungal species strains exist which differ in virulence for a given insect species (Ferron, 1967). Although virulence is difficult to measure, only highly virulent strains should be considered in microbial control attempts.

In general, fungal pathogens of insects act as imperfectly density-dependent mortality factors (Milne, 1958). When given suitable weather conditions, they infect a greater proportion of insects as the host population density increases. If we apply this axiom to pest management systems where subeconomic populations of a pest species are acceptable, one could assume that fungi will cause greater mortality and perhaps epizootics will occur more quickly and/or more frequently where the host carrying capacity is higher.

Unusually high populations have often been regarded as the primary factor favoring the development and spread of fungal epizootics. However, certain fungi such as strains of _B. bassiana_ and some _Entomophthora_ species are capable of initiating epizootics at relatively low host population densities (MacLeod, et al., 1966). Unfortunately, there is very little quantitative data to demonstrate whether some fungi are regulatory agents or if a threshold density exists above which a major epizootic will occur.

Because fungi have a cuticular mode of host invasion, host susceptibility is influenced by factors such as nutrition, crowding, molting, etc., that affect the composition of the insect cuticle. Therefore, host susceptibility to fungus invasion may vary with host behavior (e.g., Jaques, et al., 1968), stage of development and age of the host.

Temperature, humidity, and light appear to be the critical environment factors affecting fungus survival (MacLeod, et al., 1966). These parameters must be measured in the micro-climate around the host to achieve a meaningful understanding as to how environment influences fungal epizootics.

Much information is available on the effects of temperature on germination and/or growth of entomogenous fungi in vitro (e.g., Muller-Kogler, 1965). In general, the limits for growth range between 5 and 35°C and optima fall between 20 and 30°C. Optimum temperatures for spore germination are similar, but more restricted than those for growth.

The effect of relative humidity on fungal epizootics is more significant than temperature. High humidity is needed by most fungi for spore germination and spore dispersal to spread and heighten the epizootic. Although there are reports of insects becoming infected at low humidities, even as low as 46% (e.g., Madelin, 1963), most fungi germinate only at very high relative humidity, i.e., usually 90% or higher. It is widely held that the spores of many pathogenic fungi will germinate only in a water film and not in saturated air (MacLeod, et al., 1966). It is likely that the differences found in published information concerning humidity requirements for spore germination have developed because of the difficulty involved in measuring accurately micro-humidity. However, these difficulties should be minimal in the future, with improved meterological instrumentation (i.e., micro-sensors, etc.) now available.

Light, presumably ultraviolet radiation, appears to affect both longevity of spores and germination (e.g., Muller-Kogler, 1965) and neglect of this fact may be responsible for some of the microbial control failures recorded with fungi. Proper formulation and/or application in the evening should reduce this problem.

Possible Approaches to Microbial Control

Theoretically, entomogenous fungi can be used in three ways for insect and mite control depending upon the characteristics of the fungus and host involved. These ways are by colonization, as a microbial insecticide, and in integrated control.

Colonization is analogous to the classical introduction of parasites and predators, where an exotic organism is placed into a new environment where its action supplements the existing natural control to the point of economic benefit. Few attempts have been made in the past to introduce beneficial fungi through colonization; the best example, being the transfer of *Coelomomyces* sp. into mosquito populations free of this parasite (e.g., Laird, 1967).

To my knowledge, no fungal pathogen is being produced commercially for crop protection today. It is generally agreed, however, that inundative applications of fungi may be useful in initiating epizootics before they would normally occur or in increasing the amount of inoculum in areas where insufficient amounts are present to start epizootics (Baird, 1958). Here again, moisture is still of paramount importance. Time of application must be synchronized with optimum weather conditions or cultural methods such as irrigation, must be available to augment environmental conditions. Dunn and Mechalas (1963) and Schaerffenberg (1964) suggest that infection can occur with low humidity, if large numbers of spores are used. Further research is needed to validate this work.

If fungal applications can be synchronized with environmental conditions and still supply the necessary crop protection, they still must compete with insecticides in availability, cost of production, ease of application, etc. This can be accomplished most efficiently in commercial production by adapting fungus production to deep-tank fermentors with liquid medium. This production method has numerous advantages, but creates an immediate problem, that is, many fungi will not sporulate in liquid culture. In view of these difficulties, McCoy, et al., 1971, by-passed the sporulation phase in the laboratory production of *Hirsutella* and disseminated the pathogen in the field as fragmented mycelia. Since the fungus mycelia began to sporulate within 48 hr after exposure to air, infective spores could be produced in the immediate environment of the target organism. In this method of mycelia application as a foliar spray, the actual number of spores produced in the field was relative to weather conditions after application. Although post-treatment spore levels have not been determined using this method of application, it would appear that the viable spore reservoir may be higher and persist for a longer time than a spore powder inoculum. Although the elimination of the sporulation phase would make commercial production of fungi cheaper, difficulties with storage and formulation develop when using mycelia as a field inoculum. The vegetative stage is also less resistant than spores to environmental conditions, so formulations would have to be patterned after virus preparations and field application would have to be confined to the evening. Anyway, many advantages do exist and with more research in storage and formulation of the field inoculum, consistent success may be achieved with this method of application.

One of the basic components of pest management or integrated control concerns selective pesticides. These are materials that reduce the pest population, but cause a minimum of harm to the natural enemies. Fungicides used to control plant-attacking fungi and nutritional sprays used to supply minor elements to citrus trees have been considered limiting factors in the development of fungal epizootics in pest populations in Florida (Griffiths and Fisher, 1950). Some fungicides appear to have less effect on naturally occurring epizootics than others. Jaques and Patterson (1962) have shown that *E. sphaerosperma* killed more apple sucker when orchards were treated with glyodin rather than Captan or Ferbam. Also, Shands, et al. (1962) were still able to find more than 11% of the potato aphids infected with *Entomophthora* spp. after weekly applications of Zineb, Nabam, and zinc sulfate.

Inhibition of entomogenous fungi by commercial insecticides and fungicides has been demonstrated in the laboratory. Hall and Dunn (1959) found that wettable sulfur, Dithan 2-78, Ferbam, Bordeaux 5-5-10 and Captan restricted the vegetative growth of the entomophthoraceous fungi attacking the spotted alfalfa aphid. Also, some chlorinated hydrocarbons and organic phosphates retarded the germination of *E. virulenta* resting spores. Yendol (1968) showed inhibition of conidial germination of *E. coronata* by malathion and fungicides.

The effect of fungi on other beneficial insects must be considered in pest management. Strains of even broad spectrum fungi are known which are much more virulent with one insect as compared to others (e.g., Veen, 1968).

As a concluding remark, I would like to say that many fungi we have talked about this afternoon do have potential as microbial agents and can be used in pest management systems of the future, if the organism is highly virulent, amenable to mass production, reasonably stable in the field, and is used in habitats where environmental conditions are optimum.

References Cited

Allen, G. E., G. L. Greene and W. H. Whitcomb. 1971. Florida Entomol. 54: 189-191.

Baker, J. R. and H. H. Neunzig. 1968. J. Econ. Entomol. 61: 1117-1118.

Baird, R. B. 1958. Proc. Intl. Congr. Entomol., 10th, Montreal, 1956, 4: 689-692.

Behnke, C. N. and J. D. Paschke. 1966. J. Invertebr. Pathol. 8: 103-108.

Berger, E. W. 1942. Florida Entomol. 25: 26-29.

Brown, A. H. S. and G. Smith. 1957. Trans. Brit. Mycol. Soc. 40: 17-89.

Clerk, G. C. and M. F. Madelin. 1965. Trans. Brit. Mycol. Soc. 48: 193-209.

Couch, J. N. and C. J. Umphlett. 1963. In Insect Pathology, An Advanced Treatise (E. A. Steinhaus, ed.) 2: 149-188. Academic Press, New York.

Couch, J. N. 1968. Proc. Joint U.S.-Japan Semin. Microbiol Control Insect Pests, U.S. Committee Sci. Cooperation. Panel 8. pp. 93-105. Fokuoka, Japan.

David, W. A. L. 1968. In Insects and Physiology (J. W. L. Beament and J. E. Treherne, eds.), pp. 17-35. American Elsevier, New York.

Dunn, P. H. and B. J. Mechalas. 1963. J. Insect Pathol. 5: 451-459.

Fawcett, H. S. 1908. Florida Agr. Exp. Sta. Bull. 76: 1-9.

Ferron, P. 1967. In Insect Pathology and Microbial Control (P. A. van der Laan, ed.) pp. 204-209. North-Holland Publ. Co., Amsterdam.

Fisher, F. E., J. T. Griffiths, Jr. and W. L. Thompson. 1949. Phytopathology 39: 510-512.

Gabriel, B. P. 1968. J. Invertebr. Pathol. 11: 70-81.

Garnham, P. C. C. and D. J. Lewis. 1959. Trans. Roy. Soc. Trop. Med. Hyg. London 53: 12-40.

Griffiths, J. T. and F. E. Fisher. 1950. J. Econ. Entomol. 43: 298-305.

Gustafsson, M. 1969. Lantbrukshogsk Ann. 35: 235-274.

Hall, I. M. 1954. Hilgardia 22: 535-565.

Hall, I. M. and P. H. Dunn. 1958. J. Econ. Entomol. 51: 341-344.

Hall, I. M. and P. H. Dunn. 1959. J. Econ. Entomol. 52: 28-29.

Hirst, J. M., O. J. Stedman and G. W. Hurst. 1967. J. Gen. Microbiol. 48: 357-377.

Jaques, R. P. and N. A. Patterson. 1962. Canadian Entomol. 94: 818-825.

Jaques, R. P., H. T. Stultz and F. Huston. 1968. Canadian Entomol. 100: 813-818.

Laird, M. 1967. Chron. World Health Organ. 21: 18-26.

Lipa, J. J. 1971. In Microbial Control of Insects and Mites (H. D. Burges and N. W. Hussey, eds.) pp. 357-373. Academic Press, New York.

Lysenko, O. and M. Kucera. 1971. In Microbial Control of Insects and Mites (H. D. Burges and N. W. Hussey, eds.) pp. 205-227, Academic Press, New York.

MacLeod, D. M. 1954. Ann. New York Acad. Sci. 60: 58-70.

MacLeod, D. M. 1959. Canadian J. Bot. 37: 695-714.

MacLeod, D. M. 1963. In Insect Pathology, An Advanced Treatise (E. A. Steinhaus, ed.) 2: 189-231. Academic Press, New York.

MacLeod, D. M., J. W. Cameron and R. S. Soper. 1966. Rev. Roumaine Biol. Ser. Bot. 11: 125-134.

Madelin, M. F. 1963. In Insect Pathology, An Advanced Treatise (E. A. Steinhaus, ed.) 2: 233-271. Academic Press, New York.

Madelin, M. F. 1966. Ann. Rev. Entomol. 11: 423-448.

Madelin, M. F. 1968. In The Fungi (G. C. Ainsworth and A. S. Sussman, eds.) 3: 227-238. Academic Press, New York.

Mains, E. B. 1951. Mycologia 43: 691-718.

Martignoni, M. E. 1964. In Biological Control of Insect Pests and Weeds (P. DeBach, ed.) pp. 579-609. Chapman and Hall, London.

McCoy, C. W. and R. F. Kanavel. 1969. J. Invertebr. Pathol. 14: 386-390.

McCoy, C. W., A. G. Selhime, R. F. Kanavel and A. J. Hill. 1971. J. Invertebr. Pathol. 17: 270-276.

McEwen, F. L. 1963. In Insect Pathology, An Advanced Treatise (E. A. Steinhaus, ed.) 2: 273-290. Academic Press, New York.

Milne, A. 1958. Nature 182: 1251.

Muller-Kogler, E. 1965. Pilzkrankheiten bei Insekten. Parey, Berlin and Hamburg.

Muller-Kogler, E. 1967. Entomophage 12: 429-441.

Muma, M. H., A. G. Selhime and H. A. Denmark. 1961. Florida Agr. Exp. Sta. Tech. Bull. 634: 1-39.

Prasertphon, S. and Y. Tanada. 1968. J. Invertebr. Pathol. 11: 260-280.

Pristauko, W. P. and V. M. Goral. 1967. In Insect Pathology and Microbial Control (P. A. van der Laan, ed.) pp. 118-119. North-Holland Publ. Co., Amsterdam.

Roberts, D. W. and W. G. Yendol. 1971. In Microbial Control of Insects and Mites (H. D. Burges and N. W. Hussey, eds.) pp. 125-149. Academic Press, New York.

Schaerffenberg, B. 1964. J. Insect Pathol. 6: 8-20.

Shands, W. A., I, M. Hall and G. W. Simpson. 1962. J. Econ. Entomol. 55: 174-179.

Steinhaus, E. A. 1949. Principles of Insect Pathology. McGraw-Hill, New York.

Steinhaus, E. A. 1954. Hilgardia 23: 197-261.

Tanada, Y. 1963. In Insect Pathology, An Advanced Treatise (E. A. Steinhaus, ed.) 2: 423-475. Academic Press, New York.

Telenga, N. A. 1958. Trans. 1st. Insertz Conf. Insect Pathol. Biol. Control. Prague: pp. 155-168.

Veen, K. H. 1968. Mededel. Landbouwhogeschool Wageningen. 68-5, 1-77.

Yendol, W. G. 1968. J. Invertebr. Pathol. 10: 116-121.

York, G. T. 1958. Iowa State Coll. J. Sci. 33: 123-129.

NEMATODE PARASITES

S. R. Dutky
Insect Physiology Laboratory
Beltsville, Maryland

Introduction

Many nematodes are known to be parasitic to insects and some are capable of a high degree of natural control of their hosts. Some that are adapted to one host and its special ecology are so highly specialized that they are not easily manipulated or artificially propagated. With additional knowledge and effort, they may one day have a place in insect pest management. Other less highly specialized forms that have wide host ranges and can be mass produced readily, lend themselves to more immediate application in control.

The terrestrial mermithid parasite of grasshoppers, Agamermis decaudata Cobb, Steiner, Christie, 1923 - Figure 1, is a good example of a nematode that can give a high degree of natural control (over 80% of the grasshoppers of some areas were infested with nematodes), but that is difficult to manipulate. The infested insects are killed on emergence of the large postparasitic juvenile nematodes that then take a year or more to maturate, mate, ovulate, and produce young that can again attack the host insect. Not all of the terrestrial forms of this important family take this long to maturate but all do require a period long or short after emergence from the host.

Aquatic mermithid parasites maturate much more rapidly after emergence from the host and thus offer less difficulty in their propagation. One of these, Reesimermis nielseni Tsai and Grundmann (1969), which parasitizes a number of species of mosquitoes of some 14 genera (Figure 2), has been shown to have considerable promise as an effective biological control agent. This species maturates, mates, and oviposits within 2 weeks after emergence and has been propagated in large numbers on laboratory-reared mosquito hosts (Figure 3). More information on mass rearing will greatly improve its potential since sex ratios and hence yields of preparasitic nematodes are greatly influenced by environmental conditions--parasite to host ratio, host species, level of nutrition (Petersen, 1972).

Another aquatic mermithid that is also readily propagated is Diximermis peterseni, Nickle (1972). This mermithid is parasitic only in certain anopheline mosquitoes, and thus has somewhat less potential for control (Figure 4).

Some nematodes sterilize their insect hosts. A good example of this is the sphaerulariid parasite, Heterotylenchus autumnalis, Nickle (1967), of the face fly, Musca autumnalis DeGeer. This nematode has a rather complex life history, with an alternation of gametogenetic and parthenogenetic generations. The female fly is sterilized; the ovaries become filled with packets of male and unmated female nematodes that develop in the sites normally occupied by eggs. During mock oviposition, the female fly deposits the packets of nematodes in manure. The deposited nematodes mate, the males die, and the impregnated young female nematodes enter the body cavities of fly maggots developing in the manure. Once inside the maggot, the small female nematode develops into the adult parasitic stage that lays eggs in the hemocoele. These eggs develop into parthenogenetic females that lay large numbers of eggs that in turn develop into small males and females. When they grow to a length of about one millimeter, they penetrate the insect ovaries, completing the life cycle. Figures 5, 6 and 7 illustrate this interesting nematode (Stoffolano, 1970).

A good example of the less specialized nematodes that have wide host ranges can be mass-produced readily, and have an immediate potential for control of many pest insects is DD-136 nematode, Neoaplectana dutkyi Jackson (1965) (Turco, et al., 1971). This nematode is presently under study by many investigators in many countries throughout the world. I will discuss this species in some detail to exemplify how nematodes can be manipulated to insect control.

Biology of the DD-136 Nematode and its Associated Bacterium

The bacterial pathogen associated with the DD-136 nematode causes a bacterial septicemia that kills the host. The bacterium also serves as food for the nematode. Also, it elaborates a wide-spectrum antibiotic that prevents putrefaction of the cadaver and the growth of microorganisms inimical to the development of the nematode (Figure 8). The infective stage of the nematode is the ensheathed second-stage larva. The nematode seeks out the host insect, enters (usually by way of the mouthparts), exsheaths, penetrates the intestinal wall, and injects the bacterium contained in its esophagus and intestinal tract into the body cavity of the host. This injection initiates the septicemia that kills the host. At 30°C., the entire process from exposure to the nematodes to death takes less than 24 hr. After the death of the host, the invading nematodes maturate and become adults in 3 days (Figures 9, 10). If both males and females are present (in this species the sex ratio is 100/100), they mate, giving rise to young.

The young are born matricidally--that is, fertile ova produce embryos within the gravid female's ovaries. When the eggs hatch, the young feed on the tissues of the mother. They escape after her death as second-stage larvae. Some (about 80%) are ensheathed and do not develop further. Others maturate, mate, and produce young until the host cadaver is filled with ensheathed larvae (Figures 11, 12). A generation is complete in about 40 hr. Ensheathed larvae then emerge from the crowded cadaver in search of a new host. At the most favorable temperature, 23-28°C, the cycle from infective stage to infective stage is 8 days. The cycle lengthens at lower temperatures. If the cadaver is not in contact with free water, the ensheathed larvae may remain inside the partially dried cadavers for at least 2 months without injury. Emergence of larvae from such cadavers begins within minutes after they are in contact with water.

The size of the adults is variable, it depends on the size of the host and the amount of nutrient remaining in the host. However, the size of ova, newly hatched larvae, and infective stage larvae are fixed, regardless of the size of the host or the female. The largest females are nearly 15 mm. long and contain more than 1,000 ova; the smallest females are less than a tenth this length and contain as few as 8 ova. The infective stage larvae are about 25 μm wide and 600 μm long. Newly hatched larvae are about 390 μm long. Female nematodes grown on restrictive artificial media used to define the sterol requirements of the species usually deposited their ova, which hatched outside the female, on richer artificial media, the development was similar to that observed in the insect host.

The nematode-bacteria complex is also most interesting because of the light it throws on host-parasite relationships. The sterol requirement of the nematode is particularly notable. Our studies showed that sterol is essential for growth, development, and reproduction of the nematode (Dutky, Robbins and Thompson 1967), and that the nematode derives its essential sterol from the cholesterol of the insect host, and converts a part of this to Δ^7-choleste (Dutky, et al., 1967). The average amount of sterol in infective larvae grown on insect hosts was 95 picograms per nematode, enough for maturation of the nematode to the adult stage, but not enough for the adult to reach its maximum size or for the females to produce fertile ova if the diet was a sterol-deficient medium (Figures 13, 14). When cholesterol was added to the medium, the female reached its maximum size and produced fertile ova (Figures 15, 16). Also, of 10 highly purified sterols tested, all except stigmasterol and ergosterol could replace cholesterol in supporting growth and reproduction. These two exceptions are similar in structure and differ from the other 8 in having a double bond at the 22,23-position (Figure 17).

More recently, in this laboratory, we have used 4-^{14}C cholestanol and ^{3}H-β-sitosterol to gain further insight into the mode of utilization of sterols by the nematode. When 4-^{14}C cholestanol was added to the medium, it was converted mainly to Δ^7 cholestenol with only a trace of cholesterol, an indication of rapid desaturation of the 7,8-position. The ^{3}H-β-sitosterol was dealkylated and converted to cholesterol which was then further metabolized, saturating the 5,6-position, desaturating the 7,8-position, and producing Δ^7 cholestenol.

Laboratory and Field Testing of the DD-136 Nematode

The DD-136 nematode can be propagated in enormous numbers in several easily reared insect hosts, some of which are presently available from commercial sources as fish bait. It can also be propagated on artificial media. The choice is largely one of convenience and should depend on the availability of help trained in the proper disciplines; with entomologists, insect hosts would usually be most convenient. The nematode can be stored in the infective stage for long periods before use. The detailed procedure for propagation and storage has been published and is available (Dutky, Thompson and Cantwell, 1964).

The DD-136 nematode has a wide host range. Nearly all the hundreds of insect species tested to date, including many important insect pests, are acceptable to this nematode and are quickly entered and killed by it. Since beneficial insects may also be attacked, some judgment must be used in selecting the site of the application, but this hazard is probably considerably less than that associated with the use of insecticides. Adults, pupae, and larvae of most insect species are equally susceptible to the nematode though the adults of some species may be more susceptible than the larvae and the adults of others may be quite resistant. To date, no evidence of extensive attack on eggs has been noted. The susceptibility of the different stages can be easily assessed in a simple laboratory test, and such tests are recommended before field testing.

Laboratory tests of candidate hosts are useful also, even if a species is reported as highly susceptible, because they provide an opportunity to train the personnel who will conduct the field tests to recognize the appearance of insects killed by nematodes, and also instill confidence in the efficacy of the agent. The yield of infective-stage larvae from the parasitized hosts should also be determined. If field test data for an insect species are already available, laboratory tests could be set up at the same time as the field tests. The effect on the nematode of pesticides (fungicides, insecticides) used on the crops in the test area should be evaluated.

The success of the nematode in field applications depends in large part on the ingenuity and skill of the investigator. The nematode is not resistant to drying and is quickly killed by desiccation. Moreover, it requires a moist surface when it migrates in search of a host. Proper timing is therefore necessary when the nematode is applied to exposed surfaces. Early morning or late evening applications of spray may produce satisfactory conditions that will last long enough for the nematode to complete its search and attack. When nematodes are sprayed on the trunks of trees protected by the canopy of leaves, conditions are much less critical, and satisfactory applications can be made at any time of day in spite of hot, dry weather. The nematode is sensitive to high temperatures and is quickly killed at 42°C, and above. This limitation must be considered when the nematodes are transported to the field, and when they are sprayed at high pressures (Figure 18). The nematode is resistant to shear and is not likely to be damaged by this stress even when small nozzles and high pressures (up to 1,000 pounds per square inch) are used, but spraying at high pressures involves a considerable production of heat in the pumping equipment, and unless care is taken, the resultant temperature rise may bring the nematode suspension above the thermal death-point. The suspension to be sprayed should be cold enough so that the heat developed during spraying will not bring the temperature above the safe limit (not higher than 35°C).

The available spray equipment should be tested to make sure that the nematodes are delivered from the nozzle in good condition, and without loss of numbers. If clean equipment and clean suspensions are used, the screens can be removed and the nematodes can be delivered freely to the nozzle. Nozzles with openings less than 2/3 the length of the nematode (400 μ) will deliver nematodes satisfactorily without blockage.

The type of pumping equipment is not critical; even gear-pumps will deliver nematodes without excessive damage. However, the suspension may recycle through the pump many times before delivery, and damage may be cumulative. Pumps should thus not be allowed to idle longer than necessary.

The infective-stage larvae of the nematode are resistant to many pesticides. Chlordane, DDT, endrin, lindane, methoxychlor, toxaphene, azinphosmethyl, Chlorthion (0-(3-chloro-4-nitrophenyl)0,0-dimethyl phosphorthioate), hexaethyl tetraphosphate, methyl parathion, paraoxon, parathion, tepp, and trichlorfon were among the insecticides tested in water solution or suspension without effect.

The miticides tested had no effect. These included Aramite (2-(p-tert-butylphenoxy) isopropyl 2-chlorethyl sulfite), dicofol, and ovex. The fungicides captan, copper sulfate, Dyrene (2,4-dichloro-6-(0-chloranilino)-triazine), ferbam, folpet, maneb, sulfur and zineb had no effect. However, two fungicides, dodine and glyodin, had considerable and limited toxicity, respectively. Two herbicides, ammonium sulfamate and monuron, were without effect, and even two nematicides tested had little or no effect on the ensheathed larvae. This resistance to so many pesticides makes it easy to use the DD-136 nematode in an integrated control program.

The large numbers of the DD-136 nematode that can be released to control an existing insect population, the high reproductive potential of the nematode, and its longevity in the infective stage make the effectiveness of the nematode independent of the population level of the host; thus this agent can virtually eliminate a population (Figures 19, 20).

Summary

Numerous nematodes are known that are parasitic in insects and capable of a high degree of natural control of their hosts. Some are highly adapted to one host and its special ecology, but these specialized forms are often not easily manipulated or artificially propagated. However, other less specialized forms that have wide host ranges and could be mass-produced easily would lend themselves to manipulation for insect control. The DD-136 nematode and its associated bacterial insect pathogen is a good example of this less specialized group.

The DD-136 nematode can be propagated in enormous numbers in several easily reared insect hosts and on artificial media. Also, it can be stored for long periods before use. Nearly all the hundreds of insect species tested to date are acceptable as hosts to this nematode, and are quickly entered and killed by it. Because of the large numbers of the nematode that can thus be released to control an existing insect population, the high potential of reproduction of the nematode, and because of the longevity of the nematode in the infective stage, the effectiveness of the DD-136 nematode is independent of the level of the host population, and can virtually eliminate a population. Since the nematode is resistant to most chemicals used commonly as pesticides (insecticides, fungicides, and herbicides), this agent can also be used in an integrated control program. The nematode can be applied with the same equipment used to apply pesticides.

Literature Cited

Cobb, N. A., G. Steiner and J. R. Christie. 1923. *Agamermis decaudata* Cobb, Steiner and Christie: a nema parasite of grasshoppers and other insects. J. Agr. Res. 23: 921-926.

Dutky, S. R., J. V. Thompson and G. E. Cantwell. 1964. A technique for the mass propagation of the DD-136 nematode. J. Insect Pathol. 6: 417-422.

Dutky, S. R., J. N. Kaplanis, M. J. Thompson and W. E. Robbins. 1967. The isolation and identification of the sterols of the DD-136 insect parasitic nematode and their derivation from its insect hosts. Nematologica 13: 139.

Dutky, S. R., W. E. Robbins and J. V. Thompson. 1967. The demonstration of sterols as requirements for the growth, development and the reproduction of the DD-136 nematode. Nematologica 13: 140.

Jackson, G. J. 1965. Differentiation of three species of *Neoaplectana* (Nematoda: Rhabditida), grown axenically. Parasitol. 55: 571-578.

Nickle, W. R. 1967. *Heterotylenchus autumnalis* sp. n. (Nematoda: Sphaerulariidae) a parasite of the face fly, *Musca autumnalis* DeGeer. J. Parasitol. 53: 398-401.

Nickle, W. R. 1972. A contribution to our knowledge of the Mermithidae (Nematoda). J. Nematol. 4: 113-146.

Petersen, J. J. 1972. Factor affecting sex ratios of a mermithid parasite of mosquitoes. J. Nematol. 4: 83-87.

Stoffolano, J. G., Jr. 1970. Parasitism of *Heterotylenchus autumnalis* Nickle (Nematoda: Sphaerulariidae) to the face fly, *Musca autumnalis* DeGeer (Diptera: Muscidae). J. Nematol. 2: 324-329.

Tsai, Y. H. and A. W. Grundmann. 1969. *Reesimermis nielseni* gen. et sp. n. (Nematoda: Mermithidae) parasitizing mosquitoes in Wyoming. Helminth. Soc. Washington Proc. 36: 61-67.

Turco, C. P., W. H. Thames, Jr. and S. H. Hopkins. 1971. On the taxonomic status and comparative morphology of species of the genus *Neoaplectana* Steiner (Neoaplectanidae: Nematoda). Helminth. Soc. Washington Proc. 38: 68-79.

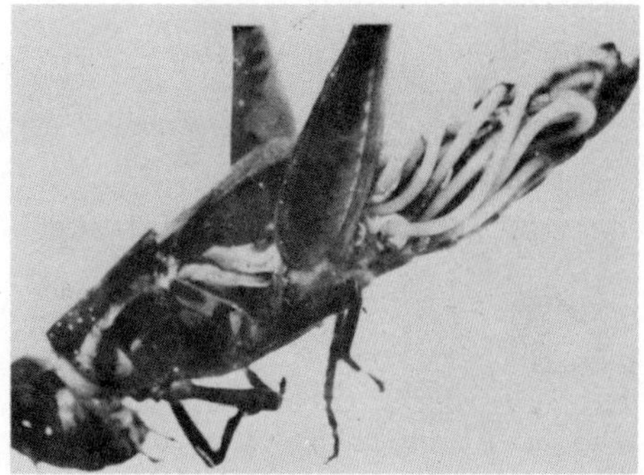

Figure 1.--Grasshopper with abdomen dissected to show mermithid parasite, Agamermis decaudata. Photograph from USDA files furnished by Dr. W. R. Nickle, USDA, Nematology Laboratory, Beltsville, Maryland.

Figure 2.--Reesimermis nielseni wound around thoraxes of fourth instar Culex pipiens quinquefasciatus. Photograph furnished by Dr. James J. Petersen, USDA, Gulf Coast Mosquito Research Laboratory, Lake Charles, Louisiana.

Figure 3.--Mass production of R. nielseni using C. p. quinquefasciatus as the rearing host. Photograph furnished by Dr. Petersen.

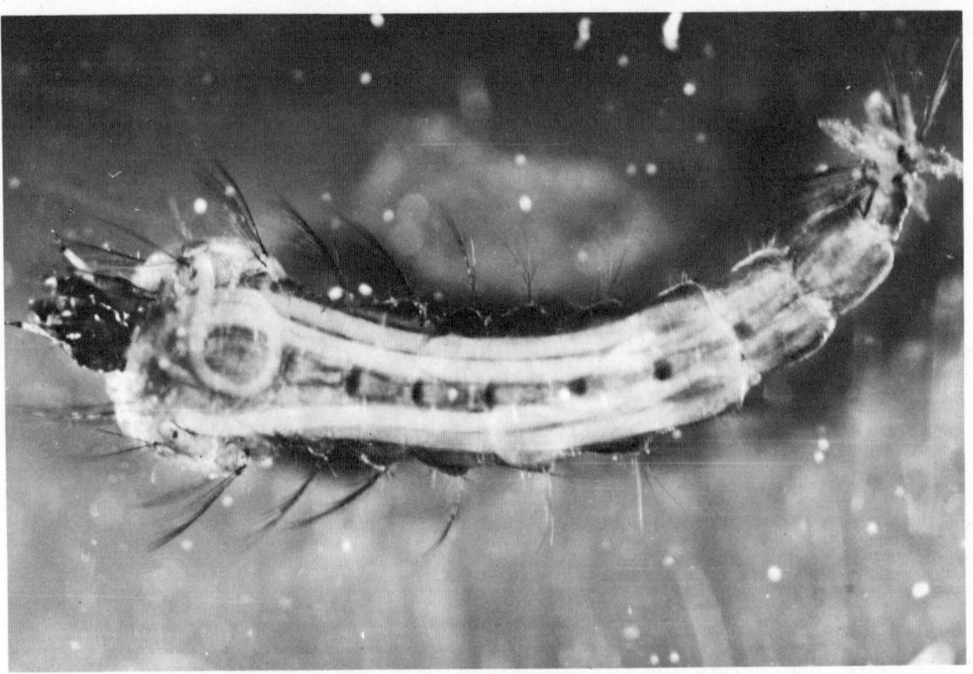

Figure 4.--Diximermis peterseni in fourth instar larva of Anopheles crucians (ventral view). Photograph furnished by Dr. Petersen.

Figure 5.--Ovaries of non-infected face fly containing several individual eggs each terminated by a dark respiratory mast (11.6X). Photograph furnished by Dr. John J. Stoffolano, Jr., Department of Entomology, University of Massachusetts, Amherst, Mass., 01002.

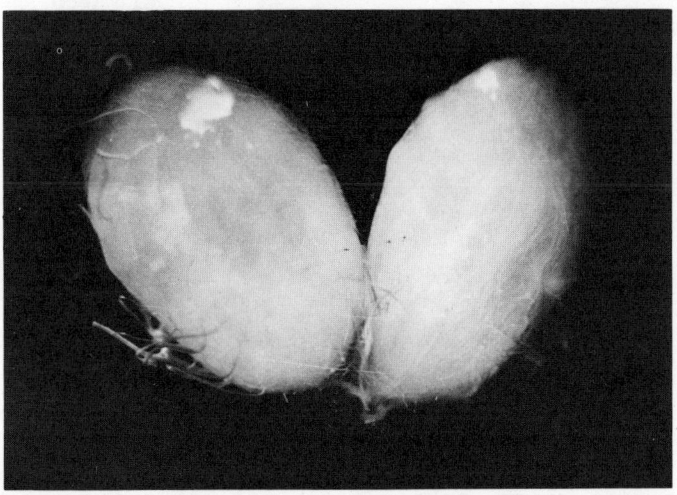

Figure 6.--Ovaries of a face fly infected with *Heterotylenchus autumnalis* and containing thousands of male and female gamogenetic larvae instead of eggs (11.6X). Photograph furnished by Dr. Stoffolano.

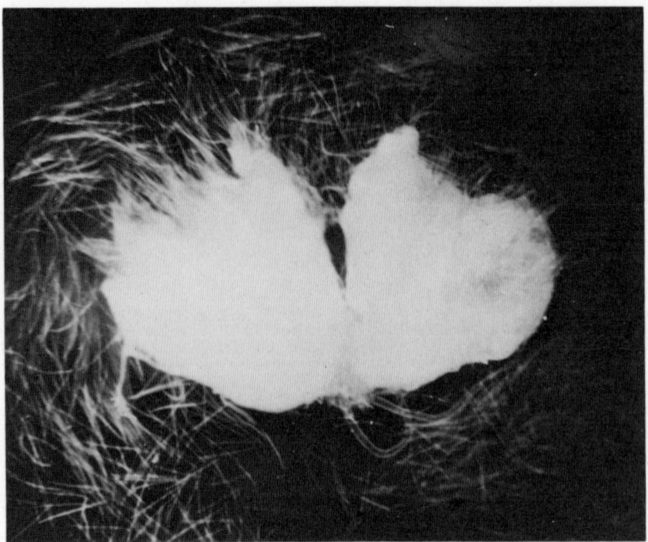

Figure 7.--Same ovaries as in Figure 6, but with the ovary sac ruptured to release the nematode larvae (11.6X). Photograph furnished by Dr. Stoffolano.

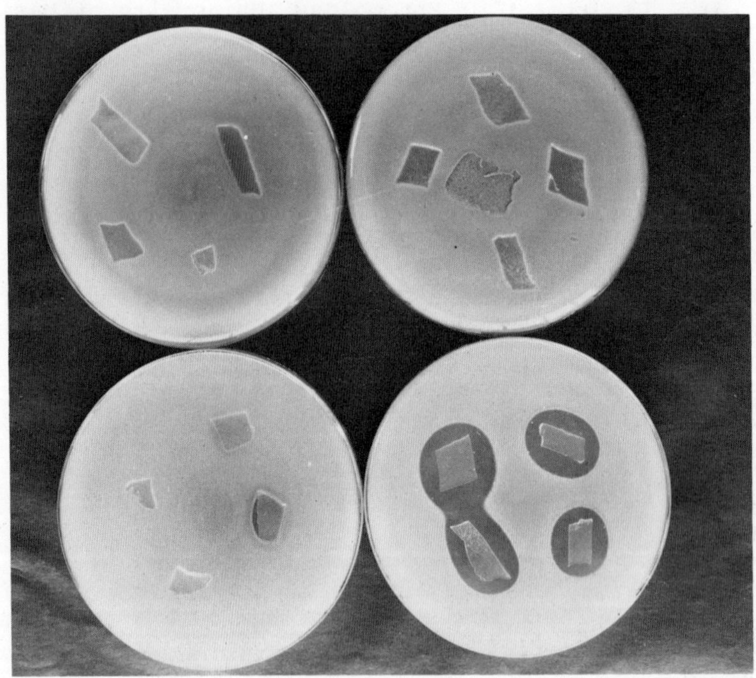

Figure 8.--Sterile agar block technique for demonstrating antibiotic production by nematode associated bacteria. Upper left: agar blocks from uninoculated plate. Upper right: agar blocks cut 30 mm away from colony on Tryptose-phosphate agar. Lower left: agar blocks cut 20 mm away from colony. Lower right: agar blocks cut within 10-15 mm from colony. Note clear zones surrounding each block. Test medium Tryptose-phosphate agar seeded with *Bacillus subtilis* spores (Mag. 0.6 X).

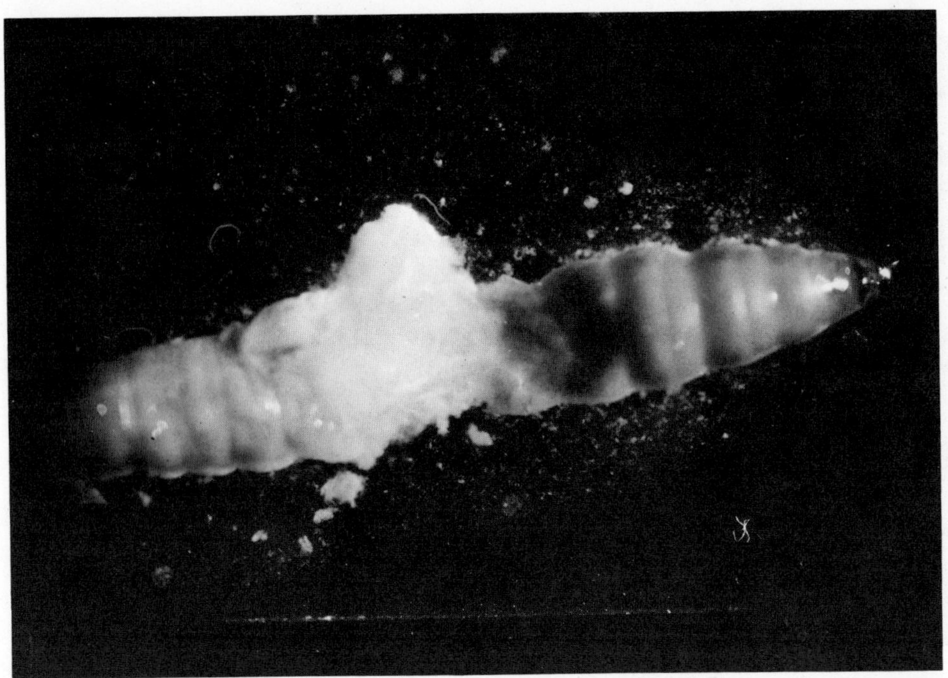

Figure 9.--Development of N. dutkyi within the dead wax moth host. Insect integument ruptured to show nematodes 2nd day after exposure. Three preadults are seen in the saline solution above host (Mag. about 3X).

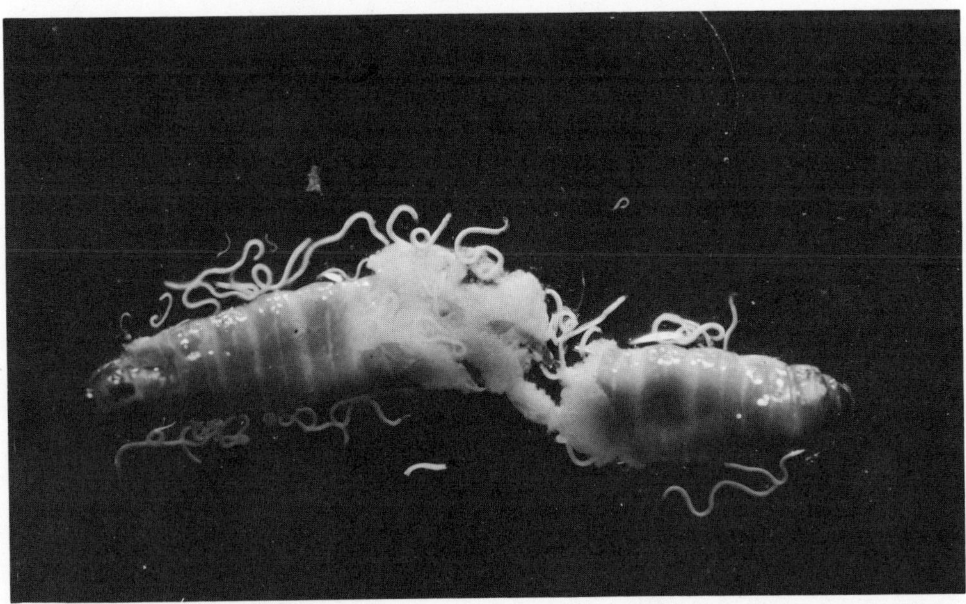

Figure 10.--DD-136 nematode development 3rd day after exposure. Note the very large size of these adult females and the very much smaller adult males (Mag. about 3X).

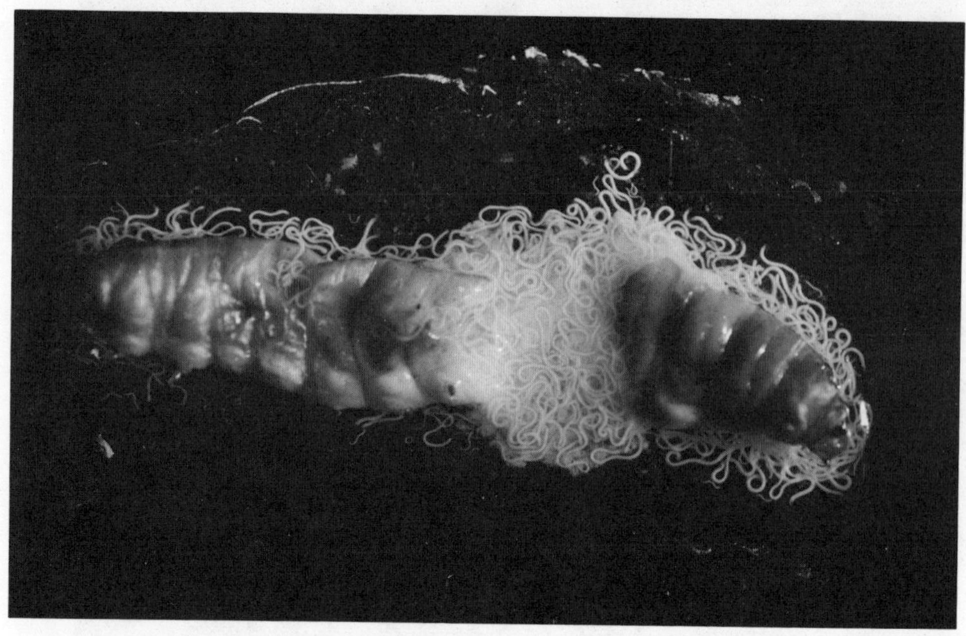

Figure 11.--DD-136 nematode development 4th day after exposure. These smaller adults are progeny of those seen in Figure 10 (Mag. about 3X).

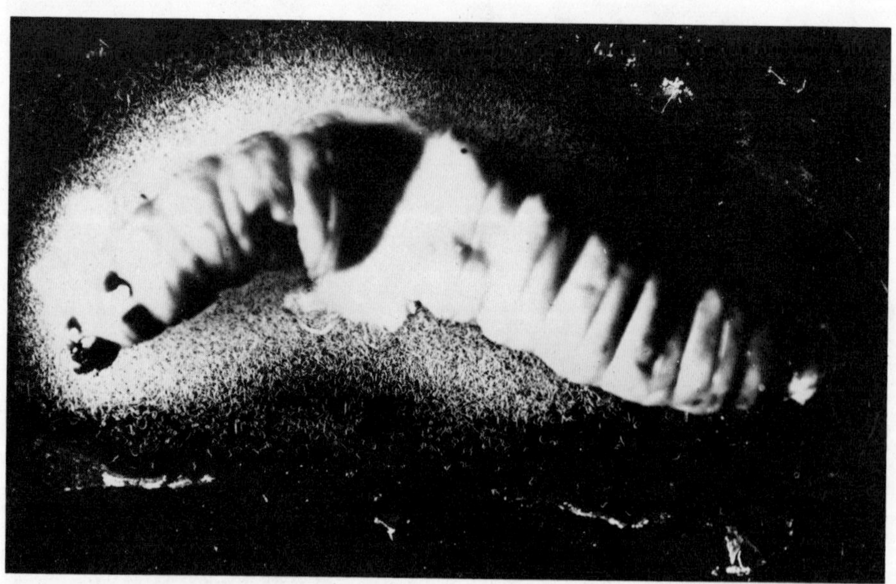

Figure 12.--DD-136 nematode development 8th day after exposure. Cadaver filled with ensheathed larvae (Mag. about 3X).

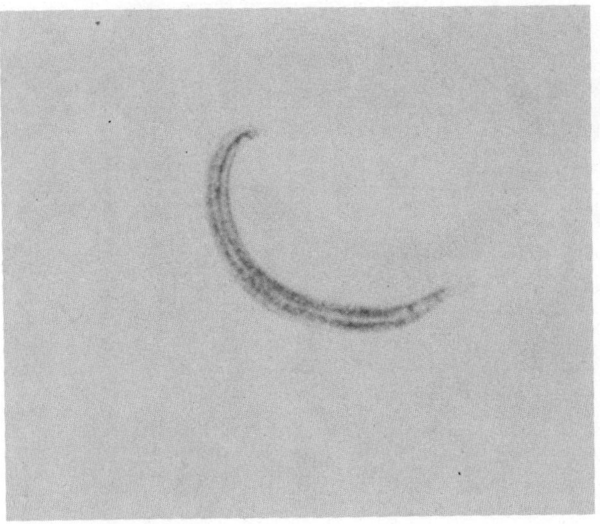

Figure 13.--Adult male N. dutkyi on sterol-deficient medium (Mag. 55X, picture width 2 mm).

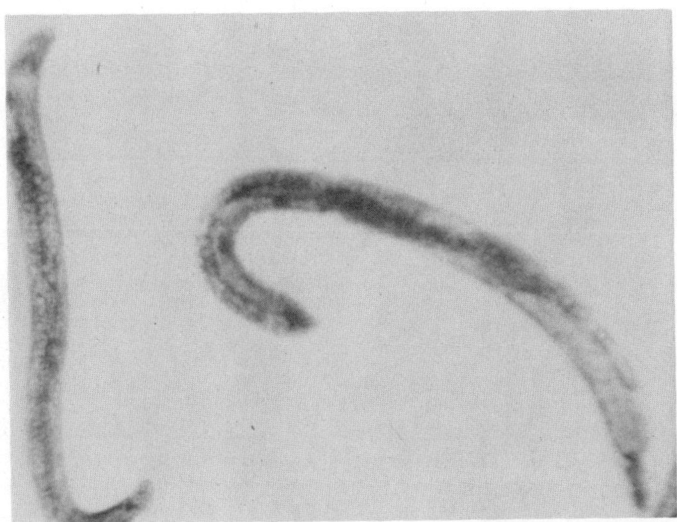

Figure 14.--Adult females N. dutkyi on sterol-deficient medium (Mag. 55X). Ova present are not fertile.

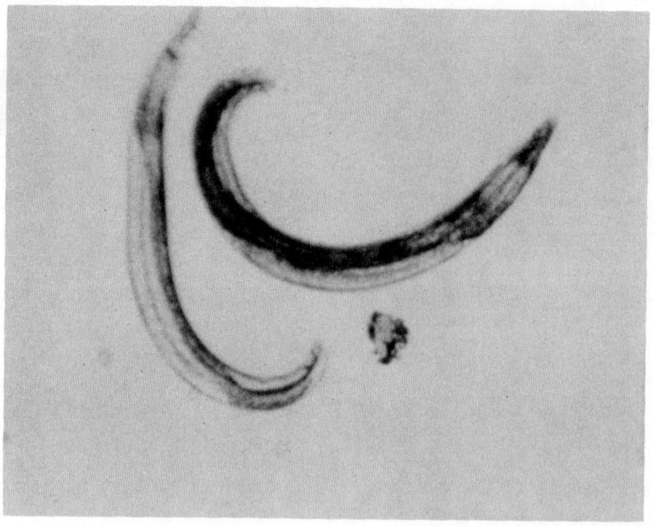

Figure 15.--Adult males N. dutkyi on sterol-enriched medium (Mag. 55X).

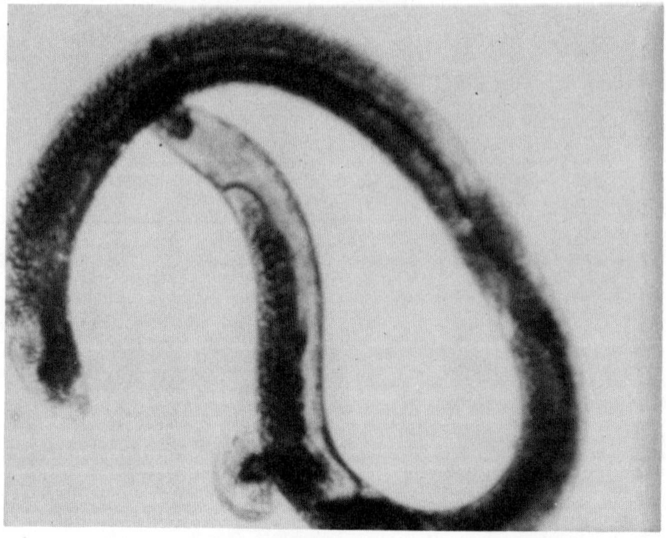

Figure 16.--Adult female N. dutkyi on sterol-enriched medium (Mag. 55X).

Sterols tested for support of nematode growth and reproduction

Sterol Tested	Structure: Double Bonds No.	Structure: Double Bonds Position	Structure: Alkyl at C-24	Utilization
3β-Hydroxy sterols				
Cholesterol	1	5	-	+
Cholestanol	0	-	-	+
7-Dehydrocholesterol	2	5, 7	-	+
Δ^7-Cholestenol	1	7	-	+
22-Dihydrobrassicasterol	1	5	Methyl	+
Ergosterol	3	5, 7, 22	Methyl	-
β-Sitosterol	1	5	Ethyl	+
Stigmastanol	0	-	Ethyl	+
Stigmasterol	2	5, 22	Ethyl	-
Ketosteroid				
4-Cholesten-3-one	1	4	-	+

Figure 17.--Sterols tested for support of growth and reproduction of _N_. _dutkyi_.

Figure 18.--Nematode application in apple orchard at White Swan, Washington. B&W copy of color transparency furnished by B. A. Butt, In Charge, Arid Areas Deciduous Fruit Insects Laboratory, Yakima, Washington.

Fate of Overwintering Codling Moth Population Under Bands

Hill Orchard

	Number of trees	Living Total	Living Per tree	No. Nematode killed Total	No. Nematode killed Per tree
9/22/60	20	33	1.65	54	2.70
11/2/60	20	30	1.50	107	5.35
5/10-11/61	37	16	0.432	71	0.19
5/23/61	28	5	0.178	14	0.50
					8.74

Overwintering brood nematode killed before emergence =

$$\frac{8.74}{8.74 + 0.178} = \frac{8.74}{8.92} = 98\%$$

Figure 19.--Results of codling moth tests at Hill Orchard, Kearneysville, West Virginia. Number of trees treated - 178 in 7 rows. Trunks and main branches sprayed with 2 million nematodes per tree 9/8/60 and 5/11/61. Rainfall frequent before and after applications.

Fate of Overwintering Codling Moth Population Under Bands

O'Neil Orchard

	Number of trees	Living Total	Living Per tree	No. Nematode killed Total	No. Nematode killed Per tree
9/14/60	12	763	63.6	143	11.9
9/28/60	7	644	92.1	229	32.7
11/9/60	9	59	6.6	614	68.2
5/3/61	10	56	5.6	37 (10% new)*	3.7
5/16/61	10	16	1.6	38 (10% new)**	3.8
					120.3

* total nematode killed 373 of which 10% were recently infected
** total nematode killed 381 of which 10% were recently infected

Overwintering brood nematode killed before emergence =

$$\frac{120.3}{121.9} = 98.7\%$$

Figure 20.--Results of codling moth test at O'Neil Orchard, Vincennes, Indiana. Number of trees treated - 118. Trunks and main branches sprayed with 4 million nematodes per tree 9/1/60 and 5/3/61. No rainfall between 8/15/60 and 10/1/60 at this location. Rainfall on 10/8/60 totaled 1.13 inches.

DISPERSION OF PATHOGENS

Irvin M. Hall
Professor
Division of Biological Control, Department of Entomology
University of California, Riverside, California

The dispersion of pathogens of insects has been touched on by many authors in their treatments of specific host-pathogen relationships, but the subject has received very little in-depth attention since the inception of modern insect pathology, with the result that there is a paucity of information in the literature on the varied mechanisms that bring about such movement. In this presentation, there will be an effort to bring together the bits of evidence concerning the means of natural spread of entomogenous microorganisms, followed by a brief discussion of manipulated dispersion, as entered into by man, including the application of pathogens as microbial insecticides.

Natural Dispersion

Any microorganism that displays pathogenicity for an insect has attained an interrelationship with its host which is regulated by the balance between the virulence of the microorganism and the degree of susceptibility of the host, augmented by the presence of an efficient means of transmission of infective stages of the pathogen from one individual or population of the host to another. The overall effect of these factors that influence the spread of any specific infection at a given place and time has been termed "dispersibility."

The ability of an entomogenous microorganism to spread or distribute itself throughout the range of its susceptible host or hosts is designated as the capacity of the pathogen to disperse. The importance of dispersion has been emphasized by Tanada (1963) in his statement that a pathogen lacking a high dispersal capacity may have only a low potential of developing an epizootic even though it may possess high virulence for the host and efficient survival capacities. The dispersal capacity may be the result of an adaptation to permit redistribution over relatively short distances, such as from soil to foliage when pathogen and host have become separated within an ecosystem due to normal environmental changes. Or, it may result from an established or fortuitous means by which the pathogen can follow its host great distances from its native habitat to a new area where environmental factors are favorable for development.

Although there are certain exceptions, such as some of the entomophilic nematodes, the motile-zoospore possessing primitive Phycomycete fungi and the higher phycomycetous Entomophthorales fungi, which can project their conidial spores, entomogenous pathogens, in general, lack meaningful means of locomotion, so that their movement from host to host within a population, or from population to population, must be accomplished through the actions of other factors, physical or biotic, in the environment. The physical factors considered to be important are wind and water.

The feasibility of wind transmission was demonstrated by Thompson and Steinhaus (1950) in their study of the nuclear polyhedrosis virus of the alfalfa caterpillar, Colias eurytheme Boisduval, in California over 20 years ago. They determined in laboratory trials that low-velocity air movement of 2 to 4 miles per hour could dislodge and splatter the fragile remains of freshly dead larvae to adjacent foliage and move contaminated dust several feet. It takes little imagination to conceive the effect of the 20 to 40 mph (or more) winds that blow frequently in the agricultural areas of California and other parts of the southwest in the distribution of this virus. Since other types of pathogens that have fallen to the soil and become attached to dust particles could be transported in this manner from one area to another, it is quite possible that wind transmission may play a major role in creating apparent new outbreaks of disease and resultant epizootics that appear to spring up suddenly in populations of many of our pest insect species. Wind, also, is a factor in the movement of alate insects, with resultant effect on the spread of any pathogens that they may be carrying.

The importance of water movement in the form of rain, irrigation water and flowing streams in the dispersion of pathogens has been suggested by a number of workers. According to Jaques (1970), the splashing action of rain is a prime contributor in the redistribution of viruses of the cabbage looper, Trichoplusia ni (Hubner), and the imported cabbageworm Pieris rapae (Linnaeus), from contaminated soil to the foliage of young cabbage plants. Moreover, Thompson and Steinhaus (1950) observed the development of epizootics in populations of the alfalfa caterpillar within two weeks after fields were irrigated by flooding, and they suspected that the thorough wetting of the new plant growth by that type of irrigation was responsible for the even distribution of the virus on the foliage, resulting in uniform epizootics in the developing Colias populations.

The role played by flowing streams has not been documented, but it is conceivable that pathogens could be washed from the foliage or leached from the soil by rain and transported by water movement to a new location where application of the water through irrigation practices could deposit viable infective stages where they could be contacted by susceptible hosts.

The biotic agents affecting the dispersion of entomogenous microorganisms are quite diverse. Of great importance is the role played by the primary host itself, since it has the closest association with the particular pathogen concerned. In any host-pathogen relationship, there is variation in susceptibility within the host population, and even when the percentage of susceptible hosts is high enough to permit an epizootic to occur, from few to many individuals may possess low susceptibility or even be immune. Thus, the diseased individuals not yet immobilized, the still uninfected susceptibles and those that are immune may serve to transmit infective material either externally on their body surfaces, to be deposited during contact with new foliage, within their digestive tracts following ingestion with food until excreted with feces, or within infected cells and tissues, to be passed on to progeny via transovarian transmission or deposited on the substrate when the host dies. Thompson and Steinhaus (1950) and Martignoni and Milstead (1962) cite the importance of the trans-ovum transmission of virus via the egg of the insect host which becomes contaminated externally during the process of deposition by a contaminated or infected parent.

Examples of diseased hosts acting to disperse their pathogens may be observed readily in certain aphid-fungus relationships. This author has repeatedly seen in the field recently killed fungus-infected alate female spotted alfalfa aphids surrounded by their first-stage young that were born just before the mothers succumbed from the disease. With the female being held to the foliage by rhizoids, the entomophthoraceous fungus pathogen was able to project conidial spores which were carried by air currents until some made contact with susceptible hosts, either the adjacent new-born young or other members of the population that were nearby.

This is only one example of the adaptation by pathogens of means of attaching hosts to a substrate in position to assure maximum opportunity for the infective stages to make contact with other susceptible individuals. In this case, as is common with many of the entomogenous fungi, the host is passive and the action of attachment by rhizoids is a function of the pathogen itself. Some fungi that attack aphids, however, do not form rhizoids, and the retention of the host in an elevated position on the foliage is by the chance holding effect of the proboscis inserted in the substrate. Obviously, this method is not as effective in assuring the localized distribution of the pathogen because many of the hosts become dislodged and fall to the ground.

Other types of entomogenous pathogens are not as well organized as are the Entomophthorales fungi, and they are forced to remain passive and rely on the activity of their hosts to assure their dissemination. In this regard, it is most interesting that diseased insects often undertake characteristic "migrations" or movements from one position to another as they reach advanced stages of disease. As discussed by Steinhaus (1949), in the case of certain virus infections, the infected caterpillars climb to the upper portions of the host plants, on which they are feeding, where they hang by their prolegs and die. A classic example of this host response is the _Wipfelkrankheit_ or high place nuclear polyhedrosis virus disease of the nun moth, _Lymantria monacha_ Linnaeus, in Europe, and other examples are common among the NPV infections of many of our noctuid pests. What prompts this host response remains unknown, but Steinhaus speculates that diseased insects may be seeking more suitable humidity and temperature conditions to relieve their discomfort. Whatever the cause, the host action does facilitate dissemination of the infectious agent, which may fall, be washed, or be blown from the disintegrating host remains onto adjacent foliage to increase the chance of contact with a healthy individual.

Nonsusceptible insect species which coexist in mixed population relationships on the same plants with insect species that are subject to disease may act as mechanical carriers of the pathogen. Most certainly, the migrations of the western yellow-striped armyworm, _Prodenia praefica_ Grote, from fields where the alfalfa caterpillar has been decimated by a virus epizootic will serve to spread the virus to adjacent fields. Similar situations undoubtedly exist in the case of migratory grasshoppers and other insects, and these nonhost species may be more effective in their spreading of a particular pathogen, to which they are not susceptible, than the previously mentioned immune individuals because of sheer numbers.

The effect of insect parasites in dispersing pathogens of their hosts was first touched on by Payne (1933) who provided evidence that a hymenopterous parasite, _Microbracon hebetor_ (Say) was capable of transmitting infective stages of the microsporidan _Thelohania ephestiae_ Mattes from one larva of the Mediterranean flour moth, _Anagasta kuehniella_ (Zeller), to another via its contaminated ovipositor. The overall effectiveness of this type of transmission is unknown, but it is quite possible that the activity of parasites in seeking out and stinging healthy hosts may play an important role in the dissemination of some pathogens. However, with contamination of the ovipositor a result of chance contact with a diseased host rather than a certainty, and the knowledge that in some cases, at least, the ovipositor may become free of contamination following one or two stings of disease-free hosts, it is possible that parasite transmission may have definite limitations in some situations.

A noted example of parasites acting as intermediaries with man's help in the introduction of a pathogen has been documented by Bird (1961) in his report on a study of the nuclear polyhedrosis virus of the European spruce sawfly, _Diprion hercyniae_ (Hartig). This sawfly, which invaded Canada early in this century, spread rapidly and became a serious pest in the spruce forests of eastern North America by 1938. Initial control of the sawfly was attained by the development of epizootics of a naturally occurring virus that appeared following the importation and colonization of parasites. The origin of the virus never has been determined with certainty, but it is considered to have been introduced accidentally into Canada with the parasites which were collected in large numbers in the native habitat of the pest in Europe and mass-propagated for release into Canadian forests. Following the rapid spread of the virus and the establishment of the introduced parasites, both agents have continued to exercise major roles in maintenance of the pest at acceptable low levels.

Predaceous insects, insectivorous small mammals and birds act much in the same manner in the dispersion of entomogenous pathogens. As they search for prey, their body surfaces may become contaminated by infectious agents in abundance in their environment, and these materials may be deposited on clean foliage as they continue to forage. In addition, they may ingest diseased insects and subsequently deposit the pathogens in new areas with their droppings. Such actions were documented during the studies on the spread of the milky disease bacilli of the Japanese beetle, *Popillia japonica* Newman, in the eastern United States in the 1940's (Hawley, 1952), and such carriers have been termed "facultative vectors" by Franz (1971).

It must be noted that man, by chance or design, can and does play an important role in the dispersion of pathogens of insects. At the local level, the worker moving within a field or from field to field may mechanically transport infective stages of pathogens, such as nuclear polyhedrosis viruses that, as part of larval remains, become smeared on clothing or tools. Moreover, in the movement of agricultural products from one area to another, man unknowingly may move viable infective materials in the remains of diseased insects on the plants, or actually transport infected living insects on or in the products being shipped or the vehicle carrying the shipment.

The effect of man in the movement of pathogens for great distances, such as from one continent to another, has not been documented to any great extent in the literature, but there are several examples that can be mentioned that would suggest that such movement might be more common than one would think.

The first example is that of the milky disease bacteria of the Japanese beetle. The history of the invasion of the eastern seaboard of the United States by this insect in the early part of this century, the continued spread of the pest up to the present time, and the extensive investigations by Federal and State agencies leading to the discovery of the milky disease bacteria and their development into effective microbial control entities, have been well recorded, but practically nothing has been postulated about the origin of the pathogens. About all that is known is that they have very close and complex relationships with their host and were not known to exist in the United States prior to the invasion by the beetle. Although they have not been detected in Japan and Korea, the native home of *P. japonica*, it probably is safe to assume that the two bacteria, *Bacillus popilliae* Dutky and *B. lentimorbus* Dutky, must have entered this country with the initial insect introduction, either within infected host insects or as resistance spores in the soil in which the grubs were living. Of importance is the strong possibility that man, although ignorant of his action, actually moved via sea shipment enough infective stages of the pathogens to have them become established in a new location far around the world from their probable home.

The second example of man's actions in long-distance dispersion already has been mentioned with respect to the transmission of an entomogenous pathogen by parasitic insects. Bird (1961, 1962) reported that the nuclear polyhedrosis virus from the European spruce sawfly, which was discovered shortly after the initiation of biological control efforts, is considered to have been accidentally introduced into Canada with shipments of parasites collected in Europe where the sawfly is native, and it became established very rapidly following propagation and release of large numbers of the parasites. Although not planned, man's effort in moving the parasite apparently was reponsible for introduction of the virus.

The third example is that of the accidental introduction of two species of entomophthoraceous fungi that attack the spotted alfalfa aphid, *Therioaphis maculata* (Buckton), into the United States in the 1950's. As the history of the invasion of this country by the pest has been determined, it first appeared on alfalfa near an Air Force base in New Mexico in the early part of that decade. With the frequency of flights by Air Force planes to and from the Mid-East home of the aphid, it may be concluded that one or more living alate female aphids must have been transported in a protected niche on a plane, arriving in good enough condition to be able to move to suitable nearby host plants to initiate a new population. The finding of the two species of pathogenic fungi, *Entomophthora exitialis* Hall and Dunn and *E. virulenta* Hall and Dunn, attacking only *T. maculata* in southern California within a few months of the time the pest reached the West Coast led to subsequent determinations that they attacked the aphid throughout its native range of the Mid-East through to India, as well as in other areas where the aphid had become established. This has brought forth the belief that one or more planes may have flown through flights of alate aphids, thereby picking up and transporting relatively large numbers of the pest, some of which were infected by one or the other of the two fungi. Since the fungi worked side-by-side in creating epizootics within a host population, it is most likely that they came in together.

Manipulated Dispersion

Although not practiced widely over the years, particularly because of the limited state of knowledge of insect pathology and microbial control and the centering of interest in other areas of biological control which were thought to offer greater chance of success, a major method of utilization of entomogenous microorganisms to control insect pests involves the direct introduction or colonization of pathogens for the purpose of establishing them into close association with the pest population, following which they can disperse throughout the range of the host aided by the varied applicable physical and biotic factors to bring about long-term reduction and possible permanent control of the pest species.

Probably the most successful program of this type has been the use of preparations of the milky disease bacteria in the control of grubs of the Japanese beetle on turf in the eastern United States since 1940. As discussed by Hawley (1952) in his review of the effort, because spores of these bacilli proved to be difficult and costly to produce, thorough-coverage application rarely has been attempted. To conserve materials, dry spore preparations generally have been applied to turf by spot treatment using equipment as simple as a rotary hand corn planter modified to inject about 2 grams of spore powder each time it is tripped. With injections made every 10 feet, checkerboard fashion, a treatment of 1.75 pounds of bacillus preparation per acre was found to be adequate. The acreage treated also varied, and it was found that only two 1/2-acre plots per square mile were needed in open agricultural areas, and smaller plots on at least one block in 10 in urban areas. The rate and degree of subsequent spread of the bacteria were found to be dependent on the size of the initial grub population and the actions of the physical and biotic factors. Hawley states that spread of the milky disease bacteria from the points of introduction is accomplished by (1) any natural or artificial movement of topsoil containing spores of the bacilli, (2) the dispersion of diseased grubs through the soil, and (3) the activity of birds and mammals that feed on diseased grubs.

Other classic examples of introduction and colonization of a pathogen have been the applications of entomogenous viruses for the control of a number of insect pests in the forests of Canada. Particularly well documented are the studies of some of the viruses of sawflies, one of which already has been mentioned. As stated by Bird (1955, 1961, 1962) and Bird and Burk (1961), the nuclear polyhedrosis virus of the European spruce sawfly was applied by spraying to a small number of trees. Following the initiation of disease in the populations of young larvae on the trees, the virus spread very rapidly. Rain appeared to be the chief agent of spread within a tree, while natural enemies, particularly parasites, and infected sawfly females appeared to be responsible for transmitting virus from tree to tree. It was found that little, if any, virus remained on the foliage over winter, and that transmission from one year to the next was mainly transovarian through eggs from infected adults, with epizootics starting from foci of diseases in less than 10% of the overwintering egg masses.

The possibility of practical use of transovum transmission, which is transmission occuring outside the ovary, has been suggested by the report of Martignoni and Milstead (1962) on their study of the dissemination of the nuclear polyhedrosis virus of the alfalfa caterpillar through release of artificially contaminated adult female butterflies. As with practically all efforts at colonization followed by natural spread, this method of virus dispersal is not conducive to the attainment of rapid control of populations of the pest, since the natural development of virus infection following eclosion cannot assure quick kill of the larvae. However, the advantages of the need for only small amounts of infective material, the simplified nature of the procedure for contamination of the adults and the lack of a need for precise timing of application would suggest that this type of release program might be useful in the future.

In the past few years, the principal interest in microbial control has been in the utilization of entomogenous microorganisms or their toxins in the manner of chemical insecticides for the quick control of economic pest infestations. This has meant the bypassing of the factors involved in the natural spread of pathogens, and the development, where feasible, of techniques for thorough coverage of plant surfaces to induce rapid mass infection of the pest populations and their subsequent quick reduction to levels approaching localized extinction in order to protect the crop. Moreover, practical utilization of microbial insecticides containing _B. thuringiensis_ has been limited by economic and residue factors to the control of insect pests on short-term vegetable or field crops, such as head lettuce and cotton, with the result that to a great extent post-treatment dispersion and extended persistence of the pathogen on a crop following initiation of disease in the pest population have not been matters of concern.

Because of the need for thorough coverage to assure rapid infection of the susceptible pests, the early testing of the _B. thuringiensis_ preparations some 15 years ago against the cabbage looper on crucifers (Hall and Andres, 1959) indicated the feasibility of using dust formulations which would give thorough coverage and place the toxic materials on the undersides of the foliage where they could be ingested by the young larvae and bring on cessation of feeding and death before the plant suffered severe damage. In the western parts of the United States, the use of bacillus dusts as the best means of mechanical dispersion for control of the cabbage looper became commonplace during the 1960's. In other areas, where weather conditions and/or habit precluded the use of dusts, spray formulations of the first-generation bacillus products often gave poor control, and sprayable BT preparations only came into their own with the development within the past three years of the much more toxic strains of the bacillus, in which the 20-fold increase in potency was enough to make up for the inherent poor coverage from spray application.

It is interesting to note that _B. thuringiensis_ is an entomogenous pathogen that has shown almost no ability to disperse on its own or through the help of environmental factors. As stated by Tanada (1963), it is a microorganism possessing high virulence (for some hosts) and efficient survival mechanisms, but it lacks a high dispersal capacity and, thus, has only a low potential for developing epizootics in populations of susceptible hosts remote from the spot where mechanically applied.

In summing up our knowledge on the subject of dispersion, it is apparent that the dispersibility of an entomogenous microorganism is the result of interactions of many factors, from the makeup of the pathogen itself, its particular relationship with its host or hosts, and the effects of the varied physical and biotic pressures in the immediate and not so immediate environment. With the varied nature of these influences, it is understandable that pathogens will have their own established patterns of spread, giving them more-or-less definable capacities to disperse. Most notable of those with apparent high dispersal capacities are the nuclear polyhedrosis viruses, which, in general, show marked levels of virulence for their hosts and are readily and rapidly transmitted great distances to cause widespread epizootics. From the standpoint of microbial control of pest insects, these viruses and other types of pathogens with similar capacities may be considered to be very effective and efficient parasites, and they may be amenable to utilization in easy to set up introduction and colonization programs. In contrast, many entomogenous pathogens do not possess such high levels of dispersion, and some, such as *B. thuringiensis* and other crystalliferous bacteria, show almost no ability to spread from the point of application. These, it is obvious, may be usable only as microbial insecticides where thorough coverage application is practiced to place the infective or toxic materials in contact with the pest species.

References

Bird, F. T. 1955. Virus idseases of sawflies. Canadian Entomol. 87: 124-127.

Bird, F. T. 1961. Transmission of some insect viruses with particular reference to ovarial transmission and its importance in the development of epizootics. J. Insect Pathol. 3: 352-380.

Bird, F. T. 1962. The use of viruses in biological control. Coll. Int. Pathol. Insectes, Paris, 1962, pp. 466-473.

Bird, F. T. and J. M. Burk. 1961. Artificially disseminated virus as a factor controlling the European spruce sawfly, *Diprion hercyniae* (Htg.) in the absence of introduced parasites. Canadian Entomol. 93: 228-238.

Franz, J. M. 1971. Influence of environment and modern trends in crop management on microbial control. *In* Microbial Control of Insects and Mites (H. D. Burges and N. W. Hussey, eds.). pp. 407-444. Academic Press, New York.

Hall, I. M. and L. A. Andres. 1959. Field evaluation of commercially produced *Bacillus thuringiensis* Berliner used for control of lepidopterous larvae on crucifers. J. Econ. Entomol. 52: 877-880.

Hawley, I. M. 1952. Milky diseases of beetles. *In* Insects--The Yearbook of Agriculture. USDA, pp. 394-401.

Jaques, R. P. 1970. Application of viruses to soil and foliage for control of the cabbage looper and imported cabbageworm. J. Invertebr. Pathol. 15: 328-340.

Martignoni, M. E. and J. E. Milstead. 1962. Trans-ovum transmission of the nuclear polyhedrosis virus of *Colias eurytheme* Boisduval through contamination of the female genitalia. J. Insect Pathol. 4: 113-121.

Payne, N. M. 1933. A parasitic hymenopteran as a vector of an insect disease. Entomol. News 44: 22.

Steinhaus, E. A. 1949. Principles of Insect Pathology. 757 pp. McGraw-Hill, New York.

Tanada, Y. 1963. Epizootiology of infectious diseases. *In* Insect Pathology, An Advanced Treatise, Vol. 2. (E. A. Steinhaus, ed.). pp. 423-475. Academic Press, New York and London.

Thompson, C. G. and E. A. Steinhaus. 1950. Further tests using a polyhedrosis virus to control the alfalfa caterpillar. Hilgardia 19: 411-445.

PRODUCTION OF MICROBIAL CONTROL AGENTS

H. T. Huang
International Minerals and Chemical Corporation
Libertyville, Illinois

In going over the program for this Institute, I could not help noticing that I am probably one of only two speakers who has not been trained as an entomologist. I am certainly the only speaker from an industrial organization. These observations do not necessarily confer a sort of distinction on me personally. Rather, they are indicative of the fact that microbial control has indeed come of age. It has graduated from the stage of laboratory research and development to the realm of practical utilization, or if you prefer, commercialization.

Since the hour is late and we have all had a long day, I shall be as brief as possible. I am going to discuss only the production of microbial agents on a commercial scale and leave aspects of the subject still under development for a separate presentation at another time.

As shown on Slide 1, the production of microbial control agents, as opposed to chemical agents, are currently accomplished by some type of biological procedure; viz. fermentation in the case of bacteria and the use of the actual target host animal in the case of viruses. There are today, three microbial agents produced commercially in this country (Slide 2). I would assume this is a resonably complete list. I have no direct knowledge of the situation in Eastern Europe and other Communist countries, but as far as we know, the same list is probably applicable worldwide.

Bacillus popillae and *Bacillus lentimorbus* are produced by one manufacturer only. Its use is limited to control of the Japanese beetle on turf. Attempts are being made to expand its usage to include forage crops, but registration has not yet been obtained (Heimpel, personal communication). The second agent, *Bacillus thuringiensis*, can qualify as a full-fledged commercial microbial insecticide. It is registered for use on a wide variety of vegetables, fruits and field crops (Federal Register, 1971), and also shade trees, forests and ornamentals. Slide 2 shows quite a long list of producers. Actually, the list includes all producers that have at one time or another been interested in and worked on the commercial production of *B. thurgingiensis*. Today, many have abandoned this business in the U.S., and we have only three producers in actual operation. Biotrol is produced by Nutrilite Products, Inc., Dipel by Abbott Laboratories, and Thuricide by International Minerals and Chemical Corporation. Bactospeine is commercial in France and perhaps in some other countries in Europe.

We now have a temporary registration for the third agent, *Heliothis* nuclear polyhedrosis virus (Federal Register, 1970). We hope that it will receive permanent registration in the near future.

I do not know too much about the production of Bacillus popilliae except that it is prepared from the whole living organism. The hosts are actually collected from the field so the possibility of very large-scale production is rather low. If its usage expands, improvement in production technology would be desirable. Slide 3 shows what a preparation of B. popilliae may look like under the microscope. You will be interested to compare it to a similar slide which will be presented later on B. thuringiensis. Please note the parasporal body being formed concurrently with the spore.

B. thuringiensis is produced by two methods. Both are fermentations using artificial media. One is the so-called surface culture. Slide 4 gives a schematic representation of a typical surface culture operation (Dulmage and Rhodes, 1971).

The inoculum is provided by a couple of stages in submerged culture. It is then spread on a semi-solid type of medium and allowed to incubate for a given length of time. When the number of spores reaches maximum level then the whole mass is collected, dried and ground to a fine powder. It is a simple procedure. The only disadvantage is that in terms of cell yield it is not as efficient as the submerged culture procedure. It is generally known that Nutrilite, Inc., produces its product by a surface culture method.

The second and more widely used procedure is submerged culture. The flow diagram on Slide 5 shows the procedure used at International Minerals and Chemical Corporation. A liquid medium in a large fermentor is sterilized and seeded with an inoculum which is itself derived from several different stages of culture in increasing volume. After optimum growth and sporulation are achieved, the material is screened, centrifuged, standardized and finally formulated into the commercial product.

A successful fermentation is based on two key factors. First, an optimum culture environment; in this case it means particularly a good medium that is conducive to production of high levels of spore and crystals.

You are already familiar with the fact that the B. thuringiensis products that are on the market today are a combination of spores and endotoxin or crystal toxin. Both entities are needed in the product for maximum activity. Slide 6 provides an example of the type of medium that is conducive to the production of spores and crystals. You will note that the protein level is quite high. B. thuringiensis is highly protolytic organism and can hydrolyze large amounts of protein during a growth cycle. Evidently these are necessary for optimal sporulation and crystal formation.

The second important factor for a successful commercial fermentation is a good culture. Of course, the medium and the culture have to be compatible with each other. One culture may work very well in one medium but it may do poorly in another medium.

Slide 7 shows actual bioassay values obtained with different B. thuringiensis cultures (Dulmage and Rhodes, 1971). These are based on products that were recovered from submerged fermentations of different strains of B. thuringiensis tested. To the best of my knowledge all three producers in the U.S. today use variety alesti. The French are still using variety thuringiensis.

One way of recovering the spore crystal complex is described on Slide 8. It is quite a useful method for producing material of high potency. After centrifugation, the spore crystal complex is suspended in a solution of lactose, and precipitated with acetone. The precipitate is dried in air to give a highly concentrated product.

Slide 9 shows the *B. thuringiensis* culture at the stage when it is almost ready for harvesting. You can see the crystals as dark spots with a diamond-shaped form in outline.

On Slide 10 is an electron micrograph of a single crystal. The striations on the surface presumably have something to do with the fine structure. I won't go into detail as to what makes the crystal tick. That is in itself a topic that requires extensive discussion. I do want to emphasize, however, that the commercial product that one buys on the market is a mixture of the spore and the crystal.

I would like to go on now to the production of the *Heliothis* virus as practiced at this time at International Minerals and Chemical Corporation (Greer, Ignoffo and Anderson, 1971). You will see on Slide 11 that we have actually two operations proceeding simultaneously. On the left side we have an insect rearing cycle and on the right side an infection cycle for the production of virus. In summary, what we are doing is to use the young bollworm larvae as fermentors to produce a virus which only propagates in the tissue of the host. The rearing cycle begins with the adult, both male and female (Slide 12). These are placed together in a mating jar (Slide 13) some 50 to 100 pairs. The jar is so constructed and furnished that the females deposit their eggs on a strip of paper that can be easily removed to collect the eggs (Slide 14). For a successful process we need first a suitable artificial diet to rear the bollworm larvae. Secondly, because the bollworm is cannabalistic, we have to find a way to rear each larva individually.

Several artificial diets have been successfully used for rearing the bollworm. One example of such a diet is shown on Slide 15.

In order to isolate individual larva during the growth cycle, we place them in cavities that are familiar to us for the packaging of small portions of jams and jellies. These come in trays of a convenient size (Slide 16). After cooking the diet for the required length of time, it is dispensed in individual cavities. The entire tray is then covered with a piece of plastic film (Slide 17).

Newly hatched larvae are introduced into these cavities and allowed to grow to the pupa stage. These emerge as adults and start on a new cycle. It is interesting to see how much a bollworm grows in size in a period of say about 9 to 12 days (Slide 18). The newly hatched larva is a speck, barely visible to the naked eye, as contrasted to the hypodermic needle. A one-day-old larva is shown to the left. To the right are the 3-day-old, 5-day-old and full-size larva. Thus, the bollworm larva makes an attractive fermentor in that one can conduct a fermentation, propagate the virus, while the fermentor itself is expanding continuously.

In the operation of the infection cycle (Slide 11), what we do is simply break the life cycle of the bollworm at about 5 days of age and infect it with the nuclear polyhedrosis virus (Slide 19). Stacks of trays of infected larvae in individual cups are incubated for a suitable number of days (Slide 20) until the larvae are mature and the disease is fully developed. As shown on Slide 21, they are either dead or they look very sick. At this stage they are harvested and processed to give the actual commercial product.

Slide 22 is an electron micrograph of polyhedral inclusion bodies (PIB) that are produced. Slide 23 shows a section of the PIB's and the individual virions within. On Slide 25 is a virion under higher magnification.

Any discussion of the production of microbial insecticides would not be complete without a few words about product standardization. In the early days of microbial control, many of you have probably had to work with products that varied in activity from batch to batch, whether it was produced in your own laboratory or by a commercial company. This was a major problem in the early days of commercialization of *B. thuringiensis*.

Obviously, one purpose of standardization is to ensure that the product will have a uniform potency from batch to batch. But another equally important purpose is to enable the product to be registered. In order to register a pesticide in the U.S., we must be able to declare the amount of active ingredient on the label in a manner consistent with the requirements of the Federal Insecticide, Fungicide and Rodenticide Act (FIFRA). The FIFRA as you probably know, was written explicitly with the aim of regulating chemical insecticides.

When the first microbial, a *B. thuringiensis* product, was submitted for registration, the USDA and FDA had to determine how it should be standardized. They had to answer the question, "What is the active ingredient in the *B. thuringiensis* product? Eventually the Government and the producing companies agreed on a system for declaring the active ingredient based on a convention which we will examine shortly.

Let us look at the next slide (No. 25) which shows three typical microbial agents. The active entity in the case of the mold will be the spore or a toxin or perhaps both. In the case of the *Heliothis* virus, it is undoubtedly the polyhedral inclusion body (PIB). In the case of *B. thuringiensis* it was thought to be spores but now we know it is a combination of spores and crystals.

If the active entity is the spore or PIB, the simplest approach would be to declare the number of spores or PIB's on the label, for example, X spores per gram or X PIB's per gram. If you look carefully at the slide of our *Heliothis* NPV label (Slide 26), you will see the statement: "Active ingredient - polyhedral inclusion bodies of the *Heliothis* virus - 0.4%." Lower down it further states the product contains 4 billion polyhedral inclusion bodies per gram. The percentage was arrived at by agreement with the Government, on the convention that one billion polyhedral inclusion bodies weigh 1 mg.

In the case of *B. thuringiensis*, the original registration was based on the assumption that spores were the only active ingredient and that one billion spores weigh 10 mg. Thus a typical label would have a declaration as shown on Slide 27. By the middle 1960's, we were well aware that the spore count by itself was a poor indicator of the potency of the material. The crystal is at least equally important. But we were unable to develop a chemical analysis of the crystal which would reflect its biological activity. We, therefore, petitioned the Government to adopt some kind of bioassay unit for declaring the active ingredient. To cut a long story short, it took more than a year's continuous deliberations to finally arrive at the currently approved label declaration based on both a unit of bioassay activity and spore count.

On Slide 28 you will see a new label for our product, Thuricide HPC. The declaration now states "*B. thuringiensis* Berliner, potency of 4000 International Units per milligram" of product. At the same time there is a guarantee regarding minimal number of spores per mg.

In any case, *B. thuringiensis* has reached a stage where it can be manufactured and distributed on a commercial scale and registered in a manner that meets with the approval of regulatory agencies of the federal government. We hope the *Heliothis* NPV will also become fully commercial shortly. If so, it will be the forerunner of a new series of viral insecticides, which will greatly increase the role of microbial agents in our unending battle against noxious insect pests.

References

Heimpel, A. H. Personal communication.

Federal Register, Vol. 36, No. 228, 180.1011, November 25, 1971.

Federal Register, Vol. 35, No. 238, p. 18690, December 9, 1970.

Dulmage, H. T. and R. A. Rhodes IN Microbial Control of Insects and Mites, (H. D. Burges and N. W. Hussey, eds.), pp. 507-538. Academic Press, New York, 1971.

Greer, F., C. M. Ignoffo and R. R. Anderson. Chemtech, June 1971, pp. 342-347.

Chemical versus microbial insecticides

Property	Chemical	Microbial
Production Technology	Chemical Synthesis	Biological Fermentation Mass Rearing of Insects
Activity Spectrum	Broad, Non-specific	Narrow, Specific
Type of Activity	Contact and Stomach poison	Stomach Poison
Rate of Action	Fast	Slow

Slide 1. Chemical versus microbial insecticides.

Slide 2.--Commercial Microbial Pathogens

Pathogen	Product Designation	Manufacturer
B. popilliae and *B. lentimorbus*	Doom	Fairfax Biological Labs., U.S.
	Japidemic	Ditman Corp., U.S.
B. thuringiensis	Agritrol	Merck and Co., U.S.*
	Bakthane L69	Rohm and Haas Co., U.S.*
	Bactospeine	Pechiney Progil Lab., Roger Bellon, France
	Bathurin	Chemapol, Biokrma, Czechoslovakia
	Biospor 2802	Farbewerke Hoechst, Germany*
	Biotrol BTB	Nutrilite Products, Inc., U.S.
	Dendrobacilin	Moskors. zarod. bakt. prepar atod., U.S.S.R.
	Dipel	Abbott Laboratories, U.S.
	Entobakterin 3	All-Union Institute Plant Protection, U.S.S.R.
	Parasporin	Grain Processing Corp., U.S.*
	Sporeine	Laboratoire L.I.B.E.C., France
	Thuricide	International Minerals & Chemical Corp., U.S.
Polyhedrosis Virus of *Heliothis*	Viron/H	International Minerals & Chemical Corp., U.S.
	Biotrol VHZ	Nutrilite Products, Inc., U.S.

*No longer in production.

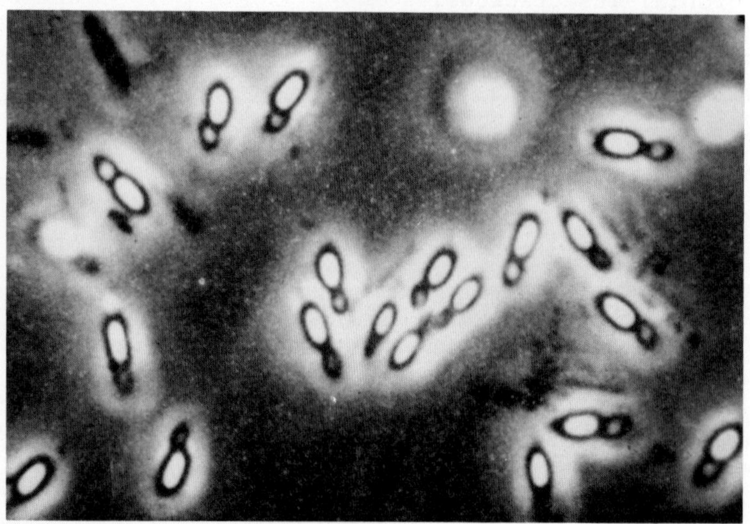

Slide 3.--Spores and parasporal bodies of *Bacillus popilliae*.

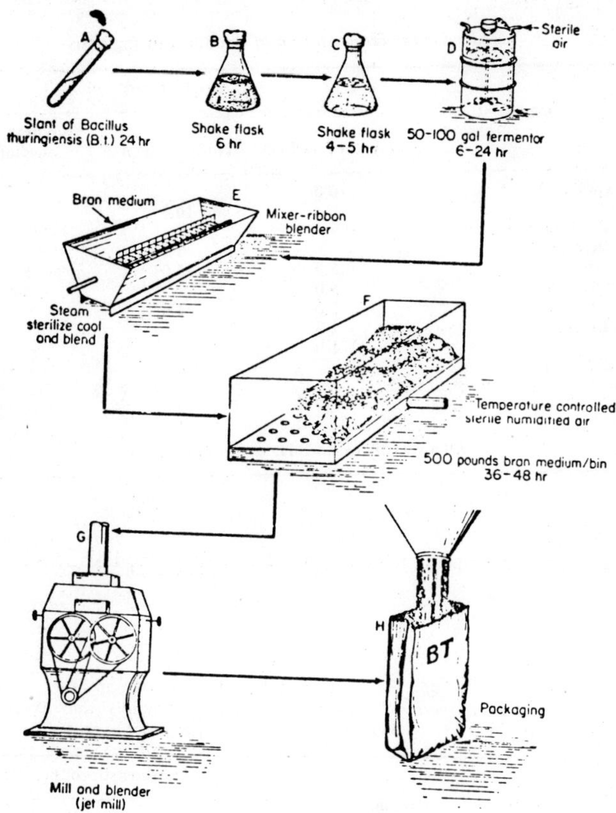

Slide 4.--Semi-solid fermentation of B. thuringiensis.

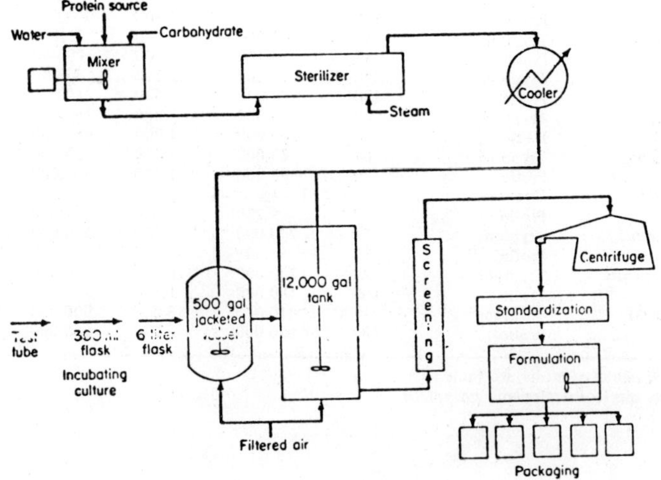

Slide 5.--Submerged fermentation of B. thuringiensis.

Media for the fermentation of Bacillus thuringiensis

Ingredient	Level (g/liter)		
	Tryptone medium	Proflo medium	Soybean meal medium
Tryptone	10.0		
Proflo		10.0	
Soybean meal			15.0
Bacto-peptone®		2.0	
Dextrose	5.0	15.0[a]	5.0
Corn starch	5.0		5.0
Yeast extract	2.0	2.0	
K_2HPO_4	1.0		
KH_2PO_4	1.0		
$MgSO_4.7H_2O$		0.3	0.3
$FeSO_4.7H_2O$		0.02	0.02
$ZnSO_4.7H_2O$		0.02	0.02
$CaCO_3$		1.0	1.0

All media diluted to volume with distilled H_2O

Slide 6.--Media for fermentation of B. thuringiensis.

Characteristics of the insecticidal activities of powders recovered from fermentation beers of different variants of Bacillus thuringiensis *grown on tryptone-based and Proflo-based media*

Variant	Medium[a]	Characteristics of powder			
		Spore count $\times 10^9$/g	DDU_{50}/g	$DDU_{50}/10^9$ spores[b]	DDU_{50} recovered/ liter beer[b]
aizawai	Tryptone	76	76,000	1,000	150,000
	Proflo	120	34,000	280	230,000
alesti	Tryptone	53	<3,700	<70	<5,000
	Proflo	22	<2,200	<100	<5,800
dendrolimus	Tryptone	95	43,000	450	86,000
	Proflo	250	<2,500	<10	<15,000
entomocidus	Tryptone	36	26,000	720	61,000
	Proflo	110	240,000	2,200	1,000,000
finitimus	Tryptone	42	<2,900	<70	<5,800
	Proflo	41	4,100	100	29,000
galleriae	Tryptone	46	17,000	370	33,000
	Proflo	120	44,000	370	240,000
kenyae	Tryptone	9	63,000	7,000	180,000
	Proflo	60	100,000	1,700	650,000
morrisoni	Tryptone	69	83,000	1,200	250,000
	Proflo	70	98,000	1,400	600,000
sotto	Tryptone		Poor growth		
	Proflo	7	<280	<40	<1,600
subtoxicus	Tryptone	31	220,000	7,100	620,000
	Proflo	85	53,000	620	340,000
thuringiensis	Tryptone	22	230,000	10,500	420,000
	Proflo	160	180,000	1,100	1,100,000
tolworthi	Tryptone	14	810,000	58,000	3,000,000
	Proflo	130	100,000	770	580,000

[a]For formulae of media, see Table IV.
[b]Assay species *Pectinophora gossypiella*.

Slide 7.--Activity of variants of B. thuringiensis.

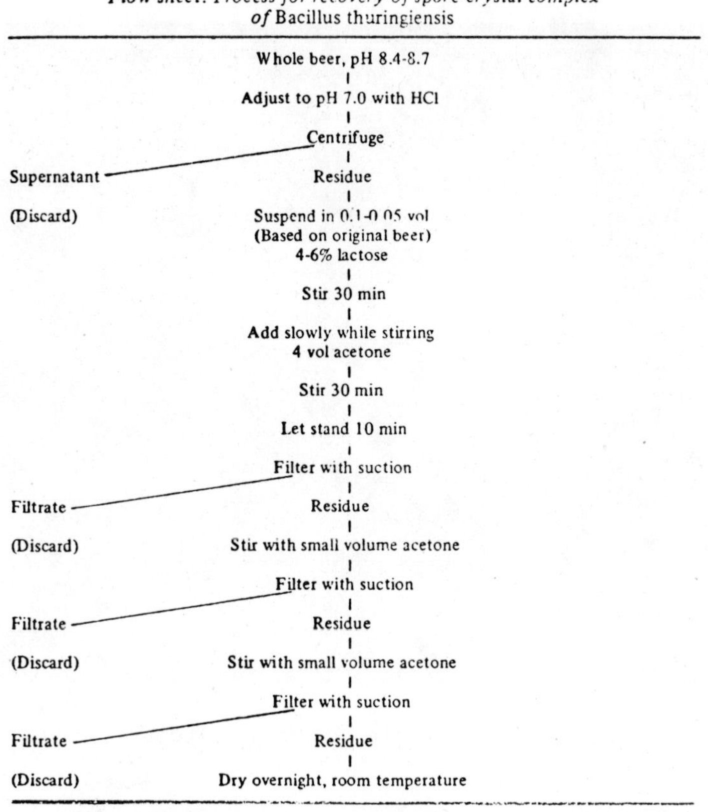

Slide 8.--Recovery of spore-crystal complex of B. thuringiensis.

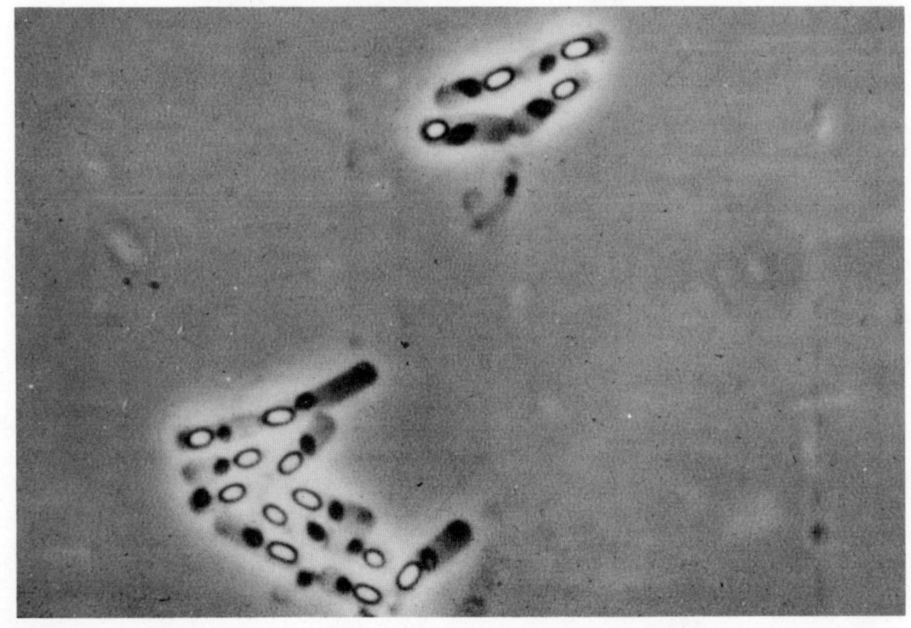

Slide 9.--Spores and stystals of B. thuringiensis.

Slide 10.--Electron micrograph of single B. thuringiensis crystal.

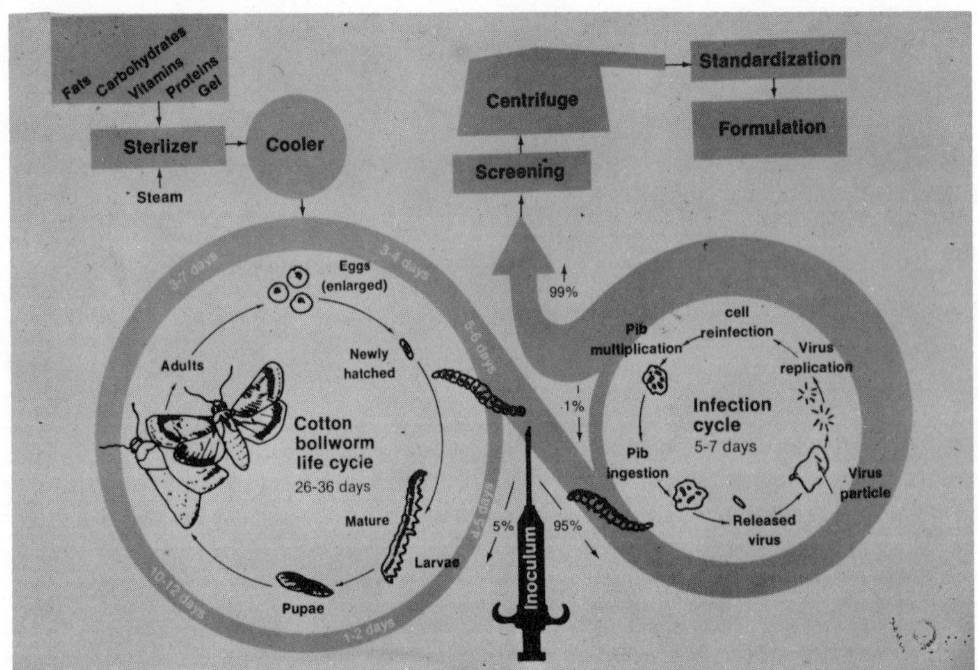

Slide 11.--Schematic diagram of process for propagation of Heliothis NPV.

Slide 12.--Male and female moths of Heliothis zea.

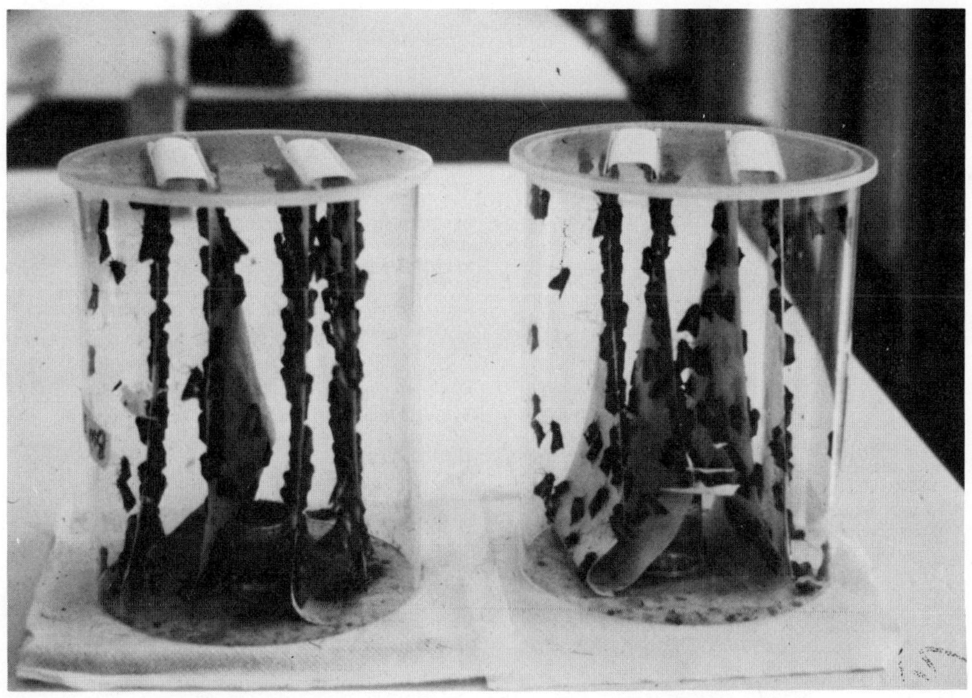

Slide 13.--Egg laying jars for Heliothis zea.

Slide 14.--Removal of eggs on paper strips.

Group	Ingredient	Quantity
I	Distilled H_2O (heated to boiling)	1200 ml
	4 M KOH	18 ml
	Casein	126 g
	Alfalfa meal (entomological grade)	54 g
	Sucrose	126 g
	Wesson salts mixture	36 g
	Alphacel	18 g
	Wheat germ	108 g
	Ascorbic acid	14.5 g
	Aureomycin (250 mg/capsule)	0.50 g
	Sorbic acid	4 g
	15% Methyl-*p*-hydroxybenzoate solution	35 ml
	10% Choline chloride solution	36 ml
	10% Formaldehyde solution	13 ml
II	Agar, granulated, dissolved in 2,500 ml boiling distilled H_2O	90 g
III	Vitamin solution A[b]	12 ml
IV	Vitamin solution B[c]	12 ml

[a] Prepare Group I in a 1-gal Waring Blendor equipped with a Powerstat; add the components in the order listed with the blendor operating at very slow speed. Add group II, cool to about 60–65°C, and add groups III and IV. Adjust Powerstat to maximum output and continue blending at slow speed for 2 min. Final *p*H of diet will be about 5.2.

[b] Vitamin solution A: nicotinic acid amide, 12.0 g; calcium pantothenate, 12.0 g; thiamine hydrochloride, 3.0 g; pyridoxine hydrochloride, 3.0 g; biotin, 0.24 g; vitamin B_{12}, 0.024 g; distilled H_2O, 1000 ml.

[c] Vitamin solution B: riboflavin, 6.0 g; folic acid, 3.0 g. Dissolve in solution of 2.24 g of KOH in 1000 ml distilled water.

Slide 15.--Artificial diet for rearing <u>Heliothis zea</u>.

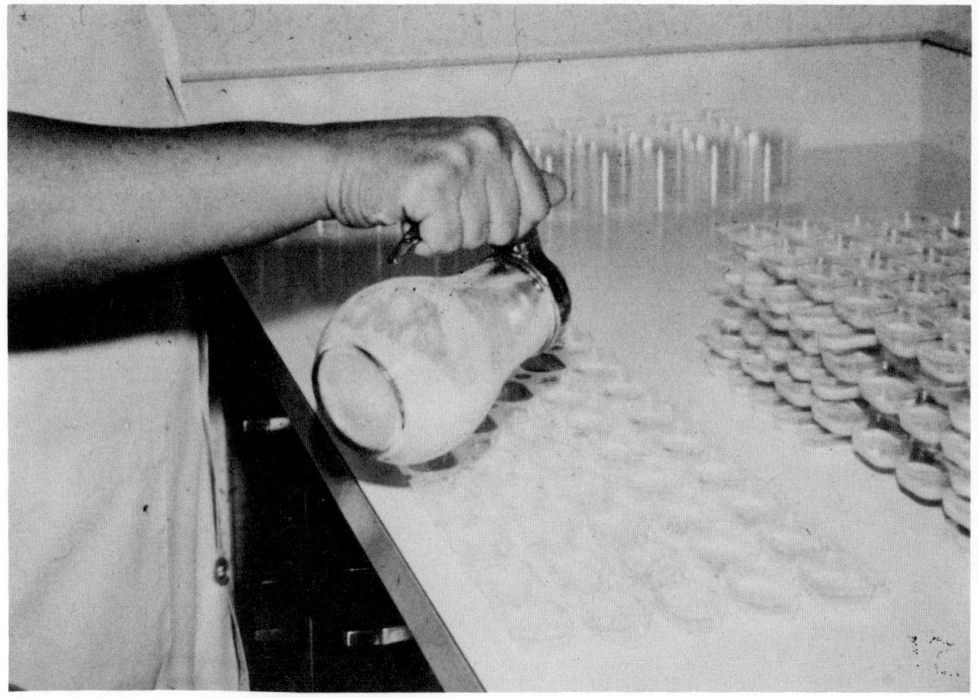

Slide 16.--Dispensing diet into individual cavities on tray.

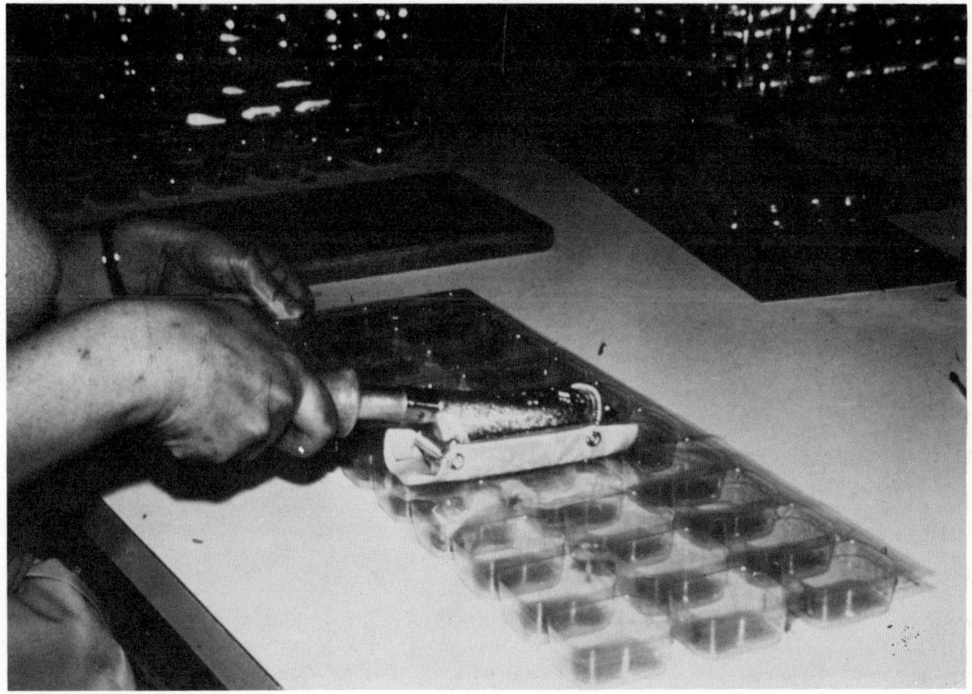

Slide 17.--Covering tray of cavities with plastic film.

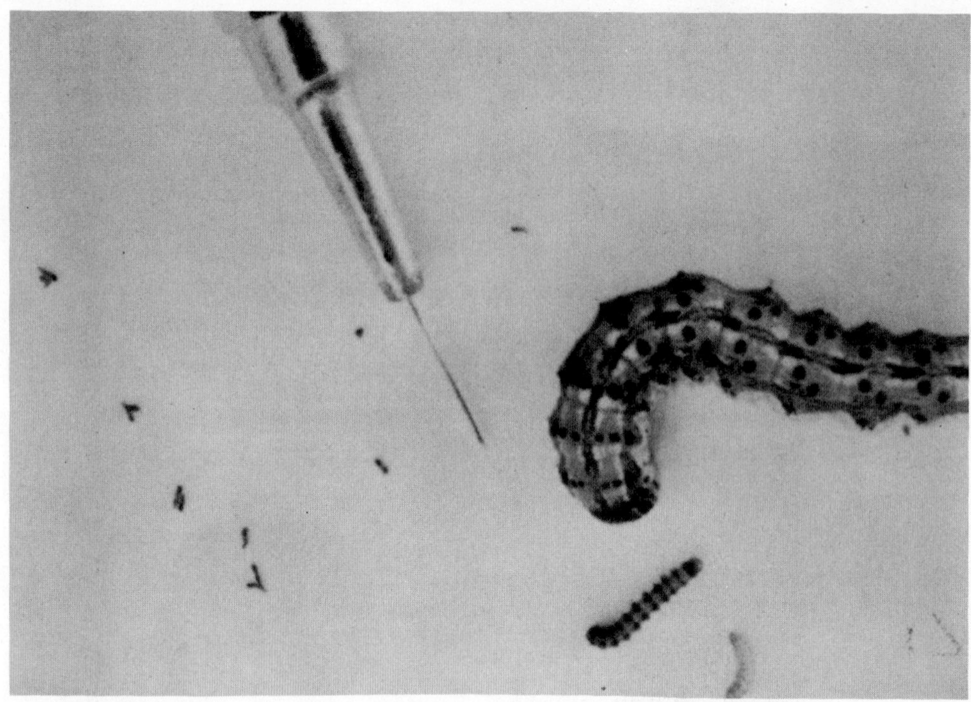

Slide 18.--Spectacular growth of Heliothis zea, one-, three-, five- and nine-day old larvae.

Slide 19.--Five-day-old larvae ready for infection with virus.

Slide 20.--Incubation chamber for trays of *Heliothis zea*.

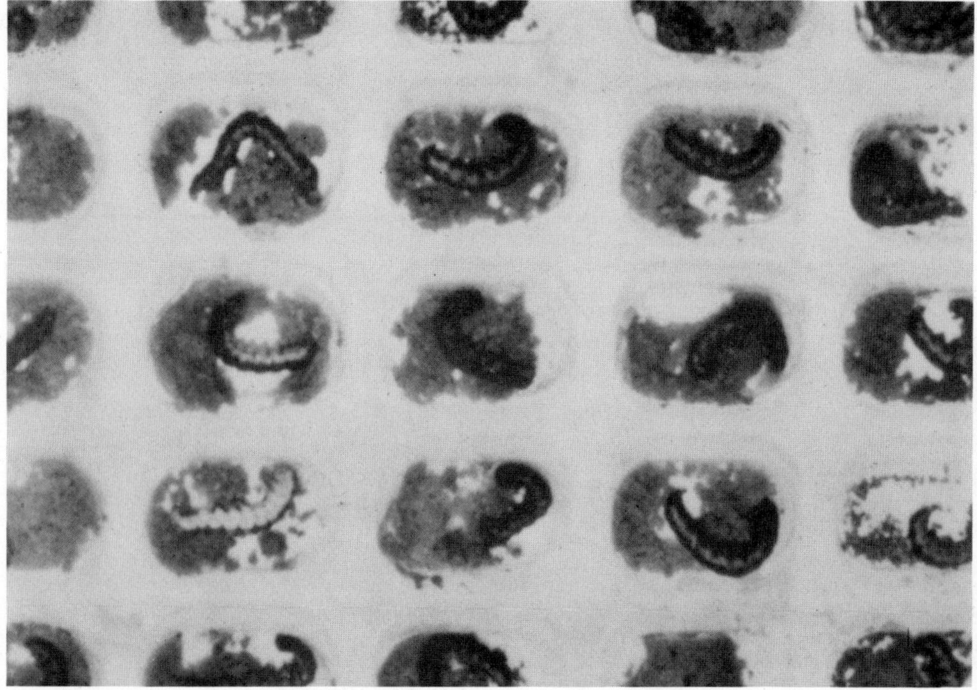

Slide 21.--Diseased larvae ready for harvest.

Slide 22.--Carbon replicas of PIB's, 15,000X.

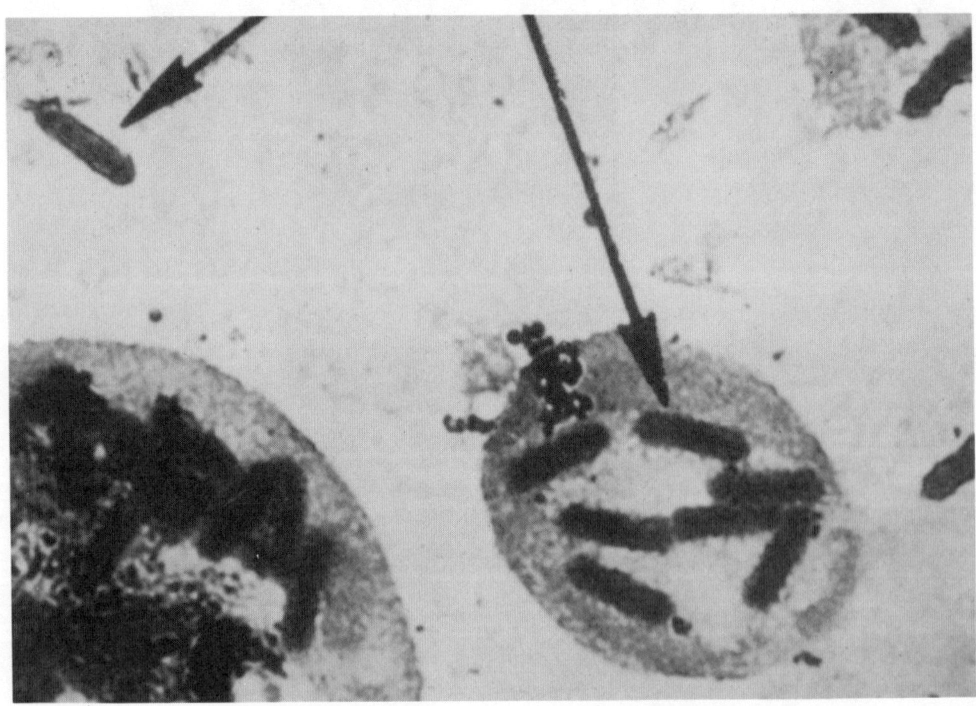

Slide 23.--Partially dissolved PIB's showing release of virions, 25,000X.

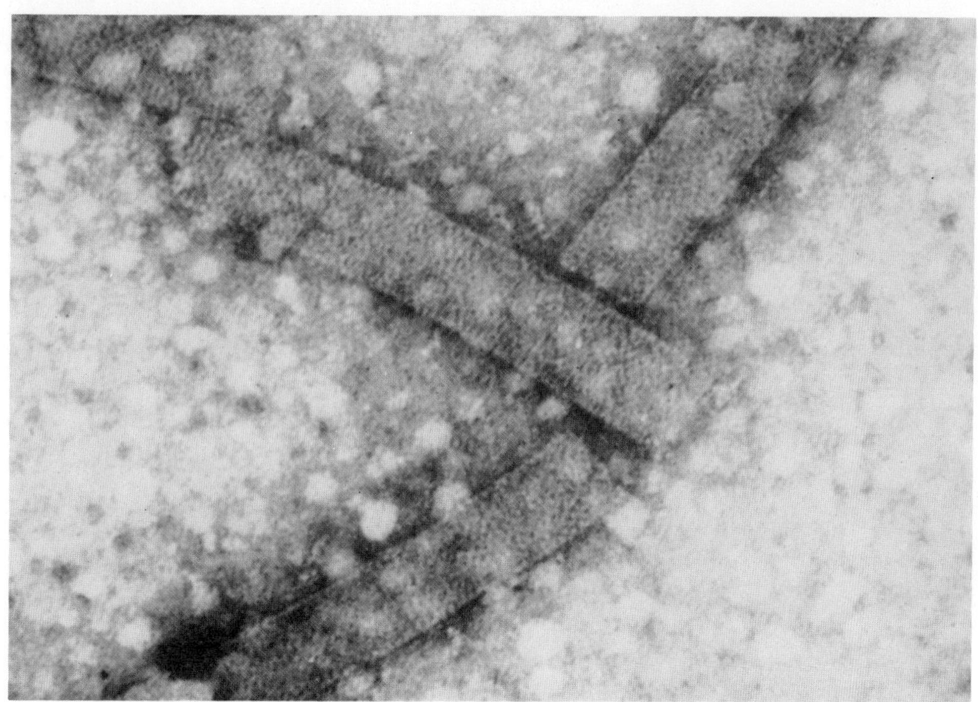

Slide 24.--Ultrastructure of Virion, 170,000X.

FOR EXPERIMENTAL USE ONLY USDA Temporary Permit 4456—Exp—11G

BIOLOGICAL INSECTICIDE FOR HELIOTHIS SPECIES...

- Cotton Bollworm,
- Tobacco Budworm

SOLUBLE POWDER

ACTIVE INGREDIENTS:

Polyhedral Inclusion Bodies of Heliothis Nuclear-Polyhedrosis Virus*	0.4%
INERT INGREDIENTS	99.6%
	100.0%

CAUTION: KEEP OUT OF REACH OF CHILDREN

Activity may be impaired by storage at temperatures above 90°F.

*Contains at least 4 Billion Polyhedral Inclusion Bodies per gram of product.

1200 GRAMS NET (1.2 Kilograms) International Minerals & Chemical Corporation

Slide 26 --Front panel - label for Viron/H.

MICROBIAL AGENT	STATUS	ACTIVE ENTITY	STANDARDIZATION
Beauvaria bassiana	Research	Spores, Toxin	Spores
Heliothis NPV	Development	Polyhedral Inclusion Body	PIB
Bacillus thuringiensis	Commercial	Crystals & Spores	Spores

Slide 25.--Active entities of three microbial agents.

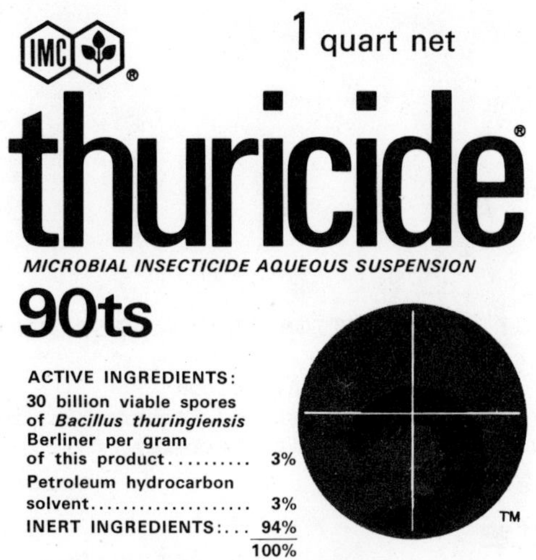

Slide 27.--Front panel - label for Thuricide 90TS.

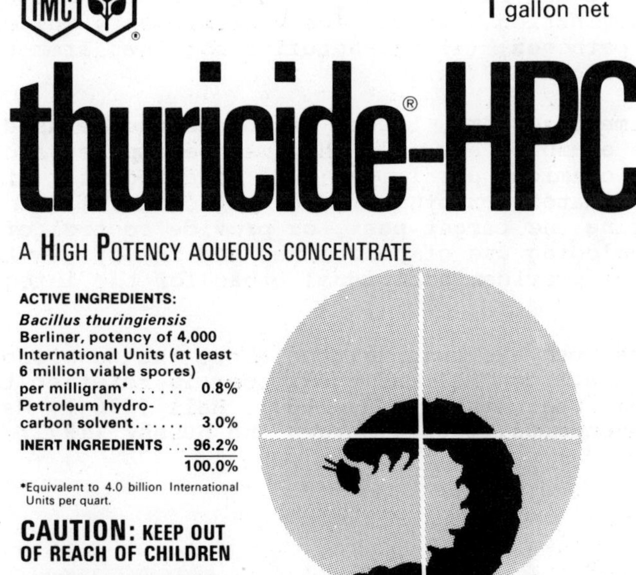

Slide 28.--Front panel - label for Thuricide HPC.

INSECT PATHOGENS: INTEGRATION INTO A PEST MANAGEMENT SYSTEM[a]

L. A. Falcon
Associate Insect Pathologist and Lecturer in Entomology
University of California, Berkeley, California

I. Pest Management System

An effective pest management system is one in which realistic economic injury levels are used to determine the need for insect control. In such a system all possible is done to protect and preserve naturally-occurring biotic mortality agents such as parasites, predators and pathogens. When artificial controls are needed they are employed in a selective manner and only when their use is economically and ecologically justified. The ultimate objective of the system is to produce maximum return (crop, comfort, recreation) at minimum cost with the least possible deleterious effect to the environment (Falcon, 1972).

The effective integration of insect pathogens into pest management systems requires many considerations: (1) thorough knowledge of the biology, ecology, phenology and behavior of the target insect to determine when a pathogen can be employed for maximum effectiveness; (2) the pathogen selected must be safe, easy to use, reasonably selective and sufficiently virulent to effectively control its host; (3) the method of dissemination must provide a persistent, uniformly distributed lethal deposit of the pathogen; (4) the benefits obtained from using the pathogen must justify its use (Falcon, 1971a).

In pest management systems, insect pathogens can be used in almost every conceivable way. Some examples are: (1) they may be applied alone, before, after, or in combination with chemical pesticides; (2) used together with released or native parasites and predators or other insect pathogens; (3) in sterile male programs, to aid in reducing the target pest, or provide control of other pests in the treated area. The developing use of pheromones and insect growth regulating hormones for insect control provides additional areas for the integration of insect pathogens.

In recent years much has been written about the development and use of insect pathogens for insect control and the reader is referred to these for additional information: Steinhaus (1963, 1964); Hall (1964); Martignoni (1964); Tanada (1964, 1967); Anonymous (1969); Burges and Hussey (1971) and Falcon (1971a).

---

a/In this paper, pest management and integrated control are used synonymously.

II. Practical Application

Since the pest management approach has only recently started its rise, there are few good case histories of such systems employing insect pathogens. Currently though, there is much interest and activity in this area in the United States and other parts of the world (see Burges and Hussey, 1971).

A. Bacillus thuringiensis in California.--A successful operation in California involves the integration of commercial products containing the bacterium *Bacillus thuringiensis* (Bt), into pest management systems for the control of Lepidoptera which defoliate trees and shrubs in roadside areas that are managed by the California State Division of Highways (Pinnock, 1971; Pinnock, et al., 1971). The major species are California oakworm, *Phryganidia californica* (Dioptidae); red-humped caterpillar, *Schizura concinna* (Notodontidae); and fruit tree leafroller, *Archips argyrospila* (Tortricidae). Other lepidopterous pests which have also been controlled with Bt as part of this program include tent caterpillars, *Malacosoma californica*, *M. constricta* (Lasiocampidae), and Western grapeleaf skeletonizer, *Harrisina brillans* (Zygaenidae).

A good example of the types of pest management programs being developed under the Highway project is the one against the red-humped caterpillar. This insect has a wide host range and several generations per year. Formerly, when chemical insecticides were used in its control, the Division of Highways applied many applications per year of Sevin® or Diazinon®. However, in developing the pest management program, Pinnock (1972) found that red-humped caterpillar larvae are heavily parasitized by two species of indigenous hymenopterous parasitoids, *Hyposoter fugitivus* (Ichneumonidae) and *Apanteles schizurae* (Braconidae). He also found that (1) the populations of these parasitoids were greatly reduced in the spring when chemical insecticides were applied for the control of aphids feeding on pyracantha; (2) due to the destruction of the parasitoids, the red-humped caterpillar population increased rapidly and required repeated treatments with chemical pesticides during the spring and summer months; and (3) this resulted in continuous suppression of parasitoids and other natural enemies throughout the period when red-humped caterpillar was active.

In the pest management program, a mixture of green soap and water is used to control aphid on pyracantha in the spring. The soap mixture greatly reduces the aphid population, but enough survive to attract and serve as food for predaceous insects such as ladybugs (*Coccinellid* spp.), and lacewing (*Chrysopa* sp.). Once established, these predators efficiently regulate the aphid populations and further controls are not required until summer. At this time the natural enemy populations decline in abundance and the red-humped caterpillar larval population increases rapidly. At this time the one and only application of Bt is made and the red-humped caterpillar population is reduced below damaging levels.

More recently studies have shown that the parasitoid adults need to feed on flower nectar to enhance mating, reproduction, and longevity. While nectar sources are abundant in the spring, they disappear in the summer, and it is this factor which leads to the disappearance of the parasitoids. A search is now underway to locate plant species which flower during the summer and thus maintain parasitoids through this period.

B. Metarrhizium anisopliae in Brazil.--An integrated control program in sugarcane and pastures is under development in northeast Brazil (Guagliumi, 1971). The program is directed at the control of nymph and adult populations of several species of froghoppers (*Mahanarva posticata*, *M. fimbriolata* in sugarcane; and *Deois schach* and *Zulia entreriana* in pasture) and it involves (1) chemical insecticide applications of benzene hexachloride (BHC) or aldrin; (2) cultural control by burning sugarcane straws and pastures; (3) biological control with mass-rearing and liberations of a hymenopterous egg-parasite, *Acmopolynema hervali* (Mymaridae); and dipterous nymph-predators (*Salpingogaster nigra* and *S. pygophora*) (Syrphidae) and (4) microbial control with mass-culture and dissemination of the entomogenous fungus, Metarrhizium anisopliae.

The fungus is obtained by culturing it on rice which was previously autoclaved at 120°C. The spores thus produced are collected and sprayed on froghopper infested sugarcane at 500-1000 grams in 500 liters water per hectare. One to 2 hectare areas in a plantation are treated with the objective of establishing a foci from where the fungus can spread to the untreated areas. The fungus is spread into surrounding areas by wind, infected froghopper adults, parasites, predators, insectivorous birds and field workers. According to Guagliumi (1971), M. anisopliae has become the most important mortality factor regulating populations of froghoppers in sugarcane in northeast Brazil.

C. The Role of Entomogenous Fungi in Nicaragua.--Cotton pest management programs have been under development in Nicaragua since 1968 (Falcon, 1971b). In studies conducted to determine the components of the cotton ecosystem, entomogenous fungi were found to be the dominant insect mortality factor in the months of October and November (Figure 1). The species thus far identified are: Aspergillus flavus, Entomophthora sp., Penicillium sp. and Spicaria rileyi. S. rileyi appears to be the most abundant.

The initial appearance of the fungi coincides with the end of the rainy season, the disappearance of beneficial insects, and the start of the period of most intensive chemical pesticide usage. Unknowingly then, chemical pesticides have been integrated with the naturally-occurring entomogenous fungi to regulate pest insect numbers. Now that there is an awareness of what occurs in this ecosystem it is possible that the naturally-occurring fungi can be made more effective by (1) disseminating field-collected or laboratory cultured fungi earlier in the year (August - September), thus accelerating their period of activity and increasing effectiveness; or (2) the application of pathogens such as Bt, and insect viruses to augment or otherwise enhance the effectiveness of the fungi.

Of additional interest is the possibility that the fungi are responsible for the cataclysmic reductions which occur in the populations of parasitoids and predators. This may result either through direct infection of the insects by the fungi, or indirectly, as the fungi infect and destroy the hosts used as food by the parasites and predators.

D. Granulosis Virus of Codling Moth.--Several years of field evaluations in California with a granulosis virus of codling moth (Laspeyresia pomonella) have shown it to be effective in reducing populations of this cosmopolitan pest of deciduous fruits (Figure 2) (Falcon, et al., 1968; Falcon, 1971a). Similar results have been obtained with this virus tested in Victoria, Australia (personal communication, D. S. Morris, Victoria Plant Research Institute). Currently a major disadvantage to using the granulosis virus for control of the codling moth is the need of very frequent applications (4-to-5-day intervals).

However as pest management programs are developed there will be many opportunities to integrate the virus for control of the codling moth. One such example would be to integrate it with the sterile male technique. In such a program the virus could be used to lower populations of the target insect, thus reducing the quantity of sterile insects required to obtain control. Unlike sterile male insects, the virus can be produced in the laboratory at any time and stored until needed.

In the field studies conducted in California it was found that the virus interfered with the survival of parasitoids of the codling moth, these being a Dipteran, Lixophaga variabilis (Tachinidae) and a Hymenopteran, Ascogaster quadridentatus (Braconidae). The major interference caused by the virus appeared to be destruction of the host before the parasitoids were able to complete their development, resulting in marked reduction in the abundance of these parasitic species (Figure 3).

E. Nuclear Polyhedrosis Viruses.--Considerable effort has been devoted to the integration of insect pathogens into cotton pest management programs in California (Falcon, 1971a, 1972; Van den Bosch, et al., 1971).

The respective nuclear polyhedrosis viruses of the cabbage looper, the beet armyworm and the western yellow-striped armyworm occur naturally in cotton-growing areas of California and are important in the population regulation of their respective hosts. Collected in the field, or mass-cultured in the laboratory, and disseminated at the proper time, the viruses have provided consistent, effective control of their respective hosts. Commercial formulations of these viruses are urgently needed at this time.

In an effort to encourage the use of these viruses, the 1972 Pest and Disease Control Program for Cotton, published by the University of California, provides specific instructions for collecting and using the viruses:

> Polyhedroses are naturally-occurring insect diseases which can be important in suppressing populations of cabbage looper, beet armyworm, and western yellow-striped armyworm. In evaluating need for control, every effort should be made to determine if disease is present in the field. If diseased worms are found, they may signal an impending general epidemic and chemical controls may not be necessary. Apparently each insect species is susceptible to a specific polyhedrosis and therefore a disease which infects one species will not infect another. Infected larvae usually show few distinctive symptoms until about 2 or 3 days before death at which time they become somewhat sluggish and turn yellowish or pale; they also may swell slightly and become limp and flaccid. Shortly before and after death the integument is fragile and easily ruptured; when ruptured it emits the liquefied body contents composed of polyhedra and disintegrating tissue. Dead larvae eventually dry to dark brown or black and are commonly found hanging from the host plant. Polyhedroses may kill larvae of any size but are most lethal to the younger ones--occasionally, death may not occur until the pupal stage. The polyhedra can be spread by infected larvae or moths of affected species and by wind, rain, birds, other insects, and activities of man.
>
> Polyhedroses may also be spread by using diseased caterpillars from naturally infested fields. To prepare infectious material for field distribution diseased worms are suspended in water (pH 6.0 to 8.0) and the suspension is homogenized in a blender or allowed to putrefy under cool, dark conditions.
>
> This preparation can be stored (45°F) in the dark until needed. For application, the material is coarse-filtered to remove large insect parts and a wetting agent added (again being certain the pH of the mixture is between 6.0 and 8.0). The following equivalents of diseased caterpillars should be applied per acre: cabbage looper, 10 large worms; beet armyworm and yellow-striped armyworm, 25 large worms. Any type of conventional spray equipment can be used. For best results, applications should be made when larvae are small.

In contrast to the aforementioned viruses the nuclear polyhedrosis virus of *Heliothis* (bollworm virus) does not appear to occur naturally in the San Joaquin Valley of California. However, laboratory-cultured experimental and commercial formulations have been tested in the field since 1964 with promising results (Falcon, 1971a). The virus has provided reductions in worm populations comparable to chemical insecticides (Table 1), none however (virus or chemical insecticides) give economic control of bollworm on cotton in the San Joaquin Valley. Consequently the University of California 1972 Pest and Disease Control Program for Cotton does not recommend treatments of any kind for bollworm.

Only in recent years with the efforts to develop microbial, macrobial and chemical controls, has the actual status of bollworm been determined. Some of the major factors which appear to influence bollworm abundance and damage to cotton include: (1) climatic conditions, especially temperatures; (2) the moon; (3) numbers of fertile egg-laying female moths present; (4) abundance and effectiveness of parasitoids and predators; and (5) growth and fruiting patterns of the cotton plant.

When moths are present and active, mating and oviposition are greatly reduced during the light of the full moon (Figure 4). This effect appears to begin one night before the moon is brightest and ends two nights later. After full moon, oviposition increases and usually reaches a peak around new moon. The intensity of each period and the actual times when they occur may be influenced by many factors; temperatures, host plant condition and the effectiveness of predators appear to be the most important. In the San Joaquin Valley, predators may destroy 80% or more of the bollworm eggs and hatching larvae (Van den Bosch, et al., 1969).

Knowledge of the growth and fruiting pattern of the cotton plant is important to developing an understanding of the bollworm situation. In the San Joaquin Valley, Acala SJ-1 variety cotton is in the fruit formation period for about nine weeks from early June to early August (Figure 5 (Falcon, 1972). The squares formed June 1 to July 10 produce the flowers of June 25 to August 6, and the plants normally have set 80% or more of their boll-carrying capacity by early August and 100% by mid- to late August. Squares, flowers and small bolls, however, continue to be produced into late September, but these late-produced fruiting parts do not contribute to the harvest.

Putting together the plant growth and moon information, it is readily apparent that the San Joaquin Valley bollworm populations first appear after the full moon which occurs sometime in mid to late July or early August, depending on the year. By this time, much of the harvestable cotton crop is set. However, an abundance of squares, flowers and small bolls are still being produced and the developing bollworm population feed almost totally on these extra fruiting parts and not on the harvestable bolls. In this system it appears important to maintain vigorous plant growth and bud formation in the bollworm period, because without this, the larvae might attack and destroy a high percentage of the harvestable bolls.

Research undertaken to integrate bollworm virus into the cotton pest management system has resulted in showing that the bollworm seldom needs to be controlled in the San Joaquin Valley. Although this has not led to the use of bollworm virus it demonstrated the benefits which can be derived from undertaking such studies, and has set the stage for future work.

III. Current Research

In the microbial control research program in California much attention is being given to the application of insect pathogens in an effort to find simple, efficient, effective and inexpensive ways to apply them.

One approach, designated the "auto-contamination" method, is to attract adults of a target species to a source where they are contaminated with a pathogen. The adults are then released and spread the pathogen to (1) untreated adults in the process of mating; (2) eggs as they are oviposited; and (3) the plant or other substrates where the immature stages feed. Nontarget insects which are attracted to the auto-contamination site also pick up and aid in disseminating the pathogen but with no harm to them because the pathogens used are selective in their host range. This procedure is being tested in cotton in the San Joaquin Valley using specially adapted ultraviolet light traps and the nuclear polyhedrosis viruses of the bollworm, cabbage looper and beet armyworm.

Another method of application under investigation is the use of ultra-low volume, wind-drifted, aerosol applications of insect pathogens delivered by cold fogger-generator type equipment. The key to success appears to be the use of non-evaporative carriers which provide necessary buoyancy to the small droplets, adhesion on impact to leaf surfaces, and protection of the pathogen from desiccation and solar irradiation. Crude cottonseed oil appears to be an excellent carrier for nuclear polyhedrosis viruses and Bacillus thuringiensis preparations.

Table 1.--Effectiveness of various mortality agents in reducing populations of small-sized bollworm larvae on cotton in the San Joaquin Valley of California[1 2 3]

Agents	Percent worm kill
I. Beneficial insects[4]	
Geocoris pallens	50
Nabis americoferous	78-89
Chrysopa carnea	41-84
Natural complex	66-93
II. Chemical insecticides[5]	
carbaryl (4 lb.)	25-40
methyl parathion (1 lb.)	45-50
Gardona (1 lb.)	50-65
Lannate (0.5 lb.)	40-56
III. Microbial pathogens[6]	
Thuricide HPC (1.5 qt)	20-40
Viron H + UV filter (40-200 le)[7]	21-50[8]
Biotrol VHZ + UV filter (50-200 le)	20-50

[1] Compiled from information given in 1972 Pest and Disease Control Program for Cotton, published by California Agricultural Experiment Station and Extension Service;

[2] Heliothis zea;

[3] Small larvae are under 1/2" in length;

[4] Conducted in replicated small cage tests;

[5] Pounds of active ingredients in parenthesis;

[6] Thuricide contains Bt; Viron H and Biotrol VHZ contain nuclear polyhedrosis virus of Heliothis;

[7] le stands for larval equivalent (6 x 10^9 PIB);

[8] Addition of 1% Buffer-X to the mixture in the spray tank significantly improved worm kill.

Literature Cited

Anonymous. 1969. Microbial control of insects. pp. 165-195. In Principles of Plant and Animal Pest Control. Vol. 3. Insect Pest Management and Control Nat. Acad. Sci. Publ. No. 1695. Washington, D.C. 508 pp.

Burges, H. D. and N. W. Hussey (eds.). 1971. Microbial Control of Insects and Mites. Academic Press, New York. 861 pp.

Falcon, L. A. 1971a. Microbial control as a tool in integrated control programs. Chap. 15 In Biological Control (C . B. Huffaker, ed.). Plenum Press, New York. 511 pp.

Falcon, L. A. 1971b. Progreso del control integrado en el algodon de Nicaragua. Revista Peruana de Entomologia 14(2): 376-378.

Falcon, L. A. 1972. Integrated control of cotton pests in the far west. Proc. 1972 Beltwide Cotton Res. Conf. Natl. Cotton Council of America: 80-82.

Falcon, L. A., W. R. Kane and R. S. Bethell. 1968. Preliminary evaluation of a granulosis virus for control of the codling moth. J. Econ. Entomol. 61: 1208-1213.

Guagliumi, P. 1971. Lucha integrada contra las cigarrhinhas (Homopt.: Cercopidae) en el noreste del Brazil. Revista Peruana de Entomologia 14(2): 361-368.

Hall, I. M. 1964. Use of microorganisms in biological control. Chap. 21 In Biological Control of Insect Pests and Weeds (P. DeBach, ed.). Reinhold Publ. Co., New York. 844 pp.

Heimpel, A. M. 1971. Safety of insect pathogens for man and vertebrates. Chap. 22 In Microbial Control of Insects and Mites (H. D. Burges and N. W. Hussey eds.). Academic Press, London. 861 pp.

Martignoni, M. E. 1964. Mass production of insect pathogens. Chap 20. In Biological Control of Insect Pests and Weeds (P. DeBach, ed.). Reinhold Publ. Co., New York. 844 pp.

Pinnock, D. E. 1971. Microbial control of lepidopterous larvae. Research Rept. RH 71-8. State of California Business and Transportation Agency, Department of Public Works, Division of Highways, Sacramento.

Pinnock, D. E. 1972. Microbial control of *Schizura concinna* (Lepidoptera: Notodontidae) on ornamental trees in California. Paper delivered at 14th International Congr. of Entomol., Canberra, Australia. 12 pp.

Pinnock, D. E. and J. E. Milstead. 1971. Control of the California oakworm with *Bacillus thuringiensis* preparations. J. Econ. Entomol. 64(2): 510-513.

Steinhaus, E. A. (ed.). 1963. Insect Pathology, An Advanced Treatise. Vols. 1, 2. Academic Press, New York. 661 pp and 689 pp, respectively.

Steinhaus, E. A. 1964. Microbial diseases of insects. Chap. 18 In Biological Control of Insect Pests and Weeds (P. DeBach, ed.). Reinhold Publ. Co., New York. 844 pp.

Tanada, Y. 1964. Epizootiology of insect diseases. Chap 19 In Biological Control of Insect Pests and Weeds (P. DeBach, ed.). Reinhold Publ. Co., New York. 844 pp.

Tanada, Y. 1967. Microbial pesticides, pp. 31-88. In Pest Control: Biological, Physical and Selected Chemical Methods (W. W. Kilgore and R. L. Doutt, eds.). Academic Press, New York. 477 pp.

van den Bosch, R., T. F. Leigh, L. A. Falcon, V. M. Stern, D. Gonzales and K. S. Hagen. 1971. The developing program of integrated control of cotton pests in California. Chap. 17 In Biological Control (C. B. Huffaker, ed.) Plenum Press, New York. 511 pp.

van den Bosch, T. F. Leigh, D. Gonzales and R. E. Stinner. 1969. Cage studies on predators of the bollworm in cotton. J. Econ. Entomol. 62: 1486-1489.

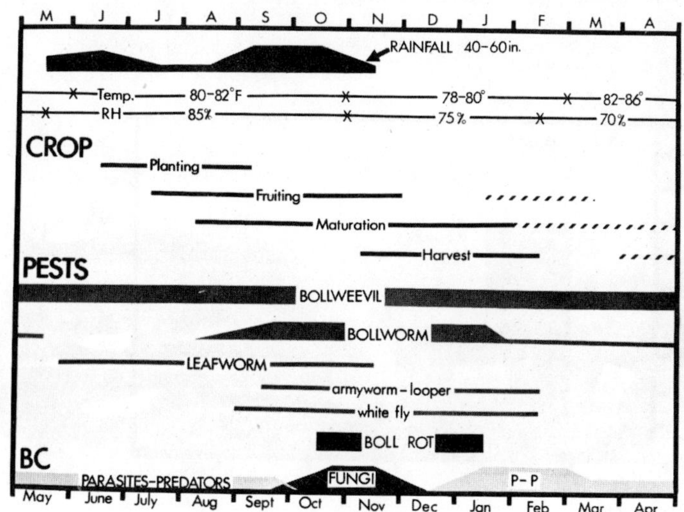

Figure 1.--A general scheme of the cotton agroecosystem in Nicaragua depecting the periods of activity, some interactions and the relative importance of the major components. "BC" refers to Biological Control and shows the periods of activity of insect parasitoids and predators and the entomogenous fungi discussed in the text.

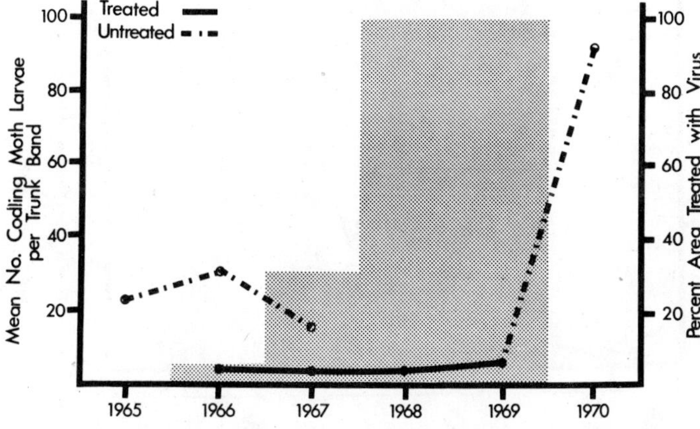

Figure 2.--Impact of a granulosis virus on survival of codling moth (*Laspeyresia pomonella*) larvae as determined by the trunk band sample method.

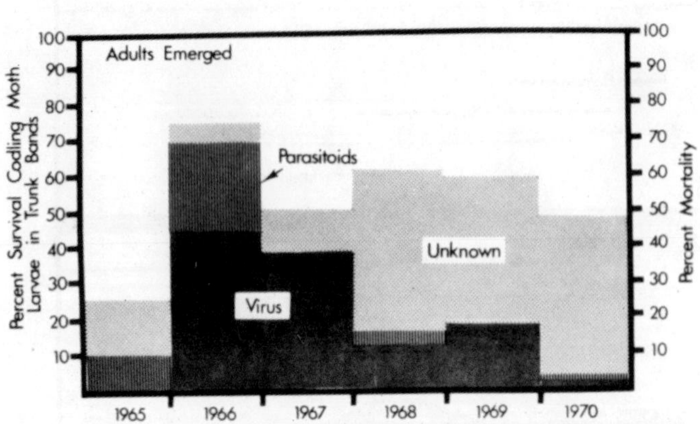

Figure 3.--Fate of codling moth (Laspeyresia pomonella) larvae collected from trunk bands and held under laboratory conditions until death of adult emergence.

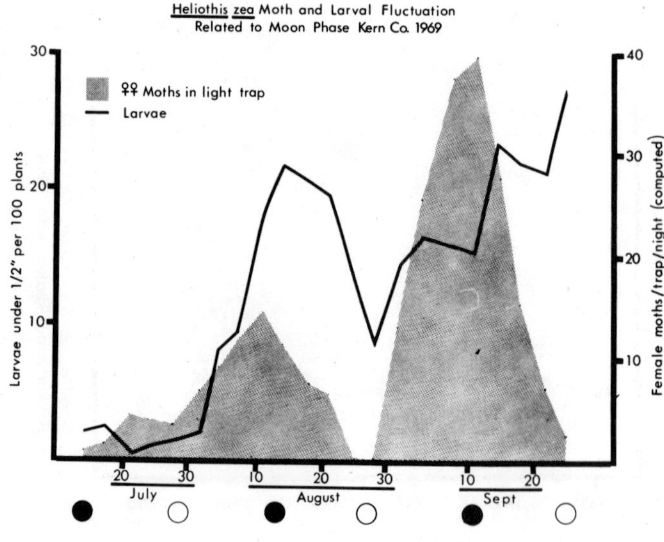

Figure 4.--Bollworm (Heliothis zea) moth flight activity and larval density pattern relative to the lunar cycle in the San Joaquin Valley of California during 1969.

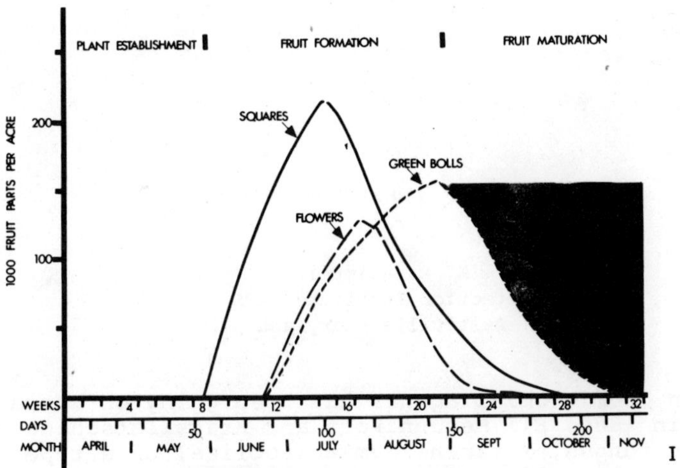

Figure 5.--A general scheme of the growth and fruiting patterns for Upland cotton variety Acala SJ-1 in the San Joaquin Valley of California.

SAFETY OF INSECT PATHOGENS

A. M. Heimpel
Plant Protection Institute, ARS, USDA
Beltsville, Maryland

The production of microbial agents has been going on sporadically for some time past. Starting in 1880, Eli Metchnikoff and Krassilstschick produced a wide-spectrum, entomogenous fungus, Metarrhizium anisopliae, on a type of beer mash for control of the sugarbeet curculio, Cleonus punctiventris. There is no record of any safety tests done by these scientists and I believe that such experiments never occurred to them.

Another product under the tradename "Sporeine," produced by a French firm Laboratoire Libec, 26 Rue d'Allerary, 15 Arrondissement, Paris, France, was used to control the meal moth, Anagasta kuehniella, in ground grains. To my knowledge no extensive safety experiments were carried out on this product.

Bacillus popilliae, a pathogen of the Japanese beetle, was isolated and described by Dutky (1937, 1940) and this scientist took out a public patent describing a method of control of the Japanese beetle using B. popilliae in 1941. In 1942, Dutky applied for a second public patent on a method of production of a product based on the bacterium.

Before launching into extensive dissemination programs using the bacterium, Dutky contacted the Federal Bureau of Entomology and Plant Quarantine, the Public Health Service, the Bureau of Plant Industry, the Bureau of Animal Industry and the Apiculture Division. His presentation to these organizations included his feeding tests using starlings and chickens, showing that the bacterium did not germinate in the bird gut and remains viable during passage through the gastro intentinal passage. He described the highly specific conditions required to germinate and grow the bacterium and showed that the human body temperature, 98.6°F is higher than the top growth temperature, 96°F, for this bacterium. Of course the body temperatures of farm animals are higher than man's normal temperature.

His arguments proved sufficient for those times and there ensued a federal-states cooperative work that disseminated some 200,000 pounds of this bacterial product on 194,000 sites in 14 states on the eastern seaboard and the District of Columbia.

The product is now made commercially by the Fairfax Biological Laboratory in Clinton Corners, New York, and registered under the tradename Japanese Beetle DOOM.

In 1968, the Pesticide Regulation Division withdrew the registration for B. popilliae for use on pastures and since then safety tests and health studies of production personnel have been carried out. Petition for registration should be submitted this year.

From this point on, the registration of a microbial product became more formal, governed by the regulations laid down under the Federal Insecticide, Fungicide and Rodenticide Act signed into law on June 25, 1947. This law is now enforced mainly by two Divisions in the relatively new Environmental Protection Agency (EPA) namely, the Pesticide Regulation Division (PRD) and the Pesticide Tolerance Division (PTD). Other government organizations are involved as consultants and these are the various EPA groups, Microbiology, Pharmacology and Toxicology Pesticide Registration, etc., and others including the ARS, Department of Interior, National Cancer Institute, Universities and State experimental stations, and so forth.

It might be useful here to review the various requirements for registration of a microbial agent. This information is partly drawn from Ignoffo's paper presented at the National Meeting of the Entomological Society of America in Los Angeles, 1972. Although there are general requirements, each case may differ, and this refers particularly to safety tests; protocols are still being developed and are in evolution as it were.

A petition for registration of a microbial agent should consist of several sections:

Section A contains the following information:
1. The product's tradename, identity and composition.
2. A description of the organism.
3. A description of the production method.
4. A list of the target insects and plants to be protected.

Section B contains information of the efficacy of the product:
1. Recommended doses on each target insect on each crop.
2. Efficacy data.
3. Since exemption from tolerance is required the data on residue limitations and cut off dates supplied in this section for a chemical pesticide are usually not required.

Section C
1. Data concerning specificity and safety of the microbial agents for animals and plants.

Section D - standardization
1. Methods of bioassay.
2. Counting methods.
3. Serological techniques for identification of the organism and for residue studies.

Section E deals with levels of residue and should be included; however, a statement that there is no intention or need to remove residues should be included.

Section F contains a statement on the tolerance sought, as well as a listing of crops for which tolerance is sought.

Section G contains an explanation of the potential usefulness and advantages of the microbial agents over other pesticides and the reasons for developing them.

The purpose of this paper is to review the work done on the safety of microbial agents. However, let us digress long enough to study the list of required protocols for establishing exemption from tolerance for microbial agents as of this date, bearing in mind that these protocols are not necessarily fixed and may be altered to accommodate for the type of microorganism.

The following list may apply to nuclear polyhedrosis viruses (see Table I).

A standard dose has been tentatively approved by EPA for use in the tests given in Table 1. The dose is based on the amount of the microbial product applied to 100 acres as it applies to man (average weight, 70 Kgms). For example, suppose that the dose per acre is 35 larval equivalents (L.E. - 6 x 10^9 polyhedra) for good insect control. Then a 100-acre dose would be 100 x 35 x 6 x 10^9 = 21 x 10^{12} polyhedra. Based on a man's weight the dose per kilogram would be 21 x 10^{12}/70 = 3 x 10^{11} polyhedra per Kg. The dose for a 20 gm mouse would then be 3 x 10^{11}/1000 x 20 = 6 x 10^9 polyhedra per mouse. Should the technical virus be too much for a 20 gm mouse to handle by stomach incubation the dose could be reduced to a 10-acre dose/70 Kg man and this would be .6 x 10^9 polyhedra/mouse or *circa* 30 x 10^9 virions per mouse. This is still a respectable dose.

Included in this presentation are the protocols that have been approved by EPA to test the safety of the *Autographa californica* multiple embedded virus (MEV). It is incumbent upon me to emphasize that in each case, the protocols must be reviewed by PTD and PRD of EPA and are subject to revision.

Please note in Table 2, that the requirements to test for storage and replication of the virus in treated animals is fulfilled by bioassaying tissues taken from 2 males and 2 females on days 0, 1, 3, 7, 14 and 21. Blood samples taken from this test prepared as a sera is to be used to detect viral antibodies in test animals.

In Tables 3-6, please note that treated tissues will also be examined for implication of replication of the virus. Table 7 gives the protocol for the 90-day feeding test and serological examination of the blood of the chronically exposed animals after 3 months, aids in requirements to test for the presence of antibody against the virus in treated animals.

After these tests are completed officials in PRD and PTD of EPA will review the results and indicate any further tests necessary. Of course, basic protocols for sub-acute tests and long-term tests are now available but should be discussed with scientists in PRD and PTD before starting testing.

With this introduction to the requirements for registration of a microbial control agent, we can now review our knowledge of the safety of microbial pathogens for plants and animals. I will not go into extreme detail here since these are to be found in several recent presentations by Heimpel (1971); Ignoffo (1972) and Cantwell, et al. (1972).

We have briefly covered the status of *Bacillus popilliae* previously. One of the first insect control agents registered in the USA is *Bacillus thuringiensis*. Fermentation experiments with this organism were initiated in 1956 by the Pacific Yeast Products Company (later to become Bioferm Corporation). Other companies that also registered products were Rohm and Hass, Merk, Sharpe and Dohme, Grain Processing Company and Nutrilite Products, Incorporated. Recently, Abbott Laboratories has registered a new *B. thuringiensis* product. In 1958, a temporary exemption from tolerance was granted by FDA and PRD issued an experimental permit for testing *B. thuringiensis* on vegetable and forage crops. Finally on 14 April 1960 the FDA granted full exemption from tolerance and registration was accomplished.

This registration certainly set a precedent, however, despite extra tests on a variety of animals, the requirements were virtually the same as previously described.

Acute inhalation, allegicity and oral tests were carried out on mice, guinea pigs and rats, respectively. Acute inhalation and oral feeding tests were also carried out on human volunteers. Both acute and chronic feeding studies in chicks, laying hens, young swine and hogs, fish, adult and larval honey bees. In addition, mice and guinea pigs were injected with massive doses of *B. thuringiensis* vegetative cells. In all cases mentioned above the organism was proven harmless.

Since B. thuringiensis resembles Bacillus cereus very closely, Dr. Steinhaus was concerned that the organism might contain "masked pathogens" such as isolated by Brown, et al. (1958) from populations of B. cereus. Accordingly, Brown and his colleagues tested varieties thuringiensis alesti; sotto; entomocidus and subtoxicus using identical isolation techniques to those used with B. cereus and could not isolate any strains pathogenic to mice.

The first virus Heliothis zea nuclear polyhedrosis (NPV) petitions to PRD were presented by International Minerals and Chemical Company by Nutrilite Products, Incorporated in 1968. The safety data used in these petitions included the results from the six basic acute toxicity pathogenicity tests previously described, the sub-acute toxicity pathogenicity tests listed in Section B of Table 1, including the teratogenicity and carcinogenicity tests. Tests listed in Sections E and F were also completed. These were the required tests.

There were several other tests carried out on H. zea NPV as well as other viruses between 1965 and 1971 that I am sure provide supporting evidence. These included skin sensitization tests on guinea pigs, intracerebral, intravenous, and intraperitoneal injection of virus into mice, inhalation and acute oral tests of mice and guinea pigs without any adverse effects on the test animals (Ignoffo, and Heimpel, 1965; Ignoffo, 1968). Human feeding tests were conducted; however, these were not acceptable to FDA for technical reasons (Heimpel and Buchanan, 1967). Test of four NPV's from Trichoplusia ni, Spodoptera exigua, Spodoptera frugiperda, H. zea and a granulosis virus from the salt marsh caterpillar Estigmene acrea, were carried out under USDA contract at Rosner-Hixson laboratories in 1966-1967. These were briefly reported by Heimpel (1971). The NPV from H. zea was also tested in several mammalian tissue culture lines without damage to or cytological changes in the cells (Ignoffo and Rafajko, 1972). Chautani, et al. (1968) showed that 2 hours incubation of H. zea NPV in human gastric juices killed all the virus. After temporary exemption from tolerance was granted, IMC carried out a study on monkeys injected with the virus. Although there was some swelling of the lymph nodes in the treated animals, it can be assumed that the reaction may have been the result of subclinical infections of the monkeys. No teratogenic responses were obtained in mammals tested with Heliothis nuclear polyhedrosis virus (Ignoffo, et al., 1972). Tests with the Heliothis polyhedrosis virus in other systems including plants, vertebrates and invertebrates are summarized by Ignoffo (1968, 1972) and by Greer, et al. (1971).

Other viruses tested, along with those mentioned above, were the NPV from T. ni (Heimpel, 1966) and the non-inclusion, densonucleosis virus from Galleria mellonella (Giran, 1966); these tests included intraperitoneal injection of live virus suspensions into adult mice, rabbits and newborn mice, subcutaneous injections in rabbits and intracerebral injections in newborn mice. The treated animals were unharmed by the virus.

Dr. D. W. Roberts (personal communication) has tested the Entomopox virus from Asmacta moorei (produced in the salt marsh caterpillar, E. acrea) in mice by intracerebral and intraperitoneal injection. All of the treated animals are still normal after one year's observation.

Apparently the insect virus types tested are safe for other life forms.

Protozoa

The microsporidia affecting insects will likely be the most numerous of protozoa agents selected for insect control. However, microsporidia pathogenic for insects apparently don't harm vertebrates.

The microsporidia have been found causing diseases in rodents (Nosema and Thelohania); in the common toad, Bufo bufo (Plistophora) in fresh and salt water fish (Nosema, glugea and Plistophora); and in Crustacea (Plistophora); and in man (Encephalitozoon) (Heimpel, 1971).

Fungi

Of all the fungi isolated from insects there are relatively few groups or species that have been seriously considered as microbial agents for insect control, these are the Entomophthoraceae, the Laboulbeniales, Beauveria bassiana, Metarrhizium anisopliae and Hirsutella thompsonii.

Entomophthora coronata has been detected causing phycomycosis in man and horses (Heimpel, 1971).

Georg, et al. (1962) reported B. bassiana the causative agent of pulmonary mycosis in the giant tortoises Tesudo elephantopus and T. gigantea elephantina and B. bassiana has also been isolated from vertebrates (MacLeod, 1954).

Fairly extensive feeding inhalation intravenous and subcutaneous injection experiments have been carried out using rats by Schaerffenberg (1968) with B. bassiana and M. anisopliae. No adverse effects were detected in non-stressed animals. The tests reported by Schaerffenberg would not be adequate to prove safety for establishing exemption from tolerance in the United States. Indeed, the carcinogenicity tests are required for fungi as previously mentioned.

There have been several allergic reactions to B. bassiana spore dust reported (Muller-Kogler, 1965). These reactions were never severe but are a warning to people who are especially sensitive. The wearing of masks and protective clothing is highly recommended.

Rickettsiae

Rickettsiae affecting insects will likely never be used for insect control for two reasons--the Rickettsia kill very slowly and they have been proven pathogenic for mammals (Heimpel, 1971).

Conclusions

Apparently, microbial agents, particularly the viruses and spore forming bacteria can be manufactured in adequate quantities, cheaply enough to compete with chemical pesticides, both from the efficacy point of view and from the aspect of quality control.

Furthermore, these microorganisms are completely safe for distribution in the environment without harm to any other life form except the target insect(s). This is a remarkable improvement over the toxic chemical and should excite the environmentalist. Why then have we not made progress over the last three decades? The answer is somewhat complicated. First, the financial support has been borderline when compared to investments in certain chemical insecticides. Secondly, the academic support has been very poor although it is improving. It is my contention that an entomologist is not properly trained until he has had an elementary microbiology course as a prerequisite to a course in insect pathology. One of the first major problems a graduate entomologist encounters in his first job is disease in his insect cultures.

Indeed, many entomologists know so little about insect diseases and asceptic techniques in rearing that they usually do not recognize chronic disease and may ignore frankly diseased insects. Many of the past studies in physiology, fecundity, etc., have been carried out using diseased populations of insects.

With enthusiastic support from the academic world a large number of jobs in insect pathology would be available and graduate research would help supply some of the much needed background to develop microbial control.

Table 1.--Requirements for safety evaluation of microbial agents.

A. Required in all cases:
Acute Toxicity
1. Acute oral (rats or another rodent).
2. Acute dermal (rabbits).
3. Inhalation (rats).
4. Eye irritation (rabbits).
5. Primary skin irritation (rabbits).
Sub-acute Toxicity Test
1. 90-day feeding (rats and one other species).

B. These requirements depend on the results of acute testing and the nature of the formulation.
Sub-acute Toxicity.
1. Twenty-one-day dermal (rabbits).
2. Fourteen-day inhalation (rats).
3. Skin sensitization (Landsteiner allergenic test - guinea pigs).
4. Respiratory sensitization - allergenicity test (guinea pigs).
5. Reproductive and teratogenicity studies (rats).

C. Long-Term Studies (same stipulation as in B).
1. Carcinogenicity studies (These will definitely be required for fungi).

D. Other Related Studies (Required).
1. Storage and replication of the biological agent in mammalian species.
2. Antibody studies.
3. Mutation studies.
4. Chemical and biological controls to ensure uniformity of product.

E. Beneficial Insect and Wildlife Toxicology (Required).
1. Acute toxicity - honey bees
2. LC_{50} - fish (rainbow trout and bluegills).
3. 8-day LC_{50} birds (quail and mallards).
4. Sub-acute and/or chronic bird toxicity (including reproduction).

Table 2.--Acute oral toxicity - rats.

Duration: 21-day observation period.
Animals: One hundred and twenty weanling albino rats (60 males and 60 females) will be randomly assigned to the experimental groups listed below. The animals will weigh between 150 and 250 grams at initiation of the test. All animals will be individually housed with food and water available ad libitum.

Group No.	No. of animals		Test material
	Male	Female	
1	20	20	saline control
2	20	20	lactose carrier + saline
3	20	20	*A. californica* + lactose + saline

Pathogen Administration: Acetone precipitated lactose-virus polyhedra slurries, dried *in vacuo*, will be resuspended in 0.8% saline at the rate of 40×10^9 polyhedra per ml.

This dosage will be administered to each animal in Group No. 3 at the rate of 1.0 ml per rat by gastric intubation. The remaining animals will receive 1.0 ml of the appropriate control by gastric intubation.

Table 2.--continued.

Observations: Daily observations for appearance and mortality will be made. Body weights and food consumption determinations will be made weekly. Body temperature of each animal will be recorded twice daily for the duration of the experiment.

Clinical Studies: Blood glucose and serum glutamic-pyruvic transaminase will be determined on five males and five females from each group at days 0 and 21. The following hematological parameters will be determined on five males and five females from each group at days 0, 14 and 21:

Hemoglobin	erythrocyte count
microhematocrit	total leukocyte count
coagulation time	differential leukocyte count
prothrombin time	thrombocyte count

Interval Sacrifice: Four animals from each group (two males and two females) will be sacrificed on the following days: day 0 (immediately after dosing, day 1, day 3, day 7, and day 14. Six animals from each group (three males and three females) will be sacrificed on day 21. One ml of serum from each animal will be frozen for pickup by the sponsor. Additionally, the following tissues will be removed, stored in sterile containers, and reserved for the sponsor:

mesentery	mesenteric lymph node
liver	gastro-intestinal contents from various sections along the GI tract.

Note: These tissues must be removed with sterile instruments. The same instrument must not be used to remove more than one of the above organs. Possible cross-contamination by the instruments must be prevented.

Termination: The study will be terminated at 21 days and necropsies will be performed. The following procedure will be followed for any animal which dies during the experiment and on all survivors at termination:

Organ weights: for each rat

heart	adrenals
lung	thyroid
liver	prostate
kidneys	uterus
gonads	pituitary

Portions of these tissues plus portions of the following will be frozen and delivered to the sponsor:

duodenum	cervical lymph node
intercostal muscle	bronchial lymph node
pancreas	bone marrow
urinary bladder	stomach
mesenteric lymph node	brain

Additionally, eyes will be taken, grossly observed, preserved in Zenker's fixative, and delivered to the sponsor.

Histopathological examination - The following tissues will be examined from any animal that dies and from three males and three females from each group sacrificed at day 21:

fat	spleen	cervical lymph node
heart	lung	mesenteric lymph node
liver	salivary gland	bronchial lymph node
kidneys	small intestine	

All slides will be delivered to the sponsor at completion of the study.

Report - At termination a report will be submitted, giving:

experimental design	hematology data
body weight data	clinical chemistry data
food consumption data	gross necropsy findings
mortality data	microscopic examination results

Table 3.--Acute dermal toxicity - guinea pigs.

Duration: 14-day observation period.
Animals: thirty albino guinea pigs (15 males and 15 females) will be randomly assigned to the experimental groups listed below. The animals will be individually housed with feed and water freely available.

Group No.	No. of animals		Test material
	Male	Female	
1	5	5	Lactose control
2	5	5	*A. californica* virus polyhedra
3	5	5	*A. californica* ME virus (released rods)

Pathogen Administration:
Group No. 1: The vehicle control (2% lactose in 0.8% saline) will be administered in the manner described for the experimental groups.
Group No. 2: Acetone precipitated lactose-virus polyhedral slurries, dried *in vacuo*, will be resuspended in 0.8% saline at the rate of 40 x 10^9 polyhedra per ml. This dosage will be applied to the shaved backs of 10 guinea pigs on two areas of intact skin and two areas of abraded skin on each animal at the rate of 0.10 ml per application (Draize's technique).
Group No. 3: Freed virus rods (by the carbonate technique), prepared by the sponsor will be used. The free virus from 40 x 10^9 clean polyhedral (43.5 mgms) will be suspended per 1.0 ml of saline and applied to the test animals in the same manner as the polyhedra (using Draize's technique).
Observations: Animals will be examined daily for the first week and at termination on the 14th day. Animals will be observed daily for mortality. Body weight and food consumption will be determined weekly. Body temperature will be recorded for each animal daily for the first 7 days and on the 14th day.
Termination: The study will be terminated on the 14th day and gross necropsies performed. The following will be performed on all animals that die during the experiment and on all survivors at termination:

Organ weights: for each animal

heart	adrenals
lung	thyroid
liver	prostate
kidneys	uterus
gonads	pituitary

Portions of these tissues plus portions of the following tissues will be frozen and shipped to the sponsor:

duodenum	cervical lymph node
intercostal muscle	bronchial lymph node
urinary bladder	bone marrow
pancreas	stomach
mesenteric lymph node	brain

In addition, eyes will be taken, grossly observed, and preserved in Zenker's fixative prior to delivery to the sponsor.

Histopathological Examination: The following tissues will be examined from any animal which dies during the study and from three males and three females from each group sacrificed on day 14:

fat	lung	bronchial lymph node
heart	salivary gland	small intestine
liver	cervical lymph node	mesenteric lymph node
kidneys	spleen	

All slides will be delivered to the client at completion of the study.

Report: At completion of the study a report will be submitted giving:

experimental design	results of dermal exposure
physical appearance	mortality data
body weight data	gross necropsy observations
food consumption data	results of microscopic examination

Table 4.--Eye irritation test - rabbits.

Duration: 14-day observation period.

Animals: Twenty albino rabbits (10 males and 10 females) will be randomly assigned to the following groups. The animals will be individually housed with food and water freely available. Prior to initiation of the study all animals will be examined for corneal injury by instillation of 5% fluorescein solution and flushing the eye with distilled water 20 seconds after application. The animals will then be examined with a band slit lamp. Any animal with damaged cornea will be excluded from the study.

Group No.	No. of animals		Test material
	Male	Female	
1	5	5	ME virus in saline suspension
2	5	5	Freed virus in saline suspension

Pathogen Administration:

Group No. 1: Acetone precipitated lactose-virus polyhedral slurries, dried *in vacuo*, will be resuspended in 0.8% saline at the rate of 40 x 10^9 polyhedra per animals. This dosage will be applied to the left eye of each animal at the rate of 0.1 ml per animal (4.35 x 10^6 polyhedra per eye). The right eye will serve as a control and will receive 0.1 ml of a 0.8% saline solution containing 2% lactose.

Group No. 2: Freed virus rods (by carbonate technique), prepared fresh by the sponsor, will be used. The free virus from 43.5 mg of clean polyhedra will be suspended in 1.0 ml of saline. The left eye of each rabbit will receive 0.1 ml of this suspension. The right eye will serve as the control and receive 0.1 ml of the 0.8% saline solution.

Observations: Examination for injury or irritation will be made at 24, 48, and 72 hours after treatment, and at 7 and 14 days. Eye irritation will be graded and scored according to the method of J. H. Draize. Body temperature will be recorded for each animal daily for the first 7 days and at weekly intervals thereafter.

Necropsy: Should damage be present at 14 days, the animal should be sacrificed and the eye and conjunctivae preserved in 10% neutral buffered formalin for histopathological examination. All slides should be sent to the sponsor at completion of the study.

Report: At completion of the study, a report will be submitted giving:

experimental design	tabulation of eye scores
ocular findings	microscopic findings.

Table 5.--Primary skin irritation - rabbits.

Duration: 72-hour observation period.

Methods: Six albino rabbits, clipped free of hair, will be divided into two equal groups.
The exposure area of one group will be abraded and the skin of the remaining group will remain intact. Acetone precipitated lactose-virus polyhedra (*A. californica*), dired *in vacuo*, will be resuspended in 0.8% saline at the rate of 40 x 10^9 polyhedra per animal; 0.5 ml of this material will be introduced under a 1" x 1" gauze patch which will be secured in place with adhesive tape.
The animals will be immobilized in stocks and the entire trunk of the animal will be wrapped with a non-absorbent binder for 24 hours.

Observations: After the 24-hour exposure, the patches will be removed and the skin reactions will be evaluated. A second evaluation will be made 48 hours later (72-hour observation). The reaction will be scored according to the method of J. H. Draize. Body temperatures will be recorded for each animal daily.

Report: The report will include the following:
- details of experimental design
- rating of the irritancy according to the Federal Hazardous Substances Act.
- tabulated scoring of skin reactions

Table 6.--Acute inhalation toxicity - rats.

Objective: The purpose of this test is to determine the toxicity and pathogenicity of a single one-hour exposure of rats to a single concentration of *A. californica* ME virus.

Animals: Forty SPF albino rats obtained as weanlings (23-26 days of age) from Charles River Breeding Laboratories will be used after a 7-day acclimation period at HLT in an air-conditioned holding room.

Groups: Two groups, each consisting of 10 males and 10 females, will be formed by random selection from the original 40. Group No. 1 will be exposed to the experimental material at 40 x 10^9 polyhedra per animal and Group No. 2 will be exposed to 2% lactose powder as a control.

Exposure Conditions: Both groups will be exposed to their respective materials in 100-liter glass and stainless steel inhalation chambers of cubical design with pyramidal tops and bottoms. The materials will be drawn into the chambers from the top by vacuum at the bottom at a rate of 10 liters per minute of airflow. The rats will be individually housed in stainless steel mesh cages in two layers during exposure. A Wright dust feeder situated at the top of each chamber will feed the material to be generated into filtered and dehumidified makeup air at 72°F to provide a constant concentration of material in air during the one-hour exposure after equilibration.

Materials Needed: A quantity of pathogen (suitably mixed with the carrier) to provide the desired dosage/hour will be required from the sponsor. Assuming a delivery concentration in air of 10 mg/liter and a flow rate of 10 l/minute through the chamber, 12 gm of material will be generated in one hour. Thus, approximately 15 gm of material should be supplied. Also, 15 gm of control material will be needed.

Post-Exposure Observations: The two groups will be housed in separate rooms in individual metal cages for 14 days after the termination of exposures. The animals will have continuous access to water. Each animal will be observed daily for signs of toxicity and any moribund animal will be sacrificed for necropsy immediately. Total food consumption will be determined by weight for individual animal weekly by preweighing and post-weighing the food remaining in the food bin.

Necropsies: All animals will be necropsied after 14 days (sacrificed with barbiturate overdose) or upon death. The following organs will be weighed:

heart	lung	liver
kidneys	gonads	adrenals
fat	thyroid	prostate or uterus
pituitary		

Portions of these organs plus portions of the following tissues will be grossly examined and preserved in 10% formalin:

duodenum	intercostal muscle
urinary bladder	pancreas
mesenteric lymph node	cervical lymph node
bronchial lymph node	bone marrow
stomach	brain
spinal cord	

In addition, eyes will be taken, grossly observed and preserved in Zenker's fixative. All preserved tissues shall be sent to the Insect Pathology Laboratory, Entomology Building A, ARC, Beltsville, Maryland 20705.

Histopathology: The following tissues will be examined from any animal dying and from three males and three females from each group on days 4 and 21:

brain	spinal cord
fat	heart
liver	kidneys
spleen	lung
salivary gland	small intestine
cervical lymph node	mesenteric lymph node
bronchial lymph node	

All slides shall be sent to the Insect Pathology Laboratory, Beltsville, Maryland.

Report: A final report will be submitted, describing the methodology and results in detail, within 30 days after completion of the histopathological examination.

Table 7.--Sub-acute dietary administration - rats.

Duration: 13-week observation period.

Animals: Forty healthy, young albino rats (Sprague-Dawley) will be selected and placed in the following experimental groups by stratified randomization:

Group No.	No. of animals		Test material
	Male	Female	
1	10	10	2% lactose
2	10	10	*A. californica* ME virus in 2% lactose

The animals will be individually housed in screen bottom cages with feed and water available ad libitum.

Pathogen Administration: Acetone precipitated lactose-virus polyhedral slurries, dried in vacuo, will be mixed into the diet at a level specified by the sponsor. The pathogen will be administered in the diet for the 13-week observation period. Fresh diets will be prepared weekly.

Observations: Body weights and food consumption will be recorded weekly. Daily observations for mortality will be made and weekly records will be maintained of appearance, behavior, and signs of toxic or pharmacologic effects. Body temperature will be recorded for each animal daily for the first 7 days, and at weekly intervals thereafter.

Clinical Studies: The following observations will be made on five males and five females from the control and test groups:

Hematology: at 45 and 90 days:
- erythrocyte counts
- leukocyte counts
- hemoglobin

Urine Analysis (pooled samples) - at 45 and 90 days:
- specific gravity
- glucose
- pH
- blood (presence or absence)
- total protein

Interval Sacrifice: Four animals (two males and two females) from each group at 45 days.
Termination: The study will be terminated at 90 days.
Necropsy Procedure: The following will be performed at each sacrifice:
Gross examination: on all animals
Organ weights: all animals:
- liver
- kidney
- heart
- gonads
- spleen

Histopathological Examination: The following tissues should be processed:

brain	colon	kidney
thyroid	adrenal	heart
lung	pancreas	stomach
salivary gland	liver	spleen
duodenum	uterus	bone marrow
jejunum	prostate	ovaries
lymph nodes	testes with epididymis	

Report: The report will include the following:
- experimental design
- general appearance and behavior
- effects on growth, body weight, food consumption, and survival
- signs of toxic or pharmacologic effects
- clinical findings
- gross and microscopic necropsy findings.

References

Brown, E. R., M. D. Moody, E. L. Treece and C. W. Smith. 1958. J. Bateriol. 75: 499-509.

Cantwell, G. E., T. Lehnert and J. Fowler. 1972. Amer. Bee J. (in press).

Chautani, A. R., D. Murphy, D. Claussen and C. S. Rehnborg. 1968. J. Invertebr. Pathol. 12: 145-147.

Dutky, S. R. 1937. Ph.D. Thesis. 113 pp., Rutgers Univ., New Brunswick, N. J.

Dutky, S. R. 1940. J. Agri. Res. 61: 57-68.

Dutky, S. R. 1941. U.S. Patent 2, 258, 319.

Dutky, S. R. 1942. U.S. Patent 2, 293, 890.

Georg, L. K., W. M. Williamson, E. B. Tilden and R. E. Getty. 1962. Sabouradia, J. Int. Soc. Hum. Anim. Mycol. 2: 80-86.

Giran, F. 1966. Entomophaga 11: 405-407.

Greer, F., C. M. Ignoffo, and R. F. Anderson. 1971. Chemtec, June. 342: 347.

Heimpel, A. M. 1966. J. Invertebr. Pathol. 8: 98-102.

Heimpel, A. M. 1971. In Microbial Control of Insects and Mites (Burges and Hussey, eds.), Chap. 22 - 269-489. Academic Press, London.

Ignoffo, C. M. 1968. Bull. Entomol. Soc. Amer. 14: 265-276.

Ignoffo, C. M. 1972. Bull. Entomol. Soc. Amer. (in press).

Ignoffo, C. M. and R. R. Rafajko. 1972. J. Invertebr. Pathol. (in press).

Ignoffo, C. M. and A. M. Heimpel. 1965. J. Invertebr. Pathol. 7: 329-340.

MacLeod, D. M. 1954. Canadian J. Bot. 32: 818-890.

Muller, Kogler, E. 1965. In Pilzkrankheiten bei Insekten, p. 130. VerlagPaulParey, Hamburg, Berlin.

Schaerffenberg, B. 1968. Entomophaga 13: 175-182

INFORMATION AND ASSISTANCE IN THE USE OF MICROBIAL AGENTS

John D. Briggs
Entomology Department
The Ohio State University
Columbus, Ohio

Introduction

The infectious diseases of insects caused by one or a combination of the major groups of microorganisms (fungi, bacteria, nematodes, protozoa, and the viruses) have been specifically treated both as entities and in terms of their utilization by other contributors in the Institute. Constructive critics state that biological control efforts based on the use of microorganisms are precariously balanced on too few successes to provide firm guidelines to the person working in crop protection. I agree that the number of practical, viable, regularly employed, and consistently operating biological control programs utilizing microorganisms is limited to no more than you can itemize on one hand. This situation does not reflect the amount of information available and applicable to a microbiological control effort. The lack of frequent field use is often the result of a conventional, insect control decision-making process, where the criteria for the selection of a control agent are those used for chemical application, e.g., short-term results, plot-by-plot killing of insects for "control," and local action rather than biotype or regional crop protection planning.

We have benefitted for more than a quarter of a century from the superb, spectacular and effective research, development, production and sale of chemical insecticides. I should add at this point, in order that the interpretation of my remarks are correct, that the usefulness and the value of chemical pesticides should not be diminished for service to society because their uses have been criticized and the consequences of some uses were not predictable.

Earlier speakers in this Institute have referred to four commercial products that are based upon microorganisms and labeled for sale. These products represent the efforts of industry capitalizing upon the characteristics of two kinds of bacteria and a virus. There are other products based upon virus pathogens and fungus pathogens of insects that will be available for use. Now we are limited to registered products for use that are formulated upon *Bacillus popillae*, *Bacillus thuringiensis* and a "nuclear polyhedrosis virus." The microorganisms serving as the active ingredients for those products were not synthesized by a biologist or a chemist standing at a bench. On the contrary, they were detected by a biologist working in the field with the particular host insect population. Beyond that point the commercial development and subsequent availability to the public of microbial agents is not different from those circumstances that assure the development, production, and availability of a conventional chemical insecticide. I must emphasize the significance of this fundamental difference between microbiological and conventional chemical insecticidal products. For the microbial agents that may be candidates as active ingredients in a biological control product, you and I as teachers, field biologists, and entomologists who have contact with insect populations are the individuals who can detect and bring attention to the biological agent as it occurs in insect populations. We will be effectively initiating the development of a product.

Diagnosis and Information

Concerning the diagnosis of insect diseases, it is evident from the literature that to 1940 approximately, the principal efforts devoted to the microbiological agents affecting insects were directed to the protection of beneficial insects from disease, particularly silkworms and honey bees. Governments, commercial, and non-profit organizations concerned with beneficial insects employed the skills of microbiologists, particularly bacteriologists or other scientists to assist in meeting the challenge of infectious diseases of beneficial hosts. A prominent example was the enlistment of Dr. Louis Pasteur for the silkworm industry in France. In the United States, the Department of Agriculture provided opportunities for bacteriologists in investigations of the diseases we recognize as "foulbroods" and other infectious diseases of bees. We can consider the pre-1940 period as "the past" which was characterized by information sources and assistance in the detection and diagnosis of infectious diseases of insects centered upon individuals, generally microbiologists, who were specifically charged with the protection of a particular kind of beneficial insect.

The "present" or post-1940 period was first identified with the detection and subsequent commercial development and availability of the Japanese beetle milky disease organisms (*Bacillus popillae* and *B. lentimorbus*). Additional industrial/commercial sources of information have developed in a manner similar to that which we expect for chemical insecticides. Technical and popular literature is available from four companies in the United States.

Fairfax Biological Laboratory
Clinton Corners
New York 12514 (*Bacillus popillae*, *B. lentimorbus*)

Crop Protection Department
P.O. Box 1489 Nuclear Polyhedrosis Virus,
Homestead, Florida 33030 (*Bacillus thuringiensis*)

Agricultural and Veterinary Products Division
Abbott Laboratories
North Chicago, Illinois 60064 (*Bacillus thuringiensis*)

Thompson-Hayward Chemical Company
P.O. Box 2383
Kansas City, Kansas 66110 (*Bacillus thuringiensis*)

The rules "read the label," and follow label recommendations, with cognizance of state and federal regulations pertinent to your area are as important for microbiological agents as they are for chemical insecticides in answering: "What can be used and how?"

In contrast to the period prior to 1940, the diagnoses of insect diseases are not predominantly from individuals who specialize in the diseases of a single species of insect. The diagnoses are from laboratories in land-grant universities, federal agencies, and non-profit private institutions (e.g., Boyce-Thompson Institute), where there are groups of individuals principally trained as entomologists who serve as general practitioners for sick insects of a variety of species. The region in which the insect pathologists are located may limit the kinds of host insects that they examine; however, it is expected by those who seek their services that they will recognize the nature of an infectious agent that is inevitably believed to cause the death of the insect(s). I use the terms "inevitably believed" because animals that die of starvation or dehydration, are paralyzed by venom of Hymenoptera, are run over by automobiles or stepped on, suffer insecticidal poisoning or are victims of trauma, are often expected by those who submit the specimen(s) to have as cause of death a specific and conspicuous microorganism. Without taking into consideration all of the physical factors that may have brought the life of an insect to an end, we should appreciate that the single "pure-culture" infection of a host may be rare. The host-parasite association existing at the peak of an epizootic disease in an insect population is the result of the "upper-hand" achieved by a microorganism that may cause the death of many individuals and consequently the collapse of the population. An example is the effect of a nuclear-polyhedrosis virus infection on cabbage loopers. For the individual insect pathologist/biologist, the results of his observations on a disease may be published for our information in a matter of months or years following the study of the materials brought to his attention. Today the conventional written communication frequently forces us to accept this delay.

In the present period of our history, entomologists have successfully employed the general diagnostic services of several laboratories around the world. In the United States, the Pioneering Insect Pathology Laboratory in the USDA Agricultural Research Center at Beltsville, and the Laboratory of Insect Pathology at the University of California at Berkeley have until recently provided two distinctive focal points for the diagnosis of insect diseases. Without active diagnostic services on a regional, national, and international scale, the individual cannot be assured of those services necessary to help himself, or society. Without a continuing source of new infectious agents that affect insects, commercial development, and distribution of microbial based insecticides will not be stimulated.

With the foregoing brief description of the past, and the present situation with respect to sources of information and assistance for use of microbiological agents, prognostications are in order. Provided that the present industrial/commercial efforts can remain viable we should expect, based on past performance, the same outstanding high quality information and assistance from them concerning specific microbiological insecticidal agents. The availability of diagnostic services for determining what is wrong with a population of insects based on individual specimens, as I have indicated in my earlier discussion, is not satisfactory and we must take steps immediately to improve the situation. A single individual or group of individuals in a laboratory located on a continent the size of North America cannot be expected to handle the volume of materials that should be processed, if we are to take seriously the use of microorganisms for the biological control of insects. Regionalization of large or faunistically diverse geographical areas for purposes of insect disease diagnoses is a direction in which we should move to efficiently and effectively provide immediate information on the microbiological agents which are affecting insect populations. Regions should provide service on an international basis with the appropriate inter-communication. Regionalization can be implemented for biotypes of host plants or distribution of animals rather than according to geographic or political boundaries. The World Health Organization (WHO) has designated an International Reference Center for Diagnosis of Diseases of Vectors (IRC) essentially a global diagnostic center. Diagnostic activities at the Center, in the Ohio State University, Columbus, Ohio, are associated directly with Vector Biology and Control in WHO and
7 WHO research units throughout Asia and Africa. The International Rererence Center is therefore specialized for diagnostic activities with respect to biological agents affecting vectors. To date the accessions are principally mosquitoes and other biting flies parasitized by microorganisms. The responsibilities of the IRC include solicitation of specimens from all parts of the globe, purposefully limiting the diagnostic activities to invertebrate vectors of human disease. The information generated by the IRC is provided directly to the unit or individual initiating accession of the specimens and the compiled information is published on an annual basis by the World Health Organization (Briggs, 1965-

1970). The lag time for public availability of the results of diagnosis is evident, although much shorter than we generally experience in the conventional process of a publication. Regardless of the way in which diagnostic activities are organized, the global concern today for knowing what kind of biological agents have been detected, and are currently being detected in hosts (including both agricultural and public health invertebrates) necessitates the development of an information network, of which the diagnosis of invertebrate diseases is the central element. The network should draw upon a data bank which can be updated with information as it is available, and that may be queried by individuals in any part of the world. A data bank must include notice of those detections of biological agents that are not generally evident because they are never published, in addition to advance notice of information intended for publication. The occurrence of a virus infection in insects in Tupelo forests in Alabama could be of importance to scientists in Ceylon; or the knowledge of protozoan parasites in a species of noctuid in Rhodesia should be available to scientists in Mexico with a minimum of lag time. The Society for Invertebrate Pathology, an international organization, is taking steps to implement the coordination and computer processing of information to assure its availability throughout the world. The first projects include virus diseases of insects, and biological agents affecting populations of mosquitoes. The technology for implementing a System for the Pathology of Invertebrates (SPI) is available by electronic data processing, and access to data banks by telephone lines and satellite communications.

Accurate information and assistance is available for the use of microbiological agents when employing a registered product. With the profit incentive clearly in evidence, the manufacturer will be interested in the results of the application of the product, and the user can depend upon the quality of the products and the results of the applications. It is this direct manufacturer-user relationship that has been well established for the chemical pesticides, particularly herbicides and insecticides. The essential element to assure the availability to society of certain microbiological agents is a strong industrial/commercial contribution. These contributions can be accelerated by the opportunity for industry to develop effective microbiological agents perhaps found by you. I am not advocating the industrial/commercial institutions as the only basis for the availability of effective microbiological control products and information to meet minimum governmental regulations, however, a competitive capitalistic system provides the incentives to apply the best industrial skills to research, development, production and marketing. A parallel effort must be the on-site exploitation of infectious diseases of insects in specific ecosystems, for example, in forests (from tropical to alpine) or forage crops, where an industrial development of a microbiological agent is not technically or economically feasible.

Detection of Diseased Insects

Field biologists or entomologists are faced with detecting or recognizing abnormal individuals in an insect population. Once recognized we must take avenues available to us for disease diagnosis, and determination of the microbiological agent (or agents) responsible for the abnormal condition. Sufficient data must be provided upon which a decision can be made for the abandonment or further investigation of the microbiological agent. What are the routes available to the field biologist for diagnostic services, what can the biologist do to capitalize upon infectious microbiological agents in insect populations?

At the present time, recalling my brief review of the history of information sources, we do not have a conspicuous series of centers for diagnosis of insect diseases, nor is an information network available that would permit an insect biologist to make an inquiry concerning a particular kind of agent or agents. We lack adequate services that will help him determine the nature of the abnormality in the host population in question. Two simple steps can be identified at this moment for assisting the individual who detects an abnormal condition in an insect population that is thought to be caused by an infectious agent. First, it is essential that you bring the circumstances concerning the abnormality in the insect population to the attention of the individual or group of individuals who have research responsibility for insects of that particular crop. Second, attempt to deliver specimens of diseased and normal insects to a diagnostic facility. For example, in central Mississippi, the entomologists in the Boll Weevil Research Laboratory or Department of Entomology would be the group to whom you would bring the abnormal specimens of insects. Often in land-grant institutions there are entomologists who have specific charges for "insect control," on particular kinds of crops, e.g., forage crops, small grains, ornamentals, etc. They are key people!

With your name and address, abnormal living and/or dead specimens of the insect in question for examination should be accomplished with concise and accurate information:

a. Identity of affected insects;
b. Host of insect, or habitat;
c. Where was it found on the host or in the habitat;
d. When was it found;
e. % estimate of host insect affected;
f. Crop protection history of collection site.

Specimens should be handled in the following manner:

a. Single dead individuals in closed glass or plastic shell vials, without preservative
b. Protect from heat;
c. Transfer by efficient route, hand carry or air mail;
d. Observe legal restrictions on movement of living insect.

Small containers, e.g., glass or plastic shell vials, are satisfactory for living or dead individuals. Do not place them in alcohol or other preservatives. Paper sacks with the mouth securely fastened with paper clips or rubber bands are satisfactory for groups of dead or dying specimens on samples of foliage. The important point is to get specimens and information to an entomologist who is familiar with and is working with the insects in question. This action accomplished two things--it brings the insect disease condition to the attention of the responsible individual and permits him or her the opportunity to attempt infection of other specimens under experimental conditions. Laboratory screening for detecting infectious agents is a valuable step to tentatively isolating the microbial agent primarily responsible for the disease. Particularly when it is masked by secondary invasion of the specimens by other microorganisms, e.g., saprophytic fungi and bacteria.

I have mentioned two simple steps to assist the individual who detects a possible microbial infection in an insect. The first is to deliver the specimens to a responsible entomologist. The second for both you and the entomologist who seeks a primary or confirming diagnosis of an infectious condition, is to put the specimens into the hands of a specialist and insect pathologist. At this moment I can commit only one laboratory and individual to this task. I trust that we will be able to identify others. For the moment please send one or several dead specimens and information to me:

John D. Briggs
OSU/Entomology
Columbus, Ohio 43210

I will provide a tentative or confirming diagnosis and refer it to an individual who should be able to act and serve your needs.

This group assembled by the conference can provide the beginning of a self-help system that we can refer to as PATHOGEN ALERT. Pathogen alert should be the community effort to reveal who is finding what pathogens, where, and on which crops. What microbial agents do I have? Can someone else use the agents? The "someone else" may be your neighbor, or an industrial group in St. Louis, or a colleague in Zambia or Brazil. Each of us that has contact with insect populations, in the field or in the laboratory, can contribute to the Pathogen Alert system which should become an integral part of all insect reporting services--in states, regionally, nationally, and internationally. A rapid reporting system cannot be specified now, however, the need is evident and it will come to pass. In 1972 the USDA Cooperative Economic Insect Report does not carry references to insect diseases. Two states, Illinois and Wisconsin, in their weekly insect survey reports, record the occurrence of insect diseases.

How can we increase our sensitivity to the detection of diseased insects? This conference and similar activities is one way to assure continued professional development. How are we preparing the students in colleges and universities who will soon carry the responsibility? Most of us have not been doing a good job when it comes to insect diseases. Frankly, until recently it hasn't been necessary. Courses in apiculture have faced disease problems in honey bees, whereas the diseases of other insects were a curiosity--unless it was our armyworm culture that is dying of "natural causes"--then it is a calamity! For an entomology curriculum, instructing in what is "wrong" with an insect is as important as specifying what is "right" with an insect.

The microbial parasites of insects can no longer be the responsibility of the parasitologist, the microbiology department or the burden of the extension apiculturist. I understand that as a faculty member there is one thing over which the faculty has control--the curriculum. We expect students in entomology to take an insect physiology course, why not provide them with a course in insect microbiology or insect pathology? If not as a full course, then as a seminar (either undergraduate or graduate) in entomology and zoology. For your action, please refer to the library Pathfinder for insect pathology and microbial control, it may be useful in academic and/or political interactions in your institution for curricular revision.

LIBRARY PATHFINDER INSECT PATHOLOGY/MICROBIAL CONTROL

Scope: The infectious diseases affecting insects, the causal agents are certain microorganisms, some species of which are subject to production *in vivo* and/or *in vitro*, and can be used for insect population management and in the formulation of microbiological insecticides.

An Introduction to this Topic appears in:

McGraw-Hill Encyclopedia of Science and Technology (1971) under the entry "Insect Pathology."

and

Steinhaus, E. A. Living insecticides, Scientific American, 195: 96-104. August 1956.

Journals devoted to investigations of the pathology of insects and other invertebrates.

Journal of Insect Pathology
Vol. 1 (1959)-Vol. 6 (1964).

Journal of Invertebrate Pathology
Vol. 7 (1964)-

Proceedings of International Congresses of Entomology

Proceedings of meetings devoted to or including material on insect pathology and microbial control.

Transaction of the First International Conference of Insect Pathology and Biological Control. Prague, Czechoslovakia, 1958.

Colloque International sur la Pathologie des Insects et la Lutte Microbiologie, Paris, France,

Books dealing with insect pathology are listed in the subject card catalog. Look for the subjects:
- Insect Pathology
- Insect Microbiology
- Microbial Control
- Biological Control

Frequently mentioned texts include:

Steinhaus, E. A.
Principles of Insect Pathology (1949)
McGraw-Hill, Inc., New York, N. Y.

Steinhaus, E. A., ed.
Insect Pathology (2 vol.) (1963)
Academic Press, Inc., New York, N. Y.

Burges, H. D. and Hussey, N. W., eds.
Insects and Mites (1971)
Academic Press, Inc., London

Bibliographies which contain material on insect pathology and microbial control include:

Annual Review of Entomology
Vol. 1 (1956)
Annual Review, Inc., Palo Alto, California, USA
Biological Control (general)
Pathology (specific)

1962. Entomophaga. Memoire No. 2. pp. 566, 1964.

Proceedings of the International Colloquium on Insect Pathology and Microbial Control Wageningen, The Netherlands, 1966.

Proceedings of the Fourth International Colloquium on Insect Pathology. College Park, Maryland, USA, 1970.

References Cited

Briggs, J. D. 1965-1970. Activities of the World Health Organization International Reference Center for Diagnosis of Diseases of Vectors, World Health Organization, Vector Biology and Control Series, Geneva, issued annually.